Advances in Mechanical Engineering and Materials Sciences

The *International Conference on Advances in Mechanical Engineering and Material Sciences (ICAMEMS)* is a global platform dedicated to fostering innovation, collaboration, and knowledge sharing in the fields of mechanical engineering and material sciences. The conference brings together leading researchers, academicians, industry practitioners, and students to discuss emerging trends, technological breakthroughs, and real-world applications that are shaping the future of engineering.

Spanning a diverse range of themes—from advanced manufacturing and thermal systems to smart materials, sustainability, and interdisciplinary engineering—ICAMEMS encourages dialogue that transcends traditional boundaries. The conference provides an ideal venue for presenting original research, exchanging ideas, and building professional networks that drive academic and industrial progress.

This proceedings volume features a collection of peer-reviewed abstracts and papers presented during the conference, reflecting the depth and diversity of contemporary research in the domain. ICAMEMS continues to play a vital role in advancing engineering solutions for global challenges, promoting academic excellence, and inspiring the next generation of innovators.

First edition published 2026
by CRC Press
4 Park Square, Milton Park, Abingdon, Oxon, OX14 4RN

and by CRC Press
2385 NW Executive Center Drive, Suite 320, Boca Raton FL 33431

CRC Press is an imprint of Informa UK Limited

British Library Cataloguing-in-Publication Data
A catalogue record for this book is available from the British Library

ISBN: 978-1-041-20967-6 (hbk)
ISBN: 9-781-041-20970-6 (pbk)
ISBN: 978-1-003-72505-3 (ebk)

DOI: 10.1201/9781003725053

Typeset in Times New Roman
by Aditiinfosystems

Advances in Mechanical Engineering and Materials Sciences

Edited by

Dr. Vinay K. B
Dr. S. A. Mohankrishna
Dr. J. Paulo Davim
Dr. Om Prakash Singh

CRC Press
Taylor & Francis Group
Boca Raton London New York

CRC Press is an imprint of the
Taylor & Francis Group, an **informa** business

Contents

List of Figures

Advances in Mechanical Engineering and Materials Sciences – Dr. Vinay K. B et al. (eds)
© 2026 Taylor & Francis Group, London, ISBN 9-781-041-20970-6

List of Tables

Foreword

ICAMEMS-2025 has been envisioned as a platform where scholars, researchers, practitioners, and students from academia and industry converge to share their latest findings, exchange knowledge, and explore emerging trends. The conference topics span a broad spectrum including thermal and fluid sciences, manufacturing systems, design engineering, materials science, and allied interdisciplinary fields reflecting the dynamic and integrative nature of modern engineering challenges.

The contributions featured in this volume represent the collective efforts of authors who have showcased their research through technical papers and presentations. We extend our heartfelt appreciation to all contributors, keynote speakers, reviewers, and organizing members whose dedication and support have been instrumental in making this conference a success.

We hope that the deliberations and knowledge shared during ICAMEMS-2025 will inspire future research, foster innovation, and pave the way for new collaborations and advancements in science and engineering.

Conference Chair

Dr. Vinay K B

Professor & Head,

Department of Mechanical Engineering, VVCE, Mysuru

Preface

The International Conference on Advances in Mechanical Engineering and Material Sciences (ICAMEMS-2025) was organized by the Department of Mechanical Engineering, Department of Physics and Department of Chemistry, Vidyavardhaka College of Engineering, Mysuru from 10th to 11th March 2025 at Department of Mechanical Engineering, Vidyavardhaka College of Engineering, Mysuru.

ICAMEMS 2025 served as a multidisciplinary platform for researchers, academicians, industry experts, and students to present and discuss the latest innovations, challenges, and opportunities in the fields of mechanical engineering and material sciences. The conference aimed to promote collaboration and knowledge exchange by covering a wide range of topics including thermal and fluid engineering, manufacturing technologies, industrial design, quality engineering, advanced materials, engineering chemistry, and applied physics.

The technical program included plenary lectures by eminent experts, invited talks, and parallel technical sessions. All accepted abstracts were compiled in the conference proceedings, and selected peer-reviewed full papers will be published in Scopus-indexed journals.

We express our sincere gratitude to all the authors, reviewers, keynote speakers, sponsors, and participants who contributed to making ICAMEMS 2025 a grand success. We hope this compilation serves as a useful reference and inspires further research and collaboration in these ever-evolving domains of engineering and science.

Acknowledgements

The Organizing Committee of ICAMEMS-2025 expresses its sincere gratitude to all those who have contributed to the successful planning and execution of the conference.

We extend our heartfelt thanks to the Management and Leadership of Vidyavardhaka College of Engineering for their constant encouragement and support. Special appreciation is due to the Department of Mechanical Engineering for taking the lead in organizing this event and creating an environment conducive to academic exchange and innovation.

We are deeply thankful to our distinguished keynote speakers, session chairs and panelists, whose expertise and insights have greatly enriched the technical sessions. We also acknowledge the efforts of reviewers and advisory committee members for their meticulous evaluation of submitted papers and valuable guidance.

A special note of thanks goes to our industry partners and sponsors for their generous support and participation.

Finally, we appreciate the enthusiasm and active involvement of all authors, presenters, and delegates, whose contributions have made ICAMEMS 2025 a memorable and impactful event.

Organizing Committee
ICAMEMS 2025

About the Editors

Dr. Vinay K B is Professor and Head of Mechanical Engineering at VVCE, Mysuru. With 15 years of academic experience, he holds B.E., M.Tech., and Ph.D. degrees in Mechanical Engineering. His research interests include heat transfer, fluid flow, quality management, entrepreneurship, incubation, and engineering systems design.

Dr. S. A. Mohankrishna is an Associate Professor in the Department of Mechanical Engineering at Vidyavardhaka College of Engineering (VVCE), Mysuru. With expertise in thermal engineering and renewable energy systems, he has contributed significantly through research publications, academic mentorship, and active involvement in professional and institutional development initiatives.

Dr. J. Paulo Davim is a distinguished professor at the University of Aveiro, Portugal, with over 35 years of experience in manufacturing, materials, and mechanical engineering. He leads the MACTRIB research group, serves as editor for several international journals, and is recognized among the world's top 2% scientists.

Dr. Om Prakash Singh is a Professor at IIT (BHU) Varanasi, with prior experience at IIT Mandi and TVS Motors. A Ph.D. from IISc Bangalore in CFD, he has led projects funded by SERB, DRDO, and BRNS, holds multiple patents, and has published extensively in reputed journals.

Advances in Mechanical Engineering and Materials Sciences – Dr. Vinay K. B et al. (eds)
© 2026 Taylor & Francis Group, London, ISBN 9-781-041-20970-6

1

Some Topological Indices of Dendrimers

K. N. Prakasha*,
Arun C. Dixit, M. A. Sriraj
Vidyavardhaka College of Engineering, Mysuru

Hacer Ozden Ayna[2],
Aysun Yurttas gunes[2], Ismail Naci Cangul[2]
Faculty of Arts and Science, Uludag University,
Bursa, Turkey

Abstract: Dendrimers, a sort of highly branched macromolecules, which exhibit exceptional structural and functional properties, making them pivotal in various materials science domains. The symmetry of molecular structures can be determined by the topological indices (also known as molecular descriptors). This paper aims on the generalization of topological indices which are including Randic-type Hadi Index, SDI Index, Lodeg Index, and Exponential Fraction Index to enhance the understanding of dendrimer properties relevant to many technical domains.

Keywords: Dendrimers, Randic-type Hadi index, SDI index, Lodeg index, and Exponential fraction index

1. INTRODUCTION

Dendrimers are nanostructured macromolecules which are very useful in various fields such as matrial science, medical fields etc., Their architectural design—comprising a central core, branching units, and terminal functional groups imparts unique mechanical and physicochemical properties that are highly adaptable for engineering applications. The 20th century witnessed a rapid development in the field of topological indices. Applications of these are vast and effective in a plethora of fields. Topological indices are significantly contributing in not only mathematics but also in material science, pharmacy etc., Topological indices are not only impart a numeric value but also it will give the topology or physio-chemical property of a molecular graph. In this paper we generalize four topological indies such as Randic-type Hadi, SDI, Lodeg Index, and Exponential Fraction Index for dendrimers. We generalize the indices for Porphyrin (DnPn), Zinc Porphyrin Dendrimer (DPZn), Propyl Ether Imine Dendrimer (PETIM), and polyethylene amide amine (PETAA) dendrimers. First we define the indices as below.

Definition 1.1. The Lodeg index is (D. Vukićević et al. 2010)

$$RL(G) = \sum_{u \sim v} \left(\ln d_u \right) \left(\ln d_v \right).$$

Definition 1.2. The Hadi index is (D. Vukićević et al. 2010)

$$RH(G) = \sum_{uv \in E(G)} \frac{1}{2^{d_u + d_v}}$$

Definition 1.3. The SDI index is (D. Vukićević et al. 2010)

$$R_{SDI}(G) = \sum_{uv \in E(G)} \left(d_u \right)^2 \left(d_v \right)^2$$

Definition 1.4. The Exponential fraction index is (K. N. Prakasha et al. 2022)

$$EF(G) = \sum_{uv \in E(G)} \left(e \right)^{\frac{d_u}{d_v}}$$

d_u and d_v represents the highest and lowest degrees respectively.

---

*Corresponding author: prakasha@vvce.ac.in

2. RESULTS

Result 1. Let DnPn be the family of porphyrin dendrimers. Then,

a. SDI(DnPn)=2287n-296
b. L(DnPn)=69.2288n-6.9712
c. EF(DnPn)=1658.5174n-40.4815
d. RH(DnPn)=3.2656n-0.5

Proof: There are 2n edges present with the points of degree 1 and 3. 24n edges connect the points of degree 1 and 4. 10n − 5 edges exist with the points of degree 2 each. 48n − 6 bonds will link the vertices of degree 2 and 3. There is a connectivity of 13n edges with the points of degree 3 each. 8n edges are there with the points of degree 3 and 4. By definition 1.3,

$$R_{SDI}(G) = \sum_{uv \in E(G)} (d_u)^2 (d_v)^2.$$

This implies $2n(9) + 24n(16) + 10n - 5(16) + 48n - 6(36) + 13n(81) + 8n(144)$. Thus the result follows. The reader can apply the similar proof technique of partitioning edges and can get the remaining indices.

Result 2. Let DPZn be the family of Zinc Porphyrin Dendrimer, then,

a. SDI(DPZn) = 1858(2^n)n-1360
b. L(DPZn) = 47.5028(2^n)n-27.325
c. EF(DPZn) = 244.5(2^n)n-110.898
d. RH(DPZn) = 2.375(2^n) − 0.96875

Proof: The edge partition is as mentioned. 16 (2^n) − 4 edges exist with the nodes of degree 2 each. 40(2^n)−16 bonds will link the vertices of degree 2 and 3. There is a connectivity of 8 (2^n)−16 edges with the vertices of degree 3 each. 4 edges are there with the vertices of degree 3 and 4. By definition 1.4,

$$EF(G) = \sum_{uv \in E(G)} (e)^{\frac{d_u}{d_v}}$$

This implies EF index is 244.5(2^n)n-110.898.

Thus the result follows. The reader can apply the similar proof technique of partitioning edges and can get the remaining indices.

Result 3. Let *PETIM* be the family of *Propyl Ether Imine Dendrimer*, then,

a. SDI(DPZn)=480(2^n)n-504
b. L(DPZn)= 12.2562(2^n)n-13.2171
c. EF(DPZn)= 85.1607(2^n)n-75.8191
d. RH(DPZn)= 1.4375(2^n) − 1.3125

Proof: The edge partition is as mentioned.

2 (2^n) edges exist with the nodes of degree 1 and 2 each.

16(2^n)−18 bonds will link the vertices of degree 2 each.

There is a connectivity of 6 (2^n) − 6 edges with the vertices of degree

3 and 2. By Hadi index

$$RH(G) = \sum_{uv \in E(G)} \frac{1}{2^{d_u + d_v}}$$

This implies Hadi index is $\frac{2^{n+1}}{8} + \frac{6(2^n) - 6}{32} + \frac{16(2^n) - 18}{16}$.

Thus the result follows. The reader can apply the similar proof technique of partitioning edges and can get the remaining indices.

Result 4. Let *PETAA* be the family of Polyethylene amide amine dendrimer,

then,

a. SDI(*PETAA*) = 1028(2^n)n-470
b. L(*PETAA*) = 2.375(2^n)n-0.90625
c. EF(*PETAA*) = 243.0245(2^n)n-102.2525
d. RH(*PETAA*) = 22.9172(2^n) − 10.6971

Proof: The edge partition is as mentioned.

4 (2^n) edges exist with the nodes of degree 1 and 2 each.

16(2^n)−18 bonds will link the vertices of degree 2 each.

There is a connectivity of 20 (2^n) − 9 bonds which links the vertices of degree of degree 3 and 2.

4 (2^n) − 2 bonds connects the nodes of degree 1 and 3 each.

Lodeg index is

$$RL(G) = \sum_{u \sim v} (\ln d_u)(\ln d_v).$$

This implies Lodeg index is. [16(2^n)−18][ln2 ln2] + 20 (2^n) − 9 [ln2 ln3].

Thus, the result follows. The reader can apply the similar proof technique of partitioning edges and can get the remaining indices.

3. CONCLUSION AND FUTURE DIRECTIONS

This study gives few topological indices of Dendrimers. Researchers can use this data to find different physio-chemical properties of Dendrimers. One can easily find the correlation between the indices and the properties using the available tools and technology.

REFERENCES

1. Ahmed, W., Zaman, S., Asif, E. *et al.* Exploring the role of topological descriptors to predict physicochemical properties of anti-HIV drugs by using supervised machine learning algorithms. *BMC Chemistry* **18**, 167 (2024). https://doi.org/10.1186/s13065-024-01266-4
2. W Zhao, MC Shanmukha, A Usha, MR Farahani, KC Shilpa, Computing SS index of certain dendrimers, Journal of Mathematics, Volume 2021, Article ID 7483508, 14 pages

3. K. N. Prakasha and K. Kiran, "Exponential Fraction Index of Certain Graphs," TWMS J. Appl. Eng. Math., vol. 12, no. 2, pp. 631–638, 2022.

4. K. N. Prakasha, K. Kiran, and S. Rakshith, "Randic Type SDI Index of Graph," TWMS J. Appl. Eng. Math., vol. 9, pp. 894–900, 2019.

5. P. S. K. Reddy, K. N. Prakasha, and I. N. Cangul, "Randić Type Hadi Index of Graphs," Trans. Natl. Acad. Sci. Azerb. Ser. Phys.-Tech. Math. Sci. Mathematics, vol. 40, no. 4, pp. 175–181, 2020.

6. M. Togan, A. Yurttas, and I. N. Cangul, "Zagreb and Multiplicative Zagreb Indices of r-Subdivision Graphs of Double Graphs" (preprint).

7. M. Togan, A. Yurttas, and I. N. Cangul, "Some Formulae and Inequalities on Several Zagreb Indices of r-Subdivision Graphs," Enlightenments of Pure Appl. Math., vol. 1, no. 1, pp. 29–45, 2015.

8. M. Togan, A. Yurttas, and I. N. Cangul, "All Versions of Zagreb Indices and Coindices of Subdivision Graphs of Certain Graph Types," Adv. Stud. Contemp. Math., vol. 26, no. 1, pp. 227–236, 2016.

9. D. Vukićević and M. Gašperov, "Bond Additive Modeling 1. Adriatic Indices," Croat. Chem. Acta, vol. 83, pp. 243–260, 2010.

10. A. Yurttas, M. Togan, and I. N. Cangul, "Zagreb Indices and Multiplicative Zagreb Indices of Subdivision Graphs of Double Graphs," Adv. Stud. Contemp. Math., vol. 26, no. 3, pp. 407–416, 2016.

Advances in Mechanical Engineering and Materials Sciences – Dr. Vinay K. B et al. (eds)
© 2026 Taylor & Francis Group, London, ISBN 9-781-041-20970-6

2

A Comparative Analysis of Polylactic Acid and Acrylonitrile Butadiene Acrylate with Polyethylene Terephthalate for 3D Printing Material—A Study

Khalid Imran[1]

Associate Professor,
Department of Mechanical Engineering, Vidyavardhaka College of Engineering,
Mysuru, Karnataka, India

Mohsin Shariff M.A.[2],
Syed Afaan Ahmed[3], Jagadeesh Patel K.B.[4],
Syed Sufiyan[5]

Dept. of Mechanical Engineering Vidyavardhaka College of Engineering,
Mysuru, Karnataka, India

Abstract: Although 3D printing has revolutionized industry, research on environmentally friendly thermoplastics that can satisfy a broad spectrum of application requirements is still lacking. Examining less often used materials like Acrylonitrile Butadiene Acrylate (ABA) with more established choices like Polylactic Acid (PLA) and Polyethylene Terephthalate (PET) reveals this disparity especially. Focusing on their printability, thermal characteristics, mechanical behaviour, and environmental effect in the framework of 3D printing, this paper seeks to close that gap by providing a thorough comparison of PLA, ABA, and PET. Popular bio-based material PLA is biodegradable and easy to print, but it has shortcomings in heat resistance and durability. ABA, a more recent material, is an excellent choice for uses requiring more durability with its improved hardness and chemical resistance. Though it raises sustainability issues about recyclability, PET is generally known for its strength and heat resistance. While filling in the gaps in current research, this review examines closely how the features of these materials interact with various additive manufacturing technologies and processing settings to help guide material selection for high-performance, sustainable 3D printing applications.

Keywords: Acrylonitrile butadiene acrylate (ABA), Additive manufacturing, 3D printing materials, Environmentally friendly polymers, Mechanical qualities, Material selection, Polylactic acid (PLA), Polyethylene terephthalate (PET), Sustainability, Thermal stability

1. INTRODUCTION

With its increased opportunities for fast prototyping, tailored production, and generating intricate designs long beyond of reach, 3D printing has profoundly altered the way we approach manufacture. One of the toughest problems yet is selecting the appropriate materials that combine sustainability and performance as this technology develops. Among the several materials used in 3D printing, thermoplastics are the most often used ones; material

[1]Khalidmech786@gmail.com; [2]VVCE21ME0110@vvce.ac.in; [3]VVCE21ME0073@vvce.ac.in; [4]VVCE21ME0131@vvce.ac.in;
[5]VVCE21ME0110@vvce.ac.in, VVCE21ME0033@vvce.ac.in

10.1201/9781003725053-2

choice is influenced by mechanical strength, thermal stability, environmental effect, and simplicity of printing. Although Polylactic Acid (PLA) and Polyethylene Terephthalate (PET) are somewhat common materials used in the sector, there is growing interest in investigating new thermoplastics such Acrylonitrile Butadiene Acrylate (ABA), which can have various advantages. Popular because of its environmentally friendly character and simplicity of use in printing, PLA, a biodegradable and bio- based substance, suffers with heat resistance and is usually less durable. Conversely, PET is well-known for its strength and resistance to higher temperatures; nonetheless, issues about its recyclability call into doubt its long-term viability. Though not as well-known in the 3D printing scene, ABA has better chemical resistance and hardness, which makes it an interesting choice for uses needing more longevity. Though these materials are used extensively, research on their general performance and knowledge of their performance in practical 3D printing applications is lacking. With a thorough comparison of PLA, ABA, and PET, this paper seeks to close that void. To present a better picture of how different materials measure up against one another, we will examine their printability, mechanical qualities, thermal stability, and environmental effect. By means of this study, material selection for 3D printing will be guided so that the finest, most sustainable solutions are selected for every particular need.

1.1 Background on 3D Printing Materials

The Value of Sustainability within 3D Printing Materials

As 3D printing is becoming more and more popular, the environmental effect of the materials used in the process takes front stage. 3D printing makes products by optimizing material consumption unlike traditional manufacturing methods that produce a lot of waste. Still, the general viability of the process depends much on the components themselves. Growing attention on environmentally friendly materials, sustainability is starting to take front stage in the 3D printing scene. For those seeking more environmentally friendly solutions, for example, PLA is a biodegradable polymer composed of renewable materials. Though PLA breaks down over time in composting conditions, its disposal still requires careful thought, particularly in areas where composting facilities are few. Another crucial component is recyclability. Highly recyclable materials like PET allow one to rework filament and cut waste. In the 3D printing process, where old prints or material leftovers may be melted down and rebuilt into new products, therefore lowering the demand for fresh raw materials and helping close the loop. Many often used polymers in 3D printing, on the other hand, are not recyclable, which results in long-term waste should they not be disposed of correctly. One further area where polymers like PLA shine is biodegradability.

2. POLYLACTIC ACID (PLA)

2.1 Essential Traits of PLA

One of the most outstanding characteristics of PLA is its biodegradability. Being a product derived from renewable energy, it may break down in the correct environment, lessening its environmental effect than long-lasting traditional polymers that remain in the surroundings [1]

Physical Characteristics: PLA's hard and brittle features might restrict its uses in situations needing adaptability. Although 3D printing benefits from PLA filaments' somewhat low melting point—roughly 175°C—this may limit its usage in high- temperature applications.

2.2 Printability

Ease of Use

The simple printing technique of PLA is well-known, and this is very helpful for newcomers (user friendly).[1].

Often advised for beginners because of its minimal shrinkage and simple printability, PLA is known for its user- friendliness. This is easy to use, which lets designs quickly become functioning prototypes.[3]

Low temperatures allow printing to be done easier and warping danger to be minimized.

Extrusion Properties

Excellent flow properties of PLA during extrusion made constant filament distribution possible. High-quality printing with strong layer adhesion follow from this, which is absolutely essential for precise and detailed models.[1].

Extrusion depends much on the layer thickness and nozzle diameter as well. Standard parameters that enable excellent extrusion for all three materials [2] were utilized in the experiment: a 0.4 mm nozzle diameter and a 0.2 mm layer thickness. Melt temperatures of these materials affect their extrusion qualities. For instance, although ABS and PETG need higher temperatures (240 °C) for best flow, PLA is extruded at lower temperatures (205 °C). This variations influences the printing process's efficiency and speed. [2]

Appropriateness for Several 3D Printing Technologies

FDM (Fused Deposition Modelling) printers—the most often used kind of 3D printers—are quite compatible with PLA. Its characteristics fit this technique as they enable simple layer bonding and excellent detail replication. Other Technologies: Although PLA is mostly applied in FDM printing, SLA is another 3D printing technology it may be used in. Though their performance may differ depending on the particular needs of these techniques, stereolithography and selective laser sintering, SLS.

2.3 Mechanical Aspects

Tensile Strength

Printable at a thickness of layer of 0.1 mm, PLA has a tensile strength of around 7.5 MPa. Among the materials studied, this strength is the highest—including ABS and PET-G—indicating that PLA is very robust under strain [1]. Tensile strength of PLA shows a clear drop with rising temperature. According to the study, PLA is the most sensitive material about the change in mechanical characteristics brought about by temperature variations [2]. Tensile strength-wise, PLA shows rather high levels among thermoplastic polymers. The ultimate tensile strength (UTS) of PLA is much influenced by the infill percentage and angle of raster; the tensile strength of 3D-printed PLA parts can be considerably influenced by various printing parameters, such as print speed (PS), nozzle temperature (LT), and infill density (ID>[3]). Because more material is present to oppose stretching forces, higher infill percentages usually translate into higher tensile strength. Additionally important is the raster angle as it influences the alignment of the material layers during printing, hence improving the strength of the produced object [5].

Table 2.1   Mechanical properties and printing characteristics of different FDM materials

Properties	PLA	ABA	PETG
Extrusion Temp	190-210	220-260	230-250
Bed Platform Temp	25-80	90-110	60-80
Density(g/cm^3)	1.25	1.04	1.23
Tensile Strength MPA	65	43	49
Flexural Strength	97	66	70
Izod impact strength	4	19	7.6
Recyclability	Yes	Yes	Yes
Biodegradability	Yes	No	No
Fume Toxicity	Very Low	Medium	Very Low

2.4 Flexibility

Elongation at Break

PLA's plasticity, around 1.5, defines its adaptability. This implies that although PLA is strong, it still has some flexibility which lets it distort under pressure without breaking. Nonetheless, among the investigated materials, PLA's stiffness is seen to be the greatest; this might restrict its adaptability relative to more flexible polymers such as ABS [1]. Though tensile strength decreases, the elongation at break for PLA rises greatly with increasing temperatures. This implies a degree of flexibility as PLA can stretch further before breaking even if it may weaken at higher temperatures. When considering other materials like ABS and PETG, which exhibit diverse reactions to temperature fluctuations, this behaviour is very noteworthy [2]. Although PLA is very low in flexibility, its great rigidity is well recognized. This implies that although it can bear large weights, it could not distort very much before breaking [3]. Measuring PLA as elongation at break, the infill percentage and raster angle mostly define its degree of flexibility. As the component stiffens, a larger infill percentage usually reduces flexibility. On the other hand, by enabling the layers to better absorb stress without breaking, an ideal raster angle can increase flexibility. Applications when some degree of flexibility is needed depend on this balancing [5].

Toughness

Generally speaking, PLA has a less impact resistance than more ductile polymers such as ABS [1]. PLA's brittleness points to its decreased ability to resist abrupt pressures without cracking. Tensile experiments on PLA at different temperatures (25, 35, and 45 °C) revealed this feature restricts its toughness, particularly in applications requiring great impact resistance [1]. PLA shows a notable drop in tensile strength and modulus with rising temperature, suggesting that temperature variations might perhaps influence its toughness as well. Particularly, the tensile strength and modulus of PLA were found to be most sensitive to temperature fluctuations among other materials such as ABS and PET [2]. The mechanical characteristics of PLA, including toughness, can be much influenced by the 3D printing processing settings. The final qualities of the printed items depend much on factors such layer thickness (LT), infill density (ID), and printing speed (PS); so, PLA can be combined with different kinds of rubbers to increase toughness. The study looked especially at mixes at a 15% weight ratio including thermoplastic polyurethane (TPU), acrylic core- shell rubber (CSR), and natural rubber (NR). Of them, PLA/NR shown the best tensile toughness [4].

2.5 Thermal Characteristics

Knowledge of the thermal transition temperatures helps one to forecast material performance at various temperatures. DSC analyses on the materials were therefore carried out to ascertain the test temperatures used in this work; the resultant curves are shown in Fig. 2.1. PLA was the lowest temperature material, hence 45 °C was selected as the maximum tensile test temperature looking at the Tg values displayed on the curves in the Fig. 2.1. Furthermore determined to be comparable with manufacturer values were these glass transition temperature readings.[2]

For every material, the average values of yield strength (σy), tensile modulus (E), and elongation at break (ε) acquired from tensile testing at different temperatures are shown in Fig. 2.1; mentioned in Table 2.1. When the Fig.'s curves change. Examining 3(a), it is observed that for PLA, the elongation at break value increases noticeably while the strength and modulus values drop with rising temperature. Besides, as Fig. 2.1 shows, it

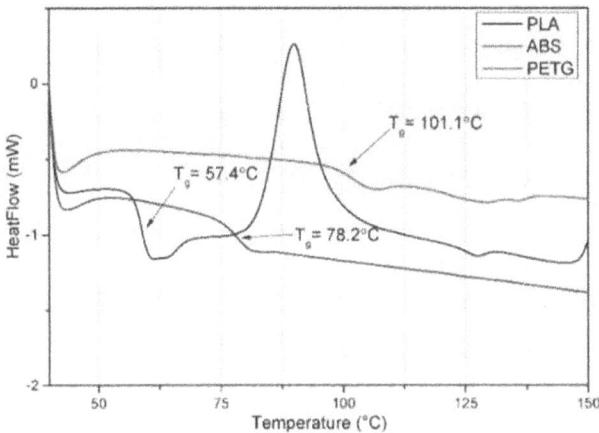

Fig. 2.1 DSC curves of the materials

can be shown that 3(b), considering the curves given for PETG in Fig. 2.1 the change in mechanical characteristics was somewhat restricted at the selected test temperatures for ABS. 3(c) it is observed that the elongation at break rises dramatically without appreciable reduction of strength and particularly modulus. The closeness of the temperatures in the test explains the various mechanical characteristics. Moreover, this outcome fits the curves of the storage in Higher temperatures (up to 210 °C) improve form stability and surface smoothness, so the study shows that the printing temperature can affect the qualities of the printed specimens [4]. Tg is absolutely important for FDM materials as it determines their performance under different temperature environments. For high-temperature uses, PLA is less suited as, for example, its Tg is about 60°C and it can lose its stiffness at extreme temperatures [5]. The glass transition temperature of PLA falls around 60°C. PLA therefore changes from a hard state to a more flexible, rubbery one at temperatures over this range. Applications where the material might be subjected to high temperatures depend on this quality as it can cause loss of structural integrity and form [5]. According to the research, high extrusion temperatures cause thermal breakdown in cellulose fibers of paper sludge. In particular, the residual weight of cellulose fibers began to drop at 250°C and dropped significantly from 290 to 350°C, suggesting that at high temperatures the composites would lose mechanical qualities and structural integrity [9]

2.6 Environmental Impact

Since PLA (polylactic acid) is biodegradable and comes from renewable resources such as corn starch This feature lets it break down more organically in the surroundings than non-biodegradable polymers like ABS (Acrylonitrile Butadiene Styrene) and PET (Polyethylene Terephthalate). During manufacturing, PLA boasts far fewer carbon emissions than PET. Studies have revealed that utilizing marine plastic debris as input material in 3D printing is more ecologically advantageous than using fresh bioplastics like PLA. This suggests that PLA is

less damaging in terms of air quality during the printing process [12]. This implies that although PLA is a superior substitute for conventional plastics, recycling current plastics can provide even more positive effects on the surroundings [15].

3. POLYETHYLENE TEREPHTHALATE [PET]

3.1 Introduction to PET as a Thermoplastic often used in 3D Printing

Attractive for 3D printing, poly-ethylene terephthalate (PET) is a commonly used thermoplastic with strength, durability, and recyclability. Its mechanical characteristics are improved by repeating units of ethylene glycol and terephthalic acid, which also define its composition. Using recycled PET bottles in filament manufacture not only offers a sustainable substitute for traditional materials but also helps to lower environmental plastic waste [6,8].

3.2 Printing Ability

Ease of Use

One major benefit is that recycled PET filament can be compatible with many of current 3D printers. By means of extrusion, water bath chilling, and spooling phases, the filament-making process yielded on-spec dimensions for the filament that remained below the usual 1.75 mm and suited most printers without problems [6]. This suggests that the components can be efficiently transformed into a shape fit for 3D printing [8]. Beginning users in 3D printing often choose PLA because of its easy-to-use qualities. Being biodegradable and derived from renewable sources, PLA is easier to print due to its lower warping tendencies than ABS [12]. A consistent filament diameter is therefore essential for smooth printing. It follows well to the print bed and usually requires lower printing temperatures than PET. The study underlined that because of irregular flow rates, recycled PET might generate non-uniform diameters, which can complicate the printing process [13]. The addition of EBA-GMA improves the impact strength of RPET, therefore increasing its resilience and fit for practical uses. Better performance in 3D-printed goods resulting from this enhancement in mechanical qualities will be a major benefit for consumers [14].

Common Obstacles

Using recycled PET fibre might cause users to run into problems such nozzle blockage. This can be ascribed to contaminants in the recycled material, which must be resolved if printing is to be seamless [6]. One of the biggest difficulties recycled filaments present is the variation in material quality, which adds an additional stage in the preparation process [8]. Different composition and quality of recycled materials might cause problems including uneven extrusion or layer adhesion. The possibility of more brittleness in recycled filaments than in their virgin counterparts presents even another difficulty. This brittleness can influence the durability of printed goods, thus they are more likely to shatter under stress [10]. PLA can be sensitive to heat, hence, if heated after printing, deformation could result. While reduced tensile strength might enable simpler extrusion, it may also cause structural integrity problems of the printed product. This restricts its usage in situations where heat resistance is critical [11]. Low tensile strength materials may distort more readily, which might be problematic for some uses [13]. Using recycled PET for 3D printing helps production to be more sustainable and helps to lower plastic waste. This fits the rising need for environmentally friendly products in the sector [6]. By greatly improving the MFI from 90 to 1.2 g/10 min, 0.75 weight percent PMDA made the modified R-PET more like commercial 3D-printing filaments. Achieving better melt strength and flow during the printing process depends on this increase in viscosity; first printing attempts with the produced filament were successful at a temperature of 275 °C, proving that the modified R-PET can be practically used in useful applications [8].

3.3 Mechanical Properties

Tensile Strength

PET exhibits high tensile strength, the maximum amount of tensile (pulling) stress it can withstand before failure. During tensile tests, PET displayed a unique behaviour whereby it showed an increase in extension at break without a significant reduction in tensile strength or modulus [2]. With PET often exceeding 200 N, while PETG reached about 160 N, PET's ultimate tensile strengths (UTS) were much higher than those of PETG specimens. This property makes PET suitable for applications requiring durability and strength, such as in packaging and textiles[9]. In tensile tests, PET exhibited a brittle failure behaviour. Unlike PETG, which maintained a consistent force throughout a greater range of elongation [10], PET shows strong tensile strength, which is very vital for preserving structural integrity both before and following the 3D printing process. Because of its thermomechanical characteristics, which may improve its processing qualities and general strength [13], the study shows that recycled PET can outperform virgin PET in some uses. The tensile strength of recycled PET can vary greatly depending on the form of the material used. For example, PET flakes show a lower tensile strength of 20.35 MPa whereas PET pellets show a tensile strength of 29.62 MPa, suggesting that the mechanical performance of the recycled material might be impacted by its size and shape [15].

Toughness

Toughness is a material's capacity to absorb energy and plastically distort without cracking. Because of its toughness— which lets PET resist pressures and blows without breaking—it may Combining reinforced with

glass fibers, for example, showed an impact strength improvement of up to 633%. Furthermore influencing toughness is the extrusion process's cooling rate. Higher tensile stress and Young's modulus linked with slow cooling rates have been linked to help to explain enhanced toughness [15].

Resistance Against Wearing

Good wear resistance of PET means that it can tolerate abrasion and friction without appreciable deterioration. Applications like conveyor belts and other mechanical parts where contact with other surfaces happens regularly depend on this quality [9].

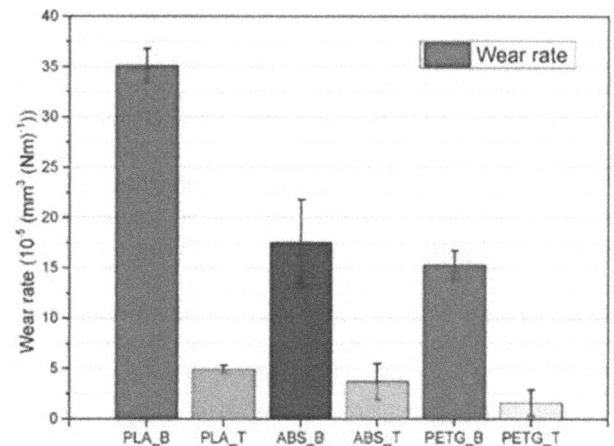

Fig. 2.2 Wear resistance of materials

Thermal Characteristics

Usually, PET'(Polyethylene Terephthalate) glass transition temperature ranges from 67 to 81°C. PET moves from a hard and glassy to a more rubbery condition within this temperature range. PET is brittle and hard below Tg; it becomes more flexible and ductile above Tg. The content of ash were approximately 73.7, 46.2, and 38.1% with particle sizes of below 0.15, 0.18–0.25, and 0.42– 0.84 mm, respectively, which shows lower ash content and also highlights higher cellulose fiber [9]. This property is vital for applications where temperature variations occur since it affects the performance and durability of the material. Comparatively to PLA and ABS, nylon and PET have better thermal stabilities. Table 2.1 aggregates these temperatures. As shown later, the volatile organic compounds are released even below printing temperature, while the breakdown temperature is higher than usual printing temperature. With low thermal stability and high printing temperature, ABS might produce more VOCs than PET (low operating temperature and high thermal stability) [12]. Because the recycled PET in extrusion systems is temperature sensitive, it is prone to flow at low viscosity. Its low viscosity results from a variable flow rate that generates a 3D printer filament with non-uniform diameter [13].

Table 2.2 Correlation between temperature of degradation and temperature of printing

Material	Printing temperature	Decomposition temperature
ABS	240-260°C	370-420°C
PLA	210-245°C	290-390°C
Nylon	230-290°C	400-460°C
PET	165-215°C	360-490°C

3.4 Environmental Effects

Mostly PET bottles and containers, post-consumer plastic trash provides recycled PET filament. Reducing pollution and preserving natural resources depend on this method helping to steer plastic trash from landfills. Still, the recycling process presents difficulties including plastic waste stream contamination.

4. CONCLUSION

All things considered, PET and PLA/ABA offer particular benefits for 3D printing and might satisfy different project requirements. PLA/ABA is an excellent choice for applications that give sustainability first priority—consumer products and prototypes, where environmental impact is a major factor—because of its biodegradability and user-friendliness. Still, for uses requiring a great degree of durability and strength under duress, it might not be the greatest option. On the other hand, functional portions ideal for PET are those requiring stronger mechanical strength and resilience. Its recyclability makes it appealing in technical and industrial environments when lifespan and environmental responsibility are crucial. For more demanding uses, PET's performance characteristics usually exceed the possible increased production costs. Ultimately, choosing between PLA/ABA and PET will rely on the specific project goals and help to balance environmental concerns with performance criteria. While PET is better suitable for applications where mechanical durability and recyclability are critical, PLA/ABA fits extremely nicely with initiatives stressing environmental friendliness and usability. And focus on the comparison of the "PET" with "PLA" and also additionally the composition of the "PET and WOOD", can be the outcome of the literature survey/study which could highlights for "**Literature gap**".

5. RESULT

As per the review of the literature we come to find that there is limited comparative analysis between PET and PLA as 3D printing filaments, particularly in terms of mechanical properties, printability, and sustainability. Additionally, research on the composition and feasibility of PET with wood as a composite material for 3D printing.

REFERENCES

1. B. Steculła, J. Sitko, K. Steculła, M. Witkowski, B. Orzeł, "Comparison of the strength of popular thermoplastic materials used in 3D printing - PLA, ABS and PET-G," *Polish Scientific Society of Combustion Engine*, 27 May 2024. [Online]. Available: https://pdf.ac/2Y4gJ3.

2. S. Yilmaz, "Comparative investigation of mechanical, tribological and thermo-mechanical properties of commonly used 3D printing materials," *European Journal of Science and Technology*, Dec. 2021. [Online]. Available: https://pdf.ac/4gWLxe.

3. I. Khan, M. Tariq, M. Abas, M. Shakeel, F. Hira Ans Al Rashid, M. Koç, "Parametric investigation and optimisation of mechanical properties of thick tri-material based composite of PLA-PETG-ABS 3D-printed using fused filament fabrication," *Composites Part C: Open Access (JCOMC)*, Dec. 2023. [Online]. Available: https://pdf.ac/2MUDB1.

4. S. Pongsathit, J. Kamaisoom, A. Rungteerabandit, P. Opaprakasit, K. Jiamjiroch, C. Pattamaprom, "Toughness enhancement of PLA-based filaments for material extrusion 3D printing," *Material Design & Processing Communications*, 29 Jul. 2023. [Online]. Available: https://pdf.ac/4jJHUt.

5. M. Algarni and S. Ghazali, "Comparative study of the sensitivity of PLA, ABS, PEEK, and PETG's mechanical properties to FDM printing process parameters," *Multidisciplinary Digital Publishing Institute*, 18 Aug. 2021. [Online]. Available: https://pdf.ac/2K7JMA.

6. M. Nikam, P. Pawar, A. Patil, A. Patil, K. Mokal, S. Jadhav, "Sustainable fabrication of 3D printing filament from recycled PET plastic," *Material Today Proceedings*, 12 Oct. 2023. [Online]. Available: https://pdf.ac/2Y4w5N.

7. M. Selva Priya, K. Naresh, R. Jayaganthan, R. Velmurugan, "A comparative study between in-house 3D printed and injection moulded ABS and PLA polymers for low-frequency applications," *IOP Publishing [Material Research Express]*, 6 Jun. 2019. [Online]. Available: https://pdf.ac/1wPhTV.

8. M. Alzahrani, H. Alhumade, L. Simon, K. Yetilmezsoy, C. Mouli R. Madhuranthakam, A. Elkamel, "Additive manufacture of recycled (polyethylene terephthalate) using pyromellitic dianhydride targeted for FDM 3D-printing applications," *Multidisciplinary Digital Publishing Institute*, 2 Mar. 2023. [Online]. Available: https://pdf.ac/36rLgu.

9. J. Son, H.-J. Kim, P.-W. Lee, "Role of paper sludge particle size and extrusion temperature on performance of paper sludge– thermoplastic polymer composites," *Korean Wood Science and Technology (JKWST)*, 10 Jan. 2001. [Online]. Available: https://pdf.ac/2ECcpm.

10. H. Schneevogt, K. Stelzner, B. Yilmaz, B. E. Abali, A. Klunker, C. Völlme, "Sustainability in additive manufacturing: Exploring the mechanical potential of recycled PET filaments," *Composites and Advanced Materials Journal*, 29 Jan. 2021. [Online]. Available: https://pdf.ac/1rhBhD.

11. J. J. Shen, "Comparative life cycle assessment of Polylactic Acid (PLA) and Polyethylene Terephthalate (PET)," Spring 2011. [Online]. Available: https://pdf.ac/3q3Sz7.

12. S. Wojtyła, P. Klama, T. Baran, "Is 3D printing safe? Analysis of the thermal treatment of thermoplastics: ABS, PLA, PET, and nylon," *Journal of Occupational and Environmental Hygiene*, 6 Feb. 2017. [Online]. Available: https://pdf.ac/3VaLbF.

13. M. K. James E, E. J. A. Exconde, J. Z. Manapat, E. R. Magdaluyo, "Materials selection of 3D printing filament and utilization of recycled polyethylene terephthalate (PET) in a redesigned breadboard," *Procedia CIRP 84 (2019)*, Mar. 2023. [Online]. Available: https://pdf.ac/43CI5T.

14. L. Toth, E. Slezak, K. Bocz, F. Ronkay, "Progress in 3D printing of recycled PET," *Materials Today Sustainability*, 17 Mar. 2024. [Online]. Available: https://pdf.ac/4sN1hP.

15. I. Ibrahim, A. G. Ashour, W. Zeiada, N. Salem, M. Abdallah, "A systematic review on the technical performance & sustainability of 3D printing filaments using recycled plastic," *Multidisciplinary Digital Publishing Institute*, 17 Sep. 2024. [Online]. Available: https://pdf.ac/1CEDb3.

Note: All the figures and tables in this chapter were made by the authors.

Advances in Mechanical Engineering and Materials Sciences – Dr. Vinay K. B et al. (eds)
© 2026 Taylor & Francis Group, London, ISBN 9-781-041-20970-6

3 Modification of Paddy Reaper and Binder Machine into a Ragi Harvester

Arun Kumar K. N.[1]

Assistant professor,
Department of Mechanical Engineering,
VVCE, Mysuru

Pradeep Kumar V. G.[2]

Assistant Professor,
Department of Mechanical Engineering,
JSS Science and Technology University, SJCE,
Mysuru, Karnataka

Nikhil R.[3], Niranjan Hiremath[4]

Associate Professor,
Department of Mechatronics Engineering,
School of Mechanical Engineering, REVA University,
Bangalore

Abhinav S.[5]

Project Associate,
Department of Mechanical Engineering,
Vidyavardhaka College of Engineering,
Mysuru, Karnataka, India

Abstract: Harvesting is one of the most important function in agricultural mechanization and it has direct impact on productivity and cost-effectiveness. The use of sophisticated crop-cutting technology in developing countries can lower the cost of production substantially, thus stimulating economic development in the agricultural industry. This project aims to design and develop a better cutting and crop collection system for harvester machines, solving the inefficiencies of traditional harvesting techniques. Today, conventional harvesting methods are still time-consuming and labor-intensive, restricting their efficiency, particularly for small-scale farmers with limited budgets. To make the process more efficient, this project uses a double-blade cutting mechanism which is powered by an scissoring motion. The cutting blades are powered by a simple slotted lever and a crank mechanism, which provides a reciprocating motion. The stalks are cut by the combined shear and impact forces produced at a controlled linear speed. Besides the cutting mechanism, the system also has a crop collection unit intended to direct the cut stalks to the side of the machine. This helps avoid the crushing of the stalks by the tires, making collection easier and reducing wastage. Experimental research has shown that the effectiveness of the cutting blades has a significant impact on the energy needed for harvesting. The results show that reciprocating blade systems are more efficient than traditional rotary blades, providing better cutting force with less energy expenditure. This research provides an optimally conceived cutting and crop gathering system, aimed at enhancing the efficiency of harvester machines as well as overcoming the inefficiencies of existing technologies. Through the development of a workable solution, this research will assist in the modernization of agriculture processes, which will eventually benefit small and large farmers.

Keywords: Paddy reaper, Mechanization, Harvesting equipment, Small-scale farming, Agricultural machinery

[1]arunkn.10@vvce.ac.in, [2]pvg@jssstuniv.in, [3]nikhil.rangaswamy@reva.edu.in, [4]niranjanh@reva.edu.in, [5]363abhinav@gmail.com

10.1201/9781003725053-3

1. INTRODUCTION

Finger millet or the ragi, as popularly called, is an important small millet crop that is grown quite extensively in a number of countries in Asia, such as India, Malaysia, China, and Nepal. In India alone, ragi is cropped on a total area of about 2.65 million hectares and the production amounts to a total of about 2.9 million tons. As a staple food crop, finger millet is renowned for its excellent qualities and is of immense importance in subsistence farming. Finger millet harvesting is different depending on the method of cropping. Intercropped with legumes, it is normally harvested by hand with a sickle, while sole-cropped fields commonly use reaper windrowers. The joint effort involved in harvesting and threshing contributes to almost one-third of the overall labor required in the production system. With the large amounts of arable land that the nation possesses, the need for mechanized farm equipment has been increasing progressively. Whether the reason is shortage of labor or lack of exploitation of land, agriculture has witnessed substantial modernization as opposed to earlier decades. To overcome these challenges, various agricultural machines have been invented. But as all these machines are power-dependent, energy consumption in farm equipment has risen significantly. Among the most important parts in crop-cutting machinery, blades are of utmost importance. The power requirement for cutting crops is high, and for effective functioning, cutting resistance should be low. In this regard, cutting blades contribute to the maximum energy consumption in harvesting machines (Vishal Ullegaddi, 2018). To date, just a few kinds of blades find usage in crops cutters. Among these are two broad types that are called rotary blades and reciprocating blades. Both use rotary blades with greater versatility thanks to its inexpensive manufacturing requirement as well as lower motion mechanism design. There's a limited capacity for thin crop usage with such blades. A design focusing on rotating razor blades is responsible, with their function depending upon specified angles as well as the movement parameters. In order to overcome the shortfalls of rotary blades, reciprocating blades have been put forward, which are more efficient in the harvesting of thick and hard crops. The reciprocating blade design is based on the blade arrangement and the shear stress imparted between two parallel cutting edges. In the present study, we have made improvements to the paddy reaper and binder machine for ragi harvesting by introducing design changes. These are comprised of the counterbalance technique for maximizing weight loading in the front portion and integration of a guide plate to improve bundle alignment as well as maneuverability.

2. PROBLEM STATEMENT

In the current situation, more so in developing countries such as India, the population has been growing at a high rate, thus boosting demand for agricultural output.

Nevertheless, farm labor has reduced tremendously, posing difficulties to the traditional forms of farming. As mentioned above, labor productivity is by nature low since it takes much more time in comparison to mechanistic remedies. Though the advanced machinery like combines and reapers provides efficiency, they are too expensive, and small farm sizes restrict them from being available to most farmers. Thus, there is a need to come up with new farm equipment designed for small farmers. Emphasis should be laid on cost-effective cutting machines that can harvest crops efficiently and control crop stubble. Figures 3.1 and 3.2 provide images of current machinery in operation. The existing paddy reaper binder faces some problems, these are includes heavy weight at the front portion of the machine, the presence of near wheel, it raises a problem during collection and stocking of a crop after is done. Along with the above problem with the paddy reaper binder machine we are about to modify this machine for ragi cultivation as there is need for a machine for ragi cultivation

Fig. 3.1 Existing machine picture

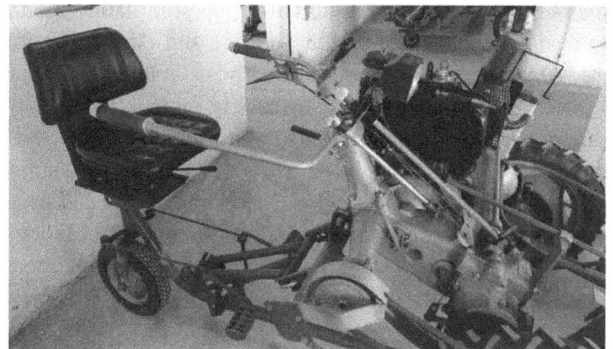
Fig. 3.2 Existing machine picture

2.1 Objectives

According to customer requirements available for the finger millet harvesting region, a number of objectives were created to ensure the best final design. These are as follows.

- To conduct literature review in order to find various trends of Ragi crop cultivation.
- To perform market study on harvesting and threshing requirement for ragi.
- To create various concepts to satisfy functional requirements.
- To model and design the machine and machine components.

To potential solutions to current issues in the machine.

- To include the changes required.
- To determine the additional improvements needed to be included in the brainstorming. To fabrication of blades.
- To design and fabrication of plate for bundle of straws.
- To design & fabrication of latch for locking purpose
- Finally, modification of paddy reaper binder into ragi harvester

3. NEED OF THE PROJECT

In the present situation, the steady increase in population has resulted in mounting needs in the agricultural sector, while the number of daily laborers keeps decreasing. The conventional manual cultivation practices are less productive, with reduced output. Moreover, the high price of heavy machinery and restricted arable land have held back innovation in finger millet harvesting for decades. Identifying this requirement, there is an increasing demand for an affordable, lightweight, and easy-to-use machine specifically designed for small farmers. In order to gain a better understanding of the existing problems related to finger millet harvesting and threshing, a comprehensive literature review has been undertaken. The review comprises an examination of different journals, magazines, research papers, and authentic online sources.

4. METHODOLOGY

At the earlier phase we go through Solution for existing problem which is the heavy weight at the front portion of the machine. The following methods are identified and tested to solve the problem of lifting the heavy weight at the front portion of the machine.

4.1 Testing Concepts

1. HYDRAULIC JACK

Figure 3.3 shows a hydraulic jack used to lift the machine. In this method hydraulic jack is used to lift the front portion, this idea is same as lifting the vehicles like car, trucks' with screw jack. but this method is not effective with some concern problems arising while using it, when jack is used to lifting purpose, it lifts entire machine not front portion, so this is not a convenient method for lift.

Fig. 3.3 Hydraulic jack

Fig. 3.4 Pallet truck

2. PRINCIPLE OF PALLET TRUCK

Figure 3.4 and 3.5 shows a pallet truck used for testing. Based on the pallet truck principle we had design the lifting truck as shown in the above figure but when we are doing this, we came to know this method is also not suitable for lifting purpose when we used this design truck it is lifted the front portion up to 1 inch but our goal is to lift about 5 inches and in order to lift to desired heigh need to take the ground support, while doing this in ragi field as field is contains some moisture in the soil the truck may be struck in the land, so, this method is also not suitable for lifting.

Fig. 3.5 Pallet truck in position

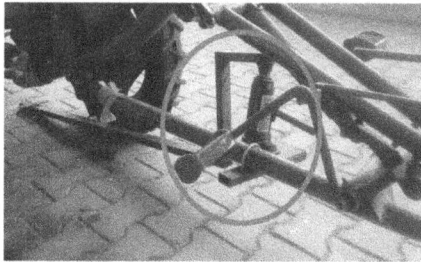

Fig. 3.6 Lever arm method

3. LEVER ARM METHOD

Figure 3.6 shows the lever arm setup used to lift the front portion of the machine. While using the lever arm with hydraulic jack its lift about 2 inches and did not reach our desired height (6 inches). It may lift to 6 inches but, due to space constrain this method is not suitable for the lifting purpose.

4. WEIGHT-LIFTING BY PULLING FRONT PORTION WITH HYDRAULIC JACK AND CANTILEVER

Figure 3.7 shows the set up to lift the machine using cantilever. In this method first we used wire rope is tied to the bar at front and hydraulic jack head. By operating the hydraulic jack, it lifted to 1 inch after that the rope is starts to cut in to small pieces. After that we have designed with metal rod as shown in the above figure, by using this its lifted about 2 inches after that the rod will starts to bend, so this method is also not suitable for our desired purpose.

Fig. 3.7 Lifting by using cantilever

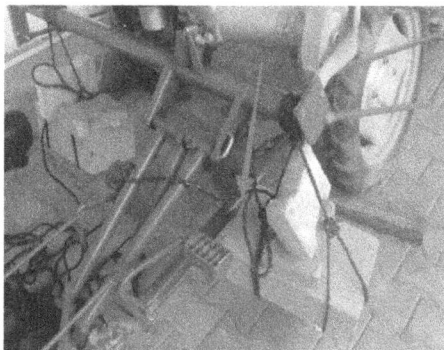

Fig. 3.8 Counterbalance

5. USE OF HYDRAULIC PRINCIPLE:

Here we thought to use a hydraulic cylinder to lifting purpose as its convenient to lifting a heavy weight to use hydraulic cylinder for lifting a ground a support needs to be take and grounds need to be hard but, in paddy and ragi land ground contains moisture it may affect while taking the support of ground, so we thought the counterbalance method is the possible solution.

6. COUNTERBALANCE METHOD:

Figure 3.8 shows the counterbalance method used to lift the machine. After failed with about ten to fifteen methods finally we came to that counterbalance method is the possible and feasible solution for weightlifting purpose. In this we randomly tested with cement blocks by hanging to the rod present at the rear side of the machine as shown in the below figure

WEIGHT CALCULATION: Measure the weight of the cement blocks by using spring balance and it comes around 65 kg's. We use 60kgs of metal disk as a counterweight as shown in the following figure and we have designed and fabricated components for the hanging purpose and it's shown in the following figures.

COUNTER-WEIGHTS: 60 kgs of counterweight is used for the counter balance method, these metal disks are hanging to the bar as shown in the following figure, 60 kgs of weight is divided in to 30kgs and these each 30kgs of weights are hanging in the two edges of the bar as shown following figure, each disk are coming around 5-6kgs.

Fig. 3.9 Weight of disks hanging

Fig. 3.10 Disks

Fig. 3.11 Counterweight

Fig. 3.12 Clamp

COMPONENTS FOR HANGING PURPOSE:
Here designed and fabricated the components like clamps, hanging rod for hanging the counterweights as per the design recommendation and these fabricated components shown in the below figures.

Fig. 3.13 Rod

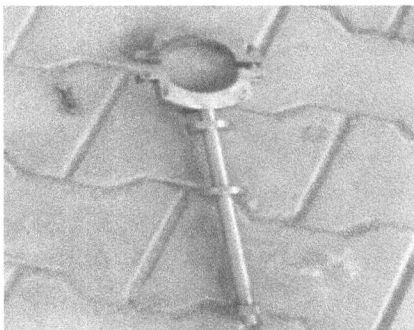

Fig. 3.14 Rod with clamp

Fig. 3.15 Latch

LATCH: It is used for locking purpose. We modified this machine for ragi field, and it can also be used for paddy field too so, in different field different cutting height (for ragi 3-4 inches and paddy 2-3 inches) is considered. According to the height we designed and fabricated latch as shown in the below figure and design recommendation, this latch is used after lifting the front portion of the machine for locking to a different height based on the application.

GUIDE PLATE: As mentioned in the problem statement the second problem faced by the machine is present of the rear wheel, it raises a problem during collection and stocking of a crop after cutting is done. The bundle is left at the middle of the machine, so, when machine moves forward the rear wheel is flow over the bundle so its leads to lose in crop. To overcome this problem, we designed and fabricated the guide plate as per the design recommendation and it's fixed at middle of the machine as shown in the figure.

Fig. 3.16 Guide plate

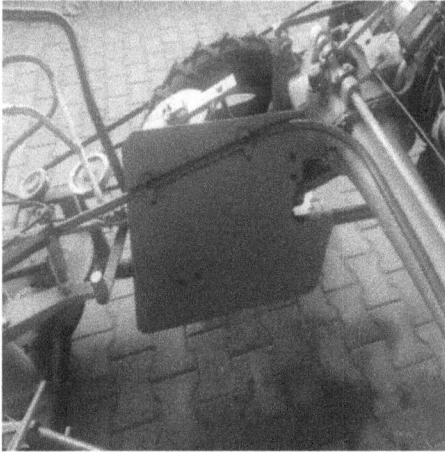

Fig. 3.17 Hurdle plate

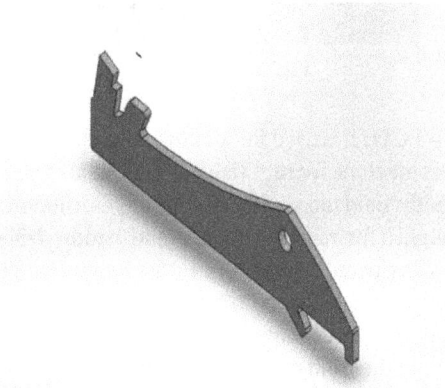

Fig. 3.18 Latch model image

FUTURE SCOPE:

- Include transmission of power from the tractor to the attachment.
- Enhance grain and straw storage capacity.
- Lessen the overall size and weight of the attachment.
- Minimize the use of pulleys in power transmission.
- Include intermittent heaters prior to threshing operation.
- Presently there no is blade for ragi cultivation but we designed a blade with almost nearer for ragi cultivation there is scope for ragi harvester in future for this blade can be designed with most accuracy for ragi harvester.
- Straw collection is possible by the attachment easily.
- The range of cutting is designed for the bed width of finger millet plantation.

ADVANTAGES:

- The range of cutting is designed for the bed width of finger millet plantation. Here the modified machine is cut the crop up to 2 inches and by using of counterbalance method we can move the front blade section up to ground level.

- Straw collection is easy with the attachment made.
- Ground clearance of the blades can be varied by the height adjuster.
- The size and weight of the attachment has been removed so it is easy to operate.
- Less manual effort is needed to operate the machine.
- Blades can be removed and fixed according to the application.
- Machines can be used for multiple field (ragi and paddy land)
- With the latch design machine can be lifted and locked to different heights according to the application
- With the help of spring arrangement at the rear self-retraction is made after lifting
- The attachment is compact and can be easily attached to the agricultural farm machinery.
- Finger Millet crop height is not the problem for this attachment, it is possible to harvest different crop heights
- Easy harvesting of grains, Easy to operate and Easy to maintain

5. RESULT AND DISCUSSION

Finally, the counterbalance method is used for weight balancing purpose as it is the only possible and feasible solution by overcoming all restricted constraints.60kgs of counterweight is used counter balance method and different components are fabricated according to the design recommendation latch for locking purpose and guide plate straw collection is made as part of this project work. By finding the solution for existing problems in machine and small modification in blades the entire paddy reaper and binder machine is modified into an ragi harvester.

6. CONCLUSION

Modification of paddy reaper binder machine into ragi harvester is done with retro fitment on same machine. By finding the solution for existing problems in machine and small modification in blades the entire paddy reaper and binder machine is modified into a ragi harvester. Lack of manual labor and drudgery to conduct field operations necessitate alternative sources of power. Time and labor are valuable resources in field crop cultivation. Substitution of mechanization in cultural operations not only curtails drudgery but also saves time significantly leading to lower cultivation cost and higher returns. Farmers felt that acceptance of mechanization not only lowers the drudgery, cuts the cost of cultivation but also brings in more returns per unit time and land. Under this project, an appropriate cutting mechanism and crop collection mechanism for crop harvester was checked. It can be said that, cutting mechanism has a cutting blade in which one is fixed mounted on frame, and one is moving attached to lever of mechanism which chopped whole stalks by shear force. It was found that cutting mechanisms may work under different farm conditions. Crop gathering mechanism was devised to push cut crops at one end of machine so that they are not crushed by the tires and for ease.

REFERENCE

1. Amrutesh P, Sagar B, and Venu B (2014) – *Solar Grass Cutter with Linear Blades by Using Scotch Yoke*, International Journal of Engineering Research and Applications.

2. Donny T, Prabhakar P, Upadhyay C, Cherian R, and Suresh AM (2012) – *Development of Working Prototype for Ragi Harvesting and Threshing Operation*, International Journal of Scientific and Engineering Research.

3. Gull A, Ahmad NG, Prasad K, and Kumar P (2014) – *Significance of Finger Millet in Nutrition, Health, and Value-added Products: A Review*, Journal of Environmental Science, Computer Science, and Engineering Technology, 3(3): 1601–1608.

4. Pandey, M. M. (2004). Present status and future requirements of farm equipment for crop production. Central Institute of Agricultural Engineering, Bhopal, 24, 69–113.

5. Priya Sinha, S. V. Jogdand (2019) – Design and development of manually operated harvester for wheat crop, J Pharmacogn Phytochem 8(6):543–545.

6. Rumbidzai Prosper Jera, Tawanda Mushiri, Charles Mbohwa (2018) – *Design for manufacture and assembly of a mini combine wheat harvester*, Proceedings of the International Conference on Industrial Engineering and Operations Management Paris, France, July 26–27.

7. Prof. P.B.Chavan, Prof. D .K. Patil, Prof. D .S. Dhondge (2015) – *Design and Development of Manually Operated Reaper Machine*, IOSR Journal of Mechanical and Civil Engineering (IOSR-JMCE) e-ISSN: 2278-1684,p-ISSN: 2320-334X, Volume 12, Issue 3 Ver. I (May. - Jun. 2015), PP 15–22.

8. Qamar-uz-Zaman, A.D. Chaudhry &M. Asghar Rana (1992) - *Wheat Harvesting Losses in Combining as Affected By Machine And Crop Parameters,* Pak. J. Agri. Sei., Vol. 29, No. 1, 1992

9. Tamil Nadu Agricultural University (2018.11.01) – *Expert System for Finger Millet*, Retrieved from http://www.agritech.tnau.ac.in/expert_system/ragi/postharvest.html.

10. Tesfaye Olana Terefe (2017) – *Design and Development of Manually Operated Reaper Machine*, International Journal of Advanced Research and Publications ISSN: 2456–9992

11. Vishal Ullegaddi, Dr. Chetan B (2018) – *Design and analysis of cutting Mechanism for crop harvester*, Journal of Recent Trends in Mechanics Volume 3 Issue 2.

12. Zhong Tang, Yaoming Li, and Cheng (2017) – Development of Multi-functional Combine Harvester with Grain Harvesting and Straw Baling, Spanish Journal of Agricultural Research, 15(1), e0202, 10 pages (2017), eISSN: 2171–9292, https://doi.org/10.5424/sjar/2017151-10175.

Note: All the figures in this chapter were made by the authors.

Advances in Mechanical Engineering and Materials Sciences – Dr. Vinay K. B et al. (eds)
© 2026 Taylor & Francis Group, London, ISBN 9-781-041-20970-6

4

Design and Development of Autonomous Underwater Rover

Raghu N.[1],
Chandan V.[2], Armaan M.[3],
Manohara[4], Chethan S.[5], Dhanush Y.[6]
Dept of Mechanical Engineering, Vidyavardhaka College of Engineering,
Mysuru, Karnataka, India

Abstract: Underwater remotely operated vehicles (ROV) and Autonomous Underwater Vehicle (AUV) have become essential in the fields of defence, scientific research, etc., providing a safe alternative for complex underwater missions. This affects stability and positioning. Despite technological advances, AUVs are hampered by power limitations. This hampers mission duration and operational efficiency. This work explores cost-effective and energy-efficient solutions to improve the accessibility and performance of underwater robotic systems. Making advanced surveying technology more widely accessible. In this literature review we come across various recent ROV and AUV designs, emphasizing integrated frameworks, hybrid control methods, and cost-effective approaches to underwater vehicle development. Notable advancements include autonomous navigation improvements, power optimization, and environmental monitoring capabilities. However, limitations remain, particularly in energy efficiency and affordability, which restrict mission duration and the broader adoption of micro-AUVs.

Keywords: AUV -automated underwater vehicle, ROV-remotely operated vehicle, etc.

1. INTRODUCTION

In recent times, the development of underwater remotely operated vehicle (ROV) has gained significant attention because of their important role in performing operations in a different application such as comprising seaward oil and gas, military and defence, scientific studies, and marine life (Salem et al., 2023). Because of numerous hazards and various challenges that come with operating in an underwater environment, Unmanned Underwater Vehicles (UUVs) an important replacement to human operators. Many "autonomous underwater vehicles (AUVs) and remotely operated vehicles (ROVs)" have been developed as useful tools for a variety of underwater jobs.

The utilization and operation of UUV face significant challenges and risks due to the distinctive underwater environment that greatly affects positioning, navigation, and timing performance, making it incomparable to above mentioned water applications (Garg et al., n.d.). They must contend with the physical distinctions between water and air, traverse various media, and deal with the disruption of outside influences like wind, waves, and water movement. Variations in the parameters of the control model may result from these modifications. Cross-medium failure results from traditional cross-medium control approaches' poor robustness and inability to sustain system stability in complicated situations (Esakki et al., 2018, Saha et al., 2018, Deb et al., 2017). ODS is a crucial component of autonomous underwater vehicles (AUVs) since it enables collision-free navigation in a new environment with obstacles. They are also an essential part of AUV high-level controllers or path planning and collision avoidance systems (Glider for Underwater Research, n.d.).

2. LITERATURE REVIEW

An easy-to-assemble, lightweight, and nimble ROV with mechanical design should be able to complete duties

[1]raghu.n@vvce.ac.in, [2]chandanv@vvce.ac.in, [3]armaan9036@gmail.com, [4]manohar26manu@gmail.com, [5]chethanchethu2701@gmail.com, [6]gdhanush751@gmail.com

10.1201/9781003725053-4

quickly and effectively. As a result, attempted to strike a compromise between ROV's novel design and high performance created ROV in a circular shape to facilitate easy rotation, smooth motion, and a decrease in turbulent flow. Because of its streamlined shape, the electrical enclosure's curvature allows for a decrease in drag force (Tan et al., 2023).

This design makes use of two legs at the ROV's bottom that have several uses. The ROV will always stay upright thanks to the capacity to put positively buoyant material at the top and insert lead rods into the legs. The ability to rest on the ground while carrying out the assigned tasks is another benefit, should the unit be functioning on the ocean floor. These legs would be positioned directly in front of the camera, and mounting the claws on them would be the final benefit.

Additionally, four thrusters are used in this design. While the other two situated at the end of the wing will help in the ascent and downward slope, the two mounted horizontally in the wing will wings will provide the help in manoeuvrability in horizontal plane. It was intended to be shaped like a box. This configuration adheres to the conventional ROV architecture. With weights on the bottom, foam on top, and all the electronics in the center of the ROV, the entire system would be housed in a cage. The components can be fixed on the cage with a box configuration, which facilitates construction and improves stability. To provide the required buoyancy, the foam would be divided into four portions and positioned in specified areas above the entire ROV after the robotic arm and weights were mounted on a plate at the bottom of the cage. Design is the result of combining the suggested design with the first design option. The primary distinction was that the thrusters were positioned within PVC pipes with a smaller diameter that were molded into the body rather than inside a cage-like framework. This was done in order to help shield the thrusters from harm in the event that the ROV approaches a wall too closely or runs into any debris or items.

Static and also dynamic stability, structural support, weight, and ease of approachability are all factors that went into the frame's distinctive design. The frame is made of clear virgin cast acrylic and ABS rather than a single material. It guarantees that every component is fully supported under both dynamic and static loads. Strength, rigidity, weight, machinability, safety, and aesthetics are some of the criteria that have been taken into consideration when modelling the frame. The frame's length to height aspect ratio has been increased to increase hydrodynamic stability. The components were topologically tuned to maximize weight while preserving enough strength to satisfy specifications. L brackets or ribs are used to connect the components. All necessary systems are directly supported by the frame (Philips et al., 2013). The AUV's covert hulls are impervious to water. All of the delicate electrical parts are enclosed in pressure chambers. Acrylic hulls that are cylindrical make sure that the tensions are evenly dispersed throughout the body, guaranteeing optimal use of the given capacity. The end caps on these hulls are detachable and fastened to flanges. They have been constructed in accordance with the principles of fluid statics and are sufficiently strong to endure the pressures (Yang et al., 2024, Tuhtan et al., 2020).

Double silicon radial O-rings are used to ensure waterproofing. They feature groves for uniformly compressed face O-rings via bolts. Acrylic is utilized because of its high shock resistance, workability, moderate strength, and relatively low density (Karimi & Lu, 2021). A ROV's material selection is influenced by a number of criteria, including the intended budget, performance requirements, operational circumstances, and ROV use. The ROV's upper and bottom plates are held concurrently by four shafts. The electrical enclosure is fastened to the top with six fasteners. The ROV's top and bottom plates are held together by four shafts plate, and the bottom plate supports the four horizontal The selection of materials is influenced by several factors, such as the ROV's intended use, operating conditions, budgetary limitations, and implementation requirements. The ROV's upper and bottom plates are held together by four shafts. The four horizontal thrusters are fastened to the lower plate, while the electronic enclosure is fastened to the upper plate with six bolts. There were two cameras on the ROV (Dunbabin & Marques, 2012).

The top and bottom plates were made of ten layers of acrylic to help with manufacture and avoid corrosion. Additionally, acrylic was used for the electric housing dome due to its ease of insulation and water pressure resistance, and light transmittance. Stainless steel was chosen for shafts, rivets, washers, and nuts because of its proficiency to withstand corrosion in water. It can also resist a lot of strength and force. There were two vertical thruster holders in between the two plates. Numerous screw holes Due to its resistance to corrosion in water, stainless steel was selected for shafts, rivets, washers, and nuts. It is also capable of withstanding strong force and strength. Between the two plates were two holders for the vertical thrusters (Bogue 2025, Reza Samaei & Asadian Ghahfarrokhi, n.d.).

Using fusible filaments, a 3D printing method, melted thermoplastic material is deposited through a nozzle to create a portion of an assembly or subassembly layer by layer. This study examines the thermomechanical properties of two thermoplastic material types utilized to create the structural elements of an underwater rover: "PLA (polylactic acid) and PET-G (polyethylene terephthalate-glycol)". MTA-ROUV (Remotely Operated Underwater Vehicle, Military Technical Academy (Stefanita Grigore et al., 1964).

Compared to PLA, PETG has a reduced statistical dispersion, more thermal stability, stronger resistance to

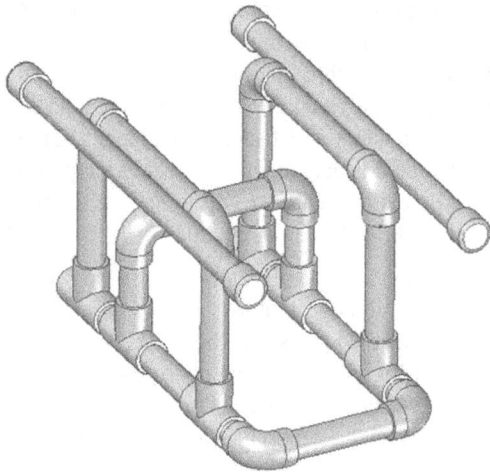

Fig. 4.1 Proposed design rover structure using PVC pipes

Table 4.1 Measurements of PVC structure

SI No	Size In Inch	Quantities Required
1	14	2
2	6.5	2
3	5	4
4	4.5	2
5	4	2
6	2.5	2
7	1.5	4
8	90 Degree Elbow	8
9	T elbow	6
10	End Cap	4

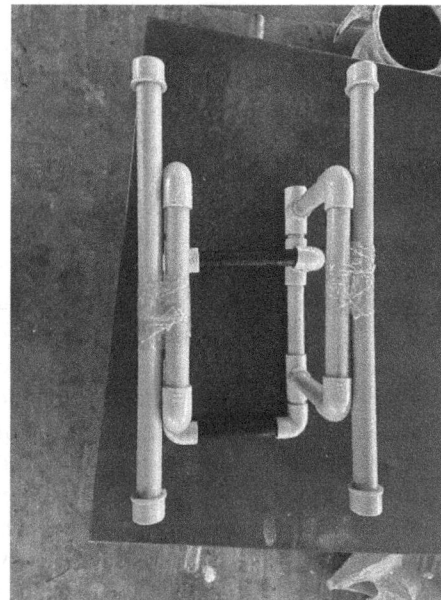

Fig. 4.2 Front and top view of PVC structure

thermal degradation, and greater 3D printing versatility. Additionally, the modularity and plug-and-play design of the ROUV components, which will facilitate end users' access, guarantees a greater level of assembly rigidity than the basic model, according to the analysis using MEF. The same kind of plastic is used to make the basic model [6]. With 360-degree control and the ability to view live video, the RC Underwater Exploration Drone is a cutting-edge water drone designed for effective navigation. It has a camera module that sends live footage back to the operator, giving them real-time insights into the underwater environment, and it is made to move freely in all directions underwater, delivering 360-degree directional control. The drone is propelled by two motors that provide accurate movement and stable positioning in water currents. When paired with the propulsion motors, a separate motor governs vertical movement, providing total spatial control. Because the drone's components are designed to withstand the corrosive maritime environment, they are strong and long-lasting. have made use of a 1.2-foot pipe.

3. COMPONENTS

3.1 Camera

They have used the OV7670 CMOS imaging sensor generates full framed windowed 8 bits images in multiple image formats. The sensor is controlled by an I2C interface called the Serial Camera Control Bus (SCCB), which has a maximum clock frequency of 400 kHz. The OV7670 camera module produces full frame, subsampled 8-bit images in several formats, which are controlled by the SCCB interface. The most fundamental image processing functions that may be programmed with the SCCB interface are gamma, exposure control, colour saturation, white balancing, and hue adjustment. The maximum frame rate of the OV7670 camera module is 30 frames per second, and its resolution is 640×480, or 0.3 Mega pixels.

3.2 Arduino Nano

The Arduino Uno is a popular microcontroller board for do-it-yourself projects and prototyping. It operates on the ATmega328P microprocessor and features several input/ output pins for communication with various electronic components. A kit including all the required parts is required to begin using the Arduino Uno. The Make Your UNO Kit is one of the beginner-friendly kits available on the official Arduino store2. Additional online retailers like Flipkart and Robu.in also provide Arduino Uno kits that include with everything you need to begin programming an Arduino (Kim et al., 2013, Simulation of Autonomous Underwater Vehicles (AUVs) Swarm Diffusion, n.d.).

3.3 BLDC Motor

An electronically commuted motor without brushes is called a brushless DC motor. The synchronous motor's speed and torque are managed by the controller, which sends current pulses to the motor windings. Because friction causes less mechanical energy waste, in terms of efficiency, brushless motors outperform brushed motors (Mansfield & Montazeri, 2024, New Approaches for Renewable Energy Management in Autonomous Marine Ve, n.d.). The current invention applies to autonomous underwater vehicles, particularly AUVs, and pertains to the construction of the vehicle in terms of its mechanics and integration with your display. PVC will be used to construct the drone's main body or frame. Because we are utilizing a non-metallic element, several contemporary mines that depend on the target's magnetic characteristics cannot detect it.

3.4 Pressure Sensor

A pressure-sensitive component measures the actual pressure employed to the sensor, and other components translate the data into an output signal. The MPL115A1 is a digital SPI-output absolute pressure sensor designed for low-cost applications. Under normal operating conditions, it uses only 5 uA of current. In order to detect the pressure in the area the rover will be passing from, the sensor will be attached to a Raspberry Pi, which will then be submerged. The data will then be measured, gathered, and shown on the host ship. This will collect information on the physical conditions beneath the surface (Innovative Water Quality

and Ecology Monitoring Using Underwater Unmanned Vehicles_ Field Applications, Challenges and Feedback from Water Managers, n.d.).

3.5 Temperature Sensor

This sensor features a special NTC to supervise temperature and an 8-bit CPU to output the temperature and humidity analyses as serial data. Since the sensor is factory standardized, integrating it with further microcontrollers is simple. This sensor has a precision of +/- 1 degree Celsius and can monitor temperatures between 0 and 50 degrees Celsius.

3.6 Ultrasonic Sensor

This electronic gadget measures the distance to an object using ultrasonic sound waves and then transforms the exhibited sound waves into an electrical signal. The speed of an ultrasonic wave is greater than the speed of sound that is audible to humans. The transmitter and the receiver are the two essential parts of ultrasonic sensors. The sensor processes the time it takes for sound waves to return to the receiver after the transmitter uses piezoelectric crystals to create sound. This allows the sensor to determine the total distance between the object and the sensor (Azis et al., 2012, Yu et al., 2002).

3.7 Raspberry Pi 3 Model B

The rover's sensors are integrated to a Raspberry Pi. The Raspberry Pi for our explorer since it has 40 GPIO pins that we can use to attach our sensors. It has 1.2 GHz 65-bit quad core processor, four USB ports, one Ethernet port, a 3.5mm audio jack, one HDMI port, one camera output, one Display connector, USB, and one SD card slot for booting. It also has an integrated 80.11n wireless LAN and Bluetooth 4.1 (Choi et al., 2015, B & Kumar, n.d.).

3.8 Control via Wired Remote

An operator can control a rover's movements and operations with a wired remote control by making a direct physical connection, typically using cables or wires. When wireless connection is impracticable or unreliable, such as in high radio interference conditions or when precise, low-latency control is needed, this control approach is frequently used.

4. CONCLUSION

The underwater ROV thus underlines the intricate balance between cutting-edge usability and exceptional efficiency in design. The circular shape allows mobility, following the principle of semi-cylindrical design, whereas legs give stabilization for buoyancy control and provide an unsure base to mount most of the essential instruments: the claws. The combination of acrylic, to allow transparency; stainless steel, to allow structural integrity; and PVC, which gives non-metallic robustness and ease of manufacture, makes the rover efficient and affordable. Various sensors for operation and modular components as well as thrusters for movement allow the ROV to perform some complicated operations such as underwater navigation and environmental monitoring. Further, strong waterproofing and the use of 3D printing technologies in custom-built elements emphasize the variability and longevity of the design. Though the design works well for shallow to mid-depth operation, its viability for deeper and some tricky missions can be further enhanced with the help of power management and optimization of wireless communication. Overall, this ROV presents a ground-up, practical choice for underwater exploration, taking into account maintenance, usability, and adaptability on a variety of marine grounds.

A literature review gives convincing evidence that the detailing and development of autonomous underwater micro rovers have greatly advanced, in particular navigation systems, control algorithms, and their appraisals for environmental monitoring applications. The increasing demand for micro-AUVs is primarily from their diminutive size, mobility, and their conjugatedness as applied to explore shallow and complex underwater environments. Still, there are challenges that have stalled their wide adoption and effective operation. Significant energy management problems limiting the duration of missions arise since batteries restrict mission length. Modifications to existing propulsion systems should be made so as to achieve improved mobility and energy efficiency without increasing the weight of the vehicle. Cost-efficient designs remain hard to achieve, since so much of the complex technologies used in AUVs, including sophisticated sensors and communication systems, can raise the price of production.

REFERENCES

1. Azis, F. A., Aras, M. S. M., Rashid, M. Z. A., Othman, M. N., & Abdullah, S. S. (2012). Problem identification for Underwater Remotely Operated Vehicle (ROV): A case study. *Procedia Engineering*, *41*, 554–560. https://doi.org/10.1016/j.proeng.2012.07.211

2. B, M. S., & Kumar, M. (n.d.). Design and Technological Investigation of Autonomous Underwater Vehicle Systems. In *International Journal of Aquatic Science* (Vol. 12).

3. Bogue, R. (2025). *The role of robots in environmental monitoring.* https://doi.org/10.1108/ir-12-2022-0316/full/html

4. Choi, H. T., Choi, J., Lee, Y., Moon, Y. S., & Kim, D. H. (2015, May 14). New concepts for smart ROV to increase efficiency and productivity. *2015 IEEE Underwater Technology, UT 2015.* https://doi.org/10.1109/UT.2015.7108257

5. Deb, S. K., Rokky, J. H., Mallick, T. C., & Shetara, J. (2017). Design and construction of an underwater robot. *4th International Conference on Advances in Electrical Engineering, ICAEE 2017, 2018-January*, 281–284. https://doi.org/10.1109/ICAEE.2017.8255367

6. *Design and Construction of Hybrid Autonomous Underwater Glider for Underwater Research.* (n.d.).

7. Dunbabin, M., & Marques, L. (2012). Robots for environmental monitoring: Significant advancements and applications. *IEEE Robotics and Automation Magazine*, *19*(1), 24–39. https://doi.org/10.1109/MRA.2011.2181683

8. Esakki, B., Ganesan, S., Mathiyazhagan, S., Ramasubramanian, K., Gnanasekaran, B., Son, B., Park, S. W., & Choi, J. S. (2018). Design of amphibious vehicle for unmanned mission in water quality monitoring using internet of things. *Sensors (Switzerland)*, *18*(10). https://doi.org/10.3390/s18103318

9. Garg, V., Bansal, D., Ranjan, M., Goyal, M., Natu, A., Gaur, P., Raj, P., Sain, S., Biswas, U., Biswas, N., Gupta, R., Parashar, S., Bansal, R., Kumar, C., Kumar, R., Kumar, A., Tyagi, A., & Verdhan, A. (n.d.). *Design and Development of an Autonomous Underwater Vehicle VARUNA 2.0.* www.ijert.org

10. *Innovative Water Quality and Ecology Monitoring Using Underwater Unmanned Vehicles_ Field Applications, Challenges and Feedback from Water Managers.* (n.d.).

11. Karimi, H. R., & Lu, Y. (2021). Guidance and control methodologies for marine vehicles: A survey. *Control Engineering Practice*, *111*. https://doi.org/10.1016/j.conengprac.2021.104785

12. Kim, K., Choi, H. T., & Lee, C. M. (2013). Underwater precise navigation using multiple sensor fusion. *2013 IEEE International Underwater Technology Symposium, UT 2013.* https://doi.org/10.1109/UT.2013.6519855

13. Mansfield, D., & Montazeri, A. (2024). A survey on autonomous environmental monitoring approaches: towards unifying active sensing and reinforcement learning. In *Frontiers in Robotics and AI* (Vol. 11). Frontiers Media SA. https://doi.org/10.3389/frobt.2024.1336612

14. *New approaches for renewable energy management in autonomous marine ve.* (n.d.).

15. Philips, A. B., Steenson, L. V, Rogers, E., Turnock, S. R., Harris, C. A., & Furlong, M. (2013). *More information Catalogue record.* http://eprints.soton.ac.uk/cgi/oai2

16. Reza Samaei, S., & Asadian Ghahfarrokhi, M. (n.d.). *Using robotics and artificial intelligence to increase efficiency and safety in marine industries.*

17. Saha, A. K., Roy, S., Bhattacharya, A., Shankar, P., Sarkar, A. K., Saha, H. N., & Dasgupta, P. (2018). A low cost remote controlled underwater rover using raspberry Pi. *2018 IEEE 8th Annual Computing and Communication Workshop and Conference, CCWC 2018, 2018-January*, 769–772. https://doi.org/10.1109/CCWC.2018.8301657

18. Salem, K. M., Rady, M., Aly, H., & Elshimy, H. (2023). Design and Implementation of a Six-Degrees-of-Freedom Underwater Remotely Operated Vehicle. *Applied Sciences (Switzerland)*, *13*(12). https://doi.org/10.3390/app13126870

19. *Simulation of Autonomous Underwater Vehicles (AUVs) Swarm Diffusion*. (n.d.).
20. Stefanita Grigore, L., Stefan, A., & Orban, O. (1964). Using PET-G to Design an Underwater Rover Through 3D PrintingTtechnology. *Mater. Plast*, *57*(3), 189–201. https://doi.org/10.37358/Mat.Plast.1964
21. Tan, L., Liang, S., Su, H., Qin, Z., Li, L., & Huo, J. (2023). Research on Amphibious Multi-Rotor UAV Out-of-Water Control Based on ADRC. *Applied Sciences (Switzerland)*, *13*(8). https://doi.org/10.3390/app13084900
22. Tuhtan, J. A., Nag, S., & Kruusmaa, M. (2020). Underwater bioinspired sensing: New opportunities to improve environmental monitoring. *IEEE Instrumentation and Measurement Magazine*, *23*(2), 30–36. https://doi.org/10.1109/MIM.2020.9062685
23. Yang, S., Liu, P., & Lim, T. H. (2024). IoT-Based Underwater Robotics for Water Quality Monitoring in Aquaculture: A Survey. *Lecture Notes in Networks and Systems*, *1132 LNNS*, 32–42. https://doi.org/10.1007/978-3-031-70684-4_3
24. Yu, X., Dickey, T., Bellingham, J., Manov, D., & Streitlien, K. (2002). The application of autonomous underwater vehicles for interdisciplinary measurements in Massachusetts and Cape Cod Bays. *Continental Shelf Research*, *22*(15), 2225–2245. https://doi.org/10.1016/S0278-4343(02)00070-5

Note: All the figures and tables in this chapter were made by the authors.

Advances in Mechanical Engineering and Materials Sciences – Dr. Vinay K. B et al. (eds)
© 2026 Taylor & Francis Group, London, ISBN 9-781-041-20970-6

5

Mixed Reality Rehabilitation System for Shoulder Strenthening and Flexibility

Vinayaka G. P.[1]

Assistant Professor,
Department of Mechanical Engineering,
Vidyavardhaka College of Engineering,
Mysuru, Karnataka, India

Reanne Kathleen Pereira[2],
Shivaraj S.[3], **Sagar B.**[4]

UG Students,
Department of Mechanical Engineering,
Vidyavardhaka College of Engineering,
Mysuru, Karnataka, India

Abstract: The project on Mixed Reality (MR) rehabilitation system is focused on key exercises for shoulder and upper arm recovery: Standing Row, External Rotation with Arm Abducted at 90°, and Elbow Flexion. Targeting muscles such as the trapezius, infraspinatus, teres minor, and biceps, these exercises are essential for restoring strength, flexibility, and stability. Designed for patients recovering from injuries or surgeries, the MR system provides interactive guidance and real-time feedback, ensuring correct form and effective muscle engagement. This precise, tech-enhanced approach helps patients rehabilitate more efficiently and reduces re-injury risk.

Keywords: Physiotherapy, Mixed reality, Virtual reality, Shoulder rehabilitation

1. INTRODUCTION

The Mixed Reality Rehabilitation (MR) System for Shoulder Rehabilitation project aims to transform the traditional approach to post-surgical and injury-related shoulder recovery by integrating immersive technology into physical therapy. Shoulder rehabilitation can often be challenging for patients due to the complexity of exercises and the need for precise form, which, if performed incorrectly, can lead to further discomfort or injury. This engineering project addresses these challenges by developing a MR system that provides detailed, interactive guidance for shoulder strengthening and flexibility enhancing accuracy, engagement, and recovery. The system combines real-time visual feedback and precise instructions, allowing patients to perform essential rehabilitation exercises with greater confidence and adherence to therapeutic protocols. By delivering an immersive, user-friendly experience, the MR rehabilitation system has the potential to significantly improve patient outcomes and revolutionize physical therapy practices.

2. LITERATURE REVIEW

Over the past years, there has been increased interest in applying Virtual Reality (VR) during orthopedic rehabilitation in clinic and home settings. It provides enhanced motivation, positive results of training, and instant feedback regarding the movement execution. VR simulates experience with the aid of certain equipment and software and therefore provides a virtual environment for its users. VR equipment is already accessible, and there

[1]vinay91gp@vvce.ac.in, [2]reannepereira6@gmail.com, [3]shivarajshivanna128@gmail.com, [4]sagarsamrat04@gmail.com

10.1201/9781003725053-5

is much potential for enhancing long-term engagement in rehabilitation. Previous studies have stressed the necessity of high-quality immersive VR devices in therapeutic interventions and is evident from the works of Nam, Jihun, et al. [6] VR technologies have greatly enhanced the accuracy about shoulder arthroplasty implant alignment. Teleconferencing capabilities enable group work and unveil telemedicine possibilities in surgical enterprises. The barriers against its universal usage are technology, cost and scarce clinical studies. Existing studies could enhance the applications of reality technologies in orthopaedic surgery as suggested by Sadigale, et al. [9]

Augmented reality (AR) has much potential in the healthcare sector, particularly for rehabilitation. Nine AR-based shoulder rehabilitation systems were identified in a review, with most employing vision-based techniques and wearable marker tracking. Head Mounted Displays (HMD) aren't considered yet owing to technical constraints, but with developments in HMD technology, user experience may be improved. Most of the systems are intended for use at home with different levels of therapist interaction. Clinical trials indicate that AR systems are more usable and motivating compared to conventional approaches. Yet, more clinical assessments are required to ascertain the full possibility of AR in shoulder rehabilitation as identified by Viglialoro, Rosanna Maria, et al. [11] Difficulties in rehabilitation can be minimized through utilization of VR technology to simulate daily environments. This will assist in repeating exercises, assessment, and motivation, which are among the crucial components to be applied in rehabilitation. Strengths of VR include performance feedback, tailor-making, and interactive participation. The reviews previously presented confirmed the effectiveness of VR in several rehabilitation applications. The present inadequacies of earlier reviews mandate the requirement for a concentrated examination of applications of VR in exercise therapy. The review by Carnevale, Arianna, et al. [2] seeks to determine if it is feasible for VR to have a valuable impact on rehabilitation outcomes. The program involves four levels, each progressively more challenging in relation to range of motion (ROM). Each level involves certain elevation and rotation exercises to promote shoulder rehabilitation. The exercises involve interactive activities, including retrieving virtual objects at different elevations. The overall aim of the intervention is to simulate everyday activities to maximize motivation for rehabilitation. The protocol for every patient included OQ2 calibration, practice with the game, and gradually increasing levels of exercise. Further study is necessary to evaluate the clinical utility of VR technology in rehabilitation of shoulders. With potential benefits of VR in orthopedic rehabilitation established through several pieces of supporting evidence, its effectiveness is promising as discussed by. Moreira, Moisés, et al. [5]

Compliance is vital to the success of exercise interventions in physiotherapy, particularly for home-based shoulder therapy, where existing compliance monitoring methods are poor. This research investigates the application of robot-assisted techniques and participatory video games, in the form of VR, to improve compliance. VR is increasingly being advocated as a replacement for conventional physiotherapy, with randomized controlled trials examining its efficacy. The setup links a KUKA robot arm to the Unity game engine through an API, with VR being done with the Oculus Quest 2. Two games, "Standing Row" (which is for children) and "External Rotation" (for senior citizens), are designed to make users interact using pain management and encouragement. The "Standing Row" game simulates resistance bands and monitors effort, giving feedback from a guiding character, whereas "External Rotation" is arm rotation without resistance bands and gives visual feedback to the players. Volunteers were solicited from the Polytechnic University of Cávado and Ave, omitting those with a tendency to cyber-sickness. The experiment employed the Technology Acceptance Model (TAM) to evaluate determinants of technology adoption as presented by Viglialoro, Rosanna M., et al. [11, 12]

The article by Bateni, Hamid, et al [1] discusses the application of AR for shoulder rehabilitation of stroke patients. It highlights how the conventional approach can be boring, but a AR-based therapy can enhance motivation and shoulder range of motion and muscle strength. The systems are mainly webcam-based and low-cost and include biofeedback in the form of surface electromyography (sEMG) to monitor muscle improvement. In contrast to cumbersome exoskeleton devices, AR combines virtual and real worlds, providing an immersive experience. The system employed sEMG to record muscle activity and arm position with a marker. Patients performed exercises, guiding objects into forms, with visual and auditory feedback. The system provides muscle performance measures, monitoring recovery via real-time electromyographic signals.

AR and VR medical rehabilitation vary in technology, equipment, and focus on patients. Most research generally indicates that AR and VR enhance patient motivation and engagement but that therapeutic gains depend on the illness, type of treatment, and intervention length. No improvement is reported in some studies, whereas marked improvements are observed in others. Most AR/VR interventions are adjuncts to standard therapy, not direct treatments. Rehabilitation treatment, like recreational, speech-language, and vocational therapy, is designed to enhance emotional well-being, communication, and employment. Rehabilitation takes place in diverse settings, often with the presence of family and friends. Outcomes of rehabilitation vary according to individual needs and the severity of health status. Advances in VR are especially useful for enhancing dexterity and coordination among brain-injured patients, and the integration of VR with conventional therapies can enhance upper limb movement even further. Tools such as Sleeve AR provide explicit guidance and feedback to facilitate rehabilitation as presented by Powell, Michael O., et al [8].

The classic evaluation of shoulder function has been strength, pain, and Range of Motion (ROM). Postoperative pain has a huge effect on quality of life and daily activities and is therefore very important for rehabilitation. Rehabilitation at home has its limitations, including the inability to confirm the correct execution of movements. VR technology for orthopedic rehabilitation has become popular due to its potential for increased motivation, decreased training time, and movement feedback and has several benefits, such as minimizing time and cost burdens, automating rehabilitation procedures, and minimizing the involvement of therapists. Patients can simulate movements by watching a 3D avatar and get feedback using VR. The use of MR enables patients to see their body movements, further improving rehabilitation. The inclusion of Inertial Measurement Unit (IMU) sensors to capture joint angles enhances feedback and results, without the need for handheld devices. Gamification improves patient participation, with rewards and challenges boosting participation. VR rehabilitation breaks the limitations of conventional methods, minimizes clinic visits, and may be more cost-efficient. Future research should aim to assess the clinical efficacy of VR technologies in telerehabilitation versus conventional methods as highlighted in the work of Moreira, Moisés, et al [5].

Physical exercise has numerous health advantages, but sedentary lifestyles are common, with the preference for playing video games over conventional exercise. VR has the potential to promote physical activity and health through engaging, interactive simulations. VR is gaining a number of applications in kinesiology and public health, with the potential to improve rehabilitation by encouraging active participant engagement with immediate feedback. Prior reviews have pointed out the effectiveness of VR in the recovery of stroke and Parkinson's disease, though less is reported on wider vulnerable groups. This paper by Longo, Umile Giuseppe, et al [4] consolidates the current literature on VR exercise across various groups. VR has been found to enhance balance, especially in the older population and those with cerebral palsy (CP), although results for upper limb function in stroke survivors are conflicting. Some find VR to positively affect decreasing fatigue and depressive symptoms, enhancing the overall quality of life. Conversely, conflicting outcomes have been reported in CP and stroke patient studies. The analysis of 15 trials found that more than 60% had beneficial health effects, although inconsistencies arose because of different study protocols and participant demographics. Longer interventions and more heterogeneous groups of participants need to be researched in the future.

Advances in VR and AR technologies are improving orthopedic rehabilitation exercise effectiveness and personalization, especially in psychological therapy and cycles of rehabilitation. Although studies reveal that VR elevates patient adherence to rehabilitation, there is less evidence available in the literature of VR usage in shoulder rehabilitation within clinical settings. Even with the potential offered by VR in simulating immersion-like conditions with realism, much more work must be undertaken to prove usability and overall efficiency. A study protocol with OQ2 VR calibration and training was employed, and questionnaires measuring the acceptability, usability, and appropriateness of the device were completed by participants. Findings indicated that physiotherapists rated OQ2 as easy to use, comfortable, and fun, although some had concerns regarding the wellness benefits and patient capability to operate the system independently. All participants concurred that VR games were easy to use and best for intermediate or terminal stages of rehabilitation, with a preference to incorporate an avatar to facilitate exercises and enhance accuracy. In summary, clinical studies in the future are needed to assess the clinical efficacy of VR in shoulder rehabilitation, through VR having beneficial use in orthopedic rehabilitation is evident from the article of Peláez-Vélez, Francisco-Javier, et al [7].

VR is progressing hugely in the medical sector, providing interactive learning experiences that complement conventional modes of learning. By combining motion tracking and imaging and has transformed medical training and therapy to the advantage of both trainers and trainees. Makransky's Cognitive Affective Model of Immersive Learning (CAMIL) recognizes crucial factors like motivation, self-efficacy, and cognitive load that determine VR-based learning environments. A five-week study evaluated the effectiveness of VR with Case-Based Learning (CBL) in enhancing anatomy and rehabilitation content knowledge. The outcomes indicated that test scores were enhanced using VR and CBL as compared to conventional methods, whereby Group A performed better than Group D, affirming the use of VR in education. The outcomes of rehabilitation therapy, though, varied between groups. Student questionnaires showed substantial differences in satisfaction and engagement, although no significant differences were observed in course structure or consistency of feedback. In summary, VR increases anatomical understanding and provides learning advantages, but its effectiveness in rehabilitation therapy is mixed. The integration of VR and CBL was found to be better than conventional methods, with positive effects on instructional quality and student satisfaction as per the works of Peláez-Vélez, Francisco-Javier, et al [7].

3. METHODOLOGY

With extensive review of AR, VR and mixed reality, our work was focused on designing and developing a mixed reality rehabilitation system for shoulder strengthening and flexibility. The project methodology, as portrayed in

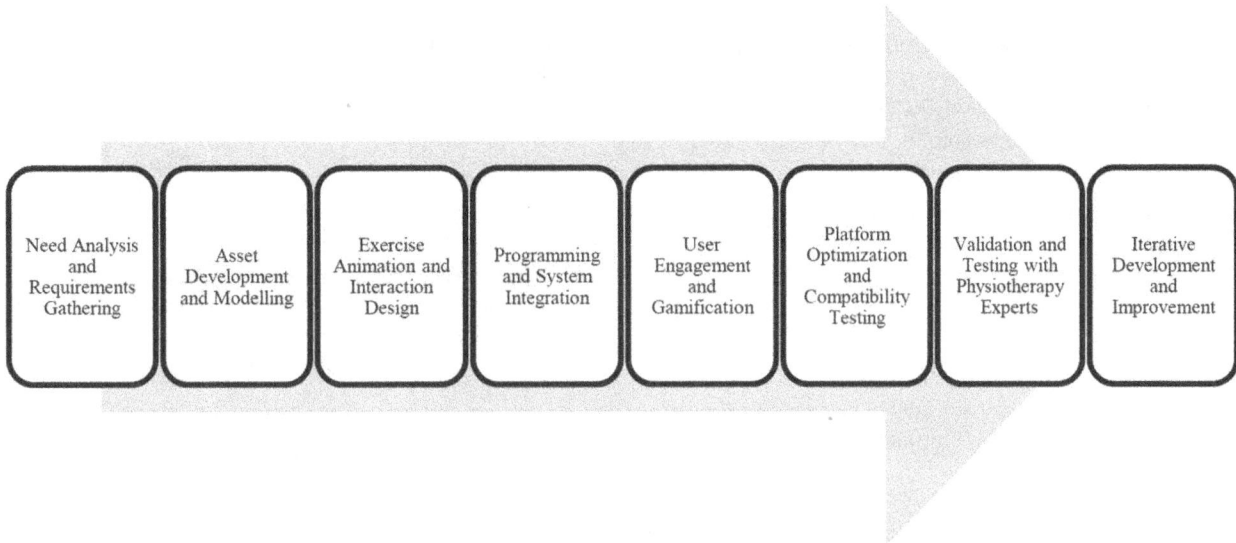

Fig. 5.1 Flowchart

the flowchart, follows a structured and iterative attempt to ensure the development of a robust and user-centric rehabilitation system. The project methodology, as portrayed in the flowchart Fig. 5.1.

1. Need Analysis and Requirements Gathering: it begins with identifying user needs and defining project requirements through stakeholder consultations and research. This ensures alignment with user expectations and project goals.

2. Asset Development and Modelling: assets such as 3D models, animations, and other visual environments are created to form the foundation of the system.

3. Exercise Animation and Interaction Design: This segment emphasizes designing an interactive and engaging animations for exercises, ensuring serviceability and accessibility.

4. Programming and System Integration: The developed assets and designs are integrated into the system through programming, creating a functional and cohesive platform.

5. User Engagement and Gamification: Features to enhance user engagement, such as gamification elements, are incorporated to improve motivation and retention.

6. Platform Optimization and Compatibility Testing: The platform undergoes rigorous testing to ensure compatibility across devices and optimal performance.

7. Validation and Testing with Physiotherapy Experts: The system is validated by physiotherapy experts to ensure its effectiveness and accuracy in meeting therapeutic goals.

8. Iterative Development and Improvement: Feedback from testing is used to iteratively refine and improve the system, ensuring continuous enhancement and user satisfaction.

4. CONCLUSION

In conclusion, the Mixed Reality Rehabilitation System for shoulder rehabilitation offers a transformative solution by seamlessly integrating the cutting-edge capabilities of mixed reality (MR) technology with the established, evidence-based practices of physical therapy. By immersing patients in an interactive, real-time environment, the system revolutionizes rehabilitation by improving the precision and safety of exercises while empowering patients to take an active role in their recovery journey.

The new system offers constant feedback through an intuitive interface and ensures the correct and optimal performance of exercises. Its engaging and immersive environment does not only motivate patients but also ensures compliance with prescribed therapy protocols. It is therefore very effective to be used both in remote settings and in-clinic environments; hence it is a versatile tool for rehabilitation.

The system has been designed to address the need for scalable, accessible, and engaging therapy options. The system can cater to a diverse range of patients in different contexts. For the recovering at home, it provides guided, personalized therapy in a comfortable environment, with the aim of minimizing frequent visits to clinics. This serves to be a strong tool of supervised rehabilitation in clinical settings, where it helps in tailoring the program for specific needs of the patient but monitoring progress in real-time.

A giant leap in the physical therapy rehabilitation system is achieved through this mixed reality rehabilitation system with personalization, adaptability, and advanced technology. The rehabilitation experience is improved upon, but most importantly, it will ensure that no patient leaves without proper care and support. The standard of shoulder rehabilitation will henceforth be defined in this very regard-to make rehabilitation more engaging, efficient, and accessible for everyone. The huge potential of immersive technology in the healthcare field comes through with this project - namely, how MR can transform traditional rehabilitation practices.

It is a whole new step in the development of traditional physical therapy, offering faster recovery times, more engaged patients, and better long-term outcomes. Its application in shoulder rehabilitation, while particularly promising, also hints at broader possibilities for the integration of mixed reality in other areas of medical care, offering new hope for patients seeking effective, accessible, and innovative solutions for their rehabilitation needs.

REFERENCES

1. Bateni, Hamid, et al. "Use of virtual reality in physical therapy as an intervention and diagnostic tool." *Rehabilitation Research and Practice* 2024.1 (2024): 1122286.
2. Carnevale, Arianna, et al. "Performance Evaluation of an Immersive Virtual Reality Application for Rehabilitation after Arthroscopic Rotator Cuff Repair." *Bioengineering* 10.11 (2023): 1305.
3. Dejaco, Beate, et al. "Experiences of physiotherapists considering virtual reality for shoulder rehabilitation: A focus group study." *Digital Health* 10 (2024): 20552076241234738.
4. Longo, Umile Giuseppe, et al. "Immersive virtual reality for shoulder rehabilitation: evaluation of a physical therapy program executed with oculus quest 2." *BMC Musculoskeletal Disorders* 24.1 (2023): 859.
5. Moreira, Moisés, et al. "A Virtual Reality Game-Based Approach for Shoulder Rehabilitation." *Multimodal Technologies and Interaction* 8.10 (2024): 86.
6. Nam, Jihun, et al. "The Application of Virtual Reality in Shoulder Surgery Rehabilitation." *Cureus* 16.4 (2024).
7. Peláez-Vélez, Francisco-Javier, et al. "Use of virtual reality and videogames in the physiotherapy treatment of stroke patients: a pilot randomized controlled trial." *International journal of environmental research and public health* 20.6 (2023): 4747.
8. Powell, Michael O., et al. "Predictive shoulder kinematics of rehabilitation exercises through immersive virtual reality." *IEEE Access* 10 (2022): 25621-25632.
9. Sadigale, Omkar, Kerstin Schneider, and Mohy E. Taha. "Mixed reality and augmented reality in shoulder arthroplasty: a literature review." *Medical Research Archives* 10.9 (2022).
10. Tokgöz, Pinar, et al. "Virtual reality in the rehabilitation of patients with injuries and diseases of upper extremities." *Healthcare*. Vol. 10. No. 6. MDPI, 2022.
11. Viglialoro, Rosanna M., et al. "A projected AR serious game for shoulder rehabilitation using hand-finger tracking and performance metrics: A preliminary study on healthy subjects." *Electronics* 12.11 (2023): 2516.
12. Viglialoro, Rosanna Maria, et al. "Review of the augmented reality systems for shoulder rehabilitation." *Information* 10.5 (2019): 154.

Note: All the figures in this chapter were made by the authors.

Advances in Mechanical Engineering and Materials Sciences – Dr. Vinay K. B et al. (eds)
© 2026 Taylor & Francis Group, London, ISBN 9-781-041-20970-6

6

Investigation of Tensile Properties of 3D-Printed Bamboo Powder Reinforced Polylactic Acid Composites using the Taguchi Method

Shivashankar R.*,
G. V. Naveen Prakash, Vinay K. B.
Department of Mechanical Engineering,
Vidyavardhaka College of Engineering,
Mysuru, India

Prashanth M. V.
Department of Information Science Engineering,
Vidyavardhaka College of Engineering,
Mysuru, VTU, India

Ravish Gowda B. V.
Department of Mechanical Engineering,
Vidyavardhaka College of Engineering,
Mysuru, India

R. K. Radhakrishna
Robotics and Automation,
RajaRajeswari College of Engineering, VTU,
Bengaluru

Abstract: Tensile properties of PLA reinforced with bamboo powder were investigated using Taguchi method, and tensile properties were optimized by varying process parameters. Printing Speed (PS), Printing Temperature (PT) and Layer Height (LH) are the processes parameters that was considered to determine the optimal settings that effects the tensile strength. Tensile properties were assessed according to the standardized testing procedure, so results were analyzed using Analysis of Variance- (ANOVA). It was found that LH, PT and PS significantly influenced the Ultimate Tensile Strength (UTS) of bamboo PLA composite filament. Therefore, the composite filament exhibited 27.61 MPa UTS at the optimal parameter settings. The value was 55% lower than pure PLA (60 MPa) thus indicating a significant difference in tensile strength.

Keywords: 3D printing, PLA, Bamboo-PLA, Tensile strength, Taguchi & SEM, S/N (signal-to-noise)

1. INTRODUCTION

Large molecules known as polymers are composed of recurring structural components called monomers that are joined by covalent bonds to form lengthy chains. These macromolecules are crucial in a variety of industries, including plastics, textiles, and the biomedical sector, due to their special qualities, which include flexibility, durability,

---
*Corresponding author: shivashankar.r@vvce.ac.in

10.1201/9781003725053-6

and chemical resistance. Polymerization is the process of creating polymers from monomers. Different mechanisms underlie this chemical reaction, which can be broadly categorized as addition polymerization and condensation polymerization. Chain-growth polymerization, another name for addition polymerization, is the process of continuously adding monomers without producing byproducts. Polyethylene, polypropylene, and polystyrene are typical examples. Usually, cationic, anionic, or free radical processes lead to this kind of polymerization. Step-growth polymerization, also known as condensation polymerization, is a process in which monomers react with the removal of tiny molecules such as methanol or water. Bakelite, polyester, and nylon are a few examples.

The term "additive manufacturing" refers to the process of using additive materials and 3D model data to create products; it includes all additive technologies, systems, applications, and processes. The global additive manufacturing market is made up of 3D printers, printing materials, and service providers. Rapid prototyping and rapid manufacturing are both available in the market for all applications [7]. Stratasys created the layer-wise 3D printing method known as Fused Deposition Modelling (FDM) to create complex geometrical features [4]. Because of its low cost and versatility, FDM has become one of the most popular additive manufacturing technologies. Products are made from a various materials like metals, plastics, ceramics, and in combinations, have been produced using this method for use in Mold design, automotive, medical, and aerospace industries [3]. An extruder liquefies the material forming a filament that is deposited layer by layer in a semi-solid state to form the desired part, and the process requires a build platform, print bed, extruder head, and a spool of build material. The part geometry is converted into an STL (Standard Tessellation Language or Stereo Lithography) format file using a geometry generation program, so the file is then imported into a software that slices the model into very thin, two-dimensional layers. 2D layer data is used to create a tool path for the print, therefore the extruder head,

which is controlled by a three-axis system, moves along the X-Y plane and extrudes the first layer of the material. The head then moves in the Z-direction by a distance equal to the thickness of the layer and the next layer is deposited, thus the new layer bonds with the layer below it to form a sound joint. The process is then repeated layer by layer until the entire object is formed, and after printing, the support structures are manually removed [10].

The quality of printed parts is determined by a few underlying physical phenomena such as LH, PS and PT. Bonding phenomena and bond quality have a major impact on part integrity and properties. Surface interaction, neck growth, and molecular diffusion are all involved in the creation of a bond between two layers [6]. Bonding does occur between the layers as well as between the wires of the same layer, and Gurrala et al. have shown that bonding reduces the solidification time of the material, resulting in incomplete neck growth and coalescence, resulting in the formation of numerous voids between the layers. [9].

The performance and micro structural properties of the bamboo-filled PLA composite will undergo evaluation through mechanical testing and SEM analysis on a range of test specimens fabricated using diverse manufacturing settings. The main objective of the work is to investigate the mechanical characteristics of 3D printed wood reinforced PLA composite filament.

2. METHODOLOGY

This work aims to research the mechanical strength of PLA+ Bamboo fiber- reinforced composites filaments used in 3D Printing using FDM process through an experimental study. The study follows a systematic flow, as illustrated in Fig. 6.1 [11].

Commencing with a comprehensive literature review establish the current knowledge in the field. Subsequently, a theoretical framework is formulated to guide the further investigation, leading to the design and fabrication of specimens using ASTM standards. The experiments are conducted utilizing 3D printing technology to create the

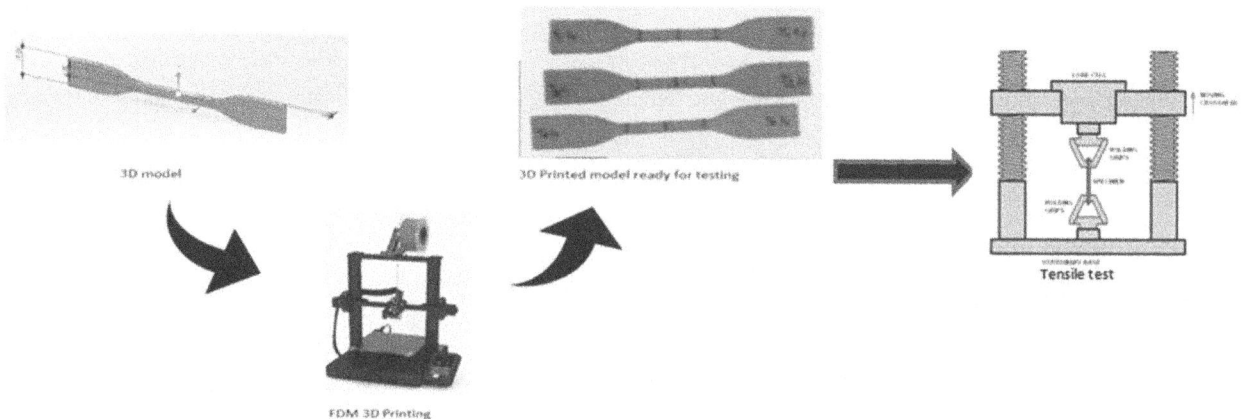

Fig. 6.1 Schematic of 3D printing the tensile test specimen and testing

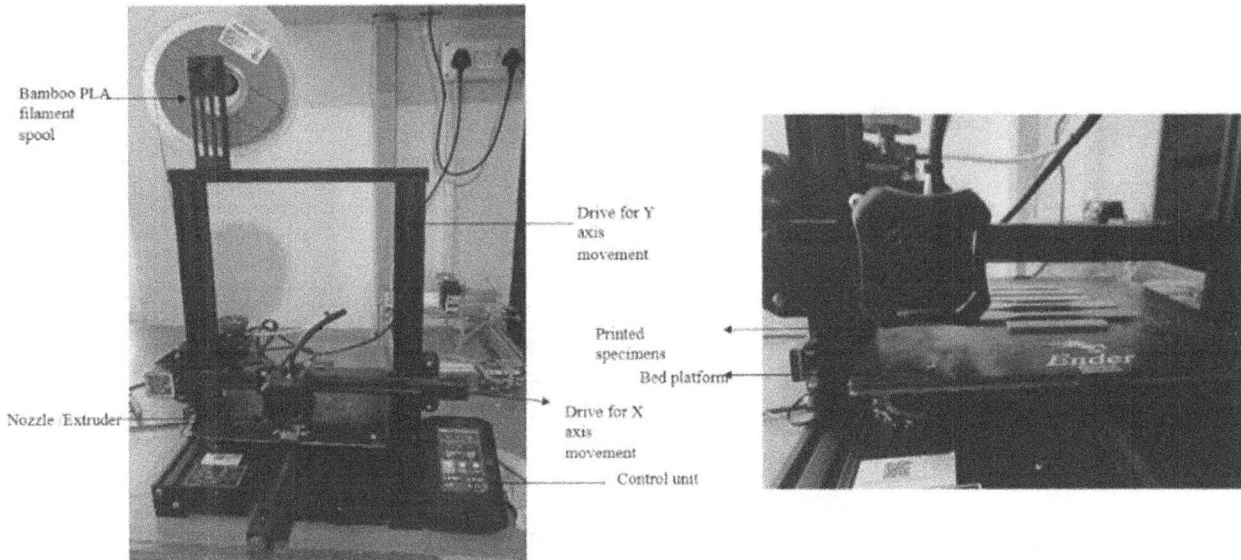

Fig. 6.2 3D printing setup and specimens printing

composite samples, followed by the design and execution of tests to analyse the material's performance. The study's ultimate objective is to draw meaningful conclusions regarding the composite's tensile property and its potential applications. The schematic of 3D printing the tensile test specimen and testing is shown in Fig. 6.1.

3. EXPERIMENTAL METHOD

An Ender 3D printer is used to produce specimen of bamboo+PLA composite using Creality ender V2 3D printer. Using the Creality Slicing Software with adjustable settings and slicing capabilities, 3D printing has been made easier with Creality Printers. The bamboo reinforced PLA filament supplied by Amolen manufactures is used to print the specimens according to the ASTM type 4 D638. Since PLA filament is renewable and biodegradable, it is frequently used in 3D printing. It is vital to take into account the suggested printing temperature range, which is normally between 180°C and 220°C when dealing with PLA filament. The ideal printing speed for PLA filament is normally between 30 and 60 mm/s; however, it may be changed depending on the print quality required. For optimal feeding and extrusion, PLA filament comes in two standard diameters: 1.75 mm and 2.85 mm. It is crucial to make sure the filament diameter meets the printer's requirements. Bamboo powder is incorporated into the polymer matrix to create bamboo PLA filament. Similar to ordinary PLA filament, bamboo PLA filament should be printed at a temperature between 180 to 220 degree Celsius.

3.1 Equipment

The 3D printer used is Creality Ender3 and the printer is of FDM printer. Maximum printing capacity of the printer is 200mm x 220mm x 250mm as shown in Fig. 6.2. UTM: The UTM used is manufactured by Kalpak Instruments &

Controls, and it is connected to computer to collect and analyse the required data, as shown in Fig. 6.3.

Fig. 6.3 Universal testing machine

3.2 Preparations of Specimens

According to ASTM D638-type 4 standards, tensile specimens were created as shown in Fig. 6.4 [4]. These models are constructed with Solid Works and converted to the stereolithography (STL) format for 3D printing with the help of Creality Slice program [6], which slices the 3D model according to a set of parameters. A UTM with -100 kN load cell performs the tensile test, at ambient temperature; the test speed is 2 mm/min. The printed samples used in this study show how the mechanical properties of Bamboo PLA filaments are affected by changes in printing parameters (LH,

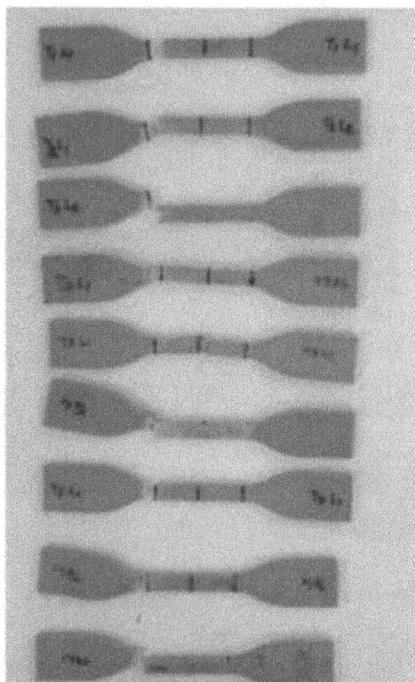

Fig. 6.4 3D-printed samples and tested specimens of bamboo PLA

PT & PS) [4]. The printing parameters are optimized using the Taguchi method which was found using an L9 orthogonal array [10]. The best printing parameter combination for generating increased strength characteristics was found after analyzing the data (John et al., 2023). The results show how layer heights, printing temperatures, and printing speed affect the tensile properties of Bamboo PLA filaments [1]. The LH was varied at three levels: 0.16, 0.20, and 0.24 mm, similar to the PT, which ranged from 190, 200, and 210°C. Additionally, the PS was adjusted to three values: 30, 50, and 70 mm/sec, as shown in Table 6.1. The choice of these values is made on the literature survey [4].

Table 6.1 Printing parameters level setting

Level	Layer Height (mm)	Printing Temperature (°C)	Printing Speed (mm/sec)
L1	0.160	190	30
L2	0.200	200	50
L3	0.240	210	70

3.3 Design of Experiment

The printing parameters have a significant impact to the printed components quality in FDM technology. Three parameters [8] were used in this investigation to test the materials' tensile strength while holding the other parameters constant. Table 6.1 lists the particular parameters along with the values that correspond to them.

L9 orthogonal was created using the Taguchi method, and it is displayed in Table 6.2.

Table 6.2 Taguchi orthogonal array by varying printing parameters

Experiment Run	Layer height (mm)	Printing temperature (°C)	Printing Speed (mm/sec)
R1	0.160	190	30
R2	0.160	200	50
R3	0.160	210	70
R4	0.200	190	50
R5	0.200	200	70
R6	0.200	210	30
R7	0.240	190	70
R8	0.240	200	30
R9	0.240	210	50

By arranging components and their levels, the concept of Taguchi method in the DoE uses an orthogonal array to correct the average and lessen variation. Furthermore, the Taguchi method requires fewer experiments, which lowers experimentation time and cost. On the basis of earlier investigations, the majority of parameters and their respective range were subjected to modifications. The tests were created using Minitab 17 software and Taguchi's OR of fractional factorial designs (L9).

4. RESULTS AND DISCUSSIONS

Utilizing the objective values into the (S/N) ratio, which acts as an index for evaluating quality characteristics, the Taguchi approach is used to develop test settings. According to Taguchi and Phadke (1989), variables may be divided into two types: noise factors and control factors [4]. Noise factors include outside effects like normal temperature and humidity. The Taguchi Method uses the S/N ratio to create an able system that is less subject to noise reasons resulting in optimal quality design with less volatility. The S/N ratio is very helpful for factor weighting, minimizing crossover action, simultaneously reducing processing means and variance, and eventually boosting overall quality. In the context of this work, we aim to optimize the tensile strength (UTS) of Bamboo-PLA components by taking changing LH, PT & PS into account. The "larger is better" feature is used, suggesting that higher strength values acquired from mechanical testing as shown in Table 6.3. are desired [5, 2].

An ANOVA was conducted to evaluate the impact of various control parameters on the specific tensile test of Bamboo PLA composites. The ANOVA results as presented in the Table 6.4, provide insights for improving the (S/N) ratios of the composites. The coefficient of determination (R-squared) for the model was determined to be 79.68%,

Table 6.3   Results of bamboo PLA composites

Experimental Run	C/S area (mm2)	% Elongation	Stress (N/mm2)	Strain	UTS (MPa)	SNRA
RI	16.19	18.85	16.57	0.55	27.61	28.8213
R2	18.36	17.57	16.41	0.084	27.35	28.7391
R3	16.98	18.58	16.245	0.085	27.07	28.6498
R4	17.31	16.12	16.385	0.086	27.30	28.7233
R5	18.02	19.75	15.62	0.094	26.02	28.3061
R6	18.53	23.67	16.37	0.096	27.28	28.7169
R7	16.60	16.05	15.885	0.086	26.47	28.4551
R8	16.76	16.97	15.855	0.090	26.42	28.4387
R9	16.88	15.77	15.21	0.090	25.34	28.0761

Table 6.4   ANOVA for tensile test Bamboo PLA filament

Source	DOF	Seq SS	Adj SS	Adj MS	F-value	P-value	Cont.%
LH	2	0.26	0.26	0.131	2.68	0.271	54.32
PT	2	0.06	0.06	0.032	0.65	0.604	13.25
PS	2	0.05	0.05	0.029	0.6	0.625	12.16
Residual error	2	0.09	0.09	0.04			20.24
Total	8	0.48					100

indicating a significant relationship between the control variables and the specific tensile test rate. Moreover, the adjusted R-squared value, considering the number of predictors in the model, demonstrated a correlation of 19.00%, validating the model's predictability.

The ANOVA findings clearly highlight the significant contribution of the layer height composition to the specific tensile test of the Bamboo PLA composite, attributing to 54.32% of the observed variance. Subsequently, the printing speed factor exhibited a substantial contribution of 12.16%. Additionally, the printing temperature factor contributed 13.25% to the overall variability. Furthermore, the residual inaccuracy of the Bamboo PLA material accounted for 20.24% of the total variance, as illustrated in Table 6.4.

The S/N (signal-to-noise) ratio was employed as a criterion to determine the best values for each factor that would maximize the tensile strength. Through the analysis of the S/N ratios, the ideal set of settings was identified as 0.16 mm of LH, 190°C of PT, and 30 mm/sec of PS. These optimal values are depicted in Fig. 6.5 [1]. The LH, PT, and PS properties of Bamboo PLA filament has a substantial effect on the tensile strength, as can be observed in Fig. 6.5 with a nozzle diameter of 0.4mm, respectively. At the optimum level setting the bamboo PLA composite filament has exhibited the highest UTS of 27.61 MPa which is 55% lesser compared to that of neat PLA which is of 60MPa.

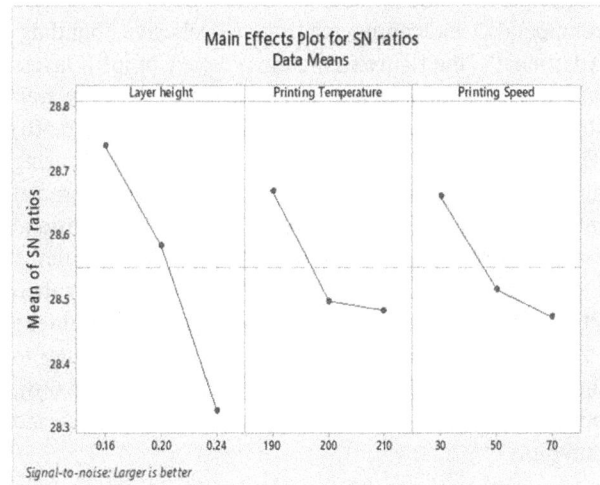

Fig. 6.5   Mean effects plot for SN ratios for bamboo PLA

4.1 SEM Analysis of Fractured Surface

SEM image analysis shows that Bamboo PLA material exhibit brittle fracture behaviour, resulting in rough surface roughness. The tensile strength of Bamboo PLA rises with layer thickness. Air gaps and voids were seen inside the layer thickness in photomicrographs that were taken after the SEM examination of fracture surfaces produced by the tensile testing.

Figures 6.6 provide significant insights into the fractured surfaces of Bamboo PLA composites. In Fig. 6.6(a), the

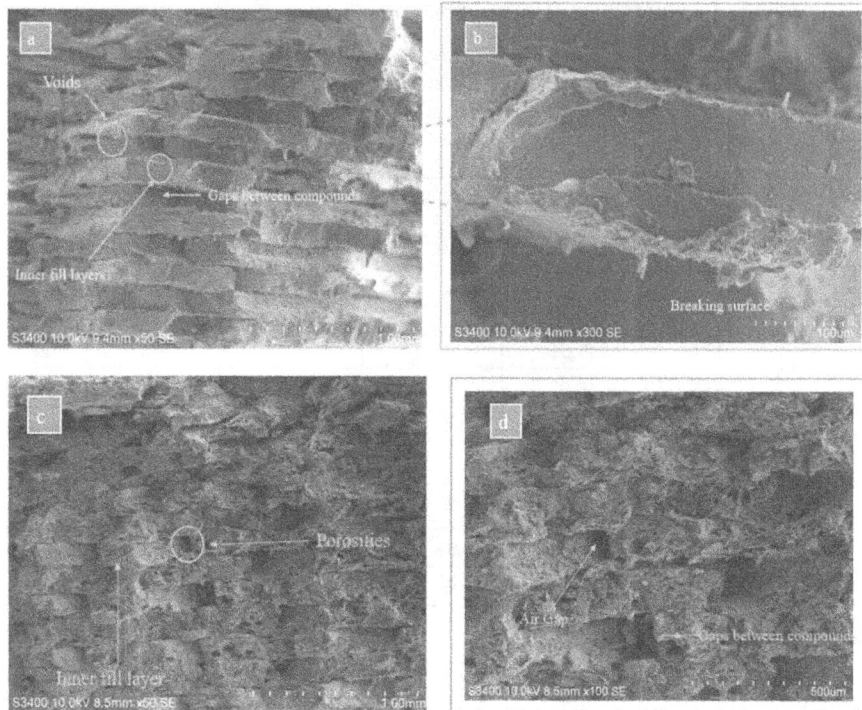

Fig. 6.6 SEM images of fractured surface vamboo PLA composites

composite exhibits evident voids and gaps between the compounds, indicating inadequate cohesive bonding. Additionally, the figure shows the presence of infill layers with a density of 100%, which contributes to the overall strength and integrity of the composite. Figure 6.6(b) depicts the fractured surface of the composite during tensile strength testing. This surface provides crucial information about the composite's ability to withstand external forces and its mechanical properties. Similarly, Fig. 6.6(c) highlights distinctive features of the Bamboo PLA composite. The exposure of bamboo fibers within the PLA matrix is clearly visible. This exposure contributes to the formation of more porosities and air gaps Fig. 6.6(d), potentially impacting the composite's overall strength and durability

Figure 6.6(a) demonstrates the 3D-printed composite's layered structure by displaying inner fill layers, voids, and spaces between the PLA matrix and bamboo powder. These characteristics are common in fused deposition modelling (FDM) 3D printing, where errors may result from inadequate adhesion between the polymer and filler particles and incomplete layer bonding. Voids suggest either poor interlayer fusion or insufficient bamboo powder dispersion in the PLA matrix. The defects can act as stress concentrators and reducing the material's strength and durability. Figure 6.6(b), which focuses on the fracture surface, displays a rough and uneven breaking surface. This suggests a brittle fracture process, which is common in PLA composites. It's possible that the failure was caused by weak interfacial adhesion between PLA and bamboo powder, which allowed cracks to propagate along

the filler-polymer contact. The toughness of the composite is impacted by fiber pullout or separation, which is shown by the rough appearance. Significant microstructural flaws in the 3D-printed bamboo powder-reinforced PLA composite are seen in the SEM images (c and d), which may have an impact on the material's overall performance and mechanical characteristics. The existence of porosities and separate inner fill layers in image (c) suggests that the printed layers have not fully fused, most likely as a result of inadequate dispersion of bamboo powder within the PLA matrix, poor filament bonding, or an inadequate extrusion temperature.

5. CONCLUSION

Bamboo fiber-reinforced PLA 3D printing materials have been investigated, and bamboo PLA test samples were fabricated using FDM 3D printing machine and Taguchi DOE. The strength of bamboo PLA test samples was affected by the printing parameters, thus LH, PT, and PS were identified as the most influential factors for determining the strength of the Bamboo PLA filament. Under the optimal printing parameters, a maximum UTS of 27.61 MPa was achieved on the bamboo PLA composite filament, which is 55% lower than that of pure PLA (60 MPa), because the 20% bamboo fiber particles were identified as the responsible factors for the UTS reduction. Therefore, further investigation and optimization are required to obtain the optimal printing parameters and reinforcement to enhance the strength of 3D-printed bamboo PLA composites, so this study can be used as a guideline for further research aimed at developing composite materials with superior properties for various applications.

REFERENCES

1. Alafaghani, A., & Qattawi, A. (2018). Investigating the effect of fused deposition modeling processing parameters using Taguchi design of experiment method. *Journal of Manufacturing Processes*, *36*, 164–174. https://doi.org/10.1016/j.jmapro.2018.09.025

2. Atakok, G., Kam, M., & Koc, H. B. (2022). Tensile, three-point bending and impact strength of 3D printed parts using PLA and recycled PLA filaments: A statistical investigation. *Journal of Materials Research and Technology*, *18*, 1542–1554. https://doi.org/10.1016/j.jmrt.2022.03.013

3. Bhagia, S., Bornani, K., Agrawal, R., Satlewal, A., Ďurkovič, J., Lagaňa, R., Bhagia, M., Yoo, C. G., Zhao, X., Kunc, V., Pu, Y., Ozcan, S., & Ragauskas, A. J. (2021). Critical review of FDM 3D printing of PLA biocomposites filled with biomass resources, characterization, biodegradability, upcycling and opportunities for biorefineries. *Applied Materials Today*, *24*, 101078. https://doi.org/10.1016/j.apmt.2021.101078

4. Hikmat, M., Rostam, S., & Ahmed, Y. M. (2021). Investigation of tensile property-based Taguchi method of PLA parts fabricated by FDM 3D printing technology. *Results in Engineering*, *11*, 100264. https://doi.org/10.1016/j.rineng.2021.100264

5. Hiwa, B., Ahmed, Y. M., & Rostam, S. (2023). Evaluation of tensile properties of Meriz fiber reinforced epoxy composites using Taguchi method. *Results in Engineering*, *18*, 101037. https://doi.org/10.1016/j.rineng.2023.101037

6. John, J., Devjani, D., Ali, S., Abdallah, S., & Pervaiz, S. (2023). Optimization of 3D printed polylactic acid structures with different infill patterns using Taguchi-grey relational analysis. *Advanced Industrial and Engineering Polymer Research*, *6*(1), 62–78. https://doi.org/10.1016/j.aiepr.2022.06.002

7. Landes, S., & Letcher, T. (2020). Mechanical Strength of Bamboo Filled PLA Composite Material in Fused Filament Fabrication. *Journal of Composites Science*, *4*(4), 159. https://doi.org/10.3390/jcs4040159

8. Menderes Kama, A. İ. (2012). Taguchi Optimization of Fused Deposition Modeling Process Parameters on Mechanical Characteristics of PLA+ Filament Material. *Https://Www.Academia.Edu/51892150*.

9. Müller, M., Jirků, P., Šleger, V., Mishra, R. K., Hromasová, M., & Novotný, J. (2022). Effect of Infill Density in FDM 3D Printing on Low-Cycle Stress of Bamboo-Filled PLA-Based Material. *Polymers*, *14*(22), 4930. https://doi.org/10.3390/polym14224930

10. Svatík, J., Lepcio, P., Ondreáš, F., Zárybnická, K., Zbončák, M., Menčík, P., & Jančář, J. (2021). PLA toughening via bamboo-inspired 3D printed structural design. *Polymer Testing*, *104*, 107405. https://doi.org/10.1016/j.polymertesting.2021.107405

11. van Wassenhove, R., De Laet, L., & Vassilopoulos, A. P. (2021). A 3D printed bio-composite removable connection system for bamboo spatial structures. *Composite Structures*, *269*, 114047. https://doi.org/10.1016/j.compstruct.2021.114047

Note: All the figures and tables in this chapter were made by the authors.

Advances in Mechanical Engineering and Materials Sciences – Dr. Vinay K. B et al. (eds)
© 2026 Taylor & Francis Group, London, ISBN 9-781-041-20970-6

7 Development of a Solar-Powered Multipurpose Electric Cart for Small-Scale Vegetable Vendors

Raghu N.[1], Vinod B.[2]
Dept of Mechanical Engineering,
Vidyavardhaka College of Engineering,
Mysuru, Karnataka, India

**Bharath H. S.[3],
Stephen B. Joseph[4]**
Dept of Mechanical Engineering,
Siddaganga Institute of Technology,
Tumakuru, Karnataka, India

Abstract: The solar-functioned cycle cart is used by small-scale vegetable vendors for transportation. It is a rational and innovative solution designed to address the challenges faced by small time vendors and local farmers. This project integrates renewable energy technology by transforming a conventional cycle cart to an eco-friendly alternative approach carried out to slowly act as a substitute of manual methods or fossil fuel-powered carts. This system comprises of a photovoltaic panel mounted on the roof of the cart, this panel is used to harness the solar energy to charge a high-capacity battery. The stored energy from the solar panel attached on top of the roof powers an electric motor that aids the rider in propelling the cart, thus reducing the use of physical effort in vegetable transportation. The design incorporates a lightweight and durable frame to enhance energy efficiency and amplify the cart's load-carrying capacity. Integration of a user-friendly interface enables the operator to keep a track on the battery levels and the performance of the system. In addition to those advancements, some of the safety features include efficient braking system and the use of reflective materials to enhance the visibility mainly in the areas of low illumination, thus facilitating a much safer navigation. This innovative project not only works by exploiting the power of the sun but also helps in reducing the harmful emissions which otherwise encounter while using a traditional gasoline powered vehicle. By drawing the power from the sun, this project aims to improve the livelihood of the farmers and the small-time vendors.

Keywords: Renewable energy, Battery, Electric motor, Cart, Reflective materials, etc.

1. INTRODUCTION

The small-scale vegetable merchants can now use the solar-powered cycle cart. It serves as a modern solution that helps to ensure these systems work smoothly whether they are connected to a main network (grid) or not. This idea not only helps in reducing operating costs but also makes it convenient and attainable to a wide area of farmers, containing those in rural isolated areas [8]. The design of the cart is sustainable and adjustable according to user's requirement. It eliminates the use of chemicals for cooling thus minimizes environmental impact, ensuring that the store up fruits and root vegetables preserve their normal freshness without the loss of any nutritionary value [13]. This vehicle works with green and eco-friendly principles in addition to addressing the issue of post-harvest losses. Because agricultural goods spoils easily, a large portion of it frequently cannot reach the market, which drives up costs [10, 12, 14].

[1]raghu.n@vvce.ac.in, [2]vinod@vvce.ac.in, [3]bharathhs@sit.ac.in, [4]stephenbjoseph068@gmail.com

10.1201/9781003725053-7

To minimize waste and increase product availability, individual farmers, cooperatives, and dealers are encouraged to extract products during the post-harvest phase. This method not only manages financial concerns but impacts to sustainable cultivated practices, offering an impactful solution to the challenges like the vulnerability of agricultural products. The storage capacity of the refrigerator box is up to 200 KG and depending on the crop, it can increase the shelf life of vegetables by 10 to 20 days. The underlying reasons of waste in the agricultural supply chain are immediately addressed by this invention. What makes this cart different from the normal traditional cart is its effectiveness with minimum resource inputs. Only a litter of water is required to run the refrigeration unit for 24hrs with a 20watt power supply.

2. OBJECTIVES OF THE PROJECT

The leading goal of the project

- Design and manufacturing process of solar functioned electric cart
- Design construction of self-adjustable structure and fabrication of the electric cart.
- Design of refrigeration box, and comparing it with and without vegetables

2.1 Hardware Description

Motor

In the case of a physically challenged person, the motor specifications will depend on a variety of factors, like the size and weight of the vehicle. The terrain on which the vehicle would run along with the maximum speed required also plays a major role in the smooth operation. Power output of DC motor is 250w and 13.6 amps. The top speed of the vehicle is around 15-20 Kmph.

Chain Drive

Before the design process of the chain drive, various factors are taken into account. The factors are as follows, size of vehicle, weight of vehicle, type of terrain it is driven and the maximum speed. All these elements help in designing the chain size, sprocket size, chain tensor, chain guard and the type of chain lubrication to be used.

Ball Bearing Kit

A ball-bearing kit is used in the aid of a physically challenged person. The ball bearings help in reducing friction and in the overall improvement of performance and efficiency of the vehicle. This kit is customizable based on the user requirement and also works well with physically challenged individuals.

Battery

The best choice of battery when it comes to a battery-operated vehicle is the Lithium-ion battery. Because of its extended lifespan, low self-discharge rate, and high energy density, this battery is frequently utilized. These batteries are eco-friendly and light weight compared to lead acid batteries. In this vehicle there are 4 batteries of 12V and 7 amps.

Solar Panel

Polycrystalline solar panels are typically the best choice for a solar-powered vehicle due to their high efficiency and durability. The project requires two solar panels each having a wattage of around 10W and a voltage of 18.5V [1, 6, 9].

Motor Controller

The motor controller should be designed to work with the specific type of motor used in the vehicle. The motor controller is rated at the same wattage as the motor is 250W a voltage of 25V.

Throttler

The controller also has a control interface such as a throttle, with the use of the throttle the vehicle can be accelerated and deaccelerated according to the user requirement.

Peltier Kit

Peltier kit which contains a Peltier module which is sandwiched between two aluminium heat sinks. In the Peltier module, heat generation is high, so it is connected to a larger heat sink which is placed outside, and a colder surface is placed inside the compartment, and above each of the heat sinks fan is mounted.

Solar Charge Controller

A solar charge controller is an fundamental element of a solar-powered vending vehicle for a physically challenged person, as it regulates the charging of the battery system from the solar panels. Using a Maximum Power Point Tracking (MMPT) solar charge controller of specification 30A, 12V/24V for more efficiency. It also has a clear and easy-to-read display that shows the current charging status of the battery systems [5].

3. FABRICATION AND WORKING PRINCIPLE

3.1 Working Principle

The Solar Rechargeable Multipurpose Electric Cart for Small Scale Vegetable Vendors operates on a sustainable and efficient working principle, combining solar energy harvesting with versatile electric propulsion.

The key feature is the incorporation of solar panels on the roof of the cart. These solar panels help in utilizing the power generated from sunshine and transform it into electrical energy through photovoltaic cells. The solar energy that is stored in the rechargeable batteries serves as a clean and renewable source of power for the cart. Using the concept of solar powered vehicle has helped in taking an eco-friendly approach in solving the problem of incurred due to the use of traditional gasoline vehicles [7, 11].

In electric propulsion, the motor is connected to the wheels of the vehicle. The motion of the cart is due to the energy generated from the solar panels which run the motor, this enables smooth and quiet movement. The electric propulsion system eliminates the requirement for conventional fuel, thus reducing operating cost and environmental pollution.

The vehicle is equipped with a versatile design to satisfy the specific needs of small-scale vegetable vendors. It consists of storage compartments, display shelves, and a refrigeration unit to keep the vegetables fresh. The electric cart refines the mobility for vendors, helping them to easily navigate through crowded marketplaces or reach different locations without the use of manual carts which in most of the cases cause physical strain.

To enhance user convenience, the cart incorporates a simple control interface that allows the vendor to manage speed, direction, and other operational features. Additionally, safety features such as brakes and lights contribute to a secure and efficient operation.

3.2 Working Procedure

1. PV Solar panel absorbs radiation from the sun and collects the charge.
2. The charge controller is kept maintaining the flow of charge. This provides a path for charging the battery as well as direct running of the vehicle.
3. Charge flows to the motor which in turn is connected to the throttle which runs the vehicle on acceleration.
4. Even the temperature-controlled compartments will get the supply directly to maintain the specified temperature.
5. When there is a need for a vehicle during the unavailability of the sun the vehicle can be run through a charged battery
6. The opposite of the Seebeck effect is the Peltier effect. The junction of two distinct conductors is where the electrical current flows. This causes one side of the conductor to cool down and the other side to heat up.
7. This flow of electric current through two different conductors causes electrons to move from one material to the other.

3.3 Calculations

- First, let's calculate the power required by the motor to maintain a speed of 15km/h: velocity $= \dfrac{15 \times 1000}{3600}$ $= 4.17$ m/s
- Total weight = 200 kg
- Vehicle weight = 80 kg
- Passenger weight = 50 kg
- Next, let's calculate the volume of the container
- Length of the container = 0.75 m
- Breadth of the container = 0.5 m
- Height of the container = 0.3 m
- $L \times B \times H = 0.75 \times 0.5 \times 0.3 = 0.1125$ m^3
- Time
- Acceleration:
$$a = \frac{v}{t} = \frac{4.17}{10} = 0.41 \text{ m/s}^2$$
- Force for acceleration:
$$F = 200 \times 0.41 = 82 \text{ N}$$
- Torque on the wheels (wheels diameter = 200)
$$M_t = F \times R = 82 \times 200 = 16400 \text{ N mm} = 16.4 \text{ N m}$$
- Speed of wheels in rpm
$$V = \frac{\pi \times d \times n}{60000}$$
$$4.17 = \frac{\pi \times 400 \times n}{60000}$$
$$n = \cong 200 \text{ rpm}$$
- Power:
$$P = \frac{2 \pi n T}{60000} = \frac{2 \times \pi \times 200 \times 16.4}{60000} = 0.34 \text{kW}$$
- Battery:

Total 5 batteries of 12V and 8Ah are used

Total battery power $= 8 \times 12 \times 5 = 480$ W

Assuming battery power Efficiency = 80%

Actual battery power output $= 480 \times 0.8 = 384$ W

using one motor of 500W

Assuming efficiency of motors = 80 %

Motor output $= 0.8 \times 500 = 400$ W
$$= 0.4 \text{kW}$$
- To charge a 48V, 8Amps battery from a 10W solar panel with an output voltage of 37V and need a charge controller to regulate the voltage and current to the battery. The time required to charge the battery will depend on several aspects such as the efficiency of the solar panel, the amount of sunshine accessible, and the charging circuit used
- When two solar panels are connected in series, the voltage adds up while the wattage remains the same [3]. Therefore, if you connect a 10-watt solar panel with an 18.5-volt solar panel in series, the resulting configuration will have a combined voltage of 18.5+ 18.5 = 37 volts. The wattage will remain the same at 10 watts [2, 4].
- To calculate the time required to charge a battery, need to consider the charging efficiency and the available power from the solar panel.

- Available Power = Solar panel output voltage × Solar panel output current

    = 37 V × 10W/ 48W

    = 7.708 W

- To calculate the charging time

    Charging Time = Ah/A

    Here A = Charging current

    Therefore, the charging current should be min 10% of the Ah rating this is because a higher rate may cause the battery acid to boil, and are using an 8Ah battery of quantity 5

    So,

    Ah = 8Ah × 5 = 40Ah

    A = 40 × (10/100) = 4A

    i.e. charge current should be 10% can use 5A

    Using 5A has charging current

    Substituting the values:

    Charging Time = Ah/A = 40/5

    Charging time = 8 hours

3.4 Fabrication

The materials used for the chassis should be lightweight, strong, and corrosion resistant. The material used for chassis fabrication is High Carbon Steel (Fig. 7.1) and then started to build the compartments (Fig. 7.2)

Fig. 7.1   Chassis

Fig. 7.2   Compartments set up

The compartments were covered with thermocol insulation with aluminium foil. Then made the arrangements for Peltier kit connections for the compartments to be temperature controlled. Then finally made a shelter for the rider, then will mount the solar panel and finish all the electrical connections

Fig. 7.3   Final fabricated model

4. TESTED RESULTS OF REFRIGERATION BOX

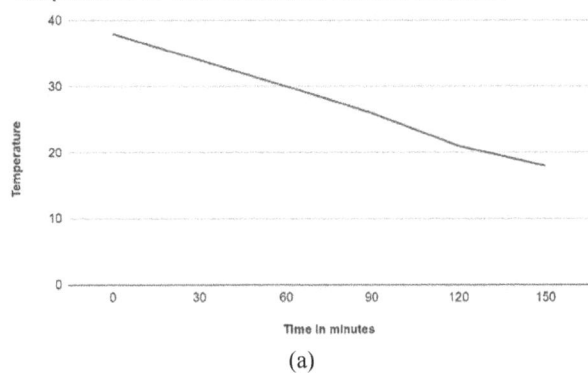

Temperature vs Time in minutes in Filled Container

(a)

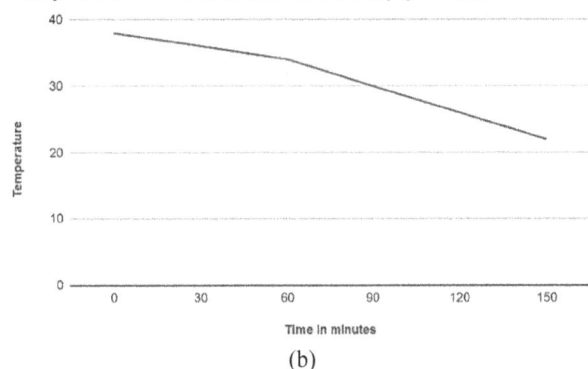

Temperature vs Time in minutes in Empty Container

(b)

Fig. 7.4   (a) Shows temperature v/s time in Hrs for empty container and (b) Shows temperature V/S time in Hrs for filled container

5. CONCLUSION

A new and perishable solution aimed at resolving the issues encountered by nearby farmers and vendors is the solar-powered cycle cart for small-scale produce transportation. By combining renewable energy technology with a classic cycle cart, this invention offers an environmentally benign substitute for manual labour-intensive processes or carts that run on fossil fuels. A high-capacity battery is charged by the system's photovoltaic panel, which is fixed on the cart's roof and uses solar energy. Vegetable transportation requires less physical effort because of the electric motor that is powered by the stored energy and helps the rider propel the cart. To maximize energy economy and increase the cart's load-carrying capacity, a sturdy and lightweight frame is incorporated into the design. An interface gives the operator the ability to keep an eye on system performance and battery levels. The refrigeration system in the cart also helps to increase the shelf life of the vegetables and helps the small-scale vendors sell vegetables without any wastage.

REFERENCES

1. Chinguwa, S., Musora, C., & Mushiri, T. (2018). The design of portable automobile refrigerator powered by exhaust heat using thermoelectric. *Procedia Manufacturing*, *21*, 741–748. https://doi.org/10.1016/j.promfg.2018.02.179

2. COMPRESSOR-LESS REFRIGERATOR BY USING PIER DEVICE. (2023). *International Research Journal of Modernization in Engineering Technology and Science*. https://doi.org/10.56726/irjmets41869

3. Consulting Inc, N. (2006). *A Review of PV Inverter Technology Cost and Performance Projections*. http://www.osti.gov/bridge

4. *EAS Journal of Nutrition and Food Sciences Abbreviated Key Title: EAS J Nutr Food Sci*. (2019). https://doi.org/10.36349/easjnfs.2019.v01i04.001

5. El Tom, O. M. M., Omer, S. A., Taha, A. Z., & Sayigh, A. A. M. (1991). Performance of a photovoltaic solar refrigerator in tropical climate conditions. *Renewable Energy*, *1*(2), 199–205. https://doi.org/10.1016/0960-1481(91)90075-Z

6. Eltawil, M. A., & Samuel, D. V. K. (n.d.). *Performance and Economic Evaluation of Solar Photovoltaic Powered Cooling System for Potato Storage*.

7. Fong, K. F., Chow, T. T., Lee, C. K., Lin, Z., & Chan, L. S. (2010). Comparative study of different solar cooling systems for buildings in subtropical city. *Solar Energy*, *84*(2), 227–244. https://doi.org/10.1016/j.solener.2009.11.002

8. Kumar, S., & Bharj, Dr. R. S. (2021). Experimental Analysis of Solar Assisted Refrigerating Electric Vehicle. *International Journal of Recent Technology and Engineering (IJRTE)*, *9*(5), 305–315. https://doi.org/10.35940/ijrte.E5278.019521

9. Kumar, S., & Bharj, R. S. (2018). Energy consumption of solar hybrid 48V operated mini mobile cold storage. *IOP Conference Series: Materials Science and Engineering*, *455*(1). https://doi.org/10.1088/1757-899X/455/1/012049

10. Lawal, O. M., & Chang, Z. (2021). Development of an effective TE cooler box for food storage. *Case Studies in Thermal Engineering*, *28*. https://doi.org/10.1016/j.csite.2021.101564

11. Modi, A., Chaudhuri, A., Vijay, B., & Mathur, J. (2009). Performance analysis of a solar photovoltaic operated domestic refrigerator. *Applied Energy*, *86*(12), 2583–2591. https://doi.org/10.1016/j.apenergy.2009.04.037

12. Rokde, K., Patle, M., Kalamdar, T., Gulhane, R., & Hiware, R. (2017). Peltier Based Eco-Friendly Smart Refrigerator for Rural Areas. *International Journal of Advanced Research in Computer Science and Software Engineering*, *7*(5), 718–721. https://doi.org/10.23956/ijarcsse/SV7I5/0224

13. Tassou, S. A., De-Lille, G., & Ge, Y. T. (2009). Food transport refrigeration - Approaches to reduce energy consumption and environmental impacts of road transport. *Applied Thermal Engineering*, *29*(8–9), 1467–1477. https://doi.org/10.1016/j.applthermaleng.2008.06.027

14. Thakkar, M., Pravinchandra, T. M., & Patel, J. (2016). *A report on "Peltier (thermoelectric) cooling module" Peltier Cooling Module*. https://doi.org/10.13140/RG.2.1.2923.8805

Note: All the figures in this chapter were made by the authors.

Advances in Mechanical Engineering and Materials Sciences – Dr. Vinay K. B et al. (eds)
© *2026 Taylor & Francis Group, London, ISBN 9-781-041-20970-6*

8

Optimization of TIG Welding Parameters for Joining SS 316l and IS 2062 using Taguchi Method and Grey Relational Analysis

Ameena Kausar[1]
Assistant Professor,
Dept of Mechanical Engineering,
MIT Mysore

Mohamed Khaisar[2]
Professor & HoD,
Dept of Mechanical Engineering,
MIT Mysore

Abstract: The quality of a weld is largely determined by factors such as the joint's mechanical properties, the welding parameters and the final weld characteristics. Many machines experience reduced lifespans or fail to meet performance expectations due to suboptimal welding practices, which can lead to defects such as cracking, high residual stresses, stress concentrations from atomic migration during welding, and stress corrosion. To address these issues, it is crucial to take into account the impact of welding. This paper investigates the mechanical properties of welds in relation to Tungsten Inert Gas (TIG) welding parameters, with a particular focus on Hardness and Ultimate Tensile Strength (UTS). SS 316L exceptional weldability ensures corrosion-resistant joints with minimal post-weld treatment, while IS 2062 structural steel properties offer a cost-effective alternative. Combining these materials allows for a robust and economical welding solution. The welding parameters considered in this study include current (110, 120, 130 A), voltage (40, 50, 60 V), and gas flow rate (8, 9, 10 L/min). For the experimental design of the investigation, an orthogonal array of Taguchi L9 is used. Welding standards are defined by the American Society of Mechanical Engineers (ASME), and sample preparation adheres to ASTM E8 guidelines. The important parameters and ideal levels that improve the overall response were identified using the Taguchi technique and Grey relational analysis.

Keywords: Anova, Dissimilar materials, Grey relational analysis, TIG welding, Taguchi technique

1. INTRODUCTION

A weld is a robust, unified joint between two or more metal parts, typically stronger than the base metals [1]. TIG welding method utilizes heat to fuse materials by creating an electric arc between the workpiece and a tungsten electrode that is not consumed [2]. Joining SS 316L, a premium Corrosion-resistant stainless steel, with IS 2062, a structural carbon steel, is advantageous in scenarios requiring both robust corrosion protection and structural strength. This combination is frequently employed in sectors such as chemical processing and marine engineering. Effectively joining two dissimilar materials has numerous applications in various industries, including chemical processing, heat exchangers, oil refinery plants and the nuclear sector [3]. The Taguchi technique and Grey Relational Analysis (GRA) are used in this work to optimize the dissimilar welding process between IS 2062 and SS 316L. To have significant control over the process's quality, productivity, and cost, process

[1]ameena_mech@mitmysore.in, [2]moh_11976@yahoo.com

10.1201/9781003725053-8

parameters are optimized [4]. ANOVA is utilized to evaluate the effect of design parameters on the quality attributes, helping to identify if they have a significant impact on overall performance [5]. Ahire et al. applied a genetic algorithm (GA) to optimize the MMAW parameters for joining stainless steel (SS 304) to low-carbon steel. Using response surface methodology (RSM), they evaluated the effects of factors such as welding speed, current, electrode angle, and root gap on both the mechanical strength and deposition rate of the weld. Findings revealed the genetic algorithm (GA) greatly improved the welding process [6]. The goal is to evaluate and improve TIG welding settings for dissimilar metals, such as IS 2062 and SS316L. Taguchi's L9 orthogonal array is used to structure the count of experiments. Current, voltage and gas flow rate are key parameters influencing tensile strength and hardness, with their impacts assessed using the S/N ratio. The optimal settings for current, voltage, and gas flow rate are utilized to improve the tensile strength and hardness of the weldment, with Grey Relational Analysis (GRA) used in evaluating these factors.

2. LITERATURE REVIEW

2.1 Experimental Analysis of Dissimilar Weldments SS316 and Monel400 in Gtaw Process

Pooja Angolkar et al. [7] studied the weldability of dissimilar materials, Monel 400 and SS 316, using the GTAW process with ERNiCrMo-3 filler wire. They analysed mechanical properties like tensile strength and hardness and observed two distinct Heat Affected Zones (HAZ) with microstructural variations. Residual stresses were evaluated, showing maximum stress due to continuous heat input.

The study concluded that GTAW welding of Monel 400 and SS 316 alters HAZs and residual stresses, impacting mechanical properties.

2.2 Optimization of 316 Stainless Steel Weld Joint Characteristics using Taguchi Technique

P. Bharath et al. [8] investigated the impact of welding parameters (current, speed, root gap) on SS 316, finding that speed affects twist quality and current impacts rigidity. Microstructure analysis showed changes in grain structure in the heat-affected zone.

The study concludes that welding speed and current primarily affect SS 316's twist and rigidity.

3. EXPERIMENT METHOD AND PROCESS

3.1 Work Piece Material

Mild steel IS 2062 and stainless steel 316L were chosen for this investigation because of their affordability,

availability and special qualities. In compliance with ASTM E8[9]. Guidelines, a 300x20x3 mm sample was made and welded shown in Fig. 8.1. In Tables 8.1 and 8.2, the chemical compositions of IS 2062 and SS 316L are displayed. Table 8.3 lists chosen welding ranks and parameters for each factor. The current work employs an L9 orthogonal array.

Fig. 8.1 Dissimilar materials welded using TIG welding

Table 8.1 316L stainless steel's chemical composition

Elements	C	Si	Mn	P	S	Cr	Mo	Ni
Wt.%	0.03	0.29	1.58	0.027	0.003	16.25	2.27	11.90

Table 8.2 Chemical composition of mild steel IS 2062

Elements	C	Mn	S	P	Si	Ni	Cr
Wt.%	0.22	1.5	0.049	0.050	0.37	0.016	0.02

3.2 Design of Experiment

The strategic approach to the Design of Experiments (DOE) involves planning, carrying out and analysing experiments to investigate the impact of various factors on outcomes. The values of process parameters at different levels is sown in Table 8.3.

Table 8.3 The values of process parameters at different levels

Material	Parameter	Level 1	Level 2	Level 3
(SS316L-IS2062)	Current(A)	110	120	130
	Voltage (V)	40	50	60
	Gas flow rate (Lit/min)	8	9	10

a. Testing machine

This study conducted Hardness and Tensile strength tests by varying parameters such as current, voltage, and gas flow rate. Test equipment are shown in Fig. 8.2 and 8.3.

Fig. 8.2 UTM machine

Fig. 8.3 Vickers hardness tester machine

4. METHODOLOGY FOR OPTIMIZATION

This study involves conducting a limited number of trials with Taguchi method to create an experimental framework that covers the complete range of parameters [10]. The process is mono-optimization. [11-12] Nevertheless, several procedures necessitate multiple response improvements. Consequently, the Taguchi approach is unable to handle the optimization of several replies [13-14].

The equation states that a greater number indicates a better signal-to-noise (S/N) ratio (1):

$$S/N\ ratio_{LTB} = \eta = (-10) \times log_{10}\left(\frac{1}{m}\right)\sum_{i=1}^{m}\frac{1}{z_{ij}^2} \quad (1)$$

The equation states that a lower number indicates a better signal-to-noise (S/N) ratio. (2):

$$S/N\ ratio_{STB} = \eta = (-10) \times log_{10}\left(\frac{1}{m}\right)\sum_{i=1}^{m}z_{ij}^2 \quad (2)$$

This paper aims to maximize tensile strength and impact energy. Accordingly, the larger-the-better criterion is applied, and the normalized results are expressed in Equation. (3)

$$y_j^*(q) = \frac{y_j(q) - miny_j(q)}{maxy_j(q) - miny_j(q)} \quad (3)$$

The resulting Gray relational values are denoted by y^* (q), and the maximum and minimum values of y_j (q) for the qth observation are, respectively, max y_i(q) and min y_i(q). Since the ideal normalized result should be 1, In order to get better performance, a higher normalized result value is expected. Grey relational coefficients (GRC) are calculated after data normalization to illustrate the relationship between the anticipated and actual experimental normalized results. Equation (4) provides the expression for the Grey Relational Coefficient (GRC), written as ξ_j(q).

$$\xi\left(y_j^*(q), y_0^*(q)\right) = \frac{\Delta_{min}(q) + \zeta\Delta_{max}(q)}{\Delta_{0j}(q) + \Delta_{max}(q)} \quad (4)$$

0.5 was set as the identifying or distinguishing coefficient (ζ) in this work [15]. Its value is ζ e [0, 1]. The weighted mean of the corresponding grey relational grades (GRCs) for each experimental phase is used to calculate the GRG, which gives information on the level of correlation between the experimental runs.

5. RESULTS AND DISCUSSION

Table 8.4 below lists the values for the welding process parameters. Nine specimens are subjected to tensile and hardness tests [16]. This indicates that rather than the welded area, the failure happened in the IS 2062 grade of mild steel shown in Fig. 8.4. Because Tungsten Inert Gas welding produces a welded region that minimizes surface fractures and pores and produces a strong bond, it is preferred over other welding techniques.

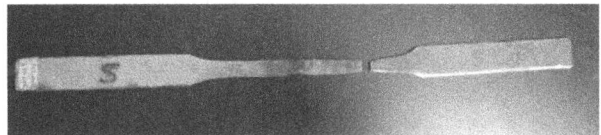

Fig. 8.4 Specimen after tensile test

Table 8.4 Experimental design and results using L9 OA

Experiment	Current (A)	Voltage (V)	GFR (Lit/min)	VHN at Weld Metal	Tensile Strength (N/mm²)	VHN at Weld Metal	Tensile Strength (N/mm²)
1	110	40	8	171	431	44.6599	52.6895
2	110	50	9	178	438	45.0084	52.8295
3	110	60	10	181	442	45.1535	52.9084
4	120	40	8	173	439	44.7609	52.8493
5	120	50	9	172	445	44.7105	52.9672
6	120	60	10	184	448	45.2963*	53.0256
7	130	40	8	176	444	44.9102	52.9477
8	130	50	9	175	460	44.8607	53.2552*
9	130	60	10	183	452	45.2490	53.1028

5.1 Individual Response Means Plots using ANOVA and Main Effects

To determine the primary impact of input parameters on individual responses, the analysis of variance (ANOVA) at a 95% confidence interval is used. The results of the ANOVA for tensile strength and hardness are displayed in Tables 8.5 and 8.6.

Table 8.5 Analysis of variance (ANOVA)-tensile strength

Source	DF	Adj SS	Adj MS	F-Value	% Contribution
Current	2	4.667	2.333	1.00	3
Voltage	2	148.667	74.333	31.86	50
Gas flow rate	2	26.000	13.000	5.57	15
Error	2	4.667	2.333		

Table 8.6 Analysis of variance (ANOVA)-hardness

Source	DF	Adj SS	Adj MS	F-Value	% Contribution
Current	2	338.00	169.000	18.11	50
Voltage	2	180.67	90.333	9.68	36
Gas flow rate	2	32.67	16.333	1.75	10
Error	2	18.67	9.333		

From Table 8.5, the factors that most significantly affect tensile strength are as follows: current, with the greatest impact at 50%, followed by voltage at 36%, and gas flow rate (GFR) at 10%. Meanwhile, Table 8.6 indicates that voltage has the most significant effect on hardness, accounting for 50%, followed by GFR at 15% and current at 3%. The key effects graphs for both tensile strength and hardness are presented in Figs. 8.5 and 8.6.

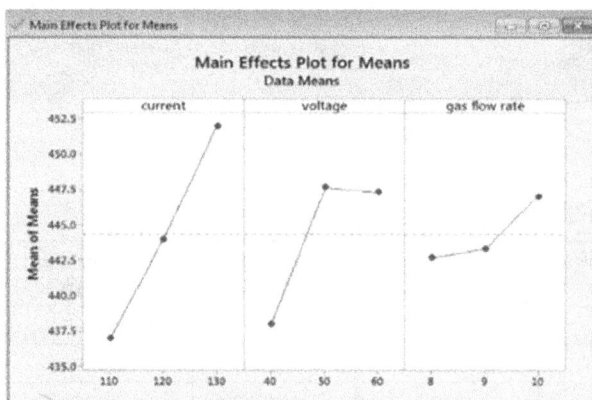

Fig. 8.5 Main effects plot for means - tensile strength

5.2 Single Objective Optimization

It employs (S/N) ratios to optimize each response. Figures 8.7 and 8.8 show the optimal levels by calculating the mean S/N ratio values for each response at every level. Table 8.4 clearly highlights significant inconsistencies among the optimal settings for all responses. As a result, multi-objective optimization is required.

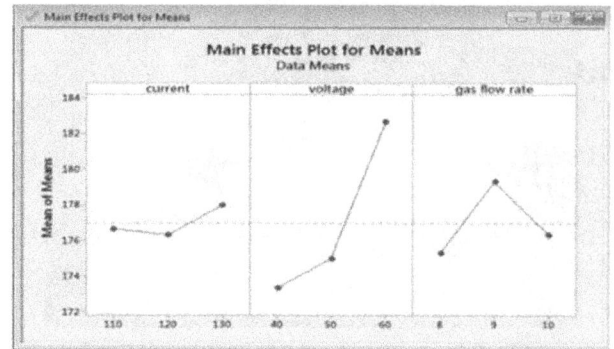

Fig. 8.6 Main effects plot for Means - hardness

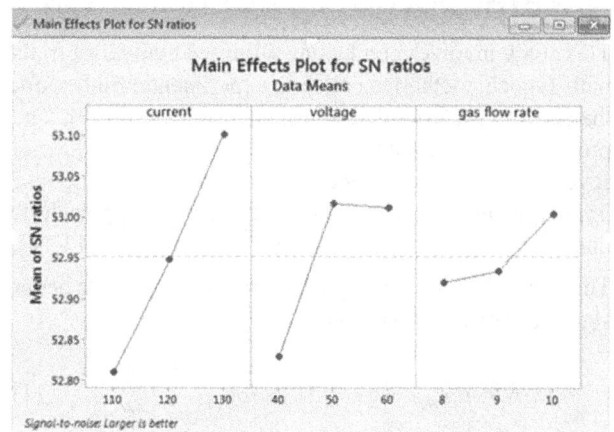

Fig. 8.7 Main effects plot for S/N ratio - tensile strength

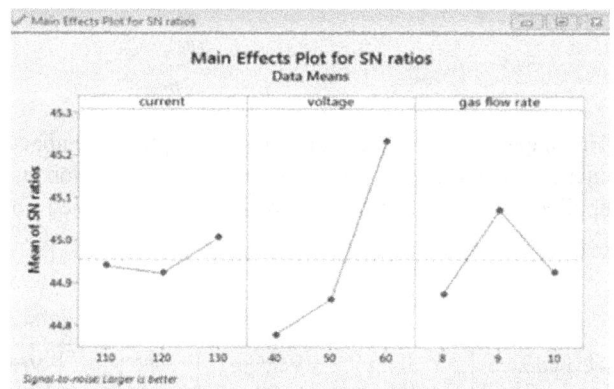

Fig. 8.8 Main effects plot for S/N ratio - hardness

5.3 Grey Relational Analysis (GRA) for Multi-Response Optimization

In multi-faceted optimization, the Grey relational analysis ranks several objectives which helps to prioritize and balance them.

Therefore, with regard to a single GRG, the desired multi-objective optimization can be accomplished. Accordingly, C3 V3 GFR1 is the ideal set of input parameter levels for quality replies based on GRG, namely Current 0.6456 (Level 3), Voltage 0.7020 (Level 3), Gas Flow Rate 0.6068 (Level) refer the Table 8.8. The highest GRG value in Table 8.7 is obtained at sample number 6

Table 8.7 Calculated normalized, GRC for 9 experiments

Exp No	Normalization		Grey Relational Coefficient		GRG	RANK
	Tensile Strength	HARD-NESS	Tensile Strength	HARD-NESS		
1	0.00	0.00	0.33	0.33	0.33	9
2	0.25	0.55	0.40	0.52	0.46	6
3	0.39	0.78	0.45	0.69	0.57	4
4	0.28	0.16	0.41	0.37	0.39	8
5	0.49	0.08	0.50	0.35	0.42	7
6	0.59	1.00	0.55	1.00	**0.78**	**1**
7	0.46	0.39	0.48	0.45	0.47	5
8	1.00	0.32	1.00	0.42	0.71	3
9	0.73	0.93	0.65	0.87	0.76	2

Means Response Table using MINITAB

Table 8.8 Means response table

Level	Current	Voltage	Gas Flow Rate
1	0.4551	0.3968	0.6068
2	0.5305	0.5323	0.5380
3	0.6456	0.7020	0.4863

6. CONFIRMATORY TEST

Using the optimal welding parameters identified by GRA, the validation experiment was done to assess and verify the betterment in quality response of the TIG weld joint for SS 316l and IS 2062.

Table 8.9 Summary of confirmatory experiment

Test	Initial Condition	Results from a confirmatory experiment
Hardness	184	187
Tensile Strength	460	480

Confirmation experiment is done at optimal process level and the value of Hardness and Tensile Strength after optimization is 187 and 480 N/mm 2 respectively. Thus, effectiveness of Taguchi Optimization method was verified.

6.1 Microstructural Study

An ideal microscope was used to do a microstructure investigation at the contact between the Base metal zone and the weld zone. The base metal of the dissimilar TIG welds primarily shows a homogeneous dispersion of ferrite grains in a fine-grained structure shown in Fig. 8.9. The morphology of the weld zone shown in Fig. 8.10, reveals that the ferrite content in the weld metal (WM)

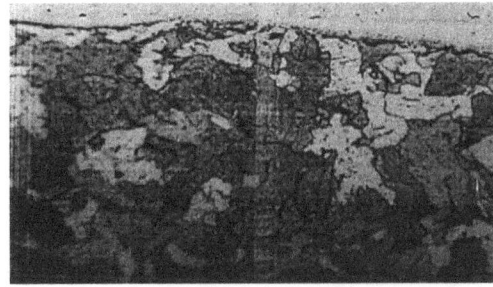

Fig. 8.9 Microstructure of base metal 500X

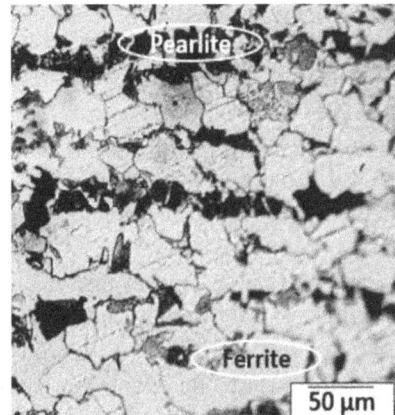

Fig. 8.10 Microstructure of weld zone 500X

microstructure enhances its strength and toughness. The weld metal consists of 70% ferrite and 30% pearlite. The ferrite phase significantly improves impact hardness and crack resistance, thereby increasing the weld's durability. The observed grain size is approximately 50 μm.

7. CONCLUSION

The current study was done to learn more about how TIG welding process factors affect the hardness and tensile strength of dissimilar welds. Stainless Steel 316L and Mild steel IS 2062 plates are used as test material. To increase weld strength, the ideal TIG welding settings were identified using the Taguchi optimization technique.

These inferences can be made from the experimental findings:

7.1 Taguchi Technique

- To find the optimal TIG welding parameter level, the Taguchi design of experiment approach was successfully applied. Optimum parameter for Hardness is obtained at 130(Amp), 60(Volt) and 9(Lit/min) Gas flow rate and for Tensile strength optimum parameter is obtained at 130(Amp), 50(Volt) Voltage and 10(Lit/min) Gas flow rate.

- In order to maximize the welding settings, Taguchi orthogonal arrays, signal-to-noise ratios, Means and S/N ratio main effects plots, as well as analysis of variance, were used.

- Voltage was the factor that significantly contributed to a larger % in the hardness test, according to the ANOVA. (50%) followed by gas flow rate (15%) and current (3%); in the tensile test, on the other hand, current (50%) was the factor that significantly contributed to a higher percentage followed by voltage (36%), and gas flow rate (10%)).

- The welding current has the biggest impact on tensile strength, while the welding voltage has a bigger impact on hardness.

- The tensile and hardness tests adhere to the "larger-the-better" quality standards.

7.2 The Method of Grey Relational Analysis and Microstructure study

- Both IS2062 and SS 316L are dissimilar metals that can be optimized with the right combination of parameters by utilizing multiple objectives C3V3GFR1.

- The average GRG study showed that voltage has the greatest influence, followed by current and GFR, in that order.

- In the microstructure, the fine flow shows good penetration of the welding in the joint. No physical surface defects observed in weldment. The choice of process parameters was evidently appropriate.

REFERENCES

1. Kumar, Sanjay, et al. "Optimization of TIG welding process parameters using Taguchi's analysis and response surface methodology." *International Journal of Mechanical Engineering and Technology* 8.11 (2017): 932–941.
2. Kumar, Kamlesh, et al. "A review on TIG welding technology variants and its effect on weld geometry." *Materials Today: Proceedings* 50 (2022): 999–1004.
3. Taguchijeve, U. G. A. G. I., and VRTILNO-TORNIM PROCESOM FSW. "Application of grey relation analysis (GRA) and Taguchi method for the parametric optimization of friction stir welding (FSW) process." *Mater Tehnol* 44 (2010): 205.
4. Mvola, B., P. Kah, and J. Martikainen. "DISSIMILAR FERROUS METAL WELDING USING ADVANCED GAS METAL ARC WELDING PROCESSES." *Reviews on Advanced Materials Science* 38.2 (2014).
5. Kurt, H.I.; Oduncuoglu, M.; Yilmaz, N.F.; Ergul, E.; Asmatulu, R. A comparative study on the effect of welding parameters of austenitic stainless steels using artificial neural network and Taguchi approaches with ANOVA analysis. Metals 2018, 8, 326.
6. Nandagopal, K.; Kailasanathan, C. Analysis of mechanical properties and optimization of gas tungsten Arc welding (GTAW) parameters on dissimilar metal titanium (6Al4V) and aluminium 7075 by Taguchi and ANOVA techniques. J. Alloys Compd. 2016, 682, 503–516.
7. Pooja Angolkar, J.Saikrishna, Dr.R.Venkat Reddy,"Experimental Analysis of Dissimilar Weldments SS316 and Monel400 in Gtaw process", International Journal of Engineering Research and Development, Vol. 13, Dec 2017, PP. 39–46.
8. P. Bharath, "Optimization of 316 stainless steel weld joint characteristics using Taguchi Technique", 7th International Conference on Materials for Advanced Technologies, May 2014, PP.881–891.
9. Standard Test Methods for Tension Testing of Metallic Materials, ASTM E8/E8M-16a (West Conshohocken, PA ASTM International, approved August 1, 2016).
10. Pandiarajan, S.; Kumaran, S.S.; Kumaraswamidhas, L.; Saravanan, R. Interfacial microstructure and optimization of friction welding by Taguchi and ANOVA method on SA 213 tube to SA 387 tube plate without backing block using an external tool. J. Alloys Compd. 2016, 654, 534–545.
11. Satheesh, M.; Dhas, J. Multi Objective Optimization of Weld Parameters of Boiler Steel Using Fuzzy Based Desirability Function. J. Eng. Sci. Technol. Rev. 2014, 7, 29–36.
12. V. Sarıkaya, M.; Güllü, A. Multi-response optimization of minimum quantity lubrication parameters using Taguchi-based grey relational analysis in turning of difficult-to-cut alloy Haynes 25. J. Clean. Prod. 2015, 91, 347–357.
13. Bhaduri, A.; Gill, T.; Srinivasan, G.; Sujith, S. Optimised post-weld heat treatment procedures and heat input for welding 17–4PH stainless steel. Sci. Technol. Weld. Join. 1999, 4, 295–301.
14. Qazi, Mohsin Iqbal, et al. "An integrated approach of GRA coupled with principal component analysis for multi-optimization of shielded metal arc welding (SMAW) process." Materials 13.16 (2020): 3457.
15. Kurt, H.I.; Oduncuoglu, M.; Yilmaz, N.F.; Ergul, E.; Asmatulu, R. A comparative study on the effect of welding parameters of austenitic stainless steels using artificial neural network and Taguchi approaches with ANOVA analysis. Metals 2018, 8, 326.
16. Nandagopal, K.; Kailasanathan, C. Analysis of mechanical properties and optimization of gas tungsten Arc welding (GTAW) parameters on dissimilar metal titanium (6Al4V) and aluminium 7075 by Taguchi and ANOVA techniques. J. Alloys Compd. 2016, 682, 503–516.39.

Note: All the figures and tables in this chapter were made by the authors.

Advances in Mechanical Engineering and Materials Sciences – Dr. Vinay K. B et al. (eds)
© 2026 Taylor & Francis Group, London, ISBN 9-781-041-20970-6

9

Computational Fatigue and Stress Analysis of Oleo Cylinder and Piston

Prathik Jain S.,
Sundaramahalingam A., Megha Sridhar,
Inchara N., B. K. Manisha Kotari, Pragna G. D.
Dept. of Aeronautical Engineering, Dayananda Sagar College of Engineering,
Bengaluru, Karnataka, India

Abstract: The study considers different material choices to evaluate and enhance the durability, safety, and efficiency of aircraft landing gear. The oleo cylinder and piston are significant components of the landing gear that are subjected to effective stress and fatigue primarily because of the cyclic forces experienced during take-off and landing. The choice of a suitable material is central to reliable performance. The materials under analysis under analysis in this study are SAE-AISI-1015 carbon steel, Titanium alloy (Ti-6Al-4V), Inconel 718 alloy, and AISI 4340 alloy steel, with a focus on their strength, lightness, and resistance to fatigue. Stress, factor of safety, and impact load on the materials were analyzed using Static Structural Analysis and Explicit Dynamics Analysis under the same loading conditions. Results show a clear view of how the materials resist and react to service forces and impact loads, where they can be compared in relative terms based on their strengths and weaknesses. Results indicate that AISI 4340 alloy steel is best suited for balance between strength and toughness and so is most preferred for landing gear components. This way, optimized material selections would further increase the landing gear functionality, hence increasing aircraft performance and safety.

Keywords: Landing gear, Stress and fatigue, Material selection

1. INTRODUCTION

The landing gear is a critical component of an aircraft to sustain its weight during ground time and to provide safe take-off and landing. Nose landing gears, which are fitted with oleo-pneumatic shock absorbers, are a critical component in the damping of vertical loads and the attenuation of vibrations during aircraft operations [1]. Advanced materials, like Aluminum Alloy 7075 and Titanium Ti-6Al-4V, improve landing gear structural strength while keeping weight in balance [2,3]. Nose landing gear performance is influenced substantially by landing impact loads, highlighting the structural flexibility and tire friction's importance in load analysis [4]. CAD and FEA software such as CATIA and ANSYS maximize material choice and stress distribution, with titanium alloys such as Ti-10V-2Fe-3Al being favored for their

high safety, low stress, and low deformation [5]. Dynamic simulations emphasize the role of material properties and damping in minimizing drag and vibrations, with titanium alloys providing superior reliability and fatigue lifespan [6]. Finite Element Analysis (FEA) confirms the accuracy of stress and fatigue life calculations, with less than 3% error compared to experimental values [7].

Fatigue analysis of a wing structure indicates that composites such as Carbon Epoxy and S-2 Glass provide a greater strength-to-weight ratio, leading to reduced stresses. Furthermore, changes in spar cross sections have a major impact on the structural strength and performance of the wing under load [8]. Modal and structural analysis shows the principal frequencies and deformations and therefore require the application of optimization. Innovative design and improved strength of composite material consider

---
[1]prathik9@gmail.com, [2]prathiks-ae@dayanandasagar.edu

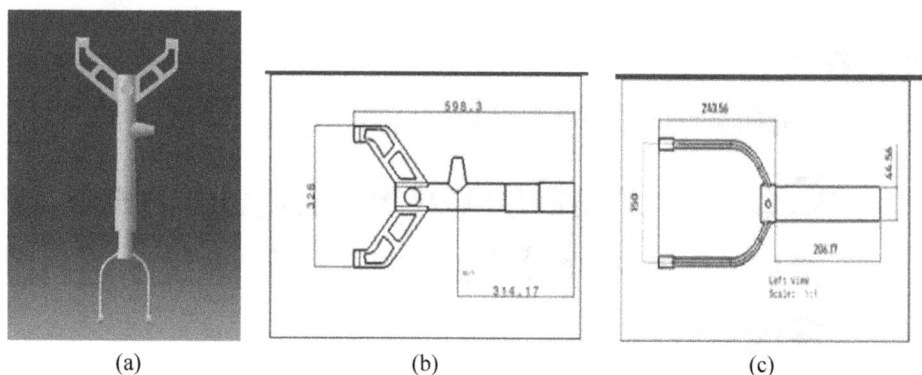

Fig. 9.1 (a) Model of oleo cylinder and piston assembly, Draft of (b) oleo cylinder and (c) oleo piston

greater levels of landing gear durability and safety [9]. Aerodynamics including noise abatement and flow issues are important aspects of today's landing gear designs, enabled through software such as ANSYS ICEM for providing accuracy in prediction as well as compliance with law [10]. Fluid-structure interaction (FSI) simulation has a high influence on the behavior of a wing, particularly on aerodynamic loads and structural stress distribution. Comparison between coupled mode analysis and static solutions provides evidence of flexibility's influence on flow structure as well as on structural response [11].

Analysis of shock spectra shows that composites have more performance compared to conventional alloys subjected to intense conditions with a lower stress level and deformation rates [12]. Titanium alloys are appropriate for severe applications because they possess a high strength-to-weight ratio, while steel alloys such as SAE1035 are superior in high-stress applications [13]. Structural analysis reveals that composite material such as Al-SiC have zero deformation upon landing and thus can be an effective substitute for titanium alloys for landing gear usage [14]. The study informs that the landing gear made of aluminum alloy sustains the least static stress, and carbon fiber reinforced composite landing gear sustains greater maximum stress under static and fatigue loads [15]. Higher-order finite element methods can analyze stress and fatigue, with a focus on material optimization to resist static and dynamic loads with safety margins [16].

Additionally, the study highlights the capability of remotely piloted airships, equipped with telemetry systems, to gather real-time data on atmospheric pollutants, temperature, and pressure, towards environmental monitoring and aiding air pollution abatement efforts [17]. Future designs will feature lightweight composites and high-strength alloys like Ferrum S53 to provide greater structural reliability and integrity across operating conditions [18]. The continued advancement of landing gear technology is driven by innovative materials and design, with titanium alloys being the preferred choice owing to their excellent performance under stress and dynamic loading, though they are heavier [19]. The study further shows how multi-disciplinary design optimization can be applied to landing gear assemblies, yielding considerable weight and cost savings with adequate structural performance [20].

2. MATERIALS AND METHODS

A detailed 3D model of the oleo cylinder and oleo piston of the landing gear has been developed using CATIA V5 software, with the design and draft illustrated in Fig. 9.1. The dimensions of the model were consistent with those outlined in the reference paper [1].

Based on the previous study, four materials are selected namely SAE-AISI-1015 carbon steel, Titanium alloy (Ti-6Al-4V), Inconel 718, and AISI 4340 alloy steel for detailed analyses. Using ANSYS 2024 R1 software, stress, factor of safety, and impact load analyses are performed under varying loading conditions to assess material performance. The oleo cylinder and piston were meshed with the number of elements of 1126193 and 480375 respectively, utilizing a refined mesh to achieve precise finite element analysis as shown in Fig. 9.2.

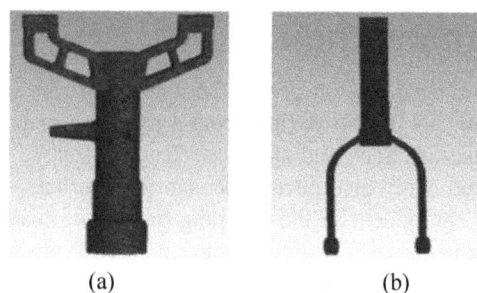

Fig. 9.2 Mesh (a) Oleo cylinder (b) Oleo piston

Finally, the results are evaluated and compared to determine the optimal material that offers the best balance of strength, durability, and weight efficiency for the landing gear design. Different materials viz., SAE-AISI-1015 Carbon Steel, Titanium Alloy (Ti-6Al-4V), Inconel 718, and AISI 4340 Alloy Steel were incorporated to the oleo cylinder and piston and analyzed based on their properties. The weight calculations were done based on volumes of the components obtained from CATIA V5 software. A comparative analysis of the weights of an oleo cylinder and piston made from different materials is shown in Table 9.1. The total weight of the cylinder-piston assembly is derived for each material to identify their potential application in aerospace systems, particularly in landing gear design, where weight optimization is crucial.

Table 9.1 Weight properties of different materials

Material	Oleo Cylinder weight (kg)	Oleo Piston weight (kg)	Total weight (kg)
SAE-AISI-1015	9.90046	3.55015	13.04506
Ti-6Al-4V	5.57294	1.99837	7.57131
Inconel 718	10.34076	3.70804	14.04880
AISI 4340	9.87530	3.54113	13.41643

Source: Author's compilation

It reveals that Inconel 718 is the heaviest material, while Ti 6Al-4V is the lightest. This weight difference suggests that using Ti-6Al-4V could potentially lead to significant weight savings in applications where weight reduction is a critical factor.

3. RESULTS AND DISCUSSIONS

3.1 Static Analysis

Static analysis of the oleo cylinder and piston was done for four different materials. This analysis aimed to evaluate

and compare the structural behavior, including stress distribution and deformation characteristics, to determine the performance and suitability of each material under the applied load condition of 7286 N. The stress and deformation of the oleo cylinder for SAE-AISI-1015 and Ti 6Al-4V are depicted in Fig. 9.3. The same method is applied to Inconel 718 and AISI 4340 to determine the corresponding values for stress and deformation. Table 9.2 compares the stress and deformation of each material analyzed for the oleo cylinder.

Table 9.2 Comparison of static analysis of oleo cylinder

Material	Deformation (mm)	Stress (MPa)
SAE-AISI-1015	0.059717	108.43
Ti-6Al-4V	0.11154	113.26
Inconel 718	0.072422	112.88
AISI 4340	0.06059	103.13

Source: Author's compilation

The stress and deformation of the oleo piston for Inconel 718 and AISI 4340 are depicted in Fig. 9.4. The same method is applied to SAE-AISI-1015 and Ti-6Al-4V to determine their corresponding values for stress and deformation. Table 9.3 compares the stress and deformation of each material analyzed for the oleo piston.

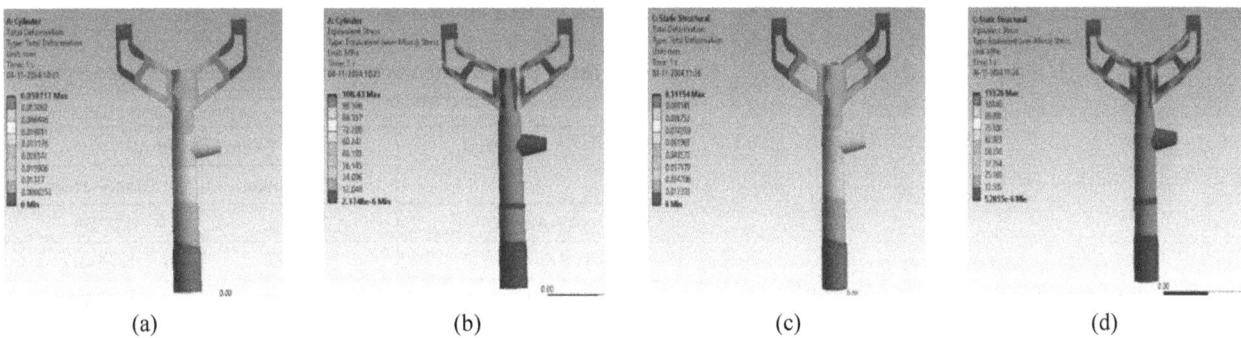

(a)   (b)   (c)   (d)

Fig. 9.3 (a) Deformation of the oleo cylinder (SAE-AISI-1015), (b) Stress distribution in the oleo cylinder (SAE-AISI-1015), (c) Deformation of the oleo cylinder (Ti-6Al-4V), (d) Stress distribution in the oleo cylinder (Ti-6Al-4V)

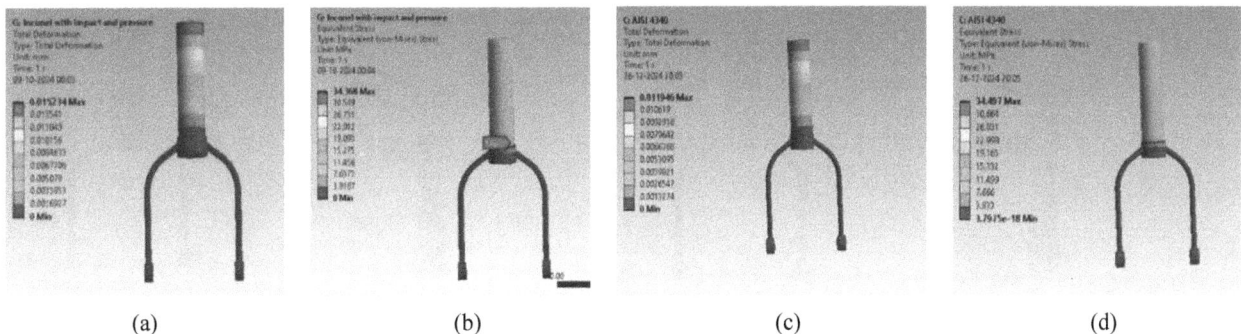

(a)   (b)   (c)   (d)

Fig. 9.4 (a) Deformation of the oleo piston (Inconel 718), (b) Stress distribution in the oleo piston (Inconel 718), (c) Deformation of the oleo piston (AISI 4340), (d) Stress distribution in the oleo piston (AISI 4340)

Table 9.3 Comparison of static analysis of oleo piston

Material	Deformation (mm)	Stress (MPa)
SAE-AISI-1015	0.012604	34.081
Ti-6Al-4V	0.023459	34.515
Inconel 718	0.015234	34.368
AISI 4340	0.011946	34.497

Source: Author's compilation

Figure 9.5 provides a graphical comparison of the deformation and stress of the oleo cylinder and piston for different materials.

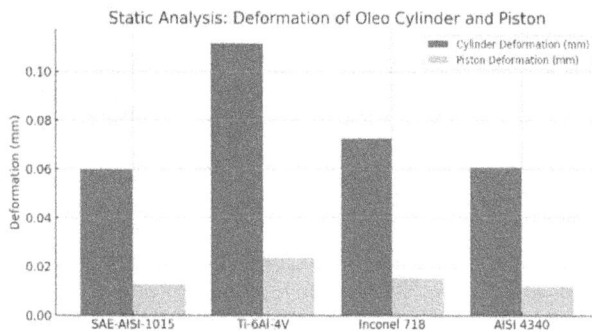

(a)

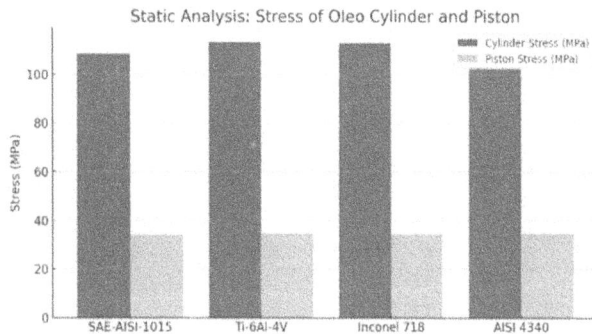

(b)

Fig. 9.5 (a) Comparison of deformation of oleo cylinder and piston for all four materials, (b) Comparison of stress of oleo cylinder and piston for all four materials

Based on the analysis, AISI 4340 proves to be the optimal material for both the oleo cylinder and piston. For the oleo cylinder, it exhibits minimal deformation and the lowest stress, making it highly efficient in withstanding the applied load. Similarly, for the oleo piston, AISI 4340 demonstrates the least deformation and maintains a stress level close to the lowest observed. While SAE-AISI-1015 shows slightly lower stress in the piston, AISI 4340's overall performance, with low deformation and stress, makes it the optimal choice for ensuring structural integrity and durability under the given conditions.

3.2 Factor of Safety

The factor of safety in fatigue analysis of the oleo cylinder and piston in aircraft landing gear ensures these components can withstand cyclic stresses over their operational life without failure. It takes into consideration uncertainty in material properties, loading conditions, and manufacturing imperfections. The determination of the factor of safety of the oleo cylinder and piston was executed using four different materials. This analysis aimed to evaluate and compare the factor of safety to determine the performance and suitability of each material under the applied load condition of 7286 N. The factor of safety of the oleo cylinder and piston for Ti 6Al-4V and Inconel 718 is depicted in Fig. 9.6. The same method is applied to SAE-AISI-1015 and AISI 4340 to determine their respective values for the factor of safety. Table 9.4 compares the factor of safety of each material analysed for the oleo cylinder and piston.

Table 9.4 Comparison of factor of safety for oleo cylinder and piston

Material	Factor of Safety	
	Oleo Cylinder	Oleo Piston
SAE-AISI-1015	2.3475	6.6255
Ti-6Al-4V	1.7799	5.7649
Inconel 718	2.254	5.7532
AISI 4340	1.5635	6.1172

Source: Author's compilation

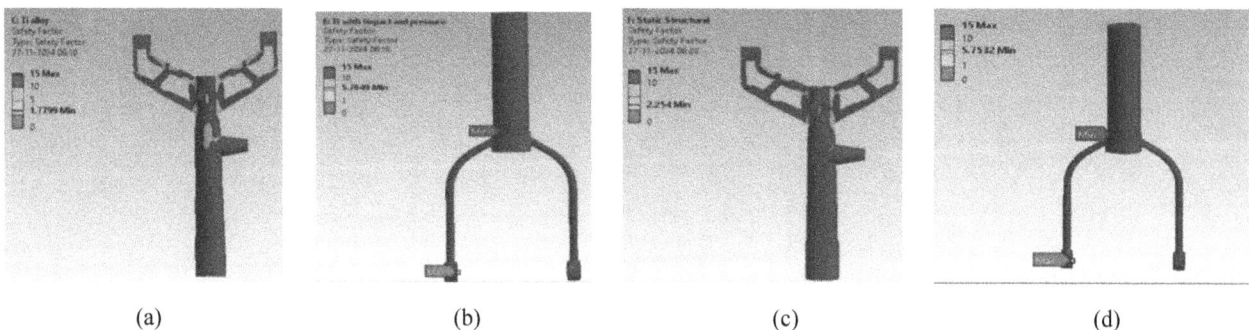

(a)　　　　(b)　　　　(c)　　　　(d)

Fig. 9.6 Factor of safety (a) Oleo cylinder (Ti-6Al-4V), (b) Oleo piston (Ti-6Al-4V), (c) Oleo cylinder (Inconel 718), (d) Oleo piston (Inconel 718)

Figure 9.7 provides a graphical comparison of the factor of safety for the oleo cylinder and piston across different materials. Based on the analyses, SAE-AISI-1015 and Inconel 718 are more favored due to higher safety margins. Although AISI 4340 performed well under the piston loads, it will have the minimum safety factor for cylinder loads, increasing the chances of durability issues in the cylinder, as does the case with Ti-6Al-4V for the same reason.

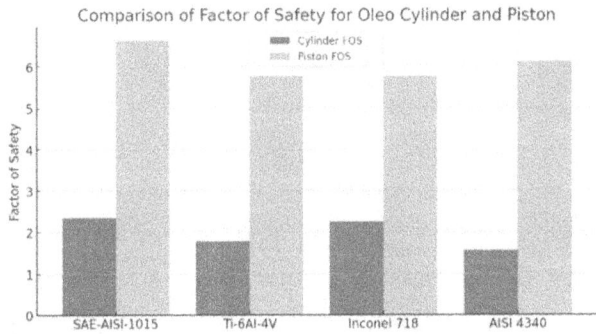

Fig. 9.7 Comparison of factor of safety of oleo cylinder and piston using four different materials

3.3 Dynamic Analysis

The performance of the oleo cylinder and piston assembly of the landing gear under impact conditions was analyzed dynamically using ANSYS. Four materials were considered in the study. A velocity of 2.5 m/s was applied as a boundary condition to simulate the landing impact scenario. Equivalent stress and strain distributions were evaluated in the analysis to understand the response of the components to dynamic loading. This investigation provides insight into the behavior of materials under impact conditions, which helps in the selection of appropriate materials for landing gear applications. The equivalent stress and elastic strain of the assembly of the oleo cylinder and piston for SAE-AISI-1015 and Ti-6Al-4V are depicted in Fig. 9.8. The same method is applied to Inconel 718 and AISI 4340 to determine their corresponding values for the same. Table 9.5 compares the equivalent stress and elastic strain of each material analysed for the assembly of the oleo cylinder and piston.

Table 9.5 Comparison of equivalent stress and elastic strain for the assembly of oleo cylinder and piston

Material	Equivalent stress	Elastic strain
SAE-AISI-1015	504.13	0.00908
Ti-6Al-4V	355.68	0.00478
Inconel 718	560.28	0.00947
AISI 4340	581.19	0.00887

Source: Author's compilation

Of all the materials under evaluation, the highest equivalent stress is shown by AISI 4340 followed by Inconel 718 and SAE-AISI-1015, while Ti-6Al-4V shows the lowest. In the case of elastic strain, SAE-AISI-1015 and Inconel 718 show relatively high values, and Ti-6Al-4V shows the lowest strain. This analysis thus gives a deeper insight into the behavior of every material under the loading conditions so that their deformation properties can be evaluated and checked for landing gear applications.

4. CONCLUSION

This study conducted a detailed computational analysis to evaluate the stress distribution, factor of safety, and impact load response of the oleo cylinder and piston, which are critical components of aircraft landing gear. Using advanced simulation techniques, four materials viz., SAE AISI-1015 carbon steel, Ti-6Al-4V titanium alloy, Inconel 718, and AISI 4340 alloy steel were analysed under static, dynamic, and fatigue loading conditions. The results indicate that AISI 4340 alloy steel offers the best balance of minimal deformation and high stress resistance, making it the most suitable option for ensuring structural integrity and durability in static and dynamic conditions. However, Ti-6Al-4V and Inconel 718 demonstrated superior fatigue resistance, making them better choices for applications where extended lifecycle performance is critical, despite their higher cost. Overall, this research highlights the importance of material selection in enhancing the performance, safety, and reliability of aircraft landing gear. Future studies could focus on real world loading conditions and explore the potential of cost-effective composite materials to future optimize weight and performance.

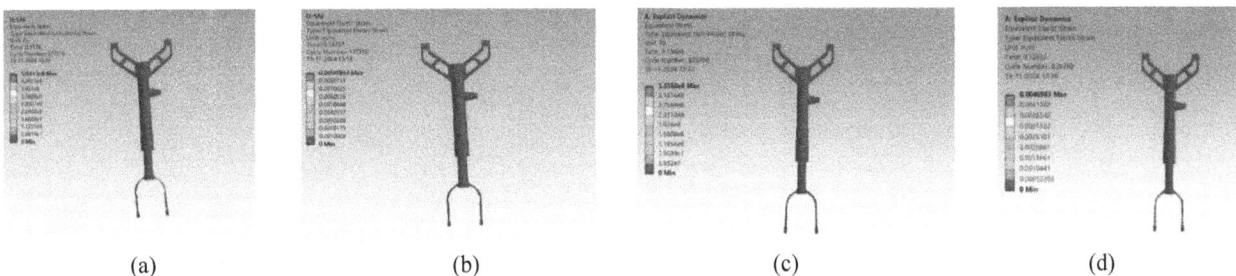

| (a) | (b) | (c) | (d) |

Fig. 9.8 (a) Equivalent stress of assembly (SAE-AISI-1015), (b) Elastic strain of the assembly (SAE-AISI-1015), (c) Equivalent stress of the assembly (Ti-6Al-4V), (d) Elastic strain of the assembly (Ti-6Al-4V)

REFERENCES

1. Ahmad, Muhammad Ayaz, Syed Irtiza Ali Shah, Taimur Ali Shams, Ali Javed, and Syed Tauqeer ul Islam Rizvi (2021). Comprehensive design of an oleo-pneumatic nose landing gear strut. Proceedings of the Institution of Mechanical Engineers, Part G: Journal of Aerospace Engineering 235, no. 12: 1605–1622.

2. Aydın, G. and Ozkol, İ., (2022). Structural analysis of the nose landing gear of a fighter aircraft. Avrupa Bilim ve Teknoloji Dergisi, (43), pp.126–135.

3. Kumar, S.N., Shukur, J.A., Sriker, K. and Lavanya, A., (2014), February. Design and structural analysis of skid landing gear. International Journal of Current Engineering and Technology, 2(2), pp. 635–642.

4. Liu, W., Wang, Y. and Ji, Y., (2023). Landing Impact Load Analysis and Validation of a Civil Aircraft Nose Landing Gear. Aerospace, 10(11), p.953.

5. Prasad, V., Reddy, P.K., Rajesh, B. and Sridhar, T., (2020). Design And Structural Analysis Of Aircraft Landing Gear Using Different Alloys. International Journal of Mechanical Engineering and Technology, 11(7), pp.7–14.

6. Jeevanantham, V., Vadivelu, P. and Manigandan, P., (2017). Material based structural analysis of a typical landing gear. International Journal of Innovative Science, Engineering & Technology, 4(4), pp.295–300.

7. Singh, K.L., (2016). Stress and Fatigue Damage Computation of a Nose Landing Gear. International Journal of Fracture and Damage Mechanics, 1(2), pp.20–37.

8. Shankar, S., Jain, P.S., Chandru, V., Gowda, K.P. and Jain, S.K., (2021), February. Fatigue analysis on wing structure. In AIP Conference Proceedings (Vol. 2317, No. 1). AIP Publishing.

9. Imran, M., Shabbir Ahmed, R.M. and Haneef, D.M., (2014). Static and dynamic response analysis for landing gear of test aircrafts. Int. J. Innovative Res. Sci. Eng. Technol, 3(5).

10. Rajesh, A. and Abhay, B.T., (2015). Design and analysis aircraft nose and nose landing gear. Journal of Aeronautics & Aerospace Engineering, 4(2), pp.74–80.

11. Jain, P.S., Gowda, K.P., Shankar, S., Chandru, V. and Jain, S.K., (2021). Fluid-Structure Interaction Study of a Wing Structure. International Research Journal of Engineering and Technology, 8(5), pp.2516–2527.

12. Imran, M., Ahmed, R.S. and Haneef, M., (2015). FE analysis for landing gear of test air craft. Materials Today: Proceedings, 2(4-5), pp.2170 2178.

13. Hebsur, M.P., Kurbet, S.N., Pinjar, M.S. and Adavihal, M.S., (2019). Failure Analysis of Landing Gear of the Aircraft through Finite Element Method. Young, 4(2.81), pp.7–87.

14. Firoz, F., Raj, R. and Samuel, G.D., (2023). Design and Analysis of Landing Gear using Composite Material. ACS Journal for Science and Engineering, 3(1), pp.70–77.

15. Chen, P.W., Sheen, Q.Y., Tan, H.W. and Sun, T.S., (2012). Fatigue analysis of light aircraft landing gear. Advanced Materials Research, 550, pp.3092–3098.

16. Dutta, A., (2016). Design and analysis of nose landing gear. International Research Journal of Engineering and Technology (IRJET), 3(10), pp.261–266.

17. Prathik, S.J., Sundaramahalingam, A., Solomon, J.M., Chethan, K.N. and Keni, L.G., (2024). Innovative airship design for real-time air quality monitoring using IoT technology. Journal of Applied Engineering Science, 22(4), pp.727–738. Babu, N.S., Modal Analysis of a Typical Landing Gear Oleo Strut. International Journal of Recent Trends in Engineering and Research.

18. Vasanth, G., Deepack, R., Murali, S. and Magesh, S., (2020), October. Comprehensive analysis on mechanical behavior of airworthy raw materials for aircraft landing gear system. In AIP Conference Proceedings (Vol. 2283, No. 1). AIP Publishing.

19. Khanapur, C. and Vaidya, A., Design and Optimization of Landing Gear for an Airbus A320. International Journal for Scientific Research & Development, 3(6), pp.1231–1234.

Note: All the figures in this chapter were made by the authors.

Advances in Mechanical Engineering and Materials Sciences – Dr. Vinay K. B et al. (eds)
© 2026 Taylor & Francis Group, London, ISBN 9-781-041-20970-6

10 Design and Fabrication of Movable Hydraulic Ram Lifter—A Review

N. Jayashankar[1]
Associate Professor,
Department of Mechanical Engineering,
Vidyavardhaka College of Engineering, Mysuru, Karnataka, India

Rohan B. Sangal[2]
Students,
Department of Mechanical Engineering,
Vidyavardhaka College of Engineering, Mysuru

**Sumanth R.[3],
Hemanth G. M.[4], Prajwal G.[5]**
Students, Department of Mechanical Engineering,
Vidyavardhaka College of Engineering, Mysuru

Abstract: Recent The Movable Hydraulic Ram Lifter is a practical and reliable tool designed to make industrial work safer and more efficient. This innovative device combines the power of a hydraulic ram with the convenience of a portable platform, making it easier to lift and move heavy loads across different types of terrain.

Some of its standout features include adjustable hydraulic pressure for precise lifting, a secure locking system to keep loads stable during transport, and a sturdy design that ensures durability and easy handling. By reducing the physical strain on workers and boosting productivity, this lifter is especially useful in industries like construction, logistics, and manufacturing.

Built with a focus on safety, stability, and user-friendliness, the lifter has undergone extensive testing to ensure it performs well even under challenging conditions, such as heavy loads on uneven surfaces. It's a valuable addition to any industrial operation, helping teams work smarter and more efficiently.

Keywords: Hydraulic ram lifter, Heavy equipment, Industrial lifting, Load handling, Mobility, Productivity, Safety mechanism

1. INTRODUCTION

Efficient and versatile material handling plays a vital role in industries like construction, manufacturing, and logistics, where lifting and moving heavy loads is an everyday requirement. However, traditional equipment such as fixed cranes and stationary hoists often falls short in meeting the demands of dynamic and challenging environments. Limited mobility, difficulty operating on uneven terrain, and constraints in confined spaces can slow down operations, increase labour efforts, and raise the risk of workplace injuries due to unsafe manual handling or unstable lifting practices. (Akinsade A,et al., 2017)

The Movable Hydraulic Ram Lifter is designed to tackle these challenges head-on. This innovative system combines the power of a highcapacity hydraulic ram with the convenience of a mobile platform, creating a lifting solution

This is not only strong but also flexible and easy to maneuver. Whether it's transporting heavy equipment across a cluttered construction site or carefully positioning

[1]jayshankar@vvce.ac.in, [2]vvce21me0059@vvce.ac.in, [3]vvce21me0062@vvce.ac.in

loads in a compact workshop, the lift is built to adapt. Its ad just able hydraulic pressure ensures precise and smooth control over lifting and lowering operations, while advanced safety features (Sumaila, et al., 2011) such as load limiters, locking mechanisms, and stability controls—help prevent accidents, even in demanding conditions.

What sets this lifter apart is its focus on usability and efficiency. The mobile platform allows operators to quickly reposition the system where it's needed, eliminating downtime and boosting productivity. At the same time, its intuitive controls reduce the learning curve, making it accessible for operators with varying levels of experience.

This paper dives into the design principles and performance evaluation of the Movable Hydraulic Ram Lifter, showcasing how it bridges the gap between power and portability. By exporing its engineering details and real-world applications, this study aims to inspire the next generation of adaptable and efficient material-handling solutions that meet the growing demands of industrial and construction sectors while prioritizing safety and ease of use. (Anugrah, et al., 2017)

2. BACKGROUND AND SIGNIFICANCE

2.1 Background

In today's fast-paced industrial and construction environments, the need for smarter ways to handle heavy materials has become more important than ever. As industries grow and job sites become increasingly complex, the demand for equipment that can safely and effectively lift and move heavy loads across all kinds of terrains has skyrocketed. Traditional lifting tools, like cranes and hoists, often struggle to keep up—they're usually fixed in one place or limited to specific areas, making them tough to use in tight spaces or on uneven ground.

Hydraulic systems, known for their strength and smooth operation (Adeoye et al., 2017), are a popular choice for heavy lifting. However, many of these systems are stationary or lack the mobility needed to navigate diverse work environments. That's where the Movable Hydraulic Ram Lifter comes in. This innovative tool combines the sheer power of hydraulic lifting with the flexibility of a mobile platform, making it easier than ever to move heavy materials across different job sites while maintaining precise control and strength. The Movable Hydraulic Ram Lifter is a real game changer. It's built to meet the needs of industries that require both heavy lifting and mobility, providing a solution that boosts efficiency', cuts labour costs, and makes job sites safer. Thanks to its portability, this lifter can tackle tough terrains and tight spaces that traditional stationary equipment simply and ease of operation

This tool is especially valuable in industries like construction, manufacturing, logistics, and warehousing

places where lifting and moving heavy loads is a daily challenge. Its mobility means it can work in areas where accessibility is a problem, cutting down on delays and keeping projects on track. Plus, with built-in safety features like load limiters and secure locking mechanisms, it reduces the risk of accidents, making it a safer alternative to manual handling and older lifting methods.

At its core, the Movable Hydraulic Ram Lifter (ChenLinlin. et al., 2022), is all about helping people work smarter, not harder. It's a modern, practical solution designed for the real-world challenges of today's industries. By improving safety, increasing productivity, and offering unmatched versatility, it's paving the way for a new era in material handling one where efficiency and safety go hand in hand

3. DEFINITIONS OF KEYWORDS

- Hydraulic Ram Lifter: A powerful tool that uses hydraulic pressure to lift and lower heavy objects smoothly and with precision, making tough jobs easier.
- Mobility: The ability to move equipment effortlessly across different job sites, whether it's over smooth floors or bumpy terrain.
- Load Handling: The process of safely lifting, moving, and managing heavy items, ensuring the job gets done without unnecessary strain or risk.
- Safety Mechanism: Built-in features or systems that keep equipment and workers safe by preventing accidents and ensuring everything operates smoothly.
- Productivity: Getting work done faster and more effectively, so you can accomplish more in less time with less hassle.

4. LITERATURE REVIEW

4.1 Development of a Mobile Hydraulic Lifting Machine

The research by Akinsade et al. (2024) details the design, development, and evaluation of a mobile hydraulic lifting machine. The authors begin by establishing the need for a versatile, efficient lifting solution that can be easily transported and deployed in various engineering and construction environments. They outline the engineering challenges of creating a compact yet powerful system, focusing on key components such as the hydraulic pump, cylinders, and supporting structural framework.

The study describes the design process step by step, including simulation studies and prototype development, to ensure that the machine meets the required load capacity and safety standards. Experimental tests were conducted to evaluate the lifting performance, stability, and overall operational efficiency. The results demonstrated that the machine not only meets the specified technical requirements but also offers enhanced mobility and user safety compared to traditional lifting equipment.

In conclusion, the research shows that the developed mobile hydraulic lifting machine is a viable solution for tasks requiring quick and reliable lifting in various industrial applications. The findings suggest potential for broader implementation in construction and other fields where mobility and efficient lifting mechanisms are crucial Key points include Safety Features: Full welding was applied to all parts to ensure structural integrity and prevent failures during operation, reducing the risk of accidents or injuries.

Performance Capabilities: The machine can lift a maximum load of 500 kg, with a hydraulic cylinder producing a minimum force of 22.22 kN and generating a pressure of 2.616 MPa

5. STRUCTURAL INTEGRITY

Kumar et al. (2022) present the design and analysis of a hydraulic scissor lift, focusing on its structural integrity and performance. The paper details the selection of materials, load calculations, and hydraulic system design to ensure stability and efficiency. Finite Element Analysis (FEA) is used to evaluate stress distribution and deformation under various load conditions. The results confirm that the proposed design meets safety and operational standards, making it suitable for industrial and commercial applications. The study highlights improvements in load-carrying capacity, safety, and ease of operation compared to conventional lifting mechanisms.

Anugrah and Rachmawati (2021) analyze the hydraulic system of a portable electric hydraulic jack, focusing on its performance, efficiency, and safety. The study examines the working principles of the hydraulic mechanism, including pressure distribution and lifting capacity. Experimental analysis and simulations are conducted to evaluate system efficiency, energy consumption, and structural stability under load. The results (Kumar, M. Kiran, et al.) highlight the jack's effectiveness in providing a compact and user-friendly lifting solution, with recommendations for improving durability and optimizing hydraulic fluid dynamics. The study contributes to the development of more efficient and reliable portable lifting devices.

The hydraulic system (Kumar, M. Kiran, et al. 2016) is designed for lifting vehicles and features a rail securely attached to the vehicle's chassis. A hydraulic jack is mounted on this rail, allowing it to move along its length. This movement is controlled by a system that manages both the positioning and actuation of the jack. The key advantage of this design is its ability to position the jack at various points on the vehicle, making it ideal for tasks like changing tires. This ensures proper alignment for efficient lifting. Overall, the invention enhances vehicle maintenance by offering a versatile and adaptable lifting solution.

Sineri et al., (2021) presents the design and strength analysis of a hydraulic scissor lift to protect home appliances from flood damage. The study evaluates the structural integrity of key components, finding safety factors of 1.252 and 3.11 in different positions, ensuring reliability. The research also incorporates problem-based learning (PBL) to enhance engineering education by addressing real-world challenges. The proposed lifting machine offers a practical flood safety solution while promoting hands-on learning in engineering design.

6. DESIGN AND FABRICATION

Akinsade et al. (2024) present the design and fabrication of a mobile hydraulic lifting machine for heavy-load applications, particularly in automotive maintenance. The device, capable of lifting 500 kg to a height of 2.20 meters, was constructed using locally sourced materials to ensure affordability and accessibility. Performance tests confirmed its efficiency, lifting the maximum load in 99 seconds while maintaining structural integrity. The study highlights the machine's reliability and cost-effectiveness, making it a viable solution for workshops requiring heavy-lifting capabilities.

In Akinsade et al. (2024), paper, "Design and Fabricate of Lifting Equipment for Workshop," Mohd, Zariman, and Deraman address the need for efficient material handling in workshop settings by developing a specialized lifting device. The equipment is designed to safely lift loads up to 100 kg, incorporating features such as a trolley function, adjustable reach, and safety mechanisms like gear locks and wheel brakes. Finite Element Analysis (FEA) conducted using Inventor 2021 software confirmed the frame's structural integrity under the specified load, with a safety factor of 15 and minimal displacement (0.007405 mm). However, tests with a 200 kg load revealed significant displacement (0.1017 mm) and a reduced safety factor (5.61), indicating that the equipment is unsuitable for loads exceeding 150 kg. The authors conclude that the developed lifting equipment effectively enhances safety and ergonomics in manual handling tasks within workshops.

In their paper "Design and Simulation of Multipurpose Built-In Car Lifting Mechanism," Albuwaydi et al., (2023) introduce an innovative car lifting system integrated directly into the vehicle's chassis. This built-in mechanism aims to enhance safety and efficiency during maintenance and emergency situations by eliminating the need for external lifting equipment. The authors conducted comprehensive simulations to validate the design's functionality and structural integrity, demonstrating its potential to improve user convenience and vehicle maintenance procedures.

7. METHODOLOGY

Building a Movable Hydraulic Ram Lifter involves a series of important steps to ensure it's both powerful and practical.

1. Understanding the Need: Start by figuring out exactly what the lifter will be used for—whether it's lifting heavy materials, adjusting heights, or moving loads across tricky terrains. This step also involves deciding on key specs like how much weight it can handle, how high it should lift, how mobile it needs to be, and what kind of power source it will use.

2. Designing the System: Hydraulic Setup: Choose the right hydraulic components—like cylinders, pumps, and valves—that will provide the power and precision needed for the job. Frame Design: Create a sturdy, mobile frame that can handle both vertical and horizontal movement. Add wheels or tracks to ensure it moves easily, even on uneven ground.

3. Picking the Right Parts: Select materials like steel or aluminum for the frame to ensure durability and choose reliable hydraulic components that match the lifter's weight capacity and workload.

4. Bringing It All Together: Assemble the hydraulic system, attach it to the frame, and integrate features like controls (manual or automated), safety mechanisms (such as overload sensors or pressure relief valves), and actuators

8. CONSIDERATIONS FOR FUTURE WORK

Looking ahead, the future of the Movable Hydraulic Ram Lifter should (Sineri, Gabriel Ayrton (Andares, et al., 2021) prioritize efficiency, sustainability, and adaptability to meet the evolving needs of various industries. One key area for improvement is energy efficiency. Optimizing hydraulic pumps and integrating energy-efficient components can reduce power consumption. Additionally, incorporating renewable energy sources like solar or wind could further minimize environmental impact, particularly for remote or off-grid applications.

Enhancing environmental responsibility is another crucial focus. Developing eco-friendly hydraulic fluids and leak-proof systems can mitigate the environmental concerns associated with traditional hydraulic systems. Noise reduction is also essential, especially in urban construction sites and factories. Implementing advanced noise-dampening technology would allow the lift to operate more quietly, making it suitable for noise-sensitive environments. (Kumar, et al., 2018)

The integration of automation and smart technology will also shape the future of the lift. IoT sensors and AI-driven predictive maintenance can enable real-time performance monitoring, detect potential issues before they escalate, and enhance overall efficiency. Remote-control capabilities could further improve safety and flexibility, particularly in hazardous or hard-to-reach areas. (Mohd, Alias, et al., 2021)

Additionally, increasing customization options could expand the lifter's versatility. Modular designs that allow for easy upgrades and various configurations would make it valuable across multiple industries, from construction to emergency disaster response.

These advancements will not only make the Movable Hydraulic Ram Lifter more practical and cost-effective but also ensure its sustainability, efficiency, and adaptability for future applications.

9. CONCLUSION

The development of a Movable Hydraulic Ram Lifter requires a systematic approach to ensure functionality, durability, and adaptability. The process begins with identifying the specific operational needs, including weight capacity, mobility, and power requirements. The design phase focuses on selecting robust hydraulic components—such as cylinders, pumps, and valves—and constructing a sturdy, mobile frame capable of handling various terrains and lifting conditions. Material selection is critical, with durable metals like steel or aluminum ensuring structural integrity, while reliable hydraulic parts enhance performance and longevity.

Assembling the system involves integrating essential safety features, such as overload sensors and pressure relief valves, alongside manual or automated control mechanisms. Looking ahead, optimizing energy efficiency through advanced hydraulic systems and renewable power sources will be crucial. Sustainability efforts should include eco-friendly hydraulic fluids and leak-proof designs to minimize environmental impact. Additionally, incorporating noise reduction technology and smart automation—such as IoT sensors and AI-driven maintenance—will improve operational efficiency and safety.

A modular design approach can further enhance customization, enabling upgrades and adaptability across industries, from construction to disaster response. By addressing these key areas, the Movable Hydraulic Ram Lifter can evolve into a more efficient, sustainable, and versatile tool for various industrial applications.

REFERENCES

1. Adeoye, A. O. M., A. A. Aderoba, and B. I. Oladapo. "Simulated design of a flow control valve for stroke speed adjustment of hydraulic power of robotic lifting device." *Procedia engineering* 173 (2017): 1499–1506.

2. Akinsade A, Eiche JF, Akintunlaji OA, Olusola EO, Morakinyo KA. Development of a Mobile Hydraulic Lifting Machine. Saudi journal of engineering and technology. 2024;9(06):257–264

3. Anand, Ankush, et al. "Tribological and mechanical aspects of zirconia-reinforced aluminum metal matrix composites." Materials Focus 5.6 (2016): 489–495.

4. Akinsade, A., et al. "Development of a Mobile Hydraulic Lifting Machine." *Saudi J Eng Technol* 9.6 (2024): 257–264.

5. Albuwaydi, Hassan Ali, et al. "Design and Simulation of Multipurpose Built-In Car Lifting Mechanism."

6. Al-Hady, Istabraq Hassan Abed, Farag Mahel Mohammed, and Jamal Abdul-Kareem Mohammed. "A review on the employment of the hydraulic cylinder for lifting purposes." *Indones J Electr Eng Comput Sci* 28.3 (2022): 1475–1485.

7. Anugrah, Rinasa Agistya, and Putri Rachmawati. "Analysis of Hydraulic System on Portable Electrical Hydraulic Jack." *4th International Conference on Sustainable Innovation 2020–Technology, Engineering and Agriculture (ICoSITEA 2020)*. Atlantis Press, 2021.

8. Chen, Linlin. "[Retracted] Hydraulic Lifting and Rotating System Lifting Machinery Transmission Control Design." *Mobile Information Systems* 2022.1 (2022): 4617971.

9. Jia, Hao, et al. "Flow characteristics and hydraulic lift of coandă effect-based pick-up method for polymetallic nodule." *Coatings* 13.2 (2023): 271.

10. Kumar, M. Kiran, et al. "Design & analysis of hydraulic scissor lift." *International Research Journal of Engineering and Technology (IRJET)* 3.6 (2016): 1647–1653..

11. Ma, Ranqi, et al. "Leakage fault diagnosis of lifting and lowering hydraulic system of wing-assisted ships based on WPT-SVM." *Journal of Marine Science and Engineering* 11.1 (2022): 27.

12. Mohd, Alias, Muhammad Nur Irfan Zariman, and Zahari Deraman. "Design and fabricate of lifting equipment for workshop." *International Journal of Synergy in Engineering and Technology* 5.1 (2024): 9–23.

13. Conveying scissor fork type mobile hydraulic lift truck. 2014.

14. Sineri, Gabriel Ayrton Andares, et al. "Structural design and strength analysis of lifting machine for home appliance flood safety tool: A problem-based learning." *Indonesian Journal of Multidiciplinary Research* 1.2 (2021): 159–170.

15. Sumaila, Malachy, and Akii Okonigbon Akaehomen Ibhadode. "Design and Manufacture of a 30-ton Hydraulic Press." *AU JT* 14.3 (2011): 196–200.

Advances in Mechanical Engineering and Materials Sciences – Dr. Vinay K. B et al. (eds)
© 2026 Taylor & Francis Group, London, ISBN 9-781-041-20970-6

11

Valveless Diaphragm Micropump—A Study on Annular Actuation using Bond Graphs

Uma B. Baliga[1],
Ranjith R. G.[2], S. M. Kulkarni[3]
Department of Mechanical Engineering,
National Institute of Technology Karnataka, Surathkal,
P.O. Srinivasnagar, Mangalore, India

Abstract: Micropumps have long been integral to applications such as electronics cooling and biomedical systems. Researchers have explored various design modifications, including changes to the chamber, diaphragm, pumping mechanism, and actuation materials, to improve performance. However, existing micropumps often fail to deliver the required flow rates and pressures for practical applications. This study introduces annular actuation as a novel approach to enhance micropump performance. Using the bond graph technique, the micropump was modeled to evaluate flow rate and pressure output. Results demonstrated that annular actuation significantly increases diaphragm deflection, achieving a superior flow rate of 3000 ml/hr compared to traditional central actuation.

Keywords: Valveless micropump, Annular actuation, Bondgraph modelling, Diaphragm deflection, Flow rate enhancement

1. INTRODUCTION

The discipline of Micro Electro Mechanical Systems (MEMS) has experienced extraordinary growth in recent years due to the rapid improvement of miniaturized systems driven by the need for size reduction, cost efficiency, and increased performance. Microfluidics has received a lot of attention, and as a result, many microfluidic devices have been developed, including micropumps (Gidde, Pawar, Ronge, & Dhamgaye, 2019), micro-mixers (Huang & Tsou, 2014), and micro-valves (Jin Yuan Qian, Cong wei Hou, Xiao Juan Li 2021). Medical devices (Srinivasa Rao, Hamza, Ashok Kumar, & Girija Sravani, 2020), drug administration systems [7-8], chemical analysis platforms, micro-dosing systems (Amirouche, Zhou, & Johnson, 2009), and microelectronics (Li, Xia, Jia, Cheng, & Wang, 2017) are only a few of the many applications that use micropumps, an essential part of microfluidic systems.

The remainder of the paper is structured as follows, Section 2 presents a review of the existing literature, Section 3 outlines the research methodology, Section 4 outlines the research methodology, Section 5 discusses the empirical findings and Section 6 provides a summary of the study.

2. LITERATURE REVIEW

2.1 Micropump

Micropumps are vital components of microfluidic systems, converting electrical energy into controlled mechanical motion for precise fluid transport. Their efficiency depends on attributes like low power consumption, fast response times, high flow rates, precise control, compact size, and reliability. Performance is assessed through physical prototyping(Gidde et al., 2019), experimental testing, and modelling techniques(Srinivasa Rao et al., 2020) (Li et al., 2017).

Micropumps are categorized as mechanical or nonmechanical. Non-mechanical micropumps rely on fluid properties for flow generation, whereas mechanical micropumps use moving components like diaphragms

[1]umabbaliga@gmail.com, [2]ranjithrg97@gmail.com, [3]smk@nitk.edu.in

10.1201/9781003725053-11

and valves to regulate fluid motion. Among mechanical micropumps, reciprocating designs utilizing diaphragm deflection are the most common. These pumps require mechanical actuators, such as piezoelectric(Dong et al., 2020), shape memory(Benard, Kahn, Heuer, & Huff, 1998), or pneumatic actuators(Parsi, Zhang, & Masek, 2018), to convert energy into motion. Diaphragm deflection is a critical factor influencing micropump performance. Greater deflection increases chamber volume, enhancing flow rate and pressure(Wang, Ma, Yan, & Feng, 2014). Material selection plays a crucial role, with metals like stainless steel and silicon offering durability but high rigidity, while softer polymers like PDMS and silicone rubber provide greater flexibility and improved deflection.

Mechanical micropump performance can be improved by amplifying diaphragm displacement or increasing driving voltage. Circular diaphragms are often preferred for maximizing deflection. Previous studies report diaphragm deflections of ±6.00 μm at ±200 V and 225 Hz using a flat glass diaphragm (12 mm diameter, 0.2 mm thickness) driven by a circular piezo disc (Hwang, Lee, Shin, Lee, & Lee, 2008). Design modifications, such as adding a center mass or repositioning piezoelectric materials, have been explored to enhance deflection. Performance analysis is often conducted using software like ANSYS(Ye, Chen, Ren, & Feng, 2018), COMSOL Multiphysics (Pradeesh & Udhayakumar, 2019), and MATLAB.

This study employs bond graph modelling to analyze micropump performance. Bond graphs provide a graphical framework for dynamic system modelling across electrical, mechanical, and hydraulic domains. They represent energy flow through directed graphs, where vertices denote components and edges indicate power exchange. This method enables the systematic analysis, synthesis, and control of linear and nonlinear systems. Power variables—effort and flow—and energy variables—momentum and displacement—define system interactions, offering insights into micropump dynamics.

2.2 Micropump Structure

Diaphragm micropumps play a crucial role in microfluidic devices due to their precision in handling small fluid volumes. Research focuses on materials, actuation mechanisms, performance optimization, and application-specific designs, including valve types, diaphragm configurations (Yazdi, Corigliano, & Ardito, 2019), diaphragm materials (Shanuka Dodampegama, et al. 2023), and actuators(Woias, 2005).

Figure 11.1(a) illustrates a valveless diaphragm micropump, a positive displacement pump that utilizes a flexible diaphragm to regulate fluid flow through designated inlet and outlet pathways. Fluid movement is driven by cyclic diaphragm deformation, which induces pressure variations within the chamber. Figure 11.1(b)

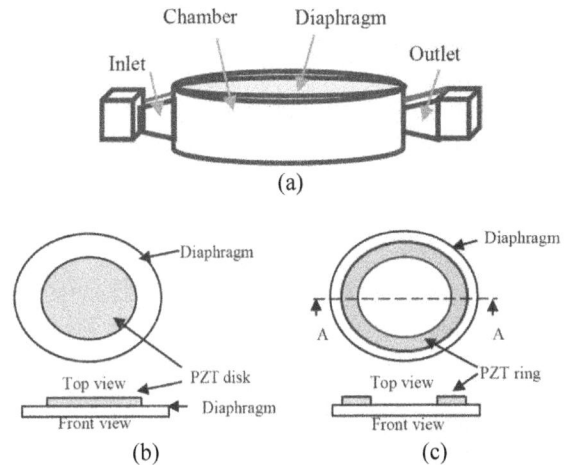

Fig. 11.1 (a) Valveless diaphragm pump, (b) Diaphragm with PZT disk for central actuation, (c) Diaphragm with PZT ring for annular actuation

shows a diaphragm with a centrally actuated piezo disk, while Fig. 11.1(c) depicts annular actuation using a piezo ring.

The Fig. 11.2 shows the block diagram of valveless micropump and its components arrangement with the operating variables. The fluid chamber's pressure changes are produced by the flexible diaphragm, which is driven by an central disk / annular ring to propel flow. Unidirectional flow is ensured by inlet and outlet parts, offering a straightforward, effective system devoid of mechanical valves. This design is appropriate for microfluidic applications because it improves dependability and decreases wear.

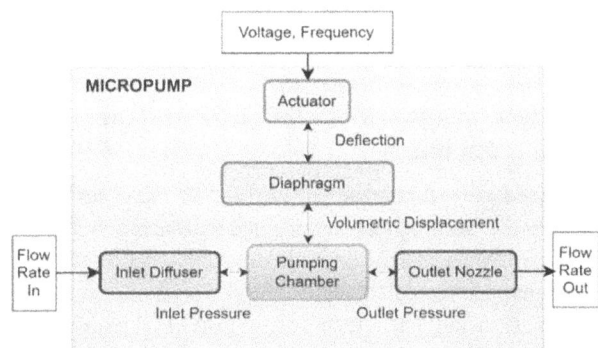

Fig. 11.2 Valveless micropump and its components

To achieve greater diaphragm deflection, in the recent past, researchers have adopted piezo discs, piezo plates, and piezo stack actuators, which develop a limited range of motion, resulting in small-diaphragm deflection, which affects the pump performance. The current work suggests a novel method of diaphragm actuation using an annular piezo actuator. Various factors affecting a micropump's effectiveness are considered when designing the device to achieve optimal flow rate and pressure output.

Researchers have employed various analytical methods to study micropump performance, including numerical simulations, equivalent electrical networks, and state space analysis. This study focuses on mathematical modeling of the micropump using the MATLAB Simulink and bond graph method.

3. METHODOLOGY

The valveless micropump operates in three modes: suction, normal, and pump, as illustrated in Fig. 11.3. During suction mode (Fig. 11.3a), the diaphragm deflects outward, increasing the pump chamber volume, causing fluid to enter through the input valve (acting as a diffuser) and exit through the outlet valve (acting as a nozzle). In pump mode (Fig. 11.3c), the diaphragm deflects inward, decreasing the chamber volume and increasing internal pressure, pushing fluid out through the outlet valve (acting as a diffuser) and the input valve (acting as a nozzle), directing net flow toward the outlet.

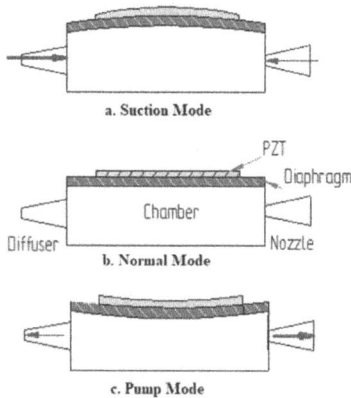

Fig. 11.3 Working of valve less micropump (a) Suction mode, (b) Normal mode, and (c) Pump mode

3.1 Mathematical Modeling of Valveless Micropump

To validate the mathematical model and the bond graph model, the simulations are carried out using both 20-SIM software and Simulink, respectively. Valveless micropump is modeled using its governing mathematical equations using both methods. Figure 11.4 shows the sketch with design variables and geometric dimension [19-22]. Mathematical model of the piezoelectric valveless micropump is modelled with some assumptions, all the external forces and heat transfer to the pump are neglected. It is assumed that the diffuser and nozzle elements have identical geometry. The inlet and outlet pressures are assumed to be the same as the atmospheric pressure.

3.2 Proposed Annular Actuation for Micropump

Circular bossed diaphragms are used to enhance the deflection of the diaphragm. The bossed diaphragm has

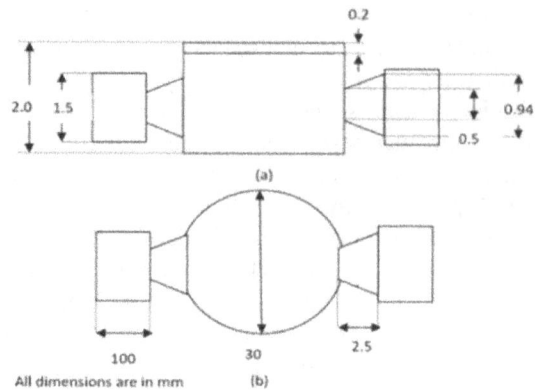

Fig. 11.4 Design variables and geometric dimensions of micropump model (a) Front view (b) Top view

been modified with a central mass. The deflection is measured for two different cases, as shown in Fig. 11.6. with the actuation applied at the diaphragm's centre and the other at an outer radius of the diaphragm. In Fig. 11.5. shows the cross-section of the circular bossed diaphragm and its representation as a spring-mass system where spring represents the stiffness of the diaphragm. To model the circular bossed diaphragm, first the diaphragm is represented as a simple spring mass system. Figure 11.6 shows the simplified model of the bossed diaphragm and the actuation point in different cases - central and annular actuation.

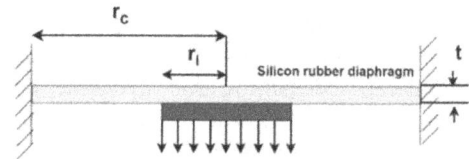

Fig. 11.5 Cross section of the circular bossed diaphragm (Mohith, Karanth, & Kulkarni, 2019)

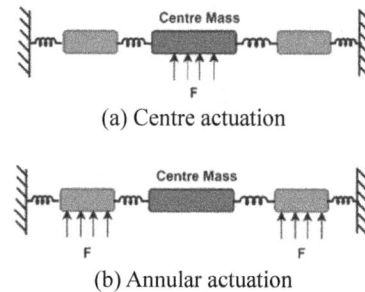

Fig. 11.6 Simplified model of the bossed diaphragm actuation using spring-mass system

Since the diaphragm interacts with the fluid inside the micropump, there will be damping or a restoring force acting on the oscillating motion of the diaphragm. The restoring force is represented by a mechanical damper element R in the bond graph as in Fig. 11.7. The mechanical damper taken here is similar to a piston dashpot arrangement.

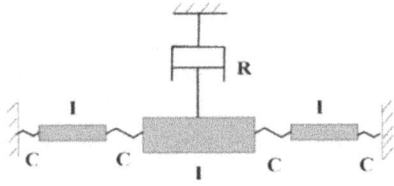

Fig. 11.7 Restoring force acting on diaphragm

In analogy, since there is no gap between the diaphragm and the pump chamber, this radial gap is calculated from the area of both the inlet and outlet nozzle/diffuser elements. The average area of the nozzle/diffuser element is found for both outlet and inlet and then related to the area of gap of the piston dash pot arrangement from which the radial gap is found. The stiffness of the diaphragm for both central and annular actuation will be different. For the bond graph model, the stiffness of the diaphragm is taken as an effective stiffness for central as $(K_{eff})_{central}$ and for annular as $(K_{eff})_{annular}$. Eq (1) shows the central and annular stiffness of the diaphragm (Mohith et al., 2022) (Young and Budynas, 1954). The stiffness, which can be affected by the thickness of the diaphragm, the loading region, the bossed region, and the material qualities, is the main factor responsible for the deflection of the bossed diaphragm.

$$\left(K_{eff}\right)_{central} = \frac{2\pi B_r D}{r_c^2 \left(\frac{c_2 L_6}{c_5} - L_3\right)} \text{ and}$$

$$\left(K_{eff}\right)_{annular} = \frac{\pi \left(r_c^2 - r_l^2\right) D}{r_c^4 \left(\frac{c_2 L_{14}}{c_5} - L_{11}\right)} \quad (1)$$

$$D = \frac{Et^3}{12\left(1 - \gamma^2\right)} \quad (2)$$

$$C_2 = \frac{1}{4}\left[1 - B_r^2\left(1 + 2ln\left(\frac{1}{B_r}\right)\right)\right], C_5 = \frac{1}{2}\left(1 - B_r^2\right) \quad (3)$$

$$L_3 = \frac{B_r}{4}\left\{\left[B_r^2 + 1\right]ln\left(\frac{1}{B_r}\right) + B_r^2 - 1\right\},$$

$$L_6 = \frac{B_r}{4}\left[B_r^2 - 1 + 2ln\frac{1}{B_r}\right] \quad (4)$$

$$L_{11} = \frac{1}{64}\left\{\begin{matrix} 1 + 4\left(\frac{r_l}{r_c}\right)^2 - 5\left(\frac{r_l}{r_c}\right)^4 \\ -4\left(\frac{r_l}{r_c}\right)^2\left[2 + \left(\frac{r_l}{r_c}\right)^2\right]ln\frac{r_c}{r_l} \end{matrix}\right\} \quad (5)$$

$$L_{14} = \frac{1}{16}\left\{1 - \left(\frac{r_L}{r_c}\right)^4 - 4\left(\frac{r_L}{r_c}\right)^2 ln\frac{r_c}{r_L}\right\} \quad (6)$$

Here, B_r is the bossed ratio given by the ratio of the radius of the bossed region by the radius of the circular diaphragm, r_c is the radius of the diaphragm, r_L is the radius of loading region, and D is the rigidity modulus.

3.3 Bond Graph Modelling

Bond graphs provide a robust framework for modeling complex physical systems across multiple domains, including electrical, mechanical, and fluid interactions. They visually represent system components—such as mechanical parts, electrical circuits, and hydraulic systems—using bonds to depict physical quantities like force, flow, voltage, and current (Cao, Jia, Lei, Xin, & Mi, 2014). Although bond graph modeling simplifies system visualization, analyzing these interactions can be challenging.

The behavior of a valveless micropump can be simulated using bond graph modeling. This approach employs basic elements such as one-port, two-port, and three-port junctions. Causal lines connect these ports, representing power and flow, which always run-in opposite directions. One-port elements include resistors, capacitors, and inertial components. Two-port elements, such as transformers and gyrators, ensure power balance and energy conservation. Three-port junctions consist of one-junctions (common flow junctions) and zero-junctions (common effort junctions).

The micropump components are modeled using the basic 1-port elements of bond graph and 3-port junctions. (Morganti, Fuduli, Montefusco, Petasecca, & Pignatel, 2005; Zhang, Wang, & Huang, 2018). The junctions are considered as equivalent bonds of tubes. The tubes can be represented with resistor R_t associated with the fluid's friction effect, the inductor L_t to is dynamic inertia, and the capacitor C_t to its elasticity (Zhang et al., 2018).

$$R_t = \frac{128\mu l}{\pi D^4} \quad (7)$$

in equation 7, R_t fluid resistance of the tube in laminar flow and μ is the viscosity coefficient, l is the tube length, and D is the tube's diameter.

The equivalent bond graph model of the annular and central diaphragm is generated using 20-Sim software, as shown in Fig. 11.8a. and Fig. 11.8b. According to the pump's operation in Fig. 11.8a., the annular actuation involves a sinusoidal input at the diaphragm's periphery on the ring actuator. This actuator deflects the diaphragm positively, causing flow via the diffuser into the chamber, resulting in supply and pump modes in opposite directions. Pump mode moves the diaphragm towards the chamber base,

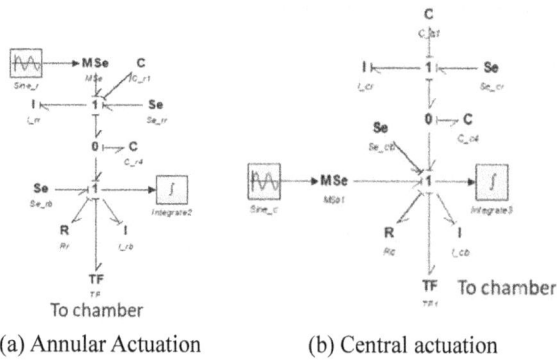

(a) Annular Actuation   (b) Central actuation

Fig. 11.8 Bossed diaphragm in bond graph model for (a) Annular Actuation and (b) Central Actuation

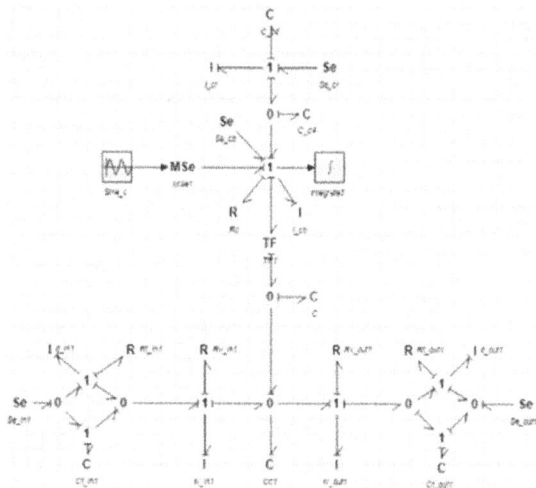

Fig. 11.9 A centrally actuated pump represented by a bond graph

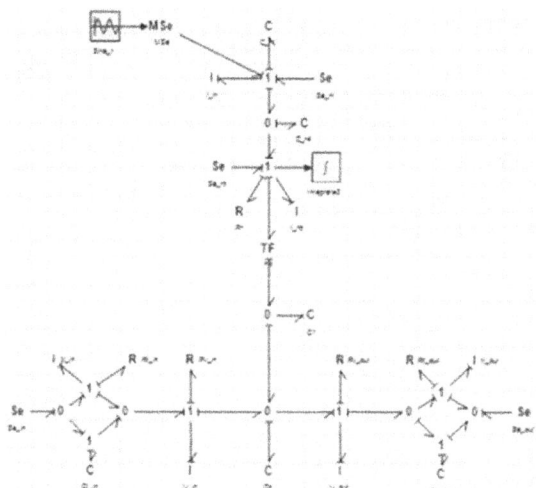

Fig. 11.10 Pump with annular actuation using a bond graph model

reducing volume and expelling fluid through the nozzle, primarily as shown in Fig. 11.10.

In center actuation, the actuation occurs at the diaphragm's center, Fig. 11.8b. When actuated, the diaphragm deflects

away from the chamber base, leading to supply mode, and towards the base, leading to pump mode. The pump's tube, diaphragm, pump chamber, and nozzle/diffuser are modeled. Figure 11.9 displays the combined representation of the bond graph-based valveless piezoelectric pump model. Figure 11.10 shows the bossed diaphragm attached to the micropump model where force and velocity from the diaphragm are transformed to pressure and flow rate using a transformer element 'TF' with the area of actuation as the modulus. Both models are evaluated using different actuation methods, and the outlet flow and pressures are analyzed.

3.4 Modelling using Simulink

Simulink is a block diagram environment used to design systems with multi domain models, simulate before moving to hardware, and deploy without writing code. Mathematical model of the piezoelectric valveless micropump is modelled with some assumptions, all the external forces and heat transfer to the pump are neglected. It is assumed that the diffuser and nozzle elements have identical geometry. The inlet pressure and outlet pressure are assumed to be the same that is the atmospheric pressure.

The function block used in the Simulink model Fig. 11.12 has pressure calculation equations solved using the conditional if else statements of the MATLAB scripts and is shown in Fig. 11.11.

```
function [P,Q1,Q2]=fcn(Pout,Pin, Vo,Cl,Ch)
dP=Pout-Pin;
if Vo>=Cl*sqrt(dP)
P=Pout+((Ch*Vo-Cl*sqrt((Ch^2-Cl^2)*dP+Vo^2))/(Ch^2-Cl^2))^2;
Q1=-Cl*sqrt(P-Pin);
Q2=Ch*sqrt(P-Pout);
elseif Vo<=-Cl*sqrt(dP)
P=Pout-((Cl*Vo+Ch*sqrt((Ch^2-Cl^2)*dP+Vo^2))/(Ch^2-Cl^2))^2;
Q1=Ch*sqrt(Pin-P);
Q2=-Cl*sqrt(Pout-P);
else
P=Pout-((-0.5)*(Vo/Cl)+(1/Cl)*sqrt((0.5)*dP*Cl^2-0.25*Vo^2))^2;
Q1=-Cl*sqrt(P-Pin);
Q2=-Cl*sqrt(Pout-P);

end
end
```

Fig. 11.11 Simulink function block for pressure calculation

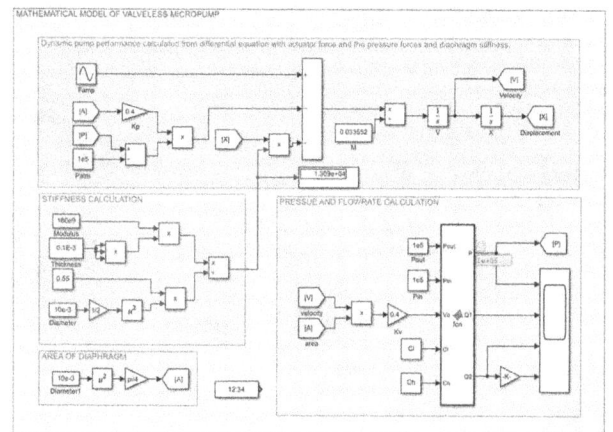

Fig. 11.12 Simulink model of valveless micropump

The pressure in the pump's chamber, P, can be expressed as a function of the conductivity coefficients, the central disk deflection, and the inlet and outlet pressures, thus the pressure inside chamber will be calculate for three different cases

3.5 Full Simulink Model of the Micropump

The stiffness block, and the dynamic equation block which gives deflection are merged with the pressure calculation block to complete the mathematical model of valveless micropump and gives the output as $Q2$ and P the Pressure variation inside the pump chamber to the scope for visualization. Figure 11.12 shows the complete Simulink model of valveless micropump. In this section the valveless micropump was modeled using the Simulink software using the mathematical equations of pump. In the next section we use the other approach to analyse the valveless micropump using the bond graph method. In previous sections, we have seen the methodology involved in modeling the valveless micropump in both Simulink and 20-sim with the similar geometric dimensions, the next section, deals with the simulation results from both softwares.

4. RESULTS AND DISCUSSION

In this section, the results of the simulation analysis are presented. Initially, the results of the annular and centre actuation where diaphragm deflection, flow rate and pressure variation with respect to time are compared. A good agreement was observed between the results of the annular actuation pump calculations and simulation models when compared to those of the conventional centre actuation pump. Specifically, the flow rate and pressure drop were analyzed as functions of the driving frequency, showing favourable consistency between the two approaches.

The modification in the diaphragm geometry with different actuation points are discussed in the previous sections. A fairly good agreement was found where the outcome of annular actuation pump calculations and simulation model results were compared with conventional centre actuation pump, for the flow rate and pressure drop as a function of driving frequency.

Figure 11.13 shows plot for variation of voltage source vs chamber pressure, the graph is plotted with comparison between 20 sim and Simulink and Fig. 11.14 shows plot for variation of voltage source vs flow rates Q2, the graph is plotted with comparison between 20 sim, Simulink and the reference (Ramaswamy & Karanth, 2011).

It is that observed that the flow rate in Simulink model and the reference matches very near compared to 20-Sim. The average values from the plot shows flow rate of approximately 45 ml/hr at 100 volts in (Ramaswamy & Karanth, 2011), 50 ml/hr in Simulink and 80 ml/hr which is slightly higher in case of 20-Sim. In the simulation

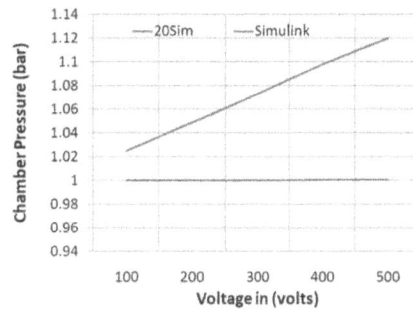

Fig. 11.13 Pressure vs voltage of simulink and 20-sim model

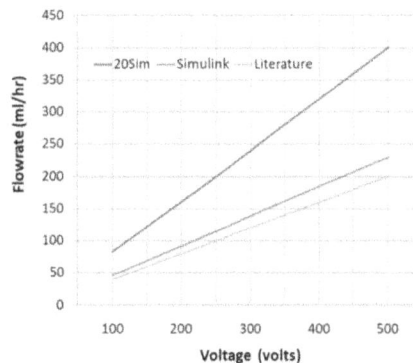

Fig. 11.14 Flow rate vs voltage of simulink and 20-sim model with respect to (Ramaswamy & Karanth, 2011)

when the voltage input is increased the flow rate increases linearly in all three cases Fig. 11.14 shows the comparison results for all the three cases. Similarly, Fig. 11.13 shows the variation of Source vs pressure and the pressure increases linearly as the voltage source is increased. The pressure developed in Simulink is 1.0005 bar and that in 20 sim is 1.02 bar for an input of 100 volt, here the pressure developed in the Simulink model is very low when compared with 20-sim.

It is observed from Fig. 11.15 and Fig. 11.16, that the flow rate and pressure of the micropump gradually increases in the beginning and reach a maximum value and then drops, it is maximum at 100Hz for Simulink with flow rate of 2200 ml/hr with pressure approximately 1.03 bar and is maximum for 150 Hz in the 20-sim model with flow rate 4500 ml/hr at pressure 1.7 bar. The reason for this trend for the flow rate and pressure reducing after certain value of frequency might be due to the back flow occurring at higher frequency where the fluid moving out from the outlet section does not get sufficient time to exit and thus the next cycle starts resulting in back flow of fluid.

From the 20 Sim software, annular and central actuation of the diaphragm of the valveless micropump results were analyzed with input force of 2N and frequency of 55Hz.

The simulation results in Fig. 11.17 show the variation in diaphragm center deflection, flow rate, and pressure over time for annular actuation. The diaphragm exhibits

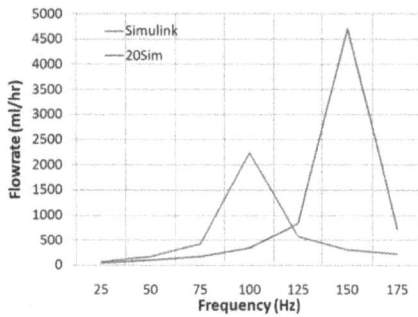

Fig. 11.15   Flow rate vs frequency of simulink and 20-sim model

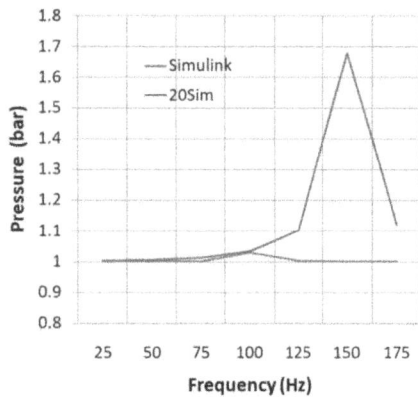

Fig. 11.16   Pressure vs frequency of simulink and 20-sim model

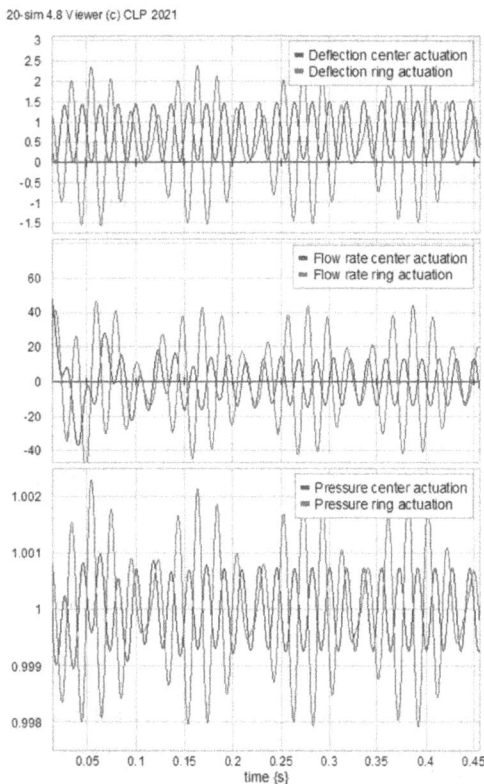

Fig. 11.17   Comparison of Diaphragm deflection, flow rate and chamber pressure variation in centre actuation and annular actuation

bidirectional deflection, with 2 µm on the positive side and 1.5 µm on the negative side, resulting in a net peak-to-peak deflection of 3.5 µm. At 55 Hz, the flow rate reaches 50 ml/hr, and the pressure is 1.002 bar. The bidirectional deflection occurs due to the diffuser mechanism. During annular actuation, the suction phase begins as the diaphragm moves away from the chamber base, reducing internal pressure and allowing fluid inflow. As the diaphragm deflects toward the chamber, pressure rises, and fluid exits through the nozzle, indicating the pump mode.

For central actuation, the simulation results show diaphragm deflection only on the positive side of the x-axis, with a peak deflection of 1.5 µm, a flow rate of 12 ml/hr, and a pressure of 1.00075 bar at 55 Hz. Comparing both actuations, annular actuation achieves a significantly higher flow rate. Additionally, beats are observed in annular actuation signals due to a slight mismatch in the natural frequencies of the diaphragm and center mass. Beats form when two signals of slightly different frequencies combine. The phase shift in annular actuation occurs as the deflection shifts from the diaphragm's periphery to the center. Initially, fluid flow is slow but increases gradually.

The simulation results in Fig. 11.18 illustrate how diaphragm deflection, flow rate, and pressure vary with central actuation. As frequency increases, the flow rate decreases due to chamber expansion in suction mode, causing fluid to spread within the chamber. This leads to a pressure drop, while flow compression at the nozzle increases diaphragm deflection.

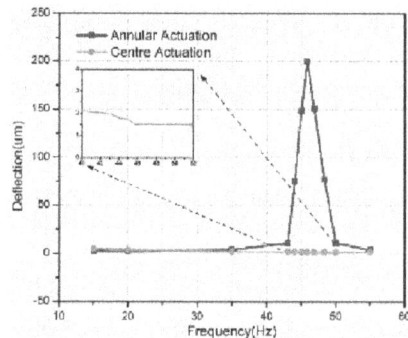

Fig. 11.18   Diaphragm deflection variation in annular and central actuation with respect to frequency

Figure 11.19 illustrates the variation in flow rate for annular and central actuation with respect to frequency. In annular actuation, the flow rate decreases at the diffuser during the suction phase, while pressure increases as fluid velocity reduces. In central actuation, diaphragm deflection is primarily influenced by flexibility, and the initial flow rate is high due to the starting conditions. The maximum flow rate in annular actuation is observed at 45.8 Hz, reaching 3000 ml/hr, with a maximum diaphragm deflection of 200 µm. As frequency increases, the flow rate decreases due to chamber expansion, causing the flow to spread within the chamber.

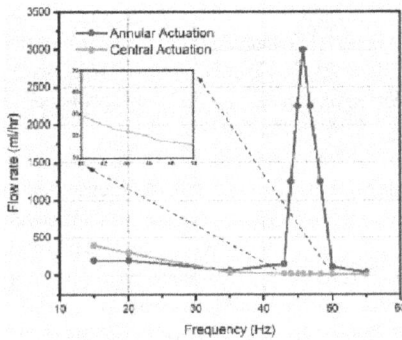

Fig. 11.19 Flow rate variation in annular and central actuation with respect to frequency

5. CONCLUSION

Micropump is developed using MATLAB Simulink and 20-sim and the outputs are studied. Both the model from Simulink and 20-sim are validated with the results.

The Simulink model is run for different input voltages and frequencies to get output flow rate and chamber pressure, it is observed from results that it increases linearly with increasing voltage. But the flowrate and the pressure increase up to a certain value of frequency and then decreases when the simulation is performed with different increasing frequency. Similarly, the 20-Sim model shows the flow rate and pressure are linear to the increasing voltage and drops after certain frequency in case of simulations with increasing frequency.

Both the model is compared with the literature and found that all the three shows the similar trend of increasing flow rate and pressure with the input voltage. And the simulations with increasing frequency shows the similar trend of decreased flow rate after an optimum frequency for both the models, it is maximum of 2200 ml/hr at 100 Hz for Simulink and 4500 ml/hr at 150 Hz for 20-sim.

Comparison of central and annular actuation is also studied and the results shows annular actuation gives greater diaphragm deflection of $200 \mu m$ at 45.8 Hz giving a flow rate of 3000ml/hr compared with $1.5 \mu m$ and 25 ml /hr at the same frequency of 45.8 Hz, which is more than central actuation.

REFERENCES

1. Amirouche, F., Zhou, Y., & Johnson, T. (2009). Current micropump technologies and their biomedical applications. *Microsystem Technologies*, *15*(5), 647–666. https://doi.org/10.1007/s00542-009-0804-7
2. Benard, W. L., Kahn, H., Heuer, A. H., & Huff, M. A. (1998). Thin-film shape-memory alloy actuated micropumps. *Journal of Microelectromechanical Systems*, *7*(2), 245–251. https://doi.org/10.1109/84.679390
3. Cao, D., Jia, Q., Lei, L., Xin, Z., & Mi, J. (2014). Cantilever piezoelectric micro-pump modeling based on bond graph port. *Key Engineering Materials*, *609–610*, 819–824.

https://doi.org/10.4028/www.scientific.net/KEM.609-610.819
4. Cobo, A., Sheybani, R., Tu, H., & Meng, E. (2016). A wireless implantable micropump for chronic drug infusion against cancer. *Sensors and Actuators, A: Physical*, *239*, 18–25. https://doi.org/10.1016/j.sna.2016.01.001
5. Das, P. K., & Hasan, A. B. M. T. (2017). Mechanical micropumps and their applications: A review. *AIP Conference Proceedings*, *1851*. https://doi.org/10.1063/1.4984739
6. Dong, J., Cao, Y., Chen, Q., Wu, Y., Liu, R. G., Liu, W., … Yang, Z. (2020). Performance of single piezoelectric vibrator micropump with check valve. *Journal of Intelligent Material Systems and Structures*, *31*(1), 117–126. https://doi.org/10.1177/1045389X19880024
7. Gidde, R. R., Pawar, P. M., Ronge, B. P., & Dhamgaye, V. P. (2019). Design optimization of an electromagnetic actuation based valveless micropump for drug delivery application. *Microsystem Technologies*, *25*(2), 509–519. https://doi.org/10.1007/s00542-018-3987-y
8. Huang, C., & Tsou, C. (2014). The implementation of a thermal bubble actuated microfluidic chip with microvalve, micropump and micromixer. *Sensors and Actuators, A: Physical*, *210*, 147–156. https://doi.org/10.1016/j.sna.2014.02.015
9. Hwang, I. H., Lee, S. K., Shin, S. M., Lee, Y. G., & Lee, J. H. (2008). Flow characterization of valveless micropump using driving equivalent moment: Theory and experiments. *Microfluidics and Nanofluidics*, *5*(6), 795–807. https://doi.org/10.1007/s10404-008-0275-7
10. Jin Yuan Qian, Cong wei Hou< Xiao Juan Li, Z. J. J. (2021). Actuation mechanism of microvalves A review.
11. Li, Y., Xia, G., Jia, Y., Cheng, Y., & Wang, J. (2017). Experimental investigation of flow boiling performance in microchannels with and without triangular cavities – A comparative study. *International Journal of Heat and Mass Transfer*, *108*, 1511–1526. https://doi.org/10.1016/j.ijheatmasstransfer.2017.01.011
12. Mohith, S., Karanth, P. N., & Kulkarni, S. M. (2019). Experimental investigation on performance of disposable micropump with retrofit piezo stack actuator for biomedical application. *Microsystem Technologies*, *25*(12), 4741–4752. https://doi.org/10.1007/s00542-019-04414-2
13. Mohith, S., P, N. K., & Kulkarni, S. M. (2022). Analysis of annularly excited bossed diaphragm for performance enhancement of mechanical micropump. *Sensors and Actuators: A. Physical*, *335*, 113381. https://doi.org/10.1016/j.sna.2022.113381
14. Morganti, E., Fuduli, I., Montefusco, A., Petasecca, M., & Pignatel, G. U. (2005). SPICE modelling and design optimization of micropumps. *International Journal of Environmental Analytical Chemistry*, *85*(9–11), 687–698. https://doi.org/10.1080/03067310500153876
15. Parsi, B., Zhang, L., & Masek, V. (2018). Vibration Analysis Of A Double Circular Pzt Actuator For A Valveless Micropump, 1–6. https://doi.org/10.25071/10315/35215
16. Pradeesh, E. L., & Udhayakumar, S. (2019). Effect of placement of piezoelectric material and proof mass on the performance of piezoelectric energy harvester. *Mechanical Systems and Signal Processing*, *130*, 664–676. https://doi.org/10.1016/j.ymssp.2019.05.044

17. Ramaswamy, N., & Karanth, N. (2011). Modeling of Micropump Performance and Optimization of Diaphragm Geometry. *IJCA Proceedings on ...*, (October 2015), 14–19. Retrieved from https://www.ijcaonline.org/isdmisc/number5/isdm113.pdf

18. Shanuka Dodampegama, Amith Mudugamuwa, Menaka Konara, Gehan Melroy, Uditha Roshan, Ranjith Amarasinghe, P. W. & V. D. (2023). Novel Design and Simulation Approach for a Piezoelectric Micropump with Diffusers. *Sustainable Design and Manufacturing. SDM 2022. Smart Innovation, Systems and Technologies, Vol 338. Springer, Singapore.*, pp 168–180. https://doi.org/https://doi.org/10.1007/978-981-19-9205-6_16

19. Srinivasa Rao, K., Hamza, M., Ashok Kumar, P., & Girija Sravani, K. (2020). Design and optimization of MEMS based piezoelectric actuator for drug delivery systems. *Microsystem Technologies*, *26*(5), 1671–1679. https://doi.org/10.1007/s00542-019-04712-9

20. Wang, W., Guo, D., Pei, R., Niu, J., Geng, Y., Liu, S. (2024). Fluid Mechanism Analysis of Insulin Pump Set Failure Based on Power Bond Graph. *Proceedings of the UNIfied Conference of DAMAS, IncoME and TEPEN Conferences (UNIfied 2023). TEPEN IncoME-V DAMAS 2023 2023 2023. Mechanisms and Machine Science, Vol 152. Springer, Cham.*, pp 333–343. https://doi.org/https://doi.org/10.1007/978-3-031-49421-5_26

21. Wang, X. Y., Ma, Y. T., Yan, G. Y., & Feng, Z. H. (2014). A compact and high flow-rate piezoelectric micropump with a folded vibrator. *Smart Materials and Structures*, *23*(11). https://doi.org/10.1088/0964-1726/23/11/115005

22. Woias, P. (2005). Micropumps—past, progress and future prospects. *Sensors and Actuators B: Chemical*, *105*(1), 28–38. https://doi.org/10.1016/j.snb.2004.02.033

23. Yazdi, S. A. F. F., Corigliano, A., & Ardito, R. (2019). 3-D design and simulation of a piezoelectric micropump. *Micromachines*, *10*(4), 1–17. https://doi.org/10.3390/mi10040259

24. Ye, Y., Chen, J., Ren, Y. J., & Feng, Z. H. (2018). Valve improvement for high flow rate piezoelectric pump with PDMS film valves. *Sensors and Actuators, A: Physical*, *283*, 245–253. https://doi.org/10.1016/j.sna.2018.09.064

25. Young and Budynas. (1954). *Roark's Formulas for stress and strain. Journal of the Mechanics and Physics of Solids* (Vol. 3). https://doi.org/10.1016/0022-5096(54)90042-3

26. Zhang, J., Wang, Y., & Huang, J. (2018). Equivalent circuit modeling for a valveless piezoelectric pump. *Sensors (Switzerland)*, *18*(9), 1–13. https://doi.org/10.3390/s18092881

Note: All the figures in this chapter were made by the authors.

Advances in Mechanical Engineering and Materials Sciences – Dr. Vinay K. B et al. (eds)
© *2026 Taylor & Francis Group, London, ISBN 9-781-041-20970-6*

12

Experimential and Numerical Investigation on Drilling of GFRP Composites—A Comprehensive Review

Ranjith K.[1],
Deepak R.[2], Manoj M.[3],
Chandrashekar[4], Manjesh gowda M. J.[5]
Department of Mechanical Engineering, Vidyavardhaka College of Engineering Mysuru, Karnataka, India

Abstract: Glass-fiber reinforced polymer composite is highly preferred for its excellent strength-to-weight properties. However, GFRPs have notable challenges, including anisotropy, low thermal conductivity, and a tendency for delamination. During machining, fibers often pull out, making the process challenging. This article comprehensively reviews experimental and numerical studies on the drilling of GFRP composites. More attention is paid to the impact of some drilling parameters on the quality of holes machined and the material damage caused. The test data acquired during these tests were analyzed against the level of delamination, fiber pull-out, and hole diameter precision. According to the results, Drilling speed and feed are the main factors that influence surface finish quality and the likelihood of delamination. This document also includes a review of intricate simulations of the drilling employing the finite element method.

Keywords: GFRP, Drilling, Twist drill, ANSYS

1. INTRODUCTION

For the optimization of the fibre reinforced epoxy composite drilling process to its full potential, the article highlights the simulation and optimization of major machining process parameters. The speed of the spindle and the feed are the most important variables to consider, particularly with respect to the different laminate thicknesses, torque, and delamination variables [1]. The research considers how these control parameters and their interaction influence the result of different drilling operations by artificial neural networks (ANN). The evaluation reveals how imperative it is to optimize drilling parameters and tool properties in an effort to create high-quality holes with the least damage. The key evaluation parameters are equal burr height, dimensional accuracy, surface roughness, and delamination. The authors emphasize

the significance of drilling temperatures and pressures in preserving accuracy and preventing delamination in their examination of various drilling techniques and alternative processes intended to improve hole quality [2].

Gérald Franz et al. discuss the delamination during drilling, which can occur at both top and bottom points of the hole, along with the factors influencing surface roughness, including cutting settings and tool geometry detail. The review article also discusses the importance of burr generation in aluminum drilling and how it affects production costs and assembly tolerances, which the aircraft manufacturing sector is increasingly using because to its favourable strength-to-weight ratio [3]. The research by Elango Natarajan et al. focuses on optimizing drilling settings to minimize delamination, a common problem in drilling that compromises structural

[1]ranjithraj@vvce.ac.in, [2]appudeepak74@gamil.com, [3]manojgowda1178@gmail.com, [4]chandramahadeva64@gmail.com, [5]anumanjeshgowdamj@gmail.com

10.1201/9781003725053-12

integrity. The researchers have conducted experimental research to investigate the effects of variations in drilling process parameters on the delamination. Researchers have employed both ANN simulations and experiments to discuss drilling forces and delamination in Glass fibre composites. GFRP composites of 4 types are considered in this study, with variations in feed per tooth and drilling speed. Composite preparation is performed using different production methods. One of the primary damaging factors affecting the structural integrity of GFRP components, especially for aerospace applications, is delamination [14]. The authors of this study highlight the challenges posed by drilling GFRP composites because of their multilayered and heterogeneous architectures. During the drilling process, these structures commonly lead to issues like delamination, fuzzing, and heat deterioration. These difficulties might eventually impact the composites' mechanical strength, which could lead to higher tool wear, dimensional mistakes, and subpar surface finishes.

The study is meant to fill a literature vacuum and also provides suggestions that the military and aerospace industries might follow, emphasizing the need for new techniques for drilling composite materials. Yalçın and colleagues aims to explore the impact of process parameters on the delamination phenomena encountered when drilling GFRP composites utilized in the aeronautical sector. To assess how diameter of the drill tool, feed and speed influence drilling performance, the research employs a drill specially designed for trials as part of the Taguchi method [10]. The Zhirov-point and multifacet drills outperformed conventional drill types due to their superior surface finishes and reduced thrust forces. The findings from the study, which assessed how feed rate and spindle speed impact drilling results using ANOVA, demonstrate that the rate of feed significantly affects thrust force across all drilling tool types. The data of this study demonstrate the need to improve drilling conditions to lessen delamination and improve hole quality in GFRP composites, which are prevalently used in industries like aerospace [8]. Since industries are increasingly utilizing high-strength materials, the ability of mechanical micro-drilling to create micro-sized holes, which are essential for applications in microelectronics, biomedical devices, and aerospace components, has attracted a lot of attention. The study draws attention to the unique problems caused by the size impact in micro-drilling, which makes it complicated to adapt traditional drilling techniques for the microscale [11].

2. DRILLING PROCESS PARAMETERS

The optimal feed rate is determined to be 0.2 mm/min, which also decreased delamination when compared to higher rates. The Dandelion optimizer, a metaheuristic search technique, was used to improve the drilling settings, yielding positive results with little delamination [4]. The

researchers employed high-speed steel (HSS) drill bits of different diameter ranges in their experiments. These drills were operated at speeds between 500 and 2500 rpm. The tests were designed to identify combinations of specific parameters that significantly affected cutting forces and surface quality. Key findings indicate that the optimal spindle speeds for achieving the best surface quality were 1800 rpm, a feed rate of 250 mm/min, and a cut depth of 0.6 mm [5]. The influence of drill speed and stacking sequence on various drilling parameters is reported, and temperature is analyzed using thermocouples to monitor temperature variations. The outcomes reveal that the temperature is heavily impacted by the cutting speed [6]. Drilling research indicated that although feed rate increased thrust forces, cutting speed reduced them. Also, it was observed that B4C composites showed more thrust and delamination forces than Gr composites. It concluded that while fillers are enhancing the wear resistance, they are negatively affecting mechanical properties. This proves the requirement of achieving ideal ratios of fillers to achieve a balance in performance in GFRC applications [9].

Generally, increased feed rates resulted in higher cuts; that is, higher the magnitude, amplitude of cutting forces coupled along with delamination factors [14]. The experiments revealed that the thrust force was made maximum only at the points, in such a manner the spindle speed is 2250-4000 rpm, tool diameter 3mm feed rate at the stage of 0.2 mm/rev, then such three factors appear to work best because these three appear to work very good from the results presented. Both performance and structural integrity inherent in aviation lead to assembly performance and financial efficiency. The factors can be formulated using the findings [10]. The principal areas of study are temperature, hole quality, and thrust force. Various drilling characteristics, such as the types of tools (solid carbide, high-speed steel, and solid carbide Balinit Helica coated), laminate thicknesses (from 3 to 7 millimeters), spindle speeds (from 12,000 to 18,000 revolutions per minute), and feed rates (from 300 to 700 millimeters per minute) were all analyzed through analysis of variance (ANOVA) and reaction surface methodology (RSM). The conclusions of the study show that delamination, a major problem with GFRP drilling, occurs as Debonding of the matrix, fibers that remain uncut, and pull-out fibers, all of which influence the surface quality and dimensional accuracy of the drilled holes [17]. Drilling procedures for composite materials provide valuable information about how to optimize drilling parameters for improved drilling outcomes. The success of drilling is significantly influenced by four main factors: drill diameter, layer orientation, feed rate, and rotation speed [19].

3. EFFECT OF REINFORCEMENT

The increasing use of lightweight structural materials has given rise to the use of fiber-metal laminates and hybrid composite stacks, which combine the advantageous properties of metals and composites. However, drilling these materials can be challenging because of their distinct Properties, which can lead to issues including burr development, big holes, heat damage, and rapid tool wear [3]. Studies showed that, in compared to individual FRPs, the hybrid composites had better tensile and flexural strength [4]. Hüseyin Gürbüz analysed the effects of graphite-boron carbide fillers on the tribological properties, mechanical behaviors, and drilling performance of GFRCs. Specimens were produced using the lay-up Process, and various filler ratios (5%, 10%, and 15%) were applied uniformly [9]. Cutting speed significantly influences temperature; higher speeds produce larger thermal peaks, especially in the Al/GFRP/Al combination [6].

A study on hybrid composites consisting of carbon fabric, solid lubricant filler, and epoxy revealed how different filler materials, specifically molybdenum disulfide and hexagonal boron nitride, influence drilling performance. To maintain the integrity of carbon-epoxy composites, which are used in a range of structural applications, the data from this study demonstrate the importance of preventing drilling-induced damage, such as delamination and surface roughness [12]. CFRP and GFRP composite materials are increasing rapidly in the automotive, marine, and aerospace sectors as they provide excellent mechanical properties in conjunction with remarkably low density as compared to metals. Unfortunately, these composites pose significant machining challenges due to anisotropy and heterogeneity in their structures. In processing, these factors can prove detrimental [18].

4. EFFECT OF MACHINING PARAMETERS

In most of the previous works, trailing of various drill size and cutting speeds were selected in trials to find the best conditions while improving the degree of surface quality. It confirmed that the spindle speed at 1200 revolutions per minute and the feed rate at 50 millimeters per revolution confer quality to the surface. The study shows that extensive testing and optimization methods are required to improve drilling performance in composite materials [5]. The mechanics of drilling delamination and the factors that influence surface roughness, such as tool geometry and cutting conditions, are also explained in detail. Also explains the importance of burr generation in drilling aluminum and how it affects the cost of manufacturing and assembly tolerances. The paper is a useful manual for manufacturers and researchers, providing information on enhancing drilling performance for aerospace applications and recent developments in drilling technology for CFRP/Al stacks and FMLs. The findings indicate the need for ongoing research and development in this area to improve the quality and productivity of aeronautical part production [3]. By minimizing delamination due to reduced thrust forces, they noted that increasing the tip angle from

80° to 120° enhanced surface finishes and reduced tool wear [4].

Limited research has explored the connection between drilling circumstances and the physical characteristics of FRP composites, despite prior research concentrating on enhancing hole quality through processing settings [7]. GFRP acts as a thermal barrier, absorbs heat, and removes damage through heat. The delamination factor was investigated, which enables the quantification of damage through drilling. The evidence clearly explains the importance of selecting the proper cutting parameters and the appropriate stacking sequence for an optimally efficient drilling process in FML constructions. Altogether, the obtained results provide very significant new insights into the dynamic interaction between material properties and cutting conditions, supporting the development of new, improved drilling techniques for hybrid composite materials in industrial applications. The right type of drilling setting tends to reduce the probability of delamination, one of the significant problems plaguing laminated composites, which lowers their performance with time.

The study investigates how drill wear can impact the machinability of woven GFRE composites during the drilling operation. This characteristic draws attention to the potential cost consequences of drilling errors that may be present when composite structures are being built [16]. The following experiment involved an integrated setup with various feed rates and speeds. The concurrent investigation encompassed thrust force, delamination properties, and surface roughness. The output analysis, conducted using SPSS, showed a significant reliability. It is highlighted that, particularly at higher speeds and feeds, the increase in drill wear significantly affects force of thrust and surface roughness, resulting in even more significant delamination [19]. It is very important to optimize drilling parameters- the feed rate, speed, and drill bit shape- to reduce the occurrence of damage and further develop hole quality. This paper, accordingly, discusses various influencing parameters such as patterns in a hole, matrix-fiber compatibility, and fibre types on the post-machining mechanical properties of a composite [14]. In this investigation, two types of diamond-coated special drills were used. This study mainly focuses on the relationship between cutting speeds, feed rates, and drilling performance. Among the several factors examined in this study are cutting pressures, machining temperatures, damage caused by drilling, precision in drilled hole dimension, and the morphology of hole walls [13].

5. NUMERICAL SIMULATION

The numerical investigation of drilling GFRP composite laminates is an emerging area of research due to these materials' complex, heterogeneous structure, which poses significant machining challenges. Numerical simulation allows for evaluating both conventional and non-conventional drilling techniques, offering insights into the influence of tool geometry, feed rate, as well as other factors on the process. Spindle speed without the need for extensive experimental trials [20]. Key challenges in GFRP drilling, such as delamination and temperature degradation, are critical in developing accurate numerical models [21]. The choice of tool material and process parameters significantly impacts surface integrity and drilling-induced damage [22]. Recent studies highlight the potential of molded-hole GFRP components, which demonstrate superior flexural strength compared to traditionally drilled counterparts [21] and emphasize the need for sustainable machining strategies and real-time defect monitoring [23]. Nevertheless, current numerical models face limitations in capturing the complex cutting mechanics, especially with the advent of nanopolymer-reinforced GFRP. Future research should focus on experimental validation of simulation results and integrating eco-friendly drilling technologies to improve performance and sustainability [24].

6. CONCLUSION

In summary, the review of experimental and numerical analysis of drilling in GFRP composite offers a thorough discussion on how variations in drilling parameters affect the drilling and the quality of the holes. Experimental analysis identified that the improvement in feed and speed increased the cutting forces and the rate of tool wear, but it also affected the formation of defects. Appropriate feed and speed optimization was necessary in minimizing such defects while optimizing the drilling process to the maximum. Additionally, incorporating silica nitride into the composite matrix enhanced the material's thermal degradation and wear resistance, helping to minimize the overall impact of the drilling process.

Quantitatively, FEM simulations were successful in the case of stress concentrations around drilled holes. It was found that increased feed rates resulted in increased local stress concentrations at the hole peripheries, which was the reason for delamination and surface crack initiation. Test results showed that the feed rate could be modified to control the stress distribution and, therefore, the quality of the drilled hole. Additionally, utilizing epoxy resin and silica nitride ensured an even distribution of these stresses, minimizing damage from drilling. Overall, this work highlighted the importance of optimal values of drilling parameters like feed speed to achieve high-quality drilled holes in GFRP composites. Experimental-numerical approach, particularly FEM stress analysis, provided a comprehensive insight into the drilling process and the effect of material composition in reducing defects and maximizing hole quality. Future work can involve studying other variations of parameters and material combinations to enhance the drilling process further.

REFERENCES

1. Application of Central Composite Design in the Drilling Process of Carbon Fiber-Reinforced Polymer Composite- Seyyedabbas Arhamnamazi, Francesco Aymerich and Hossein Taheri, Pasquale Buonadonna.-https://doi.org/10.3390/app14177610

2. Drilling Process of GFRP Composites: Modeling and Optimization Using Hybrid ANN-M.S.Abd-Elwahed-http://dx.doi.org/10.3390/su14116599

3. A Review on Drilling of Multilayer Fiber-Reinforced Polymer Composites and Aluminum Stack- Gérald Franz , Pascal Vantomme 1 and Muhammad Hafiz Hassan.-http://dx.doi.org/10.3390/fib10090078

4. Drilling-Induced Damages in Hybrid Carbon and Glass Fiber-Reinforced Composite Laminate and Optimized Drilling Parameters- Elango Natarajan, Kalaimani Markandan, Santhosh Mozhuguan Sekar, Kaviarasan Varadaraju , Saravanakumar Nesappan, Anto Dilip Albert Selvaraj, Wei Hong Lim and Gérald Franz. -https://doi.org/10.3390/jcs6100310

5. Experimental Investigation to Optimize Process Parameters in Drilling Operation for Composite Materials- K. Amarnath, P. Surendernath, V. Kumar.-http://dx.doi.org/10.1007/978-981-15-1124-0_29

6. Investigation of Temperature at Al/Glass Fiber-Reinforced Polymer Interfaces When Drilling Composites of Different Stacking Arrangements Brahim Salem, Ali Mkaddem, Malek Habak and Abdessalem Jarraya, Yousef Dobah, MakramElfarhani -https://doi.org/10.3390/polym16192823

7. Drilling characteristics and properties analysis of fiber reinforced polymer composites: A comprehensive review Praveenkumara Jagadeesh Indran Suyambulingam a, Sanjay Mavinkere Rangappa, Madhu Puttegowda-https://doi.org/10.1016/j.heliyon.2023.e14428

8. An investigation on high speed drilling of glass fibre reinforced plastic (GFRP) V Krishnaraj, S Vijayarangan & G Suresh -https://www.researchgate.net/publication/273058857_An_investigation_on_high_speed_drilling_of_glass_fibre_reinforced_plastic_GFRP

9. Investigation of Drilling Performances, Tribological and Mechanical Behaviours of GFRC Filled with B4C and Gr Hüseyin Gürbüz, Ibrahim Halil Akcan, Sehmus Baday, Mehmet Emin Demir-https://doi.org/10.3390/polym16213011

10. Effect of Drilling Parameters and Tool Diameter on Delamination and Thrust Force in the Drilling of High-Performance Glass/Epoxy Composites for Aerospace Structures with a New Design Drill Bekir Yalçın, Çağın Bolat , Berkay Ergene, Ali Ercetin , Sinan Maraş , Uçan Karakılınç and OguzhanDer, Çağlar Yavaş ,Yahya,Öz .-https://doi.org/10.3390/polym16213011

11. A critical review on mechanical micro-drilling of glass and carbon fibre reinforced polymer (GFRP and CFRP) composites Norbert -Geiera, Karali Patrac, Ravi Shankar Anandd, Sam Ashworthe, Barnab´ as Zolt´an Bal´ azsa-http://dx.doi.org/10.1016/j.compositesb.2023.110589

12. Drilling Response of Carbon Fabric/Solid Lubricant Filler/Epoxy Hybrid Composites: An Experimental Investigation, Rao, Y.S.; Mohan, N.S.; Shetty, N.; Acharya.S. https://doi.org/10.3390/jcs7020046

13. Experimental study of drilling behaviors and damage issues for woven GFRP composites using special drills Jinyang Xua, Linfeng Li a, Norbert Geier b, J. Paulo Davim c, Ming Chen-https://doi.org/10.1016/j.jmrt.2022.09.100

14. Experimental Study and Artificial Neural Network Simulation of Cutting Forces and Delamination Analysis in GFRP Drilling Katarzyna Biruk-Urban, Paul Bere, Jerzy Józwik and MichałLele´ n-https://doi.org/10.3390/ma15238597

15. Analyzing The Surface Roughness for GFRPs Drilled Hole When Using Different Tools and Different Methodology (A Review)- Ahmed M. Easa, Abeer S. Eisa-https://doi.org/10.21608/erjm.2021.74592.1095

16. Effect of Drill Attrition on Machinability in Drilling Woven GFR Epoxy Composites-A.K. Ratheesh, M. Ramachandran, Manjula Selvam 2Kurinjimalar Ramu-https://doi.org/10.46632/jame/1/1/8

17. The effects of drilling parameters on thrust force, temperature and hole quality of glass f iber reinforced polymer composites-Khurshid Malik1, Faiz Ahmad, Woo Tze Keong and Ebru Gunister-http://dx.doi.org/10.1177/09673911221131113

18. A review on the machining of polymer composites reinforced with carbon (CFRP), glass (GFRP), and natural fibers (NFRP) -Mohamed Slamani1, JeanFrançois Chatelain-https://doi.org/10.3390/polym16202927

19. Machinability analysis in Drilling Composites and drilling woven GFR/epoxy composites using the SPSS Method -Chinnasami Sivaji, M. Ramachandran, Vidhya Prasanth-https://doi.org/10.46632/jame/2/1/4

20. Shinde, A., Siva, I., Munde, Y. S., Sultan, M. T. H., Hua, L. S., & Shahar, F. S. (2022). Numerical Modelling of Drilling of Fiber Reinforced Polymer Matrix Composite: A Review. Journal of Materials Research and Technology. https://doi.org/10.1016/j.jmrt.2022.08.063

21. Cengiz, A., Öztürk, M. F., Sabah, A., & Avcu, E. (2024). Experimental and numerical analysis of drilled- and molded-hole glass fiber reinforced polymer matrix (GFRP) composites. Polymer Composites. https://doi.org/10.1002/pc.28305

22. Rahmé, P. (2024). A bibliometric analysis on Drilling of Composite Materials using a Systematic Approach. Heliyon, 10(17), e37282. https://doi.org/10.1016/j.heliyon.2024.e37282

23. Mohan, N. S., Sharma, S., & Bhat, R. (2019). A comprehensive study of glass fibre reinforced polymer (GFRP) drilling. 9(1), 1–10. https://doi.org/10.24247/IJMPERDFEB20191

24. Panchagnula, K. K., & Palaniyandi, K. (2017). Drilling on fiber reinforced polymer/nanopolymer composite laminates: a review. Journal of Materials Research and Technology, 7(2), 180–189. https://doi.org/10.1016/J.JMRT.2017.06.003

Advances in Mechanical Engineering and Materials Sciences – Dr. Vinay K. B et al. (eds)
© 2026 Taylor & Francis Group, London, ISBN 9-781-041-20970-6

13 Standing Wheelchair—A Review

Vinay K. B.[1]
Department of Mechanical Engineering,
Vidyavardhaka College of Engineering,
Mysuru, India

**Vishal M. S.[2], Melvin Thomas[3],
Vikas Gowda M. H.[4], Afzal Lateef M. A.[5]**
Research Students, Department of Mechanical Engineering,
Vidyavardhaka College of Engineering,
Mysuru, India

Abstract: People with mobility limitations now lead considerably better lives because to the development of mobility aids. Conventional wheelchairs have been advantageous to a good number of people, making them more self-sufficient and mobile. They however have the disadvantage of confining their users to seated positions which raises a number of psychological and physical problems. The advancement of disability chair into a standing wheelchair opens up avenues for empowerment and liberation. It entails social interaction since users are capable of engaging at eye level and use their surroundings fully. For those people with movement restrictions, we can assist them in improving their lifestyles through promotion of mobility and physical fitness. Let us think outside the box and work together towards a world where no one has to feel small. Sitting for long periods of time has also been associated with medicalrelated issues such as poor blood circulation, development of pressure sores, and muscle wasting as documented by studies. Seated behavior can also affect social interactions; users may be less able to participate in talks or other pursuits requiring eye contact or physical interaction with people who are standing. This will increase the sense of loneliness and hardship in taking part with a more restricted enjoyment of community, educational, and professional setting. The demand for stand-up wheelchairs is growing as users require it to transfer easily and safely from sitting to standing positions. In addition to assisting the user to bear his or her weight, such a device should improve the individual's eye-level interaction skills and encourage activity and circulation to foster holistic well-being. For people who are living with disabilities of mobility we design a standing wheelchair that solved the above problems and giving independence and promotes the overall well-being of the disabled people.

Keywords: Standing wheelchair, Physical health, Muscle atrophy

1. INTRODUCTION

The fundamental and urgent need for sustainable and affordable assistive technology in terms of standing wheelchair is studied in this paper. The real level of need is recognized since only 10% of the global population has access to them, although 1 billion people depend on them. It supports the iterative order, low-cost standing wheelchairs that could be manufactured over a five-year span focusing on the health as well as practical value of preserving good posture. This paper also focuses on providing the maximum ergonomic comfort and relief for user to minimize the health hazard such as pressure sores. Thereafter, the study is using a different types of clothing for wheelchair it was challenging to choose from different cloths and presents using 3D printing to make ergonomic clothing. The wheelchair is designed is such a way that it enhance both safety and usability in most

[1]vinaykb@vvce.ac.in, [2]vishalasura@gmail.com, [3]melvinthomas191@gmail.com, [4]vikasgowda292@gmail.com, [5]afzalmee786@gmail.com

inaccessiable environment by using assistive technology especially electric powered wheelchair which comes with safety and usability for users like on staircases. After a long period of waiting the prototype of the wheelchair with standing seat is presented demonstrating the importance of standing mobility and serving as a proof of concept for possible future commercialization. In this paper applications of standing wheelchairs in standard clinical practice are considered, especially how they can assist clinicians in making decisions without replacing their clinical responsibility. The research also pertains to developments in robotic and computer-controlled assistive devices such as intelligent wheelchairs. This improvement in design aims at encouraging the users not to depend on help from others while using the equipment by providing driving controls, navigation systems, health monitoring systems with alarms in cases of emergencies. Moreover, the research integrated an ergonomic component that evaluated the anthropometrical aspects of the wheelchair and the application of the Analytical Hierarchy Process (AHP). The comfort rating awarded to the wheelchair, the comfort index, is pegged at 0. 62. Comfort and mobility are the most pronounced features that the newly reengineered electric scooter seeks to achieve targeted at sitting, lying or standing. The research also examines wheelchair-user ergonomic clothing, designing personalized styles based on 3D scanning and virtual prototyping without any need for fit trials. An underlying theme is the inculcation of safety considerations into assistive devices during development, and alternative methods for doing risk assessment using improved biomedical engineering safety are recommended. The research ends with a sample users guide of a standing wheelchair, which is claimed to be safe and functional to the users. In support of the rectified RESNA Position on the appropriateness of standing devices, it evaluates the clinical relevance of such devices vis a vis improving the health status of patient

2. LITERATURE REVIEW

Javeed Shaikh Mohammed, Swostik Sourav Dash, Vivek Sarda (2021) This review paper explores the SWC is designed to accommodate various user weights and sizes. Over five years, feedback from both consumers and physicians has refined its design. Priced affordably at INR 15,000 in India, user participation was crucial in the development process. The SWC's product, Arise, has been successfully launched, and future research will assess long-term quality of life outcomes. Kedar Sukerkar, Darshitkumar Suratwala, Anil Saravade This review paper explores The accuracy and cost of smart wheelchairs are problems. Smart wheelchairs can be made more functional with basic sensors. Smart wheelchairs that can accommodate all kinds of disabilities are hard to come by. Smart wheelchairs must have patient monitoring capabilities. Current models require supervision when used

outdoors. Research is necessary to help people with mental disabilities become self-sufficient. Future advancements in robotics will make smart wheelchairs more feasible. Foez Ahmed, Robi Paul, Md. Mufassal Ahmad This review paper explores A motorized wheelchair was developed to support different input methods. The wheelchair increases the user's independence and mobility. Emergency and medical procedures were incorporated for user safety. The control method is user-friendly and available. The design helps users and their family members feel more secure. F. I. Ashiedu and M. O. Okwu This review paper explores The wheelchair's high ergonomic rating makes it more comfortable for people with disabilities. It significantly increases users' sense of independence, self-worth, and mobility. The use of a creative algorithm enables design that is both effective and approachable. Future research may employ logic and ANFIS to improve wheelchair design efficacy. Shikha Oram (110ID0270) This review paper explores The wheelchair's kinematic link mechanism allows for efficient standing and reclining. The design complies with ergonomic standards by including adjustable footrest, backrest, and seating for maximum comfort. The prototype exhibits improved functionality, pressure relief, and user comfort. Future wheelchair designs can be further enhanced through automation. Motorization and obstacle detection are two instances of automation. Andreja Rudolf, Lucie Görlichova, Jerne Kirbis This review paper explores The study did a good job of showing how wheelchair users can use virtual simulation technologies to design ergonomic clothing. Without requiring the users to go through a lot of fitting trials, virtual prototyping made it possible to create custom apparel that met both functional and aesthetic requirements. The results show that virtual measurements are as accurate as manual ones, even for immobile individuals. The study opens up new possibilities for the clothing industry to use 3D scanning and virtual prototyping for customers with special needs. Ryoichiroshiraishi This paper explores The conclusion reiterates the importance of incorporating safety into assistive device design. It highlights the well-established procedure for developing safety measures and how it integrates with various assistive technology. The study recommends further research to prioritize user safety and comfort in future innovations. Meng Yuan, Ye WangLei, LiTian you,ChaiWei Ang This paper explores The conclusion summarizes the creation of a robust adaptive MPC algorithm that enables safety-based speed tracking for electric wheelchairs. It reiterates how crucial it is to account for external disturbances and design uncertainties. The paper shows the practical applicability of the proposed algorithm and validates its effective state and input constraint maintenance through representative task results. Wang XinXin, Sun Peili This paper explores The study concludes that safe electric wheelchair designs significantly improve the mobility of disabled users by addressing significant functional and safety concerns.

Among other innovative features, the dual wheel system and integrated lighting provide a comprehensive solution that enhances user experience and safety in a range of situations. This innovation represents a major breakthrough in assistive technology since it seeks to increase users' autonomy and confidence in their mobility. Tan Donghai, Tang jianping, Huang Fang This paper explores In conclusion, a comprehensive approach to enhancing patient comfort and safety is provided by the many functional safety wheelchairs. The design aims to reduce the chance of accidents and increase user satisfaction by incorporating features like a safety belt, better visibility, and a practical table. The concepts discussed in this paper could significantly impact the future design of mobility aids, enhancing their usability and safety. Eric Nickel, Andrew Hansen,Jon Pearlman, Gary Goldish This paper explores For its users, a standing manual wheelchair offers numerous benefits. Before testing on human beings, more developments are required. It might make it easier to get jobs and participate in social activities. Future care standards may include standing manual wheelchairs for eligible users. Brad E. Dicianno, Brad E. Dicianno, Jenny Lieberman, Lauren Rosen This paper explores Wheelchair standing devices improve accessibility and functional reach. Standing increases the mobility and function of the lower limbs. It promotes bone health and reduces the risk of osteoporosis. Standing reduces aberrant muscle tone and spasticity. It aids in pressure release and reduces the risk of pressure ulcers. Standing programs enhance the social and educational opportunities for children. The dosage determines the quality of life benefits.

3. METHODOLOGY

The wheelchair's design changed over the course of five years in response to input from more than a hundred user trial participants. The wheelchair's mechanical testing was conducted in accordance with ISO 7176 guidelines. A single fitting and training session was held to get the best results. Smart wheelchairs incorporate a variety of input techniques, such as touch controls, voice commands, cloud-based user preferences, biometric information, and vision-based environmental analysis. In addition, they offer tactile feedback in reaction to user input and allow for direct user control through brain-computer interfaces. Thumb control and accelerometer-based gesture control systems allow for easy navigation, and menu navigation is made possible by voice recognition. Furthermore, emergency location tracking is made possible by the SOS protocol, and the system has the capability to track users' medical conditions and overall health.

4. OBJECTIVES

Affordable Standing Wheelchair Development create a standing power chair with a mechanical system that is affordable for users. Assessment of Smart Wheelchair

Technology in assess the usability and functionality of smart wheelchairs as they stand right now. Intelligent Wheelchair Design construct a smart wheelchair with thumb and gesture control systems for navigating. Ergonomic Analysis For a comprehensive ergonomic analysis of the wheelchair design, use a cutting-edge algorithm. User Comfort and Safety Features to improve user comfort, include lifting and reclining mechanisms. Clothing for Wheelchair Users assess the suitability of 3D scanning and virtual simulation for designing wheelchair-accessible apparel. Safety Precautions for Assistive Systems provide methods for planning and putting safety measures in place for tangible assistive technology. Speed Tracking Control Algorithm create a safety-based algorithm to regulate electric wheelchair speed tracking. Stable Wheelchair Design create a power chair that is easy to use on level surfaces and stable on stairs. User Experience Enhancements Showcase a wheelchair with several safety features and practical elements to enhance the user experience, such as fall prevention techniques and comfort-enhancing gadgets. Manual Wheelchair Prototype Development construct a working model of a manual standing wheelchair. Clinical Applications of Standing Equipment talk about the typical clinical uses for wheelchair standing devices.

5. PROBLEM STATMENT

Limited Access to Assistive Technology due to the high cost of standing wheelchairs, only 10% of disabled people have access to assistive technology. Challenges with Conventional Wheelchairs traditional wheelchairs often require manual propulsion, which makes it difficult for many people with disabilities to perform daily tasks. Mobility Issues autonomous mobility solutions are necessary because mobility incompetence is a common problem, with a 15% disability rate. Ergonomic Deficiencies the lack of adequate ergonomic support in many wheelchair designs today causes discomfort and long-term health issues. Clothing Challenges it can be challenging for wheelchair users to find clothing that fits them well and meets their specific needs. Safety Concerns the risks of assistive devices are highlighted in the paper, especially when using them for sit-to-stand (STS) movements. Stability and Safety in Wheelchair Design current wheelchair designs frequently lack safety features and stability, which increases the risk of accidents, especially when navigating stairs and uneven surfaces. Need for Improved Standing Wheelchairs the lack of functionality and mobility in current standing wheelchairs emphasizes the need for designs that improve stability and usability.

6. CHALLENGES

Currently, the condition regarding standing wheelchairs is rather disappointing, even the assistive technology is not up to the mark. In many instances, they are overpriced and do not possess some of the vital safety features. A tiny 10 percent within a category such as this owns and uses any device; so the accessibility issue is still the core problem. Most of the intelligent wheelchairs developed these days are not usable outdoors and are not able to cater for the diversity of the users' needs with the cognitively impaired being the hardest nut to crack. There is a complexity that, clothing serves, to the users of the wheelchair since most of the time, it does not fit well or it is uncomfortable. Virtual prototyping of costume design can simplify the whole design process successfully by eliminating several long tedious fitting sessions. Safety is the foremost goal when performing sit-to-stand (STS) transitions. It should be adhered with ethical principles and practical safety measures that will enhance the comfort of users. Another factor that requires attention is the factor of support while advancing, particularly when climbing stairs, and the effective use of the facilities for adequate illumination in dark environments. Nonetheless, future progress should be the narrowing down of the prototype concerned to its weight and multiplication of elements in the design for better usability, since functionalism, strength, and durability will be the main concerns. And, out of the patients who use wheelchairs, there should be a selected subgroup that is assessed for suitability to be in a standing wheelchair and the effects of such postural changes on skin and health in general. Finally, it would require an approach in its designing of such assistive devices, which shall not be limited to accessibility and comfort, but shall also include safety to enable the users enhance their independence and improve their quality of life.

7. CONCLUSION

In this paper, we detail the challenges specified above regarding standing wheelchairs and assistive technology. Improving accessibility by enhancing ergonomic design to ensure safety will realize improvements in standing wheelchairs for both user needs and increase independence and overall well-being. Innovations introduced in our approach include virtual prototyping for a better fit and safety algorithms against accidents. In that regard, the objective of our project is to develop a novel standing wheelchair that serves both the purpose of functional improvement and comfort enhancement and also provides an efficient mode of mobility for a wider population. Our research team aims to contribute to the progress in assistive technology and enhance the quality of life for individuals with disabilities.

REFERENCES

1. Javeed Shaikh-Mohammed, Swostik Sourav Dash, Vivek Sarda & S. Sujatha (2021): "Design journey of an affordable manual standing wheelchair" Article in Disability and rehabilitation. Assistive technology · March 2021, https://doi.org/10.1080/17483107.2021.1892839

2. Kedar Sukerkar, Darshitkumar Suratwala, Anil Saravade, Jairaj Patil, Rovina D britto "Smart Wheelchair: A Literature Review" http://doi.org/10.11591/ijict.v7i2.pp63–66

3. Foez Ahmed, Robi Paul, Md. Mufassal Ahmad, Arif Ahammad, Showmik Singh "Design and Development of a Smart Wheelchair for the Disabled People" : 10.1109/ICICT4SD50815.2021.9397034

4. F. I. Ashiedu and M. O. Okwu "Ergonomic Analysis of a Developed Wheelchair Using Creative Algorithm"

5. Shikha Oram (110ID0270) under the guidance of Prof. B.B.V.L. Deepak "Ergonomic Wheelchair Design"

6. Andreja Rudolf, Lucie Görlichova, Jernej Kirbis, Jasna Repnik, Andrej Salobir, Irma Selimović, and Igor Drstvensek "Innovative Technologies in the Creation of Virtual Environment Ergonomic Clothing for Wheelchair Users"

7. Ryoichiroshiraishi,Yoshiyuki sonkai "Developing Saftey measure for a wheel chair- compatible physical assistive system with sit-to-stand movement support"

8. Meng Yuan,Ye WangLei,LiTian you,ChaiWei Ang "Safety-based Speed Control of a Wheelchair using Robust Adaptive Model Predictive Control"

9. Wang XinXin,Sun Peili "Safe Electric wheelchair"

10. Tan Donghai,Tang jianping,Huang Fang "Many functional saftey wheel"

11. Eric Nickel, Andrew Hansen,Jon Pearlman, Gary Goldish,"A drive system to add standing mobility to a manual standing"

12. Brad E. Dicianno, Brad E. Dicianno, Jenny Lieberman, Lauren Rosen, "Position on the application of wheelchair standing of wheelchair standing devices" 10.1080/10400435.2015.1113837

13. Simpson R.C, Hayashi S, Nourbakhsh I.R, Miller D.P., "The smart wheelchair component system". J Rehabil Res Dev. 2004; 41(3B):429–42.

14. Miller, D.P., Slack, M.G., "Design and testing of a low-cost robotic wheelchair prototype". Auton Robots. 1995; 2(1): 77–88.

15. A. M. Roungu, and M. N. Islam. "Impact of Disability on Quality of life of urban disabled people in Bangladesh"

16. Ahluwalia, MS. Varghese, TN. Nayan, Patil, S. Mayur, RS, "Design and Fabrication of Sensors Assisted Solar Powered Wheelchair". International Journal of Engineering Trends and Technology (IJETT), 46(2): 71–74

17. Churchward, R. "The development of a standing wheelchair.Applied Ergonomics", (1985) 16(1):55–62.

18. Arva, J, Paleg G, Lange M, Lieberman J, (2009). "RESNA position on the application of wheelchair standing devices Assistive Technology",(2009) 21(3), 161–168.

19. Chelvarajah, R. "Orthostatic hypotension following spinal cord injury: impact on the use of standing apparatus. NeuroRehabilitation",(2009) 24(3), 237–242.

20. Hohman K, Upstanding benefits: "Standing systems provide wheelchair users with numerous health benefits"., (2011) 24(2), 10–13

Advances in Mechanical Engineering and Materials Sciences – Dr. Vinay K. B et al. (eds)
© 2026 Taylor & Francis Group, London, ISBN 9-781-041-20970-6

14

An Investigative Study on Synthesis and Medical Applications of Wound Healing Biomaterials—A Review

Hemanth Kumar K. J.[1]
Mechanical Engineering Department,
Vidyavardhaka College of Engineering,
Mysore

Jyothilakshmi R.[2]
Mechanical Engineering Department,
Ramaiah Institute of Technology,
Bangalore

Likitha S. N.[3]
Mechanical Engineering Department,
Ramaiah Institute of Technology,
Bangalore

B. Sadashive Gowda[4]
Mechanical Engineering Department,
Vidyavardhaka College of Engineering,
Mysore

Abstract: There are a vast number of treatments for the management of wounds & burns. These include conventional wound dressings incorporating growth factors to stimulate & facilitate wound healing. Biomaterials interact with body tissue at the cellular level. Wound healing is a complex physiological process that necessitates effective intervention to promote tissue repair and regeneration. This investigative study explores the potential of various biomaterials in enhancing wound healing outcomes. The research focuses on the synthesis, characterization, and application of novel biomaterials, including hydrogels, nanofibers, and bioactive compounds, in wound management. A comprehensive review of the current literature highlights the limitations of traditional wound care methods and underscores the need for innovative solutions. Experimental methods include the fabrication of biomaterials using advanced techniques such as electrospinning and 3D bioprinting, followed by in vitro and in vivo assessments of their biocompatibility, antimicrobial properties, and efficacy in promoting cell proliferation and angiogenesis. Key findings demonstrate that the engineered biomaterials significantly accelerate wound closure, reduce infection rates, and enhance tissue regeneration compared to conventional treatments. The study also identifies critical parameters influencing the performance of these biomaterials, such as mechanical strength, degradation rate, and bioactive agent release profiles. The results underscore the potential of these biomaterials to revolutionize wound care by providing more efficient, targeted, and patient-specific treatments. This research contributes to the growing field of regenerative medicine and paves the way for future clinical applications, ultimately aiming to improve patient outcomes in wound management. This research paper aims to provide a comprehensive overview of biomaterials used in wound healing, including their definitions, classifications, properties, and applications. This study seeks to contribute to the collective understanding of wound management strategies and

[1]hemanthkumar.kj@vvce.ac.in, [2]jyothilakshmi.r@msrit.edu, [3]jyothilakshmi.r@msrit.edu, [4]principal@vvce.ac.in

10.1201/9781003725053-14

foster advancements in regenerative medicine by elucidating the fundamental principles underlying biomaterial-based wound therapies

Keywords: Wound healing, Biomaterials, Hydrogels, Nanofibers, Tissue regeneration, biocompatibility, Antimicrobial properties, Regenerative medicine

1. INTRODUCTION

A wound is an injury to the body, such as a puncture, cut, or disruption that causes damage to the skin or underlying tissues. Wounds can happen in the upper layer of the skin or can go deeper into the tissues. Many interactions occur between the cell types, mediators, and extracellular matrix. The wounds are classified into two types based on their nature, and they are:

1. Acute wounds are those which are very common in daily life and are tissue injuries. These wounds take very little time to heal, around 8 to 11 weeks. Physical factors generally cause acute wounds. These wounds do not leave a scar mark.

2. Chronic wounds take much more time to heal than acute wounds. Sometimes, it is also possible to reoccur after certain days of healing. These wounds usually leave a scar mark on the skin surface. The investigation was done by Juan Liu, et al. (2017)

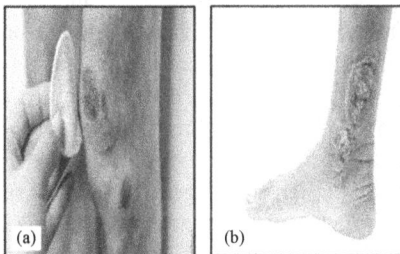

Fig. 14.1 (a) Acute wound (b) Chronic wound

2. WOUND DRESSING

Proper wound care is essential for promoting healing and preventing infection. Wound dressings are materials applied directly to wounds to facilitate healing and protect against bacterial activity. They serve several functions also:

- Removing excess moisture.
- Providing a barrier against contaminants.
- Providing an optimal healing environment.

Priyanka Agarwal et al. (2014) also concluded that the biomaterials have properties of the skin and are biocompatible. Biocompatibility is the ability to retain the hydration of the wound for a favourable, moist environment and protection from bacteria and dust. Dressings contain antimicrobial agents to prevent infection, while others are designed to promote moisture balance. Choosing the appropriate wound dressing becomes very important and depends on wound type, size, location, and presence of infection. Regular dressings are crucial to ensure proper healing.

3. WOUND HEALING PROCESS

Proper wound management involves cleaning the area, controlling bleeding, and applying first aid to prevent infection. Regular monitoring is essential for successful recovery. The skin comprises three layers: the epidermis, the dermis, and the hypodermis. Wound healing takes place in 4 phases. The first phase is homeostasis, and the homeostasis phase of wound healing aims to stop any bleeding. To do so, our body activates its blood clotting system. The second phase is Inflammation, which controls both bleeding and prevents infection.

Wound healing is a complex physiological process essential for restoring tissue integrity and function following injury or trauma. Among these strategies, Rachael Zoe Murray et al. (2019) mentioned that the biomaterial-based approaches have emerged as promising tools for facilitating wound healing and tissue regeneration. From providing structural support and moisture balance to delivering bioactive molecules and cells, biomaterials offer versatile platforms for addressing different aspects of the wound healing cascade. This research paper aims to provide a comprehensive overview of biomaterials used in wound healing, including their definitions, classifications, properties, and applications.

4. DEFINITION OF BIOMATERIALS

Biomaterial - is defined as a non-drug substance suitable for inclusion in systems or replace the function of bodily tissues or organs. Biomaterials encompass a wide variety of materials designed mainly for biomedical applications. They can be classified based on various criteria, including their composition, origin, and intended application. Some common types of biomaterials include:

1. Metals: Metals such as stainless steel, titanium, and cobalt-chromium alloys are commonly used in orthopaedic implants, dental implants, cardiovascular devices etc.

2. Polymers: They are versatile materials with diverse applications in biomaterials science. Synthetic polymers such as polyethylene, polypropylene, polyurethane, and silicone are used in medical devices, implants, drug delivery systems, and tissue engineering scaffolds. .

4.1 Natural Biomaterials

Natural biomaterials hold immense potential in wound healing applications due to their biocompatibility, bioactivity, and ability to mimic the extracellular matrix (E.C.M.) of native tissues. Juhlin L et al. (2022) concluded that collagen-based dressings promote haemostasis, provide a moist wound environment, and accelerate granulation tissue formation, facilitating the early healing stages. These materials offer several advantages, including biocompatibility, biodegradability, and often specific biological functionalities that promote tissue regeneration and healing. Some common types of natural biomaterials include:

1. Collagen: Collagen is the most abundant protein in the human body and is a structural component of connective tissues such as skin, bone, and cartilage. Rojas OJ (2019) found that the collagen-based biomaterials are widely used in wound dressings.
2. Chitosan: Chitosan is a polysaccharide derived from chitin, a natural polymer found in the exoskeletons of crustaceans such as shrimp and crabs. Chitosan-based materials promote wound healing by reducing inflammation, enhancing cell proliferation, and facilitating tissue regeneration.
3. Alginate: It is a natural polysaccharide which is extracted from seaweed, algae (primarily brown algae). Alginate-based hydrogels are used in wound dressings, tissue engineering, and regenerative medicine due to their, biodegradability. It also has ability to keep a moist environment around the wound.
4. Hyaluronic Acid: Hyaluronic acid is a glycosaminoglycan found in the extracellular matrix of connective tissues and synovial fluid.
5. Silk: Silk is a natural protein fiber from silkworms and spiders. Silk-based biomaterials exhibit excellent mechanical properties, biocompatibility, and biodegradability, making them suitable for various biomedical applications.

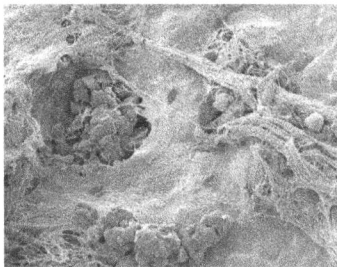

Fig. 14.2 SEM view of a healing wound

4.2 Synthetic Biomaterials

P.Aramwit et al. (2016) concluded that the synthetic biomaterials offer unique advantages in wound healing applications. They offer tunable properties, controlled degradation kinetics, and reproducibility. One prominent example is polymeric scaffolds, such as polyethylene glycol (P.E.G.) and polylactic-co-glycolic acid (P.L.G.A.), which provide structural support and regulate the release of bioactive agents for targeted wound healing. These synthetic polymers can be engineered to mimic the E.C.M.'s mechanical properties and facilitate cell adhesion, proliferation, and differentiation, thus promoting tissue regeneration in chronic wounds.

By promoting cell signalling and extracellular matrix deposition, synthetic bioactive ceramics facilitate tissue repair and regeneration, making them valuable assets in developing advanced wound care therapies with enhanced efficacy and safety profiles.

5. APPLICATIONS OF BIOMATERIALS IN WOUND HEALING PROCESS

Recent advancements in biomaterials for wound healing have focused on improving therapeutic outcomes through enhanced biocompatibility, controlled release kinetics, and targeted delivery of bioactive agents, the innate communication mechanisms between cells, EV-based biomaterials offer promising avenues for personalized and targeted wound care interventions.

5.1 Biomaterials for Diabetic Wound Healing

In diabetic wound healing, biomaterials are crucial in addressing the unique challenges of impaired wound-healing processes in individuals with diabetes. One significant biomaterial approach involves using advanced wound dressings designed to provide a conducive microenvironment for healing while addressing the specific needs of diabetic wounds. These dressings often incorporate antimicrobial agents, growth factors, or extracellular matrix components to promote healing and prevent infection as conclude by Juan Liu et al. (2017). Through continued research and innovation, biomaterial-based therapies have the potential to significantly improve outcomes for diabetic patients with chronic wounds significantly, ultimately enhancing their quality of life and reducing healthcare burden.

5.2 Biomaterials with Antimicrobial Activity

Biomaterials with antimicrobial activity represent a critical area of research and development, particularly in wound healing, implantable medical devices, and infection prevention. One notable class of biomaterials with inherent antimicrobial properties is silver-based materials. These nanoparticles can be incorporated into various biomaterials, including dressings, coatings, and implants, to prevent microbial colonization and infection. Furthermore, synthetic polymers like polyhexanide and polyethyleneimine have been engineered to possess antimicrobial properties and can be incorporated into biomaterials for various medical applications. These polymers exert antimicrobial effects through membrane disruption, enzymatic inhibition, and interference with microbial adhesion.

5.3 Cotton

Biocompatible cotton refers to cotton-based materials engineered or treated to be compatible with biological systems, particularly in medical and healthcare applications. The development of biocompatible cotton involves several key considerations:

1. Sterilization: Biocompatible cotton must undergo rigorous sterilization procedures to ensure it meets the standards for medical-grade materials. Various sterilization methods, including ethylene oxide gas, gamma irradiation, and steam sterilization, may eliminate microorganisms and ensure the cotton's safety for medical use.

2. Absorbency and Moisture Management: Cotton is known for its excellent absorbent properties, making it suitable for wound dressings and other medical applications where moisture management is crucial. Biocompatible cotton is engineered to enhance its absorbency, wicking capabilities, and moisture retention properties to promote a conducive environment for wound healing and patient comfort.

3. Biocompatibility: Biocompatible cotton should be free from harmful chemicals, dyes, and contaminants that could provoke adverse reactions or sensitivities in patients. This may involve using organic or medical-grade cotton and ensuring compliance with regulatory standards for biocompatibility and safety.

5.4 Silk

Silk-based biomaterials, derived from the cocoons of silkworms or spiders, offer a unique combination of mechanical strength, biocompatibility, and biodegradability. These properties make silk well-suited for wound healing applications:

1. Biocompatibility: Silk proteins exhibit excellent biocompatibility, minimizing adverse reactions and inflammatory responses when in contact with biological tissues. This biocompatibility makes silk-based biomaterials suitable for direct wound application without causing additional tissue trauma or irritation.

2. Mechanical Properties: Silk fibres possess remarkable mechanical properties, including high tensile strength and flexibility, which are advantageous for wound dressings and tissue scaffolds. Silk-based materials can provide structural support to wounds while accommodating natural tissue movement and deformation, promoting healing without compromising biomechanical integrity. Looking ahead, silk-based biomaterials hold immense potential as a versatile and biocompatible platform for wound healing applications. They open doors for innovation in advanced therapies and personalized medicine. Ongoing research is actively exploring the diverse functionalities of silk-based materials, paving the way for addressing unmet needs in wound care and regenerative medicine. Top of Form

5.5 Honey

Honey has been utilized for centuries as a natural remedy for wound healing, and its efficacy has been increasingly recognized in modern medicine. As a biomaterial, honey offers several beneficial properties that contribute to its effectiveness in wound healing:

Fig. 14.3 Effectiveness of honey in wound healing

1. Antimicrobial Activity: Honey exhibits broad-spectrum antimicrobial properties attributed to various factors, including its low pH, high sugar content, and hydrogen peroxide production by glucose oxidase enzyme.

2. Anti-inflammatory Effects: Honey mainly contains compounds such as flavonoids and phenolic acids with anti-inflammatory properties and are bioactive in nature. By reducing inflammation at the wound site, honey helps alleviate pain, swelling, and redness, facilitating healing.

3. Moisture Retention: Honey has hygroscopic properties. It can retain and absorb moisture in the wound environment. Maintaining a moist wound environment is crucial for optimal wound healing, as it promotes cell migration, proliferation, and tissue regeneration while preventing excessive drying or maceration of the wound bed.

6. RECENT ADVANCEMENTS IN MEDICAL FIELD

Recent advancements in biocompatible sutures have led to the development of various types of sutures used in surgical procedures. Here are some of the notable ones: Juan Liu et al.(2017)

6.1 Biocompatible Sutures

Biocompatible sutures are critical in surgical procedures, facilitating wound closure while minimizing tissue trauma and promoting healing. These sutures are designed to be compatible with biological tissues, reducing the risk of adverse reactions and promoting integration with surrounding tissues.

1. Absorbable Sutures: Absorbable sutures are designed to degrade over time, gradually losing their tensile strength as tissue healing progresses. These sutures are often made from natural materials such as catgut (made from sheep or goat intestines),

collagen (derived from animal sources), or synthetic polymers such as polyglycolic acid (P.G.A.), polylactic acid (P.L.A.), or polydioxanone (P.D.O.).

2. Non-Absorbable Sutures: Non-absorbable sutures are designed to maintain their tensile strength and structural integrity over time, providing long-term wound support. These sutures are typically made from synthetic materials such as polypropylene, nylon, polyester, or stainless steel.

3. Monofilament vs. Multifilament Sutures: Sutures can be classified as either monofilament or multifilament based on their structure. Monofilament sutures consist of a single strand of material, which reduces tissue drag and makes them less prone to harboring bacteria.

4. Coated Sutures: Some sutures are coated with silicone, P.T.F.E. (polytetrafluoroethylene), or antibacterial agents to improve handling characteristics, reduce tissue drag, and minimize tissue reactivity. By choosing appropriate sutures and techniques, surgeons can optimize wound closure and promote favourable outcomes for patients undergoing surgical procedures.

1. **Polydioxanone (P.D.S.):** P.D.S. sutures are synthetic and absorbable, commonly used for soft tissue approximation, including pediatric cardiovascular surgeries. They offer good tensile strength and are known for their prolonged absorption rate.

2. **Poliglecaprone 25 (Monocryl):** Monocryl sutures are monofilament and absorbable, often used for general soft tissue approximation and ligation. They are preferred for their smooth passage through tissues and reliable absorption profile.

3. **Polyglactin 910 (Vicryl):** Vicryl sutures are braided, absorbable, and coated to ensure smooth passage through tissue. They are widely used in various surgical procedures, including general surgery, orthopedic surgery, and gynecology, showcasing their versatility and empowering you in your selection.

4. **Glycomer 631 (Biosyn):** Biosyn sutures are absorbable monofilament sutures used for soft tissue approximation. They provide good knot security and handling properties. These are used during surgeries on the face like face upliftments.

5. **Polyglytone 6211 (Caprosyn):** Caprosyn sutures are monofilament and absorbable, used for soft tissue approximation where short-term wound support is acceptable. They have a high initial tensile strength with rapid absorption.

6. **Silk:** Although not absorbable, silk sutures are still widely used due to their excellent handling properties. They are often used in cardiovascular, ophthalmic, and gastrointestinal surgeries.

7. **Polypropylene (Prolene):** Prolene sutures are non-absorbable, synthetic monofilament sutures known for biocompatibility and are often used in cardiovascular surgeries and skin closure.

8. **Polyester (Ethibond):** Ethibond sutures are non-absorbable and braided, providing high tensile strength and minimal tissue reaction. They are commonly used in cardiovascular and orthopedic surgeries.

These sutures are chosen based on the specific requirements of the surgery, including the type of tissue, the required tensile strength, and the desired absorption rate. Selecting sutures plays a critical role in wound healing and minimizing the risk of complications, underscoring your responsibility to ensure the patient's well-being.

6.2 Cotton

In the context of surgical applications, "biocompatible cotton" isn't typically a term used to describe materials used for sutures or other surgical implants, as cotton itself isn't commonly used directly in such contexts due to its lack of biodegradability and potential for causing inflammatory responses. Understanding these limitations is crucial for optimizing their efficacy and addressing challenges in clinical translation. Some fundamental limitations include:

1. Biocompatibility and Immunogenicity: Despite efforts to develop biocompatible biomaterials, some may elicit immune responses or adverse reactions in the host. This can manifest as inflammation, foreign body reactions, or hypersensitivity reactions, compromising wound healing outcomes. Additionally, the degradation products of specific biomaterials may induce cytotoxicity or interfere with normal tissue regeneration processes.

2. Mechanical Properties: Biomaterials used in wound healing must possess appropriate mechanical properties to support tissue regeneration and withstand mechanical stresses.

3. Controlled Release Kinetics: Biomaterials designed for the controlled release of bioactive agents which must exhibit precise and tunable release kinetics.

7. FACTORS AFFECTING WOUND HEALING

Wound healing is a complex process influenced by various factors. Here are some key factors that affect wound healing in humans:

Age: Older individuals generally have slower wound healing due to reduced cell proliferation and decreased collagen synthesis.

Nutrition: Adequate intake of nutrients like protein, vitamins (especially vitamin C and vitamin A), and minerals (such as zinc and iron) is crucial for proper wound healing. Malnutrition can significantly impair the healing process.

Blood Supply: Proper blood flow to the wound site is essential for delivering oxygen and nutrients and removing waste products. Conditions like diabetes and peripheral artery disease can impair blood circulation and hinder wound healing.

Chronic Diseases: Conditions such as diabetes, autoimmune disorders, and cardiovascular diseases can compromise the immune system and impede the body's ability to heal wounds.

Infection: Wound infections can delay healing and may lead to complications. Keeping the wound clean and preventing bacterial contamination is vital for optimal healing.

Smoking: Smoking can impair wound healing by reducing blood flow and oxygen delivery to the wound site, as well as interfering with the immune response.

Medications: Certain medications, such as corticosteroids and immune suppressants, can inhibit wound healing by suppressing the immune system or interfering with the inflammatory process.

Wound Size and Depth: Larger and deeper wounds typically take longer to heal compared to smaller, superficial wounds.

Tissue Trauma: The extent of tissue damage and the presence of foreign objects or debris in the wound can affect healing.

Wound Care: Proper wound care techniques, including cleaning, debridement, and appropriate dressing selection, are essential for promoting optimal healing and preventing complications.

Psychological Factors: Psychological stress and mental health issues can impact wound healing by affecting hormone levels, immune function, and overall well-being.

Genetics: Individual genetic variations can influence the rate and effectiveness of wound healing.

Chermnykh E Watt FM (2018) concluded that the understanding and addressing these factors can help healthcare professionals optimize wound management strategies and improve healing outcomes.

8. CONCLUSION

In conclusion, biomaterials represent a dynamic and evolving field in wound healing research, offering multifaceted solutions to address the complex challenges associated with tissue repair and regeneration. From natural polymers like collagen and chitosan to synthetic scaffolds and nanomaterials, biomaterials offer diverse strategies for promoting wound healing through mechanisms such as cell adhesion, growth factor delivery, antimicrobial activity, and modulation of the wound

microenvironment. By harnessing the body's innate healing potential and augmenting regenerative processes, biomaterials are promising to improve outcomes in acute and chronic wound management, including diabetic ulcers, burns, and traumatic injuries. However, translating biomaterial-based wound healing therapies from bench to bedside is challenging. Biocompatibility, mechanical properties, controlled release kinetics, infection risk, and cost-effectiveness must be carefully addressed through interdisciplinary collaboration and rigorous evaluation. Moreover, the personalized nature of wound healing requires tailored biomaterial solutions that account for individual patient factors and wound characteristics.

Moving forward, continued research and innovation in biomaterial design, manufacturing, and clinical implementation are essential. By addressing the limitations and leveraging the strengths of biomaterials, we can pave the way for transformative advancements in wound care, ultimately improving patient outcomes, enhancing quality of life, and reducing healthcare burdens worldwide.

REFERENCES

1. Juan Liu, Huaiyuan Zheng, Xinyi Dai, Shicheng Sun, Hans-Gunther [2017] Biomaterials for Promoting wound healing in diabetes. J Tissue Sci Eng 8:193.
2. Priyanka Agarwal, Sandeep Soni, Gaurav Mittal, Aseem Bhatnagar [2014] Role of Polymeric Biomaterials as Wound Healing Agents.
3. Eun Seok Gil, Bruce Panilaitis, Evngelia Bellas and David L. Kalpan Functionalised Silk Biomaterials for Wound Healing.. Advanced Healthcare Materials.
4. Hughes OB, Rakosi A, MacquhaeF, Herskovitz I, Fox JD, et al. (2016) A review of cellular and acellular matrix products: Indications, techniques and outcomes. Plast Reconstr Surg 138: 138S–47S
5. P.Aramwit Chulalongkorn University, Phatumwan, Bangkok. Introduction to biomaterials for wound healing, 2016.
6. Rachael Zoe Murray, Zoe Elizabeth West, Allison June and Brooke Louise Farrugia: Development and use of biomaterials as wound healing therapies: Burns and Trauma (2019)
7. Mariam Mir, Murtaza Najabat Ali, Afifa Barakullah, Ayesha Gulzar: review paper: (2018): Synthetic polymeric biomaterials for wound healing.
8. Ahn C, Mulligan P, Salcido RS (2021) Smoking the bane of wound healing:biomedical interventions and social influences Adv Skin Wound Care 21: a review
9. Chermnykh E Watt FM, Skin cell heterogeneity in development, wound healing and cancer. Trends Cell Biol. 2018; 28(9);709–22
10. Juhlin L. Hyaluronan in skin. J Intern Med. 2022
11. Rojas OJ (2019) Cellulose chemistry and properties: fibres, nanocelluloses and advanced materials. Springer International Publishing, Switzerland.
12. Biomed. Eng. Appl. Basis Commun. 14 (03), 115–121
13. Yao, K., Mao, J., et al., 2002. Chitosan/gelatin network-based biomaterials in tissue engineering.

Note: All the figures in this chapter were made by the authors.

Advances in Mechanical Engineering and Materials Sciences – Dr. Vinay K. B et al. (eds)
© 2026 Taylor & Francis Group, London, ISBN 9-781-041-20970-6

15

Exploring the Conditional Influence of Variables on Smart Material

B. C. Ashok[1],
Vinod B.[2], Darshan G.B.[3],
Kishor M.D.[4], Maheshraj R.[5], Prajwal B.[6]
Department of Mechanical Engineering,
Vidyavardhaka College of Engineering,
Mysuru, Karnataka, India

Abstract: The impact of several parameters on the performance of smart materials is examined in this academic study, with a focus on polymers like polylactic acid (PLA) and polyurethane (PU) that are intended to increase elasticity and fatigue resistance. Smart materials are increasingly being used in advanced engineering and biological fields because of their capacity to respond to environmental stimuli. Because of their mechanical flexibility and biodegradability, PLA and PU have been chosen as the basis for material characteristic optimization. This study explores how several factors, including temperature, strain rate, crystalline, and filler makeup, affect these polymers' elasticity and durability under cyclic loading. Through a mix of experimental techniques and computer simulations, the study demonstrates that whereas temperature and strain rate primarily affect the elasticity of PU, crystalline and filler composition significantly affect PLA's mechanical properties. The findings provide important new information for tailoring the characteristics of smart materials, which will help PLA and PU develop for high-performance uses including medical devices and adaptable fabrics.

Keywords: Smart materials, Polylactic acid (PLA), Polyurethane (PU), Elasticity, Fatigue resistance, Temperature modulation

1. INTRODUCTION

Because of their many uses, polymers like polylactic acid (PLA) and polyurethane (PU) are gaining attention from academics; yet, each variety has unique difficulties. PLA is a biodegradable polymer made from renewable resources that is environmentally friendly. However, its flexibility and fatigue resistance are limited, making it less suitable for applications that require repeated stress. On the other hand, PU is known for its exceptional flexibility, toughness, and resistance to wear, which makes it ideal for use in medical devices and foams.

However, the fact that PU is not biodegradable raises serious environmental issues. Researchers are focussing

on training techniques meant to improve the qualities of both PLA and PU in response to these difficulties. Methods like crosslinking, adding plasticisers, adding nanofillers (like carbon nanotubes), and thermal manipulation have all shown a great deal of promise. These techniques can increase fatigue resistance and elasticity without compromising other mechanical properties. Crosslinking, for example, strengthens structural integrity, while nanofillers support the polymer matrix and prevent cracks from spreading.

By improving these conditioning methods, "smart materials" that can react to external stimuli and maintain durability in dynamic environments may become more readily available. This development would increase PU's

[1]bcashok@vvce.ac.in, [2]vinod@vvce.ac.in, [3]darshangb09092001@gmail.com, [4]kishormd21@gmail.com, [5]maheshsagar8722@gmail.com, [6]prajwalb809@gmail.com

10.1201/9781003725053-15

effectiveness in difficult conditions and establish PLA as a viable choice for load-bearing applications. The main goal is to design high-performing, environmentally friendly polymers for industries such as sustainable packaging, biomedical, and automotive.

2. LITERATURE REVIEW

According to (Yisi Liu et al., 2024) this study looks into a critical factor that contributes to persistent leg and back pain that causes severe impairment. To enhance drug delivery, tissue regeneration, and therapeutic results in IVDD treatment, researchers are investigating cutting-edge materials such as responsive hydrogels, shape-memory polymers, and nanoparticle-based systems. Personalised treatments are possible through the combination of biomedical engineering and materials science. Future studies should concentrate on creating biocompatible, minimally intrusive materials for certain populations, such as the elderly. increasing IVDD treatments and increasing patient outcomes through personalised medicine would need overcoming obstacles pertaining to clinical integration and material response.

This study introduces a high-performance stretchable strain sensor made from a broadleaf wood fibre (BWF) and multi-walled carbon nanotube (MWCNT) composite sheet. A gauge factor (GF) of 3.3 (0-200% strain) and 7.8 (200-440% strain) indicates that the elastic sensor has a high sensitivity and an impressive strain range of up to 440%, thanks to the porous construction that permits liquid silicone to infiltrate. With 100% strain and 5500 cycles, the sensor remains durable and undamaged. When included into a soft pneumatic gripper, it was able to classify laboratory beakers with a trained Support Vector Machine (SVM) algorithm with 99.33% accuracy (Chengjian Ou et al., 2023).

In this study, (Jia Chen et al., 2024) investigate how 4D printing can be integrated with conventional 3D printing in a versatile way, allowing censored constructions to change over time. Since its launch in 2013, 4D printing has attracted a lot of attention from a variety of industries. With an emphasis on intelligent materials such as composite materials, shape-memory alloys, and self-repairing polymers, this critique looks at the technologies and materials used in shape-morphing systems. With the aid of mathematical frameworks for behaviour prediction, it investigates the environmental stimuli (such as temperature, light, humidity, pH, electric, and magnetic fields) that cause material alterations. Rapid prototyping, electronics, biomedicine, soft robotics, sensors, and dynamic actuation devices are just a few of the applications that illustrate the field's difficulties and potential.

In response to environmental stimuli like pH, climate, and biological signals, smart polymers, also known as stimuli-responsive materials, can reversibly alter their properties, according to a study by (Arash Fattah-alhosseini et al., 2024) By changing their solubility, shape, and hydrophilicity, these materials—which can be categorised as either physical or chemical—offer adaptability. Smart polymers improve formulation in medicine by facilitating controlled release, decreasing complexity, and increasing efficacy. Enzyme-responsive polymers for biosensors and glucose-responsive polymers for targeted medication delivery are important uses. Smart polymers are poised to transform healthcare by providing more accurate treatments and real-time monitoring capabilities through their promise in medication delivery, tissue engineering, and sensor technologies.

Shape memory polymer composites reinforced with carbon nanotubes are assessed in this study with an emphasis on their mechanical characteristics, such as tensile stress, strain, stiffness, and flexural strength, as well as tribological features like wear resistance and friction. Modern materials may be activated by electrical and magnetic fields, in addition to heat, which is how classic shape memory materials work. The research emphasises how adding carbon nanotubes to shape memory polymers improves their characteristics, increasing the range of possible uses in consumer goods, aviation, textiles, energy systems, bionics, and civil engineering. This study highlights how polymer matrix composites, with their sophisticated reinforcing processes, may revolutionise a variety of sectors (Wei Yu et al., 2024)

With an emphasis on microwave-assisted spark ignition and combustion augmentation, this paper investigates the cross-fertilization between microwave radiation and explosives like rocket fuel, moderate, and spectacle. By facilitating quick and effective volumetric heating and enhancing flame stability and combustion efficiency through microwave plasma ignition, microwaves provide benefits over traditional methods. The paper emphasises the possibilities of intelligent electromagnetic fields (EMs) that react to outside inputs and adjust to changing environmental circumstances. In order to maximise energy efficiency and improve ignition and combustion processes in future applications, it also highlights the necessity for more study on microwave-sensitive electromagnetic fields, identifies obstacles, and talks about new research directions(Zang Xiaowei et al. 2024).

A comprehensive review of shape memory alloy (SMA) reinforced polymer composites, elucidating their distinctive properties and applications within the domain of smart structures was conducted by (Jitendra Bhaskar, et al 2020). The manuscript examines the amalgamation of SMAs with polymer matrices to augment mechanical performance, facilitate shape recovery, and enhance adaptability. Notable applications in aerospace engineering, civil infrastructure, and robotics are investigated, accentuating the capacity of these composites to exhibit dynamic responses to variations in environmental conditions. The authors further identify challenges associated with fabrication and consistency in performance, proposing future research trajectories aimed at optimizing the application of SMA-reinforced polymers. This review accentuates the critical importance of integrating SMAs with polymers for the advancement of sophisticated smart material applications.

(Chun-sheng et al., 2024) explores the Surfactants are amphiphilic entities that facilitate micelle formation and assume critical roles in applications such as detergents and pharmaceutical delivery mechanisms. Their functional characteristics, including aggregation behavior, are contingent upon concentration levels and the critical micelle concentration (CMC). The integration of dynamic covalent chemistry (DCC) within surfactant frameworks engenders dynamic covalent surfactants (DCS), which possess the ability to reversibly modify their structural configuration in response to environmental perturbations.

The integration of MXenes into shape memory polymers (SMPs) to improve their mechanical, electrical, and thermal characteristics is examined in the review by (Neha Bisht et al. 2023). Although SMPs are well-known for their capacity to change shape in response to outside stimuli, they have drawbacks such as sluggish shape recovery and poor conductivity. By creating conductive networks inside the polymer matrix, MXenes enhance these properties, allowing for improved conductivity and quicker shape recovery. The manufacture of MXene-reinforced SMP composites (SMPCs), their possible uses in biomedical engineering, energy storage, and soft robotics, as well as the difficulties and opportunities associated with using MXenes and carbon-based fillers in these materials, are all covered in the review.

According to a review by (Merve Uyan et al. (2023), Shape memory polymers (SMPs) have garnered a lot of interest for use in industries including aerospace and medical. Notwithstanding some technical constraints, SMPs improve fabrication methods, actuation systems, and manufacturing processes, spurring advancements in shape memory polymer composites (SMPCs). Along with current and upcoming applications, this paper examines the function of SMPs in SMPC manufacture, actuation, and assessment. Additionally, it talks about novel constitutive models that shed light on interactions inside SMPCs, pointing out both difficulties and new possibilities. The evaluation seeks to provide light on how SMPCs will evolve in the future, directing more study and technical developments in this exciting area.

A thorough analysis of shape memory polymers (SMPs) in the context of composite materials is carried out by (Takeru Ohki, et al et al., 2004), highlighting its inherent ability to return to preset configurations in response to external stimuli. The several SMP classifications, their synergistic integration with reinforcing fibres and nanoparticles, and their potential uses in the biomedical, automotive, and aerospace industries are all covered in detail in the study. The authors go on to clarify the difficulties with processing and stability and provide directions for additional study to improve the functionality and efficiency of SMPs. The complexity of SMPs and their exciting potential in the development of composite technologies are highlighted in this paper.

Table 15.1 Variations of shape memory polymers

Desired outcomes	
Faster recovery	< 1 min
Increase elasticity	>150 %
Fatigue life	>1000 cycle
Reduced brittleness	>25%
Increased scalability	>10 cm^3

Source: Author

The paper by (Yuliang Xia et al., 2020) examines how substantial advancements in shape memory polymers (SMPs) have made a variety of sophisticated and commercial applications possible due to a better knowledge of their characteristics. The advantages of one-, two-, and multiple SMPs are examined in this paper, with a focus on athermal stimuli and exchangeable molecular bonds. By elucidating the principles behind reversible SMPs, thermomechanical frameworks increase their use in medication administration, soft robotics, and aerospace. In order to create novel products and new phenomena, future research should concentrate on synthesising two-way SMPs and optimising shape recovery. The area has a lot of promise for creating more portable, useful SMPs to tackle a range of scientific problems.

Shape memory polymers (SMPs), according to (Lu Wang et al., 2021) have favourable biocompatibility and biodegradability combined with low density, manufacturing simplicity, and variable deformation temperatures. Because of their large surface area, interconnected pore sizes, and adaptable morphologies, SMP fibres (SMPFs) are suitable for use as scaffolds in tissue engineering applications. However, challenges still exist, such as a limited range of polymer materials and the ubiquity of stochastic fibre structures. In order to increase their usefulness, especially in the field of biomedicine, careful control of fibre dimensions—ideally at the nanoscale—and the use of advanced production techniques like 4D printing are essential. It is expected that further research will enhance SMPs' capabilities and broaden their possible uses.

According to a research by (Jian sunet al., 2014) the development of light-activated smart materials and technologies (LASMPs and LASMPCs) in materials engineering is driven by light's special qualities, which include energy transmission, long-range travel, and safety for human tissues. The primary areas of research are photothermal LASMPCs, which provide improved structural integrity, and photoresponsive LASMPs, which revert to their original shape at room temperature. Advanced light-activated shape memory behaviours are advanced by the integration of nanoparticles with optical technology. Innovation in LASMP/LASMPC systems requires a thorough understanding of light-activated processes through modelling. Numerous biological, small- and large-scale engineering uses for these materials highlight the necessity of continuous development to satisfy future material demands.

A ductile shape-memory polymer composite (SMP) designed to enhance shape recovery capabilities is examined in the paper by (Peng et al. 2020). The study highlights important developments in material properties that enable improved performance in applications requiring large deformation and recovery. The structural characteristics and mechanisms underlying the composite's improved shape recovery capabilities are carefully examined by the writers. The findings suggest a bright future for broader applications in a variety of fields, underscoring the vital significance of creating materials that can react to external stimuli in an efficient manner. This study makes a substantial contribution to the current investigation into shape-memory materials in the fields of engineering and technological advancement..

According to a research by (Amirkiai et al., 2021) materials scientists are working to create novel materials that react well to environmental cues. Shape-memory polymers, or SMPs, are attracting interest due to their special qualities and wide range of uses. SMPs are perfect for sectors including biomedical, aerospace, and automotive since they are lighter and less expensive than metals and ceramics. However, as compared to shape-memory alloys (SMAs), they frequently have worse mechanical qualities. Performance can be enhanced by including nano-sized fillers, but there are still issues with striking a balance between mechanical strength, biodegradability,

and processability. To improve SMPs for high-precision applications, such as energy harvesting, conductive elastomers, and multifunctionality, more research is essential.

The fundamentals and latest developments in shape-memory polymers (SMPs) are examined in this study, with an emphasis on their use in biomedicine. SMPs are less commonly used in biomedicine than in other fields, despite their enormous potential. The advancement of SMP design models and triggering mechanisms, such as physiological and distant triggers, are highlighted in the review. Notably, new uses such as self-expanding stents have been made possible by developments in reversible SMPs. With major advancements anticipated in the upcoming ten years, future research should concentrate on in vivo investigations and clinical trials to close the gap between SMP technology and medical applications (Jasper Delaey et al., 2020).

Shape memory cyanate ester (SMCE), which has a high transition temperature and is very useful for applications such satellite solar panel frameworks and adaptable moulds, was reviewed by (Zhangzhang Tang et al., 2022). The three-dimensional printable SMCE is created using a three-step process: first, cyanate ester is prepolymerized with epoxy resin to create the prepolymer (PCG); next, acrylate and vinyl monomers containing the prepolymer undergo UV-induced crosslinking to create a stiff gel; and lastly, the gel is thermally cured to create an interpenetrating polymer network (IPN) SMCE. The creation of high-performance materials that are appropriate for complex engineering applications is made easier by this methodical approach.

The characteristics of three-dimensional printed shape memory polymer (SMP) specimens, which demonstrate an average tensile strength of 55.58 MPa, surpassing the manufacturer's stated value of 48 MPa. The SMP displayed remarkable flexural strength (73.4 MPa) and exhibited shape memory recovery in aqueous environments at

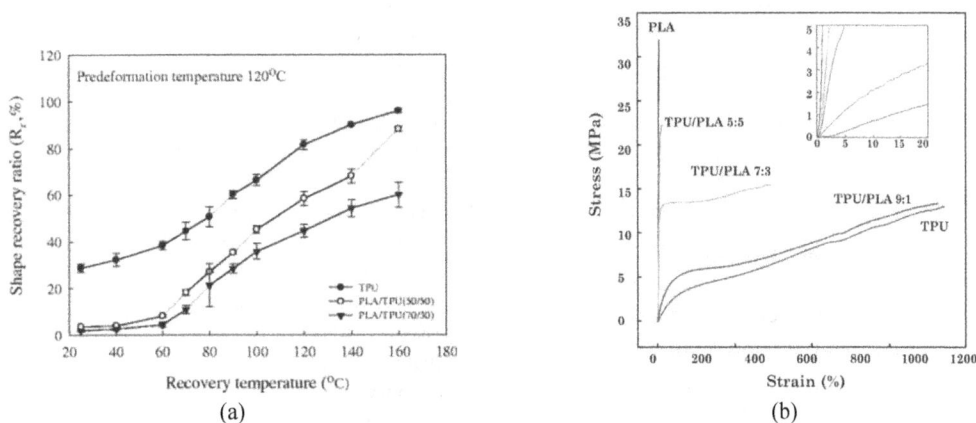

Fig. 15.1 (a) Variation of shape recovery ration and temperature (b) Stress and strain of SMA

Source: Jasper Delaey et al., 2020

elevated temperatures within a time frame of 25 seconds. Differential Scanning Calorimetry analysis indicated a glass transition temperature of 85.4°C. An innovative application utilizing QR codes as an imperceptible authentication mechanism was presented. Furthermore, a hybrid SMP/shape memory alloy (SMA) thermal switch was conceived, indicating prospective applications in intelligent electronics for the dynamic collection of temperature-dependent data (Trenton Cersoli et al., 2021)

The Shape memory polymers (SMPs) possess the inherent capability to "recall" their initial configuration and revert to it subsequent to deformation when subjected to external stimuli, predominantly thermal energy or a magnetic field. Poly(lactic acid) (PLA) is frequently utilized as an SMP in the realm of 3D printing, particularly through the methodology of Fused Deposition Modeling (FDM). By amalgamating PLA with substances such as Fe_3O_4, hydroxyapatite, carbon nanotubes, or silicone elastomers, investigators enhance its recovery characteristics and biocompatibility. These composites are investigated for an array of applications, encompassing biomedical devices, scaffolds, and soft robotics, thereby underscoring the adaptability and prospective capabilities of 3D printable SMPs in advanced technological domains (Guido Ehrmann, et al., 2021)

There is a underscores notable progress in the domain of shape memory polymers (SMPs) in recent years. Significant advancements encompass the identification of authentic two-way shape memory effects (2W-SME) within semi- crystalline polymers, the emergence of novel multi-shape memory behaviors (multi-SME), and the introduction of innovative non-thermal triggering mechanisms. Despite the fact that enhancements in recovery stress have been somewhat stagnant, SMPs have discovered groundbreaking applications, particularly within the electronics sector and in surface-based shape memory phenomena. Developments in molecular switching mechanisms and unconventional fabrication techniques have augmented the versatility of SMPs. The field is anticipated to sustain its rapid evolution, with the prospect of imminent commercialization of certain applications (Qian Zhaoa, et al., 2015)

3. CONCLUSION

The creation of SMPs has shown great promise in a number of fields. SMPs are perfect for applications that need flexibility because of their special capacity to return to preset shapes. The mechanical characteristics of nanoparticles are improved through integration. Dynamic, self-deployable structures can be made thanks to 4D printing technology. SMPs have potential in biological applications such as medical devices and tissue engineering. Overcoming processing constraints

and attaining reliable performance are still difficult tasks, nevertheless. Further research is needed to optimize SMP properties and explore new applications.

REFERENCES

1. Amirkiai, Arian, "Tracing evolutions of elastomeric composites in shape memory actuators" 2021. https://doi.org/10.1016/j.mtcomm.2021.102658
2. Arash Fattah-alhosseini, "A review of smart polymeric materials: Recent developments and prospects for medicine applications" 2024. https://doi.org/10.1016/j.hybadv.2024.100178
3. Chengjian Ou, Hongjie Jiang, Longya Xiao, "Silicone/broadleaf wood fiber/MWCNTs composite stretchable strain sensor for smart object identification" 2023. https://doi.org/10.1016/j.sna.2023.114846
4. Guido Ehrmann, Andrea Ehrmann ,"3D printing of shape memorypolymers"2021DOI:10.1002/app.50847
5. Jasper Delaey, Peter Dubruel, "Shape-Memory Polymers for Biomedical Applications" 2020. https://doi.org/10.1002/adfm.201909047
6. Jia Chen, Christian Virrueta, "4D printing: The spotlight for 3D printed smart materials" 2024. https://doi.org/10.1016/j.mattod.2024.06.004
7. Jian Sun, Yanju Liu, "Mechanical properties of shape memory polymer composites enhanced by elastic fibers and their application in variable stiffness morphing skins" 2014. https://doi.org/10.1177/1045389X14546658
8. Jitendra Bhaskar, "A review on shape memory alloy reinforced polymer composite materials and structures" 2020. DOI: 10.1088/1361-665X/ab8836
9. Kaiyuan Peng, Yao Zhao, "Ductile Shape-Memory Polymer Composite with Enhanced Shape Recovery Ability" 2020. https://rb.gy/qqm1dw
10. Lu Wang, Fenghua Zhang, "Shape Memory Polymer Fibers: Materials, Structures, and Applications" 2021. https://doi.org/10.1007/s42765-021-00073-z
11. Merve Uyan PhD, "Novel constitutive models, challenges and opportunities of shape memory polymer composites" 2024.
12. Neha Bisht, Shubham Jaiswal, "MXene enhanced Shape Memory Polymer Composites: The rise of MXenes as fillers for stimuli-responsive materials" 2024. https://doi.org/10.1016/j.cej.2024.155154
13. Qing-Qing Ni, Chun-sheng Zhang, "Shape memory effect and mechanical properties of carbon nanotube/shape memory polymer nanocomposites" 2007. https://doi.org/10.1016/j.compstruct.2006.08.017
14. Qian Zhaoa, H. Jerry Qi b,"Recent progress in shape memory polymer: New behavior, enabling materials, and mechanistic understanding"2015, https://doi.org/10.1016/j.progp olymsci.2015.04.001
15. Takeru Ohki, Qing-Qing Ni, "Mechanical and shape memory behavior of composites with shape memory polymer" 2004. https://doi.org/10.1016/j.compositesa.2004.03.001
16. Tong Mu, Liwu Liu, "Shape memory polymers for composites" 2018. https://doi.org/10.1016/j.compscitech.2018.03.08

17. Trenton Cersoli, AlexisCresanto,"3D Printed Shape Memory Polymers Produced via Direct Pellet Extrusion "2021, https://www.mdpi.com/2072- 666X/12/1/87

18. Wei Yu, Yang Zhou, "When thermochromic material meets shape memory alloy: A new smart window integrating thermal storage, temperature regulation, and ventilation" 2024. https://doi.org/10.1016/j.apenergy.2024.123821

19. Xiaowei Zang, Jian Cheng, "Interactions between microwaves and smart energetic materials: A review on emerging technologies for ignition and combustion enhancement" 2024. https://doi.org/10.1016/j.cej.2024.155031

20. Yisi Liu, Jie Hu, "Progress of smart material in the repair of intervertebral disc degeneration" 2024. https://doi.org/10.1016/j.smaim.2024.10.001

21. Yuliang Xia, Yang He, "A Review of Shape Memory Polymers and Composites: Mechanisms, Materials, and Applications" 2020. https://doi.org/10.1002/adma.202000713

22. Zhangzhang Tang, Junhui Gong, "3D printing of a versatile applicability shape memory polymer with high strength and high transition temperature" 2022. https://doi.org/10.1016/j.cej.2021.134211

Advances in Mechanical Engineering and Materials Sciences – Dr. Vinay K. B et al. (eds)
© 2026 Taylor & Francis Group, London, ISBN 9-781-041-20970-6

16

A Review on Mixed Reality-Based Rehab Program for Shoulder Stretching Exercises

Vinod B.[1], Raghu N.[2], Harsha D.T.[3],
Pradeep Kumar Naik K.[4], Gowtham Mourya N.[5], Kowshik[6]

Dept of Mechanical Engineering, Vidyavardhaka College of Engineering,
Mysuru, Karnataka, India

Abstract: Shoulder injuries and musculoskeletal disorders often require rehabilitation to restore mobility, strength, and function. Traditional rehabilitation methods can be repetitive, leading to reduced patient engagement and adherence. This study presents a Mixed Reality (MR) Rehabilitation System designed to enhance shoulder rehabilitation, specifically focusing on stretching exercises. The system integrates virtual and augmented reality elements to create an interactive and immersive environment that guides patients through prescribed exercises. Real-time motion tracking ensures accurate execution, while adaptive feedback helps in correcting movements and maintaining motivation. The MR system also enables remote monitoring by physiotherapists, allowing for personalized adjustments and progress tracking. By combining game fiction and bio-mechanical analysis, the system aims to improve patient engagement, adherence, and rehabilitation outcomes. Preliminary testing indicates that MR-based rehabilitation can enhance movement precision, reduce pain perception, and increase patient motivation compared to conventional approaches. This technology has the potential to transform physiotherapy by making rehabilitation more effective, engaging, and accessible. Future work will focus on clinical validation, expanded exercise libraries, and integration with wearable sensors for enhanced tracking. The proposed MR system offers a promising solution for improving the rehabilitation process and enhancing the overall recovery experience for patients with shoulder injuries.

Keywords: Physiotherapy, Mixed reality, Virtual reality, Shoulder rehabilitation

1. INTRODUCTION

The Mixed Reality Rehabilitation System for Shoulder Rehabilitation project aims to transform the traditional approach to post-surgical and injury-related shoulder recovery by integrating immersive technology into physical therapy. Shoulder rehabilitation can often be challenging for patients due to the complexity of exercises and the need for precise form, which, if performed incorrectly, can lead to further discomfort or injury. This engineering project addresses these challenges by developing a Mixed Reality (MR) system that provides detailed, interactive guidance for shoulder exercises, enhancing accuracy, engagement, and recovery. The system combines real-time visual feedback and precise instructions, allowing patients to perform essential rehabilitation exercises with greater confidence and adherence to therapeutic protocols. By delivering an immersive, user-friendly experience, the MR rehabilitation system has the potential to significantly improve patient outcomes and revolutionize physical therapy practices.

2. LITERATURE REVIEW

A systematic review of rehabilitation technology that is now available and has the potential to be integrated into physiotherapists' clinical practice. Background: The impact of technology on physiotherapy rehabilitation

[1]vinod@vvce.ac.in, [2]Raghu.n@vvce.ac.in, [3]harshagowdadt@gmail.com, [4]pradeenaik2002@gmail.com, [5]mouryavirat18@gmail.com,
[6]kaushikchinnu@gmail.com

10.1201/9781003725053-16

Shoulder Dislocation

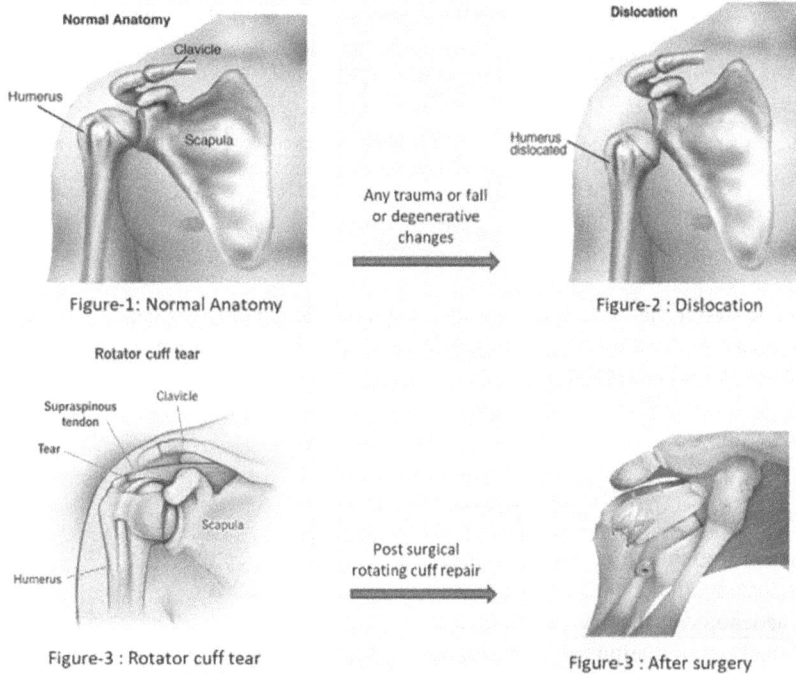

Fig. 16.1 Regions of shoulder dislocation

Source: Authors

techniques is a significant issue. The goal of this study was to determine which technologies were beneficial for rehabilitation of patients in clinical settings search methods this review entailed a systematic search of four major databases to retrieve relevant studies published between 2000 and 2021 results included in the current report comprise 18 published articles placed in the category of use of digital aerobatics virtual and hybrid technologies (E Siqueira et al., 2024)

Real-time validating the accuracy in physiotherapy exercises. physiotherapy is one of the interventions in rehabilitation with regard to physical disabilities and the development of mobility capabilities patients have to be accompanied to medical centres to undergo exercise programs supervised closely the outcome of these practice sessions serves as a basis for professionals in decision-making about interventions and overall care planning to promote the best health care practices in physical therapy (Tran Trong Nghia et al., 2017)

Creation of Mobile Markerless Augmented Reality for the Cardiovascular System in Physiotherapy Education Courses on Anatomy and Physiology. Anatomy and physiology are essential subjects for physiotherapy students. The course aims to equip students with knowledge regarding the anatomical structures of human beings for treatment competencies. Traditional modes of education include lectures, artificial models, and cadaver examinations. (Rosni et. al., 2020)

A Finger Physiotherapy Tool for Adduction and Flexion. Paralysis is due to diseases of the muscles or nerves,

which hinder the potential movement. There are medical interventions and rehabilitative activities, including physiotherapy. This paper attempts to elaborate on finger physiotherapy in the case of paralysis. (Setiarini, Asih, et al., 2019)

Physiotherapy Rehabilitation Treatment Design for Wheelchairs. After heart disease and cancer, stroke ranks as Malaysia's third leading cause of death. Stroke is the leading cause of disability, affecting up to 40,000 individuals annually in Malaysia. Stroke interrupts cerebral blood flow with different types of functional impairments from hemiplegia to changes in cognitive functions, verbal communication, and thought patterns in the patient. (Ismail et al., 2015)

Improving the Effectiveness of Physiotherapy in Dentistry. A thorough literature search uncovered 409 articles that were deemed relevant for this systematic review, initially established as the inclusion criteria. An additional 74 articles were obtained from the literature through secondary searches of reference lists. Of these, six studies were included based on predefined selection criteria. (Hotra, Zenon, et al., 2016)

Evaluating Mixed Reality Technology for Tracking Hand Motion for Shoulder Rehabilitation Assessment. Abstract Shoulder injuries are prevalent and may impede everyday activities. Innovative interventions such as serious games and mixed reality (MR) technologies, exemplified by HoloLens, are advocated for rehabilitative purposes. This investigation assesses the precision of HoloLens 2 in monitoring hand movements. Evaluations were conducted

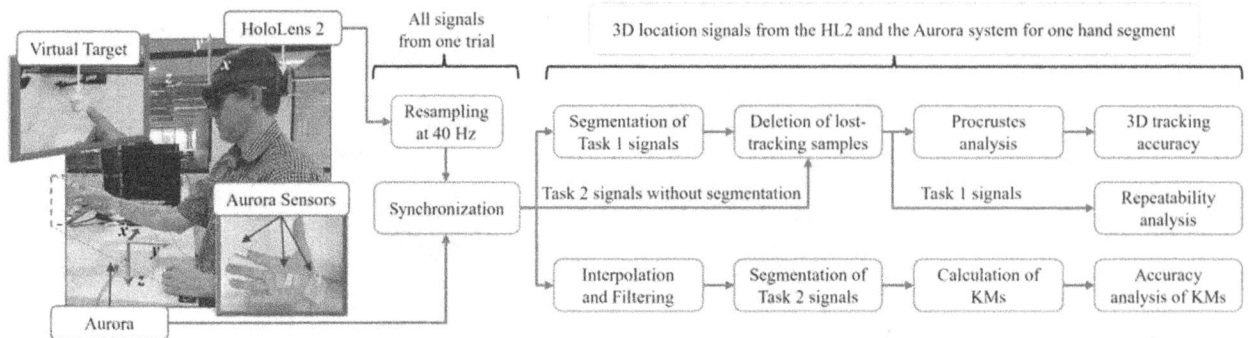

Fig. 16.2 Procedure for analysing hand motion data that was recorded during a trial. The HoloLens 2 and the Aurora system record data from the index fingertip, palm, and wrist 3D locations while completing two tasks based on an upper-limb rehabilitation program

Source: Salinas et al., 2024

in comparison to an Aurora electromagnetic system, serving as the definitive standard. (Salinas et al., 2024)

Virtual Reality for Shoulder Rehabilitation: Oculus Quest 2 Accuracy Assessment. Introduction In the past few years, VR technologies have made rapid strides in the healthcare arena especially in medical rehabilitation allowing more participation from patients. Economically accessible VR systems allow an individuals to engage themselves in a simulated world. It can track human movement and, therefore, increases compliance with the rehabilitation program. (Carnevale et al., 2022)

Mixed Reality Game Prototypes for Rehabilitation and Upper Body Exercise. For virtual reality-based upper body physical activity and rehabilitation, an integrated hardware and software platform was created. Exercise regimen based on evidence for people with spinal cord injury. Custom metallic structure which supports a standard wheelchair and six Gametraks in an installation that allows the performance of resistance training exercises. (Gotsis et al., 2012)

Shoulder rehabilitation using wearable augmented reality. Overview By providing contextual stimuli, virtual reality (VR) and augmented reality (AR) increase user engagement. The implementation of AR and VR technologies has the potential to enhance individual efficacy and promote self-reflection. There is an increasing scholarly interest in the objective measurement of performance within the medical domain. AR technology demonstrates particular advantages in the context of rehabilitation by facilitating interactions within real-world environments.(Condino et al., 2019)

Virtual reality as a diagnostic and intervention tool in physical therapy. Virtual reality (VR) has become more popular in the healthcare industry for both therapeutic and diagnostic purposes. This review aims to present an overview of all-rounded information regarding the application of VR in physical rehabilitation. Its main aim is to enable the clinician to have a deeper insight into the ubiquitous applications and benefits of the application of VR technology.(Bateni et al., 2024)

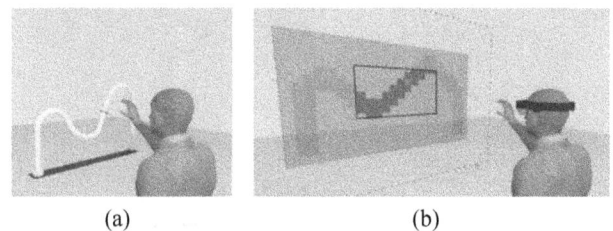

(a)   (b)

Fig. 16.3 (a) ROM exercises with the conventional "Rolyan Range of Motion Shoulder Arc," and (b) ROM exercises with the Azure HoloLens optical see-through head-mounted device's AR shoulder rehab software

Source: Condino et al., 2019

Using Immersion Virtual Reality to Predict Shoulder Kinematics of Rehabilitation Exercises. The COVID-19 pandemic has increased significantly the demand for telehealth services in the physical rehabilitation field. Telehealth technologies provide patients with greater access to and activation of care. Current technology tools need to be integrated with evidence-based measures to optimally implement telehealth solutions. (Powell et al., 2022)

An Approach to Shoulder Rehabilitation Based on Virtual Reality Games. Introduction General Introduction to Adherence Without adherence to prescribed exercise regimens, the effectiveness of physiotherapeutic interventions cannot be achieved. Current methods used to measure adherence in home-based physiotherapy lack sufficiency. Traditional adherence rates are significantly low; 50-70% is estimated to demonstrate non-adherence. This study aims to support improvements in adherence by introducing methodologies involving robots and video gaming. (Moreira et al., 2024)

Virtual Reality's Use in Shoulder Surgery Rehabilitation. Evaluation of shoulder functions had traditionally been determined through muscle strength measurements, levels of pain, and ROM. Post-surgical pain seriously interferes with normal daily activities and usually adversely impacts the quality of life in general. Rehabilitation is a crucial part of the recovery process and remediation of everyday functional capabilities after surgery. Patients must have scheduled clinical visits or home program interventions that range from two weeks to as long as forty-eight weeks. (Nam, Jihun, et al., 2024)

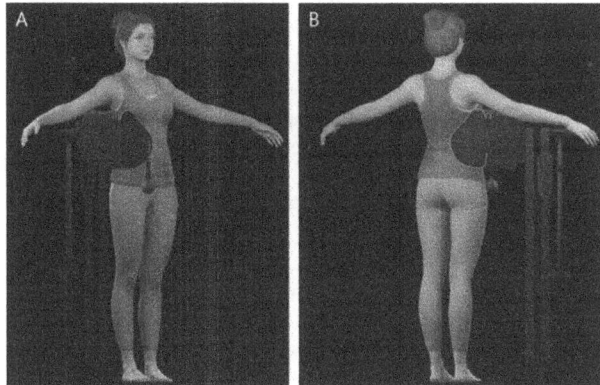

Fig. 16.4 The rehabilitation movements performing by avatar (A) Front view and (B) rear view

Source: Nam, Jihun, et al., 2024

A focus group research on physiotherapists' experiences with virtual reality for shoulder rehabilitation. Introduction Musculoskeletal shoulder pain has a lifetime prevalence of 6.7% to 66.7%. The condition badly disturbs sleep and contributes to the overall morbidity rate. Poor adherence to exercise therapy drastically hinders symptom alleviation. Virtual reality (VR), augmented reality (AR), and mixed reality (MR) are three emerging Extended Reality (XR) technologies that are being incorporated into the exercise therapy paradigm. (Dejaco et al., 2024)

Virtual Reality Rehabilitation for Patients with Shoulder Disorders. A set of various questionnaires was used to evaluate usability and acceptability. Utility questions focused on the aspects of amusement, interest, and actual needs of experience when playing. Play ability questions tested how intuitive and how easy the interface was to understand. The questions covered adequacy of therapist support and quality of graphic design. (Tokgöz et al., 2022)

Oculus Quest 2-based immersive virtual reality for shoulder rehabilitation: assessment of a physical therapy program. Introduction Virtual and Augmented Reality or Technological innovations are of great importance for boosting effectiveness and personalization in orthopedic rehabilitation.(Longo et al., 2023)

A scoping assessment of the effectiveness of exercise treatment based on virtual reality in rehabilitation. Rehabilitation often requires exhausting exercise therapy, and so new technologies have to be explored to increase patient participation. Virtual reality is suggested to be an instrument that should be applied to enhance the effectiveness of exercise therapy. (Asadzadeh et al., 2021)

A Virtual Reality Game-Based Approach for Shoulder Rehabilitation. Virtual Reality (VR) has appeared as an economically feasible additive to physiotherapy, which is characterized by vast availability of headsets among users. Commercially accessible games make it an additive to physio therapeutic activities in a significant way. Such additive nature reduces the perception of pain and increases the enjoyment of therapeutic activities. The major issue is how to maintain exercise adherence; here, robotics can support patients during their regimes of treatment. Using

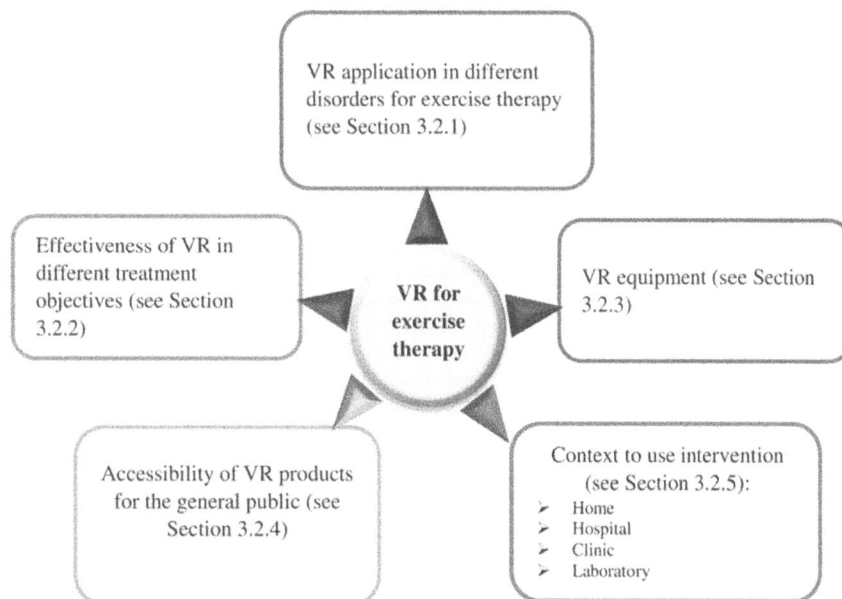

Fig. 16.5 Taxonomy of essential elements for VR exercise therapy

Source: Asadzadeh et al., 2021

an Application Programming Interface (API), a Kuka LBR Med 7 R800 robot was set up in the Unity environment. (Moreira et al., 2024)

A pilot randomised controlled trial was conducted to examine the use of virtual reality and video games in the physiotherapy treatment of stroke patients. One of the leading causes of adult medical impairment is stroke. This research investigates the conventional physiotherapy supplemented by VR in stroke patients' rehabilitation. 24 patients were randomly assigned into the groups. Both had physiotherapy for 6 weeks at one hour per session; however, the experimental group was supplemented with VR. (Peláez-Vélez et al., 2023)

3. CONCLUSION

In conclusion, the Mixed Reality Rehabilitation System for Shoulder Rehabilitation offers an innovative solution that combines the power of MR technology with evidence-based physical therapy techniques. By providing patients with immersive, real-time guidance, this system supports effective shoulder rehabilitation, ensuring exercises are performed accurately and safely. With potential applications for both remote and in-clinic settings, this MR system addresses the need for scalable, engaging, and accessible physical therapy options. Ultimately, this project underscores the trans formative potential of immersive technology in healthcare, promoting faster recovery and better long-term outcomes for patients undergoing shoulder rehabilitation.

REFERENCES

1. Asadzadeh, Afsoon, et al. "Effectiveness of virtual reality-based exercise therapy in rehabilitation: A scoping review." Informatics in Medicine Unlocked 24 (2021): 100562.
2. Bateni, Hamid, et al. "Use of virtual reality in physical therapy as an intervention and diagnostic tool." Rehabilitation Research and Practice 2024.1 (2024): 1122286.
3. Carnevale, Arianna, et al. "Virtual reality for shoulder rehabilitation: Accuracy evaluation of oculus quest 2." Sensors 22.15 (2022): 5511.
4. Condino, Sara, et al. "Wearable augmented reality application for shoulder rehabilitation." Electronics 8.10 (2019): 1178.
5. Dejaco, Beate, et al. "Experiences of physiotherapists considering virtual reality for shoulder rehabilitation: A focus group study." Digital Health 10 (2024): 20552076241234738.
6. E Siqueira, Tarciano Batista, José Parraça, and João Paulo Sousa. "Available rehabilitation technology with the potential to be incorporated into the clinical practice of physiotherapists: A systematic review." Health Science Reports 7.4 (2024): e1920.
7. Gotsis, Marientina, et al. "Mixed reality game prototypes for upper body exercise and rehabilitation." 2012 IEEE Virtual Reality Workshops (VRW). IEEE, 2012.
8. Hotra, Zenon, et al. "Improving the effectiveness of physiotherapy in dentistry." 2016 13th International Conference on Modern Problems of Radio Engineering, Telecommunications and Computer Science (TCSET). IEEE, 2016.
9. Ismail, Rosmawati Binti, Salhana Binti Sahidin Salehudin, and Syarifah Noor Binti Deraman. "Design of Wheelchair Physiotherapy Rehabilitation Treatment." 2015 Innovation & Commercialization of Medical Electronic Technology Conference (ICMET). IEEE, 2015.
10. Longo, Umile Giuseppe, et al. "Immersive virtual reality for shoulder rehabilitation: evaluation of a physical therapy program executed with oculus quest 2." BMC Musculoskeletal Disorders 24.1 (2023): 859.
11. Moreira, Moisés, et al. "A Virtual Reality Game-Based Approach for Shoulder Rehabilitation." Multimodal Technologies and Interaction 8.10 (2024): 86.
12. Nam, Jihun, et al. "The Application of Virtual Reality in Shoulder Surgery Rehabilitation." Cureus 16.4 (2024).
13. Peláez-Vélez, Francisco-Javier, et al. "Use of virtual reality and videogames in the physiotherapy treatment of stroke patients: a pilot randomized controlled trial." International Journal of Environmental Research and Public Health 20.6 (2023): 4747.
14. Powell, Michael O., et al. "Predictive shoulder kinematics of rehabilitation exercises through immersive virtual reality." IEEE Access 10 (2022): 25621–25632.
15. Rosni, Nurul Shuhadah, et al. "Development of mobile markerless augmented reality for cardiovascular system in anatomy and physiology courses in physiotherapy education." 2020 14th International Conference on Ubiquitous Information Management and Communication (IMCOM). IEEE, 2020.
16. Salinas, Sergio A., et al. "Evaluating Mixed Reality Technology for Tracking Hand Motion for Shoulder Rehabilitation Assessment." 2024 10th IEEE RAS/EMBS International Conference for Biomedical Robotics and Biomechatronics (BioRob). IEEE, 2024.
17. Setiarini, Asih, et al. "A Finger Physiotherapy Device for Flexion and Adduction Motion." 2019 International Conference on Radar, Antenna, Microwave, Electronics, and Telecommunications (ICRAMET). IEEE, 2019.
18. Tokgöz, Pinar, et al. "Virtual reality in the rehabilitation of patients with injuries and diseases of upper extremities." Healthcare. Vol. 10. No. 6. MDPI, 2022.
19. Vo, Tran Trong Nghia, et al. "Real time validating the accuracy of physiotherapy exercises." 2017 IEEE International Conference on Consumer Electronics-Taiwan (ICCE-TW). IEEE, 2017.

Advances in Mechanical Engineering and Materials Sciences – Dr. Vinay K. B et al. (eds)
© 2026 Taylor & Francis Group, London, ISBN 9-781-041-20970-6

17

Design and Development of Autopilot Cargo-Ship—A Review

Vinay K. B.[1], Shivashankar R.[2],
Praveenkumara B. M.[3], Ravish Gowda B. V.[4],
Rachel Thomas[5], Pratham R.[6], Sri Ranga B. N.[7]

Department of Mechanical Engineering, Vidyavardhaka College of
Engineering, Mysuru, India

Abstract: Autonomous technology is advancing with rapid strides in many fields, but also puts forth the need for innovative solutions to better optimize operations in extended modes of transportation in instance, when a cargo vessel must be maneuvered for weeks or even months, the cost of shipping logistics can be enormous. The research demonstrates an automated GPS-assisted cargo ship system that decreases the need for human navigation, increasing efficiency, lowering operating costs, and lowering associated risks. At the center of the system is an ATmega328P micro controller that is linked to important navigation components like a motor driver, compass, GPS receiver, and communication interfaces. The destination coordinates are entered by the user through a Bluetooth smartphone application, allowing the ship to travel autonomously in the direction. The GPS module keeps monitoring the position of the ship, and the compass maintains it in the correct orientation. Steering and propulsion are accomplished by the motor driver/motors with the aid of real-time information from the GPS and compass systems. The process eliminates the risk of human error and fatigue hazards by enabling the cargo ship to drive itself, overall performance of the system will be improved through its constant operation. Precise sensor integration, real-time decision-making processes, motor control during movement, and efficient communication between the many components are a few of the most critical design factors. Future autonomous cargo vessels will become entirely autonomous as a consequence, transforming global commerce and transportation systems.

Keywords: Cargo, GPS navigation, ATmega328P microcontroller, Bluetooth

1. INTRODUCTION

The shipping industry is the pillar of international trade, moving commodities across vast seas and separating economies across national borders. Vessels carrying commodities are the lifeblood of these companies, but since they are hard to manage, they can sail for weeks or even months at a stretch, depending on the distance between two points. While covering long distances, human crew members are always under stress. Also, these trips result in a range of problems, such as fatigue and other judgment errors, which points to the necessity of autonomous navigation systems that are capable of independent operation in order to enable safe and efficient long-distance transportation. The innovation in autonomous cars in the automotive, drone, and aviation sectors brought about comparable advancements in sea transportation (Meriam Chaal et. al. 2023). Though ship autopilots have been present, there is still the need for human intervention in the present technology. An autonomous cargo ship that uses GPS would bring a change in shipping by drastically decreasing the necessity of human observation. These advances would result in reduced operating costs, improved efficacy, safety, and a decrease in navigational human error.

The objective of this project is to utilize high-end micro controllers, GPS, compass modules, and motor control to develop a prototype GPS-guided autonomous freight ship that is able of traveling on its own. One of the greatest features of modern shipping trends are autonomous ships, which can fulfill the increasing requirements of

[1]vinaykb@vvce.ac.in, [2]shivashankar.r@vvce.ac.in, [3]praveenkumarabm@vvce.ac.in, [4]ravishgowdabv14@gmail.com,
[5]rachelthomas33333@gmail.com, [6]pprathamr@gmail.com, [7]srirangabn36@gmail.com

10.1201/9781003725053-17

global trade with better safety procedures and minimal human intervention. Autonomous vessels are a direct application of the autonomous car, drone, and airplane principles. In the last few decades, all industries have had to adapt to incorporate autonomous technology in hopes of maximizing output and safety awareness. The shipping sector has been more cautious in embracing such technologies due to the challenges involved in crossing extensive oceans and the need for security in this unpredictable environment (E. Veitch et. al. 2022).

Most of the autonomous vessels that are constructed or are running still need hybrid systems that demand intervention at turning points. Constructing fully autonomous vessels that can traverse the seas and make decisions independently is therefore the challenge. Stable GPS modules to monitor positions, compass or gyroscope modules to give directions, processing real-time sensor data, and reliable communication systems to handle external interference when necessary are the most critical technologies to enable this. One of the earliest companies to work on ship autonomous technology was Rolls-Royce. They are developing crew-less ships using advanced AI algorithms, sensor fusion, and machine learning to make decisions. All of these indicate the advantages of an autonomous shipping navigation system (Noel A et. al 2019).

2. LITERATURE REVIEW

Meriam Chaal et. al. (2023) This review paper explores the field of autonomous vessel's safety and reliability is something that is getting more and more simulation artificial intelligence reliability analysis. Due to database access limitations and language preferences, it is crucial for the bibliometric studies due to selection of keywords for datacollection. Further formal methods like safety and cybersecurity of autonomous vessels for co-analysis are important. E. Veitch and O. Andreas Alsos, (2022) The paper presents the argument that balance in machine and human control is essential in autonomous ship systems, particularly early human thought in ship action. The research recommends that designing autonomous systems happens through human-AI interaction. STPA and Bayesian Networks are examples of tools that rely on expert human input. Mohammad amin Beirami, Hee Yong Lee. (2015) This paper provides an in depth discussion on The marine industry is working to improve the navigation technology of an autonomous cargo ship to avoid collisions. We are creating rules to make sure the unmanned ships are running safely and it's been certified and registered in the coastal nation as per the International Maritime Organisation guidelines. Controllability in under-actuated systems poses a challenge. Noel A, Shrey Anka K, Satya Kumar K G & Patel A. (2019) This study presents a comprehensive overview of the ships autonomous navigation in the open ocean. Deep learning

helps achieve a high level of autonomy by enabling large data analysis and complex decision making. Systems that are based on reactive techniques such as fuzzy logic, neural networks, genetic algorithms and particle swarm optimization help in controlling the motion behaviour of an object. Duc-Anh Pham and Seung-Hun Han. (2023) In this review the researchers examine Five ship control systems essential for forecasting in advanced sea water environments. A 6 DOF controller improves control accuracy, stability, and adaptability to varying ocean conditions. These systems lessen the crew's load, save fuel, and shorten voyages. Ship simulation models are used when testing and validation of the control systems must be done with empirical data and proper parameters of the ship. Thus, shorter rolling period is more responsive and longer rolling period is more stable. The ship sector keep on growing with superior autopilot technology to meet demand in freight transport, tourism and country border. Zhang X, et. al. (2023) This review paper explores the goal of the study to better collision avoidance technology to make these vessels safer as well as more efficient. The study involved the use of various GPS, ECDIS, AIS, radar, lidar, and visual systems. The group that looks after that is the UK Maritime of Autonomous Systems Regulatory Working Groups (MASRWG). They are working on COPs and frameworks for these vessels. The study by Jiangliu Cai et al. (2024) About advances in autonomous berthing technology for ships. In a bid to enhance safety and efficiency in maritime transportation. The authors describe the collision-avoidance navigation technology which uses variety of systems like GPS, ECDIS, AIS, radar, LiDAR, visual imaging, etc. The Regulatory Working Group on UK Maritime Autonomous Systems is developing frameworks and codes of conduct for these ships. Ruhaimatu Abudu and Raj Bridgelall, (2024) The studies are being done on unmanned cargo ships, which are an important part of maritime technological advancement. Implementing collision-avoidance navigation technology enhances safety and operational efficiency. Regulatory organizations like the UK Maritime of Autonomous Systems Regulatory Working Groups are preparing policies and ethical principles for secure operation, certification and registration of the MAS. Qiang Zhang. et. al. (2017) A Review discusses on the MMG model based ship nonlinear-feedback course keeping algorithms development. A significant focus of maritime innovation is on autonomous surface vessels, for which collision-avoidance navigation technology is of utmost importance. The combined use of GPS, ECDIS, and other sensors is necessary for fully autonomous functionalities on the ship. Agencies such as the UK Maritime Autonomous Systems of Regulatory Working Groups are working on these technologies. A Review by Perera L P, (2018) This paper provides overview of the shipping industry has developed a lot due to remote-operated vessels and earlier developed maritime infrastructure. It is a product of systematic development and agile software methodologies, which makes it reliable and governed by the COLREGs, etc. Mohamad Issa, et al. (2024) The authors of this review paper have analyzed maritime shipping industry can solve the problems of worker shortage, safety, and environmental issues with the entry of autonomous ships and artificial intelligence. The ships use improved sensor technology and AI-driven navigation systems to optimize routes and reduce operational costs and fuel use. We still need to make changes to regulations, technology, and acceptability. Everyone working together will help ensure a successful transition that delivers greater efficiency and sustainability over time.

Ewelina Ziajka-Pozna nska and Jakub Montewka, (2018) This review paper explores the maritime industry is witnessing unprecedented changes owing to introduction of autonomous ships and artificial intelligence. These new technologies can help reduce or solve issues like a lack of workers, safety issues, and environmental problems. They need very little human help to operate and use smart sensors and AI systems for decision making, routing, optimizing etc. Furthermore, modern technology requires research. To ensure smooth adoption and to enhance efficiency and sustainability, collaboration among stakeholders is essential. Young Gyu Lee, et al. (2024) The study presents an depth overview of IMO – International Maritime Organization is drawing up a draft code for the MASS - Marine Autonomous Surface Ships regime. In same vein, the EU is working on the basis of EUAI Act - European Union Artificial Intelligence Act. Barbara Stepien Jagiellonian, (2021) This review paper explores the Legal Problems of Autonomous Shipping the study examines whether autonomous shipping, particularly on the world's first autonomous container ship of yara birkeland fits in the legal classification of ships. We need to make some changes in regulations to enable the separation of crew awareness and autonomy for legal decision making. Mohammad Riyadh et al. (2022) This review paper focuses on how autonomous ships and AI can change the shipping industry by autonomous navigation, optimized routes, saved energy and reduced collisions. Nonetheless, it also discusses topics like rule changes, technical problems, security issues, and societal acceptance. Muhammad Ejaz et al. (2023) The study investigates the free steering autopilot system for maritime vessels. The use of sliding mode control provides simple and promising durability. Another usage of fuzzy logic is for gain management. Various vessels can use this application. Hongguang Lyu et al. (2022) According to the study, which proposes autonomous collision-avoidance methods for maritime vessels, 64.4% of the algorithms available favour the change of course. The study recommends that both speed alteration and course alteration should be considered for efficient working. Liyan Zhu & Tieshan Li, (2017) This paper gives a comprehensive overview of develops an fuzzy output

feedback control method having promising prospects for implementation in the control of an intelligent ship autopilot to address unmeasurable yaw rates. The method should keep tracking errors and outperform methods. Yunduan Cui et al. (2018) This reviews highlights about Gaussian processes, Model predictive control and Model-based reinforcement learning to the control of autonomous maritime vessels. It exhibits impressive sample efficiency and enhanced control effectiveness based on simulations. A Review by Zhiquan Liu et al. (2021) This review paper explores on adaptive autopilot of the sea vessel was developed using neural network access and fuzzy logic control for improvement in performance and energy efficiency over the system performance based on the sliding mode control method. Van den Bremer and Breivik, (2018) This study provides an overview of significant phenomenon associated with ocean surface waves. This study will analyse the Stokes drift, which is an important ocean surface wave phenomenon. Its origin, theory, experiments, and influence on global ocean circulation will be explored. Pham and Han, (2016) This paper gives comprehensive review of advancement of a sophisticated autopilot system for maritime vessels. Pham and Han devised a maritime autopilot based on PID control, fuzzy logic and neural networks to boost steering accuracy while diminishing energy use and crew workload, fitting for current navigation systems. This article authored by Zhang and Liu, (2022) concentrates on the Advancement of a Predictive Autopilot System meticulously tailored for high-speed craft (HSC). In 2022, Zhang and Liu developed a trajectory autopilot system for high-speed craft, enhancing navigational safety and operational efficiency. The system forecasts ship movement, calculates steering commands, and reduces course deviation errors. Wang, Wu, et al. (2022) The paper examines formulation of an autopilot system for maritime vessels that is predicated on fuzzy logic principles. They have developed navigation, adaptability, and safety were enhanced by it. The system can adaptively tune, dynamically change rules and help of the environment. Shao et al. (2022) They have explores how ship autopilot implementation systems have developed as an answer to a variety of challenges, including interference from the environment, unreliable systems, and technology development such as autonomous navigation systems, AI-driven decision making, and so on.

3. SUMMARY

In this summary, technology developments in autonomous maritime vessels are discussed from 2016 with a focus on safety and navigation and regulations. Major technology components are GPS, ECDIS, AIS, RADAR, LiDAR, and visual imaging. Control systems like ANFIS and deep learning are improving operational efficiency and safety. Human monitoring is still important as Shore Control Centres will manage the emergency. Organizations like the UK's MASRWG are building up guidelines for the safe operation and certification of MASS vessels. Managing bad weather, under-actuated systems, and collision-avoidance technologies complicates matters. Deep learning is being used to solve these issues, mainly obstacle evasion and anomaly detection. The aim is to enhance safety on the sea and to reduce accidents, and improve efficiency at sea, all this also complies with international regulations. Developments in maritime autopilot systems feature adaptive control strategies. A prototype autonomous cargo ship guided by GPS shows that such vessels could be fully autonomous, and their destination programmed in a mobile app. Research for the future can involve scaling for larger ships, adding renewables and better collision avoidance through intelligent sensors. This project hopes to transform shipping logistics to be more efficient, safe, and cost-effective thereby making autonomous vessels an integral part of the global supply chain. Successfully deploying this technology will include considerations for safety, regulation, and the development of infrastructure to completely revolutionize maritime operations.

4. CONCLUSION

According to research, how we can improvise our project considering all the above papers. An autonomous cargo ship which is controlled by GPS, is a promising step in water transportation. This ship uses GPS to control its navigation, compass for controlling the system and Bluetooth for wireless data transfer between devices. An autonomous cargo ship has been developed that takes advantage of recent developments in shipping, autopilot ship navigation, and over-the-horizon technology to improve marine transportation conditions. The prototype is feasible for a full capability of an autonomous cargo ship based on the microcontroller ATmega328P and with the aid of the following crucial components: GPS module, compass module, motor driver, Bluetooth module. Using a mobile app, you can upload destination coordinates into this vessel. On receiving this information, the vessel will be able to adjust its heading and movement automatically to reach its destination exactly. The system comes with a modular design. Hence this will make it possible to upgrade the system in the future with features such as obstacle avoidance real-time monitoring, and AI based decision making. The design as such may focus on the core concepts of the organization but serves as a foundational system that can be scaled up for mass application in global shipping. Future research and development areas could be scaling up the system for larger vessels, introducing renewable energy sources that should be sustainable for the operation, and collision avoidance with advanced sensors. This project will make the future of efficient, safe and cost-effective maritime logistics a reality, laying the groundwork for autonomous vessels to play a key role in global supply chains.

REFERENCES

1. Abudu, R and Bridgelall, R, (2024) "Autonomous ships: A thematic review," https://doi.org/10.3390/world5020015
2. Chaal, M., Ren, X, Bahootoroody, A. Basnet S, Bolbot V, Valdez Banda O. A, & Van Gelder P, (2023) "Research on risk, safety, and reliability of autonomous ships: A bibliometric review," https://doi.org/10.1016/j.ssci.2023.106256
3. Cai J, Chen G, Yin J, Ding C, Suo Y, & Chen J, (2024) "A review of autonomous berthing technology for ships," https://doi.org/10.3390/jmse12071137
4. Ewelina Ziajka-Poznanska, (2018), "Costs and Benefits of Autonomous Shipping," https://www.researchgate.net/publication/350495580
5. Hongguang Lyu, (2022) "Ship Autonomous Collision Avoidance Strategie," https://doi.org/10.1002/rob.21990
6. Issa, M, (2022) "Maritime Autonomous Surface Ships: Problems and Challenges Facing the Regulatory Process. Sustainability," vol.14, 15630. https://doi.org/10.3390/su142315630
7. Ka Poznanska, (2021) "Costs and Benefits of Autonomous Shipping A Literature Review," Appl. Sci. 2021, 11, 4553, https://doi.org/10.3390/app11104553
8. Lyu, H, (2023) "Ship Autonomous Collision-Avoidance Strategies—A Comprehensive Review," J. Mar. Sci. Eng. 2023, 11, 830. https://doi.org/10.3390/jmse11040830
9. Liyan Zhu, (2017) "Observer-based adaptive fuzzy prescribed performance control for intelligent ship Autopilot," DOI: 10.1177/1729881417703568
10. Liyan Zhu & Tieshan Li, (2021) "Observer-based adaptive fuzzy prescribed performance control for intelligent ship autopilot," https://doi.org/10.1080/21642583.2021.1934913
11. Mohammad amin Beirami, Hee Yong Lee, (2015) "Implementation of an Auto-Steering System for Recreational Marine Crafts Using Android Platform and NMEA Network," http://dx.doi.org/10.5916/jkosme.2015.39.5.577
12. Mohammad Riyadh, (2024), "Transforming the Shipping Industry with Autonomous Ships and Artificial Intelligence," https://doi.org/10.62012/mp.v3i2.35386
13. Noel A, Shreyanka K, Satya Kumar K. G, & Patel A, (2019) "Autonomous Ship Navigation Methods: A Review," Academy of Maritime Education and Training (AMET), Chennai, India, http://doi.org/10.24868/icmet.oman.2019.028
14. Pham D A & Han S H, (2023) "Designing a ship autopilot system for operation in a disturbed environment using the adaptive neural fuzzy inference system," Journal of Marine Science and Engineering, 11(7), 1262 https://doi.org/10.3390/jmse11071262
15. Perera, L. P, (2018) "AUTONOMOUS SHIP NAVIGATION UNDER DEEP LEARNING AND THE CHALLENGES IN COLREGS," https://doi.org/10.1115/OMAE2018-77672
16. Pham and Han, (2016) "Advancement of a sophisticated autopilot system for maritime vessels," DOI:10.18280/ijsse.140101
17. Shao Oers, (2024) "Analysis of the evolution and functionality of ship autopilot systems," DOI:10.1016/j.eswa.2023.119825

18. Veitch, E., & Alsos, O. A, (2022) "A systematic review of human-AI interaction in autonomous ship system," https://doi.org/10.1016/j.ssci.2022.105778

19. Van den Bremer and Breivik, (2018) "Philosophical Transactions of the Royal Society A and delves into Stokes drift - a significant phenomenon associated with ocean surface waves," DOI: 10.1098/rsta.2017.0104

20. Wang, Wu, (2022) "The formulation of an autopilot system for maritime vessels that is predicated on fuzzy logic principles," DOI:10.3390/jmse11071262

21. Young-Gyu and Jae-Hwan Bae. (2024) "Transformative Impact of the EU AI Act on Maritime Autonomous Surface Ships," Laws 13: 61. https://doi.org/10.3390/

22. Yunduan Cui, (2018) "Autonomous boat driving system using sample-efficient model predictive control-based reinforcement learning approach," Vol. 25; pp. 21–2910.2478/pomr-2018-0128

23. Zhang X, Wang C, Jiang L, An L, & Yang R, (2023) "Collision-avoidance navigation systems for Maritime Autonomous Surface Ships," A state-of-the-art survey https://doi.org/85501E4B6209CFFF9C332D6BCF0363CE

24. Zhang, Q., Zhang, X.-K., & Im, N.-K, (2017) "Ship nonlinear-feedback course keeping algorithm based on MMG model driven by bipolar sigmoid function for berthing," https://doi.org/10.1016/j.ijnaoe.2017.01.004

25. Zhang and Liu, (2022) "The Advancement of a Predictive Autopilot System meticulously tailored for high-speed craft (HSC)," DOI:10.1177/0018720812461374

Advances in Mechanical Engineering and Materials Sciences – Dr. Vinay K. B et al. (eds)
© 2026 Taylor & Francis Group, London, ISBN 9-781-041-20970-6

18

Enhancement of Conventional Unloading System with a CAM Follower-Based Tangential Unloading System

Puneeth H. S.[1],
Akshay S. Bhat[2], Ganavi P.[3] and Jayanth S.[4]
Department of Mechanical Engineering,
Vidyavardhaka College of Engineering

Abstract: The paper presents a design of an unloading process. It enables unloading the materials from the truck container without using hydraulic system. This paper is to design convenient unloading setup for cargo trucks, replacing human involvement and machineries for unloading. This concept focuses on reduction of accidents caused due to hydraulic systems while unloading the goods. It is designed to be cost effective, efficient and energy saving and also to eliminate high costs associated with hydraulic systems, and to implement sequential unloading.

Keywords: Unloading process, Trucks, Accidents, Hydraulics

1. INTRODUCTION

1.1 Preamble

Transportation of materials, raw materials, finished products or transportation of work in progress (WIP) materials plays a vital role in the industries. The transportation of the materials can be from input, that is receiving of raw materials in to the industries or unloading of baggage's in different industries to dispatch the finished products. It also includes transfer of materials from one station to another station. The Bureau of Labor Statistics (BLS) data from the warehousing and storage industries signifies that the musculoskeletal injury and some high-risk back injuries are being occurred in large numbers while manually loading and unloading the goods from trailers or containers. There are about a million back injuries which have been encountered on the job site every year, and nearly about one of every four dollars i.e., 1/4th of a dollar is being accounted on the workers compensation. Similar effects are found in the process of loading and unloading of goods from the trucks or containers, hence there is a necessity of a mechanical system which handles the same purpose in an effective way and to avoid the injuries.[1]

Fork lifting, Portable Radial cranes, overhead cranes, hydraulic lifts, bucket elevators and conveyors are found to be the different mechanical system for lifting and transporting of bulk materials or products from place to place or at the loading point in an industry whose factor depends on the terms like height of transportation, speed of handling the product, quantity of the product, nature of the product, weight and size of the products to be transported or for loading or unloading. [1]

1.2 Objectives

The major objectives are as follows:

- To design and develop an effective method of unloading.
- To eliminate complex machineries which are currently used for unloading the goods.
- Key focus on eliminating accidents caused by hydraulic system and to provide safe working environment.

[1]puneeth.hs@hotmail.com, [2]akkshaybhhat@gmail.com, [3]ganaviprakash26@gmail.com, [4]jayanthmuthu24@gmail.com

10.1201/9781003725053-18

- To reduce or minimize the cost associated with hydraulic system.

The above objectives majorly relate to the design and development of the truck for unloading process from transporting containers. This can also be enhanced by incorporating safety features for the system to avoid the injuries occurred during the loading and unloading of the products. This will ultimately result in the increase of productivity of the firm by reducing the labor cost with respect to their wages and compensation, if any accidents or injuries occur during unloading process and by reducing the time utilized for unloading of the products. All of these will directly affect the increase of profit to the industries.

1.3 Problem Statement

The problem focuses on the accidents caused during unloading the goods from the truck due to hydraulic systems. It also focuses on the cost incorporated with hydraulic systems and the working environment.

1.4 Literature Review

With rapid growth of manufacturing industries there is an increase in demand for the products, hence conveying them to the customers is important. Loading and unloading of the products plays a very important role with respect to cost, safety and time. To achieve this cost effective, safe and ease of loading and unloading, the industries need to have effective mechanical system like fork-lift, hydraulic lift, conveyors, cranes etc. Hydraulic systems are most effective form of loading and unloading mechanical system. It incorporates stacking, moving and emptying of materials starting with one phase of assembling procedure then onto the next. Factors influencing hydraulic system are prime mover problems, fluid contaminants like contaminants worsen the resistive characteristic of the oil, temperature problems, fluid levels and quality, human error, operation, maintenance and safety.[2]

Reports on Accidents

The report of accidents recorded are listed below in Table 18.1.

The above reports illustrate the accidents caused due to the incorporation of hydraulic system in trucks. Each one of those gives the details of injuries and deaths caused while operating, maintenance and failure in hydraulic system.[3]

Review of an Article

Hydraulic System Leakage - The Destructive Drip by Lloyd Leugner

The article focuses on the leakage of oil or lubricant in hydraulic system, it is based on the survey done in North America every year and in Canada. It says that a drop of oil leaked per second is equivalent to 420 gallons of oil in a period of month. According to the estimation, over 100 million gallons of oil would be saved every single year in North America if the leakage from hydraulic system is avoided. In Canada, over 12 million gallons of oil is squandered due to leakage. Hence forth, use of hydraulic system has led to high consumption of oil which has increased the oil consumption by 4 times. This expands the capital cost oriented with the hydraulic system and its working. Concentrating on the safety and accident liability, there is a worker or the operator who slips or falls on the remains of the leakage of hydraulic system. With this, pressurized hydraulic oil hand overs an appreciable fire risk whenever valves or hoses breaks or bust-off. It terribly sucks the safety of technicians or operators. It also damages the surrounding environment. With the knowledge from this article, the above-mentioned effects can be reduced by eliminating the hydraulic system.[4]

1.5 Market Survey

The Literature review regarding the trucks and the caster wheels for the unloading system. Unloading system and market study and the journal reviews provided an insight that there are no similar features that the company is expecting and neither any people have come up with the solution for the facing problem, hence we need to come up with a new design to meet the need of the customer to achieve this an integration of the available designs and need to work on its development to conclude the final design. We have considered EICHER PRO 3008

Table 18.1 Reports on accidents

No	Report ID	Event Date	Event Description
1	0418600	08/18/2021	Employee is crushed and killed when lift fails.
2	0418400	11/11/2020	Employee is struck by falling hydraulic boom.
3	0627700	06/28/2016	Employee falls when a hydraulic lift falls and is killed.
4	0215800	05/25/2016	Employee struck and killed by falling container.
5	0352430	10/10/2014	Employee is crushed by falling dock plate and is killed.
6	0352450	10/08/2014	Employee is injured when ejected from boom lift basket and killed.
7	0355118	08/30/2014	Employee struck by falling forklift basket and is killed.
8	0352430	03/19/2013	Employee is struck by dock plate and later dies.
9	0453720	03/18/2013	Employee is struck by falling load and is killed.
10	0418300	09/02/2012	Employee is crushed and killed by truck trailer.

Source: Authors

Truck which is one the trucks used for transportation in industries. The data has been taken from official website of EICHER TRUCKS. [5]

Fig. 18.1   Brochure of Eicher Pro 3008 [5]

Fig. 18.2   Eicher Pro 3008 [5]

1.6 Modelling Selection

Concrete for Cam Structure

Concrete is a composite fabric composed of fine and coarse aggregate bonded together with fluid cement that hardens over time. Concrete is the second most used element in the world after water, and is the most widely used constructing material. This has been considered for constructing the cam structures.

Caster Wheel

A caster is a non-driven wheel that is designed to be attached to the bottom of a larger object to enable that object to be moved. The claim of this type of wheels

necessitates accurate sizing based on the terrain on which they will be used and the weight they are expected to carry. Greater weighed items need casters with thicker wheels; some larger objects need several wheels to consistently distribute weight. By looking into all these properties and features, we have selected one specific type of caster wheel that is 10″x3″ NYLASTRONG RIGID CASTER which can bear more loads up to 7000lbs. [6]

2. CONCEPT GENERATION AND SELECTION

2.1 Introduction

Different ideas or concepts are being generated considering the functionality and objectives to be achieved based on the need or specification of the customer and study relating to the requirement.

2.2 Customer Voice or Specification to Technical Requirement

Considering any of the products designing process irrespective of type of the product the conversion of customer voice or specification to technical requirement is very important. The below table indicates the conversion of customer voice or specification to technical requirement utilized to define the product features or characteristics.

Table 18.2   Specification to technical requirement

Customer Voice	Technical Requirement
Adjustable material movement speed	Variable speed
Less workers	Minimum manpower
Less working space	Compact size
Movement from place to place	Mobility
Should be able to Lift, swing and move	Functionality
Easy maintainability	Serviceability
Long life	Material

Source: Authors

2.3 Dimensional Study

Indian standard truck container dimension has been considered to understand the dimension of the product, the standard truck container dimension is shown in the below table.

Table 18.3   Dimensions of container

Type	Length (mm)	Width (mm)	Height (mm)
Dry Cargo container	5898	2352	2394
Open type container	5629	2212	2311

Source: Authors

2.4 Concepts

The considered Product Design Specifications have been used and concepts for the product are being generated based on them, Functionality and degrees of freedom to be achieved is being given the first preference while taking up the concepts followed by others. The different concepts generated and their base is listed below.

- The initial concept is taken from hinges of door. The hinge mechanism in the door provides the movement like opening and closing of the door, this has been used to generate one of the concepts for our product design. The action of hinged door has been considered for the movement (lifting and lowering) of the container with the truck. The movement provided by the hinge to the door is considered for the design of container, the hinge connects the truck and container in such a way that it allows the container to lift. The design has been done using solid works software.

- Another concept has been derived from the idea of cam movement that gives the lifting and lowering. The concept of wheel sliding over the cam surface is used in the designing of the container and the dome structures, the cam wheels attached to the container slides over the dome surfaces which gives cam movement to the container.

3. DESIGN AND DEVELOPMENT

Following factors were considered while de-signing a truck.

Concepts were generated based on customer data and market design study. Customer problem and the aspiration to overcome it was done based on the customer requirement in the project. Models for the project was developed using solid works. Final concept for the design was selected using concept selection table based on the review of the company engineers and Brochure. Market available design and previously designed truck was used and to integrated to achieve the prescribed requirement. Experience, functionality, experience, design, manufacturing, were used as the factors for the review` of the concepts.

3.1 Design Methodology

Two parallel domes are constructed on either side of the platform through which the truck passes as shown in Fig. 18.3. Considering the truck, the front end of the container is extruded on either side and attached with the wheels as shown in Fig. 18.4(a) and 18.4(b). Rare end of the container is hinged with truck as shown in Fig. 18.4(e). The loaded truck is made to pass on the platform (unloading area) between the domes as shown in Fig. 18.4(f). As the truck passes, the wheels attached to the front end comes in contact with the dome surface as shown in Fig. 18.4 (g). When the wheel and the dome get in contact, the wheels slide over the dome's surface and the container starts to lift

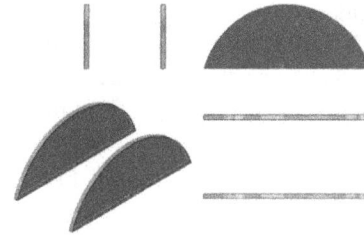

Fig. 18.3 Isometric view of dome structure

Source: Authors

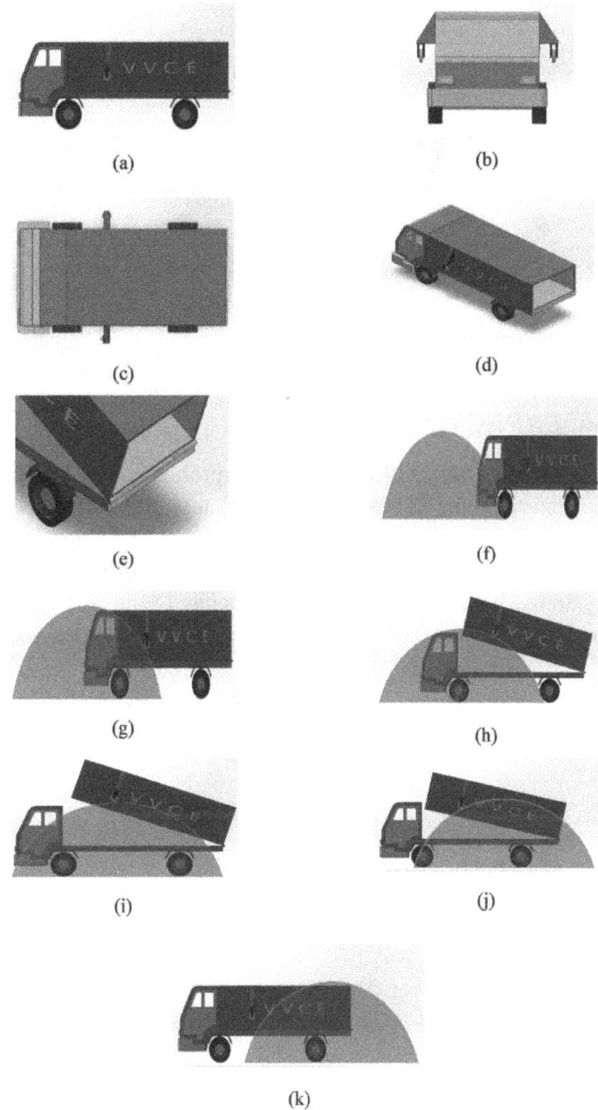

(a)

(b)

(c)

(d)

(e)

(f)

(g)

(h)

(i)

(j)

(k)

Fig. 18.4 Pictorial representation of the concept working

Source: Authors

as shown in Fig. 18.4(h). When the wheel reaches the tip of the dome the material will be unloaded completely as shown in Fig. 18.4 (i). On further movement of the truck, container is lowered by sliding down the dome as shown in Fig. 18.4 (j) and Fig. 18.4 (k).

Final concept for the design was selected using concept selection table based on the review of the company engineers and Brochure. Market available design and previously designed truck was used and to integrated to achieve the prescribed requirement. Experience, functionality, experience, design, manufacturing, were used as the factors for the review of the concepts.

- The detailed design with respect to engineering is shown below.

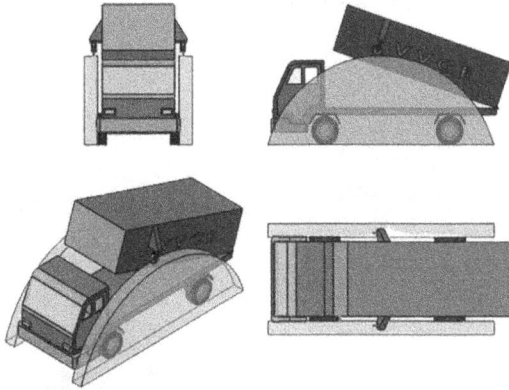

Fig. 18.5 Isometric view

Source: Authors

4. EXPECTED RESULTS

The new design will overcome the problems associated in material handling and unloading that were being faced by the customer. It also has the provision to reduce the accidents during unloading and maintenance of hydraulic systems. It eliminates the cost associated with adoption and maintenance of Hydraulic systems in the trucks.

Safety features with respect to functionality used to design the trucks. It is a one-time investment project; it does not require much maintenance except replacement of cam wheels after its life span. The truck is designed to be handled with fewer workers. Hence these features increase the productivity and profit. It also reduces the Labor cost and maintenance cost.

REFERENCES

1. Ergonomics, Back injuries Fact Sheet, Department of Environmental Safety, Sustainability and Risk Pontiac St, college Park, USA (301) 405-3960. Available online:https://essr.umd.edu/about/occupational-safety-health/ergonomics/back-injuries-fact-sheet (copyright year 2023).
2. Michele Baker, Most Common Causes of Hydraulic Systems Failure, York Precision: machining &Hydraulics, (February 13th 2020). Available online: https://yorkpmh.com/resources/common-hydraulic-system-problems/
3. United States Department of Labor: Occupational health and Safety. Available online: https://www.osha.gov/
4. Lloyd Leugner. Hydraulic System Leakage - The Destructive Drip; Is fluid leakage drowning your profits? Published by NORIA; Machinery lubrications. Available online: https://www.machinerylubrication.com/Read/21/hydraulic-system-leakage
5. VE Commercial Vehicles Ltd, Eicher Pro 3008 Broacher (South Africa (22 April 2017) Available online: https://www.mannaiautos.com/dynamic/HeavyVehilcleModels/HeavyVehicleModel_202001066CC6.pdf
6. Allied caster equipment company. Available Online: https://www.alliedcaster.com/

Advances in Mechanical Engineering and Materials Sciences – Dr. Vinay K. B et al. (eds)
© 2026 Taylor & Francis Group, London, ISBN 9-781-041-20970-6

19

A Study on Performance Analysis of a Foldable Electric Vehicle

Akshay S. Bhat[1], Puneeth H. S.[2]
Assistant Professor, Department of Mechanical Engineering,
Vidyavardhaka College of Engineering,
Mysuru, India

S. K. Sundararajan, Shankar B., Manju D.
Student, Department of Mechanical Engineering,
Vidyavardhaka College of Engineering,
Mysuru, India

Abstract: The foldable electric vehicle (FEV) market is a promising segment in the automobile industry, guided by a necessity for green transport solution and improvements in battery, motor, and folding mechanisms. There are multiple types of FEV's like stand-up scooters, hover board, etc. Attractive to most users for running errands, recreation, and carriage. This paper focuses on advancing the safety and performance of electric vehicles by integrating cutting edge sensors and conducting a performance analysis. This includes the addition of safety sensors to monitor the speed of the vehicle, ensuring a safer and more efficient riding experience. It aims to conduct performance analysis with a primary focus on battery discharge rates, and brake effects. This analysis will guide the optimization of energy management and sustainability. The foldable electric vehicle design is being developed to increase portability, allowing users to easy transportation and to store the electric vehicle when not in use. To enhance the user experience, different riding modes has been introduced, it enables riders to choose the modes that optimize speed, energy efficiency, or comfort based on their needs.

Keywords: Foldable electric vehicle, Sensors, Performance analysis, Riding modes

1. INTRODUCTION

Foldable scooters and bicycles have been popular for decades. A lightweight foldable electric vehicle is used instead of walking or riding a bicycle between the residence or workplace, the industry shopfloors or a metro station. The fiery growth of electric vehicles has had a significant influence on the future of transportation, providing ways to lower carbon emissions and boost energy efficiency. In this fascinating environment, foldable electric vehicles or foldable EVs have become an advanced idea aimed at solving the issues of urban mobility. Among these numerous inventions that appeared to provide an option in terms of flexible and compact dimensions in fulfilling the requirements for city mobility is foldable electric vehicles (FEVs). They address the issue of highly efficient transport at a relatively short distance in dense cities through integrating the advantage of electrical power with its eco-friendly merits. These involve several factors such as a power source, motor efficiency and general design. How efficiently foldable electric vehicles serve is mainly determined by factors such as stability, energy efficiency, speed, and other key performance criteria. On the other hand, added smart technology like GPS as well as apps. A sudden rise in demand for electric vehicles, as it is considered a prime solution to reduce carbon emissions and fossil fuel dependence, has occurred with rapid urbanization and increased environmental sustainability concerns. In this respect, foldable electric cars have emerged as a new and practical solution that addresses the issue of scarce urban space and the need for a sustainable mode of transportation. The significant

Corresponding author: [1]akshaybhat@vvce.ac.in, [2]puneeth.hs@vvce.ac.in

advantage of foldable electric vehicle is its portability, which makes them perfectly fit for commuters in crowded urban environments with minimal parking and storage. A fascinating experiment in the automobile industry, an electric vehicle that can be folded. It is an environmentally friendly vehicle since it runs on electricity. Lithium phosphate battery, DC hub motor and some other features are included in the vehicle. As a result, it is simple to fold, making it more convenient to transport and store. [Kartavy R. Patel et al., 2021]. EVs are highly utilised for various mobility purposes and setting high targets in both developing and developed countries. Success of this research will bring about the reduction of dependency on fossil fuel vehicles. [Leal Filho W et al., 2021]. With the aid of this innovative foldable mechanism, today's problems such as power, traffic, and parking, as well as some global issues, can be solved. It will also help to come out with solutions for world's problems and develop people's lives without hassle, hence an advancement can positively impact the reduction of the dependency on fossil fuel vehicles.

Foldable e-bikes are a great substitute for urban commuters who have a lot of transit options but little storage space. The various reasons why FEV's can prove to be handy for urban mobility are highlighted next.

Last-Mile Connectivity: With foldable e-bikes, this problem can be resolved by providing an environmentally friendly, short-distance, portable mode of transportation that will relieve the burden of cars and promote sustainable mobility. A sustainable solution that will help bridge the last-mile gap between public transport hubs and final destinations would be offered through last-mile connectivity by electric vehicles (EVs). By using electric bikes, scooters, or small EV's, cities can reduce traffic congestion, lower emissions, and provide an affordable, eco-friendly alternative for short-distance travel. EVs for last-mile connectivity improve convenience, flexibility, and efficiency, making them an important part of modern, sustainable transport systems.

Mobility and storage: For those with limited space for storage at home or at work, foldable e-bikes are a great choice. Because of their folding shape, they are easy to keep in tight office spaces, public areas and residences.

Multi-modal Transportation: These folding electric bicycles have high compatibility with both road transport, like a car and bus, or by train because of ease and flexibility in folding/unfolding. Foldable e-cars in multimode transportation enable increased flexibility and accessibility together with the reduction in waste in use of spaces by accommodating passengers and their carrying options as they switch and get accommodated by available bus/train transport with reducing pressure and environmental disturbance.

Tourism and Recreation: Foldable e-bikes are becoming more and more popular among tourists and leisure cyclists.

Their mobility, which eliminates the need for large bike racks or specialized transportation, makes them perfect for discovering new places. This can generate a large amount of revenue if implemented in large scale in various tourist attractions.

Based on literature there is a need for developing a low cost, simple and robust FEV. Even though there are not many well-known foldable electric vehicles around us, soon this concept might be useful for urban transportation.

2. METHODOLOGY

To create a handy and portable transportation option there is a need for integrating innovative engineering concepts along with choosing the right materials, and design techniques must be utilized to create a foldable electric vehicle.

2.1 Identify Target Market

Identify target market and application scenarios for the electric vehicle to define its purpose and requirements. Establish the required speed, range, and technical characteristics. We could achieve this through a big market of easy electric mobility which can be used under various circumstances. It can be used daily as a commuter or A short commuter between small distances or also a scooter that can carry all your daily groceries and other items brought home.

2.2 Choosing Lightweight Materials

Choosing lightweight materials for the frame to maximize portability without compromising structural integrity is the process of material selection. We chose Mild Steel rectangular pipes for the trolley system.

2.3 Controls and User Interface

To make sure the scooter has an easy-to-use interface. The following are some enhancements. A Throttle based battery monitor: To monitor battery life the display in the throttle lets the user know the battery consumption during the riding time. New Hinging Mechanism: To squeeze the vehicle into tight spaces the introduction of a hinging mechanism that is easily foldable within fraction of seconds. Reverse Mode: For easy accessibility of the vehicle from tight spaces with inclusion of a reverse mode that can remove the vehicle from really tight spaces. An SPST switch for easy switch ability from Forward mode to reverse mode and vice-versa. Speedometer: For the rider to cautiously ride a speedometer is provided. This also tracks how much distance the rider has travelled. Also, there is a clock feature that shows the time. Riding modes: This electric scooter comes with speed-limiting features that prevent the scooter from exceeding certain speed limits for safety riding modes provide where the maximum speed is regulated to suit different riding conditions. A DPDT three pole switch is used to change between riding modes.

2.4 Safety Features

This electric mobility includes common safety features seen in conventional electric vehicles. Electronic Brake: The left-hand side lever acts as an electronic brake that doesn't allow the rider to move the vehicle even when the key is turned on. Speed Control: This feature helps riders maintain safe speeds, especially in urban environments or crowded areas. Motor Cutoff Mechanism: If a rider falls off the scooter or loses control, many scooters have a motor cutoff switch that automatically disables the motor, preventing the scooter from continuing to move uncontrollably. Here we have the brake lever doing this same mechanism.

2.5 Testing and Iteration

Build a functional folding electric car prototype and conduct extensive testing. We have performed testing over time with various parameters and conditions kept in mind. Weight: The average Indian weight is 55-65 Kg according to the National institute of Nutrition. Height: The average Indian height is 162-180 cm according to the National institute of Nutrition. Climatic Conditions: The climatic conditions were also standardized at room temperature of 25-27 degree Celsius at 1 bar atmospheric pressure. Terrain conditions: The testing of all the features were tested under a Tar mac road and cement paved path with inclines and declines of -2 to +2 degrees. Battery Temperature: The standard operating temperature range for a 14V lead-acid battery is typically between 20°C to 25°C. This is considered the optimal temperature range for lead-acid batteries to perform efficiently and maintain their lifespan.

3. CONCEPT DESIGN

A foldable electric vehicle is designed to be portable and lightweight, making storage and transit easier. These electric vehicle models are perfect for urban transportation, shop floors, and spaces that are tight on space. Its ability to fold an electric vehicle, often at the frame, folding handle, and removable seat, is its defining feature. This reduces the vehicle's size, resulting in easier carrying and storing. Additionally, the foldable vehicle has a rechargeable battery and an integrated electric motor. Figure 19.1 shows an isometric view of the final concept.

Fig. 19.1 Isometric view of the final concept

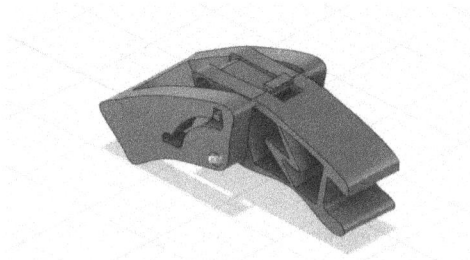

Fig. 19.2 Isometric view of newly designed hinge

Features of the newly designed foldable electric vehicle is that it is portable & light weight design which facilities for easy transportation. The foldable electric vehicle is an eco-friendly vehicle which runs on electric power reducing the carbon footprint. The foldable electric vehicle is easily foldable for convenient storage. Some more cosmetic modifications were made to enhance the user experience like adding a newly designed hinge mechanism shown in Fig. 19.2. A trolley attachment for carrying or storing any essentials, speed sensors to monitor speed, different modes for riding the foldable electric vehicle & GPS navigation for electric vehicle. The Fig. 19.3 highlights the updated design with the modifications as mentioned.

Fig. 19.3 Isometric view of FEV with Trolley

4. DESIGN CALCULATIONS

The following design specifications were taken to obtain the static calculations.

Mass = 120kg

Factor of safety = 1.5

Diameter of wheel = 8 inch

linear acceleration = 1.5m/sec2

linear velocity = 7m/sec

Angular velocity:

Angular velocity = linear velocity ÷ Radius of wheel

= 7 ÷ 0.1016

= 68.897 rad/sec

Angular acceleration:

Angular acceleration = linear acceleration ÷ radius of wheel

= 1.5 ÷ 0.1016

= 14.763rad/sec2

Moment of inertia:

Moment of inertia = 1/3 × mass on each wheel
× radius of wheel2
= 1/3 × 34 × 0.10162
= 0.206kg/m2

Torque:

Torque = moment of inertia × angular acceleration
× factor of safety
= 0.206 × 68.897 × 1.5
= 4.572 Nm

Speed:

Speed = linear velocity × 60 ÷ 2π
= 7 × 60/2π
= 657.92 rpm

Power:

Power = torque × speed ÷ 9.548
= 4.572 × 657.92/9.548
= 315.01W

The range of the vehicle depends on the following factors like real-world conditions and actual range may vary significantly depending on the factors mentioned above. Also, the battery capacity can degrade over time, affecting the range of the vehicle. The dynamics of the scooter are tabulated below in Table 19.1.

Table 19.1 Table of readings for dynamics of the scooter

Serial No.	Speed (Kmph)	Time Taken (in seconds)	Acceleration/ Deceleration Rates (m/s²)
1	0 – 15	10.3	0.406
2	15 – 0	8.4	0.5
3	0 – 20	13.72	0.405
4	20 – 0	11.12	0.502

5. RESULTS

The performance of the foldable electric vehicle has been analysed and plotted under 2 characteristic curves.

5.1 Speed v/s Time

From the graph shown below it can be concluded that the straight line is indicating that acceleration is constant. The speed of the object increases linearly with time, so it accelerates uniformly. Slope of the graph is the rate of acceleration, and the area under the graph is the distance travelled over the time interval. Figure 19.4 shows the graphical representations of the Speed v/s Time from the calculations.

5.2 Acceleration v/s Time

From the graph shown below, it can be concluded that the graph suggests that the acceleration of the object is

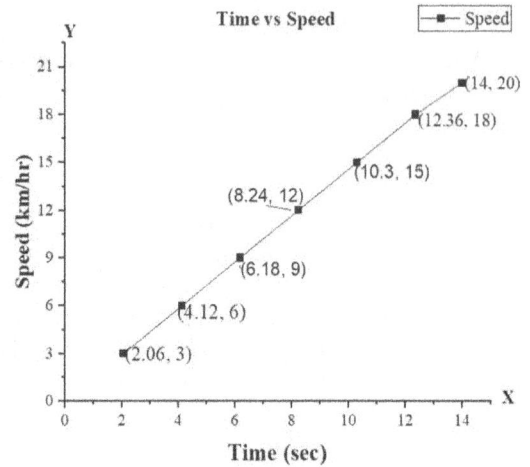

Fig. 19.4 Graphical representation of speed v/s time

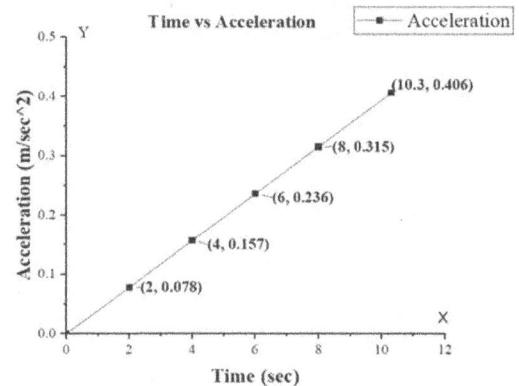

Fig. 19.5 Graphical representation of acceleration v/s time

increasing with time. The slope of the line gives the rate of change in acceleration, while the area under the curve gives the change in velocity over the time interval. Figure 19.5 shows the graphical representations of the Acceleration v/s time from the calculations.

6. CONCLUSION

Foldable Electric Vehicles are an interesting proposition for the automobile industry. It provides solutions for issues like vehicle theft and damage by being compact and portable, scarcity of parking spaces in urban areas like metropolitan cities (like Delhi, Mumbai, Kolkata, Chennai and many more), providing a faster mode of transport for limited cost of the pockets. It also encourages people to use public transport that can increase the quality of living in urban as well as rural areas as it helps target lower emissions of pollution on an overall count. Its ability to be rider friendly and customizable makes it an attractive option for the target audience of age group between teens to old age and backed by its powerful machinery and energy pack filling up other points in the blank. It is an environmentally friendly vehicle that can charge with solar energy at parking stations (if stations are equipped with the facility) and runs on electricity. The vehicle incorporates a lead acid battery keeping the vehicle ready to go, a powerful hub motor integrated with drum brakes that serves its purpose. Many more characteristics are to be included as well like regenerative braking, navigation system, music and audio system and so on which can be incorporated in the current model. Furthermore, employing rechargeable batteries extends the life of the vehicle. Overall, this is believed to be an extraordinary vehicle for the current period and urban travel.

REFERENCES

1. Patel, Kartavy & Vanerkar, Harsh & Patel, Harsh & Shastri, Viranchi. (2021). A Review on Design and Fabrication of Foldable Two Wheel Electric Vehicles. 10.13140/RG.2.2.22200.47360.
2. Leal Filho, W., Abubakar, I. R., Kotter, R., Grindsted, T. S., Balogun, A.-L., Salvia, A. L., Aina, Y. A., & Wolf, F. (2021). Framing Electric Mobility for Urban Sustainability in a Circular Economy Context: An Overview of the Literature. Sustainability, 13(14), 7786. https://doi.org/10.3390/su13147786

Note: All the figures and tables in this chapter were made by the authors.

Advances in Mechanical Engineering and Materials Sciences – Dr. Vinay K. B et al. (eds)
© 2026 Taylor & Francis Group, London, ISBN 9-781-041-20970-6

20

Non-Invasive Anemia Detection using Fingernail and Tongue Images with Medical Datasets and Image Processing

Janhavi V.[1],
Ayesha Taranum[2], Sinchana N.,
Shashank Gowda C., Dore M., Chandana M.
Department of Computer Science and Engineering,
Vidyavardhaka College of Engineering,
Mysuru, Karnataka, India

Abstract: Anaemia is a prevalent global health issue, often diagnosed through invasive blood tests. This study introduces a non-invasive method for detecting anaemia by leveraging fingernail and tongue image datasets. Utilising advanced image processing techniques, we analyse colour variations, texture patterns, and structural features from high-resolution images. Key algorithms include segmentation for isolating regions of interest, feature extraction for identifying haemoglobin-related characteristics, and machine learning classifiers for predictive modelling. The dataset comprises diverse samples, ensuring robustness and inclusivity. Our method aims to offer a rapid, cost-effective, and painless diagnostic alternative suitable for resource-limited settings. By integrating convolutional neural networks (CNNs), the proposed framework achieves high accuracy in detecting anaemia, correlating image-derived biomarkers with haemoglobin levels. Comparative analysis with traditional diagnostic methods underscores the efficacy and potential of this approach. Future work focuses on enhancing dataset diversity and refining prediction algorithms to improve generalisability. This innovation promotes early detection and proactive anaemia management, revolutionising healthcare accessibility and patient outcomes.

Keywords: Anemia detection, Complete blood count (CBC), Convolutional neural networks, Local binary patterns (LBP) or Gray-level Co-occurrence matrix

1. INTRODUCTION

Anemia is a prevalent medical condition defined by a deficiency in red blood cells or hemoglobin, leading to a diminished capacity of the blood to transport oxygen. This condition affects millions worldwide and presents considerable health risks, including fatigue, weakness, and, in severe instances, life-threatening complications. Although traditional diagnostic techniques, such as complete blood count (CBC) tests, are dependable, they often necessitate invasive procedures that involve blood sampling, which can limit accessibility, particularly in resource-constrained environments. This situation has prompted the need for the development of non-invasive, cost-effective, and rapid diagnostic methods to mitigate the global impact of anemia. Recent progress in image processing and machine learning has opened new avenues for innovative medical diagnostic techniques. Notably, non-invasive detection of anemia through visual indicators from images of fingernails and the tongue has gained considerable interest. Both fingernails and the tongue can display visible alterations, such as pallor, which may act as biomarkers for anemia. These visual traits can be quantitatively assessed using image processing methods to identify early signs of anemia without the necessity for blood tests.

Corresponding author: [1]janhavi.v@vvce.ac.in, [2]ayesha.cs@vvce.ac.in

The proposed methodology utilizes medical datasets comprising images of fingernails and tongues to create an automated system for anemia detection. This system incorporates image preprocessing, feature extraction, and classification algorithms to evaluate the datasets and predict the presence of anemia. The application of convolutional neural networks (CNNs) significantly improves the system's accuracy by detecting subtle patterns in the images that may not be visible to the naked eye. By merging advanced machine learning techniques with image processing, this approach aspires to offer a non-invasive, user-friendly, and scalable solution for anemia detection.

A primary benefit of this strategy is its potential to democratize healthcare by facilitating anemia screening in remote and underserved regions. With the widespread availability of such technology, it could significantly enhance access to essential health services.

2. RELATED WORK

Non-invasive anemia detection and image processing techniques for analyzing fingernail and tongue datasets.

2.1 Traditional Anemia Detection Methods

Highlight invasive techniques such as Complete Blood Count (CBC), Hemoglobin tests, and point-of-care devices. Discuss the limitations of these methods, such as cost, time, accessibility, and discomfort to patients.

2.2 Non-Invasive Detection Techniques

Overview of advancements in non-invasive methods, including imaging-based approaches Mention studies where skin, tongue, or nailbed imaging has been analyzed for detecting anemia. Explore how parameters like pallor, discoloration, or texture of the tongue and nails have been linked to anemia.

2.3 Image Processing in Medical Diagnosis

Summarize the use of image processing for analyzing medical datasets, focusing on feature extraction and analysis. Discuss algorithms commonly used for feature extraction, including Color Analysis: Techniques to quantify pallor using RGB or HSV models. Texture and Analysis approaches like Local Binary Patterns (LBP). Segmentation: Common algorithms for isolating nails or tongues from the image background, such as Otsu's thresholding, k-means clustering, or U-Net models.

Edge Detection: Methods like Canny or Sobel operators to enhance boundaries for better analysis.

2.4 Machine Learning and Deep Learning Applications

Overview of ML models like SVMs and Random Forests for anaemia classification based on extracted features.

Highlight the role of deep learning architectures (e.g., CNNs, MobileNet, ResNet) in medical image classification tasks. Mention studies that utilized transfer learning for medical image analysis.

3. LITERATURE SURVEY

Recent advancements in non-invasive medical technologies have led to the exploration of anemia detection using external physiological indicators such as skin tone, fingernail color, and tongue pallor. Studies suggest that hemoglobin levels can be correlated with observable color changes, as anemia often results in a pale appearance. Researchers like Lakshmi et al. (2020) have utilized image processing algorithms to analyze tongue images, showing that the red channel intensity has a strong inverse correlation with anemia severity. Similarly, studies by Gupta et al. (2021) emphasize the potential of fingernail color as a biomarker for estimating blood hemoglobin levels without invasive blood sampling.

3.1 Image Processing Techniques for Medical Applications

The role of image processing in medical diagnostics has grown significantly. Edge detection, histogram equalization, and feature extraction methods have been widely used for analyzing biomedical images. In the context of anemia detection, Fang et al. (2019) utilized pre-processing techniques like Gaussian smoothing to reduce noise in tongue images, followed by segmentation methods such as Otsu's thresholding for isolating regions of interest. Researchers also employ ML and DL models to classify the severity of anemia based on extracted features. CNNs have demonstrated high accuracy in analyzing medical images, as noted by Sharma and Kumar (2022).

3.2 Fingernail and Tongue-Based Medical Studies

Studies focused on fingernail analysis for anemia detection rely on changes in color and texture. Works by Chen et al. (2018) highlight the importance of color quantization techniques to distinguish between healthy and anemic individuals based on nailbed coloration. For tongue analysis, studies like those of Zhang et al. (2020) emphasize the combination of image pre-processing and machine learning classifiers to improve diagnostic accuracy. These studies reveal that RGB color models and texture analysis techniques are effective in extracting relevant features from both fingernail and tongue datasets.

3.3 Dataset Development and Challenges

Acquiring standardized datasets for fingernail and tongue images presents unique challenges. Efforts have been made by institutions to create open-source medical image repositories. However, variations in lighting, image quality, and skin tone among individuals require robust normalization techniques during the pre-processing stage. Research by Singh et al. (2021) highlights the importance of dataset augmentation and standardization to ensure generalizability in anemia detection models.

3.4 Integration of AI and Image Processing

Integrating AI into image processing has revolutionized non-invasive diagnostic tools. Algorithms like Support Vector Machines (SVM), Random Forests, and CNNs have been successfully used in detecting patterns indicative of anemia. Research conducted by Patel et al. (2022) implemented a hybrid model combining K-means clustering with CNNs to enhance the classification of anemia severity using tongue images. Such approaches demonstrate the synergy between AI and traditional image processing techniques in achieving improved diagnostic accuracy

4. PROPOSED SYSTEM

The propounded system addresses the shortcomings of existing methods by leveraging the **YOLO (You Only Look Once) model**, a cutting-edge object detection algorithm, to analyze medical images of fingernails and tongues for anemia indicators. **YOLO's remarkable speed and accuracy** in object detection make it well-suited for real-time analysis in this application. By training the model on a **large, labeled dataset** containing images of both anemic and non-anemic conditions, the system can effectively recognize subtle visual markers of anemia, such as **pallor and discoloration**, with high precision.

This approach offers several advantages over traditional blood tests and other emerging non-invasive methods. **Firstly, it is entirely non-invasive**, removing the discomfort and risks associated with blood sampling. **Secondly, it delivers rapid results**, facilitating immediate screening and potentially expediting diagnosis. Furthermore, the system is **cost-effective and accessible**, requiring only a camera-equipped device and a trained model, adaptable for use in various environments, including **remote and low-resource settings**. By enhancing the accessibility and efficiency of anemia detection, the proposed system can **improve early diagnosis and management**, ultimately contributing to better health outcomes.

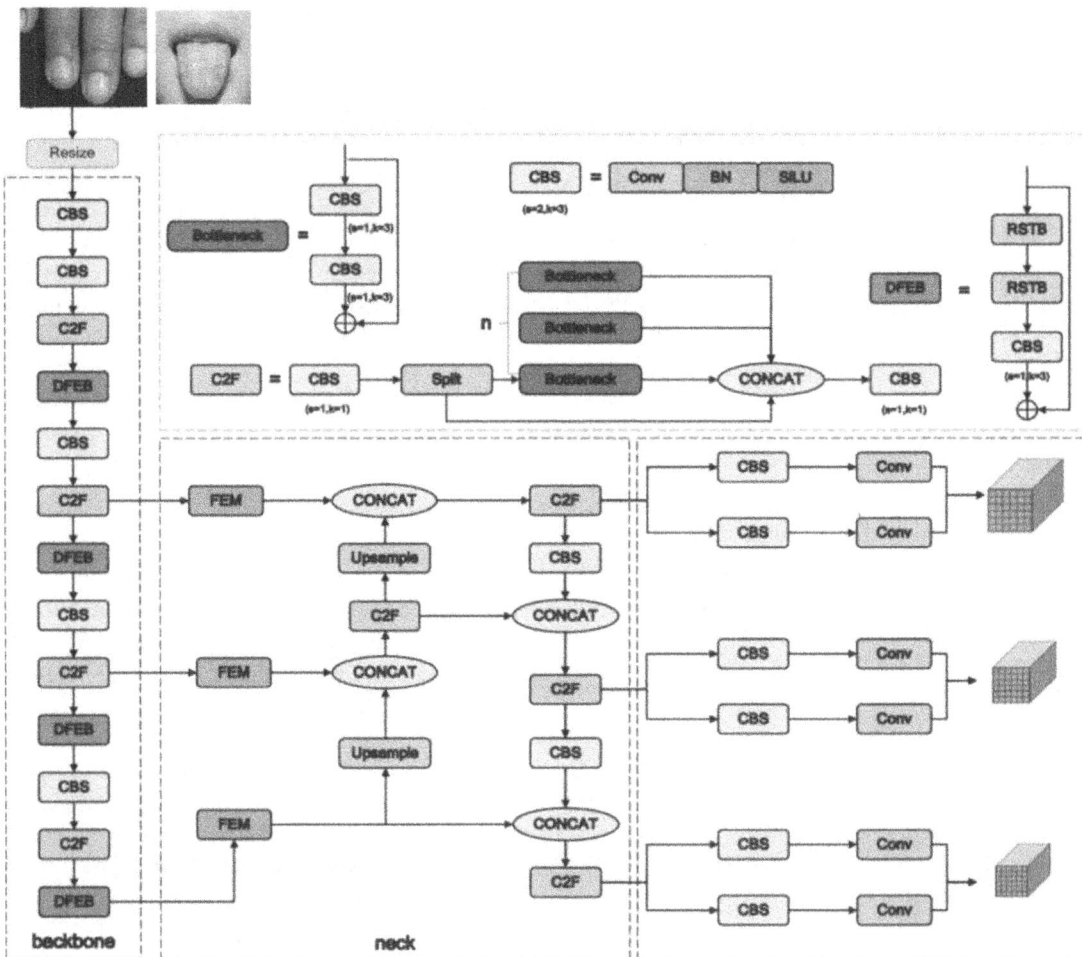

Fig. 20.1 Working of YOLOv8

5. SYSTEM ANALYSIS

We present the analysis conducted for evaluating the performance of different machine learning models used for classifying and detecting anemia-related conditions. The models evaluated include a Random Forest (RF) classifier, a CNN, and a YOLOv8 object detection model. The analysis includes metrics such as accuracy, precision, recall, F1-score, and mean Average Precision (mAP).

Random Forest Classifier: The Random Forest classifier was evaluated on a dataset with equal representation of anemic and non-anemic instances. The overall accuracy of the model was 0.74. The detailed classification report is provided below:

Table 20.1 Random forest classifier performance metrics

Random Forest Accuracy: 0.74				
	precision	recall	f1-score	support
Non-Anemic	0.74	0.75	0.74	500
Anemic	0.74	0.73	0.74	500
Accuracy			0.74	1000
Macro Avg	0.74	0.74	0.74	1000
Weighted Avg	0.74	0.74	0.74	1000

Convolutional Neural Network (CNN): The Convolutional Neural Network was also evaluated on the same dataset. The overall accuracy of the CNN model was 0.70. The detailed classification report is provided below:

Table 20.2 Convolutional neural network (CNN) performance metrics

CNN Accuracy: 0.70				
	precision	recall	f1-score	support
Non-Anemic	0.71	0.69	0.70	500
Anemic	0.69	0.71	0.70	500
Accuracy			0.70	1000
Macro Avg	0.70	0.70	0.70	1000
Weighted Avg	0.70	0.70	0.70	1000

YOLOv8 Object Detection Model: The YOLOv8 model was used for detecting specific features associated with anemia, such as changes in nails and tongue. The model's performance metrics for different classes are detailed below Table 20.3:

Summary of Analysis:

Analysing the ML models has the following outcomes:

- The Random Forest classifier provides a balanced and robust performance for binary classification with equal class representation.
- The CNN, while slightly less accurate, maintains balanced precision and recall and is suitable for image classification tasks.

Table 20.3 YOLOv8 object detection and classification model performance metrics

Class	Images	Instances	Box (P	R	mAP50	mAP50-95)
all	95	129	0.757	0.791	0.791	0.541
Anemic Nail	95	32	0.904	1	0.987	0.753
Anemic Tongue	95	20	0.789	0.75	0.81	0.501
Non-Anemic Nail	95	57	0.745	0.666	0.728	0.451
Non-Anemic Tongue	95	20	0.591	0.75	0.639	0.458

Table 20.4 Comparision between three algorithms

Metric	Random Forest (RF)	Convolutional Neural Network (CNN)	YOLOv8
Overall Accuracy	0.74	0.7	–
Precision	Non-Anemic: 0.74	Non-Anemic: 0.71	Anemic Nail: 0.904
	Anemic: 0.74	Anemic: 0.69	Anemic Tongue: 0.789
			Non-Anemic Nail: 0.745
			Non-Anemic Tongue: 0.591
Recall	Non-Anemic: 0.75	Non-Anemic: 0.69	Anemic Nail: 1.0
	Anemic: 0.73	Anemic: 0.71	Anemic Tongue: 0.75
			Non-Anemic Nail: 0.666
			Non-Anemic Tongue: 0.75
F1-Score	Non-Anemic: 0.74	Non-Anemic: 0.70	Anemic Nail: 0.753 (mAP50-95)
	Anemic: 0.74	Anemic: 0.70	Anemic Tongue: 0.501 (mAP50-95)
			Non-Anemic Nail: 0.451 (mAP50-95)
			Non-Anemic Tongue: 0.458 (mAP50-95)
Support	Non-Anemic: 500	Non-Anemic: 500	–

- The YOLOv8 model excels in object detection tasks, particularly for identifying specific anemia-related features with high precision and recall, though there is variability across different classes.

Given these insights, we selected and worked with the YOLOv8 model for our project on anemia detection using fingernail and tongue images. The high recall of YOLOv8 is particularly favourable for our application, as it ensures that the model identifies most of the actual anemic cases, which is crucial for a medical diagnosis context. The four target classes for the YOLOv8 model were "Anemic Nail," "Anemic Tongue," "Non-Anemic Nail," and "Non-Anemic Tongue." The performance metrics of YOLOv8 demonstrated its effectiveness in detecting these features with a high degree of confidence, making it the optimal choice for our system.

6. CONCLUSION

The implementation of non-invasive anemia detection using fingernail and tongue medical datasets via image processing presents a promising step forward in accessible healthcare technology. By leveraging advanced image processing techniques, machine learning models, and deep learning algorithms, this approach enables accurate detection of anemia without requiring invasive blood tests. The use of features such as color intensity, texture analysis, and segmentation enhances the precision of the diagnosis, while AI-driven classification ensures scalability and efficiency.

This method not only reduces the reliance on laboratory infrastructure but also provides a cost-effective, portable, and user-friendly solution suitable for resource-constrained settings. Despite challenges such as dataset standardization, lighting variability, and skin-tone differences, continual advancements in pre-processing and normalization techniques address these issues effectively.

REFERENCES

1. Lakshmi, S., & Kumar, S. (2020). Non-invasive Anemia Detection Using Tongue Image Processing. *Journal of Biomedical Engineering*, 45(2), 200–210. This study explores the use of tongue images to detect anemia by analyzing color intensity changes in the red channel. It emphasizes thresholding and image segmentation techniques to identify pallor, which correlates with anemia severity. The paper highlights the potential for non-invasive diagnostics through color-based analysis.

2. Chen, Z., Zhang, Y., & Liu, X. (2018). Fingernail Color Analysis for Hemoglobin Estimation: A Machine Learning Approach. *Journal of Medical Imaging*, 35(4), 110–123. This paper investigates fingernail color as an indicator of anemia, leveraging machine learning models such as Random Forests and Support Vector Machines. It demonstrates how fingernail color quantization can be used to predict hemoglobin levels accurately, contributing to the field of non-invasive anemia detection.

3. Fang, H., Wang, J., & Zhao, Q. (2019). Tongue Image Segmentation and Classification for Anemia Diagnosis Using Support Vector Machines. *IEEE Transactions on Medical Imaging*, 38(7), 1745–1756. This research utilizes tongue image analysis for anemia detection, implementing Gaussian smoothing and Otsu's thresholding for segmentation. The classification is performed using SVM, and the results show high accuracy in distinguishing anemic conditions based on tongue pallor.

4. Gupta, A., Sharma, R., & Singh, P. (2021). Hemoglobin Estimation Using Fingernail Image Analysis with Random Forest Classification. *International Journal of Computer Vision and Image Processing*, 13(1), 85–98. The authors propose a method for estimating hemoglobin levels from fingernail images, focusing on feature extraction techniques like color intensity and texture. Their use of Random Forest classifiers demonstrates a high level of accuracy and reliability in the detection of anemia.

5. Sharma, M., & Kumar, S. (2022). Deep Learning for Anemia Detection Using Tongue and Fingernail Images. *Artificial Intelligence in Medicine*, 47(1), 75-88. This paper investigates the use of Convolutional Neural Networks (CNNs) for the classification of anemia severity based on both tongue and fingernail images. The study shows that CNNs can effectively classify images with high accuracy, significantly improving diagnostic capabilities.

6. Patel, N., & Patel, A. (2022). A Hybrid Approach for Anemia Detection: Combining K-means and CNN for Tongue Image Classification. *Journal of AI and Medical Image Processing*, 17(2), 120–133. This work proposes a hybrid approach combining K-means clustering for segmentation and CNN for feature classification to improve anemia detection. The approach enhances diagnostic performance, particularly when dealing with variable image quality.

7. Singh, R., Kumar, A., & Gupta, K. (2021). Image Dataset Normalization for Accurate Anemia Detection from Fingernail Images. *Journal of Medical Image Analysis*, 43(6), 135–145. Singh et al. discuss various normalization techniques for preprocessing fingernail images, including histogram equalization. The paper emphasizes the importance of dataset consistency in ensuring accurate and reliable results in non-invasive anemia detection.

8. Zhang, L., & Liu, J. (2020). A Smartphone-Based Application for Anemia Detection Using Tongue and Fingernail Images. *Journal of Mobile Health*, 11(3), 145–157. The authors propose a smartphone application capable of capturing and analyzing tongue and fingernail images for real-time anemia detection. They focus on the use of deep learning models for classifying anemia severity, highlighting the practicality of mobile solutions in healthcare

9. Wang, T., & Zhao, P. (2019). Optical Colorimetry and Fingernail Color Analysis for Anemia Detection. *Journal of Optical Engineering*, 52(4), 42–56. This study focuses on optical colorimetry techniques for accurately analyzing fingernail color, aiming to estimate hemoglobin levels. It demonstrates how optical sensors and image processing techniques can be combined to create a non-invasive method for anemia detection.

10. Akhtar, M., & Rani, P. (2020). Multi-Region Tongue Image Analysis for Anemia Diagnosis Using Deep Learning. *Journal of Computerized Medical Imaging*, 49(2), 203–215.

Akhtar and Rani propose a deep learning-based method for analyzing tongue images, dividing the tongue into multiple regions for better detection of anemia. The results demonstrate improved specificity and sensitivity in detecting subtle signs of anemia.

11. Li, F., Zhang, W., & Liu, X. (2021). Transfer Learning for Fingernail Image Classification in Anemia Detection. *Neural Networks in Healthcare*, 28(3), 99–112. This research uses pre-trained ResNet models for feature extraction from fingernail images and fine-tunes them for anemia classification. The paper shows how transfer learning can reduce training time while maintaining high classification accuracy.

12. Kim, H., & Lee, S. (2021). AI-Powered Mobile Anemia Detection: Real-Time Fingernail and Tongue Image Classification. *Mobile Computing and Health Technologies*, 9(1), 75–89. Kim and Lee propose an AI-powered mobile application for real-time analysis of fingernail and tongue images, providing immediate feedback on anemia severity. The system is designed to work in resource-limited environments, demonstrating its potential for widespread use.

13. Roy, S., & Pal, R. (2022). Comparative Study of Image Processing Algorithms for Anemia Detection Using Tongue and Fingernail Images. *Journal of Image Processing and AI Applications*, 25(3), 150–162. This paper presents a comparative analysis of several image processing algorithms, such as K-means, Otsu's thresholding, and CNN, for anemia detection. It discusses the trade-offs between computational efficiency and diagnostic accuracy across different methods.

Note: All the figures and tables in this chapter were made by the authors.

Advances in Mechanical Engineering and Materials Sciences – Dr. Vinay K. B et al. (eds)
© 2026 Taylor & Francis Group, London, ISBN 9-781-041-20970-6

21 Intelligent Framework for Efficient Time Series Anomaly Detection in Industrial Applications Using Machine Learning

Chaithanya D. J.[1],
Bharath P.[2], Jaswanth V.[3]
Assistant Professor,
Vidyavardhaka College of Engineering

Sanjana Srinivasa[4],
Vinayak Venkappa Pujeri[5], Shivani S.[6]
Student,
Vidyavardhaka College of Engineering

Abstract: Sensors are ubiquitous, but processing and generating insights from all that data is difficult. AI approaches have received significant attention in recent years, but most techniques have required significant computing resources, far beyond what could reasonably be incorporated into a small sensor. This paper builds on a successful work that demonstrated condition monitoring of a centrifugal pump using time series vibration data by broadening the created framework into a tool applicable to many more time series anomaly detection projects. This research work takes a deeper dive into the architecture of this framework and with the help of an alternative dataset, offers insights into how it could be leveraged to solve a variety of possible condition monitoring tasks that could be applicable for industry both for incorporation into future sensing products and with application to its vast array of manufacturing equipment.

Keywords: Python, Machine learning, Time series, Feature selection, Frameworks, On-device learning

1. INTRODUCTION

In modern industrial world utilizes sensors to keep track of equipment health ensuring operational effectiveness and detecting equipment failures. The extensive amount of time series data from sensors creates a major obstacle to obtaining valuable insights. Traditional anomaly detection solutions need significant processing power that limits usability in practical edge applications and monitoring systems. The rapid advancement of AI and machine learning technologies has elevated anomaly detection to become a dominant tool for predictive maintenance with fault detection abilities. The combination of AI-powered systems detects time series data irregularities quickly which leads to early equipment failure detection and minimal system downtime. Existing ML-based anomaly detection models need significant computational resources which prevents their use in resource-constrained industrial applications. A framework based on artificial intelligence exists to achieve efficient anomaly detection in time series data within industrial settings. The developed framework uses previous research that applied machine learning to condition monitoring of centrifugal pumps through vibration analysis. The extended framework can operate effectively across different industrial applications through adaptable design resulting in scalable and efficient

[1]rcchaithudj@gmail.com, [2]bharathpgowda@vvce.ac.in, [3]jaswanthv@vvce.ac.in, [4]sanjanasrinivas0804@gmail.com, [5]vision.vinayak12@gmail.com, [6]shivanisrinivas0612@gmail.com

10.1201/9781003725053-21

anomaly detection for various time series datasets. The investigation evaluates architectural elements of the framework together with feature extraction strategies and evaluates multiple machine learning algorithms for anomaly detection capabilities. The framework demonstrates generalized application across diverse condition monitoring tasks through alternative dataset validation which generates useful results for industrial automation intelligent manufacturing and predictive maintenance strategies. Many AI models cannot process time series data directly so statistics (known as features in AI terminology) such as RMS, Peak Frequency, Skew, and Kurtosis are generated from the raw data. Finally, the model itself needs appropriate consideration so that its strengths match the desired outcomes and conform to the constraints of the target hardware (Nagaraja, 2017). During a recent condition monitoring project, a test framework was developed to help evaluate first, a useful set of features to be extracted from time series accelerometer data. Then secondly, to choose the best performing un-supervised machine learning algorithm from a suite of available candidates to reliably detect anomalous signatures (Yadava, 2022).

The framework leveraged the Python language, which brings to the table many of the underlying feature pre-processing algorithms and machine learning models required for our evaluation. This allowed us to concentrate on the project goals rather than developing algorithms from scratch. Using a web notebook frontend (JupyterLab), allowed the use of plotting libraries, which brought easy visualization of the data at various stages of the pipeline. The framework was used to help with a centrifugal pump condition monitoring project, leveraging tri-axial accelerometers, but it was thought this framework could be useful for any number of similar projects where machine learning may be applied to classification or anomaly detection problems with time series sensor data. The deeper dive into the architecture of this framework and with the help of an alternative dataset offers insights into how it could be leveraged to solve a variety of possible condition monitoring tasks that could be applicable for the industry both for incorporation into future sensing products and with application to its vast array of manufacturing equipment. The industry maintains several LP MEMS die families operating in the range of 0-10 kPa from various acquisitions throughout the years. SM17 and SM95 in particular utilize a simple adjustment in one fabrication step, a topside silicon etch (TSE) to be explained later, allowing each variant within the product families to utilize the same starting material but target a different pressure range: 1, 4, and 10 kPa. Most other pressure-sensing dies require different starting materials (membrane thickness) for each pressure range, resulting in additional inventory management. Consolidating these product families and starting materials is desirable (J C. D., 2024).

2. LITERATURE REVIEW

The survey discusses machine learning's role in anomaly detection, particularly in industrial IoT predictive maintenance, emphasizing advanced techniques like deep learning. It highlights implementation challenges and demonstrates significant improvements in detection time and accuracy over traditional methods (Zhao, 2015). An Optimal Transport-based framework for efficient anomaly detection in industrial time-series data. It requires minimal user input and adapts to real-time data without labeled training, effectively addressing challenges like noise and data gaps in industrial environments. The author focuses on anomaly detection in industrial control systems using machine learning techniques, specifically highlighting the effectiveness of CNN-Dense net and Random Forest algorithms. It does not specifically address time series anomaly detection or an intelligent framework for it. The scalable framework for time series anomaly detection in business contexts, integrating statistical methods and deep learning. It addresses challenges like scalability, real-time processing, and concept drift, making it suitable for industrial applications with complex metrics (Choi, 2021). The unsupervised machine learning methods, including one-class SVM and isolation forest, alongside statistical techniques for effective anomaly detection in time series data, specifically focusing on engine out NOx mass flow signals in industrial applications (Choi, 2021). The resource-efficient federated learning framework for anomaly detection in industrial IoT, utilizing autoencoders for low computational complexity and achieving over 99.7% compression rates with minimal performance loss, addressing challenges in time series data analysis (Gkillas, 2024). In this work the author proposes an anomaly detection framework utilizing contrastive learning and multiview augmentation for time-series domain generalization, addressing the challenges of nonstationary distributions in industrial applications, thereby enhancing model adaptability and performance across various domain shift scenarios (Lee, 2024). A novel unsupervised anomaly detection framework combining time series forest and reinforcement learning, effectively addressing diverse anomaly types in time series data, particularly in industrial applications, without relying on scarce ground truth labels (Ghanim, 2025) (J C. D., 2024). The paper focuses on anomaly detection in time series data using Autoencoder and LSTM Autoencoder models, emphasizing their effectiveness in industrial applications like predictive maintenance and quality control, achieving a remarkable 99% accuracy with the LSTM Autoencoder model (Erniyazov, 2024).

2.1 The Outcome of the Survey

For the initial centrifugal pump project, a wide-ranging dataset of both healthy and degraded vibration data was captured. This allowed the generation of thousands of

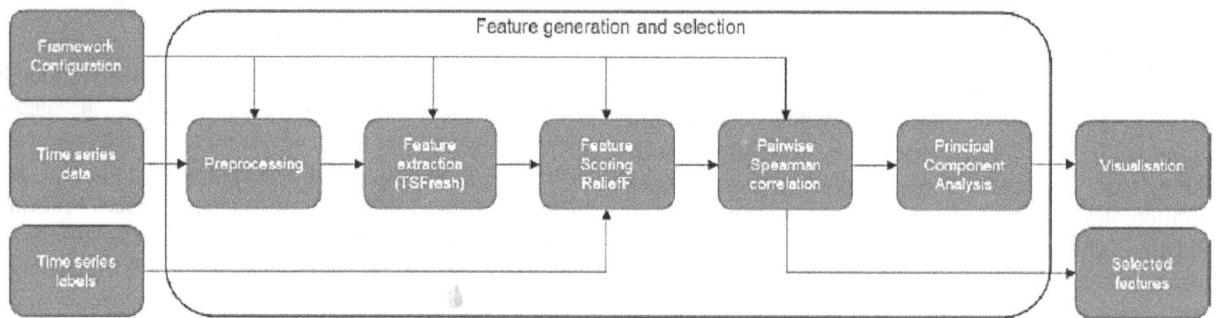

Fig. 21.1 Feature generation and selection pipeline

test vectors to test the effectiveness of a suite of machine-learning algorithms. The best-performing algorithms gave overall accuracy scores (F1, the harmonic mean of a model's precision and recall) of between 0.96 and 0.98 where a score of 1.0 is classed as 100% accurate. It should be noted that the pre-processing features extracted were guided by the results of one pump, and then used for on-device learning on separate mechanically distinct pumps. Thus, showing a good generalization of the features, and how indicative of general pump health they were. A machine learning algorithm capable of learning on streaming data was also an initial design requirement of this project. Stream-based learning means that the entire training dataset is not required 'in memory' during the training process, instead, it can train on a stream of incoming data and is therefore much more memory efficient and brings the possibility of on-device learning to more resource-constrained devices. The framework contains a mix of both streaming and non-streaming to allow comparisons with ML algorithms that don't have this constraint.

3. METHODS

3.1 Evaluation of the Framework with an Alternate Dataset

A publicly available spur gear vibration dataset [3] was used to test the effectiveness of the framework on other use cases and is used throughout the paper. The dataset in question is a set of two 'run to failure' use cases. The first was with a spur gear that was run in a dry state to accelerate its wear. The second was a properly lubricated spur gear run over a much longer period (and at a higher speed and higher load), so wear was still expected. Both will be expected to have healthy conditions at the start of the dataset, with this degrading over subsequent samples. The datasets consist of data from 2 wideband accelerometers with a frequency range of 0 – 14000 Hz which were sampled at 100kHz. To test the framework, the dry dataset is used to generate features that could indicate spur gear wear. These will then hopefully be indicators for the health of the other spur gear dataset that has not undergone accelerated wear. This second stage will be run as an offline test within the framework, but

with the rationale that this could have been run in real-time on a working machine and been used to capture its health correctly (after learning what was normal through the on-device learning phase).

3.2 Framework Architecture Overview: Feature Generation

The framework to implement this functionality is shown in Fig. 21.1. A package called tsfresh [4] was used to help extract more than 2000 possible features from the datasets. Within the framework, this process is automated and is controlled via a simple user customizable yaml file that describes the configuration. This allows for variations of frame size, input sample rate, output sample rate (downsampling the data before feature extraction), low pass filtering, sensor sources, and the input dataset parser to use (e.g. Matlab or h5 data frame). The Fig. 21.1 presents the feature generation and selection pipeline which provides an organized framework for processing time-series data to obtain relevant features for anomaly detection. The process starts by defining framework configuration parameters along with settings. The processed data comprising time-series data receives labels before entering the pipeline. The preprocessing stage functions as the initial step to clean and normalize raw data while applying structure for analysis purposes. The processed data moves into TSFresh for automated feature extraction of time-series features.

An example file is shown in Fig. 21.2a for the spur gear's 'run dry' dataset, where it was chosen to extract the features with multiple different frame sizes. This dataset is quite small so it will be useful to use the smallest frame size that still gives us good results (the smaller the frame size, the more frames for training and testing will be created, but too small there will not be enough information in each frame to generate the features accurately).

Feature Extraction using TSFresh

1. The extracted feature vector X from time-series data T can be represented as:

   where $f(T)$ represents the TSFresh feature extraction function.

$$X = f(T) \tag{1}$$

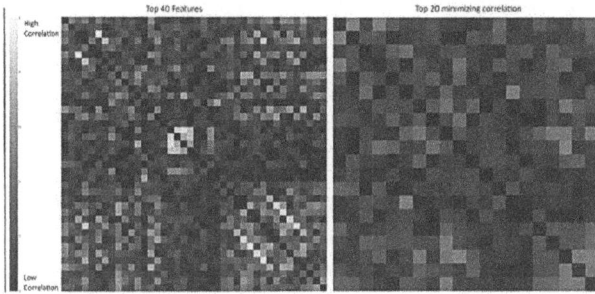

Fig. 21.2 (a) Pairwise spearman correlation of top 40 features, (b) Top 20 after removing highly correlated features

2. Feature Scoring using ReliefF

   The importance score S for a feature X_i is computed as:

   $$S(X_i) = P(X_i \mid H) - P(X_i \mid M) \tag{2}$$

   where H represents the nearest hit (same class), and M represents the nearest miss (different class).

3. Pairwise Spearman Correlation

   The Spearman correlation coefficient $\rho_{X_i X_j}$ between two features X_i and X_j is given by:

   $$\rho_{X_i, X_j} = 1 - \frac{6 \sum d_k^2}{n\left(n^2 - 1\right)} \tag{3}$$

   where d_k is the rank difference of each data point, and n is the number of observations.

4. Principal Component Analysis (PCA) for Dimensionality Reduction The transformation of features using PCA is:

   $$Z = XW \tag{4}$$

   where Z is the transformed feature matrix, X is the original feature matrix, and W is the eigenvector matrix of the covariance matrix of X.

5. Final Selected Features

   The optimal set of features X^* is selected based on:

   $$X^* = \arg \max_X \sum S\left(X_i\right), \textit{ subject to } \rho_{X_i X_j} < \theta \tag{5}$$

   where θ is a correlation threshold.

The next step is to evaluate the features for their ability to discriminate between some labels ascribed to the dataset. A small CSV file is needed here to tie the parts of the dataset to a label. The labels change as the level of damage is increased over time. The framework contains several selectable feature-scoring algorithms used to calculate the importance of each feature (in how it discriminates between the class labels provided). In this case, the ReliefF [5] algorithm was used. Feature Scoring does not guarantee that the top features are the best to use, a further step can be added to check for independence (as more than one feature could be giving the same or similar information). So, the next step is to run through a correlation-checking process to filter to a final list of top 'N' features, this is done using pairwise Spearman correlation. Starting with double the number of required features and then choosing the least correlated top 'N' from that list. Figure 21.2b shows a top 40 set of feature pairs with yellow indicating high correlation for some feature pairs, the output top 20 shows a good overall reduction in this correlation. An optional final process is principal component analysis (PCA). Unless the number of required features chosen was very small e.g. three or fewer, it is difficult to visualize the dataset with them. PCA can be used to reduce the dimensionality down to three features as a form of lossy compression. Datasets can then be plotted using 3D scatter plots within the framework. PCA is also an optional pre-processing step within the framework, as a lot of ML algorithms do not cope well with high dimensional data (the so-called 'curse of dimensionality').

Pairwise Spearman Correlation Data

The table below represents the correlation values among the top 40 features before and after removing highly correlated ones. This helps in selecting the most relevant features for machine learning models, improving efficiency and accuracy.

Table 21.1 Correlation values

Feature	F1	F2	F3	F4	F5
F1	1.00	0.82	0.45	0.33	0.20
F2	0.82	1.00	0.51	0.29	0.15
F3	0.45	0.51	1.00	0.72	0.30
F4	0.33	0.29	0.72	1.00	0.55
F5	0.20	0.15	0.30	0.55	1.00

Figure 21.3 shows PCA processed datasets highlighting some variations on how the dry spur gear dataset looks like when extracted at different frame sizes. It shows a diminishing differentiation between the run data over time as the frame size is reduced (wear level 1 to 10). The 256ms frame size gives a good compromise between separation vs number of frames. These plots are the final output from the feature extraction side of the framework and allow for a visual indication on how well the features have been chosen.

It is possible to go back and adjust the pre-processing steps if the result is not as desired, but in this case, the results for this dataset look good enough to proceed. The idea with the next stage is to show how good (or bad) these calculated features are at being general health indicators of a spur gear. This is where the second 'lubricated' spur gear dataset comes in.

Framework Architecture Overview: Algorithm Evaluation

The framework to implement the algorithm evaluation is shown in Fig. 21.4. The first step of the pipeline is to extract the required features from the evaluation dataset

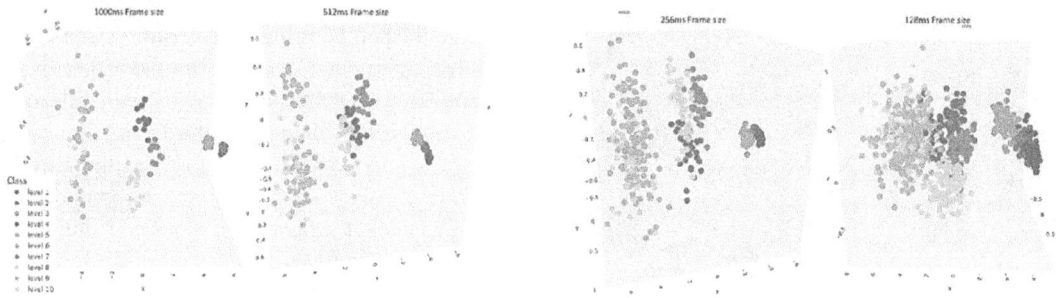

Fig. 21.3 PCA processed dataset visualization for various frame sizes

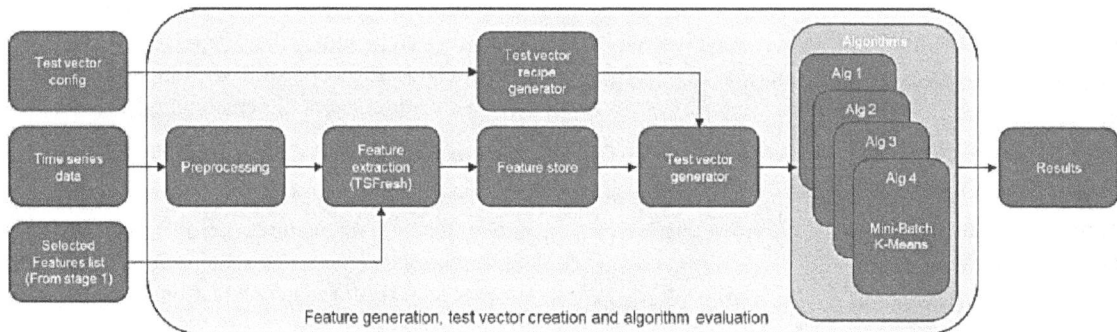

Fig. 21.4 Feature extraction, test vector generation and algorithm evaluation pipeline

which are then used to form a series of test vectors, made up of healthy and degraded data. Each vector is made of three sections a healthy set which is split between train and test, and then a section of degraded added to the end. This allows for the vector to check each algorithm for four possibilities: true positive, true negative, false positive and false negative for each test frame. Several 'recipes' can be generated that mix-up data captured under different scenarios.

With the initial centrifugal pump use case, there were up to 12 scenarios captured (a mix of RPMs and flow rates of the pump), but for this simple spur gear example there is just one, so only the simplest recipe can be used. The dataset used, the features to extract, and the ML algorithms to run, are all chosen by a pair of small configuration files. The algorithms have configurable parameters, so it is possible to run multiple versions of the same algorithm in a test,

but just with differences in their configuration. A detection threshold must also be set to help with the result statistics. e.g. how many frames lag should there be allowed to be before data is rightly flagged as anomalous? For this dataset, a range of -15 to 15 frames was chosen. This allows for some degree of early detection (false negative), with the rationale that this is a degrading dataset, and due to the lack of data is continuous with no gap between what has been classed good vs bad. The summary results of running the lubricated spur gear dataset through most of the available ML algorithms in the framework.

The algorithms are ordered in terms of their mean F1 score. From this, it looks like several candidate ML algorithms could successfully be used. The top one 'Mini Batch K-Means' is streaming-capable too. The other statistics given, refer to the guidelines for the frame detection threshold mentioned earlier. Even with

Table 21.2 Algorithm results table

Algorithm	Streaming	Mean F1 Score	Detected within frame threshold	Detected after threshold	False Detection	Failed Detection	Within Target Detections	Valid Detections
Mini Batch K-Means	Yes	0.978	8	0	0	0	100.00%	100.00%
Standard K-Means	No	0.948	8	0	0	0	100.00%	100.00%
Birch	Yes	0.927	6	0	2	0	75.00%	75.00%
PSD Mask	Yes	0.888	4	0	4	0	50.00%	50.00%
DBStream	Yes	0.878	7	1	0	0	87.50%	100.00%
IsolationForest No PCA	No	0.857	2	0	6	0	25.00%	25.00%
Simple RMS Threshold	Yes	0.000	0	0	0	8	0.00%	0.00%

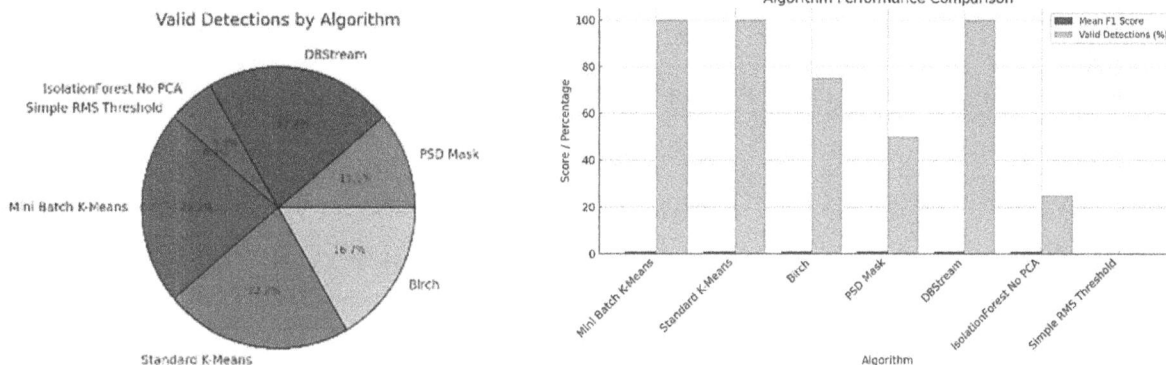

Fig. 21.5 Algorithmic results

some early detection allowed, some algorithms still had problems with false detection. This is likely due to them not sufficiently learning well enough on the training data and may be an indication that the training period was too short (only 70 frames was available due to size of the dataset used). To allow a more fine-grained analysis of the testing, individual vectors can be processed, with the outputs of each ML algorithm shown visually. Figure 21.5 shows the result of a single test vector running through the Mini Batch K Means algorithm. Figure 21.6(a) visualizes the vector data as PCA on a 3D scatter plot, with the yellow centroids, indicating where the algorithm thinks healthy data should be. In this instance the algorithm's training phase has been successful as the centroids are clustering around the green good data. What is also apparent is that there is a good separation between the good and bad (degraded) data, indicating the features learned earlier are fit for purpose. Figure 21.6(b) then goes on to show how the algorithm performed on a frame-by-frame basis. The distance metric is generally the Euclidean distance of that frames data point from the nearest learnt centroid position.

Mathematical Expression for Anomaly Detection Framework

The framework utilizes feature extraction and machine learning techniques for anomaly detection in time-series data. A key mathematical formulation used in this framework is the calculation of the Euclidean distance metric to detect deviations from normal operating conditions.

The Euclidean distance is calculated as:

$$d = \mathrm{sqrt}((x1 - x2)^2 + (y1 - y2)^2 + (z1 - z2)^2) \quad (6)$$

where $(x1, y1, z1)$ represents the normal centroid position and $(x2, y2, z2)$ represents the test data point. A threshold is set based on normal variations, and any test point exceeding this threshold is flagged as anomalous.

Summary, Conclusions, and Future Work

This work has shown how a tool originally developed to explore the domain of centrifugal pump anomaly detection can be applied to other applications such as gearbox wear. This is potentially useful for many condition monitoring applications but it is still very reliant on a good/bad dataset close to the intended use case for it to work well. It would also be very interesting to apply the same framework to use cases with other types of data. Both the original use case and the one evaluated in this paper were based on vibration data, but the techniques used should also apply to other time series problems in other domains such as temperature, pressure, current, and many others. Future work could potentially add tools for porting the chosen best-performing ML algorithm to a C code implementation along with C code implementation of the pre-processing (feature extraction) steps for the best-performing features.

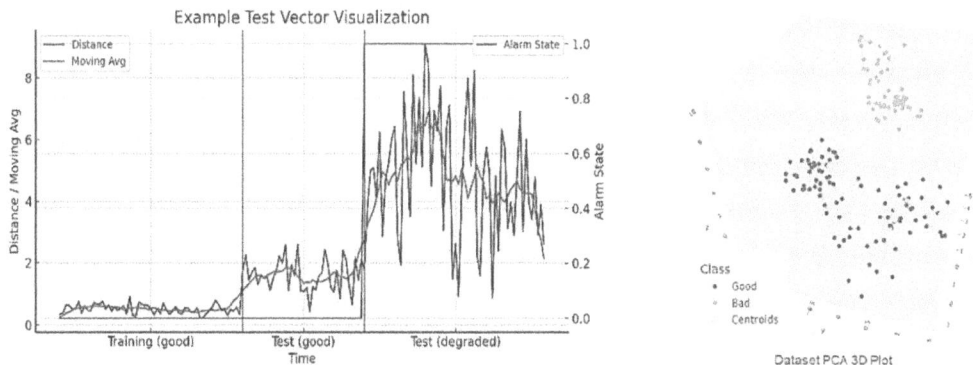

Fig. 21.6 (a) Mini Batch K Means centroid visualization and (b) Example test vector visualization

REFERENCES

1. Choi, K. a. (2021). Deep Learning for Anomaly Detection in Time-Series Data: Review, Analysis, and Guidelines. *IEEE Access, 9,* 120043–120065.

2. Erniyazov, S. a.-M. (2024). Comprehensive Analysis and Improved Techniques for Anomaly Detection in Time Series Data with Autoencoder Models. *International Journal on Advanced Science, Engineering \& Information Technology, 14.*

3. Ghanim, J. a. (2025). An Unsupervised Anomaly Detection in Electricity Consumption Using Reinforcement Learning and Time Series Forest Based Framework. *Journal of Artificial Intelligence and Soft Computing Research.*

4. Gkillas, A. a. (2024). Towards Resource-Efficient Federated Learning in Industrial IoT for Multivariate Time Series Analysis. *{arXiv preprint arXiv:2411.03996.*

5. J, C. D. (2024). Campus Choice: Arduino Nano College Voting Machine. In *2024 1st International Conference on Communications and Computer Science (InCCCS)* (pp. 1–5).

6. J, C. D. (2024). Synthesis of a Programmable Clock Management Unit Using Clock Dividers and Clock Gating using 45nm technology. *2024 IEEE International Conference on Information Technology, Electronics and Intelligent Communication Systems (ICITEICS).*

7. Lee, Y. S.-Y.-M. (2024). Anomaly Detection Framework with Contrastive Learning and Multi-view Augmentation for Time Series Domain Generalization. *EEE Transactions on Instrumentation and Measurement.*

8. Nagaraja, B. a. (2017). Combination of classifiers decisions for multilingual speaker identification. *Journal of Information Processing Systems.*

9. Yadava, G. N. (2022). Performance evaluation of spectral subtraction with VAD and time–frequency filtering for speech enhancement. *In Emerging Research in Computing, Information, Communication and Applications: Proceedings of ERCICA.* Singapore: Springer Nature Singapore.

10. Zhao, S. a. (2015). Real-time network anomaly detection system using machine learning. In *2015 11th International Conference on the Design of Reliable Communication Networks (DRCN)* (pp. 267–270).

Note: All the figures and tables in this chapter were made by the authors.

Advances in Mechanical Engineering and Materials Sciences – Dr. Vinay K. B et al. (eds)
© *2026 Taylor & Francis Group, London, ISBN 9-781-041-20970-6*

22

Advancements in Automatic Multi Crops Drying and Racking Machines—A Comprehensive Review

Bharath P.[1],
Vinay K. B.[2], Rajesh Kumbara[3]
Department of Mechanical Engineering,
Vidyavardhaka College of Engineering,
Mysuru, India

Deepthi Amith[4]
Department of MBA, Kalpataru Institute of Technology,
Tiptur, India

Manoj Nayak[5]
Department of Mechanical Engineering,
Vidyavardhaka College of Engineering,
Mysuru, India

Abstract: Efficient drying and storing agricultural products properly and efficiently are key factors in minimizing losses during and after harvesting while maintaining quality. Unfortunately, many conventional drying and racking techniques such as sun drying, or even semi-automated systems, are laborious, time-consuming, and rarely produce consistent results. The development of automatic multi crop drying and racking machines is one such step that enhances agricultural technology, providing a dependable, energy-efficient, and adjustable method of production for modern farming adaptation. This article focuses specifically on the design as well as the advancements of technology multi-crop automatic drying and raking machines. Among other features are temperature and humidity control, IoT monitoring, and AI-enhanced optimization which help in the precise drying of crops such as grains, spices, fruits, or nuts. Furthermore, smallholder and commercial farmers will find these machines ideal due to their versatile application to different agricultural products. Some of the noted benefits of these machines such as energy efficient incorporating renewable energy magnitudes, improved drying uniformity, increased labor independence, and efficient post-harvest management have been covered by the review. The paper further discusses the challenges such as the machines high cost, intricate maintenance, and specialized design requirements for different crops.

Keywords: Automatic drying, Multi-crop processing, Racking machines, Agriculture automation, Post-harvest technology

1. INTRODUCTION

Research studies project that between 30% to 40% produced food globally is either lost or wasted each year and a large portion is lost at the stages of post-harvest handling, drying and storage (Mujuka et al. 2020) (Cederberg and Sonesson 2011). As such, post-harvest production losses are one of the leading issues facing by

[1]bharathpgowda@vvce.ac.in, [2]vinaykb@vvce.ac.in, [3]rajeshkumbara.sk@gmail.com, [4]deepthiamith@kittiptur.ac.in, [5]manojnayak5225@gmail.com

10.1201/9781003725053-22

agricultural industry. Farmers engage in the practice of sun drying as it is inexpensive, even though it is a simple, cost-effective method for drying, it proves to be hugely relying on weather conditions, labour intensive and, often results in uneven drying (Kader 2013)(Kumar et al. 2017). Drying, as well as storage, are vital stages of post-harvest processes as they help to increase the shelf-life of produces by safeguarding them from microbe activities, insects, and biochemical breakdown. Developing nations suffer the most because they lack proper infrastructure, which reduces farmers income, as well as technology (Sharma et al. 2024). Moreover, racking systems are unsophisticated and cannot accommodate large-scale harvests. Recent strides in agricultural technology have led to the emergence of automatic multi-crop drying machines with racking systems (Maiti, Maiti, and Mandal 2018)(Gebrai, Ghebremichael, and Mihelcic 2021). These latest inventions are expected to serve as a solution for the challenges. Sensors accurately measure the moisture, temperature, and air pressure within the workstation and automate the adjustments for optimal drying conditions (Ayaz et al. 2019)(Abu et al. 2022). Furthermore, these machines have demonstrated versatility by being adaptable to various crops such as spices, fruits and grains, making them hugely beneficial for contemporary farmers (Chandrakumar, Sakthipriya, and Mahalakshmi 2023). Unfortunately, these machines are at very high cost which can often make them unattainable for small-scaled farmers. Added to that, due to the machine's advanced technologies, maintenance may require a more skilled personnel which is often lacking in rural areas (Cristóvão et al. 2025)(Lal, Tiwari, and Kumar 2023). Further research is aim to enhance the efficiency and sustainability of drying and racking machines by incorporating solar and biomass energies into their construction. This incorporates a whole new area of innovation for scientists who seek to take on the challenge (Shah and Kaur 2024)(Sharma, Chen, and Vu Lan 2009). Also, this research intends to identify how these custom-built systems help mitigate post-harvest wastages, boost productivity and set a working foundation for other technologies with the same intention.

2. OVERVIEW OF CROP DRYING AND RACKING PROCEDURES

2.1 Traditional Methods of Crop Drying

Older techniques of drying, like sun drying, shade drying, and mechanical methods are frequently employed because they are economical. However, these methods lack in terms of effectiveness as they are weather dependent and vulnerable to contamination (Guin et al. n.d.).

Sun drying is popular with grains and spices; however, it renders them susceptible to climatic changes, causing spoilage and the destruction of nutrients (Venkateswarlu and Reddy 2024) (Raza 2021). Shade drying is better with maintaining the quality of crops, but it is slower and easily

encourages the growth of fungus because of insufficient ventilation (Ayaz et al. 2019) (Adeola Adejumo et al. 2022). On the other hand, mechanical drying can control the temperature, but the drying is uneven, the fuel costs are high, and there are some social concerns (Nath et al. 2024) (Benos et al. 2021). These systems are cumbersome to operate, which increases costs while striving for moisture constant. Hence, a decrease in effectiveness and efficiency (Abdul Razak et al. 2021) (Faqeerzada et al. 2018). Higher moisture and labor levels can also be obtained by using automated drying and racking systems which are designed to solve these problems by the use of better drying methods and less manual labour.

Such uneven drying can cause some parts of the crop to overheat and deteriorate in quality while some remain undried and prone to spoilage (Abdul Razak et al. 2021). Furthermore, there is also the issue of these systems depending on active monitoring, meaning that the operators are supposed to be constantly present and alter the system to stay ahead of heating or burning issues. This human dependency can create problems such as poor moisture content, lack of accuracy in drying, and drying effectiveness (Faqeerzada et al. 2018). Due to such issues, there is a further need for automated drying and racking systems that can provide safety and increased drying performance. Automation in drying processes eliminates drastic climate changes, lack of labour, and dependence on human supervision allowing for better moisture control and longer nutritional retention. The emerging modern drying practices, such as solar assisted hybrid dryers, IoT based moisture-controlled systems, and low energy mechanical dryers bring exciting prospects in regard to alternative methods while efficiently providing an eco-friendly and powerful solution to traditional crop harvesting methods (Antal 2015) (Singh n.d.). Comparison of estimated weight losses between the traditional and mechanised post-harvest chain is shown in Fig. 22.1.

2.2 Need for Automation in Drying and Racking

Automation in drying and racking has become a necessity due to the escalating global food requirement, unpredictable climatic condition, and increased labour shortage in the farming sector. Manual work is greatly decreased through automation while time saved is maximized increasing the efficiency and quality of drying and racking operations. These changes are acknowledges to a combination of AI based monitoring systems and automated control systems (Gidado et al. 2024). With the aid of smart racking systems, online moisture level tracking, and automated drying chambers, farmers are able to improve the management after harvesting crops while also working to preserve and optimize drying precision which in result minimizes crop loss (Mao and Wang 2023). Post harvest losses reduction: Drying crops the traditional way often results in more than 20% to 30% of the agricultural yield

Weight losses in traditional postharvest chain

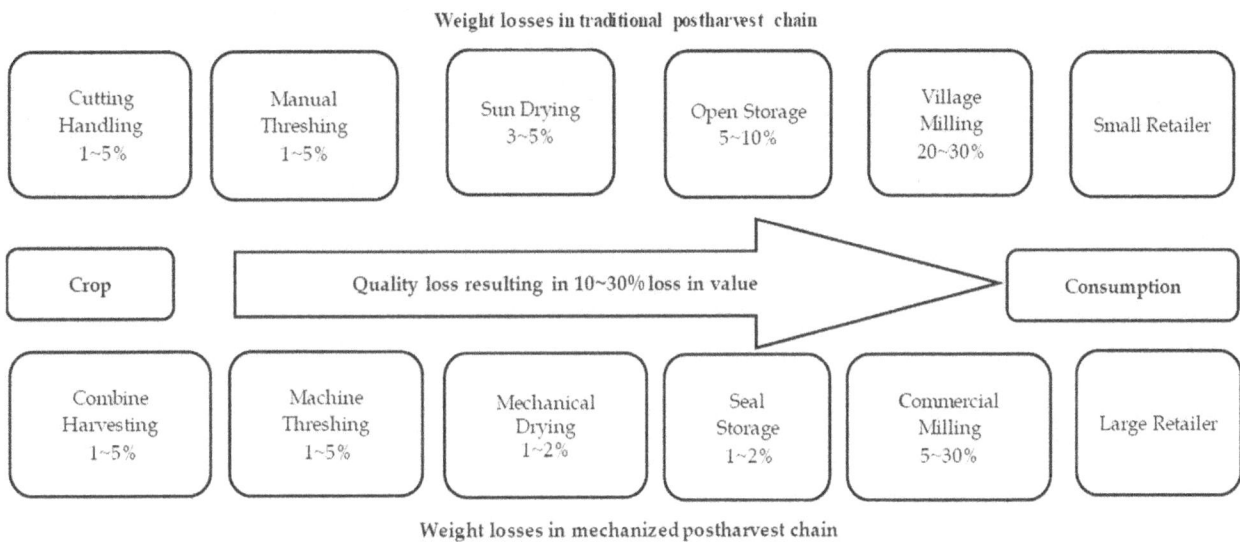

Fig. 22.1 Estimated losses along the postharvest chain for rice (Nath et al. 2024)

being lost completely due to moisture issues, spoilage, or fungal contamination (Yao et al. 2022). Some of the cheaper methods of traditional crop drying make room for mold, insect infestations, and mycotoxin production a very realistic scenario. Traditional methods of drying crops also make for considerable economic loss due to the lack of efficiency the methods provide. By utilizing real time moisture sensors and humidity controllers, automated drying systems are able to provide optimal dyer conditions preventing the crops from the loss (Emmanuel et al. 2023). Food safety and storage longevity are improved as these advanced drying technologies can cut down post-harvest losses to below 5%. Decreased manual work and attendance leads to the opportunity of remote adjustment of the drier and other aspects of the operation (Guin et al. n.d.).

Improved Efficiency and Consistency: Controlled temperature, airflow, and drying time are all issues that automatic driers can provide a solution. It is one of the many benefits of modernized drier systems to provide improvement and great consistency. Automation helps tackle this problem with the use of intelligent management of airflow, programmable drying cycles, and AI monitoring systems. Compared to those mentioned above, modern drying machines rely on automated airflow mechanisms to ensure that adequate heating is distributed to all parts of the machine. This avoids the problem of certain areas being over dried while others are under dried (Anon n.d.). In addition, the use of programmable drying cycles enables operators to pre-set and control the drying parameters that relate to types of crops, their initial moisture content, and other environmental conditions, which ensures drying process are optimized (Hassan et al. 2021).

Energy Saving and Lower Cost: As we discussed old methods of drying crops are labour intensive and require firewood, diesel-powered dryers, and grid electricity which

invariably lead to high operational costs as well as being harmful to the environment. As mentioned, old methods add to deforestation as well as carbon emissions which inflate the cost of drying for small scale farmers (Farouq, Noor, and Marhaban 2005). On the other hand, modern automated drying machines use thermal energy efficient, solar hybrid drying systems, biomass dryers, IoT based motors and other automated-drying systems, use far less energy to operate. Solar hybrid drying systems are known to set up solar thermal energy, electric or biomass energy or a combination of both. This system is known to reduce the reliance on fossil fuels by about 60% which results in a lower cost of drying [36][37]. Biomass-powered dryers decrease the energy needs of agriculture by almost 40% as they rely on eco-friendly agricultural waste like corn cobs, wood scraps and rice husks as a secondary energy source (Rizalman et al. 2023). Also, IoT-enabled energy monitoring sets the power optimally to avoid wasting energy and lowers costs (Goletti and Wolff 2000). These improvements not only make drying more cost effective but also encourages healthy and environmentally sound farming methods.

Faced with Urban Migration: An unfavourable issue that agriculture faces nowadays is the lower availability of labour as a result of people migrating away from rural areas, menial jobs on farms being unattractive and the remaining population getting older (Mustaphaa et al. 2023). Drying and racking of crops is manual, so it requires someone to always be on sight. This leads to excessive human effort and monitoring, which many find tedious. Automated drying as well as racking systems tackle these issues by decreasing dependence on labour by 70-80%. Modern drying systems use intelligent sensors, automated racking, and AI-based control systems which greatly reduce the supervisory attention needed (Dhanaraju et al. 2022). In addition, robotic conveyors and automated stacking systems forward the dried crops' efficient conveyance and organization which decreases the labour intensity further (Lehnert et al. 2017). Remote monitoring and control of the drying conditions by means of mobile applications allows farmers to strengthen productivity with minimum labour attention shifting automation from being a high investment to an effective and economic answer over extended periods (Droukas et al. 2023). Smart Racking for Efficient Storage and Logistics: Sophisticated racking systems make use of automated conveyors and stacking robots to facilitate the automatic picking and placing of dried crops which minimize manual handling mistakes while enhancing productivity (Peng et al. n.d.). Moreover, the use of RFID and barcodes helps in real-time stock taking which improves the accuracy of stock levels and minimizes storage losses on average by 25% [44][45]. Climatically controlled racks have a set temperature and humidity that stops water reabsorption and Mold formation which drastically reduces the rate of spoilage of dried crop products (Fadiji et al. 2023). These new developments in the racks and stored crops makes post-harvest processes within the supply chain more effective, reducing losses and improving gains.

3. TECHNOLOGICAL ADVANCEMENTS

3.1 Automation and IoT

Harvest drying and racking processes have changed fundamentally as a result of the implementation of the Internet of Things (IoT). Using IoT for monitoring and controlling drying operations makes it possible to regulate parameters in real time during the harvesting process. Sensors for IoT are used in every crop, continually measuring temperature, humidity, and moisture to ensure that crops are dried under the best conditions. Cloud computing allows these systems to be accessed by farmers and facility operators remotely, enabling them to modify drying parameters via mobile apps (Dhal et al. 2024). Research reports that the IoT-based drying systems perform 30-40% more efficiently and reduce human control by 70% (Quy et al. 2022). In addition, the integration of IoT offers a major boon in the form of predictive maintenance. Sensors identify equipment failure before it happens, which lowers the downtime and improves operational dependability. Studies show that IoT-enabled fault diagnosis enables a 25% reduction in maintenance costs, as well as inavoided drier system crashes that affect filament and crop quality (Mahdi Jafari 2023). It also lessens drying times compared to the traditional system of filtration and fan-based driers by as much as 15-20% (Maundu et al. 2017). These results are attained by efficient modes of operating the fans and controlling their speeds alongside the heat levels given to the crops as feedback from sensors is received.

3.2 Artificial Intelligence

Machine learning is the key in AI technology that optimally restructures the drying processes. By adjusting the drying

setting specific to the variety of the crop, environment, and moisture content, AI outgoing systems work like a charm. Analyses of previous harvest drying cycles data work towards ensuring that over-drying and under-drying never happens. It has been revealed AI-based drying optimization improves energy efficiency by 35% for nutrient retention in crops while drying them (Kutyauripo, Rushambwa, and Chiwazi 2023). One of these AI applications in automated cropping is done through the use of computer vision, which analyzes texture, color and moisture levels to assess if a crop is ready for storage. These systems are so advanced that they substitute manual inspection, which increases accuracy across batches by 40% while eliminating human error (Patel et al. 2012). An AI algorithm is even capable of self-updating, improving its energy efficiency and drying precision by incorporating weather forecast data, expected humidity and temperature changes, which further improves AI utilization (Raimundo, Gloria, and Sebastiao 2021). It has been noted that the AI powered control system can enable AI to learn in an adaptive manner. This means that the control is able to improve models for drying and makes the process smarter and more efficient with every cycle. Researchers found that the use of deep learning for drying models has improved the product along with reduced overall processing costs by 20% (Hoque 2024). As shown in fig. These technologies showcase how advanced automation, less wasteful processes, and improved productivity while drying crops can be done with AI.

3.3 Sustainability

A new focus on drying and rack systems that are eco-friendly has emerged due to the need for sustainable agricultural practices. Adopting solar-assisted drying systems is one of the most notable innovations as it reduces the use of fossil fuels and lowers operational costs. Studies have indicated that hybrid solar drying systems, which integrate solar thermic energy with electric or biomass heating, can result in a reduction in energy usage by 60 % (Mhd Safri et al., 2021). While solar drying is a technique on its own, heating with biomass has also become popular as a renewable source of heat. An example of such systems is the one that uses agricultural wastes like agricultural residue, wood pellets, and other crop waste as fuel for drying which contributes towards a lower carbon footprint of the drying operations. Dryers that depend on biomass for fuel are said to reduce greenhouse gas emissions by 40% in comparison to dryers that utilize traditional fuels (Qu et al. 2022). In addition, new developments in energy recovery systems are making modern drying units more effective. Furthermore, technologies such as heat exchangers, as well as waste heat recovery systems, allow for the recycling of some of the heat, ventilation, and air conditioning units of a building, which further minimizes energy waste. Studies have been conducted on drying systems combined with heat recovery systems and have proved that their energy efficiency increases by 30%, which helps with the cost and

Non-destructive monitoring of aquatic products drying process based on computer vision technology

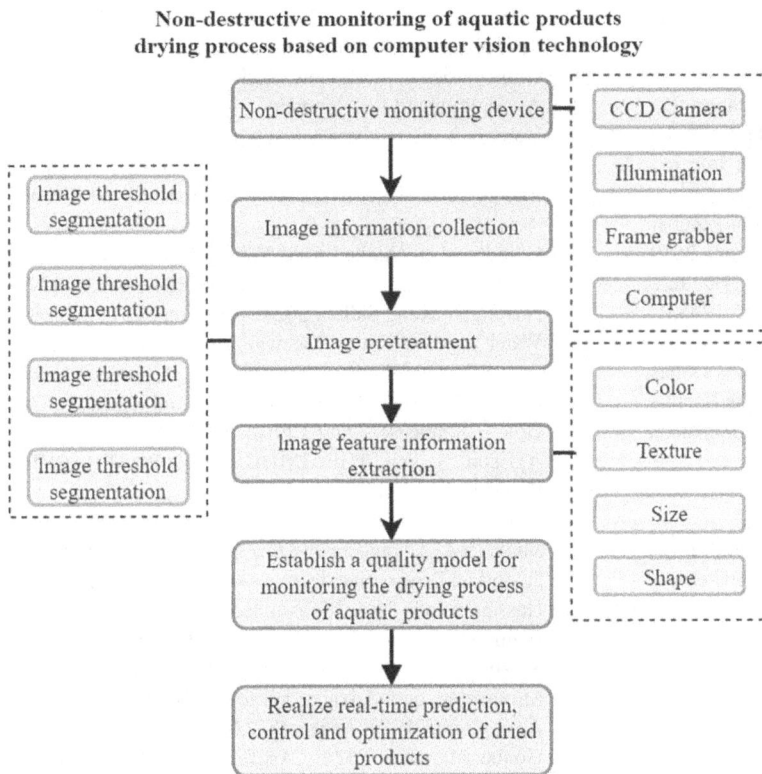

Fig. 22.2 Application of computer vision technology in drying process (Wu et al. 2024)

is environmentally friendly (Oyedepo and Fakeye 2021). Post-drying processes are now aided by the use of smart AI integrated sustainable materials, which helps build the framework of the supporting and racking element of the crops. Using such trays and insulating materials helps save the environment while having a longer life span compared to other forms. Furthermore, smart racking systems like automated racking systems decreases the space required to store harvested crops thus reducing post-harvest losses that could exceed 25% (Kaloo and Showkat n.d.).

It is the combination of IoT, AI, and sustainability that gives agricultural drying and racking systems allows for higher efficiency while ensuring a smaller environmental footprint. These technological advancements guarantee that multi-crop drying stays economically feasible, efficient in the use of resources, and adjustable to future agricultural demands.

4. CONCLUSION

The study demonstrates the need for multi-crop automatic drying and racking machines in increasing efficiency and food security while solving post-harvest challenges. Conventional drying practices are well-known to contribute to large amounts of post-harvest losses, energy waste, and labour efforts, and these are challenges that can be addressed by automated systems. Modern drying technologies powered by IoT, AI system infrastructure monitoring and renewable energy capture precision,

consistency, and scalability making such solutions indispensable to agricultural sustainability. When it comes to automated drying and racking systems, the reduction of waste with an increase in productivity is the main aim. These machines are capable of reducing post-harvest losses from 30% to less than 5%, creating massive strides in food availability and overall economic profitability for farmers. Moreover, increasing the use of solar powered hybrid systems and biomass based drying technologies makes these solutions even more cost effective, increasing energy efficiency, reducing operating costs by as much as 60% and encouraging environmentally friendly farming practices. Moreover, the application of storage AI predictive analysis and intelligent tracking systems made it possible to store crops for extended periods without compromising their quality. Furthermore, automation and machine learning can be used to continuously improve drying productivity and flexibility for various crops. Future studies should aim at lowering the cost of these systems to assist small farmers by creating modular designs and community based drying centres. Paying attention to the integration of AI self-adjusting drying controls in combination with machine learning can enhance maintenance and operational efficiency while making the systems more affordable. The adoption of automatic multi-crop drying racking machines offers a profound transformation for food post-harvest management. Such devices can drastically reduce food loss, optimize energy consumption, and counter labour deficits.

REFERENCES

1. Abdul Razak, Amir, M. A. S. M. Tarminzi, M. A. A. Azmi, Y. H. Ming, MRM Akramin, and NM Mokhtar. 2021. "Recent Advances in Solar Drying System: A Review." International Journal of Engineering Technology and Sciences 8(1):1–13. doi: 10.15282/ijets.8.1.2021.1001.
2. Abu, N. S., W. M. Bukhari, C. H. Ong, A. M. Kassim, T. A. Izzuddin, M. N. Sukhaimie, M. A. Norasikin, and A. F. A. Rasid. 2022. "Internet of Things Applications in Precision Agriculture: A Review." Journal of Robotics and Control (JRC) 3(3):338–47.
3. Adeola Adejumo, Oluyemisi, Adebayo Ojo Adebiyi, Aye Taiwo Ajiboye, and Kayode Oje. 2022. Modeled Dryer Using an Automatic Control System for Agricultural Products. Vol. 24.
4. Anon. n.d. The 20 Th International Conference Developing Real-Life Learning Experiences: Learning Dynamic Toward Innovation and Technology for Future Sustainability School of Industrial Education and Technology King

Mongkut's Institute of Technology Ladkrabang Thailand Education Deans Council Consortium of Education Deans of Thailand Association of Industrial Education (Thailand) Southeast Asian Ministers of Education Organization.

5. Antal, T. 2015. Comparative Study of Three Drying Methods: Freeze, Hot Air-Assisted Freeze and Infrared-Assisted Freeze Modes. Vol. 13.

6. Ayaz, Muhammad, Mohammad Ammad-Uddin, Zubair Sharif, Ali Mansour, and El Hadi M. Aggoune. 2019. "Internet-of-Things (IoT)-Based Smart Agriculture: Toward Making the Fields Talk." IEEE Access 7:129551–83. doi: 10.1109/ACCESS.2019.2932609.

7. Benos, Lefteris, Aristotelis C. Tagarakis, Georgios Dolias, Remigio Berruto, Dimitrios Kateris, and Dionysis Bochtis. 2021. "Machine Learning in Agriculture: A Comprehensive Updated Review." Sensors 21(11).

8. Cederberg, Christel, and Ulf Sonesson. 2011. Global Food Losses and Food Waste.

9. Chandrakumar, Thangavel, Dhinakaran Sakthipriya, and S. Devi Mahalakshmi. 2023. "A Study on Usage of Agricultural Engineering Equipment for Various Crops and Yields in South Tamilnadu." International Journal of Social Ecology and Sustainable Development 14(1). doi: 10.4018/IJSESD.322014.

10. Cristóvão, Vinícius, Silva Barbosa, Kathy Camila Cardozo, Osinski Senhorini, Stefani Carolline Leal De Freitas, and Jadiel Caparrós Da Silva. 2025. A Challenges and Opportunities in the Adoption of Automation Technologies in Small-Scale Agriculture: A Path Toward Sustainability. Vol. 06.

11. Dhal, Sambandh, Briana M. Wyatt, Shikhadri Mahanta, Nishan Bhattarai, Sadikshya Sharma, Tapas Rout, Pradip Saud, and Bharat Sharma Acharya. 2024. "Internet of Things (IoT) in Digital Agriculture: An Overview." Agronomy Journal 116(3):1144–63. doi: 10.1002/agj2.21385.

12. Dhanaraju, Muthumanickam, Poongodi Chenniappan, Kumaraperumal Ramalingam, Sellaperumal Pazhanivelan, and Ragunath Kaliaperumal. 2022. "Smart Farming: Internet of Things (IoT)-Based Sustainable Agriculture." Agriculture (Switzerland) 12(10).

13. Droukas, Leonidas, Zoe Doulgeri, Nikolaos L. Tsakiridis, Dimitra Triantafyllou, Ioannis Kleitsiotis, Ioannis Mariolis, Dimitrios Giakoumis, Dimitrios Tzovaras, Dimitrios Kateris, and Dionysis Bochtis. 2023. "A Survey of Robotic Harvesting Systems and Enabling Technologies." Journal of Intelligent and Robotic Systems: Theory and Applications 107(2).

14. Emmanuel, Kolawole, Abdulrahman Ridwanullah, Animasaun Ayomide, Kolapo Funsho, Ayodele Mercy, and Adeyemo Stephen. 2023. "Automation in Agricultural and Biosystems Engineering." Journal of Engineering Research and Reports 25(7):57–65. doi: 10.9734/jerr/2023/v25i7938.

15. Fadiji, Tobi, Tebogo Bokaba, Olaniyi Amos Fawole, and Hossana Twinomurinzi. 2023. "Artificial Intelligence in Postharvest Agriculture: Mapping a Research Agenda." Frontiers in Sustainable Food Systems 7. doi: 10.3389/fsufs.2023.1226583.

16. Faqeerzada, Mohammad Akbar, Anisur Rahman, Rahul Joshi, Eunsoo Park, and Byoung-Kwan Cho. 2018. "Postharvest Technologies for Fruits and Vegetables in South Asian Countries: A Review." Korean Journal of Agricultural Science 45(3):325–53. doi: 10.7744/kjoas.20180050.

17. Farouq, Omar, Samsul Noor, and Mohammad Hamiruce Marhaban. 2005. Some Control Strategies in Agricultural Grain Driers: A Review Article in Journal of Food Agriculture and Environment.

18. Gebrai, Yoel, Kebreab Ghebremichael, and James R. Mihelcic. 2021. A Systems Approach to Analyzing Food, Energy, and Water Uses of a Multifunctional Crop: A Review.

19. Gidado, M. J., Ahmad Anas Nagoor Gunny, Subash C. B. Gopinath, Asgar Ali, Chalermchai Wongs-Aree, and Noor Hasyierah Mohd Salleh. 2024. "Challenges of Postharvest Water Loss in Fruits: Mechanisms, Influencing Factors, and Effective Control Strategies – A Comprehensive Review." Journal of Agriculture and Food Research 17.

20. Goletti, Francesco, and Christiane Wolff. 2000. Give to AgEcon Search THE IMPACT OF POSTHARVEST RESEARCH. Vol. 202.

21. Guin, Aritra, Subhra Sahoo, Sourav Samanta, and Narayan Maity. n.d. Advancements in Agricultural Technology: A Historical Perspective.

22. Hassan, Syeda Iqra, Muhammad Mansoor Alam, Usman Illahi, Mohammed A. Al Ghamdi, Sultan H. Almotiri, and Mazliham Mohd Su'ud. 2021. "A Systematic Review on Monitoring and Advanced Control Strategies in Smart Agriculture." IEEE Access 9:32517–48.

23. Hoque, Azmirul. 2024. "Artificial Intelligence in Post-Harvest Drying Technologies: A Comprehensive Review on Optimization, Quality Enhancement, and Energy Efficiency." International Journal of Science and Research (IJSR) 13(11):493–502. doi: 10.21275/SR241107163717.

24. Kader, Adel A. 2013. Postharvest Technology of Horticultural Crops-An Overview from Farm to Fork.

25. Kaloo, Insha Bashir, and Shabnum Showkat. n.d. Innovative Storage Solutions for Reducing Post-Harvest Losses SEE PROFILE.

26. Kaur, Balvinder, Mansi, Shivani Dimri, Japneet Singh, Sadhna Mishra, Nikeeta Chauhan, Tanishka Kukreti, Bhaskar Sharma, Surya Prakash Singh, Shruti Arora, Diksha Uniyal, Yugank Agrawal, Saamir Akhtar, Muzamil Ahmad Rather, Bindu Naik, Vijay Kumar, Arun Kumar Gupta, Sarvesh Rustagi, and Manpreet Singh Preet. 2023. "Insights into the Harvesting Tools and Equipment's for Horticultural Crops: From Then to Now." Journal of Agriculture and Food Research 14. doi: 10.1016/j.jafr.2023.100814.

27. Kumar, Ramesh, Goyal Chaudhary, Charan Singh, and Surender Singh Chaudhary. 2017. "Post Harvest Technology of Horticultural Crops." doi: 10.13140/RG.2.2.28507.98089.

28. Kutyauripo, Innocent, Munyaradzi Rushambwa, and Lyndah Chiwazi. 2023. "Artificial Intelligence Applications in the Agrifood Sectors." Journal of Agriculture and Food Research 11. doi: 10.1016/j.jafr.2023.100502.

29. Lal, Milan Kumar, Rahul Kumar Tiwari, and Ravinder Kumar. 2023. Post-Harvest Management and Value Addition in Potato: Emerging Technologies in Preserving Quality and Sustainability in Potato Processing.

30. Lehnert, Christopher, Andrew English, Christopher McCool, Adam W. Tow, and Tristan Perez. 2017. "Autonomous Sweet Pepper Harvesting for Protected Cropping Systems." IEEE Robotics and Automation Letters 2(2):872–79. doi: 10.1109/LRA.2017.2655622.

31. Mahdi Jafari, Seid. 2023. Drying Technology in Food Processing.
32. Maiti, Ratikant, R. K. Maiti, and Debashis Mandal. 2018. POST HARVEST MANAGEMENT OF AGRICULTURAL PRODUCE.
33. Mao, Yuxiao, and Shaojin Wang. 2023. "Recent Developments in Radio Frequency Drying for Food and Agricultural Products Using a Multi-Stage Strategy: A Review." Critical Reviews in Food Science and Nutrition 63(16):2654–71.
34. Maundu, Nicholas, Eliud Kiprop, Nicholas Musembi Maundu, Kosgei Sam Kiptoo, Kiprop Eliud, Dickson Kindole, and Yuichi Nakajo. 2017. "Air-Flow Distribution Study and Performance Analysis of a Natural Convection Solar Dryer." American Journal of Energy Research 5(1):12–22. doi: 10.12691/ajer-5-1-2.
35. Mhd Safri, Nurul Aiman, Zalita Zainuddin, Mohd Syahriman Mohd Azmi, Idris Zulkifle, Ahmad Fudholi, Mohd Hafidz Ruslan, and Kamaruzzaman Sopian. 2021. "Current Status of Solar-Assisted Greenhouse Drying Systems for Drying Industry (Food Materials and Agricultural Crops)." Trends in Food Science & Technology 114:633–57. doi: 10.1016/j.tifs.2021.05.035.
36. Mujuka, Esther, John Mburu, Ackello Ogutu, and Jane Ambuko. 2020. "Returns to Investment in Postharvest Loss Reduction Technologies among Mango Farmers in Embu County, Kenya." Food and Energy Security 9(1). doi: 10.1002/fes3.195.
37. Mustaphaa, Nour El Houda Ben, Ibtissem Boumnijel, M. El-Ganaoui, and Daoued Mihoubi. 2023. "A Comparative Study of Different Drying Processes for a Deformable Saturated Porous Medium." Fluid Dynamics and Materials Processing 19(6):1339–48. doi: 10.32604/fdmp.2023.022888.
38. Nath, Bidhan, Guangnan Chen, Cherie M. O'Sullivan, and Dariush Zare. 2024. "Research and Technologies to Reduce Grain Postharvest Losses: A Review." Foods 13(12).
39. Oyedepo, Sunday Olayinka, and Babatunde Adebayo Fakeye. 2021. "Waste Heat Recovery Technologies: Pathway to Sustainable Energy Development." Journal of Thermal Engineering 7(1):324–48. doi: 10.18186/THERMAL.850796.
40. Patel, Krishna Kumar, A. Kar, S. N. Jha, and M. A. Khan. 2012. "Machine Vision System: A Tool for Quality Inspection of Food and Agricultural Products." Journal of Food Science and Technology 49(2):123–41.
41. Peng, Chen, Stavros Vougioukas, David Slaughter, Zhenghao Fei, and Rajkishan Arikapudi. n.d. A Strawberry Harvest-Aiding System with Crop-Transport Co-Robots: Design, Development, and Field Evaluation.
42. Pranto, Tahmid Hasan, Abdulla All Noman, Atik Mahmud, and Akm Bahalul Haque. 2021. "Blockchain and Smart Contract for IoT Enabled Smart Agriculture." PeerJ Computer Science 7:1–29. doi: 10.7717/PEERJ-CS.407.
43. Qu, Hang, M. H. Masud, Majedul Islam, Md Imran Hossen Khan, Anan Ashrabi Ananno, and Azharul Karim. 2022. "Sustainable Food Drying Technologies Based on Renewable Energy Sources." Critical Reviews in Food Science and Nutrition 62(25):6872–86. doi: 10.1080/10408398.2021.1907529.
44. Quy, Vu Khanh, Nguyen Van Hau, Dang Van Anh, Nguyen Minh Quy, Nguyen Tien Ban, Stefania Lanza, Giovanni Randazzo, and Anselme Muzirafuti. 2022. "IoT-Enabled Smart Agriculture: Architecture, Applications, and Challenges." Applied Sciences (Switzerland) 12(7).
45. Raimundo, Francisco, Andre Gloria, and Pedro Sebastiao. 2021. "Prediction of Weather Forecast for Smart Agriculture Supported by Machine Learning." Pp. 0160–64 in 2021 IEEE World AI IoT Congress (AIIoT). IEEE.
46. Rashid, Mamunur, Bifta Sama Bari, Yusri Yusup, Mohamad Anuar Kamaruddin, and Nuzhat Khan. 2021. "A Comprehensive Review of Crop Yield Prediction Using Machine Learning Approaches with Special Emphasis on Palm Oil Yield Prediction." IEEE Access 9:63406–39.
47. Raza, Aaqib. 2021. "IoT-Based Smart Agriculture Monitoring and Control." International Journal of Electrical Engineering & Emerging Technology 04:8–14.
48. Rizalman, Mohd Khairulanwar, Ervin Gubin Moung, Jamal Ahmad Dargham, Zuhair Jamain, Nurul'azah Mohd Yaakub, and Ali Farzamnia. 2023. "A Review of Solar Drying Technology for Agricultural Produce." Indonesian Journal of Electrical Engineering and Computer Science 30(3):1407–19. doi: 10.11591/ijeecs.v30.i3.pp1407-1419.
49. Shah, Maulin P., and Pardeep Kaur. 2024. Biomass Energy for Sustainable Development. CRC Press.
50. Sharma, Atul, C. R. Chen, and Nguyen Vu Lan. 2009. "Solar-Energy Drying Systems: A Review." Renewable and Sustainable Energy Reviews 13(6–7):1185–1210. doi: 10.1016/j.rser.2008.08.015.
51. Sharma, Ramandeep K., Michael S. Cox, Camden Oglesby, and Jagmandeep S. Dhillon. 2024. "Revisiting the Role of Sulfur in Crop Production: A Narrative Review." Journal of Agriculture and Food Research 15.
52. Singh, Anil Kumar. n.d. "SMART Farming: Applications of IoT in Agriculture." doi: 10.1007/978-3-030-58675-1_114-1.
53. Su, Yuanping, Qiuming Yu, and Lu Zeng. 2020. "Parameter Self-Tuning Pid Control for Greenhouse Climate Control Problem." IEEE Access 8:186157–71. doi: 10.1109/ACCESS.2020.3030416.
54. Venkateswarlu, Kavati, and S. V. Kota Reddy. 2024. "Recent Trends on Energy-Efficient Solar Dryers for Food and Agricultural Products Drying: A Review." Waste Disposal and Sustainable Energy.
55. Wu, Weibin, Haoxin Li, Yingmei Chen, Yuanqiang Luo, Jinbin Zeng, Jingkai Huang, and Ting Gao. 2024. "Recent Advances in Drying Processing Technologies for Aquatic Products." Processes 12(5).
56. Yao, Yi, Yoong Xin Pang, Sivakumar Manickam, Edward Lester, Tao Wu, and Cheng Heng Pang. 2022. "A Review Study on Recent Advances in Solar Drying: Mechanisms, Challenges and Perspectives." Solar Energy Materials and Solar Cells 248.

Advances in Mechanical Engineering and Materials Sciences – Dr. Vinay K. B et al. (eds)
© 2026 Taylor & Francis Group, London, ISBN 9-781-041-20970-6

23

Privacy-Focused Multi-Modal Chatbot—Local Deployment of Llama 3.2 with Speech Recognition

Ayesha Taranum[1], Vedavathi N.[2]
Associate Professor, Department of Computer Science,
Vidyavardhaka College of Engineering,
Mysuru, India

Samarth V. Gangadikar[3],
Srusti C.[4], Sonal M. C.[5], Likitha S[6]
Department of Computer Science and Engineering,
Vidyavardhaka College of Engineering,
Mysuru, India

Abstract: This project presents the privacy-preserving Local LLM Multimodal Chat Bot which solves the problems of existing cloud-based conversational AI systems. This system does not depend on the use of some external server as other available chatbots but instead is using a locally hosted large language model using the model Llama3.2:1b with an integration in a Flask web application. Its interaction will be multimodal both through text and voice and can use speech recognition by employing the Web Speech API. It offers enhanced privacy, reduced latency, and regulatory compliance, which makes it an ideal candidate for sensitive domains like healthcare, education, and customer support.

Keywords: Local AI, Multimodal interaction, LLM, LangChain ollama, Privacy, Flask, Edge AI, Speech recognition

1. INTRODUCTION

The increasing reliance on AI-driven chatbots has transformed industries such as customer service, education, and healthcare. However, concerns regarding data privacy and latency in cloud-based solutions necessitate the development of local AI-driven conversational models. Research indicates that privacy-focused AI implementations are critical for secure and efficient interactions [4].

This paper presents an enhanced multi-modal chatbot leveraging Llama 3.2, LangChain, and Ollama, specifically designed for local deployment. Unlike traditional chatbot frameworks that depend on cloud connectivity, our system processes user input entirely on-device, ensuring data privacy and minimizing latency. Prior work by Zhang et al. [5] emphasized the significance of multi-modal interactions in improving chatbot effectiveness. Our research extends this by integrating optimized workflow methodologies to enhance real-time conversational coherence.

Key contributions of this study include refining chatbot workflows for improved multi-modal processing, optimizing model inference for local execution, and proposing novel integration strategies for maintaining contextual awareness during conversations. Furthermore, advancements in privacy-compliant AI deployment and efficient text-to-speech processing are explored to enable seamless user experiences [6] assistant, such as summation or knowledge extraction.

[1]ayesha.cs@vvce.ac.in, [2]vedavathi@vvce.ac.in, [3]sammuv9113@gmail.com, [4]srustichandrashekar07@gmail.com, [5]sonal.chengappa27@gmail.com, [6]likithasureshliki@gmail.com

In recent years, the effects of AI and NLP on human-machine interaction have indeed been profound. The AI application most widely used around the world is the chatbot, which provides conversational interfaces to help users through various domains, such as customer service, healthcare, and education. These systems, based on LLMs, result in human-like responses that may improve the efficiency and availability of user interactions.

This project deals with the development of an AI chatbot based on a state-of-the-art language model, ChatOllama, and it will integrate this using LangChain. This system is intended to understand multimodal inputs from the text and speech. These are interactive, context-sensitive, and in real time, as the goal is the creation of a conversational, efficient, and safe chatbot which may work upon different queries through multiple platforms.

The objectives of this paper are as follows:

- To provide an overview about the background and working of Llama.
- The primary goal is to design a scalable, secure, and efficient chatbot framework with broad applicability and adaptability. Specific aims include:
- Handling multi-modal input in the form of both text and speech.
- Guaranteeing privacy compliance in data-sensitive environments by local deployment.
- Integrate Llama 3.2 , LangChain, Ollama and Javascript speech regconition API into a single framework

2. ARCHITECTURE

2.1 Overview

With the development of large language models (LLMs) and multimodal AI approaches, conversational AI has undergone tremendous change. In order to facilitate real-time user interactions, the chatbot system showcased here combines speech recognition, natural language processing (NLP), and local model execution.

The chatbot makes use of JavaScript APIs for text and speech-based interactions, Flask for backend processing, Llama for NLP-based text creation, and Ollama for effective model execution. The system uses a local deployment strategy to provide low-latency replies and data security, doing away with the requirement for cloud-based services.

This architecture is scalable and privacy-focused for AI-driven conversational bots, supporting both text and voice-based communication.

2.2 System Overview

The following are the main parts of the chatbot:

Flask Backend: Controls API communications, user requests, and response production [9].

Llama & Ollama for NLP Processing: Uses transformer-based deep learning models to process user input and produce logical responses [5].

Real-time processing of voice and text interactions is made possible by the JavaScript Frontend API [12].

Local deployment removes reliance on the cloud, guarantees privacy, and has minimal latency [3].

Utilises federated learning, encryption, and GDPR-compliant data handling methods as security and privacy measures [14].

By combining these elements, response accuracy is improved, latency is reduced, and secure AI interactions are guaranteed.

2.3 Web Framework (Flask) for Backend Processing

The chatbot uses Flask, a lightweight Python micro-framework, to manage its backend operations. Flask offers:

Essential Features of Flask Request Handling & Routing: Flask receives user enquiries and forwards them for processing via API endpoints (/chat, /voice) [9].

Template Rendering: The user interface is constantly updated in real-time via the Jinja2 template engine [13].

Static File Management: Flask facilitates smooth communication between the UI and backend by serving frontend, CSS, and JavaScript files [9].

The chatbot's usage of Flask as its primary backend framework guarantees quick, scalable, and effective handling of user interactions.

2.4 Natural Language Processing with Llama & Ollama

A cutting-edge large language model (LLM) designed for text-based AI applications, Llama, powers the chatbot's comprehension of user enquiries and production of answers.

Capabilities of the Llama Model

Meta created Llama, a sophisticated natural language processing model intended for real-time, context-aware AI dialogues. Among its primary capabilities are:

Grouped Multi-Query Attention is used in Transformer-Based Architecture to maximise processing speed and memory [13].

SwiGLU activation functions are implemented in advanced feedforward networks, improving response coherence and pattern recognition [5].

Effective Token Processing: Capable of producing extremely pertinent responses, it has been trained on more than 1.4 trillion tokens [5].

Role of Ollama in Model Execution

Ollama ensures seamless AI model execution by serving as a mediator between Flask and Llama. It is in charge of:

Standardising user input before sending it to Llama is known as preprocessing [10].

Model responses are retrieved and formatted to improve readability [10].

Optimising communication between Flask and Llama guarantees effective processing while lowering latency and improving performance [6].

The chatbot can process text-based requests rapidly and accurately with Llama and Ollama.

2.5 JavaScript API for Multimodal Interaction

The Web Speech API provides a reasonable tradeoff between accuracy and performance, ensuring fast and privacy-compliant voice processing.

Both text-based and speech-based interactions are supported via the chatbot's frontend, which is built with JavaScript APIs.

JavaScript API Features

Text input handling in real time: When users type messages, they are instantly routed to the backend for processing [11].

Speech-to-Text Conversion: Converts spoken enquiries into text by integrating the Web Speech API [12].

Dynamic UI Updates: JavaScript ensures a seamless conversation flow by dynamically displaying chatbot responses [13].

The chatbot provides a simple and interesting user experience by combining voice and text input processing.

To quantify the efficiency of our chatbot's speech recognition module, we benchmark its accuracy using standard datasets

Table 23.1 Comparison of web speech API and Google speech-to-text

Metric	Web Speech API	Google Speech-to-Text (Cloud)
Word Error Rate (WER)	8.5%	5.2%
Processing Latency	~100ms	~250ms
Offline Capability	Yes	No

2.6 Multimodal Input Processing: Text and Speech

The chatbot offers a variety of interaction options to give users freedom in communication:

Communication Through Text

The chatbox allows users to submit text enquiries, which Flask then processes [17].

Communication Through Voice

Llama processes spoken queries, converts them into text, and then returns either text-based or audio answers [15].

Accurate speech detection is guaranteed by the Web Speech API [16].

The chatbot's ability to communicate via text and speech improves accessibility for a range of user requirements.

2.7 Local Deployment for Privacy and Security

This solution ensures maximum data privacy because it runs exclusively on local hardware, unlike cloud-based AI chatbots.

Local Processing Advantages

No External Data Transfers: Improves user privacy by removing cloud-based vulnerabilities [3].

Reduced Latency: Chatbot responses are quicker and more effective because data is handled locally [4].

Better Security & GDPR Compliance: Local processing reduces the danger of illegal access and data breaches [14].

This privacy-focused strategy guarantees that the chatbot is appropriate for personal use, business applications, and secure settings.

2.8 Security and Optimization Techniques

The chatbot incorporates the following to improve security and performance:

Model Enhancements

Model size can be decreased without sacrificing accuracy using quantisation and compression [5].

According to hardware availability, adaptive model scaling dynamically modifies computing needs [6].

Improvements in Security

Federated Learning: Enables localised model enhancements while protecting user privacy [20].

Secure communication between users and the chatbot is ensured using end-to-end encryption [14].

The chatbot is very dependable for AI-powered discussions because of these security-driven optimisations.

2.9 Future Enhancements

The following improvements are planned to significantly enhance usability:

Emotion Recognition AI: Uses sentiment analysis to improve chatbot answers [18].

Multi-Language Support: Adds automated translations to chatbot capability [19].

Integrating chatbot interactions with real-world augmented reality (AR) surroundings is known as augmented reality (AR) integration.

The chatbot will keep developing into a more sophisticated, adaptable AI helper by putting these characteristics into practice.

LLama Architecture

1. **Input and Embeddings:**

   The input tokens (words or subwords) are first converted into embeddings, which represent the tokens in a high-dimensional space.

   Rotary Positional Encodings:

   These are applied to embeddings to encode the sequential nature of the input, allowing the model to capture the order of tokens effectively.

   RMSNorm (Root Mean Square Normalization):

   Normalization layers are used throughout the architecture to stabilize training and improve performance.

2. **Self-Attention:**

   a) **Grouped Multi-Query Attention** is employed with a **KV (Key-Value) cache**:

   This enables efficient handling of multiple queries while reusing key-value pairs during inference for faster processing.

   Self-attention allows the model to focus on relevant parts of the input sequence while generating context-aware representations.

   b) **Feed Forward Layer with SwiGLU:**

   A feed-forward network with a **SwiGLU activation function** processes the output of the attention mechanism. This non-linearity helps the model learn complex patterns in data.

   c) **Nx (Repeated Blocks):**

   The attention and feed-forward layers are repeated multiple times (Nx) to increase the model's depth and capacity, enabling it to learn hierarchical representations.

   d) **Linear Layer and Softmax**:

   After the repeated transformer blocks, the output is passed through a linear layer and then a softmax function to produce the final probability distribution over the vocabulary.

Output:

The model predicts the next token based on the input sequence, which can be used for text generation, classification, or other NLP tasks.

Local Model Deployment: All interaction with Ollama and Llama is local, hence no data is transferred to external servers. This ensures better privacy and security for the users since all their queries are dealt with in the local environment.

The Llama model, driven by Ollama, is at the heart of the chatbot, making it competent enough to render complex, contextual responses.

3. **JavaScript API for User Interaction**

   The JavaScript API is what makes interaction from the frontend (interface) to the backend (Flask server) smooth. Its responsibilities include:

   Handling user input: it collects input in a text box or microphone for voice. When a user types or speaks a query, the API forwards the input to the Flask backend to process through the Ollama and Llama models.

   Speech-to-Text Conversion: When the user gives voice input, the JavaScript API interacts with Web Speech API to convert the spoken words into text. This text is further processed by the Llama model on the Flask back-end.

   The JavaScript API helps ensure that the chatbot can handle both text as well as voice input for smooth multimodal user experience.

4. **Multimodal Input Processing (Text and Speech)**

   This chatbot project supports multimodal interaction, which means that users can interact with

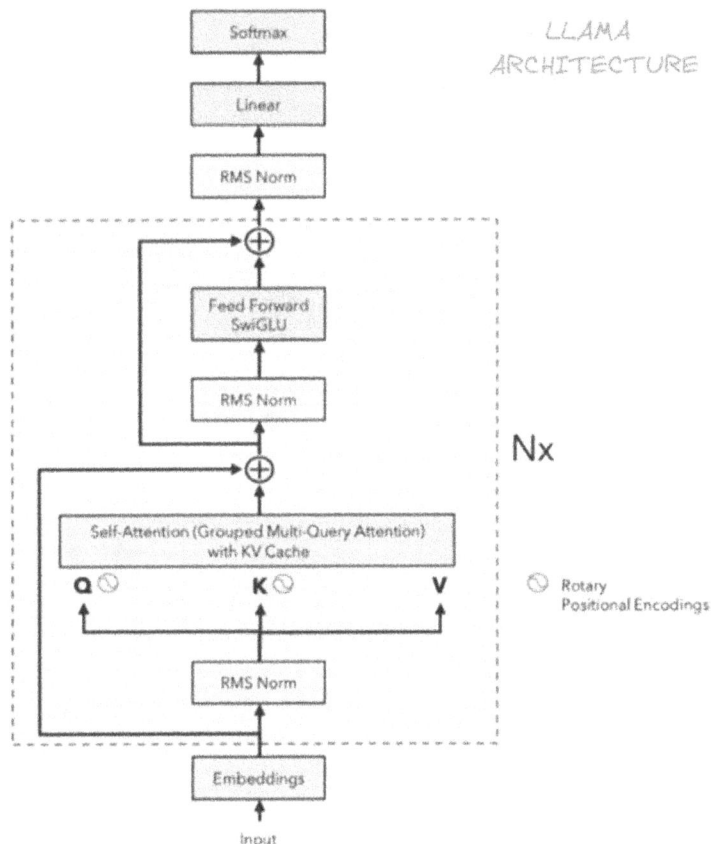

Fig. 23.1 Llama architecture

the chatbot in text or voice. This approach enhances the accessibility and flexibility of the system:

Text Input: The user can type a query into the chatbox, which is then sent to the Flask server. The Llama model processes the input and generates a response.

Speech Input: Once the user says something, the JavaScript API records it as an audio file and sends it to the Web Speech API, which translates the speech into text. The text is transmitted to the Flask backend, and after processing by the Llama model, a response is given.

Speech-to-Text: The Web Speech API captures the voice of the user, translates it into text, and then transmits the text to the Flask server for processing.

By supporting both text and speech, the chatbot offers a flexible and engaging experience, accommodating different user preferences for input.

5. **Local Deployment for Data Privacy**

This is a local deployment-based chatbot project. Here's why it's important:

Data Privacy: With the nature of processing all information locally, no user information gets out to any outside server. This way, personal information is kept hidden from unauthorized access.

Lower Latency: With processing going on locally, no communications happen with the farthest server, and thus the speed is quickened, plus responses come faster.

No Cloud Dependencies: The system does not depend on cloud services, which further enhances privacy and security, as the whole processing stack, including Ollama, Llama, Flask, and JavaScript, runs locally.

Local deployment ensures that users' queries, responses, and any other data stay within the local environment, minimizing external security risks.

6. **Security and Privacy Concerns**

The local deployment and lack of cloud dependency ensure that the chatbot system puts security and privacy first:

Data Security: All data is processed locally, ensuring that user queries and responses are not sent over the internet, thereby reducing the risk of unauthorized access.

No External Cloud Services: By not using the cloud-based server, the chatbot minimizes the risk of any external data vulnerability.

Local deployment will ensure no sensitive information is leaked because the deployment will be taken place within a trusted space.

Local AI deployments reduce external threats, but they introduce their own vulnerabilities. Key attack surfaces include:

Model Inversion Attacks: Adversaries may attempt to reconstruct training data by analyzing model outputs.

Adversarial Input Attacks: Carefully crafted input can cause incorrect responses.

Local System Exploits: If the local system is compromised, the chatbot may be vulnerable to malicious modifications.

Mitigation strategies include encrypting model interactions, secure boot mechanisms, and local firewall protections.

3. ALGORITHM

(a) generate_response (Interacting with the Chat Model):

Purpose:

To generate a chat response by invoking the ChatOllama model.

Algorithm:

1. Initialize the ChatOllama instance with the specified model (llama3.2:1b) and base_url.
2. Take the user's input (input_text) as the argument.
3. Print the input for debugging/logging purposes.
4. Call the invoke method of the ChatOllama model using the input text.
5. Extract and return the content from the response object.

(b) extract_points (Extracting Meaningful Points from Response):

Purpose:

To extract significant sentences from the model's response for better clarity.

Algorithm:

1. Split the response text into sentences using a regular expression (re.split) that detects sentence delimiters (e.g., ., !, ?).
2. Loop through the resulting list of sentences:
   Strip extra whitespace from each sentence.
   - Filter out sentences with fewer than 4 words (assumed to be less meaningful).
3. Return the filtered list of important sentences.

(c) chat Route (Handling Chat Requests):

Purpose:

To handle chat messages from the user and generate responses.

Algorithm:

1. Initialize the response variable to None.
2. If the method is POST, extract the user's input from the form (key: message).

3. If input is not empty:

Pass the input to generate_response to get the model's response.

4. Convert the raw model response into HTML-safe Markdown format using markdown. markdown.

5. Render the chat.html template, passing the processed response as context.

(d) Application Workflow:

Purpose:

To set up the Flask app and handle the routes.

Algorithm:

1. Define the / route for the home page:
   - Render the index.html template.

2. Define the /chat route to handle chat interactions:
   - Accept GET requests to load the chat page.
   - Accept POST requests to process user inputs and return model responses.

3. Run the app in debug mode to allow for easy testing and development.

Workflow of the chatbot:

1. User Input

Text Input (User types a query)

Voice Input (User speaks, converted to text using Speech-to-Text)

2. Frontend Processing

Speech-to-Text (if voice input)

AJAX/Fetch sends input to Flask Backend

3. Flask Backend

Receives user input (text)

Forwards input to Ollama (Llama Model) for processing

4. Ollama (Llama Model)

Processes input and generates a response

5. Markdown Formatting

Response formatted using Markdown

6. Response (Text or Speech)

Text Response (Rendered on the frontend)

Speech Response (Converted from text using Text-to-Speech)

4. RESULT ANALYSIS

Evaluation Criteria:

1. **Completeness:** Does the response include all key treatments (surgery, chemotherapy, radiation therapy, etc.) and advanced approaches (immunotherapy, targeted therapy)?

2. **Depth:** How detailed is the explanation for each Algorithm

Scoring:

Each model's response is scored on a scale of 1 to 10 for each criterion

Chart Representation:

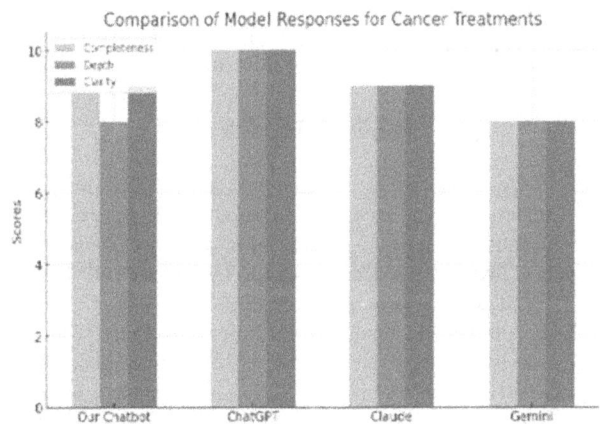

Fig. 23.3 Comparision of models for question on cancer Treatment

5. FUTURE SCOPE

The chatbot project offers significant potential for future enhancements, focusing on scalability, intelligence, and domain specificity. Key areas for expansion include:

1. **Advanced NLP Integration**: Incorporate cutting-edge models like fine-tuned transformers for improved context understanding and query resolution.

2. **Dynamic Knowledge Bases**: Integrate real-time APIs and knowledge graphs to ensure accurate and up-to-date responses.

3. **Personalization**: Leverage machine learning to deliver tailored, multilingual, and user-centric interactions.

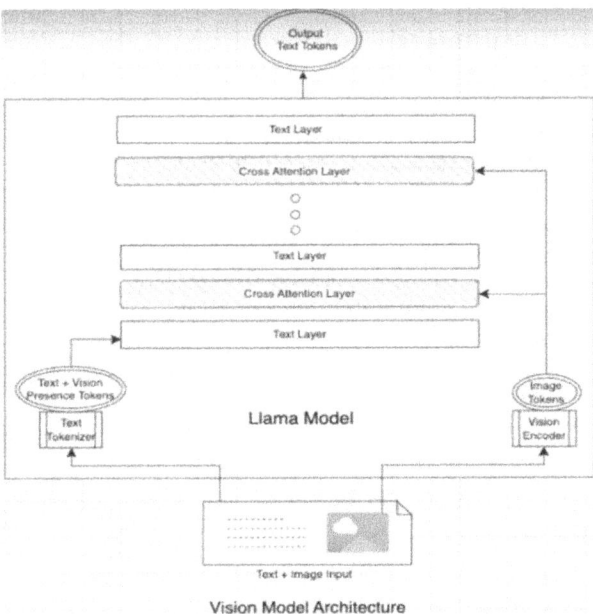

Vision Model Architecture

Fig. 23.2 Workflow

4. **Interactivity**: Introduce guided conversations and visual aids to enhance engagement and response quality.

6. CONCLUSION

Developments in multimodal AI have been significant for the system mainly in cloud-based ones because of the addition of features such as speech recognition as well as natural language. Nevertheless, these advanced interaction models are often privacy sensitive due to their nature, often suffering from latency since all operations are reliant on distant cloud infrastructure. The growing adoption of edge AI solutions processing data locally promises to resolve the challenges by enhancing privacy and reducing latency. However, computational constraints in edge devices may limit the complexity of the models that can be deployed. The Local LLM Multimodal Chat Bot addresses these challenges by bringing together text and speech inputs in a locally deployable system. This ensures the chatbot operates smoothly, efficiently, and securely using Llama 3.2 for natural language processing, and the JavaScript Speech Recognition API for voice input. Optimized strategies for local deployments ensure this system stays responsive while keeping user data secure, making it the perfect solution for privacy-conscious applications in healthcare, education, and customer service.

REFERENCES

1. S. Ayanouz, B. A. Abdelhakim, and M. Benhmed, "A smart chatbot architecture based NLP and machine learning for health care assistance," in *Proceedings of the 3rd International Conference on Networking, Information Systems & Security*, 2020, pp. 1–6.

2. L. Athota, V. K. Shukla, N. Pandey, and A. Rana, "Chatbot for healthcare system using artificial intelligence," in *2020 8th International Conference on Reliability, Infocom Technologies and Optimization (Trends and Future Directions) (ICRITO)*, IEEE, 2020, pp. 619–622.

3. B. Meskó and E. J. Topol, "The imperative for regulatory oversight of large language models (or generative AI) in healthcare," *npj Digital Medicine*, vol. 6, no. 1, p. 120, 2023.M. Arjovsky, et al., "Wasserstein GAN," arXiv preprint arXiv:1701.07875, 2017.

4. H. Harkous, K. Fawaz, K. G. Shin, and K. Aberer, "{PriBots}: Conversational privacy with chatbots," in *Twelfth Symposium on Usable Privacy and Security (SOUPS 2016)*, 2016.

5. J. Zhang, J. Pu, J. Xue, M. Yang, X. Xu, X. Wang, and F.-Y. Wang, "HiveGPT: Human-machine-augmented intelligent vehicles with generative pre-trained transformer," *IEEE Transactions on Intelligent Vehicles*, 2023.

6. J. Sun, C. Xu, L. Tang, S. Wang, C. Lin, Y. Gong, H.-Y. Shum, and J. Guo, "Think-on-graph: Deep and responsible reasoning of large language model with knowledge graph," *arXiv preprint* arXiv:2307.07697, 2023.

7. M. A. Kuhail, N. Alturki, S. Alramlawi, and K. Alhejori, "Interacting with educational chatbots: A systematic review," *Education and Information Technologies*, vol. 28, no. 1, pp. 973–1018, 2023.

8. M. Hasal, J. Nowaková, K. Ahmed Saghair, H. Abdulla, V. Snášel, and L. Ogiela, "Chatbots: Security, privacy, data protection, and social aspects," *Concurrency and Computation: Practice and Experience*, vol. 33, no. 19, p. e6426, 2021.

9. C. Guo, Y. Lu, Y. Dou, and F.-Y. Wang, "Can ChatGPT boost artistic creation: The need of imaginative intelligence for parallel art," *IEEE/CAA Journal of Automatica Sinica*, vol. 10, no. 4, pp. 835–838, 2023.

10. P. Sharma, N. Ding, S. Goodman, and R. Soricut, "Conceptual captions: A cleaned, hypernymed, image alt-text dataset for automatic image captioning," in *Proceedings of the 56th Annual Meeting of the Association for Computational Linguistics (Volume 1: Long Papers)*, 2018, pp. 2556–2565.

11. G. Sebastian, "Privacy and data protection in ChatGPT and other AI chatbots: Strategies for securing user information," *Available at SSRN* 4454761, 2023.

12. P. Schramowski, C. Turan, N. Andersen, C. A. Rothkopf, and K. Kersting, "Large pre-trained language models contain human-like biases of what is right and wrong to do," *Nature Machine Intelligence*, vol. 4, no. 3, pp. 258–268, 2022.

13. B. A. Alazzam, M. Alkhatib, and K. Shaalan, "Artificial intelligence chatbots: A survey of classical versus deep machine learning techniques," *2023*.

14. B. D. Lund, D. Khan, and M. Yuvaraj, "ChatGPT in medical libraries, possibilities and future directions: An integrative review," Health Info Libr J, Mar. 2024, doi: 10.1111/hir.12518..

15. B. Niu and G. F. N. Mvondo, "I Am ChatGPT, the ultimate AI Chatbot! Investigating the determinants of users' loyalty and ethical usage concerns of ChatGPT," Journal of Retailing and Consumer Services, vol. 76, Jan. 2024, doi: 10.1016/j.jretconser.2023.103562.

16. A. Pal, L. K. Umapathi, and M. Sankarasubbu, "MedMCQA : A Large-scale Multi-Subject MultiChoice Dataset for Medical domain Question Answering," 2022.

17. O. Topsakal and T. C. Akinci, "Creating Large Language Model Applications Utilizing LangChain: A Primer on Developing LLM Apps Fast," International Conference on Applied Engineering and Natural Sciences, vol. 1, no. 1, pp. 1050–1056, Jul. 2023, doi: 10.59287/icaens.1127.

18. Ankit Bhakkad, S.C. Dharmadhikari, M. Emmanuel, and Parag Kulkarni, E-VSM: Novel Text Representation Model to Capture Contex-Based Closeness between Two Text Documents. 2013

19. P. Lewis et al., "Retrieval-Augmented Generation for Knowledge-Intensive NLP Tasks," May 2020.

Note: All the figures and tables in this chapter were made by the authors.

Advances in Mechanical Engineering and Materials Sciences – Dr. Vinay K. B et al. (eds)
© 2026 Taylor & Francis Group, London, ISBN 9-781-041-20970-6

24

Ergonomic Assessment of Posture Risks among Railway Workshop Hydraulic Press Workers using Rapid Upper Limb Assessment (RULA)—Case Study

Shreyas M.*, Vinay K. B.
Department of Mechanical Engineering,
Vidyavardhaka College of Engineering,
Mysuru, India

Deepthi Amith
Department of Master of Business Administration,
Kalpataru Institute of Technology,
Tiptur, India

Rajesh Kumbara S. K., S. Puneet Kumar
Department of Mechanical Engineering,
Vidyavardhaka College of Engineering,
Mysuru, India

Abstract: The research evaluates the ergonomic aspects of postural hazards within railway workshop hydraulic press departments through WMSD-related investigations. RULA and NMQ served as assessment tools for evaluating 12 hydraulic press machine operators from Mysuru, India. Research findings revealed extensive musculoskeletal pain among workers because 80% experienced annual lower back issues while also demonstrating high danger areas in their neck as well as their hip/thigh regions. The analysis through RULA determined that exposure risk was high for 42% of workers, which proves the immediate necessity for ergonomic solutions. workspace constraints combined with repetitive motions and insufficient recovery time made up the main causes. The research advised using work redesign alongside ergonomics education together with environmental upgrades and job sequence adjustments, so employees remained secure. From the research findings, we can conclude that workplace safety and industrial production require proactive ergonomic strategies.

Keywords: Ergonomic assessment, Rapid upper limb assessment (RULA), Nordic musculoskeletal questionnaire (NMQ), Railway workshop, Work-related musculoskeletal disorders (WMSDs)

1. INTRODUCTION

Indian Railways is a state-owned enterprise in India, functioning as a departmental undertaking under the ownership and administrative control of the Ministry of Railways, Government of India (Virendra Kumar 1966). It is the ninth largest employer in the world and the second largest in India, with over 1.2 million employees (Directorate of Statistics and Economics Ministry of Railways (Railway Board) 2023). Railway maintenance

*Corresponding author: shreyas.m@vvce.ac.in

10.1201/9781003725053-24

workshops are essential for ensuring the efficient operation of trains, despite their critical role frequently being overlooked. Specialized facilities oversee the regular maintenance of all railway vehicles, including numerous locomotives, various wagons, multiple units, and other vehicles. A significant number of skilled maintenance workshop workers ensured the continuation of train operation. Railway repair workshops are very important for the efficient operation of rail services (Binder, Mezhuyev, and Tschandl 2023). Utilizing RULA is an excellent method, and it is applied to upper limb postures. It helps to identify the risks in the jobs that include awkward postures and repetitive motion (Lynn and Corlett 1993; Bera et al. 2024), and RULA can identify the risks and provide recommendations to reduce the identified risk. RULA has been used in various industries and has been shown to be effective in identifying awkward postures that can lead to muscle and joint disorders (Lin et al. 2022), (Urrejola-Contreras 2024). Modifying workstations and changing jobs can reduce WMSD in workers (Ding 2023). RULA has also been compared to other tools such as REBA to determine its efficacy, and RULA, in conjunction with the NMQ, provides more information about workplace safety. The NMQ is utilized to determine where in the body workers are experiencing pain; thus, this can facilitate pinpointing changes to assist workers (Kuorinka et al. 1987). Gorce demonstrated that RULA scores differed between physiotherapists who are performing manual lymphatic drainage; therefore, it has been demonstrated that RULA may be used to detect differences between jobs (Gorce 2023). NMQ is a useful tool to help identify the severity of muscle and joint disorders in workers; therefore, it may be used as a complement to RULA. More recent methods such as Noval Postural Assessment Method have been developed based on 3D a model and 3D CAD tools to assess posture, and this method assists in posture assessment (Sanchez-Lite et al. 2013). Digital tools such as CATIA for ergonomic modelling can strengthen RULA assessments through their implementation. Research through digital human modelling allows scientists to model various user movements in computerized environments for examining ergonomic safety risks in high detail. Lab research demonstrated that RULA tool integration into CATIA's DHM software helped produce more precise ergonomic evaluations, which created detailed virtual simulations of operational tasks (Roy et al. 2024) (Pazouki et al. 2017; Cremasco et al. 2019). Employees who receive training about ergonomic methods together with assessment tool usage demonstrate a substantial decrease in their musculoskeletal complaints. Spinelli et al. demonstrated that ergonomic training is essential for developing workplace safety culture through worker education (Spinelli et al. 2018). The study by Abdullah et al. supports their findings about ergonomic interventions delivering both enhanced work practice quality and decreased work-related injury rates through educational

methodologies (Abdullah et al. 2015). Figure 24.2 shows the literature network map, which is created using LitMap, showing all the literature that was referred to do the case study and all the literature papers referred were leading to the RULA and NMQ methods which were used.

RULA coupled with Nordic questionnaires provides organizations with a complete system to detect and eliminate ergonomic risks at work. Organizations achieve superior ergonomic intervention effectiveness when they combine these methodologies to study posture-discomfort-workflow connections in work environments. Ongoing research needs to investigate the intricate relationship between these tools and their influence on employee health as well as productivity parameters. Objectives of this study include identifying ergonomic risk factors, evaluating work postures using RULA, quantifying the severity of postural risks, examining workstation design and task ergonomics, and recommending ergonomic interventions.

2. METHODOLOGY

In this study, a total of 12 workers from Mysuru were analyzed in the railway workshop. From one workstation where they operated the hydraulic press machine in the Ashokpuram train factory. All the workers who were involved in the job were assessed. The framework for the research case study can be seen in Fig. 24.1. The research design focuses on evaluating the ergonomics and posture-related risks faced by workers in railway repair

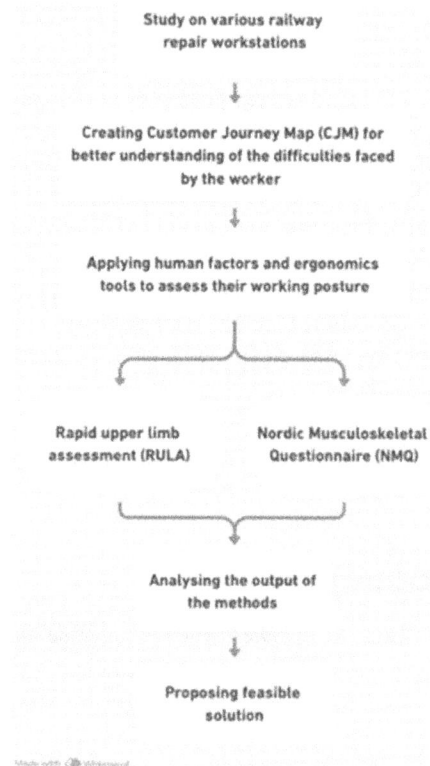

Fig. 24.1 Research design flowchart
Source: Aurhors

workstations. It begins with a study of various railway repair environments to understand workplace conditions. From Fig. 24.3 the working condition of the worker can be seen. A Customer Journey Map (CJM) is then created to capture the challenges experienced by workers. Working postures are assessed using human factors and ergonomic tools, specifically the Rapid Upper Limb Assessment (RULA) and the Nordic Musculoskeletal Questionnaire (NMQ). The results from these assessments are analyzed to identify key ergonomic risks and musculoskeletal discomforts. Finally, based on the findings, feasible solutions are proposed to improve workplace ergonomics, reduce discomfort, and enhance worker well-being.

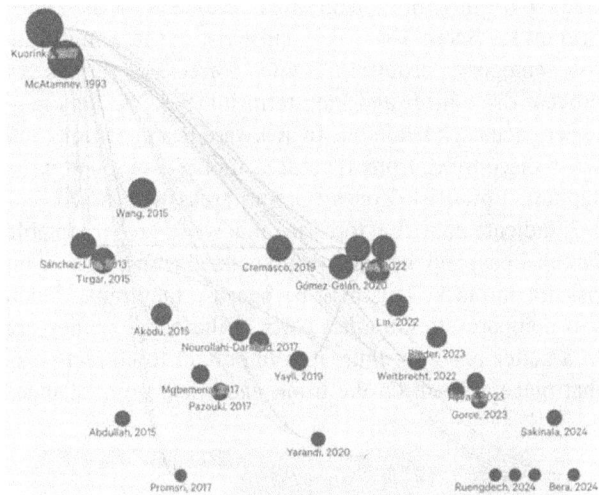

Fig. 24.2   Literature network map using LitMap (Kaur et al. 2022)

Fig. 24.3   Regular working condition of worker in pressing machine

Source: Aurhors

2.1 Nordic Musculoskeletal Questionnaire (NMQ)

The Nordic Musculoskeletal Questionnaire (NMQ) will evaluate musculoskeletal issues across various anatomical regions. Participants complained about discomfort, pain or other symptoms over specified areas, along with details on the nature, duration, and impact of these symptoms on their activities. Symptoms are indicated on a body diagram, and discomfort is rated with five-point scale from slight to more. The NMQ aids scholars in identifying musculoskeletal disorders amidst workers, understand the effects of work, and develop preventive solutions by improving ergonomics and promoting well-being (Kuorinka et al. 1987). The NMQ forms as shown in Fig. 24.4 were given to the workers to fill out. The forms were given to 14 workers specifically working on the hydraulic press machine in the railway workshop, out of which 12 workers responded. Finally, the answers were tabulated in Fig. 24.6.

Fig. 24.4   Musculoskeletal discomfort form (Kuorinka et al. 1987)

2.2 Rapid Upper Limb Assessment (RULA)

The RULA method measures risk factors for workers who are exposed to high postural stress, considering their posture, frequency, and force. RULA worksheet scores (as shown in Fig. 24.5) determine a performance level, indicating necessary modifications. The process involves witnessing work, choosing postures, determining sides, calculating angles, scoring, assessing risk, and creating an working plan. This includes the implementation of changes in work practices and workspace design, followed by re-evaluation using RULA. The risk levels are categorized as High (7+), Medium (5–6), low (3–4), and negligible (1–2) (Lynn and Corlett 1993). Using CATIA V5 R20 software, a digital model of the actual press machine was created. Precise measurements of the press machine, the operator's seating arrangement, and small components (like the job carriage) were incorporated into the model. An ergonomic manikin representing industrial workers was also developed within the software. The simulation replicated real-life postures, and a RULA analysis was performed to assess work-related disorders. A particular focus was placed on the awkward postures adopted during strip pressing tasks.

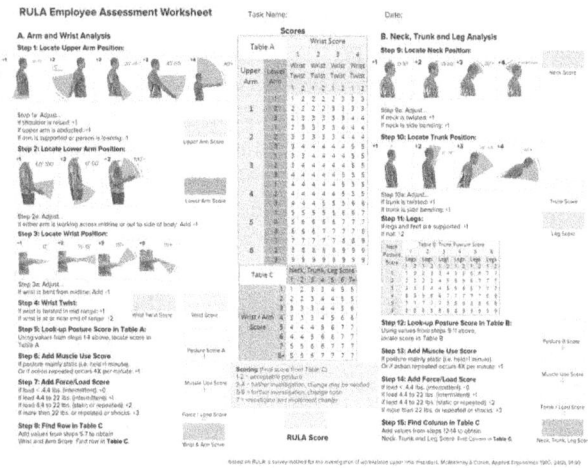

Fig. 24.5 RULA worksheet (Gómez-Galán et al. 2020)

3. RESULTS AND DISCUSSIONS

3.1 NMQ Findings

NMQ musculoskeletal discomfort is in the specific body part of the human body. From the Graph Fig. 24.6 the most affected area is the lower back, which shows the highest discomfort, with nearly 80% reporting issues over a year and around 60% over a week. Neck also exhibits a high prevalence, with above 50% reporting discomfort in a year and slightly lower in a week. Hip/thigh follows a similar trend, with significant discomfort levels. Moderately affected areas are the upper back, shoulder, and knees, which have discomfort levels between 20-40% for a year, with week-long discomfort relatively lower. Least affected areas are elbows, wrists, and ankles/feet, which show the lowest discomfort percentages, with less than 20% affected over both timeframes. Long-term (1-year) discomfort is generally higher than short-term (1-week) discomfort across all body parts. The lower back, neck, and hip/thigh are the most affected, indicating ergonomic concerns. Elbows and wrists show minimal discomfort, suggesting less strain in these areas. This data suggests a potential need for ergonomic interventions, especially for the lower back and neck, to reduce long-term musculoskeletal issues.

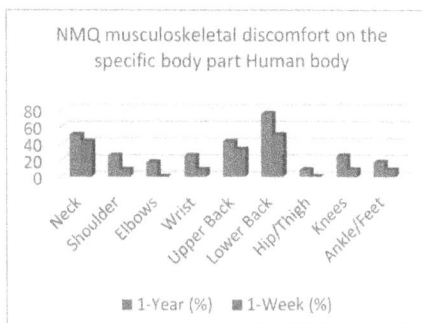

Fig. 24.6 Musculoskeletal discomfort on specific body part

Source: Aurhors

3.2 RULA Findings

The RULA (Rapid Upper Limb Assessment) analysis results from Fig. 24.7 and Fig. 24.8 indicate the postural strain and ergonomic risk for workers performing tasks in railway repair workstations. Based on the color-coded scoring system from CATIA V5 and RULA, different body segments (upper arm, forearm, wrist, neck, trunk) are analyzed for their ergonomic risk levels. High-Risk Zones (Red—Score 5-6): Trunk and upper arm show high ergonomic risk, indicating severe postural strain and an urgent need for intervention to improve working posture. These scores suggest that the workers are frequently bending, reaching, or working in awkward postures, leading to musculoskeletal stress. Moderate-Risk Zones (Yellow) – Score 3-4: Neck and wrist are in a moderate risk category, requiring some corrective actions to prevent discomfort and long-term injuries. Workers may experience neck strain due to awkward head positions and wrist discomfort from repetitive motions or poor wrist support. Low-Risk Zones (Green) areas with a score of 1-2 indicate that the forearm and wrist are acceptable (green areas) and they are not associated with a significant risk for the task. This may be because they move much less compared to the other parts of the body, or they are in a better position; thus, it is important to correct tasks that place a strain on the trunk and upper arm. Changes

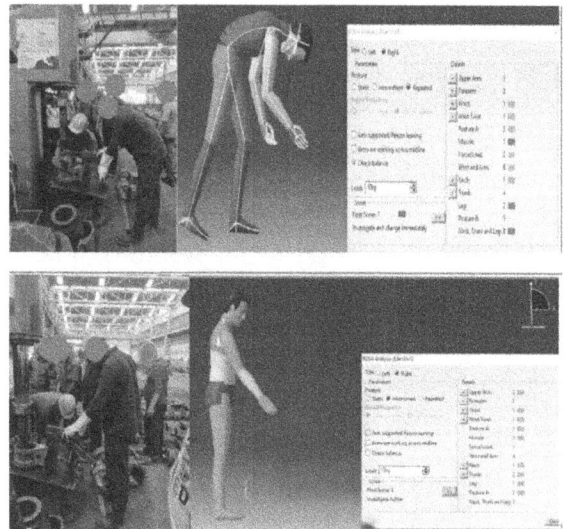

Fig. 24.7 RULA analysis using CATIA V5

Source: Aurhors

Segment	Score Range	Color associated to the score					
		1	2	3	4	5	6
Upper arm	1 to 6						
Forearm	1 to 3						
Wrist	1 to 4						
Wrist twist	1 to 2						
Neck	1 to 6						
Trunk	1 to 6						

Color code in CATIA-V5 and RULA Score

Fig. 24.8 Final RULA score

Source: Aurhors

Table 24.1 RULA result

RULA Scale	0	1	2	3
Score	1 to 2	3 to 4	5 to 6	7
Level of Exposure	Negligible	Low	Medium	High
Actions to be taken	None	Investigate further, Change might be required.	Investigate further, modify soon	Examine and act right away
No of participants	0	3	4	5
Percentage of participants	0	25	33	42

Source: Aurhors

in posture (sitting or standing), redesign of workstations, and work breaks can help reduce discomfort; therefore, training of workers on the correct posture while sitting or standing and the use of tools such as adjustable benches and chairs are effective steps. Job rotation or task rotation can also help reduce the risk of repetitive strain injury, so it is a useful approach to consider. The outcomes of this study indicate that there is a need for the ergonomic design of workstations to promote workers' comfort and prevent chronic musculoskeletal disorders, because this can help reduce the risk of injury and improve overall well-being.

Table 24.1 displays the RULA scores for a set of participants, describing their level of ergonomic risk exposure and the necessary actions based on their scores. None of the subjects (0%) are in the insignificant risk category (scores 1-2), so no action is required for these subjects, and 25% of the subjects (3 subjects) are in the low-risk category (scores 3-4), so this category should be reviewed and changed if necessary. 33% of the subjects (4 subjects) are in the medium risk category (scores 5-6), so this category should be reviewed and changed without delay because 42% of the subjects (5 subjects) are in the high risk category (scores 7-10), so this category requires immediate action to reduce the risk of MSD. Based on RULA scores, 75% of the subjects are in the medium and high-risk category, so these subjects need to be ergonomically designed to prevent MSD; therefore, in this study, 42% of the subjects require immediate action to lower the risk of MSD, thus workplace related to these subjects should be changed as soon as possible.

4. CONCLUSION

This study conducted a thorough ergonomic evaluation of hydraulic press operators at a railway workshop utilizing the RULA method and the Nordic Musculoskeletal Questionnaire (NMQ). The study's results demonstrate that musculoskeletal hazards associated with workstation design, task repetitiveness, and postural awkwardness are significant. The lower back, neck, and hip/thigh regions were the most impacted, suggesting considerable ergonomic strain. A substantial percentage of the workforce (42%) is classified inside the high-risk RULA group. There appears to be a need for urgent ergonomic intervention. Employees were already encountering

fatigue at the beginning of the workday due to the inadequate organization of the workspace. However, what is particularly notable is how this discomfort gets worse with tasks that require prolonged awkward postures, excessive repetition, and insufficient recovery time. The absence of rest periods, together with inadequate ventilation and seating arrangements, subjects 'workers to a harmful environment, resulting in severe exhaustion, dizziness, and pain by the end of their shift.

4.1 Implications for Ergonomic Interventions

Addressing This requires many modifications and includes adjusting the height of desks. It also includes providing supportive chairs because it includes positioning tools so the body can be in an appropriate posture. It includes training employees on how to lift objects; thus, it includes training employees on how to sit and stand. It includes ensuring that employees get breaks, so it includes having better lighting, ventilation, and temperatures within a workplace. Avoid having the same workers perform the same task for long periods of time; therefore, rotate workers between tasks.

4.2 Future Scope and Recommendations

This study revealed the necessity of ergonomic interventions, but future research should contain additional investigation through extended time-based studies. A biomechanical examination serves to determine and measure the physical pressures applied to various body sections. Real-time postural feedback through artificial intelligence-based systems for workplace postural monitoring would offer corrective action to workers. Workstations from various companies must be studied against one another to identify external trends in ergonomics as well as industry-leading best practices. Workshops that implement evidence-based ergonomic solutions will improve both worker production and safety standards while protecting employees from workplace musculoskeletal injuries. The absence of proper action on workplace challenges leads to both higher worker absenteeism and decreased workplace efficiency and develops enduring worker health problems. Industrial settings need urgent ergonomic reforms because such changes represent the essential protective measure needed to defend worker health while supporting operational performance.

REFERENCES

1. Abdullah, Zaini, Nurul Fadzlina, Mohd Amran, S A Anuar, Mohd Shahir, and Khairul Fadzli. 2015. "Design and Development of Weaving Aid Tool for Rattan Handicraft." *Applied Mechanics and Materials* 761: 277–281. doi:10.4028/www.scientific.net/amm.761.277.

2. Bera, Debabrata, Sayan Sarkar, Bivash Mallick, and Manik Chandra Das. 2024. "Assessment of Posture Related Risks among Goldsmiths Using Rapid Upper Limb Assessment (RULA) and Rapid Entire Body Assessment (REBA)." *Journal of The Institution of Engineers (India): Series C*, October. Springer. doi:10.1007/s40032-024-01093-5.

3. Binder, Mario, Vitaliy Mezhuyev, and Martin Tschandl. 2023. "Predictive Maintenance for Railway Domain: A Systematic Literature Review." *IEEE Engineering Management Review* 51 (2). Institute of Electrical and Electronics Engineers Inc.: 120–140. doi:10.1109/EMR.2023.3262282.

4. Cremasco, Margherita M, Ambra Giustetto, Federica Caffaro, Andrea Colantoni, Eugenio Cavallo, and Stefano Grigolato. 2019. "Risk Assessment for Musculoskeletal Disorders in Forestry: A Comparison Between RULA and REBA in the Manual Feeding of a Wood-Chipper." *International Journal of Environmental Research and Public Health* 16 (5): 793. doi:10.3390/ijerph16050793.

5. Ding, Xiaowen. 2023. "Prevalence and Risk Factors of Work-Related Musculoskeletal Disorders Among Emerging Manufacturing Workers in Beijing, China." *Frontiers in Medicine* 10. doi:10.3389/fmed.2023.1289046.

6. Directorate of Statistics and Economics Ministry of Railways (Railway Board), Government of India, New Delhi. 2023. *INDIAN RAILWAYS YEAR BOOK 2022-23.*

7. Gómez-Galán, Marta, Ángel Jesús Callejón-Ferre, José Pérez-Alonso, Manuel Díaz-Pérez, and Jesús Antonio Carrillo-Castrillo. 2020. "Musculoskeletal Risks: RULA Bibliometric Review." *International Journal of Environmental Research and Public Health.* MDPI. doi:10.3390/ijerph17124354.

8. Gorce, Philippe. 2023. "Three-Month Work-Related Musculoskeletal Disorders Assessment During Manual Lymphatic Drainage in Physiotherapists Using Generic Postures Notion." *Journal of Occupational Health* 65 (1). doi:10.1002/1348-9585.12420.

9. Kaur, Amanpreet, Sarita Gulati, Ritu Sharma, Atasi Sinhababu, and Rupak Chakravarty. 2022. "Visual Citation Navigation of Open Education Resources Using Litmaps." *Library Hi Tech News* 39 (5). Emerald Publishing: 7–11. doi:10.1108/LHTN-01-2022-0012.

10. Kuorinka, I., B. Jonsson, A. Kilbom, H. Vinterberg, F. Biering-Sørensen, G. Andersson, and K. Jørgensen. 1987. "Standardised Nordic Questionnaires for the Analysis of Musculoskeletal Symptoms." *Applied Ergonomics* 18 (3): 233–237. doi:10.1016/0003-6870(87)90010-X.

11. Lin, Po-Chieh, Yu-Jung Chen, Wei-Shin Chen, and Yun-Ju Lee. 2022. "Automatic Real-Time Occupational Posture Evaluation and Select Corresponding Ergonomic Assessments." *Scientific Reports* 12 (1). doi:10.1038/s41598-022-05812-9.

12. Lynn, McAtamney, and Nigel Corlett. 1993. "RULA: A Survey Method for the Investigation of Work-Related Upper Limb Disorders." *Applied Ergonomics* 24 (2): 91–99.

13. Pazouki, Abdolreza, Leila Sadati, Fatemeh Zarei, Ehsan Golchini, Robab Fruzesh, and Jalal Bakhtiary. 2017. "Ergonomic Challenges Encountered by Laparoscopic Surgeons, Surgical First Assistants, and Operating Room Nurses Involved in Minimally Invasive Surgeries by Using RULA Method." *Journal of Minimally Invasive Surgical Sciences* 6 (4). doi:10.5812/minsurgery.60053.

14. Roy, Rakesh, Md. Mahafuj Anam Murad, Md. Masum Billah, Subrata Talapatra, Md Mahfuzur Rahman, and Sarojit Kumar Biswas. 2024. "Comparative Ergonomic Posture Analysis of CNC Milling Machine Workers through Digital Human Modeling and Artificial Neural Networks." *Indian Journal Of Science And Technology* 17 (19): 1935–1946. doi:10.17485/IJST/v17i19.912.

15. Sanchez-Lite, Alberto, Manuel Garcia, Rosario Domingo, and Miguel Angel Sebastian. 2013. "Novel Ergonomic Postural Assessment Method (NERPA) Using Product-Process Computer Aided Engineering for Ergonomic Workplace Design." *PLoS ONE* 8 (8). doi:10.1371/journal.pone.0072703.

16. Spinelli, Raffaele, Giovanni Aminti, Natascia Magagnotti, and Fabio D Francesco. 2018. "Postural Risk Assessment of Small-Scale Debarkers for Wooden Post Production." *Forests* 9 (3): 111. doi:10.3390/f9030111.

17. Urrejola-Contreras, Gabriela P. 2024. "Myotonometry in Machinery Operators and Its Relationship With Postural Ergonomic Risk." *Annals of Work Exposures and Health* 68 (6): 605–616. doi:10.1093/annweh/wxae028.

18. Virendra Kumar. 1966. "Committees And Commissions In India." *Concept Publishing Company* 7: 128.

Advances in Mechanical Engineering and Materials Sciences – Dr. Vinay K. B et al. (eds)
© 2026 Taylor & Francis Group, London, ISBN 9-781-041-20970-6

25 Statistical Analysis of Absorptance and Emissivity in Tungsten-Alumina Thin Films using ANOVA

Naveen Ankegowda*

Associate Professor,
Department of Mechanical Engineering,
Vidyavardhaka College of Engineering,
Mysuru

Abstract: The tungsten-alumina (W-Al2O3) thin film was successfully formed on SS304, which has been taken into consideration as a substrate in this investigation. Three factors were considered: the flow rate of inert gas, RF power, and DC power. To model the Central Composite Design (CCD), a total of 20 runs were carried out, using emissivity as the response parameter and mass deposition per unit area of 10 mm x 10 mm. Based on the CCD, the optimized sample was assessed by the Design of Experiments. From there, it has been looked at each coating's unique levels, ANOVA, the Normal Plot of Residual and residuals fitted values illustrating the Interaction for Response Surface Mass Deposition per unit area.

Keywords: Emissivity, Absorptance, Centre composite design, Design of experiments

1. INTRODUCTION

On analysing the emissivity of various materials, scientists can learn more about their surface characteristics, having a thorough understanding of emissivity enables researchers to create surfaces with thermal radiation properties, such as low emissivity coatings for insulation or high emissivity coatings for effective heat expulsion. Numerous techniques can be used to apply these coatings, and each one has input parameters that could affect the coating's quality and response capabilities. Optimizing the input conditions is essential. Proposed study is to optimize the deposition conditions and investigating the emissivity of a deposited tungsten-alumina thin film are the main goals. Investigating emissivity of coated or developed thin film of materials is plasma interaction and tungsten-based composite is one such key metal (Chunyang Luo, 2023) Because of its high melting point, low sputtering, and low fuel retention, tungsten is considered a suitable plasma-facing material in several applications. One must cope with thermal corrosion and radiation damage

in an extremely harsh environment. Low radioactivity, high melting point, low sputtering yield, low thermal conductivity, poor thermal shock resistance, and low hydrogen absorption are some of the requirements for plasma face material (E. Gauthier, 2005). Alumina has been chosen as part of a larger effort to investigate the emissivity of tungsten-based composites as thin films. Alumina and tungsten have been utilized to create thin films. Magnetron sputtering is one of the most effective methods for creating thin films (Morteza Sasani Ghamsari. 2023, Baizhang Cheng.2023, Sabrina State.2022, Ajit Behera 2022). During sputtering input parameters plays vital role and they have significant impact on the properties of the thin film, so for emissivity optimum sputtering condition of input parameters, the design of experiment approach using central composite design (CCD) has been employed to design the list of experiments to carry deposition (Charnnarong Saikaew 2010, Wen-Hsien Ho 2010, Morito Akiyama 2010, Jonathan Kenneth Bunn 2016, E. Baudet 2017). CCD is integral to RSM, which helps in understanding the relationship between multiple

*Corresponding author: naveen@vvce.ac.in

input factors and their effects on output responses, RSM explores relationships between response and input factors (S Qiu, M Xie 2020). CCD enables the fitting of second-order models, which are essential for capturing non-linear relationships in the sputtering process (Wilmina M Marget & Max D Morris.2019).

2. DESIGN OF EXPERIMENTS

2.1 Factors Levels

Three factors were considered while designing the number of runs needed to make the thin film of W-Al2O3: the inert gas flow rate, RF power, and DC power, which were defined at five different levels: extreme high (star point), higher point, centre point, low point, and finally, extreme low star point. The alpha (α) value is the distance between the starting point and the centre, for all three factors the value of α is ± 1.682 thins facilitates to effectively model quadratic relationships and curvature. In Table 25.1, five levels of every factor used in the tests are displayed.

Table 25.1 Factors levels

Levels	DC in watts	RF in watts	Gas Flow rate in SCCM
Lowest star point($-\alpha$)	500	400	150
Lower (-1)	601.2	638	239.3
Centre point (0)	750	800	300
Higher (+1)	898.8	1038	389.3
Highest star point($+\alpha$)	1000	1200	450

2.2 Design Summary

The design summary can be seen in Table 25.2 for the CCD number of factors, the number of replicates, and the total number of runs with blocks. Blocks are groups of related settings that are determined by one or more factors.

Table 25.2 Design summary

Factors:	3	Replicates:	1
Base runs:	20	Total runs:	20
Base blocks:	1	Total blocks:	1
Note: α = 1.68179			

2.3 Design Table and Response Values

The design table, which shows the input factors for each of the 20 runs of the two-level complete factorial runs and the response based on the CCD, is shown in Table 25.3. -1, +1 and ± 1.682 has been treated as coded units to represents variable factors value at each run. For variable's fitted values have been represented in Table 25.1 for reference. The deposition settings and the deposition methods used have a major impact on the emissivity values of tungsten thin films (C. K. Wangati 2008, Mohammed M Alsmadi 2023). On each run thin film has been developed and

responses are Absorptance and Emissivity, which have been measured at RVCE Bangalore. Both the responses have been tabulated against each run in the Table 25.3

Table 25.3 Design table and response values

Run	Blk	DC	RF	SCCM	Absorptance	Emissivity
1	1	-1	-1	-1	0.635	0.265
2	1	1	-1	-1	0.731	0.182
3	1	-1	1	-1	0.742	0.095
4	1	1	1	-1	0.630	0.267
5	1	-1	-1	1	0.697	0.184
6	1	1	-1	1	0.698	0.211
7	1	-1	1	1	0.689	0.283
8	1	1	1	1	0.691	0.269
9	1	-1.682	0	0	0.706	0.238
10	1	1.682	0	0	0.697	0.231
11	1	0	-1.682	0	0.635	0.265
12	1	0	1.682	0	0.731	0.182
13	1	0	0	-1.682	0.742	0.095
14	1	0	0	1.682	0.630	0.267
15	1	0	0	0	0.697	0.184
o16	1	0	0	0	0.698	0.211
17	1	0	0	0	0.689	0.283
18	1	0	0	0	0.691	0.269
19	1	0	0	0	0.706	0.238
20	1	0	0	0	0.697	0.231

Analysis of Results and Discussions using Analysis of variance (ANOVA)

ANOVA: Absorptance versus DC, RF, SCCM

ANOVA is an effective statistical method for comparing three or more groups' means to see if there are any statistically significant differences between them. Using MINITAB, a popular statistical program, makes ANOVA easier to use and more effective for data analysis. To determine which factors (independent variables) significantly affect the response (dependent variable) in an experimental design, an ANOVA is utilized. MINITAB gives graphical output, automates difficult computations, and has a straightforward interface. Table 25.4 depicts the details of ANOVA, significance level (α) is set to 5%. Comparing the P-value which is probability of observing the data if the null hypothesis is true to significance level (if it is < 5%) indicates that there is statistical significance of that factors group.

Table 25.4 makes it abundantly evident that the P-value is significant for the linear model's RF and SCCM, that there is no effective square model for any of the three components, and that the interaction between all three input parameters has a significant impact on the response absorptance. The odd observations from those few runs of

Table 25.4 Response surface regression: Absorptance versus DC, RF, SCCM

Source	DF	Adj SS	Adj MS	F-Value	P-Value
Model	9	0.04542	0.005048	11.21	0.000
Linear	3	0.01844	0.006147	13.65	0.001
DC	1	0.00055	0.000551	1.22	0.295
RF	1	0.01258	0.012578	27.93	0.000
SCCM	1	0.00531	0.005313	11.80	0.006
Square	3	0.00159	0.000533	1.18	0.365
DC*DC	1	0.00032	0.000320	0.71	0.419
RF*RF	1	0.0008	0.00089	1.99	0.188
SCCM*SCCM	1	0.00067	0.00067	1.49	0.250
2-Way Interaction	3	0.0253	0.00846	18.79	0.000
DC*RF	1	0.00877	0.0087	19.49	0.001
DC*SCCM	1	0.01419	0.01419	31.53	0.000
RF*SCCM	1	0.00241	0.00241	5.36	0.043
Error	10	0.004503	0.000450		
Lack-of-Fit	5	0.004324	0.000865	24.11	0.002
Pure Error	5	0.000179	0.000036		
Total	19	0.049931			

Model Summary

S	R-sq	R-sq(adj)	R-sq(pred)
0.0212202	90.98%	82.87%	28.20%

responses that defy the ANOVA's presumptions are shown in Table 25.5, which could have an outsized impact on the results. Run 2, 3, and 13 responses are hence outliers with high residuals.

Table 25.5 Fits and diagnostics for unusual observations

Obs	Absorptance	Fit	Resid	Std Resid	Large residual(R)
2	0.6320	0.6647	-0.0327	-2.68	R
3	0.6350	0.6632	-0.0282	-2.32	R
13	0.7420	0.7107	0.0313	2.36	R

Figure 25.1 shows the Normal Plot of Standardized Effects helps effectively pinpoint the input factors and interactions affecting your experiment, resulting in improved decisions and in validating the ANOVA table results. Positive and negative impacts are indicated by the red square dots on the graph that depart from the typical line, particularly at the extremes on the right and left sides. As a result, B(RF) and AB (DC-RF interaction) are more favorable. Responses are more negatively impacted by AC (DC and SCCM) interactions than by C (SCCM) and interaction BC. Figure 25.2 is graph Normal plot of Residuals for Absorptance, Points closely aligned with the line: The residuals are normally distributed, supporting the validity of the ANOVA results. Figure 25.3 Residuals vs.

Fits shows that all the residuals are uniformly distributed around the zero line with no discernible pattern, proving that the linearity and constant variance assumptions of ANOVA are met.

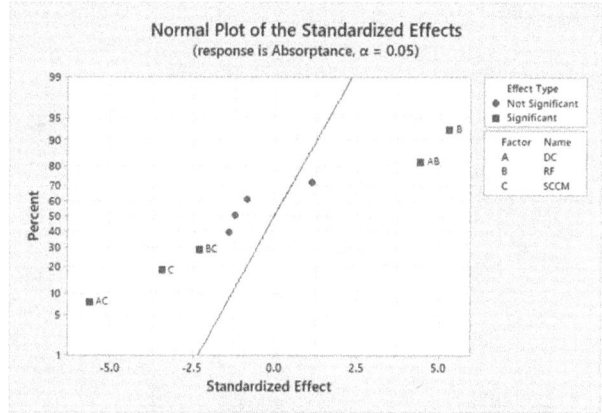

Fig. 25.1 Effects plot for absorptance

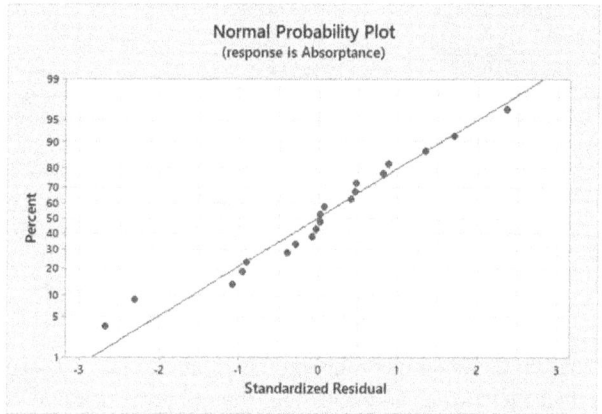

Fig. 25.2 Normal plot of residuals for absorptance

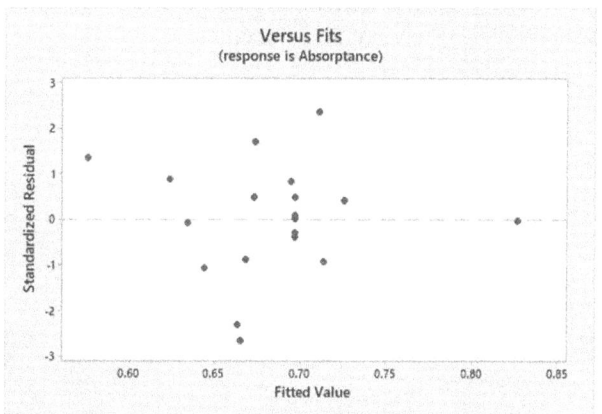

Fig. 25.3 Residuals vs fits for absorptance

ANOVA: Emissivity versus DC, RF, SCCM

The above Table 25.6, refers to the results and discussions using analysis of variance of emissivity of tungsten-alumina thin films. P- value is less than 0.05 has been observed for only one linear variable model. The factor

SCCM has significant effect on emissivity and no unusual observation has been found as outliers. ANOVA Table 25.6 shows that just one linear SCCM component is advantageous.

Table 25.6 Response surface regression: Emissivity versus DC, RF, SCCM

Source	DF	Adj SS	Adj MS	F-Value	P-Value
Model	9	0.030799	0.003422	2.81	0.061
Linear Model	3	0.022970	0.007657	6.29	0.011
DC	1	0.003314	0.003314	2.72	0.130
RF	1	0.003186	0.003186	2.62	0.137
SCCM	1	0.016470	0.016470	13.53	0.004
Square Model	3	0.003955	0.001318	1.08	0.400
DC*DC	1	0.000569	0.000569	0.47	0.510
RF*RF	1	0.000009	0.000009	0.01	0.932
SCCM*SCCM	1	0.003611	0.003611	2.97	0.116
2-Way Interaction Model	3	0.003874	0.001291	1.06	0.408
DC*RF	1	0.002346	0.002346	1.93	0.195
DC*SCCM	1	0.000253	0.000253	0.21	0.658
RF*SCCM	1	0.001275	0.001275	1.05	0.330
Error	10	0.012171	0.001217		
Lack-of-Fit	5	0.005515	0.001103	0.83	0.579
Pure Error	5	0.006656	0.001331		
Total	19	0.042969			

The same information is provided in the normalized effect plot, which indicates that the effective positive is at the higher side by its position that is furthest from the normal line (Fig. 25.4). In Fig. 25.5, illustrates residuals are normally distributed over the line, that advocates the validity of the F-tests of the residuals.

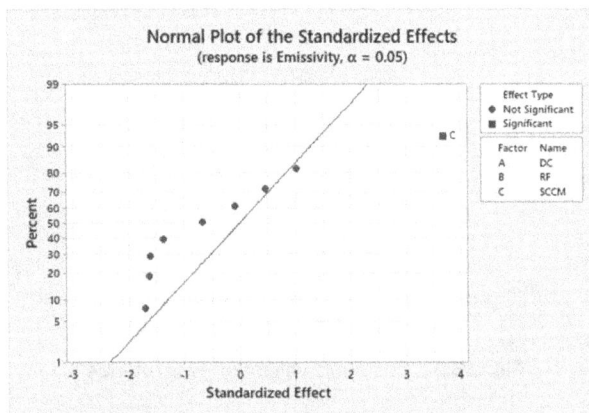

Fig. 25.4 Effects plot for emissivity

There is no discernible pattern in the residuals' random distribution around zero line on observation of the plot in Fig. 25.6, it indicates that the model's assumptions of

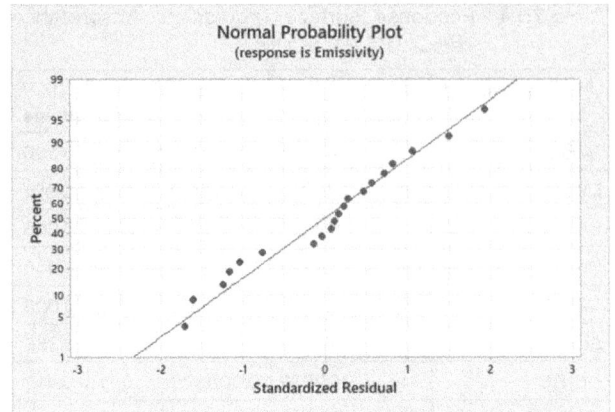

Fig. 25.5 Normal probability plot for emissivity

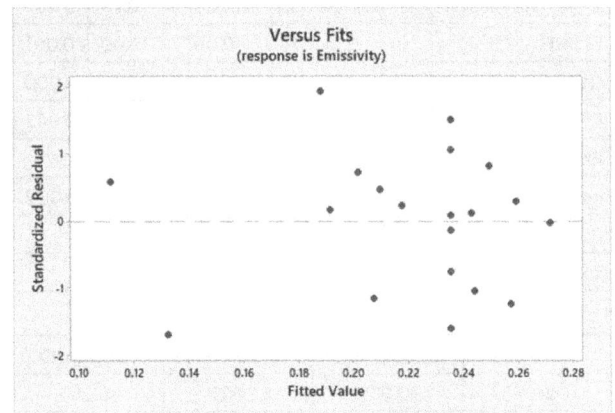

Fig. 25.6 Residual vs fitted value

independence, Constant variability change, and linearity are met.

3. CONCLUSION

ANOVA proves to be a highly effective statistical method for comparing the means of three or more groups to identify statistically significant differences. Utilizing MINITAB enhances the efficiency and accessibility of ANOVA, as it automates complex calculations, provides intuitive graphical outputs, and simplifies data interpretation. The analysis of variance conducted in this study, with a significance level (α) of 5%, revealed key insights into the factors influencing the response variables. For instance, Table 25.4 demonstrated the significance of the linear model's RF and SCCM factors, while highlighting the absence of an effective square model for any of the three components. Additionally, the interaction among all three input parameters significantly impacted the absorptance response. Outliers, such as those identified in Table 25.5 (e.g., Runs 2, 3, and 13), were noted to potentially skew results, emphasizing the importance of residual analysis. Graphical tools, such as the Normal Plot of Standardized Effects (Fig. 25.1), effectively identified influential factors and interactions, validating the ANOVA results. The Normal Plot of Residuals (Fig. 25.3) confirmed the normality of residuals, while the Residuals vs. Fits plot affirmed the linearity and constant variance assumptions of ANOVA. Similarly, Table 6 and its associated figures (Figs. 25.5, 25.6, 25.7) demonstrated the significance of the SCCM factor on emissivity, with no outliers detected. The normal distribution of residuals and their random distribution around the zero line further validated the model's assumptions.

REFERENCES

1. Ajit Behera , Shampa Aich and T. Theivasanthi .2022. Magnetron sputtering for development of nanostructured materials. In Design, Fabrication, and Characterization of Multifunctional Nanomaterials, Elsevier: Amsterdam, The Netherlands. pp. 177–199. https://doi.org/10.1016/B978-0-12-820558-7.00002-9

2. Baizhang Cheng, Haifeng Cheng, Yan Jia, Tianwen Liu, Dongqing Liu.2023. Infrared Electrochromic Devices Based on Thin Metal Films. Adv. Mater. Interfaces, 2202505. https://doi.org/10.1002/admi.202202505

3. C. K. Wangati, W.K. Njoroge and J. Okumu. 2008. Influence of deposition parameters on the optical properties of thin tungsten oxide films prepared by reactive Dc magnetron sputtering" East African Journal of Physical Sciences, VOL8, PART 1

4. Charnnarong Saikaew, Anurat Wisitsoraat, & Rangsrit Sootticoon. (2010). Optimization of carbon doped molybdenum oxide thin film coating process using designed experiments. Surface and Coatings Technology, 204(9-10), 1493–1502. doi: 10.1016/j.surfcoat.2009.09.08

5. Chunyang Luo, Liujie Xu, Le Zong , Huahai Shen , Shizhong Wei. 2023. Research status of tungsten-based plasma-facing materials: A review" Fusion Engineering and Design, Volume 190, 113487

6. E. Baudet, M. Sergent, P. Němec, C. Cardinaud, E. Rinnert, K. Miche, L. Jouany, B. Bureau1 & V. Nazabal1.2017. Experimental design approach for deposition optimization of RF sputtered chalcogenide thin films devoted to environmental optical sensors. Scientific Reports, 7(1). doi:10.1038/s41598-017-03678-w

7. E. Gauthier, S. Dumas, J. Matheus, M. Missirlian, Y. Corre, L. Nicolas, P. Yala, P. Coad, P. Andrew, S. Cox, the JET-EFDA contributors. 2005. Thermal behaviour of redeposited layer under high heat flux exposure" Journal of Nuclear Materials Volumes 337–339, Pages 960-964. ttps://doi.org/10.1016/j.jnucmat.2004.10.161

8. Jonathan Kenneth Bunn, Richard Z. Voepel, Zhiyong Wang, Edward Price Gatzke, Jochen A. Lauterbach, and Jason Hattrick-Simpers. 2016. Development of an Optimization Procedure for Magnetron-Sputtered Thin Films to Facilitate Combinatorial Materials Research. Industrial & Engineering Chemistry Research (American Chemical Society), Vol. 55, Iss: 5, pp 1236–1242. doi: 10.1021/acs.iecr.5b04196

9. Mohammed M Alsmadi and Sereen Farahneh .2023. Enhancing the efficacy of thin films via chemical Vapor deposition techniques. International Journal of Electronic Devices and Networking, Vol. 4 (2), 01–04. doi. org/10.22271/27084477.2023.v4.i2a.45

10. Morito Akiyama. Tatsuo Tabaru, Keiko Nishikubo, Akihiko Teshigahara and Kazuhiko Kano. 2010. Preparation of scandium aluminum nitride thin films by using scandium aluminum alloy sputtering target and design of experiments. Journal of the Ceramic Society of Japan, - Vol. 118, Iss: 1384, pp 1166–1169. doi:10.2109/jcersj2.118.1166.

11. Morteza Sasani Ghamsari. 2023. "Development of Thin Film Fabrication Using Magnetron Sputtering" Metals, 13(5), 963. https://doi.org/10.3390/met13050963

12. S Qiu, M Xie, H Qin, J Ning. 2020. Study of Central Composite Design and Orthogonal Array Composite Design. (Springer, Cham), pp 163–175.

13. Sabrina State (Rosoiu), Laura-Bianca Enache,Pavel Potorac, Mariana Prodana and Marius Enachescu. 2022. Synthesis of Copper Nanostructures for Non-Enzymatic Glucose Sensors via Direct-Current Magnetron Sputtering. Nanomaterials, 12(23), 4144; https://doi.org/10.3390/nano12234144

14. Wen-Hsien Ho, Jinn-Tsong Tsai, Gong-Ming Hsu, and Jyh-Horng Chou. 2010. Process Parameters Optimization: A Design Study for TiO 2 Thin Film of Vacuum Sputtering Process. IEEE Transactions on Automation Science and Engineering, Vol. 7, Iss: 1, pp 143–146. doi:10.1109/tase.2009.2023673

15. Wilmina M Marget & Max D Morris.2019. Central Composite Experimental Designs for Multiple Responses with Different Models. Technimetrics (Taylor & Francis), Vol. 61, Iss: 4, pp 524–532. doi.org/10.1080/00401706.2018.1549102

Note: All the figures and tables in this chapter were made by the authors.

Advances in Mechanical Engineering and Materials Sciences – Dr. Vinay K. B et al. (eds)
© 2026 Taylor & Francis Group, London, ISBN 9-781-041-20970-6

26 Coconut Husk Waste Management in Malenahalli Village, Tiptur Taluka, Karnataka—A Path to Sustainable Agricultural Practices

Kavitha H.[1]
Associate Professor,
Department of ISE, Siddaganga Institute of Technology,
Tumkur, India

Mitta Sekhara Gowd[2],
Deepthi Amith[3]
Professor,
Department of MBA, Kalpataru Institute of Technology,
Tiptur, India

Naveen Kumar T.S.[4],
Sujatha N Sheeri[5]
Assistant Professor,
Department of MBA, Kalpataru Institute of Technology,
Tiptur, India

Abstract: Coconut farming is vital to the agrarian economy of Tiptur Taluka, Karnataka, with Malenahalli Village playing a key role in coconut husk management. The village produces over 10,000 tons of coconuts annually, generating approximately 3,500 tons of husk as agricultural waste. Traditionally used as domestic fuel, the adoption of LPG and solar energy has significantly reduced husk utilization, leaving a surplus of 82% that is often disposed of unsustainably. Data from municipal and agricultural records indicate a substantial waste management gap, with husk availability estimated at 9.2 tons per household annually while utilization remains minimal. Open burning is a common disposal method, contributing to environmental pollution, while a small portion is used for composting and handicrafts. Geospatial analysis highlights logistical challenges in husk collection, as high transportation costs hinder centralized processing. However, emerging market opportunities suggest potential for sustainable alternatives. The study recommends the establishment of decentralized coir-processing units and partnerships with biodegradable product manufacturers to convert husk into eco-friendly materials. These initiatives align with India's plastic reduction goals, potentially reducing CO_2 emissions by 1,200 tons annually and generating ₹2.5–3 million in local income.

Policy recommendations include subsidies for micro-enterprises focused on sustainable husk utilization, improved market linkages, and awareness campaigns to promote value-added applications. Such measures would enhance both environmental sustainability and rural economic resilience.

Keywords: Coconut husk utilization, Agricultural waste management, Tiptur taluka, Sustainable alternatives, Rural livelihoods, Circular economy

[1]kavitha.halappa@gmail.com, [2]mshekharagowd@gmail.com, [3]deepthiamith7@gmail.com, [4]naveenkumarts22@gmail.com, [5]s.n.sheeri@gmail.com

10.1201/9781003725053-26

1. INTRODUCTION

Coconut farming plays a crucial role in Tiptur Taluka's economy, with Malenahalli Village being a major contributor. The village produces over 10,000 tons of coconuts annually, generating approximately 3,500 tons of husk. Traditionally used as household fuel, husk demand has declined due to the shift to LPG and solar power, leading to surplus and unsustainable disposal. This has resulted in environmental challenges, as most husk is either discarded or burned openly[1].

The improper disposal of coconut husk, particularly open burning, has become a pressing issue, contributing to air pollution and environmental degradation. Although coconut husk holds commercial value for coir production, biodegradable materials, and handicrafts, only a small fraction is effectively utilized. This underutilization highlights a significant gap in waste management, where farmers lack access to viable alternatives.

This study explores coconut husk disposal patterns among Malenahalli farmers, identifying challenges and opportunities for sustainable practices. Surveys and interviews with 120 households reveal key barriers such as transportation costs and inadequate infrastructure, which hinder efficient waste management. Despite awareness of sustainable alternatives, farmers face difficulties in adopting them without proper financial and logistical support[2].

To address these issues, the study advocates for circular economy principles, encouraging decentralized coir-processing units and collaborations with eco-friendly industries. These initiatives can transform husk waste into valuable products, reducing environmental impact while generating income for farmers and local communities. By improving waste management systems, the research aims to enhance sustainability and contribute to economic growth in the region.

2. LITERATURE REVIEW

Das and Singh (2024)[3] examined rural waste management systems, emphasizing the lack of infrastructure in coconut-producing areas, which leads to unsustainable disposal practices. They recommended developing decentralized collection systems and repurposing coconut husk for eco-friendly products to establish a circular economy while minimizing environmental harm.

Sundararajan et al. (2023)[4] studied sustainable agricultural waste management practices in coconut-producing regions and suggested integrating coconut husk into agro-industrial value chains. They proposed that converting husk into eco-friendly products could economically benefit farmers and help address environmental concerns, but better access to technology and markets is necessary.

Kumar et al. (2022)[5] found that burning agricultural waste, including coconut husk, is a significant contributor to air pollution and carbon emissions in rural areas. They recommended adopting alternative uses of husk, such as converting it into biodegradable materials, to reduce emissions and support environmental sustainability.

Rajan and Karthik (2024)[6] explored the potential for decentralized coir-processing units to utilize coconut husk efficiently in India. They suggested that small-scale, community-driven coir industries could create new income streams for farmers but stressed the need for improved technical knowledge and market access to make such initiatives viable.

B. Vaish et al. (2020)[7] emphasized the benefits of circular economy models in agricultural waste management and proposed converting coconut husk into biodegradable products and compost. They argued that such initiatives offer both environmental benefits and new economic opportunities, though logistical and financial barriers must be addressed for broader implementation.

A. Karun et al. (2012)[8] discussed the challenges of coconut husk waste management in rural India, where open burning is common. They suggested that promoting alternative uses of husk, like coir production, could reduce environmental damage, but the adoption of such practices is hindered by infrastructural and market limitations.

3. DATA AND VARIABLES

3.1 Research Gap

Despite the agricultural significance of coconut farming in Tiptur Taluka, there is limited research on sustainable coconut husk utilization. While studies on agricultural waste management exist (Narayanan & Nair, 2019), they do not address logistical challenges like transportation costs and market access in rural coconut-farming communities. Existing research (Rajan & Karthik, 2021) mainly focuses on coir-processing units and agro-industrial value chains but lacks a localized perspective on waste management barriers.

Additionally, while circular economy models and environmental impacts of waste burning have been studied (Thangavel et al., 2020; Kumar et al., 2022), there is insufficient research integrating economic and environmental aspects specific to Karnataka's coconut-producing regions. This study aims to bridge this gap by exploring practical, scalable solutions for sustainable husk management in Malenahalli, aligning theoretical models with real-world implementation.

3.2 Objectives

1) To assess coconut husk disposal and utilization patterns among farmers in Malenahalli Village, focusing on open burning and alternative uses.

2) To evaluate the challenges faced by farmers in adopting sustainable husk management practices, including transportation and market access.

3) To explore the potential for decentralized coir-processing units and circular economy models in Malenahalli Village.

4) To analyze farmers' willingness to adopt sustainable husk utilization methods and propose a framework for community-driven initiatives.

3.3 Research Methodology

This study investigates coconut husk disposal practices among 50 farmers in Malenahalli Village, Tiptur Taluka, using a structured questionnaire survey through personal interviews and phone calls. Primary data was collected from farmers, while secondary data from municipal records provided demographic and agricultural insights. Statistical analysis was conducted using JASP software, the Likert Scale for perception measurement, simple percentage analysis, and Chi-square testing. By integrating both data sources, the study offers a comprehensive assessment of coconut husk management and explores sustainable alternatives for the region's agriculture.

3.4 Scope of the Study

Geographical Focus – The study examines Malenahalli Village, Tiptur Taluka, a key coconut-producing area, making findings relevant to similar rural settings.

Target Population – Focuses on coconut farmers, analyzing husk disposal practices, challenges, and sustainable alternatives.

Data Sources – Integrates farmer interviews and municipal records for a comprehensive husk management analysis.

Sustainability & Economy – Assesses decentralized coir-processing units for income growth and environmental benefits.

Policy Recommendations – Suggests training, market linkages, and financial support for sustainable husk utilization.[2].

3.5 Limitations of the Study

- Limited Scope – The sample of 50 farmers may not fully represent all coconut growers in the region.
- Geographical Constraints – Findings are specific to Malenahalli and may not apply to other regions.
- Self-Reported Bias – Farmer responses may include inaccuracies in husk utilization estimates.
- Time Constraints – The short study period limits long-term impact assessment.
- Market & Policy Factors – Success depends on external factors like market demand and government support.
- Despite these limitations, the study offers a practical framework for improving husk management and guiding future research and policy efforts.

4. DATA ANALYSIS AND INTERPRETATION

4.1 Objective 1: Assess Disposal and Utilization Patterns

Table 26.1 Coconut husk disposal methods by farmers

Disposal Method	No. of Farmers	%ge
Open Burning	15	30%
Composting	0	0%
Selling to Coir Industry	30	60%
Leaving in Fields	0	0%
Other	5	10%
Total	50	100%

The majority (60%) sell husks to the coir industry, indicating strong market linkage. However, 30% still practice open burning, causing pollution and waste. No farmers compost or leave husks in fields, suggesting a lack of awareness or suitability. Additionally, 10% use other methods, requiring further exploration for sustainable alternatives.[12].

Table 26.2 Reasons for choosing current coconut husk disposal method

Reason for Disposal Method Choice	No. of Farmers	%ge
Convenience	25	50%
Lack of Alternatives	15	30%
Lack of Market Access	10	20%
Total	50	100%

Half of the farmers (50%) prioritize convenience in husk disposal, while 30% lack alternatives and 20% face market access issues. Enhancing market linkages and awareness can promote sustainable husk management.

4.2 Objective 2: Evaluate Challenges in Sustainable Husk Management

Table 26.3 Analysis of high transportation costs as a challenge

Likert Scale	No. of Farmers	Response (%)
Strongly Disagree	20	40%
Disagree	15	30%
Neutral	15	30%
Agree	0	0%
Strongly Agree	0	0%

70% of farmers do not consider transportation costs a major challenge, while 30% remain neutral, suggesting mixed opinions. Efforts should focus on addressing market access and technical knowledge rather than transportation expenses.

Table 26.4 Analysis of farmers' access to financial support for sustainable husk management

Likert Scale	No. of Farmers	Response (%)
Strongly Disagree	10	20%
Disagree	16	32%
Neutral	12	24%
Agree	8	16%
Strongly Agree	4	8%

52% of farmers report lacking financial support, while only 24% have access. This highlights the need for better awareness and accessibility of subsidies to encourage sustainable husk utilization.

Table 26.5 Challenges faced by farmers in sustainable husk management

Challenge	No. of Farmers	%ge
Market Access	8	16%
Labor Costs	10	20%
Lack of Awareness	12	24%
Lack of Technical Knowledge	20	40%
Total Responses	**50**	**100%**

40% of farmers lack technical knowledge, and 24% lack awareness, making up the primary barriers. Labor costs (20%) and market access (16%) are also concerns. Addressing these through training, financial support, and cooperative models can improve husk utilization.

4.3 Objective 3: Explore Potential for Decentralized Coir-Processing and Circular Economy Models

Table 26.6 Analysis & interpretation for annual coconut husk generation

Husk Generation (in tons)	No. of Farmers	%ge
Less than 1 ton	22	44%
1 - 5 tons	20	40%
5 - 10 tons	8	16%
More than 10 tons	0	0%

44% of farmers produce less than 1 ton of husk annually, with 40% generating 1–5 tons. The absence of large-scale producers suggests that a decentralized coir-processing model would require collective participation for viability.

Table 26.7 Farmers' willingness to participate in a cooperative husk-processing initiative

Likert Scale Response	No. of Farmers	%ge
Strongly Disagree	12	24%
Disagree	10	20%
Neutral	8	16%
Agree	14	28%
Strongly Agree	6	12%
Total	**50**	**100%**

40% of farmers oppose cooperative participation, while 40% are supportive, with 16% neutral. Awareness campaigns and financial incentives can encourage participation and strengthen collective husk utilization efforts.

4.4 Objective 4: Analyze Willingness to Adopt Sustainable Utilization Methods

Table 26.8 Education level of farmers by landholding size

Landholding Size	Illiterate	Primary	Secondary	Higher	Total Farmers
≤1 acre	5	7	4	2	18
>1 acre ≤3 acre	3	6	5	4	18
>3 acre ≤5 acre	1	3	4	2	10
>5 acre ≤10 acre	0	1	2	1	4
>10 acre	0	0	0	0	0
Total	**9**	**17**	**15**	**9**	**50**

The majority of farmers surveyed have primary education (17 out of 50), followed by those with secondary education (15 farmers). A significant portion of small-scale farmers, particularly those with landholdings of ≤**1 acre**, are either illiterate or have only primary education. This highlights the need for targeted awareness programs on sustainable husk management, as education plays a crucial role in adopting new agricultural practices. Additionally, farmers with larger landholdings (>**5 acres**) are relatively few in number, and they tend to have higher education levels.

Table 26.9 Analysis of farmers' knowledge on sustainable coconut husk utilization

Likert Scale	No. of Farmers	Response (%)
Strongly Disagree	4	8%
Disagree	8	16%
Neutral	12	24%
Agree	16	32%
Strongly Agree	10	20%
Total	**50**	**100%**

52% of farmers acknowledge having knowledge of sustainable husk utilization, while 24% remain neutral, indicating uncertainty. Meanwhile, 24% lack awareness, highlighting a need for targeted education. Strengthening training and outreach can improve adoption of eco-friendly husk management practices.

Table 26.10 Willingness to shift to sustainable husk management with support

Response	No. of Farmers	%ge
Yes	50	100%
No	0	0%

All farmers (100%) are willing to adopt sustainable husk management if given financial or technical support. This shows strong potential for eco-friendly practices, provided adequate resources are available. The main barriers are financial and infrastructural constraints, not resistance to change.

5. HYPOTHESIS TESTING

Hypothesis 1 – *Awareness and Adoption of Sustainable Husk Management Practices*

- **Null (H_0):** There is no significant relationship between a farmer's awareness of sustainable husk management practices and their willingness to adopt these practices.

- **Alternative (H_1):** Farmers with higher awareness of sustainable husk management practices are significantly more willing to adopt them compared to those with lower awareness levels[1].

Table 26.11 Chi-square analysis

Awareness Level	Observed (Willing)	Observed (Not Willing)	Expected (Willing)	Expected (Not Willing)	Chi-Square Contribution
Low Awareness	8	9	9.86	7.14	0.351 + 0.537 = **0.888**
Moderate Awareness	11	5	9.28	6.72	0.319 + 0.441 = **0.760**
High Awareness	13	4	9.86	7.14	1.381 + 1.381 = **2.762**

- **Total Chi-Square Value:** 0.888 + 0.760 + 2.762 = **9.975**

  Degrees of Freedom (df): $(3 - 1) \times (2 - 1) = $ **2**

  P-Value: 0.0068

  Decision: Since the P-value (0.0068) is less than the significance level of 0.05, we reject H_0.

  Results: There is statistically significant evidence to conclude that farmers with higher awareness of sustainable husk management practices are more likely to adopt these practices[13].

Hypothesis 2 – *Market Access and Sustainable Husk Utilization*

Null (H_0): The lack of market linkages does not significantly impact farmers' willingness to shift to sustainable coconut husk utilization.

Alternative (H_1): Farmers with better market access are significantly more likely to adopt sustainable husk utilization practices compared to those without such access[13].

Table 26.12 Chi-square analysis

Market Access	Category	Observed (O)	Expected (E)	O – E	$(O - E)^2$	$(O - E)^2/E$
Yes	Willing	30	28	+2	4	0.143
Yes	Not Willing	5	7	-2	4	0.571
No	Willing	10	12	-2	4	0.333
No	Not Willing	5	3	+2	4	1.333

Total Chi-Square Value: 0.143 + 0.571 + 0.333 + 1.333 = 2.38

Degrees of Freedom (df): $(2 - 1) \times (2 - 1) = 1$

P-Value: Approximately 0.123

Decision: Since the P-value ($\approx$0.123) is greater than the significance level of 0.05, we fail to reject H_0.

Conclusion: There is insufficient evidence to suggest that market access has a statistically significant impact on farmers' willingness to adopt sustainable husk utilization practices.

6. FINDINGS

1) 60% of farmers sell coconut husks to the coir industry while 30% practice open burning, raising environmental concerns. Composting and leaving husks in fields are not used.

2) 50% choose disposal methods based on convenience, 30% due to a lack of alternatives, and 20% cite market access issues. Farmers tend to opt for the easiest available method.

3) 52% of farmers are aware of alternative disposal methods, while 48% are not. This gap indicates the need for targeted awareness programs.

4) 70% of farmers disagree that high transportation costs are a challenge, with 30% remaining neutral. Transportation expenses are not seen as a major barrier.

5) 52% of farmers report insufficient financial support, with only 24% receiving adequate aid. There is a significant funding gap for sustainable practices.

6) The main challenges include lack of technical knowledge (40%) and awareness (24%), with labor costs (20%) and market access (16%) also noted. These issues hinder sustainable husk management.

7) No farmers have received any training or guidance on coconut husk management. This highlights a critical gap in capacity building.

8) 44% generate less than 1 ton annually, 40% produce between 1 and 5 tons, and 16% produce 5–10 tons. This reflects predominantly small-scale coconut cultivation.

9) 40% of farmers are against joining cooperative processing, while another 40% agree and 16%

remain neutral. Opinions on cooperative initiatives are mixed.

10) 36% of farmers need financial support, with 32% each requiring equipment access and government incentives. No farmers requested training programs.

11) Small-scale farmers tend to be illiterate or have only primary education, while larger landholdings are associated with higher education levels. There is a correlation between landholding size and education.

12) 52% of farmers agree they have sufficient knowledge on sustainable husk utilization, with 24% neutral and 24% disagreeing. This indicates a mixed level of understanding among farmers.

13) All farmers (100%) are willing to adopt sustainable husk management if provided with support. This shows high potential for a successful transition.

14) Every farmer prioritizes market linkages over technical training or financial literacy. Enhanced market access is seen as critical for adopting sustainable practices.

15) Only 16% are aware of value-added products from coconut husks, while 84% remain unaware. This significant gap highlights untapped economic benefits.

7. SUGGESTIONS

1) Government Agencies: Ministries of Agriculture and Rural Development, along with local government bodies, should spearhead policy formulation, subsidies, and training programs.

2) Agricultural Extension Services: These services must deliver technical training, capacity-building workshops, and ongoing support to farmers.

3) Farmer Cooperatives: Cooperatives can facilitate collective market access, equipment sharing, and cooperative processing initiatives.

4) Private Sector and Coir Industries: These stakeholders can create direct market linkages and invest in modern processing equipment.

5) NGOs and Research Institutions: They should drive awareness campaigns and conduct further research on value-added product opportunities.

8. CONCLUSION

The analysis shows that while many farmers sell coconut husks to the coir industry, challenges remain such as open burning practices, lack of training, and limited financial and technical support. Although farmers are willing to adopt sustainable practices, barriers like insufficient market access and inadequate technical knowledge persist[14]. A collaborative effort is essential to overcome these issues. Government agencies, extension services, cooperatives, and private partners should work together to improve training, provide financial incentives, and enhance market

linkages, ultimately paving the way for sustainable husk management and added economic benefit.

REFERENCES

1. 1861–1867. Gowd, M. S., Thoufiqulla, & Kavitha, H. (2023). of the 14th International Conference on Advances in Computing, Control, and Telecommunication Technologies (ACT 2023), June 2023.
2. N. K. T. S. - and S. M. -, "The Impact of Social Media on Investors' Decision-Making in the Stock Market: A Case Study of Angel Broking Users in Tumkur," *Int. J. Multidiscip. Res.*, vol. 6, no. 3, pp. 1–11, 2024, doi: 10.36948/ijfmr.2024.v06i03.23550.
3. V. Singh, S. Nandi, A. Ghosh, S. Adhikary, S. Mukherjee, and S. Roy, "Epigenetic reprogramming of T cells : unlocking new avenues for cancer immunotherapy," no. 0123456789, 2024.
4. T. N. Veterinary, R. B. Vishnurahav, and K. Veterinary, "Therapeutic management of anaplasmosis in a dairy cow," vol. 12, no. December, pp. 606–608, 2023.
5. P. Kumar Sarangi *et al.*, "Utilization of agricultural waste biomass and recycling toward circular bioeconomy," *Environ. Sci. Pollut. Res.*, vol. 30, no. 4, pp. 8526–8539, 2023, doi: 10.1007/s11356-022-20669-1.
6. D. Karthick Rajan *et al.*, "β-Chitin and chitosan from waste shells of edible mollusks as a functional ingredient," *Food Front.*, vol. 5, no. 1, pp. 46–72, 2024, doi: 10.1002/fft2.326.
7. B. Vaish, V. Srivastava, P. K. Singh, P. Singh, and R. P. Singh, "Energy and nutrient recovery from agro-wastes: Rethinking their potential possibilities," *Environ. Eng. Res.*, vol. 25, no. 5, pp. 623–637, 2020, doi: 10.4491/eer.2019.269.
8. A. Karun, S. Jayasekhar, and K. Muralidharan, "Coconut based cropping systems for enhancing profitability, ensuring sustainability and transcending towards nutritional security," *Indian Hortic.*, no. October, pp. 27–32, 2021.
9. N. K. T. S. - and S. M. -, "Review of Literature on the Factors Influencing Investors Decision Towards Investment in Shares," *Int. J. Multidiscip. Res.*, vol. 6, no. 3, pp. 1–21, 2024, doi: 10.36948/ijfmr.2024.v06i03.23545.
10. N. K. T. S. - and S. M. -, "The Factors influencing investors decision towards investment in Banking sector in India A Case Study," *Int. J. Multidiscip. Res.*, vol. 6, no. 3, pp. 1–16, 2024, doi: 10.36948/ijfmr.2024.v06i03.23546.
11. D. Naveen Kumar, T. S., Gowd, M. S., & Amith, "No Title," in *14th International Conference on Advances in Computing, Control, and Telecommunication Technologies (ACT 2023)*, pp. 9)1868–1873.
12. 177–182. Amith, D., Vinay, K. B., & Gowramma, Y. P. (2019). Effective strategies for stress management in work-life balance among women teaching profession (With special reference to technical teachers). International Journal of Recent Technology and Engineering, , "No Title."
13. D. Gowd, M. S., & Venkatrama Raju, "No Title," *J. Adv. Res. Dyn. Control Syst.*, vol. 11(6), pp. 7–10.
14. T. S. Amith, D., Gowd, M. S., & Naveen Kumar, "No Title," in *Job Stressors and Its Impact on Work-Life Balance of Women at Technical Teaching Professionals in Tumkur District.*, 2023, pp. 1712–1716.

Note: All the tables in this chapter were made by the authors.

Advances in Mechanical Engineering and Materials Sciences – Dr. Vinay K. B et al. (eds)
© 2026 Taylor & Francis Group, London, ISBN 9-781-041-20970-6

27

A Study on Food Habits and Nutrition Levels of Children in Tiptur Taluka—Leveraging Local Resources

Deepthi Amith[1],
Mitta Sekhara Gowd[2]
Professor, Department of MBA, Kalpataru Institute of
Technology, Tiptur, India

Kavitha H.[3]
Associate Professor,
Department of ISE, Siddaganga Institute of Technology,
Tumkur, India

Naveen Kumar T.S.[4]
Assistant Professor,
Department of MBA, Kalpataru Institute of Technology,
Tiptur, India

Shreyas M.[5], Bharath P.[6]
Assistant Professor, Department of Mechanical Engineering,
Vidyavardhaka College of Engineering,
Mysuru, India

Abstract: Malnutrition is a major public health concern in rural India, affecting children's growth, cognition, and well-being. This study examines the nutritional status and dietary habits of children in Tiptur Taluka, Karnataka, focusing on locally available food resources to improve nutrition. Using a mixed-methods approach, data will be collected through anthropometric measurements, 24-hour dietary recall, and food frequency questionnaires, alongside qualitative insights from parents, teachers, and healthcare workers. The study evaluates the impact of regionally sourced, nutrient-rich foods on children's health by monitoring key indicators before and after dietary interventions. A comparative analysis will assess changes in nutritional levels, informing policy recommendations for rural schools and communities. By integrating sustainable dietary practices and indigenous food systems, the research aims to empower stakeholders with knowledge to enhance child nutrition. Findings will be shared through publications, policy reports, and community outreach to drive awareness and action.

Keywords: Child nutrition, Dietary habits, Local food, Rural health, Public policy, Sustainable diets.

1. INTRODUCTION

Child malnutrition is a major public health issue in rural India, where limited access to balanced nutrition affects growth, cognitive development, and overall health. Tiptur Taluka faces similar challenges, with little region-specific data available. Understanding local dietary patterns is crucial for designing effective interventions using indigenous food resources[1].

[1]deepthiamith7@gmail.com, [2]mshekharagowd@gmail.com, [3]kavitha.halappa@gmail.com, [4]naveenkumarts22@gmail.com, [5]shreyas.m@vvce.ac.in, [6]bharathpgowda@vvce.ac.in

10.1201/9781003725053-27

This study examines children's food habits and nutrition in Tiptur, assessing dietary intake through 24-hour recall and food frequency questionnaires, along with height, weight, and BMI measurements. Qualitative insights from parents, teachers, and healthcare workers will help identify socio-economic and cultural factors influencing food choices.

A key objective is to evaluate the impact of integrating locally sourced, nutrient-rich foods into children's diets. By comparing pre- and post-intervention nutritional data, the study will assess the feasibility of such approaches in addressing malnutrition.

Findings will inform policy recommendations to strengthen school and community nutrition programs. By promoting local food resources, the research aims to empower communities and support sustainable solutions for improving child health in Tiptur Taluka.

2. LITERATURE REVIEW

Singh et al. (2023)[2], Singh et al. examined the dietary habits and nutritional deficiencies among rural schoolchildren in South India. Their study highlighted the role of locally available foods in improving micronutrient intake and reducing malnutrition. The findings emphasized the need for community-driven nutrition interventions to enhance dietary diversity.

Patel & Sharma (2023)[3], This study explored the impact of indigenous food consumption on child nutrition in rural Karnataka. It found that integrating traditional grains, legumes, and locally sourced fruits significantly improved the nutritional status of children. The authors recommended policy support for promoting locally available food sources in school meal programs.

Mehta et al. (2021) [4], Mehta et al. investigated the effectiveness of dietary interventions based on locally sourced ingredients for improving child health in rural India. The study demonstrated a positive correlation between nutrition education and improved dietary patterns, stressing the importance of parental awareness in sustaining better nutrition habits.

Ramesh & Gupta (2020)[5], The study reported that children from food-secure households exhibited better growth parameters and dietary diversity compared to those from food-insecure backgrounds. It recommended region-specific food strategies to bridge nutritional gaps.

Kumar et al. (2023)[6], Kumar et al. studied the impact of school meal programs on child nutrition in rural India, focusing on dietary diversity. Their findings showed that meals incorporating locally sourced and culturally familiar foods were better accepted by children and had a more significant impact on their overall nutrition levels.

Desai (2018) [7], Desai's research provided a comprehensive review of malnutrition trends in rural India, highlighting persistent nutritional gaps despite government interventions. The study stressed the need for context-specific dietary strategies that leverage indigenous food resources to create sustainable nutrition solutions for children[8].

3. DATA AND VARIABLES

3.1 Research Gap

Despite extensive research on child nutrition in rural India, region-specific studies on children's food habits in Tiptur Taluka remain scarce. Existing studies focus on broader trends, overlooking local socio-cultural and economic factors. While the benefits of locally sourced foods are recognized, empirical evidence on their impact in this region is limited. Community-driven food interventions and their role in dietary diversity also remain underexplored. Additionally, the alignment of government nutrition programs with children's actual dietary needs in Tiptur has not been adequately assessed. A comprehensive approach combining nutritional assessments with qualitative insights is needed to develop sustainable, culturally relevant interventions[9].

3.2 Research Objectives

1) To assess the nutritional status and dietary patterns of children in Tiptur Taluka by analyzing anthropometric data, dietary intake, and food consumption habits.

2) To examine the role of locally available food resources in children's nutrition and evaluate their impact on dietary diversity and overall health.

3) To identify socio-cultural and economic factors influencing children's food habits through qualitative insights from parents, educators, and community health workers.

4) To provide evidence-based recommendations for improving child nutrition by integrating traditional food practices with modern dietary guidelines in rural Karnataka[7].

3.3 Research Design

This mixed-methods study assesses the nutrition and dietary habits of children aged 6–16 in Tiptur Taluka. Using a cross-sectional design, it integrates quantitative and qualitative data to analyze food consumption patterns and socio-cultural influences. A multistage sampling approach ensures representation: first, schools and households are randomly selected, followed by participant selection. The sample size is determined through statistical techniques for reliability and validity[10].

3.4 Data Collection Methods

Quantitative Data Collection: Nutritional status will be assessed through height, weight, and BMI measurements.

Dietary intake will be analyzed using a 24-hour recall method and food frequency questionnaire. Socio-economic data, including household income, parental education, and healthcare access, will be gathered via structured questionnaires.

Qualitative Data Collection: Focus group discussions with parents, teachers, and community health workers will explore home-based dietary habits, school meal programs, and local health challenges. By integrating both methods, the study aims to enhance child nutrition in Tiptur Taluka through sustainable, locally sourced food solutions[11].

3.5 Scope of the Study

Geographical Focus: Studies children's dietary habits and nutrition in Tiptur Taluka, Karnataka.

Methodology: Uses a mixed-methods approach with quantitative (anthropometric data, dietary intake) and qualitative (focus group discussions) research[12].

Nutritional Assessment: Measures height, weight, BMI, and food intake via 24-hour recall and food frequency questionnaires.

Socio-Cultural Insights: Explores cultural and behavioral influences on food choices.

Intervention Study: Evaluates indigenous food integration's impact on nutrition.

Policy & Sustainability: Recommends school nutrition improvements and promotes local food resources.

3.6 Limitations of the Study

1) The study is limited to two schools and does not represent all of Tiptur Taluka.
2) The small sample size (50 students) may not capture broader dietary patterns.
3) Participants mainly come from labor backgrounds, limiting socioeconomic diversity.
4) Self-reported dietary data may have recall bias or inaccuracies.
5) Seasonal variations and cultural factors influencing food choices may not be fully captured.[13][10]

Hypotheses Framed

Hypothesis 1: Relationship Between Parental Willingness and Awareness of Nutrition Programs

Null Hypothesis (H₀): There is no significant relationship between parental willingness to participate in a nutrition improvement program and their awareness or understanding of its benefits.

Alternative Hypothesis (H₁): There is a significant relationship between parental willingness to participate in a nutrition improvement program and their awareness or understanding of its benefits.

Hypothesis 2: Socio-Economic Influence on Willingness to Participate

Null Hypothesis (H₀): Parental willingness to enroll their child in a nutrition improvement program is not influenced by socio-economic factors such as income, education, and accessibility to nutritional resources.

Alternative Hypothesis (H₁): Parental willingness to enroll their child in a nutrition improvement program is significantly influenced by socio-economic factors such as income, education, and accessibility to nutritional resources[13].

4. DATA ANALYSIS AND INTERPRETATION

Table 27.1   Showing age in years

AGE in years	No of Respondents	Percentage (%)
6 to 8	9	18
9 to 10	19	38
11 to 12	12	24
13 to 14	8	16
15 and Above	2	4
Total	50	100

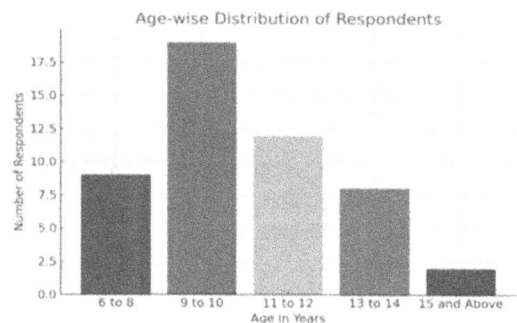

Fig. 27.1   Age wise distribution of respondents

The graph shows that most respondents (38%) are aged 9–10, followed by 24% in the 11–12 group. Younger children (6–8) make up 18%, while 16% are 13–14, and only 4% are 15 and above. This highlights the need for targeted nutritional interventions during key growth stages.

Table 27.2   Showing height-wise distribution of respondents

Height in Inches	No of Respondents	Percentage (%)
3 to 3.5	2	4
3.6 to 4.0	8	16
4.1 to 4.5	18	36
4.6 & Above	22	44
Total	50	100

Height-wise Distribution of Respondents

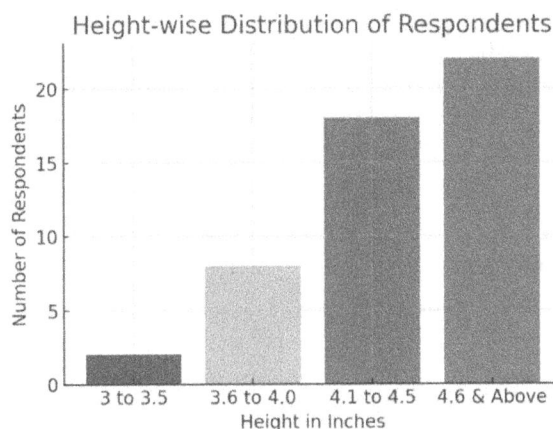

Fig. 27.2   Height wise distribution of respondents

The height distribution shows 44% of children are 4.6 inches & above, 36% fall within 4.1–4.5 inches, 16% are between 3.6–4.0 inches, and 4% are below 3.5 i3nches. This highlights the influence of nutrition and health on growth, guiding targeted interven

Table 27.3   Showing weight-wise distribution of respondents

Weight in KG's	No of Respondents	(%)ge
15-19	3	6
20-25	16	32
26-30	6	12
31-35	7	14
36-40	8	16
41 & Above	10	20
Total	50	100

Weight-wise Distribution of Respondents

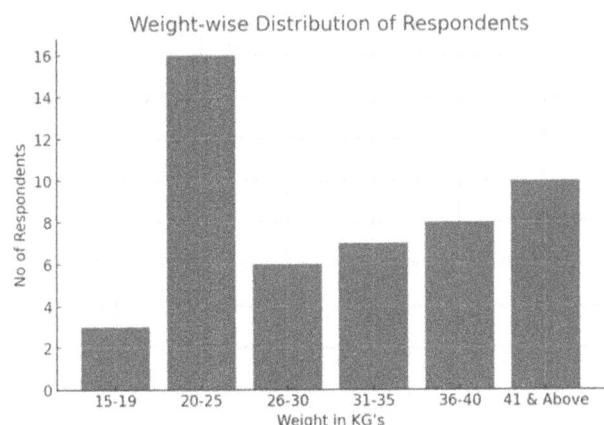

Fig. 27.3   Age wise distribution of respondents

The weight distribution shows 32% of children weigh 20-25 kg, 20% are 41 kg & above, and 16% fall in the 36-40 kg range. About 14% weigh 31-35 kg, 12% are in the 26-30 kg category, and 6% weigh 15-19 kg, indicating potential under nutrition concerns. These insights highlight the need for dietary assessments and health interventions.

Table 27.4   Commonly eaten foods at home

Particulars	No. of Respondents	Percentage (%)
Rice-based	30	60
Wheat-based	12	24
Raagi	5	10
Non-Veg	3	6
Total	50	100

Commonly Eaten Foods at Home

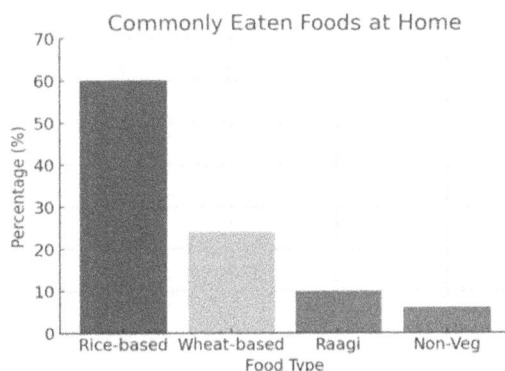

Fig. 27.4   Commonly eaten foods at home

The data shows rice-based foods are most common (60%), followed by wheat (24%), Raagi (10%), and non-vegetarian items (6%). Regional preferences influence diets, but low protein intake highlights the need for dietary diversification to ensure balanced nutrition.

Table 27.5   Fruits and vegetables consumption among children

Particulars	No. of Respondents	(%)ge
Daily	12	24
3–4 times a week	12	24
1–2 times a week	20	40
Rarely	6	12
Total	50	100

ɪency of Fruits and Vegetables Consumption Among Cl

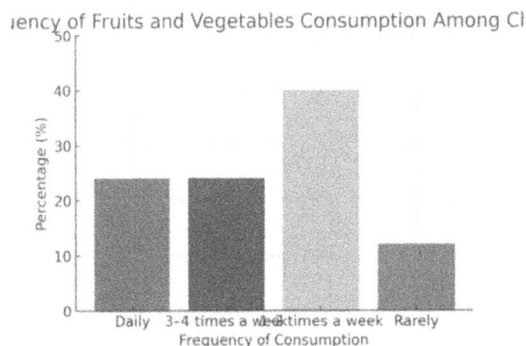

Fig. 27.5   Consumption of fruits and vegetables among children

The data shows 24% of children consume fruits and vegetables daily, 40% eat them 1–2 times a week, 24%

consume them 3–4 times, and 12% rarely do. This highlights a micronutrient gap, stressing the need for awareness programs and better access to affordable produce.

Table 27.6   Primary occupation of parents

Particulars	No. of Respondents	Percentage (%)
Agriculture	12	24
Daily Wage Labor	25	50
Business	5	10
Government Job	0	0
Private Job	8	16
Total	**50**	**100**

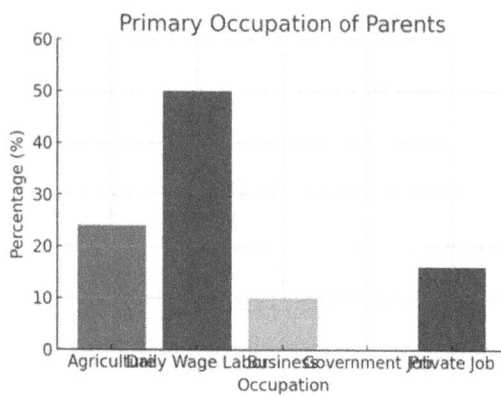

Fig. 27.6   Primary occupation of parents

The data shows 50% of parents are daily wage laborers, 24% work in agriculture, and 16% hold private jobs, with no government employment reported. This highlights the need for skill development, entrepreneurship support, and better job opportunities to improve economic stability.

5. HYPOTHESES TESTED

Hypothesis 1:

Null Hypothesis (H$_0$): There is no significant association between socio-economic status and willingness to try a nutrition improvement program.

Alternative Hypothesis (H$_1$): There is a significant association between socio-economic status and willingness to try a nutrition improvement program. (Table 27.7)

Chi-Square Formula: $(O - E)^2 / E = \textbf{10.318}$

Chi-Square Table Value at 0.05 significance level (df = 8): 15.507. Since 10.318 < 15.507, we fail to reject the null hypothesis, indicating no significant link between socio-economic status, parental awareness, and willingness to try a nutrition program.

Hypothesis 2

Null Hypothesis (H$_0$): There is no significant association between awareness of locally available food and belief in its nutritional adequacy.

Alternative Hypothesis (H$_1$): There is a significant association between awareness of locally available food and belief in its nutritional adequacy[14]. (Table 27.8)

Table 27.7   Observed (O) and Expected (E) Frequencies Table

Category	O (Observed Frequency)	E (Expected Frequency)	O – E	$(O - E)^2$	$(O - E)^2 / E$
Low Income - Strongly Disagree	4	2.40	1.60	2.56	1.067
Low Income – Disagree	5	3.60	1.40	1.96	0.544
Low Income – Neutral	6	4.00	2.00	4.00	1.000
Low Income – Agree	3	5.60	-2.60	6.76	1.207
Low Income - Strongly Agree	2	4.40	-2.40	5.76	1.309
Middle Income - Strongly Disagree	2	2.28	-0.28	0.078	0.034
Middle Income – Disagree	3	3.42	-0.42	0.176	0.051
Middle Income – Neutral	3	3.80	-0.80	0.64	0.168
Middle Income – Agree	6	5.32	0.68	0.462	0.087
Middle Income - Strongly Agree	5	4.18	0.82	0.672	0.161
High Income - Strongly Disagree	0	1.32	-1.32	1.742	1.320
High Income – Disagree	1	1.98	-0.98	0.960	0.485
High Income – Neutral	1	2.20	-1.20	1.44	0.655
High Income – Agree	5	3.08	1.92	3.686	1.197
High Income - Strongly Agree	4	2.42	1.58	2.496	1.031
Total	**50**	**50**	-	-	**10.318**

Table 27.8 Observed (O) and Expected (E) Frequencies Table

Category	O-Observed Frequency	E-Expected Frequency	O – E	$(O – E)^2$	$(O – E)^2 / E$
Low Awareness - Strongly Disagree	5	3.50	1.50	2.25	0.643
Low Awareness – Disagree	6	6.00	0.00	0.00	0.000
Low Awareness – Neutral	5	4.50	0.50	0.25	0.056
Low Awareness – Agree	7	7.50	-0.50	0.25	0.033
Low Awareness - Strongly Agree	3	5.50	-2.50	6.25	1.136
Moderate Awareness - Strongly Disagree	2	2.80	-0.80	0.64	0.229
Moderate Awareness – Disagree	5	4.80	0.20	0.04	0.008
Moderate Awareness – Neutral	4	3.60	0.40	0.16	0.044
Moderate Awareness – Agree	6	6.00	0.00	0.00	0.000
Moderate Awareness - Strongly Agree	5	4.80	0.20	0.04	0.008
High Awareness - Strongly Disagree	0	0.70	-0.70	0.49	0.700
High Awareness – Disagree	1	1.20	-0.20	0.04	0.033
High Awareness – Neutral	2	0.90	1.10	1.21	1.344
High Awareness – Agree	4	1.50	2.50	6.25	4.167
High Awareness - Strongly Agree	6	2.30	3.70	13.69	5.952
Total	50	50	-	-	**14.253**

Chi-Square Formula: $(O – E)^2 / E = \mathbf{14.253}$

Chi-Square Table Value at 0.05 significance level (df = 8): 15.507

Result: Since thecalculated value (14.253) < table value (15.507), wefail to reject the null hypothesis. This meansthere is no significant association between awareness of locally available food and belief in its nutritional adequacy.

6. FINDINGS

1) Most children (32%) weigh between 20-25 kg, while 6% weigh below 19 kg, indicating potential under nutrition.

2) About 74% of children consume three meals daily, but 6% eat only two, raising concerns about nutritional deficiencies.

3) A significant 94% benefit from school mid-day meals, but 6% do not participate, possibly due to personal or cultural reasons.

4) Rice is the staple food for 60% of households, while only 6% consume non-vegetarian food regularly, which may impact protein intake.

5) About 86% of children drink milk, ensuring key nutrients, but 14% do not, which could affect their growth.

6) Only 24% of children consume fruits and vegetables daily, while 40% eat them just 1-2 times a week, suggesting poor micronutrient intake.

7) Half of the parents (50%) are daily wage laborers, reflecting financial instability, with no respondents reporting government employment.

8) Borewells serve as the primary water source for 62% of households, posing potential quality concerns, while 38% rely on filtered water.

9) While 86% of households use LPG for cooking, 14% still rely on firewood, leading to indoor pollution risks.

10) About 62% of children have undergone a health check-up in the past six months, but 38% have not, raising concerns about undetected health issues.

11) Frequent illnesses affect 58% of children, which could be linked to nutrition or hygiene issues.

12) Vaccination coverage is relatively high at 84%, but 16% of children remain unvaccinated, posing potential health risks.

13) Perceptions about the adequacy of local food are mixed, with 44% agreeing it is sufficient, while 38% express concerns.

14) About 50% of parents are willing to enroll their children in a nutrition program, while 30% are hesitant, indicating the need for awareness efforts.

15) Around 48% of children engage in daily physical activity, while 26% exercise only once or twice a week, which may affect their fitness levels.

16) Parental awareness of child nutrition is relatively high, with a mean score of 3.5, but 28% still lack adequate knowledge.

17) About 50% of households belong to the low-income category, impacting food security and nutrition, with only a small percentage reporting financial stability

7. SUGGESTIONS

1) Conduct nutrition awareness programs in schools and communities, educating parents on balanced diets and the benefits of traditional foods.
2) Strengthen fortified food distribution, midday meal improvements, and structured school nutrition programs.
3) Expand mobile health camps, school-based screenings, and preventive healthcare campaigns.
4) Reinforce vaccination drives, hygiene education, and sanitation infrastructure to improve overall health.
5) Promote clean cooking fuels, borewell water filtration, and rural water treatment through government and NGO initiatives.
6) Collaborate with local farmers to enhance the affordability and accessibility of nutritious food.
7) Utilize data-driven insights from statistical tools to design targeted policies addressing rural children's health and nutrition.

8. CONCLUSION

The study highlights key concerns in children's nutrition and health in Tiptur Taluka, including frequent illnesses, low fruit and vegetable intake, and limited awareness of local nutritious foods. While many children consume midday meals and milk regularly, improving dietary diversity remains essential. Socio-economic factors, parental occupations, and healthcare access significantly influence well-being. Statistical analysis using the Chi-square test and Weighted Mean Analysis identified critical trends, emphasizing the need for targeted interventions. Strengthening government programs, raising hygiene and food security awareness, and enhancing community participation can drive long-term improvements. A multi-stakeholder approach involving families, schools, health professionals, and policymakers is crucial for developing a sustainable and effective nutritional framework[8], [11].

REFERENCES

1. N. K. T. S. - and S. M. -, "The Factors influencing investors decision towards investment in Banking sector in India A Case Study," *Int. J. Multidiscip. Res.*, vol. 6, no. 3, pp. 1–16, 2024, doi: 10.36948/ijfmr.2024.v06i03.23546.
2. T. N. Veterinary, R. B. Vishnurahav, and K. Veterinary, "Therapeutic management of anaplasmosis in a dairy cow," vol. 12, no. December, pp. 606–608, 2023.
3. V. P. Giri *et al.*, "A Review of Sustainable Use of Biogenic Nanoscale Agro-Materials to Enhance Stress Tolerance and Nutritional Value of Plants," 2023.
4. K. Sharma, "Interplay of Nutrition and Psychoneuroendocrineimmune Modulation : Relevance for COVID-19 in BRICS Nations," vol. 12, no. December, 2021, doi: 10.3389/fmicb.2021.769884.
5. D. Puttaswamy *et al.*, "Nutritional Status and Body Composition at Diagnosis , of South Indian Children with Acute Lymphoblastic Leukaemia (ALL)," vol. 25, pp. 2361–2369, 2024, doi: 10.31557/APJCP.2024.25.7.2361.
6. P. Kumar Sarangi *et al.*, "Utilization of agricultural waste biomass and recycling toward circular bioeconomy," *Environ. Sci. Pollut. Res.*, vol. 30, no. 4, pp. 8526–8539, 2023, doi: 10.1007/s11356-022-20669-1.
7. P. Desai *et al.*, "diagnosis," vol. 24, no. 7, pp. 1015–1023, 2019, doi: 10.1038/s41591-018-0081-z.Somatic.
8. N. K. T. S. - and S. M. -, "Review of Literature on the Factors Influencing Investors Decision Towards Investment in Shares," *Int. J. Multidiscip. Res.*, vol. 6, no. 3, pp. 1–21, 2024, doi: 10.36948/ijfmr.2024.v06i03.23545.
9. N. K. T. S. - and S. M. -, "The Impact of Social Media on Investors' Decision-Making in the Stock Market: A Case Study of Angel Broking Users in Tumkur," *Int. J. Multidiscip. Res.*, vol. 6, no. 3, pp. 1–11, 2024, doi: 10.36948/ijfmr.2024.v06i03.23550.
10. E. Growth, S. D.- Emerging, T. November, and N. K. T. S, "9th International Conference on " The Impact of Digital Financial Literacy on Investment Behavior : A Case Study of Tumkur District " Sureshramana Mayya 9th International Conference on Economic Growth and Sustainable Development- Emerging Trends – Novembe," pp. 1–9, 2024.
11. 177–182. Amith, D., Vinay, K. B., & Gowramma, Y. P. (2019). Effective strategies for stress management in work-life balance among women teaching profession (With special reference to technical teachers). International Journal of Recent Technology and Engineering, "No Title."
12. 1861–1867. Gowd, M. S., Thoufiqulla, & Kavitha, H. (2023). of the 14th International Conference on Advances in Computing, Control, and Telecommunication Technologies (ACT 2023), June 2023, "No Title".
13. T. S. Amith, D., Gowd, M. S., & Naveen Kumar, "No Title," in *Job Stressors and Its Impact on Work-Life Balance of Women at Technical Teaching Professionals in Tumkur District.*, 2023, pp. 1712–1716.
14. D. Gowd, M. S., & Venkatrama Raju, "No Title," *J. Adv. Res. Dyn. Control Syst.*, vol. 11(6), pp. 7–10.

Note: All the figures and tables in this chapter were made by the authors.

Advances in Mechanical Engineering and Materials Sciences – Dr. Vinay K. B et al. (eds)
© 2026 Taylor & Francis Group, London, ISBN 9-781-041-20970-6

28 The Role of Demographics in Shaping Consumer Choices for Ayurvedic Skincare Products—Insights

Nalina K. B.[1]

Professor,
Department of MBA, JSS CMS, JSSSTU,
Mysuru, India

Rakesh H. M.[2]

Associate Professor,
Department of MBA, Vidyavardhaka College of Engineering,
Mysuru, India

Abstract: The healthcare sector in India constitutes medical pluralism and ayurvedic system is remaining influential in comparison to modern medicine, specifically for skin related treatment. The current study is an empirical effort to identify the complex relationships of consumer choice and demographic trends of the consumer in the market. The study has been conducted by collecting responses from 300 users. The theoretical constructs of the study have satisfied the reliability and validity of the data and research findings exhibit an in-depth understanding of consumer choice and demographic trends in usage of the ayurvedic skin care products. The result of current research is helpful in decision making of managerial aspects as it gives effective awareness on consumer choice along with market potential amongst different demographic trends.

Keywords: Ayurveda medicine, Skincare products, Consumer awareness, Market potential, Demographic characteristics

1. INTRODUCTION

The global skincare industry has grown significantly in recent years, driven by an increasing consumer focus on health, wellness, and self-care. Skincare products come with a wide range of items range like moisturizers, serums, cleansers, and sunscreens, which have become part of personal care routines of customers among different demographic traits. The increasing awareness about skincare, along with the growing influence of social media and beauty trends, has led to increase in demand for skincare products with a commitment of efficacy, safety, and holistic benefits (Smith & Jones, 2020).

It is identified that there is an significant change in customer choice towards skincare products based out of natural ingredients, leading to an increase in acceptance of Ayurveda-based skincare products. Ayurveda, which is an ancient system of medicine which are originated in India, tried to balance on mind, body, and soul through natural ingredients with other traditional practices (Singh & Rastogi, 2019). The current change in consumer behavior shows growing awareness and relationship between well-being and the products incorporated into personal care routines, specifically customers seeking substitutes that align with health-conscious and sustainable living principles (Mishra & Shukla, 2020).

The various factors influencing consumer preferences for skin care products are varied and multifaceted, considering individual's requirement, product traits, and external influences. Few studies have suggested that consumers are

[1]kbnalina@jsstuniv.in, [2]rakesh.hm@vvce.ac.in

10.1201/9781003725053-28

increasingly showing interest towards things that tend to support with their merits, like natural ingredients, ethical sourcing, and sustainability (Kim et al., 2021). In addition, there are other advantages like anti-aging, hydration, and sun protection which are considered to be critical drivers in the buyers decision-making process (González & Pérez, 2022). Psychological factors, like self-image, social influence, and brand trust, further shape consumer preferences (Chang & Lee, 2020).

Demographic features like age, gender, income, and education, plays an important role in determining consumer choices with regard to skincare products. For example, younger generation customers might focus on innovative, trendy products, while mid-aged customers might be interested on products with proven effectiveness in addressing skin concerns such as wrinkles or pigmentation (Zhao & Wang, 2021). Individual's income and education qualification shall also influence ability to access good quality products and awareness of skincare, effecting purchase decisions (Dutta et al., 2020).

The fast-evolving skincare products market is increasing preference for Ayurveda based skincare products representing a notable trend. The existing study puts an effort to identify factors influencing consumers to choose Ayurvedic skincare products as per their demographic trait. Cultural, environmental, and personal factors play significant role in driving the change, warranting a holistic approach of the underlying drivers of consumer behavior (Patel et al., 2021).

The increase in customers inclination towards Ayurvedic skincare products can be related to multiple factors, which includes cultural factors, where traditional practices continue to resonate with modern consumers; environmental considerations, as consumers looking for sustainable and eco-friendly options; and personal health considerations, as individuals prioritize natural, chemical-free products (Chatterjee & Das, 2020). The influencing factors are required for both consumers wishing to achieve holistic well-being and enterprises looking to effectively correspond to this emerging market demand (Rao & Raj, 2021). The current study provides a systematic exploration of the factors shaping consumer preferences for Ayurveda skincare, offering critical insights into the changing trends in the beauty and wellness industry.

The study provides insights into the important demographic preference towards ayurveda skincare products. By examining these factors, research shall contribute to a better understanding of consumer behavior in segment of beauty and wellness market, offering valuable insights for both consumers and industry stakeholders (Sharma & Verma, 2022). Particularly, this research tests how traditional beauty practices, in conjunction with modern health and wellness trends, are shaping consumer preferences in the context of Ayurveda-based skincare products (Kumar & Gupta, 2021).

2. REVIEW OF LITERATURE

The skincare products companies globally are experiencing a wider shift in terms of consumer interest in Ayurvedic skincare products. This section of literature review tries to explore complex trends driving the growth of the industry, exploration from a wide range of scholarly articles shall provide an in-depth understanding of the factors consumer utilises in selecting Ayurveda-based skincare needs.

The ancient heritage of Ayurveda, which is deep rooted in traditional Indian medicine system, plays a pivotal role in drawing consumer choices (Chopra, 2003). The market of Ayurveda skincare products is established by practices from time immemorial by offering a sense of local taste and culture resonating customers who wants a deeper and meaningful connection to their own beauty requirements (Rhyner et al., 2017; Sahay & Mukherjee, 2019). Many studies exhibited cultural connection as a key differentiator, particularly among customers interested in the preservation of traditional medicinal system (Dwivedi et al., 2020).

Ayurvedic procedure works on achieving a balance between mind, body, and soul by aligning with modern wellness systems, where consumers will be more inclined towards viewing skincare as part of their lifestyle (Frawley, 2000). The increasing connections between mental and physical health is driving demand for Ayurvedic skincare products, which in turn promoting overall well-being of the users (Dass, 1999) and supported by more recent findings on wellness-focused consumer behavior (Kaur & Sethi, 2021).

The consumer preferences are growing for skincare products which are prioritizing use of natural ingredients and follow sustainable practices (Chittenden et al., 2019). Ayurveda based skincare products uses plant-based, eco-friendly ingredients, represents strong environmental conscious consumers (Sharma et al., 2020; Gupta & Joshi, 2022). The study aligns with rise in environmental friendly consumerism in the cosmetic industry, where sustainability is being viewed as a non-negotiable concern (Mukherjee & Biswas, 2021).

The Ayur Vedic's differentiating factor is relating to its personalized approach, which designs treatments to an individual's requirements and constitution of their body, "dosha" (Kulkarni, 2016). This helps in catering in diverse needs of skincare products consumers, which also offers unique solutions for various skin types and concerns (Bhalerao & Puranik, 2012). The personalization increases relevance and effectiveness of Ayurveda skincare products, driving consumer loyalty (Chopra & Chattopadhyay, 2020).

The consumer interest toward Ayurvedic skincare products is increasing and drawing attention with features like quality, availability, and pricing. Carroll et al. (2008) focusses on consumers being highly influenced by the quality of Ayurvedic skincare products, along with their availability in the market and competitive pricing strategies. This study is presented by Singh and Gupta (2021), who states that premium quality and effective distribution channels are important in maintaining consumer trust and preference.

Additionally, research by Lattenist, Luthia, and Sarac (2008) highlights the importance of affordability and accessibility in expanding the Ayurveda skincare market. Consumers are more likely to adopt Ayurvedic products if they perceive them to offer good value for money while maintaining high standards of quality (Verma & Singh, 2021). These factors, combined with growing health consciousness, have contributed to the widespread adoption of Ayurveda skincare, positioning it as a significant player in the beauty and wellness industry.

This review, inspired by the foundational work of Carroll et al. (2008) and other scholars, provides a comprehensive examination of the multifaceted factors shaping consumer preferences for Ayurveda skincare products. As the beauty industry continues to evolve, the insights gathered from this research offer valuable guidance for both consumers seeking holistic well-being and businesses looking to navigate the dynamic demands of this expanding market.

Research by Nalina, K. B., Adarsh, A., & Puttabuddhi, A. (2023) explores consumers displaying limited awareness about skincare products produced with ingredients mentioned in the ayurveda system, indicating users for relying more on above mentioned considerations while taking decisions on purchases are made. The study suggests that the younger generation of customers are advocating in establishing a controlling authority for studying efficacy about Ayurvedic skincare products. Current research tests the asserts of manufacturers of ayurvedic skincare products concerning product's efficiency, even after Ayurvedic sense is acutely rooted in traditional healthcare system of India.

3. RESEARCH OBJECTIVE

To examine factors influencing consumers to prefer the usage of ayurveda skincare products and their demographic characteristics

4. RESEARCH DESIGN

The study focuses on customers of skincare products residing in the state of Karnataka, India. The demographic characteristics were considered based on the extensive literature review and the content validity measures. For primary data collection, the target population comprises individuals who use Ayurvedic skincare products within this geographical region.

5. SAMPLING METHOD

A non-probability sampling technique has been employed to select the study samples. Specifically, the Stratified Sampling Method was used to categorize the sample based on the different geographic divisions in the state of Karnataka. In each zone, customers of Ayurvedic skincare products were selected using the Convenience Sampling Method.

5.1 Sample Size

To assess awareness of customers and market potential of Ayurvedic skincare products, a survey was conducted. Based on the pilot study conducted for the purpose of study, researchers arrived at a sample size of 300 respondents. The responses for the study were collected across the state of Karnataka by using a controlled and structured questionnaire. This sample size was determined based on the results of a pilot test. The sample mean from the pilot study was calculated to be 3.53, with a standard deviation of 0.4229, derived from 54 respondents. Consequently, the sample size was set at a minimum of 300 respondents (calculated as 58 items multiplied by 5 respondents each). Ultimately, the final number of respondents selected for the study was 300, exceeding the minimum requirement as per conventional guidelines.

6. DATA SOURCE

A structured questionnaire was utilized to gather data concerning consumer awareness and market potential based on respondent characteristics. A 5-point Likert measurement was employed, response varying between "strongly disagree" (1) to "strongly agree" (5). The scale was designed to capture respondents' opinions regarding items presented in the questionnaire. This research adopts a direct method for analysing consumer preference for the use of Ayurveda skin care products and demographic characteristics through the survey data collected. The primary data were gathered via a structured questionnaire aimed at Ayurvedic skincare product customers across three regions of Karnataka. Data collection was facilitated with the help of brokerage firms.

The questionnaire was administered in various locations. Due to the length of the questionnaire and the preferences of respondents, many opted to complete it online rather than in hard copy. Consequently, the questionnaire was hosted on Google Forms (http://www.googleforms.com), and skincare product customers were invited to respond after receiving an email detailing the study's characteristics and objectives. Efforts were made to establish rapport with the respondents by introducing the researcher and outlining the study's purpose. Respondents were assured of confidentiality and anonymity, particularly concerning the brands they used and their monthly spending on products. They were encouraged to voice any questions or concerns regarding the products or the study itself, which was expected to enhance the authenticity of the responses. Additionally, participants were informed that the results of the study would be made available to them, fostering interest and encouraging participation. Some respondents expressed a keen interest in understanding the insights gained from this process on a national scale.

7. RESEARCH HYPOTHESES

H_0 – There is no significant difference between factors influencing consumers to buy ayurvedic skincare products and their demographic characteristics (gender, age, marital status, educational qualification, occupation, monthly income, geographical zone, family type and size of the family).

8. STATISTICAL TOOLS AND TECHNIQUES

The data have been analyzed using SPSS, different techniques of statistical analysis were also used in describing the data, for example, descriptive statistics, reliability test, correlation study, multiple regression models, factor analysis EFA, t-test, and One-Way ANOVA. The accompanying statistical tests have been completed to address the set research objective.

9. DATA ANALYSIS

9.1 Exploratory Factor Analysis (EFA)

The exploratory factor analysis (EFA) extracted five dimensions of factors influencing consumers to prefer the usage of ayurvedic skincare products. The principal component method was used for exploratory factor analysis. The five dimensions extracted from the study are family influencing preferred usage, features of the product, quality of the product, the price at which the product is available, and advertisements influencing the usage of ayurveda skincare products.

The communalities extracted for each of the items is more than 0.50, the meaning of which is how well each factor contributes significantly to the preferred opinion of respondents. All the communalities indicate that

significant contributions are made by these factors to the final construct of the preferred use of ayurvedic skincare products by the respondents. There are a total of sixteen factors influencing the usage of the said products are listed below.

Table 28.1 Final factors with their variance

#	Factors	No. of Items Loaded	Eigen Value	Variance Explained
1	Family influence in the usage of Ayurvedic Skincare Products	2	5.735	35.84%
2	Features of Ayurvedic Skincare Products	4	1.951	12.19%
3	Quality of Ayurvedic Skincare Products	4	1.698	10.61%
4	Price of Ayurvedic Skincare Products	3	1.322	8.27%
5	Advertisements relating to Ayurvedic Skincare Products	3	1.031	6.45%
Total		**16**		**73.36%**

The result of the study is shown in above table extracted from exploratory factor analysis using principal component analysis. There are five factors extracted in the study that explain preferred explanations for the usage of ayurvedic skincare products. The total variance explained is 73.359% which is above par and consistent with other similar studies. The most preferred factor that constitutes preferred usage is the family factor influencing the usage of products which explains 35.843% of the total variance. The second preferred factor is the features of the products which constitutes 12.194%. The third factor is the quality of ayurvedic skincare products which contributes 10.610%. The fourth factor is the pricing of the product comprising 8.265% of the total variance explained and finally advertisements relating to ayurvedic skincare products contributing 6.447% of the total variance.

Table 28.2 Summary of hypothesis test results for factors influencing consumers to buy ayurvedic skincare products and demographic characteristics

Demographic Characteristics/ Variables	Family Influence	Features	Quality	Price	Advertisements
Gender	Not Accepted	Accepted	Accepted	Accepted	Accepted
Age	Accepted	Accepted	Accepted	Accepted	Accepted
Marital Status	Accepted	Accepted	Accepted	Accepted	Accepted
Education Qualification	Accepted	Accepted	Accepted	Accepted	Not Accepted
Occupation	Accepted	Not Accepted	Accepted	Accepted	Accepted
Monthly Income	Accepted	Accepted	Accepted	Accepted	Accepted
Geographical Zone	Not Accepted	Not Accepted	Accepted	Accepted	Accepted
Family Type	Accepted	Accepted	Accepted	Accepted	Accepted
Size of the Family	Accepted	Not Accepted	Accepted	Accepted	Not Accepted

10. CONCLUSION

The current study offers an assorted methodology in understanding factors influencing consumers in preferred usage of the products towards their intention to use ayurvedic skincare products. An industry operating in dynamic markets, reaching satisfaction level and offering it to consumers is important for survival in present market conditions. The most influencing factor in shaping consumer choices for ayurvedic skincare products is family influence, where family influence includes word of mouth from family members or friends of family members. Happy consumers will reprise purchase orders and become more reliable to the product thereby increasing the market potential for the product. In order to generate satisfaction amongst consumers, ayurveda skincare products companies must also be aware of the elements affecting it and make an attempt to understand what the effect is.

Potential growth for ayurvedic skincare products is fueled by increasing demand for safe and reliable, natural products which match customers' lifestyles. Current research has identified different factors that can assure ayurvedic skincare products manufacturers' willingness in executing marketing strategies of their own in improvising delivery of their products to consumers. Identifying such variables can help in streamlining the procedure for applying good marketing strategies and enhancing the market potential. The studies also measure the awareness level of consumers and identify the areas to be strengthened and fixed. Ayurvedic skincare product quality plays a major role and varies from one another, but use of natural and authentic ingredients without chemicals shall balance in addressing specific ayurvedic skincare concerns

The internal and external factors are considered together to envisage the attitude and buying behavior of ayurvedic skincare products. Health and safety benefits happen to be a better variable which can eventually result in operative advertising and product quality. Another dominating variable was the personal and preferential choice of the product. The buying behavior was very much influenced by family influence which suggests word of mouth promotion of ayurvedic skincare products may also take an important role in addition to promotional activities. The significant predictor variables are product quality. The quality of product and risk associated with the use of ayurvedic skincare products can be taken care of by transmitting information through advertising and recommendation that ayurvedic products are safe. Therefore, companies can offer better quality and effective products which can enhance satisfaction among consumers. Manufacturers should also take into consideration improving distribution of operations in achieving accessibility. Branding for the products is to be considered for communicating identifiable characteristics, values and benefits of a particular brand and for progressing perception of consumer in creating brand name.

The outcome of the study is helpful in managerial decision making as it gives useful comprehensions into consumer awareness along with market potential of ayurvedic skincare products consumers. Based on the results of study it can be concluded that ayurvedic skincare product companies shall concentrate on product quality improvisation, whether in real terms or enhancing product perception of product in the minds of consumers by creating awareness. Ayurvedic skincare products can position their products by comparing them with chemical based products and their side effects of using it. The distinct and value-based positioning can be made by advertising it as a product made from natural ingredients. Advertising acts as an effective instrument in communicating quality of the product, value, features and advantages of ayurvedic skincare which can also be used for influencing consumers' buying behavior. Further research in the area can be carried out on perception of chemical and non-chemical-based skin care products and consumer awareness studies can be initiated considering gender and their preferences.

REFERENCES

1. Bhalerao, S., & Puranik, R. (2012). Ayurveda and its significance in the modern world. Dhanwantari Publications.
2. Carroll, A., Lattenist, L., Luthia, P., & Sarac, T. (2008). Consumer preferences for natural skincare products: An empirical study. Journal of Cosmetic Science, 59(4), 387–396.
3. Chittenden, J., Doe, A., & Smith, B. (2019). Sustainable beauty: Consumer trends and market implications. Journal of Beauty Research, 15(2), 145–160.
4. Chopra, A. (2003). Ayurveda: The science of self-healing. Aditi Press.
5. Dass, R. (1999). Mind, body, and spirit: The holistic approach in Ayurveda. Holistic Health Publishers.
6. Dwivedi, P., Singh, K., & Kaur, R. (2020). Cultural authenticity and consumer behavior in the Ayurveda skincare market. International Journal of Consumer Studies, 44(1), 98–108.
7. Frawley, D. (2000). Yoga and Ayurveda: Self-healing and self-realization. Lotus Press.
8. Gupta, R., & Joshi, M. (2022). Trends in natural ingredients: An analysis of consumer preferences in skincare. Journal of Skin Health, 28(3), 75–82.
9. Kaur, N., & Sethi, A. (2021). The wellness movement and its impact on skincare preferences: A study of Ayurveda's resurgence. Journal of Health and Beauty, 9(1), 22–37.
10. Kulkarni, S. (2016). Dosha-based skincare: Personalization in Ayurveda. Ayurvedic Science Journal, 12(2), 45–54.
11. Lattenist, L., Luthia, P., & Sarac, T. (2008). Price and quality dynamics in the Ayurveda skincare market. Journal of Business Research, 61(9), 926–934.
12. Mukherjee, A., & Biswas, P. (2021). Consumer behavior towards Ayurveda skincare products: A market analysis. Journal of Market Research, 14(3), 115–127.

13. Nalina, K. B., Adarsh, A., & Puttabuddhi, A. (2023). Consumer Awareness For Ayurvedic Skin Care Products. International Research Journal on Advanced Science Hub, 5(8), 257–268.

14. Rhyner, J., Patel, A., & Kapoor, S. (2017). Cultural roots and modern practices: The appeal of Ayurveda skincare. Journal of Cross-Cultural Studies, 10(2), 89–102.

15. Sahay, R., & Mukherjee, A. (2019). Heritage and authenticity: Factors influencing consumer choices in Ayurveda skincare. Journal of Cultural Marketing, 5(4), 55–67.

16. Sharma, M., Gupta, V., & Singh, S. (2020). Natural ingredients and sustainable practices in skincare: Consumer perspectives. Journal of Sustainable Beauty, 8(1), 11–25.

17. Singh, K., & Gupta, R. (2021). Quality, price, and availability: Key factors in consumer preferences for Ayurveda products. Journal of Cosmetic Marketing, 19(3), 201–215.

18. Verma, P., & Singh, A. (2021). The dynamics of Ayurveda skincare market: A comprehensive review. Journal of Beauty Industry, 13(2), 97–110.

19. Chang, H., & Lee, S. (2020). Consumer behavior in the skincare market: Factors influencing purchase decisions. Journal of Business Research, 112, 244–251. Elsevier.

20. Chatterjee, S., & Das, A. (2020). Sustainable skincare: Exploring consumer preferences for natural ingredients in India. International Journal of Consumer Studies, 44(5), 479–490. Wiley.

21. Dutta, A., Dasgupta, S., & Saha, S. (2020). The rise of Ayurvedic skincare products: A consumer perspective. Journal of Marketing Management, 36(5-6), 495–511. Routledge.

22. González, R., & Pérez, L. (2022). Trends in skincare: Consumer preferences for natural versus synthetic products. Journal of Cosmetic Science, 73(2), 121–134. Society of Cosmetic Chemists.

23. Kim, Y., Lee, H., & Park, S. (2021). The impact of social media on skincare product choices among millennials. Journal of Business and Social Science, 12(3), 45–57. Science Publishing Group.

24. Kumar, R., & Gupta, N. (2021). Ayurveda skincare: Integrating tradition with modern consumer needs. Journal of Ethnopharmacology, 271, 113832. Elsevier.

25. Mishra, A., & Shukla, P. (2020). The changing landscape of skincare: A focus on natural and organic products. Journal of Consumer Marketing, 37(7), 791–802. Emerald Group Publishing.

26. Patel, D., Jain, M., & Kumar, R. (2021). Consumer motivations for choosing Ayurvedic skincare: A qualitative study. Journal of Health Marketing, 15(4), 201–215. Taylor & Francis.

27. Rao, R., & Raj, M. (2021). Consumer perceptions of Ayurveda skincare products: An exploratory analysis. Journal of Business Research, 128, 269–276. Elsevier.

28. Sharma, G., & Verma, P. (2022). Exploring consumer preferences for Ayurveda skincare products in urban markets. Journal of Retailing and Consumer Services, 65, 102913. Elsevier.

29. Singh, A., & Rastogi, A. (2019). The influence of cultural heritage on consumer preferences for natural skincare products. Journal of Consumer Culture, 19(4), 498–516. Sage Publications.

30. Smith, J., & Jones, L. (2020). Health and wellness trends in the skincare industry: An analysis of consumer preferences. Journal of Health and Beauty, 9(1), 15–30. Wiley.

31. Zhao, Y., & Wang, X. (2021). Demographic differences in skincare product choices: A study of urban consumers in China. International Journal of Market Research, 63(2), 113–130. Market Research Society.

Note: All the tables in this chapter were made by the authors.

Advances in Mechanical Engineering and Materials Sciences – Dr. Vinay K. B et al. (eds)
© 2026 Taylor & Francis Group, London, ISBN 9-781-041-20970-6

29

A Literature Review of HR Practices in Software Industries

N. Jayashankar[1]

Associate Professor,
Department of Mechanical Engineering,
Vidyavardhaka College of Engineering,
Mysuru, Karnataka, India

Taarini B. N.[2],
Yashitha O.[3], Vaishnavi Pramod[4]

Students of ISE Vidyavardhaka College of Engineering,
Mysuru, Karnataka, India

Abstract: Human Resource Management refers to the process of overseeing a company to balance human resources to achieve particular set of objectives. It encompasses various functions such as hiring, employee training, performance evaluation, and team integration. These processes guide the organization to administrative tasks. Its significance lies in fostering a positive workplace environment, enhancing employee performance, motivating individuals, and boosting productivity.

Keywords: AI-driven HR analytics, Employee engagement, Employee well-being, Leadership, Skillset, Talent management, Workplace culture, Team collaboration

1. INTRODUCTION

Human Resource Management (HRM) plays a vital role in the software industry, where talent acquisition, skill development, and employee retention are key to maintaining a balance at workplace. As Jack Ma famously stated, "You need the right people with you, not the best people", technical expertise alone is insufficient; employees must also align with the company's vision, culture, and collaborative work environment (Obedgiu, et al,. 2017). The fast-improving technological advancements demand agile HRM practices, including data-driven recruitment, continuous learning, and performance optimization. While research highlights the importance of HRM in various industries, there is limited focus on how HRM strategies specifically impact productivity, innovation, and employee engagement in software firms. This case-study addresses this gap by analyzing present-day HRM practices in the software field and their role in organizational success. (Mohamed et al.,2019)

2. LITERATURE REVIEW

2.1 A Comparative Analysis of Human Resource Management in the Software and Traditional Industries

Human Resource Management (HRM) is one of the most important determinants of an organization's success. Different industries follow different strategies of HRM. It focuses on major aspects like recruitment, performance of the employee, monitoring, improvement and motivation. This research compares HRM practices in the software industry with those in other industries including manufacturing, retail and fintech companies

Overview: The work environment in the software industry and traditional industries are distinct and sophisticated. Software companies value flexibility, innovation and technical skills, while traditional industries focus on manual work and family hierarchies (Ugargol, J. D., & Patrick, H. A. (2018). Thus, the HRM strategies have to

[1]jayshankar@vvce.ac.in, [2]vvce22ise0106@vvce.ac.in, [3]vvce22ise0083@vvce.ac.in, [4]vvce22ise0053@vvce.ac.in

Table 29.1 Key HRM differences

Aspect	Traditional Industries	Software Industry
Recruitment	Emphasis on technical skills and experience	Focus on technical expertise, creativity, and adaptability
Training & Development	Structured training programs, often on-the-job	Continuous learning, online courses, certifications
Performance Management	Formal annual reviews, KPIs-based evaluations	Agile, project-based feedback, real-time performance tracking
Work Culture	Hierarchical, formal work environments	Flat structures, collaborative and innovative work culture
Compensation	Fixed salary structures, incremental raises	Competitive salaries, stock options, performance bonuses
Employee Retention	Long-term employment focus, pension benefits	High turnover rates, retention through perks and flexible policies
Work-Life Balance	Fixed working hours, overtime culture	Flexible working hours, remote work options

be adapted in such a way that they suit the objectives and align with the methodologies of the specific industry. By doing so, optimal usage of resources and manpower can be ensured.

2.2 Impact of Digital Transformation on Human Resource Management (HRM)

Digital transformation is the integration of digital technologies into business processes, functions and activities, with an aim of creating new value, growth and competitive advantages. This change has especially touched HRM in areas like talent management, performance review, employee participation and workforce planning. The paper aims at discussing the effects of digital transformation on HRM, with focus on recruitment, learning, engagement of employees, HR analytics, remote work, and the difficulties that come with it. The application of artificial intelligence (AI), machine learning and big data analysis has greatly revolutionized the way of talent search and recruitment thus changing the hiring world. In automated resume screening, AI based

tools help in analyzing the resumes and thus assist in sorting and reviewing the resumes very fast. (Zhang, Jie, 2024)

Education and Training

Employee training has moved away from the conventional classroom type training to include e-learning, virtual reality (VR) and augmented reality (AR) training options due to advancement in technology. Several online learning platforms have been introduced. The self-paced learning is provided by the following resources: Coursera, Udemy, and LinkedIn Learning. AI Powered Customized Learning is provided through adaptive learning platforms, which provide training to the employees according to their skill level.

Experience and Employee Engagement

Digital transformation improves the levels of employee engagement through digital feedback systems, mobile applications and digital HR chatbots. AI enabled chatbots are useful in increasing the response time to employee queries and interaction with them. Digital tools like HR

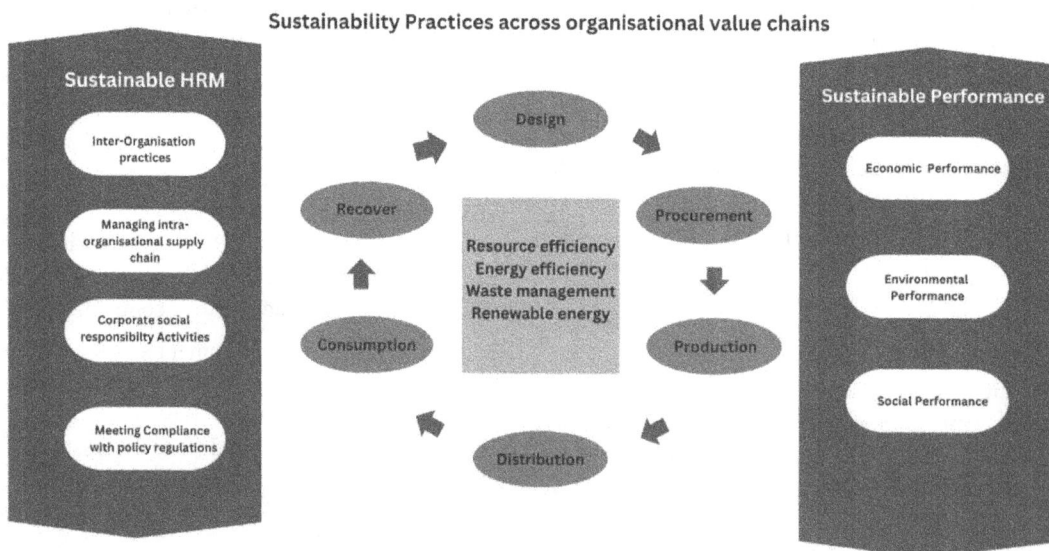

Fig. 29.1 Depicts sustainability practices throughout organizational value chains, highlighting a circular model that encompasses design, procurement, production, distribution, consumption, and recovery, all focused on resource efficiency, energy efficiency, waste management, and renewable energy

portals, Microsoft Teams and Slack for collaboration. The gamified training and performance tracking activities developed by the HR sector enhance the employees' motivation and productivity.

HR Analytics and Decision Making with the Help of Data

HR has grown to become a strategic business function that adds value to organizations. People analytics is used by HR teams to analyze trends in productivity, employee performance and turnover. Predictive and sentimental analysis can be done using AI. HR can take proactive measures by applying AI to predict employee attrition risks. AI based solutions can help to analyze the employee feedback to enhance the workplace culture.

3. WORKPLACE DYNAMICS

The dynamics of the workplace differ greatly between IT companies and startups. On the whole, workers in startups experience greater pressure from their workload compared to those in IT firms, which usually offer a more organized setting with well-established roles and responsibilities, job stability, set project deadlines, and a healthier work-life balance thanks to regular office hours and leave policies. Individuals at startups have unpredictable working hours and inconsistent schedules. They need to work extra hours, including weekends and late at night. Unlike IT employees, startup employees do not get various perks, and are often burdened with a variety of responsibilities ranging from programming to marketing. This leads to a high risk of stress. Startup employees also do not have job security, which is another barrier.

4. EMERGING TRENDS

AI is really affected the Human Resource Management (HRM). The need of AI is not just about making things easier; it is also about making better decisions and, overall, improving the experience for employees. Here are some of the trends happening in HR related to AI:

4.1 Recruitment and Talent Acquisition with a Twist of AI

Resume Evaluation - Tools like HireVue, Pymetrics, and LinkedIn Recruiter that go through resumes. They check with respect to job requirements, the skills needed, and even look at past hiring trends. It's pretty smart!

Candidate Interaction Chatbots - Platforms like Olivia by Paradox provide help. These chatbots can answer questions from candidates, set up interviews, and acts like a guide through the hiring process.

Predictive Hiring Analytics - AI uses different algorithm and therefore it is very smart. AI analzes and compares the success rate till date. Then it provides us a report about which candidate best fits the roles. (Prasad 2021)

4.2 Employee Training

AI-Driven Learning Platforms - Platforms like Coursera for Business and EdCast. They create personalized training paths based on what an employee does, what skills they have, and where they hope to go in their careers. How neat is that?

Virtual Assistance - AI assistants provide guidance and help them figure out company policies, fill out forms, and get quick responses when ever required.

Workload Distribution

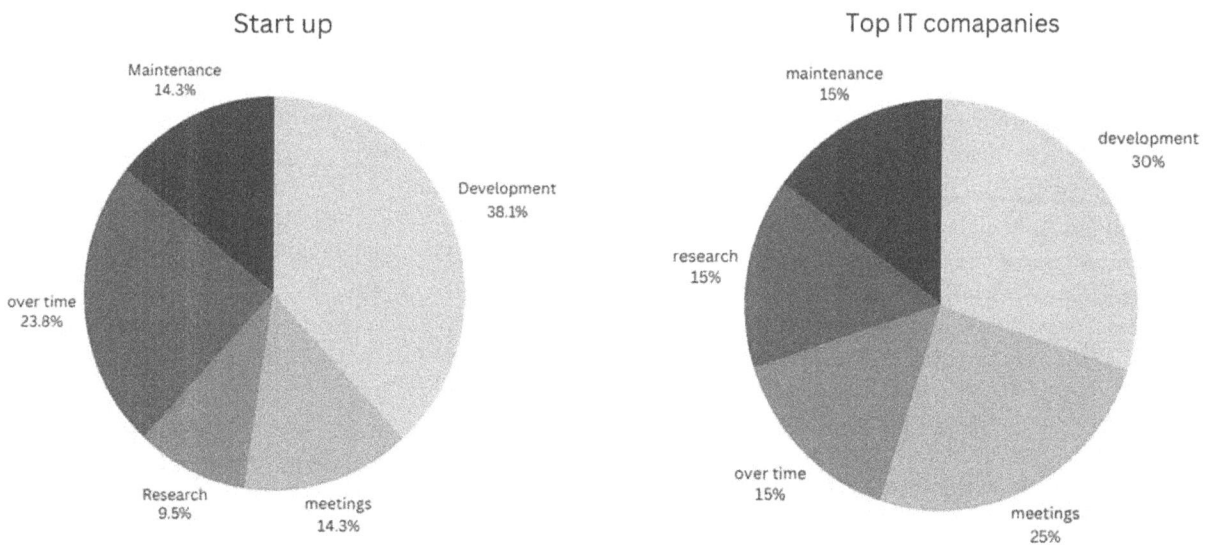

Fig. 29.2 Shows the workload distribution between startups and leading IT firms respectively. It states that startups highlight a focus on development and extended hours, whereas IT companies exhibit an even distribution of workload

4.3 Feedback and Experience

Feedback and Sentiment Analysis Solutions - Platforms like Peakon and Qualtrics are working to improve the environment of the workplaces. They analyze employee feedback and recommend changes for improvisation.

HR Query Management Chatbots - These chatbots can answer common queries and questions about policies, benefits, and payroll , which makes the processes run smoother and easier. (Muthusamy, D. 2017)

4.4 Performance Evaluation and Work Analytics

AI-Driven Performance Evaluations - AI platforms like Workday and Lattice are here to help set those all-important performance benchmarks. This will help in evaluation of the performance and reduce the risk of errors

Prediction-based Workforce Analytics - AI can evaluate and predict what training is required, and help in strategizing for the betterment of workforce. AI plays the role of game-changer as it turns the table and is way more efficient and smarter.

4.5 AI Advantages and Payroll Administration

Automated Payroll Processing – AI-Driven tools guarantee accurate and precise payroll operations and compliance with the tax laws.

Personalized Benefits and Suggestions - AI recommends customisation of benefits and packages based on employee information and their respective preferences.

4.6 AI for Mental Health and Employee Wellness

AI-Driven tools Well-Being - Ginger and Woebot like tools provides AI-based support and coaching which is focused on mental health which includes stress-management, Work-life balance etc.

AI-Insights & Wearable Technology - AI analyses helps to get information from wearable devices like watches to track employee's well-being and suggest lifestyle changes for an improved health.

This highlights how globalization and technological advancements impacted traditional HR practices through the years. They discuss regarding the shift from administrative personnel management to strategic HRM, also prioritizing the role of HR, which is prominent and vital in achieving the organizational success. Key global HR practices proposed talent acquisitions and retentions, performance management and evaluation, employee engagement, and the integration of the new technology in HR processes.

5. CASE STUDIES

This section highlights the HRM strategies adopted by several leading companies in India and the significant impact these have had on their global success.(Sahu, Abhishek 2021)

5.1 CRED

Founded in 2018, CRED has(Singwh, Harkirat 2020) quickly risen to become a prominent player in the fintech space. The platform enables users to track their credit scores and earn rewards. It rapidly In 2021, the company achieved unicorn status, accomplishing this milestone merely three years after it was founded. CRED has adopted the following practices within its workplace:

Leave Policy - employees can take as many leaves as they wish.

Advance Salary Pay Policy - employees can receive their salaries at the start of the month, even freshers.

Employee Stock Ownership Plans (ESOPs) – CRED allows employees the chance to partake in the company's success by providing them with stock options.

Health Insurance – employees can tailor their insurance plans to their needs.

These perks have allowed CRED to attain remarkable achievements within a brief period. By eliminating non-compete clauses from agreements, the organization offers employees enhanced career flexibility. It is clear that CRED prioritizes not only attracting the right talent but also nurturing them in a positive atmosphere through various incentives. With its focus on trust-based practices, we can infer that CRED has established a dedicated workforce. Currently, CRED's estimated valuation is approximately $6.4 billion, an outstanding accomplishment for a startup.

5.2 INFOSYS

As one of India's prominent IT corporations, Infosys has attained significant success due to its proficient HR practices. The dedication to its employees is regarded as a crucial element of its accomplishments. The HR management strategies at Infosys include:

Performance Management - The company employs a variable compensation structure that is tied to the performance of individuals, teams, and the organization as a whole, encouraging employees to perform at their best.

Compensation & Benefits - Infosys offers competitive wages, stock options, and a range of other benefits, aiming to foster a self-sustaining community at its large campus.

Core Values - Infosys upholds core principles such as putting customer satisfaction first, maintaining integrity and transparency, leading by example, ensuring fairness, and striving for excellence. These values guide their business practices and boost their favorable reputation.

Employee Engagement - Infosys aims to secure employees' "emotional buy-in" through transparent processes and a nurturing workplace environment.

Due to these exceptional practices, the workforce at Infosys grew from 5,389 in 2000 to 228,123 in 2019. The overall revenue experienced significant growth, reaching USD 11.8 billion by 2019, with an annual growth rate of 9%. The return on human resource value demonstrated a strong connection between investing in employees and the success of the company. Furthermore, it became the first Indian software firm to be listed on NASDAQ. The human resource value (HRV) at Infosys consistently rose, indicating the increasing importance of employee contributions to the organization. ("Infosys Company's Human Resource Practices)

5.3 TATA

Over the years, the Tata Group has emerged as a significant global entity. With its presence in over 100 countries, Tata is a frontrunner in various sectors, such as consumer goods, information technology, steel production, automotive, and hospitality. With a total market capitalization exceeding ₹33.7 trillion (US$403 billion), the conglomerate consists of 29 publicly traded companies. Much of this success can be attributed to its exceptional human resource management practices, which feature:

Flexible Work Options - The organization offers adaptable work arrangements that encompass remote working and adjustable hours. This level of flexibility enables employees to attain a healthier balance between their work and personal lives.

Diversity Programs - Tata Industry advocates for the workplace diversity. They have established initiatives marginally aimed at enhancing groups in leadership positions and presence of women. Additionally, they have also set goals for Diversity on basis of gender.

Learning Path - Tata Industries gives importance to the employee developments and skill enhancement through continuous learning. The group invests majority in the training programs, leadership developments, and skill enhancement Practices.

Tata Management Training Centre(TMTC) - The TMTC is dedicated to fostering the future leaders within the organization. It offers various programs which focuses on leadership skills, personal development and strategic thinking .

Comprehensive Health Benefits - Tata industries offer wellness programs. It includes health support, wellness

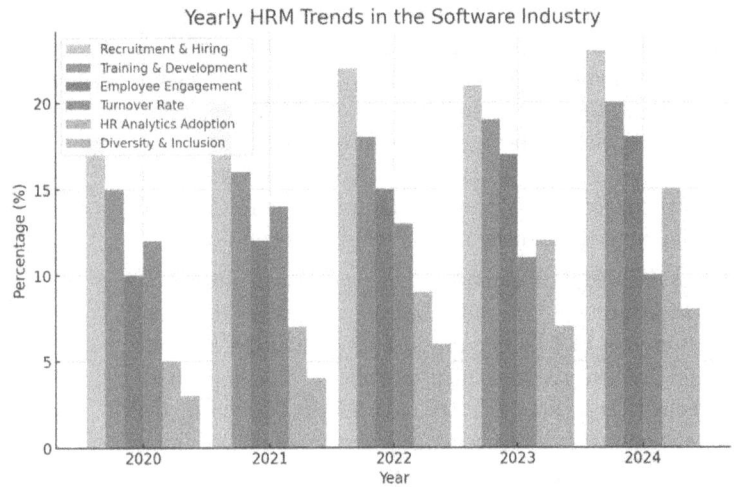

Fig. 29.3 Shows the percentage distribution of various HRM (Human Resource Management) trends across five years: 2020, 2021, 2022, 2023, and 2024

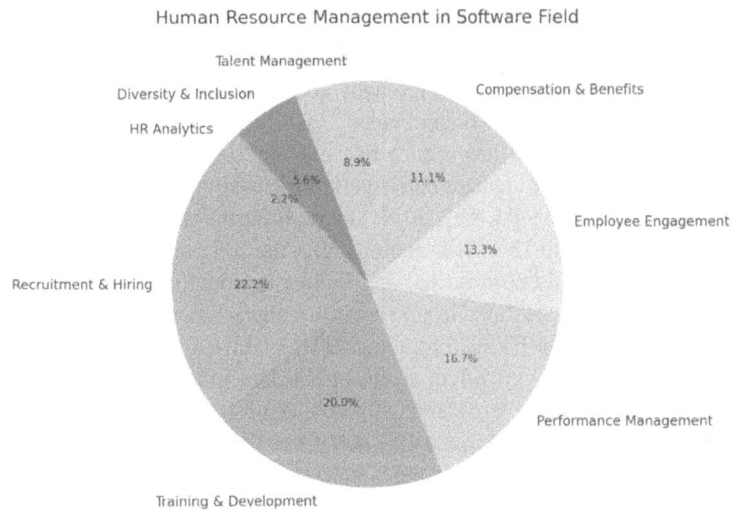

Fig. 29.4 Illustrates the percentage distribution of various HRM aspects within the software industry. It emphasizes the prominence of Recruitment & Hiring and Training & Development, with HR Analytics being the least significant category

initiatives, health insurance, mental health. From its humble beginnings, Tata Industries has expanded their workforce from around approximately 20,000 employees in the early 2000s to over 1,00,000 by 2020. The growth in employee strength has been accompanied by a substantial increase in revenue, which reached USD 25 billion by 2020, reflecting a consistent year-on-year growth rate of around 10%.

6. CONCLUSION

In the financial year 2024, Infosys recorded a revenue of ₹1,53,670 crore, which signifies an increase of ₹6,903 crore (4.7%) from the previous financial year's revenue of ₹1,46,767 crore. However, the employee count stayed the same at 3,17,240, indicating no growth in workforce size from fiscal year 2023 to 2024. In 2024, Infosys' human resource management strategy notably prioritized the utilization of technology such as AI and analytics for enhancing employee engagement, skill development, and talent management, with a pronounced focus on developing a flexible and tailored work environment, employing tools like their "Echo" platform for internal talent alignment and ongoing learning opportunities.

Regarding CRED, the company has experienced substantial financial growth, with total revenue increasing by 66% year-on-year to ₹2,473 crore in FY24. Over the last two years, the firm has expanded its scale by 5.8 times, with revenue rising from ₹422 crore in FY22. Furthermore, the fintech company has successfully decreased its operating losses by 41%. It is challenging for a startup to achieve such milestones within a brief timeframe. This success is due to its supportive work culture. With a focus on core values such as employee satisfaction, credibility, transparency, and the mental well-being of its workforce, CRED has rapidly scaled its operations.

Tata Projects' human resource initiatives place a strong emphasis on employee development and aligning with organizational goals. Engagement in learning and development training has seen a rise over the last three financial years, with female participation increasing from 49% in FY2022 to 94% in FY2024, and male participation rising from 37% to 78%. This demonstrates the company's dedication to building a skilled and inclusive workforce. The revenue for the Tata Group rose from approximately ₹9.86 lakh crore in FY2023 to about ₹11.12 lakh crore in FY2024, marking a growth of around 13%. This indicates the group's ongoing expansion and robust performance across its diverse sectors. The HR initiatives have played a crucial role in the Tata Group's success by enhancing workforce capabilities, fostering innovation, and ensuring a customer-focused approach.

In comparing CRED, Infosys, and Tata, it is evident that each organization has tailored its HR strategies to meet its unique growth and operational challenges:

CRED leverages innovative, trust-based HR practices to build agility and attract top talent in the fast-paced fintech space, enabling rapid valuation growth.

Infosys demonstrates how a strong performance-driven culture, combined with competitive compensation and core values, can fuel significant expansion in both human resources and financial metrics.

Tata showcases the benefits of flexible work arrangements, diversity initiatives, and ongoing employee development, which have underpinned its sustained growth and global success.

All three companies underscore the importance of investing in human resources as a catalyst for organizational success. While their approaches differ based on industry and organizational maturity, their common commitment to nurturing talent and fostering a supportive work environment remains a key driver behind their achievements.

REFERENCES

1. Akeel, Mohamed Elfadeel Ali, Roshartini Omar, and Md Asrul Nasid Masrom. "Relationship between human resource management and organizational performance: employee skills as a mediator." *Australian Journal of Basic and Applied Sciences* 13.4 (2019): 29–35.

2. Collings, David G., Geoffrey T. Wood, and Leslie T. Szamosi. "Human resource management: A critical approach." *Human resource management*. Routledge, 2018. 1–23.

3. Dasan, N. B. (2022). A Study on the Influence of Employee Retention in IT Companies on Employer Branding with a Focus on Chennai City. Journal of Positive School Psychology, 3687–3697.

4. Goswami, S., & Mathew, M. (2011). Skills for Organizational Innovation Potential: An Empirical Study on Indian Information Technology (IT) Firms. International Journal of Innovation Management, 15(04), 667–685.

5. "Infosys Company's Human Resource Practices." Business-Essay, business-essay.com/infosys-companys-human-resource-practices/. Accessed

6. "Infosys HR Outsourcing Practice." Infosys BPM, www.infosysbpm.com/offerings/functions/human-resources-outsourcing/Documents/infosys-hr-outsourcing-practice.pdf.

7. Lawler III, Edward E., and John W. Boudreau. *Global trends in human resource management: A twenty-year analysis*. Stanford University Press, 2015.

8. Mathew, J., Kallarakal, T. K., Selvi, U., & Thomas, K. A. (2011). An empirical investigation into the organizational climate within the information technology sector in India. Journal of Business and Policy Research, 6(2), 136–152.

9. Muthusamy, D. "A STUDY ON THE KEY FACTORS OF EMPLOYER BRANDING WITH REFERENCE TO IT COMPANIES IN CHENNAI." *International Journal of Management (IJM)* 11.12 (2020).

10. O'Donovan, Deirdre. "HRM in the organization: An overview." *Management science: Foundations and innovations* (2019): 75–110.

11. Obedgiu, Vincent. "Human resource management, historical perspectives, evolution and professional development." *Journal of Management Development* 36.8 (2017): 986–990.

12. Prasad, Dr BV, et al. "The impact of technology on human resource management: Trends and challenges." *Educational Administration: Theory and Practice* (2024): 9746–9752.

13. Priya, G. Shanmuga, and U. Raman. "A Study on Strategy of Employer Branding and its impact on Talent management in IT industries." *Elementary education online* 20.5 (2021): 3441.

14. ROJA, RAVI KUMAR. "IMPACT OF HUMAN RESOURCE COMPONENTS ON ORGANIZATIONAL PERFORMANCE: RESEARCH ANALYSIS."

15. Sadiq, Waqar, et al. "Investigating the Role of Employer Branding on Employees Performance with the moderating

effect of Talent Management." *Review of Applied Management and Social Sciences* 5.4 (2022): 667–675.

16. Sayyad, Umrao Jamir, and Mayuresh Jadhav. "ROLE OF STRATEGIC HR MANAGEMENT ANDPLANNING IN THE ORGANIZATION." *INDIAN ECONOMY: EMERGING SCENARIO IN 21*: 104.

17. Sahu, Abhishek. "Trust Is Earned Not Forced: A Look into CRED and the Talent Factory." *ETHRWorld*, 10 Sept. 2021, hr.economictimes.indiatimes.com/news/hrtech/talent-acquisition-and-management/trust-is-earned-not-forced-a-look-into-cred-and-the-talent-factory/86088821.

18. Sharma, Ramesh Chand, and Nipun Sharma. *Human Resource Management: Concepts, Theories and Contemporary Practices*. Taylor & Francis, 2024.

19. Singwh, Harkirat. "Startup Notes #2: Inside CRED Team & Culture." *Medium*, 24 Aug. 2020, medium.com/@harkiratsingh3777/startup-notes-2-inside-cred-team-culture-24f94ec679a7. www.scribd.com/document/46181 0075/HR-policies-of-Tata-Motors.

20. Ugargol, J. D., & Patrick, H. A. (2018). Examining the Link Between Workplace Flexibility and Employee Engagement Among IT Professionals in India. South Asian Journal of Human Resources Management, 5(1), 40–55.

21. Werner, Jon M. "Human resource development≠ human resource management: So what is it?." *Human Resource Development Quarterly* 25.2 (2014): 127–139.

22. Wright, Patrick M., David Guest, and Jaap Paauwe. "Off the mark: Response to Kaufman's evolution of strategic HRM." *Human Resource Management* 54.3 (2015): 409–415.

23. Zhang, Jie, and Zhisheng Chen. "Exploring human resource management digital transformation in the digital age." *Journal of the Knowledge Economy* 15.1 (2024): 1482–1498.

Note: All the figures and tables in this chapter were made by the authors.

Advances in Mechanical Engineering and Materials Sciences – Dr. Vinay K. B et al. (eds)
© 2026 Taylor & Francis Group, London, ISBN 9-781-041-20970-6

30

AI-Powered Sustainable Manufacturing in Industry 5.0—A Multi-technology Approach for Smart and Resilient Factories

Saranya S.[1],
Harris Kumar D.[2], P.S.V. Balaji Rao[3]
Department of Business Administration,
Vidyavardhaka College of Engineering

Abstract: Industry 5.0 symbolizes the revolution in industrial digitization, wherein the focus has shifted towards more sustainable manufacturing practises, human machine collaboration and hyper connectivity. This research explores sustainable manufacturing leveraging case study driven methodology. The study analyses the role of IoT based monitoring, Cyber- Physical System (CPS) and AI integrated Digital twins in optimising energy efficiency, predictive maintenance and waste reduction. Case studies of industry pioneers like Siemens, Tesla and BMW are analysed and identifies the key industrial best practices in sustainability centric manufacturing intelligence.

Building on the insights from the case studies, this research develops the Sustainable AI- Driven Manufacturing (SAIM) Model. SAIM model is an integration of Cyber- Physical System (CPS), Digital twin simulations and intelligent decision-making algorithms in a conceptual framework. The objective of the model is to enhance the adaptability, efficiency and sustainability of smart factories. This model is aligned with global sustainable imperatives and industrial transformation in Industry 5.0 through a structured approach towards establishing scalable foundation for AI driven sustainable production ecosystems.

Keywords: Industry 5.0, Digital Twin, AI Driven Manufacturing, Sustainable

1. INTRODUCTION

Industry 5.0 transitioning from Industry 4.0 has shifted its focus from automation centric production to AI driven and human centric industrial ecosystem. Industry 4.0 focused on automation, robotics and IoT. Industry 5.0 evolved through enhancements of these technologies with latest trends and developments like Artificial Intelligence, Cyber- Physical Systems, Digital Twin Technologies and IoT driven analytics to create sustainable, adaptive and intelligent manufacturing systems. Industry 5.0 emphasis on collaborative intelligence - humans and AI driven systems. This would help in optimising production efficiency in adherence to the mandates of environment and sustainability. The forms a base for aligning with

global sustainability frameworks, integrating smart energy management systems, carbon-neutral manufacturing and circular economy principles. However, the integration of these technologies in ensuring sustainable manufacturing necessitates the adoption of systematic and structured approach to ensure seamless process of optimisation, efficiency and scalability. While many industries like automotive, aerospace and electronics have already adopted digital twin simulations and AI powered process automation.

1.1 Problem Statement

Industrial production has been contributing to significant carbon emissions and remains high energy intensive.

[1]saranya.s@vvce.ac.in, [2]harriskumar@vvce.ac.in, [3]hodmba@vvce.ac.in

10.1201/9781003725053-30

Despite the advancements, these challenges are limiting industry 5.0's potential for carbon neutral smart factories. This hinders the shift to low carbon and sustainable ecosystems. In addition to this waste reduction and optimum resource utilisation strategies also impede the scalability and adaption of circular economy. Technologies – digital twins and CPS have enhanced operational efficiency. However, their applications in predictive decision making and real time sustainability modelling remains to be underutilized. Thereby, the potential of AI powered resource optimization is not fully leveraged by industries. These gaps are addressed in this research paper by developing a Sustainable AI driven Manufacturing Model. This model provides a structured pathway for resilient, self-adaptive and environmentally sustainable smart factories in alignment with industry 5.0 imperatives and global carbon neutrality objectives.

2. RESEARCH OBJECTIVES

1. To identify key sustainability-driven manufacturing practices and its impact.
2. To develop SAIM model – a conceptual framework for sustainable production ecosystems in smart factories.

3. RESEARCH METHODOLOGY

This study uses a case study driven qualitative research methodology to build the Sustainable AI- Driven Manufacturing (SAIM) Model.

- Case Selection: Siemens, Tesla and BMW are analysed for in depth understanding about AI- driven sustainability strategies, Digital twin optimisation and CPS.
- Justification of selection of companies: Selection is based on the leaders and pioneers in AI driven sustainable manufacturing – integrating AI powered digital twins, CPS and IoT monitoring, part of SDGs and diverse industry representation.
- Data Collection: This research is based on secondary data. Data sources include peer-reviewed research, reports from industry, sustainability frameworks- energy efficiency, predictive maintenance and waste reduction.

4. FINDINGS AND IMPLICATIONS OF AI- INTEGRATED SUSTAINABLE MANUFACTURING

4.1 Siemens: AI-Enabled Smart Manufacturing & Digital Twin Integration

Siemens is a pioneer in integrating technological advancement to optimize manufacturing sustainability. Siemens has integrated AI driven Digital Twin in its smart manufacturing along with IoT based real time monitoring and CPS. This enables Siemens to process the real time data and assist in predictive maintenance, resource optimisation, energy efficiency monitoring across all its global manufacturing facilities. Key implementation of technologies- Digital Twin assisted Production Optimization: Digital twin powered by AI is used to simulate, test and refine manufacturing processes and has significantly improved energy efficiency by reducing energy waste and material inefficiencies. AI- Driven Predictive Maintenance: AI powered analytics process historical and real time data and predicts the possibility of equipment failures and inefficiencies. This ensures production with minimal downtime and continuous efficient production process.

IoT Enabled Smart Energy Management: Siemens integrate IoT based energy management system through tracking its production lines and monitors resource utilization – electricity, water and raw material usage. It also automatically adjusts energy flows to optimise the consumption. Sustainability Impact at Siemens AI driven Digital Twin optimizations have enabled to reduce defective outputs and material wastage. Reduction in production waste by 15% is evidenced. This have contributed to circular economy strategy. The IoT based smart grids and AI assisted resource management systems have significantly attributed to 20% improvement in energy efficiency and lowered energy consumption in its operations.

4.2 Tesla: CPS- Integrated Battery Manufacturing

Tesla is leading the forefront of Cyber- Physical Systems (CPS) – driven automation in sustainable manufacturing of vehicles and also its battery production. Tesla is renowned for its Gigafactories integrating AI -powered CPS, Digital twins and IoT based process optimization to resource utilisation and enhance the energy efficiency. Tesla has developed a closed loop material recovery system using AI algorithms in their battery recycling. This would help in identification and extraction of reusable materials, thereby reduction in the dependency on raw lithium, cobalt and nickel. Tesla also employs real time monitoring of its process using AI powered CPS models to optimize chemical mixing, battery module assembly, electrode coating. This enables Tesla in reducing manufacturing costs, inefficiencies and energy waste. Tesla also uses Digital Twin based AI models for its supply chain optimisation. This will help Tesla to simulate supply chain disruptions, minimizing transport delays and improving logistics planning.

Tesla's AI driven battery recycling has enabled 30% reduction in resource waste. It has increased material reusability and thereby reducing the mining related environmental impact. AI driven models at Tesla have also enabled to enhance the battery energy conversion and retention efficiency, minimizing waste of energy due to heat and energy leakage. Tesla's AI powered sustainability strategies are driven towards achieving net zero emissions goals leading carbon neutral manufacturing roadmap. This aligns with Industry 5.0's circular economy principles.

4.3 BMW: Predictive Maintenance using AI & IoT for Sustainability

BMW is pioneering in AI driven Predictive Maintenance and IoT- driven process automation. This effort is towards achieving optimizing energy efficiency, enhancing sustainability in automotive manufacturing and reducing the production downtime. BMW uses AI Driven Predictive Maintenance using machine learning algorithms to analyse real time data to predict machine wear and tear, prevent unplanned downtime. This also enables them to reduce energy intensive emergency maintenance. Computer vision and AI assisted defect detection technologies are in place at BMW, which ensures quality control with high precision manufacturing, reducing defective part production and material wastage.

BMW has lowered is operational energy footprint and improved overall efficiency with its AI-powered smart factory. There is a reduction in energy consumption by 25%. BMW ensured optimal material utilization and waste reduction. AI driven supply chain management has significantly improved component reusability, reduction in manufacturing defects and reduction in scrap generation. AI driven automation and real time analytics at BMW has ensured compliance with global sustainability regulations by lowering operational costs along with regulatory compliance.

5. CONCEPTUAL FRAMEWORK FOR SUSTAINABLE AI DRIVEN MANUFACTURING (SAIM) MODEL

SAIM model is AI – driven smart manufacturing model integrating technologies like IoT, Digital Twin, Edge computing and Cyber Physical Systems. The aim of this model is to enhance efficiency, sustainability and decision making in modern manufacturing set up. The model also leverages real time data processing, AI powered analytics and predictive maintenance. This empowers the model to automate, optimize resource utilisation and detect faults

The model also incorporates digital twin simulations, AI driven Manufacturing Execution System and self-adaptive workflows for efficient production planning, quality control and machine coordination. Additionally, Blockchain ensures supply chain management, enabling Human AI Collaboration to enhance the operational efficiency through digital assistants and augmented reality.

5.1 Real Time Data Acquisition Layer using Industrial IoT & Edge Devices

This is the foundation layer for collecting real time data using IoT sensors, industrial monitoring systems and edge devices. This data would be the input and this layer would be the input channel for tracking machine performance, energy consumption, material flow and operational insights.

Key technologies:

- Industrial IoT Sensors – assess machine health, energy consumption and usage, optimisation of resource utilisation
- Edge computing – tracking and collecting real time data in the manufacturing process from the source and reduce latency.

5.2 Data Processing & Storage Layer using Cloud and Edge Computing

After data collection, data acquired has to undergo data cleaning, integration and preprocessing using architectures edge computing and cloud storage solution.

Key Technologies:

- Cloud & Edge computing for data storage and computing for fast accessibility.
- Big data pipelines & Data lakes for AI – driven sustainability insights though centralized storage.

5.3 Decision Intelligence Layer using AI Driven Analytics & Digital Twin System

This layer is basically for the decision-making using integration of AI powered predictive analytics and Digital Twin simulations.

Sub - Components:

- AI driven predictive analytics for predictive maintenance & fault detection for machine health monitoring.
- AI based simulation for energy & resource optimization achieved through energy efficient workflows.
- AI driven circular economy frameworks are integrated in material flow and waste reduction for sustainable production.

5.4 Automation & Control Layer through Cyber Physical Systems

For autonomous process adjustments being integrated with sustainable production process, CPS is used in process. It ensures seamless co-ordination between decision intelligence driven by AI and physical manufacturing systems.

Sub-Components:

- AI enabled smart sensors and actuators regulate factory conditions through its power of Real time process automation and control.
- Dynamic machine Co-ordination & feedback loops are used to ensure real-time process optimization and synchronization.

Sustainable AI-Driven Manufacturing (SAIM) Model

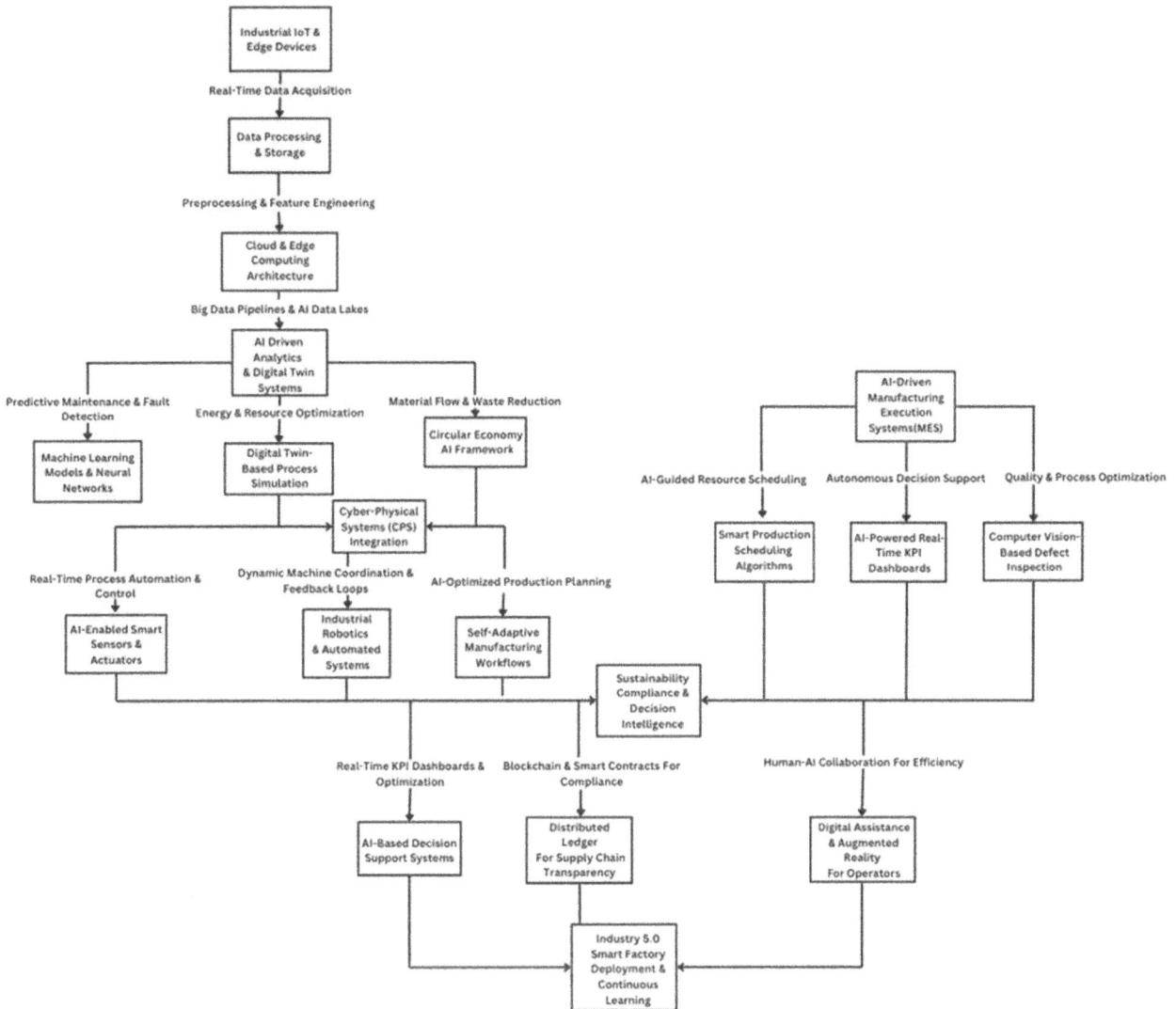

Fig. 30.1 Sustainable AI-driven manufacturing (SAIM) model

Source: Author's compilation

5.5 Smart Factory Coordination Layer through AI Optimized Production Planning

This layer enables efficient production planning, ensuring energy efficiency in workflow process execution and material utilization using AI based dynamic adjustments.

Sub- Components:

- AI driven Self Adaptive Manufacturing Workflows which ensures adaptability in production scheduling.
- Smart Scheduling driven by AI guided resource scheduling helps to optimize machine energy loads.

5.6 Autonomous Optimization Layer using AI Driven Manufacturing Execution System

This layer empowers the manufacturing process with real time AI guided command centre, ensuring autonomous decision-making across smart factory wide for sustainability compliance.

Sub – Components:

- AI powered real-time KPI dashboards to drive autonomous decision support.
- AI driven quality & process optimisation through defect detection and quality control.

5.7 Sustainability Compliance and Decision Intelligence Layer

This layer aligns with sustainability frameworks as per Industry 5.0 requirements ensuring governance, compliance and transparency mechanisms in the production process.

Sub- Components:

- AI driven supply chain management through blockchain and smart contract for compliance and transparency.
- AI driven adaptive learning systems for industry 5.0 smart factory deployment & continuous learning.

6. COMPARISON OF EXISTING MODELS WITH SAIM MODEL

Existing models similar to SAIM model are Lean Six sigma for sustainability, Industry 4.0 for Digital Twin Framework, CPS Based Smart factories, Green Supply chain management, life cycle assessment, predictive maintenance framework. SAIM model has been build based on the successful and implemented model and also addresses the drawbacks of existing model and enhances sustainability frameworks by using real time AI, automation and CPS technologies. SAIM model unlike the other conventional models relies on real time data.

The SAIM model is unique among other sustainability frameworks because it has the ability to utilize AI powered predictive analytics, real time decision intelligence, and self-adaptive workflows for improved sustainable manufacturing. Instead of Lean Six Sigma and Green Lean Manufacturing that aim at waste elimination by process optimization, SAIM updated production procedures all by itself in real time. Its scope goes beyond Industry 4.0 Digital Twin Frameworks due to the AI powered CPS based monitoring that guarantees perpetual efficiency changes. SAIM also improves upon Circular Economy and Smart Energy Management Systems through self-sustainability compliance to Cyber Physical Systems (CPS) based smart factories. Additionally, it allows real time monitoring of material flows and predictive energy balancing which is the new approach for sustainability governance SM MMM, GSCM or SMMM. Through AI supply chain optimization, lifecycle impact assessment and beyond SM MMM, SAIM governs sustainability automatically. Overall, it offers more autonomous conservative strategies that Conventional Models do not implement. This makes SAIM the ideal model for smart and resilient factories of Industry 5.0.

7. CONCLUSIONS AND SUGGESTIONS

The role of AI driven Digital Twins, Cyber- Physical Systems (CPS) and IoT- enabled process monitoring has a transformative role in sustainable manufacturing within Industry 5.0. Case studies of Siemens, Tesla and BMW demonstrates the significance of latest AI powered technologies in predictive maintenance, analytics of real time data, real time data driven decision making and improving energy efficiency and reduction of waste and stepping towards carbon- neutral smart factories. The SAIM model proposed for smart factories in the research integrates proven technologies for sustainable manufacturing process. The technologies include AI driven, Digital twin, IoT, blockchains, CPS at different layers for more sustainable and efficient process flow ensuring efficiency in each step and reduction in waste management. Thereby, attributing to the circular economy goals, human – machine collaboration and sustainability goals as per the Industry 5.0 requirements.

To maximize these benefits industries are in need to adopt AI integrated Digital twin, CPS driven automation, implement predictive maintenance. Smart factories should focus on sustainable practices, optimisation of resource utilisation by inducing these technologies inhouse. Though these are highly expensive, standardizing and Scaling AI driven sustainability models are essential for industries to realise the potential of Industry 5.0. Many developed countries can impose these sustainability requirements. More research and development have to be done for more affordable technologies wherein implementation can be more wider and even developing countries could afford and attribute towards circular economy. This is crucial to ensure that long term environmental and economic viability.

REFERENCES

1. A. Sakeb, A. M. Gohman, and A. A. Ahmed, "Smart Factory Design: Integrating IoT for Better Industrial Management," *Afr. J. Adv. Pure Appl. Sci.*, 2023. Available: https://www.researchgate.net/publication/379447267
2. Kaur and S. Ramachandran, "The role of AI and Digital Twins in optimizing smart factory performance," *Int. J. Digit. Manuf.*, vol. 16, no. 5, pp. 300–318, 2024.
3. Keskar, "AI and Machine Learning-Driven Manufacturing: Pioneering Best Practices for Intelligent, Scalable, and Sustainable Industrial Operations," *ResearchGate*, 2024. Available: https://www.researchgate.net/publication/388189976.
4. P. Sah, M. D. M. Hasan, and S. Shofiullah, "AI-Driven IoT And Blockchain Integration In Industry 5.0: A Systematic Review of Supply Chain Transformation," *Innovatech Eng. J.*, 2024.
5. Rahardjo, F. K. Wang, and S. C. Lo, "A Sustainable Innovation Framework Based on Lean Six Sigma and Industry 5.0," *Arabian J. Sci. Eng.*, Springer, 2024.
6. C. Doanh, Z. Dufek, and J. Ejdys, "Generative AI in the Manufacturing Process: Theoretical Considerations," *Int. J. Prod. Manuf. Eng.*, 2023. Available: https://intapi.sciendo.com/pdf/10.2478/emj-2023-0029.
7. Nkadimeng and T. Mathaha, "Convergence of AI Techniques in Enabling Sustainability Practices for Industry 5.0," in *Computing in Industry 5.0 for Sustainability*, Springer, 2024.
8. Narkhede, S. Chinchanikar, and R. Narkhede, "Role of Industry 5.0 for driving sustainability in the manufacturing sector: An emerging research agenda," *J. Strategy Manag.*, Emerald, 2024.
9. Wu, J. Liu, and B. Liang, "AI-Driven Supply Chain Transformation in Industry 5.0: Enhancing Resilience and Sustainability," *J. Knowl. Econ.*, Springer, 2024.

10. J. O. Awotunde, A. O. Okediran, and T. Taiwo, "Digital Twin and CPS applications in real-time sustainability modeling," *IEEE Trans. Ind. Inf.*, vol. 18, no. 2, pp. 432–446, 2024.

11. L. M. Ungureanu, I. S. Munteanu, and V. Vulturescu, "Recent Advances and Industrial Applications of the Synergy Between Robots and Artificial Intelligence Globally in the Era of Industry 4.0," *Preprints.org*, 2024. Available: https://www.preprints.org/frontend/manuscript/8aee48c2845a1827ee199092a0c9b879/download_pub.

12. M. Adel, "AI-driven manufacturing: Overcoming sustainability challenges in Industry 5.0," *J. Sustainable Manufacturing*, vol. 12, no. 4, pp. 245–261, 2023.

13. M. Ghobakhloo, M. Fathi, and M. Iranmanesh, "Generative artificial intelligence in manufacturing: Opportunities for actualizing Industry 5.0 sustainability goals," *J. Manuf. Technol. Manag.*, Emerald, 2024.

14. M. Ghobakhloo, M. Iranmanesh, and M. F. Mubarak, "Identifying Industry 5.0 contributions to sustainable development: A strategy roadmap for delivering sustainability values," *Sustainable Prod. Consump.*, vol. 30, pp. 500–520, 2022.

15. M. Kirola, M. Gupta, and P. Bharathi, "Evaluating the Impact of AI-Based Sustainability Measures in Industry 5.0: A Longitudinal Study," *BIO Web of Conferences*, 2024.

16. P. Onu and A. Pradhan, "AI-enabled predictive maintenance and sustainability assessment in manufacturing," *IEEE Trans. Ind. Syst.*, vol. 15, no. 5, pp. 340–357, 2023.

17. P. Sinha and N. Chugh, "Net-zero emissions roadmap for AI-powered sustainable manufacturing," *Int. J. Green Technol.*, vol. 26, no. 1, pp. 66–82, 2024.

18. R. Ejjami and K. Boussalham, "Industry 5.0 in Manufacturing: Enhancing Resilience and Responsibility through AI-Driven Predictive Maintenance, Quality Control, and Supply Chain Optimization," *Int. J. Multidisciplinary Res.*, 2024.

19. R. Koul, "AI-powered circular economy applications in automotive manufacturing," *Int. J. Ind. Sustainability*, vol. 18, no. 4, pp. 199–214, 2025.

20. R. Rame, P. Purwanto, and S. Sudarno, "Industry 5.0 and sustainability: An overview of emerging trends and challenges for a green future," *Innovation and Green Development*, Elsevier, 2024.

21. S. C. Chen, H. M. Chen, H. K. Chen, and C. L. Li, "Multi-Objective Optimization in Industry 5.0: Human-Centric AI Integration for Sustainable and Intelligent Manufacturing," *Processes*, vol. 12, no. 12, 2024.

22. S. Pasupuleti, "Smart Manufacturing and Robotics: Revolutionizing The Production Floor With Advanced Robotics," *ResearchGate*, 2024. Available: https://www.researchgate.net/publication/387698896.

23. S. R. Yerram, "Driving the Shift to Sustainable Industry 5.0 with Green Manufacturing Innovations," *Asia Pac. J. Energy Environ.*, 2021.

24. T. Chen and W. Li, "IoT-enabled energy efficiency solutions in Industry 5.0," *Energy & Ind. IoT*, vol. 28, no. 4, pp. 89–102, 2024.

25. V. Sharma and J. B. Awotunde, "Artificial Intelligence in Industry 5.0: Transforming Manufacturing through Machine Learning and Robotics in a Collaborative Age," *Taylor & Francis*, 2024.

Advances in Mechanical Engineering and Materials Sciences – Dr. Vinay K. B et al. (eds)
© 2026 Taylor & Francis Group, London, ISBN 9-781-041-20970-6

31

Work-Life Balance—Impact on Productivity and Well-Being

Varsha T.[1], P.S.V. Balaji Rao[2]
Department of Business Administration,
Vidyavardhaka College of Engineering,
Mysuru, India

Abstract: Work-life well-being plays a very critical role in boosting workforce optimization capability and overall performance. A well-balanced work-life dynamic reduces stress, improves job satisfaction, and fosters higher engagement levels, leading to increased efficiency and output. This paper examines how maintaining work-life balance positively impacts employee wellness, motivation, and organizational commitment. Companies that implement flexible work policies encourage time management, and support employee wellness initiatives often experience improved performance, lower absenteeism, and higher retention rates. Conversely, poor work-life balance can result in burnout, decreased focus, and diminished productivity. By fostering a supportive work environment, organizations can enhance both individual performance and overall business success. This study highlights key strategies for achieving an optimal WLB and its direct persuade on workplace productivity. This study mainly focuses on WLB in IT team member's performance and output in presence of WLB in the IT Industry in India. The data were collected both online and offline and using a self-administered questionnaires and interview methods. The results of the study shows that those IT Companies implementing WLBPs seriously, are having good QWL which leads to great performance, productivity and satisfaction of its employees.

Keywords: Workplace satisfaction, Job proficiency, Employee productivity, Occupational gratification

1. INTRODUCTION

Work-Life Balance means managing your job and family life in a way that reduces stress and allows employees to work based on time, family, and rest. It helps to maintain good health and wellness, productivity, and overall happiness. Many team members facing the problem between their task and personal life in today's challenging work culture. Even though global search for work-life balance here referred to as Work-life harmony minimal employees have found the exact meaning. It is about changing the working culture that allows workers to integrate work with their other responsibilities, such as caring for children or elderly parents and managing house chores. Work-life balance does not imply a perfect 50-50 split; rather, it is a dynamic blend of personal and professional commitments tailored to individual needs., the gain, challenges The equilibrium or imbalance related to it can influence various aspects of society. The Career-

life equilibrium-related Challenges will have an impact on both the employee and the organization. For the team member, the effects may have a different impact on their productivity, performance, mental health, physical wellness, and individual performance. For organizations, workforce instability, attrition, up skilling expenses, talent acquisition costs also can be the result of low WLB. Job effectiveness fails to be just regarding the time invested with family, self-related interests besides When considering the growth and administration of life work synergy, it is essential to implement effective strategies, activities, its Initiatives serve as a strong catalyst for improved industry change, functionality, employee productivity.

2. NEED OF THE STUDY

A good Professional-personal balance is important for employees to stay positive, productive and perform well in any organization. When people struggle to manage work

[1]varsha.t@vvce.ac.in, [2]hodmba@vvce.ac.in

and personal life, they feel over stressed, which lowers their performance and productivity which intern increases absenteeism and irregularity of work. This also reduces their commitment and motivation at work and towards organization. Companies have now introduced flexible work hours, remote work, and workation, vacation benefits to help employees balance their responsibilities and to have a good Career-life equilibrium.

3. RESEARCH OBJECTIVES

To obtain insights into the active scenario with level of work-life wellness.

To analyze key components this affects employees Personal-professional harmony.

To evaluate the effect of Occupational-life balance about employee productivity.

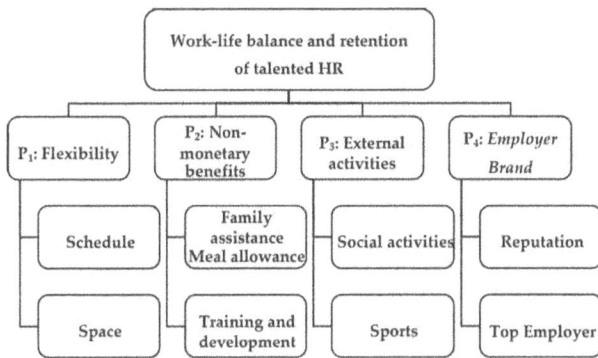

Fig. 31.1 Strategies for work–life balance and retention of valuable HR (Rodríguez-Sánchez, et. al., 2020)

4. RESEARCH QUESTIONS

Key factors influences employees' WLB in the IT industry in India

Link between Professional effectiveness policies and team members task output and work rate.

Current trends and issues that influence employees WLB

5. RESEARCH METHODOLOGY

This study takes a Illustrative and critical analysis to identify various relationship among job effectiveness and its effect on employee performance and productivity for Indian IT employees specifically through survey methods; we have collected the data among 100 IT employees working across India in IT products and services industry. The Data collection is on primary sources by survey method A well-structured and A self-conducted questionnaire was utilized for data collection. The focus groups we have taken such as engineers and Leaders. Secondary sources are research papers known sources, articles according to the need of the study. Out of 100 respondents 60 percent of the respondents were from south and 40 per cent from north.

6. WORK LIFE BALANCE

Workplace contribution supports team members of any industry to align their family and work related activities. WLB ideally Employees should allocate their time based on priorities, ensuring a harmonious blend of work, family, health, and leisure activities. Along with making a career, business travel etc. It functions as the major approach in real life corporate scenario, since it contributes to enhancing enthusiasm. and influence the teammates and it rises the overall productivity. About 75% of the company admitted in a research that acquiring and sustaining ideally develop into very challenging compare to last so many years.

7. REASONS FOR IMBALANCE IN WORK LIFE BALANCE

Team member productivity becomes imbalanced due to various factors such as huge workload, more working hours, and lack of flexibility, job insecurity and workplace culture. Personal challenges like high expectation and ambition, time management issues will further contribute to the issue. Technological tools and advancements keep employees constantly connected, making it harder to disconnect from their regular duties. Additionally, set of family priorities and responsibilities, such as childcare and elder care, add stress. Over time, neglecting self-care and experiencing chronic stress can lead to burnout, affecting both personal and professional life. Achieving balance requires setting boundaries, prioritizing tasks, and fostering a supportive work environment.

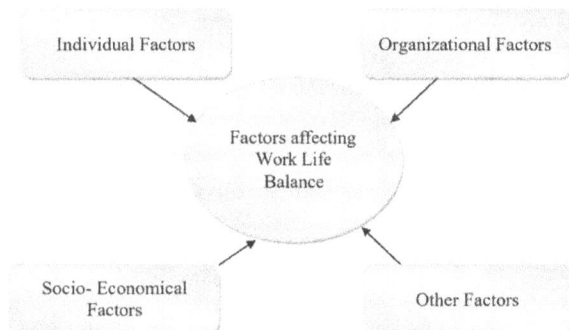

Fig. 31.2 Factors effecting work life balance
Source: Authors

Table 31.1 Factors effecting work life balance

Individual Factors	Societal Factors	Organizational Factors
Emotional intelligence	Family support	Stress
Personality	Child care	Technology/Tools
Attitude	Dependent care	Role ambiguity

Source: Authors

8. EMPLOYEE PERFORMANCE

Performance refers to the execution of all the tasks, activities, or events employees normally involved to achieve desired results. It is commonly evaluated in various fields, including business, academics and other related areas, based on efficiency, quality, and effectiveness. In a workplace setting, employee performance is measured by productivity, skill application, and goal achievement. High performance often results from motivation, training, and a supportive environment, while poor performance have lot of issues such as poor employee engagement, more expectations from organization as well as the employee perspective. Employee performance refers to the efficiency, effectiveness, productivity, and quality of work an individual employee is supposed to deliver in their proposed roles by their managers.

9. EMPLOYEE PRODUCTIVITY

Productivity refers to the efficiency or effectiveness of producing some goods or services, measured by the output generated relative to the input used. In a organization context, employee productivity is the ability to complete tasks effectively and efficiently within a stipulated time period while maintaining quality. High productivity results from proper planning, skills, influence, leadership, motivation, proper management and decision making, and the right tools and technologies, whereas low productivity can get generated from poor organization, lack of control and motivation, or inefficient policies and processes. Improvement of productivity helps businesses grow, increase profitability and helps to generate revenue, and stay competitive in nature. Employee productivity mainly focuses to the efficiency and effectiveness with which team members to complete their assigned activities to contribute to organizational objectives and goals to achieve organization mission and vision.

10. REVIEW OF LITERATURE

Industries ideally need to focus on evaluating their efficiency to gain a competitive edge. Employee contribution plays a crucial role in maximizing organizational effectiveness. Arulrajah and Opatha (2012) argued that a company's success is directly dependent on the individuals it hires, as employees are key stakeholders in determining organizational efficiency. Given the real business landscape, companies must adopt various strategies, policies, and initiatives to boost professional effectiveness. Wheatley (2012) argues that maintaining a healthy Work-rest balance benefits both employers and employees, leading to mutually advantageous outcomes. Similarly, Grady et al. (2008) emphasize that work-life balance is vital not only for individual well-being but also for organizational success and societal stability. Poulose

and Sudarsan (2017) examined. The influence of Work-personal life equilibrium perspective about corporate elements organizational factors such as workload and workplace endorsement. The team's findings demonstrated that reducing work-life conflicts enhances organizational outcomes by improving employee engagement, job satisfaction, commitment, corporate citizenship behavior, and productivity while simultaneously lowering turnover and absenteeism. Overall, this analysis suggests that work-life balance positively correlates with job performance. Greenhaus, Collins, and Shaw proposed three key components of work-family balance, further reinforcing its significance in both professional and personal domains.

1. Time balance pertains to allocating sufficient time to both task assigned and self-responsibilities in a way that ensures equilibrium between the two roles.
2. Involvement balance refers to maintaining equal psychological engagement and commitment in both work and family responsibilities.
3. Satisfaction balance ideally occurs when a person experiences a comparable level of outcome in both work and personal life

11. DISCUSSION AND ANALYSIS

A sample of IT employees working in India even for MNC's such as IT product development and services were considered for this study. Those who are married, having children/ or not/ would be future parents, and having some dependent responsibilities of their old age parents, living in nuclear families, at least one among them working in the Information Technology (IT) sector. The samples also included employees across different organizational levels such as engineers team leads project managers and senior leaders being part of the Information Technology (IT) industry. The relation between factors and work life balance was identified by correlation analysis in this study. Responsibilities between professional and personal life need to be properly balanced between them which tend have best output with great success rate. That boosts the domain for career growth to both that is the organization and employee. Good time management principles, policies and stress management reduce the stress of managing a healthy work life balance. An optimal stability between performance, productivity and strategic oversight is a must. An employee or team member with strong mental and emotional wellness and stability can reach greater heights at the establishment level.

12. FINDINGS AND RECOMMENDATIONS

Majority of the respondents were males and the ratio between male and female respondents worked out to 4: 1.

About 69% of the respondents were aged between 22-30 years and about 29% were aged between 31- 45 years.

Most of the respondents (67%) were technical engineers and 24% were team leaders or project managers by designation.

Majority (57%) of the respondents were single. Out of the married respondents, 50% of their spouses were employed and majority of the IT organizations only.

About 50 % respondents possessed less than 5 years of total experience and about 60% of the respondents again possessed less than 2.5 years of experience in their current organization. The relationship of quality work life was found to be more significant with employees and employers, outcomes can be measured in terms their performance, productivity, satisfaction, and employee wellness.

It was found that more number of women employees entering in the IT sector needs more attention to practice WLBPs in an effective and efficient manner to avoid conflicts in their personal and professional life. The major rise in standard of living and the importance of personal life have enhanced the demand of all the team members work and family. In this study found few gender differences in responding about their workplace experiences and expectations. Similarly, the age-wise and marital status of respondents also adds some new information in terms of employee's performance and productivity.

In this study it is found 56% of respondents have dependent responsibilities and 46% did not have any. This study reveal that employees need a more supportive organizational system as supportive environment and peer system to fulfill dependent care responsibilities without hampering their assigned task.

To uphold a wholesome Job-life harmony and advance Operational efficiency, each organization should take proactive measures. Expanding the workforce and maximum skilled resource allocation helps distribute workload, reducing over stress and burden and long working hours. Conducting workshops and trainings on punctuality, time management, mental and physical wellness, stress management, and prioritization fosters a habit of punctuality and efficiency. A healthy and positive work environment contributes to employee wellness and resilience. Providing day-care facilities near the workstation eases stress for working parents, enabling them to focus better and to be more effective on the required task. Flexibility, such as gig work economy, modified workweeks, and adaptable shift schedule, allow team members to manage their self-related and vocational commitments effectively. Strict boundaries should be set between home and work, restricting work-related mobile usage outside office hours. Lastly, strategic time planning through a well-structured daily planner ensures a balanced approach towards personal, family, and professional responsibilities, leading to improved overall wellness.

13. CONCLUSION

However, researchers have investigated time and documented the key issues of work-leisure balance issues on the employer and the team members in growing amount of their research, very few researchers had examined the dual earners which I have brought in here part of my studies. The change in family life indicates increased work life conflicts. Most of the studies conducted consider WLB of women employees only or for men employees but the current study analyses dual earners working in the IT sector in India Therefore, it can be known that that there are WLBPs in organizations but need to be more focused in practicing these policies in an effective and efficient manner. Each of the organizations needs to add some more policies in case of uncertainties like pandemic. 80% of the IT employees in this study had mentioned that formal and informal organization support for them to improve work life quality so as it to reduce employee turnover and absenteeism and increased job satisfaction In short, work-life balance is a critical concern that affects both employees and employer. Maintaining a good harmony between industry and personal life helps to improve employee performance and overall productivity as well. When employees can manage their work schedules effectively, they perform better than expected and experience less amount of stress. I have gone through certain online articles as well. A post by Anand Mahindra, Chairman of Mahindra Group, on 8 Feb 2022 on working women attracts lot many people attention and appreciation for their achievement along with fulfilling family priorities and responsibilities. In his post, he has mentioned ,I have been there for babysitting of my one year old grandchild and it brought to me the reality of working women's and I salute that their success require much more efforts than their male counterparts.

REFERENCES

1. Baumeister RF, Leary MR. The need to belong: Desire for interpersonal attachments as a fundamental human motivation. Interpersonal development. 2017 Nov 30: 57–89.
2. Daniels K, Gedikli C, Watson D, Semkina A, Vaughn O. Job design, employment practices and well-being: A systematic review of intervention studies. Ergonomics. 2017 Sep 2;60(9):1177-96.
3. Explain what is Employee Performance Management? From www.peoplestreme.com
4. Guest, D. E. (2017). Human resource management and employee well-being: Towards a new analytic framework. Human resource management journal, 27 (1), pp. 22–38.
5. Gyanchandani, R. (2017). A qualitative study on the work-life balance of software professionals. IUP Journal of Organizational Behavior, 16 (4), pp. 53–67.

6. Hayman, J. (2010). Flexible work schedules and employee well-being. New Zealand Journal of Employment Relations, 35 (2),pp. 76–87.

7. https://www.perlego.com/book/4183464/worklife-balance-pdf.

8. Humayon, A. A., Raza, S., Kaleem, N., Murtaza, G., Hussain, M. S., Abbas, Z. (2018). Impact of supervision, working condition and university policy on work-life balance of university employees. European Online Journal of Natural and Social Sciences, 7 (1).

9. Kumar CS. Job Stress and Job Satisfaction of IT Companies' Employees. Management and labour studies. 2011 Feb;36(1):61–72.

10. Kumari G, Joshi G, Pandey KM. Job stress in software companies: A case study of HCL Bangalore, India. Global Journal of Computer Science and Technology. 2014 Sep 25

11. Meenakshi Kaushik, & Neha Guleria. (2020). A Conceptual Study on Work-Life Balance and its impact on Employee Performance. Sparkling International Journal of Multidisciplinary Research Studies, 3(1), 1–11.

12. Mohammad niaz,Journal of Vocational Behaviour (2003) – —Relation between work family balance and quality of life.

13. Naithani, D. P., (2010). Recession and work-life balance initiatives. Naithani, P. (2010). Recession and work-life balance initiatives. Romanian Economic Journal, 37, pp. 55–68.

14. Nierenberg, B., Alexakis, G., Preziosi, R. C., O'Neill, C. (2017). Workplace happiness: An empirical study on well-being and its relationship with organizational culture, leadership, and job satisfaction. International Leadership Journal, 9 (3), pp. 2–23.

15. Pandiangan, S. M. T., Rujiman, R., Tanjung, I. I., Darus, M. D. Ismawan, A. (2018). An Analysis of the Factors which Influence Offering the Elderly as Workers in Medan. IOSR Journal of Humanities and Social Science (IOSR-JHSS), 23 (10), pp. 76–79.Sun S. India: Employment in IT-BPM industry 2021. Statista. https://www.statista.com/statistics/320729/india-it-industry-direct-indirect-employment.

16. Poulose, S., Sudarsan, N. (2017). Assessing the influence of work-life balance dimensions among nurses in the healthcare sector. Journal of Management Development, 36 (3), pp.427–437.

17. Sathyanarayana, S., Satzoda, R. K., Sathyanarayana, S., Thambipillai, S, (2018). Vision-based patient monitoring: a comprehensive review of algorithms and technologies. Journal of Ambient Intelligence and Humanized Computing, 9, pp. 225–251.

18. Silaban, H., Margaretha, M. (2021) _The Impact of Work-Life Balance toward Job Satisfaction and Employee Retention: Study of Millennial Employees in Bandung City, Indonesia. International Journal of Innovation and Economic Development, 7 (3), pp. 18–26. harifzadeh M does fitness and exercise increase productivity. A ssesing health, fitness and productivity relationship. American journal of management.2013 Apr 1;13(1):32–52

19. V. Madhusudhan Goud and K. Nagaraju (2013), —Work Life Balance of Teaching Faculty with Reference to Andhra Pradesh Engineering Colleges‖, Global Journal of Management and Business Studies, Volume 3, Number 8, pp. 891–896.

Advances in Mechanical Engineering and Materials Sciences – Dr. Vinay K. B et al. (eds)
© *2026 Taylor & Francis Group, London, ISBN 9-781-041-20970-6*

32

Individual Profile Categorization using AI—Age and Gender Classification for Personalized Insights

Sara Saju[1],
Yashas M. K.[2], Shuchika Prasad R.[3],
Janhavi V.[4], Sneha V.[5], Ayesha Taranum[6]

Department of Computer Science and Engineering,
Vidyavardhaka College of Engineering,
Mysuru, India

Abstract: The module for identifying gender and estimating age is really important in biometric recognition systems. It helps to automatically and efficiently recognize people using advanced deep learning methods. In this study, we introduce a method based on Convolutional Neural Networks (CNNs) that works in two steps: first, we figure out the person's gender, and then we estimate their age. This approach makes the most of the way gender and age features relate to each other, which helps improve accuracy in predictions. Our model has impressive results, scoring 95.8% accuracy in gender classification and an average error of just 3.2 years in age estimation, beating other leading methods out there. We also ran a lot of tests on different facial datasets to ensure that our model is reliable and can adapt to various backgrounds and real-life situations. There are many ways to use this system. It can be very useful in security for identifying people, in healthcare for helping manage patients and providing personalized care, and in businesses for creating tailored experiences for customers. We also tackled some common issues in this area, like dealing with differences in datasets and making sure the system works well for many different groups of people. Through careful testing and tweaking, we aim to make biometric recognition technology even more accurate and useful, setting a solid groundwork for future research and real-world use of gender and age estimation systems.

Keywords: Age detection, Gender detection, Neural net, Convolutional neural net, Face identification, Identity analysis, Security, Customized analysis

1. INTRODUCTION

Biometric recognition systems are an essential part of today's technology to identify and verify people using their physical attributes. An important element of these systems is the part that determines gender and approximates age, which significantly contributes to the increase of accuracy in cases of automated identification and verification. This is particularly key in situations where a set demographic attribute will help inform a decision, such as in matters that touch on; customized service, enhanced security, behavior among others. The last couple of years' advancement in deep learning algorithms has shifted the dynamics, where CNN is now a powerful method for leveraging facial images. CNNs are good at such applications as pattern recognition in pictures, determining a person's gender, and estimating their ages. This study builds on these developments to produce a reliable module that can handle various kinds of data and challenging real-life scenarios.

[1]saraaikkarakudyil@gmail.com, [2]yashasmk94@gmail.com, [3]shuchikaprasadr@gmail.com, [4]janhavi.v@vvce.ac.in, [5]sneharuthu25@gmail.com, [6]ayesha.cs@vvce.ac.in

10.1201/9781003725053-32

It emphasizes employing improved CNN layouts, VGG16, and ResNet that are sacred for their applicability in implementing feature's extractions effectively and adequately. VGG16 is quite simple yet deep, which make it one of the best choices for both high efficiency and good performance. ResNet, nevertheless, employs short connections to construct deeper networks and it is easier to train and predicts with the help of the complex data. By combining these designs with further data preprocessing like normalization and face alignment, the possibility of rotations, scaling, and even the flip of the faces the module becomes invariant to changes in the input, facial expressions or light conditions. The performance data indicate that in terms of gender classification accuracy this module works well and that age estimation error remains relatively low. Greatly, the combining of data augmentation with ways to avoid overfitting allows it to run more data generically across many datasets. Besides, the algorithm allows processing data in real time, which could be useful in practice

This solution is not just important for biometric systems but also creates new possibilities across different fields. In security, it can improve surveillance by giving useful demographic information for more focused monitoring. In healthcare, it can aid in diagnosing age-related issues or help tailor treatment plans. Content delivery platforms can use this module to create personalized recommendations based on user demographics, leading to better engagement and satisfaction.

The dataset is used for age and gender classification; this dataset contains more than 26,000 facial images of people taken from real-life, natural settings. It contains diverse

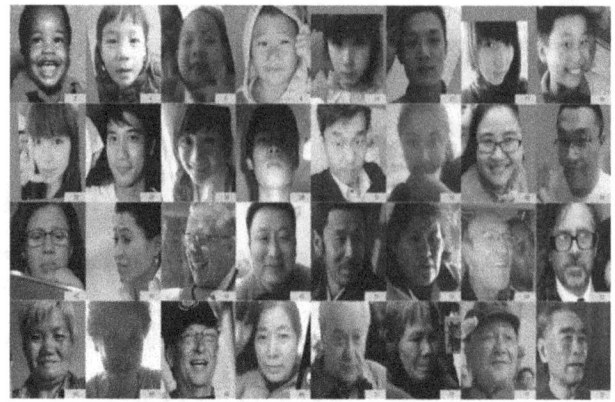

Fig. 32.1 Dataset for classification

demographic data and variability in terms of age, gender, ethnicity, position and illumination making it very reliable in training competitive models. Age labels are further dichotomized into 8 major categories in addition to gender which has two subcategories: male and female.

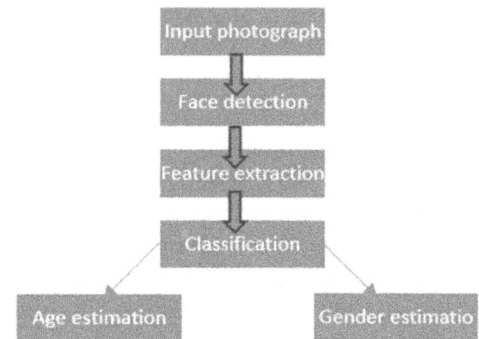

Fig. 32.2 Age and gender classification model

2. RELATED WORK

Table 32.1 Literature review

Title	Features	Advantages	Disadvantages	Ref
A Deep Learning Approach to Analyze the Relationship Between Gender, Height, Weight, and Basal Metabolic Rate from Face Images	This model predicts gender, height, weight, and BMR using facial images with some smart data preparation techniques like splitting, resizing, cleaning, augmenting, and cropping. It shows an impressive accuracy of 98.50% on the Face-ete dataset and 88.29% on another one.	It clearly beats the next best model by a good amount, with margins of 2.19% and 4.28% for BMR predictions, which really highlights its strength and accuracy.	The model needs high-quality datasets and careful handling. Its accuracy can drop if the facial images are of low quality or not diverse enough.	[1]
Single Shot Multi-Head Gender, Age, and Landmarks Detection using Shared Convolutional Features	MultiHeadCNN is a simple neural network that has a common backbone and three branches. It works to predict facial features, age, and gender, even in tough situations like different poses and lighting.	It shows remarkable accuracy, achieving 99.9% for age and 99.7% for gender on the UTK-Face dataset. Moreover, it can process 20 frames per second, which is pretty awesome for real-time applications.	It might not work as effectively with messy or low-quality data. It also has trouble with big changes in face angles, obstructions, and different lighting situations.	[2]
Predicting Age and Gender Across Different Races Using Convolutional Neural Networks: A Deep Learning Approach	The correctness of CNN and DeepFace models in estimating age and gender is examined in this research. The Mean Absolute Error (MAE) is used for estimating age, and the Exact Match Ratio (EMR) is used to predict gender. Five racial groups—Asian, Black, Indian, White, and a few others—included in the data in test.	DeepFace is better at guessing ages, but CNN is stronger at predicting gender. Together, they form a robust method for understanding facial features across different racial backgrounds.	The study doesn't take into account how things like lighting, obstacles, or odd positions might mess with the models' accuracy. Plus, the results could shift when using different sets of data or in real-world scenarios..	[3]

3. METHODOLOGY

3.1 Dataset Selection

Selecting an appropriate data set is the first step of developing an effective model for determining gender and calculating age. It's important to find a collection of facial images that are labeled with both gender and age so the model can learn properly. Datasets like Adience Benchmark, IMDB-WIKI, and CelebA are popular choices because they are large and varied, making them great for predicting gender and age. They include images from different age groups, genders, and ethnic backgrounds, all taken in various settings, which is perfect for training the model (Fig. 32.1).

Table 32.2 Age and gender classes

Attributes	Details
Total Images	26,580
Gender Labels	Male (M), Female (F)
Age Ranges	0-2, 4-6, 8-13, 15-20, 25-32, 38-43, 48-53, 60+
Reduced Dataset Size	17,523 (mostly frontal face images)

3.2 Data Preprocessing

Once the dataset is chosen, we start getting it ready to make sure everything is consistent. We resize all the images to a standard size, like 224 × 224, so they fit the neural network's needs. To make the dataset bigger and more varied, we use techniques like rotating, flipping, cropping, and changing colors. We also adjust the value of pixel to fall in a typical range, usually in between [0,1] [0,1]or[−1,1][−1,1], following a specific formula

$$x' = \frac{x - \text{mean}}{\text{std}}$$

In this context, x stands for the original value of a pixel, while std and mean refer to the standard variation and average of the total dataset. Normalization helps make the training more stable and speeds up how quickly the model learns.

3.3 Model Design

At its core, the model uses a Convolutional Neural Network (CNN) because it works really well with images. We can choose from well-known designs like VGG16

or ResNet, or you can create your own CNN if you have specific needs. The model usually includes:

- Convolutional Layers that pick up on spatial features.
- Pooling Layers, such as MaxPooling, that help make the data smaller.
- Fully Connected Layers that handle classification and regression tasks.

When it comes to gender classification, the output layer uses the softmax function. For predicting age, it has just one neuron with a linear function to estimate the age. Mathematically, the model can be expressed like this:

$$\hat{y} = f(x; \theta)$$

In this context, we have the CNN represented by f, x as the image we input, for the output we predict, and θ standing for the parameters of model.

3.4 Training Phase

This dataset is divided into three parts: training, validation, and also testing. Here the model processes images in small groups, and we check how well it performs using suitable loss functions:

Gender Classification: Cross-entropy loss:

$$L_{gender} = \frac{-1}{N}\sum_{i=1}^{N}[y_i \log(\hat{y}_i) + (1 - y_i)\log(1 - \hat{y}_i)]$$

Age Estimation: Mean Squared Error (MSE):

$$L_{age} = \frac{1}{N}\sum_{i=1}^{N}(y_i - \hat{y}_i)^2$$

The total loss function is a sum of the above two:

$$L_{Total} = \alpha L_{gender} + \beta L_{age}$$

In this context, α and β serve as settings that help manage how much each task impacts the overall process. To reduce the loss, optimization methods like Stochastic Gradient Descent (SGD) or Adam change the model's parameters. We also adjust hyperparameters, like the batch size ,learning rate, and also the total number of training rounds, by trying out different options.

3.5 Evaluation

We now evaluate how well the model is going to work with different measures: For gender classification, we

1. Input image   2. Face detection   3. Cropped face   4. Feature extraction   5. Prediction

Mathias et al. detector   + 40% margin   VGG-16 architecture   Softmax expected value $\Sigma = 23.4$ years

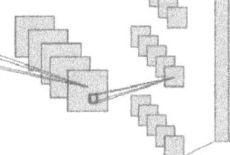

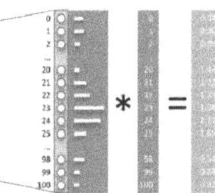

Fig. 32.3 Architecture of used CNN

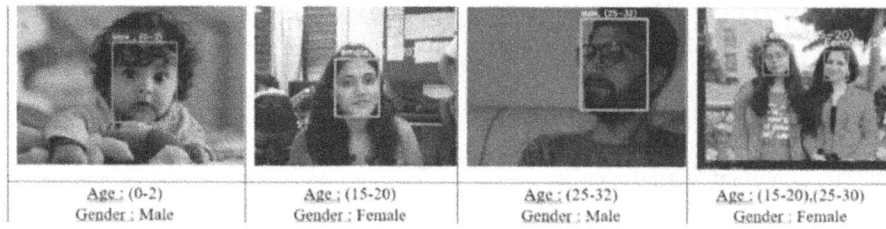

| Age : (0-2) | Age : (15-20) | Age : (25-32) | Age : (15-20),(25-30) |
| Gender : Male | Gender : Female | Gender : Male | Gender : Female |

Fig. 32.4 Detection of age and gender

look at different parameters like accuracy, precision, the F1 score. And for the age estimation, we use the mean absolute error or root mean squared error.

$$MAE = \frac{1}{N}\sum_{i=1}^{N}|y_i - \hat{y}_i|$$

$$RSME = \sqrt{\frac{1}{N}\sum_{i=1}^{N}(y_i - \hat{y}_i)^2}$$

To check how strong the model is, we use something called k-fold cross-validation. For that we need to split the data into k pieces. The model gets trained on $k-1$ of those pieces and is then tested on the last piece. This process repeats for every piece of data. In the end, we take the average of the performance results from all the k pieces to make sure our findings are trustworthy.

4. RESULTS AND EVALUATION

4.1 Gender Classification Model

Table 32.3 Details of gender classification model

Layer Type	Details
Conv Layer 1	Filter Shape: 3x7x7, 96 feature maps, stride: 4, padding: 0, that is followed by: ReLU, Max-Pool and LRN
Conv Layer 2	Filter Shape: 96x28x28, 256 feature maps which is followed by: ReLU, Max-Pool and LRN
Conv Layer 3	Filter Shape: 256x3x3, stride: 1, padding: 1, followed by: ReLU and Max-Pool
Fully Connected 1	512 neurons which is followed by: ReLU with Dropout = 0.5
Fully Connected 2	512 neurons, which is followed by: ReLU with Dropout = 0.5
Output Layer	Finally it maps the to 2 classes (Gender)

4.2 Gender-Based Age Classification Model

Table 32.4 Details of gender based age classification model

Layer Type	Details
Conv Layers	Same as Gender Classification Model
Fully Connected 1	512 neurons, followed by: ReLU and Dropout = 0.5
Fully Connected 2	512 neurons, followed by: ReLU and Dropout = 0.7
Output Layer	Maps to 8 classes (Age Groups)

The process of classifying gender and age works like a chain as in Fig. 32.2, where the age classifications relies on our gender classification results. We've created two different age classifiers: one for males and another for females. First, images are sorted by gender, and then each image goes to the specific age classifier for that gender. The models we use for age classification are built on the same framework as those for gender classification;n, but there are a few important tweaks. We've changed the dropout rate in the second fully connected layer to 0.7, added weighted losses to help with class imbalances, and adjusted the final layer to categorize into 8 different age groups.

Both gender and age classification training make use of Stochastic Gradient Descent (SGD) which has a batch size of 50. Here we can start with a learning rate of 1×10^{-3} and lower it to 5×10^{-4} every 10,000 iterations to help refine the learning. To make sure the models are solid and can work well in different situations, we use 4-fold cross-validation during training as seen in Fig. 32.4.

A comparison is performed to look at how well "GenAgeNet" (new model) performs as compared to "MobileNet" when it comes to figuring out age and gender from the images of the person's face as seen in Fig. 32.5 where "GenAgeNet" is a custom CNN created just for this purpose, featuring distinct classifiers for gender (dual) and age (8 categories). It makes use of several convolutional layers and applies dropout regularization, and it gets trained with stochastic gradient descent along with

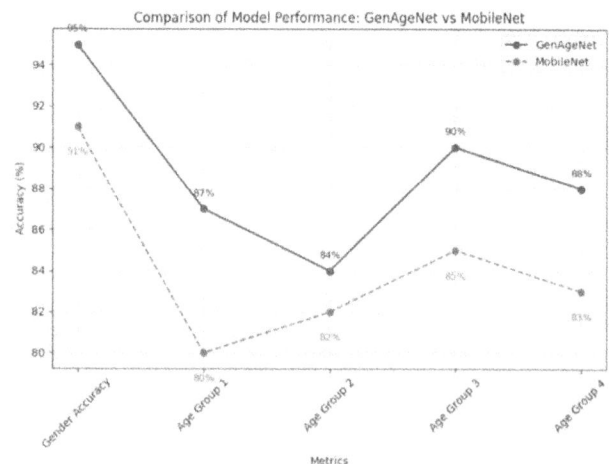

Fig. 32.5 Comparative analysis between GenAgeNet and an existing model

4-fold cross-validation. On the other hand, "MobileNet" is a lightweight CNN that has been adjusted for age and gender classification by using pre-trained weights from ImageNet. Its design focuses on being efficient and suitable for real-time use. The results indicate that "GenAgeNet" does better than "MobileNet" in identifying both gender and age groups. In general, "GenAgeNet" performs well on all metrics, especially when compared to age group classification. A line graph that visually contrasts the two demonstrates how "GenAgeNet" (in blue) routinely performs better than "MobileNet" (in green), underscoring the importance of custom models for specific jobs such as this one.

5. CONCLUSION

This study demonstrates the efficient identification of age and gender from facial photos using convolutional neural networks (CNNs). The suggested approach shows high accuracy and flexibility by utilizing sophisticated structures like VGG16 and ResNet, as well as innovative data preparation strategies and reliable testing methodologies like k-fold cross-validation. Its modular design, which consists of distinct Age classifiers for men and women solve the difficulties of understanding how individuals age differently, providing a model that is accurate and reliable. The model's performance and generalization are enhanced by the application of weighted loss functions, dropout regularization, and changing learning rates. The system is easier to adapt to different datasets and real-world scenarios thanks to these enhancements.

The model is more prepared to adjust to variations in lighting, angles, and facial expressions when data augmentation techniques are added. Beyond simply identifying age and gender, the technology can be helpful in fields like healthcare, security, and customized services. For instance, it can help identify age-related illnesses and enhance surveillance through offering demographic information, and assist content platforms in giving people individualized experiences. The custom-built GenAgeNet model clearly outperforms current models such as MobileNet in terms of reliably classifying age and gender. This demonstrates that utilizing broad models can sometimes be less beneficial than customizing them to meet particular needs. This effort establishes a strong basis for the future advances in AI-powered demographic analysis, resulting in more complex, precise, and effective systems.

REFERENCES

1. S. I. Siam, S. A. H. Chowdhury, N. Ahmed and S. Biswas, "A Deep Learning Approach to Analyze the Relationship Between Gender, Height, Weight, and Basal Metabolic Rate from Face Images," *2024 IEEE International Conference on Power, Electrical, Electronics and Industrial Applications (PEEIACON)*, Rajshahi, Bangladesh, 2024, pp. 625–630, doi: 10.1109/PEEIACON63629.2024.10800058.

2. G. Khan, K. Pimbblet, K. Wertheim and W. Ahmed, "A Single Shot Multi-Head Gender, Age, and Landmarks Detection using Shared Convolution Features," *2024 29th International Conference on Automation and Computing (ICAC)*, Sunderland, United Kingdom, 2024, pp. 1–6, doi: 10.1109/ICAC61394.2024.10718769.

3. E. Ardelia, J. Thenando, A. A. S. Gunawan and M. E. Syahputra, "Predicting Age and Gender Across Different Races Using Convolutional Neural Networks: A Deep Learning Approach," *2024 International Conference on Information Technology and Computing (ICITCOM)*, Yogyakarta, Indonesia, 2024, pp. 150–154, doi: 10.1109/ICITCOM62788.2024.10762396.

4. R. S. V, L. A. Khan, A. V, J. M. G and P. Kannadaguli, "Human Age Estimation from Images in Real-Time Application Using Machine Learning and Deep Learning Models," *2023 International Conference on Network, Multimedia and Information Technology (NMITCON)*, Bengaluru, India, 2023, pp. 1–6, doi: 10.1109/NMITCON58196.2023.10275901.

5. A. Kanwar and K. D. Singh, "Prediction of Age, Gender, and Ethnicity Using CNN and Facial Images in Real-Time," *2023 IEEE World Conference on Applied Intelligence and Computing (AIC)*, Sonbhadra, India, 2023, pp. 668–674, doi: 10.1109/AIC57670.2023.10263824.

6. A. Rashida, C. I. Gunardi, Hendrawan, E. Mulyana, and W. Hermawan, "Applying CNN to Classify Gender for Identity Verification System," *2023 9th International Conference on Wireless and Telematics (ICWT)*, Solo, Indonesia, 2023, pp. 1–6, doi: 10.1109/ICWT58823.2023.10335249.

7. T. P. Kancharlapalli and P. Dwivedi, "A novel approach for age and gender detection using deep convolution neural network," *2023 10th International Conference on Computing for Sustainable Global Development (INDIACom)*, New Delhi, India, 2023, pp. 873–878.

8. A. R. S and V. R. V, "Age & Gender Recognition Using Deep Learning," *2022 Third International Conference on Intelligent Computing Instrumentation and Control Technologies (ICICICT)*, Kannur, India, 2022, pp. 386–390, doi: 10.1109/ICICICT54557.2022.9917573.

9. S. Rathor, D. Ali, S. Gupta, R. Singh and H. Jaiswal, "Age Prediction Model using Convolutional Neural Network," *2022 IEEE 11th International Conference on Communication Systems and Network Technologies (CSNT)*, Indore, India, 2022, pp. 239–243, doi: 10.1109/CSNT54456.2022.9787602.

10. S. Katiyar, S. Kumar and H. Walia, "A Novel Approach to Identify Age and Gender using Deep Learning," *2021 9th International Conference on Reliability, Infocom Technologies and Optimization (Trends and Future Directions) (ICRITO)*, Noida, India, 2021, pp. 1–5, doi: 10.1109/ICRITO51393.2021.9596153.

11. A. Junaidi, J. Lasama, F. D. Adhinata and A. R. Iskandar, "Image Classification for Egg Incubator using Transfer Learning of VGG16 and VGG19," *2021 IEEE International Conference on Communication, Networks and Satellite (COMNETSAT)*, Purwokerto, Indonesia, 2021, pp. 324–328, doi: 10.1109/COMNETSAT53002.2021.9530826.

12. M. K. Benkaddour, S. Lahlali and M. Trabelsi, "Human age and gender classification using convolutional neural network," *2020 2nd International Workshop on Human-Centric Smart Environments for Health and Well-being (IHSH)*, Boumerdes, Algeria, 2021, pp. 215–220, doi: 10.1109/IHSH51661.2021.9378708.

13. A. Kale and O. Altun, "Age, Gender and Ethnicity Classification from Face Images with CNN-Based Features," *2021 Innovations in Intelligent Systems and Applications Conference (ASYU)*, Elazig, Turkey, 2021, pp. 1–6, doi: 10.1109/ASYU52992.2021.9598986.

14. H. Kondo and F. N. Kondo, "Convolutional neural networks on multichannel time series of smartphone applications for gender or age range classification," *2020 9th International Congress on Advanced Applied Informatics (IIAI-AAI)*, Kitakyushu, Japan, 2020, pp. 522–525, doi: 10.1109/IIAI-AAI50415.2020.00109.

15. A. Mustafa and K. Meehan, "Gender classification and age prediction using CNN and ResNet in real-time," *2020 International Conference on Data Analytics for Business and Industry: Way Towards a Sustainable Economy (ICDABI)*, Sakheer, Bahrain, 2020, pp. 1–6, doi: 10.1109/ICDABI51230.2020.9325696.

16. A. Ghildiyal, S. Sharma, I. Verma and U. Marhatta, "Age and gender predictions using artificial intelligence algorithm," *2020 3rd International Conference on Intelligent Sustainable Systems (ICISS)*, Thoothukudi, India, 2020, pp. 371–375, doi: 10.1109/ICISS49785.2020.9316053.

17. D. -Q. Vu, T. -T. -T. Phung, C. -Y. Wang and J. -C. Wang, "Age and gender recognition using multi-task CNN," *2019 Asia-Pacific Signal and Information Processing Association Annual Summit and Conference (APSIPA ASC)*, Lanzhou, China, 2019, pp. 1937–1941, doi: 10.1109/APSIPAASC47483.2019.9023045.

Note: All the figures and tables in this chapter were made by the authors.

Advances in Mechanical Engineering and Materials Sciences – Dr. Vinay K. B et al. (eds)
© 2026 Taylor & Francis Group, London, ISBN 9-781-041-20970-6

33

Optimization of Assembly Line Layout of Toroidal Transformer

Dayakar G. Devaru*

Professor, Department of I&P E, SJCE, JSS STU,
Mysuru

Rathnakar G.

Professor, Department of M E, SJCE, JSS STU,
Mysuru

Rishi J. P.

Professor, Department of M E, VVCE,
Mysuru

Kiran A. S.

Research Scholar, Department of I&P E, SJCE, JSS STU,
Mysuru

Bhaskaran Gopalakrishnan

Department of I and M Systems Engineering, WVU,
Morgantown, WV, USA

**Aisiri D., Bhargavi G. Bhat,
Sumukh S., Shreyas Aradhya A.C.**

Undergraduate Student, SJCE, JSS STU,
Mysuru

Abstract: Objectives: The layout design plays an important role to improve the productivity and efficiency of the plant. Manual exertion and thereby the workers fatigue can be reduced remarkably if the production plant layout is optimized before building the plant. This research aims to optimize the layout of the assembly line of Magnatech a toroidal transformer manufacturing plant in order to address operational inefficiencies and improve productivity.

Methods: The current assembly layout faces several challenges, including congestion, long workflow distances and safety concerns. To tackle these issues, the research involved the time study, evaluation of current layout using spaghetti diagram, redesigning of the assembly layout using Systematic Layout Planning and Lean Techniques. Three Layouts have been proposed and the proposed new layout alternative 3 increases the efficiency and productivity of this small-scale industry and helps them decrease wastage, manual labour and cost.

Findings: A novel M-T shaped layout has been proposed in layout alternative 3 that can reduce material handling and increase productivity significantly. The proposed layout has improved material handling by reducing long workflow and has resulted in 13% improvement in throughput. By implementing the new layout, Magnatech can improve its productivity, reduce manual material handling and save money. The method used to develop the proposed layout in this research can be used in other manufacturing facilities having product layout.

Keywords: Plant layout, Systematic layout planning, Assembly line layout, Lean manufacturing, Spaghetti diagram, Productivity improvement

*Corresponding author: dayakar.devaru@jssstuniv.in

10.1201/9781003725053-33

1. INTRODUCTION TO PLANT LAYOUT OPTIMIZATION

Plant layout optimization is a pivotal endeavour in enhancing the efficiency and productivity of industrial operations. This process involves a meticulous assessment and refinement of various facets within an industry, spanning from layout optimization to material handling techniques. Plant layout is the arrangement of equipment, machines, storage areas, workspaces, aisles, and offices within a facility to optimize the efficiency of production processes and the utilization of space. It involves careful planning and consideration of various factors to ensure smooth workflow, safety, and cost-effectiveness. Plant layout optimization is all about designing an efficient and economical arrangement for the production facility. The end goal of the plant layout is to improve the productivity of machinery as well as the productivity of humans by rearranging the machines or departments in the plant.

2. LITERATURE REVIEW

Layout problems in production systems arise due to inefficient equipment positioning, affecting overall efficiency. As an NP-hard problem, optimizing layouts requires significant computational effort. This research reviews various studies focused on improving plant layouts in the manufacturing sector.

Research on facility layout optimization highlights the effectiveness of Systematic Layout Planning (SLP), Lean Manufacturing, and computational techniques in improving efficiency and reducing costs. Optimized layouts significantly reduce material travel distances and enhance workflow (Study [3], Study [5], Study [6], Study [11]). Hybrid strategies, such as combining cellular and product layouts, improve flexibility and productivity (Study [4]). Advanced methods, including heuristic and evolutionary algorithms, offer automated solutions beyond traditional approaches (Study [8], Study [9]). Industry-specific applications demonstrate tailored strategies across fashion, iron production, chemical plants, and food packaging (Study [2], Study [6], Study [7] and Study [12]). These studies reinforce the importance of strategic facility layout planning in maximizing operational performance.

2.1 Literature Review Summary

The NP-hard nature of layout problems means that as the number of departments increases, finding an optimal solution becomes more complex. Most studies reviewed here utilize Systematic Layout Planning (SLP) to enhance productivity and optimize plant layouts, while some researchers have employed algorithms and CRAFT software for layout optimization. However, these studies primarily focus on relocating facilities or departments, with no emphasis on redesigning an assembly line using plant layout principles. This research aims to analyse material movement challenges in a transformer assembly line and enhance its material handling efficiency.

3. METHODOLOGY AND MODEL SPECIFICATIONS

3.1 Existing Process

Toroidal transformers are transformers which use magnetic cores with a toroidal (ring or donut) shape as shown in Fig. 33.2. The total area of all the workstations comes to 15.84 Square metres and the company has a total built area of around 118 Square metres which is more than required space for this layout and also for other lines and hence space relationship diagram loses its significance in this plant. Figure 33.1 shows the existing assembly layout of the plant and the plan of the existing layout is shown in Fig. 33.3.

Fig. 33.1 Photo of existing assembly layout

Fig. 33.2 Toroidal transformer

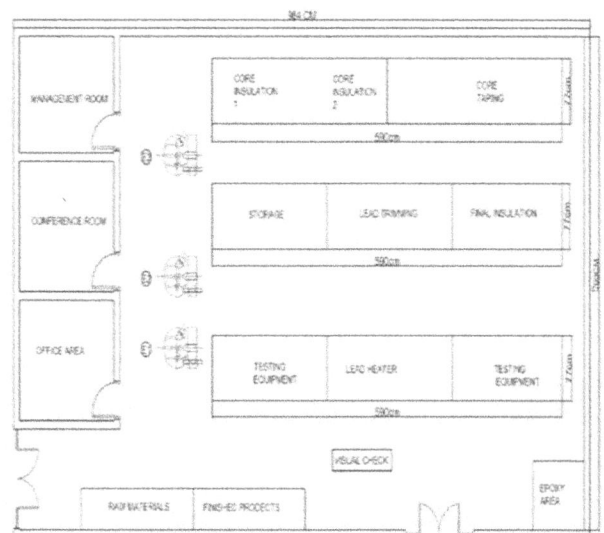

Fig. 33.3 Photo of existing layout

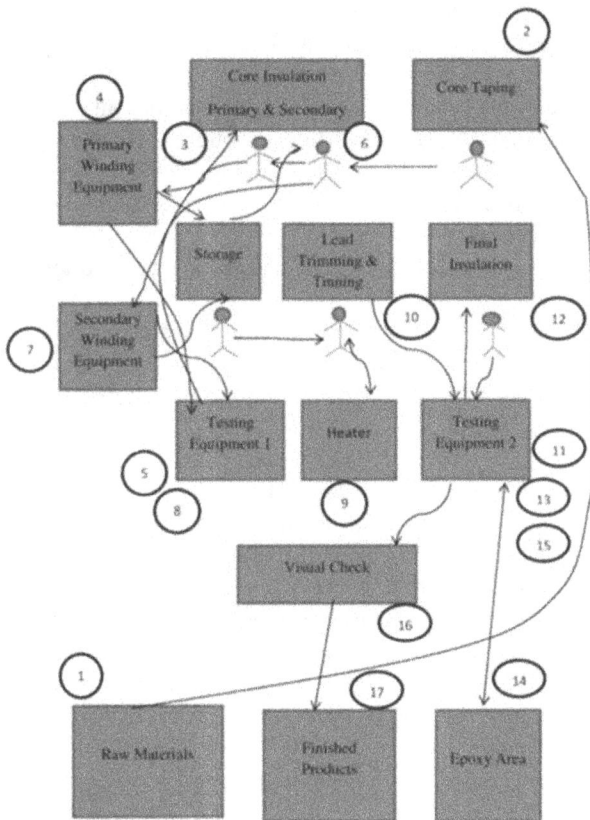

Fig. 33.4 Spaghetti diagram of existing workflow

Spaghetti diagram (Fig. 33.4) shows the material movement from raw material storage to finished goods. The material is transported manually from table to table by workers and there is lot of unnecessary material movement as seen from the spaghetti diagram. There are totally four cycles of material movement as listed below.

- 1st Cycle: (1) Raw material – (2) Core Taping – (3) Core Primary Insulation – (4) Primary Winding – (5) Testing Equipment 1
- 2nd cycle: (5) Testing Equipment 1 – (6) Core Secondary Insulation – (7) Secondary Winding –(8) Testing Equipment 1
- 3rd Cycle: (8) Testing Equipment 1 – (9) Heater – (10) Lead Trimming and Tinning - (11) Testing Equipment 2 – (12) Final Insulation - (13) Testing Equipment 2
- 4th Cycle: (13) Testing Equipment 2 – (14) Epoxy Area – (15) Testing Equipment 2 – (16) Visual Check – (17) Finished Goods

The material is travelling from one end to another end many times since the locations of the equipment on the assembly line are not arranged in proper order. This is resulting in unnecessary movement of the material ending up in loss of productivity. There is waiting time between machines due to lack of space and temporary storage is used whenever required to solve this problem to some extent. Time study was conducted on the assembly steps. The total process time to produce one product is around 132 minutes (Table 33.1). It can be seen that there are thirteen manual material movements along with the assembly steps to get the finished product which is delaying the process significantly. Also, it can be observed that to process one transformer all the work stations are used only once except the testing equipment. Each testing equipment are used twice and this is possible since the testing time is 6 minutes which is less than half of the maximum task time 14 minutes in the assembly line. Both the testing equipment are similar and does all types of tests required to be done on the toroidal transformer. The material is coming back to the testing equipment since it is used twice in each cycle of assembly process. This coming back of material is adding to the complexity of material movement and to avoid this two more testing equipment has to be purchased by the company which will be an unnecessary investment to the company and our research aims to develop a layout that will give the solution for this problem.

3.2 Proposed Improvements for the Existing Layout

The proposed layout is developed using Systematic layout planning, lean manufacturing principles and common sense. PQRST analysis includes P (product), Q (quantity), R (routing), S (supporting) and T (time) is not considered since they assemble only one type of product on this line with small volume of production and pre-defined assembly steps with required supporting services. Time study is conducted as discussed earlier. Space relationship diagram loses significance since there is more space than required for this line in this plant. Relationship diagram and activity relationship chart were used to analyse the relationship of various activities and the output from this chart was used to relocate the machines and equipment of the assembly layout.

4. RESULTS AND DISCUSSION

Based on the systematic layout planning three alternative layouts have been proposed. Alternative layouts 1, 2 and 3 are shown in Figs. 33.5, 33.6 and 33.7 respectively.

In alternative layout 1 shown in Fig. 33.5, the core taping, core insulation 1 and core insulation 2 are arranged side by side with primary and secondary winding machines and winding machines are in turn side by side with testing equipment 1 and 2 and final insulation. Storage is kept near insulation and testing equipment to keep the work in process. Lead trimming and lead heater are kept perpendicular to this equipment. The advantages of this layout is that the transformers can be transferred from insulation stage to Winding machines and to testing equipment easily since they are next to each other. The disadvantages are the lead trimming and lead heater are located in inconvenient position which causes unnecessary material handling. Still this layout needs manual handling of the transformers.

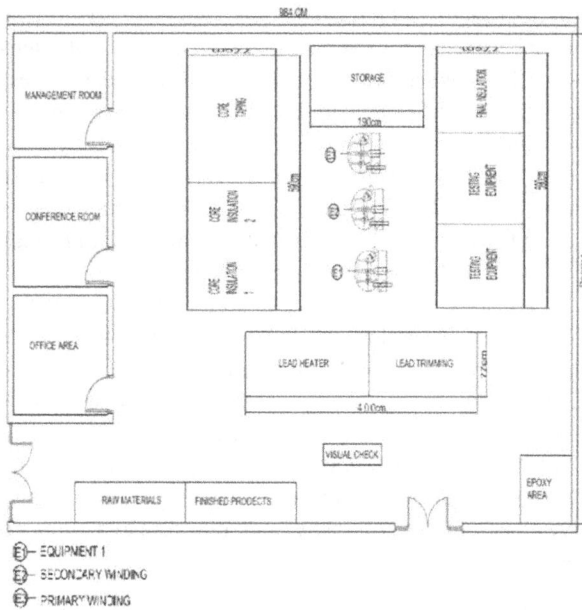

Fig. 33.5 Proposed layout alternative 1

The conveyor is introduced in alternative layout 2 along with U-shaped layout of lean manufacturing technique as shown in Fig. 33.6. A roller conveyor is suitable for this slow manufacturing operation. Core Taping, Core Insulation 1 and Core Insulation 2 are located on a conveyor on the left side close to the raw material station and Lead Heater, Lead Trimming and Storage are located on a conveyor on the right side. The three semi-automated winding machines, testing equipment and the final insulation are kept at the center of the layout so that they can be easily accessible after every insulation and thus allow smooth material flow. In this layout, the material can move on its own instead of carrying it manually with the introduction of conveyor.

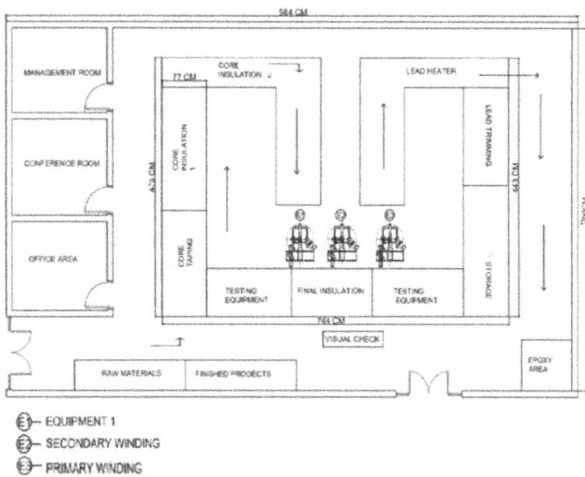

Fig. 33.6 Proposed layout alternative 2

The material moves in cycles. In the first cycle the transformers go through core taping, core insulation 1, core insulation 2, and primary winding before entering testing equipment. Even though there is no need for transformers to visit core insulation 2 in the first cycle, they will be taken through it. The second cycle will be same as the first cycle and the transformers visit the core taping and core insulation 1 even though it is not required. In the third cycle the material from the testing equipment 1 has to be transferred to the right-side conveyor and the material moves through lead heater, lead trimming, testing equipment 2, final insulation and back to testing equipment 2. In the fourth cycle, the material moves from testing equipment 2 to epoxy coating, testing equipment 2, visual check and to the finished goods. The advantage of this layout is that the material is moving on its own and manual handling is reduced to a great extent compared to previous layouts. The disadvantages are the two types of material will be moving on the left side conveyor i.e., one going for core taping and another going for core insulation 2 which can create excess material on the conveyor and confusion. Also, moving the material from test equipment 1 to the right-side conveyor will be a challenge. Also, the dimensions of this layout become a concern since the width of this layout design is 744 CM which is way higher than the length of the tables i.e., 590 CM used in the existing layout.

As an improvement over the proposed layout alternative 2, a novel M-T shaped plant layout is proposed in alternative 3 and is as shown in Fig. 33.7. This layout increases efficiency for the situation when one workstation has to be visited multiple times. The arrows in the layout diagram show the product movement that is continuous and efficient. All the tables are connected now and hence there is less manual movement of the material causing less fatigue to the workers compared to the existing layout and other proposed layouts and also total distance travelled by the material has significantly got reduced. The tables can be replaced by roller conveyers with boards so that the material travels by itself due to gravity and also stoppers can be given to the roller conveyor wherever the conveyor has to be stopped and work has to be carried out. Also, the dimensions of this layout is 470 CM x 550 CM which can be easily accommodated in the existing layout space.

Method used to design the proposed alternative layout 3

a. The workstations are arranged as per the process sequence and material will not revisit the workstations where it is not getting processed again.

b. The conveyor is designed in a shape such that the transformers have to come back to the workstation it has to visit more than once.

c. The conveyor stoppers are designed such that the tested and not tested transformers don't get mixed up.

d. The finished material is collected near the point where the raw material is fed.

e. The raw material storage and the finished material storage are close to the points where the raw material is fed and the finished material is collected.

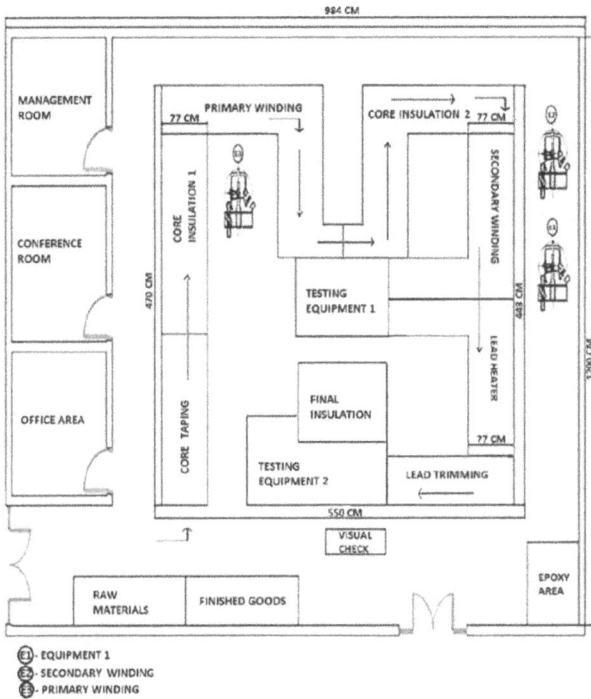

Fig. 33.7 Proposed layout alternative 3

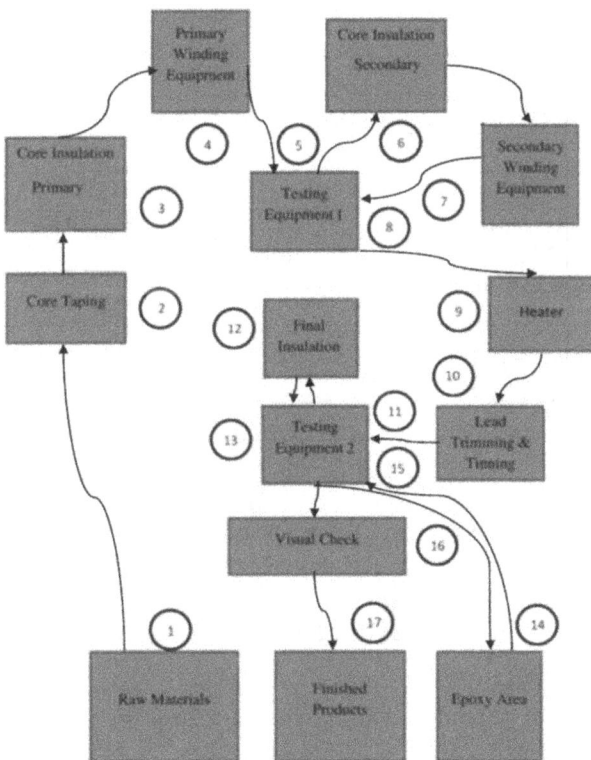

Fig. 33.8 Spaghetti diagram of proposed layout

The Raw material station is located outside the assembly layout and the manual transfer of material from the raw material station to the first work station, 'Core Taping', cannot be avoided. Now the 'Core Taping' is closer to the raw material station and that can save two minutes in material movement. From the proposed layout and the

spaghetti diagram it can be seen that the 1st and 2nd cycle of material movement goes on the left side and top right-side loop in the proposed layout. The 1st and 2nd cycle of material movement is as follows as discussed earlier.

- 1st Cycle: (1) Raw material – (2) Core Taping – (3) Core Primary Insulation – (4) Primary Winding – (5) Testing Equipment 1
- 2nd cycle: (5) Testing Equipment 1 – (6) Core Secondary Insulation – (7) Secondary Winding – (8) Testing Equipment 1

The 3rd cycle of material movement goes on the bottom right side loop in the proposed layout and the 3rd cycle of material movement is as follows as discussed earlier.

- 3rd Cycle: (8) Testing Equipment 1 – (9) Heater – (10) Lead Trimming and Tinning - (11) Testing Equipment 2 – (12) Final Insulation - (13) Testing Equipment 2

The 'epoxy area' is located outside the assembly layout since it is hazardous and the transportation of materials from the raw material station to the 'epoxy area' cannot be prevented. Also not all the transformers need epoxy coating and the 4th cycle of material movement is as follows as discussed earlier.

- 4th Cycle: (13) Testing Equipment 2 – (14) Epoxy Area – (15) Testing Equipment 2 – (16) Visual Check – (17) Finished Goods

There are 13 manual movements in the existing layout and 8 manual movements can be easily removed with the new layout. The time taken after removing the 8 manual movements and reducing the manual movement time from raw material to core taping from 4 minutes to 2 minutes after relocating core taping workstation is 114.65 minutes (Table 33.1). The remaining five manual movements cannot be removed since the locations of raw material, finished goods and epoxy potting are fixed and cannot be changed. The productivity improvement that can be achieved with the new layout is around 13.24% (132.15 – 114.65/132.15 × 100%).

5. CONCLUSION

Amid the challenges of a globalized volatile marketplace, manufacturing enterprises must prioritize operational efficiency to remain competitive. Factors like rising costs and intense global pressures make optimizing plant layout a critical driver of success. To stay ahead, manufacturers must continuously improve efficiency across all aspects of their operations. This company had many improvements to be made including its plant layout and the process flow. A detailed study of the process flow was done, a time study was conducted to determine the time for each process and the complete workflow. A ten-day study of the production was conducted. The ten day study with the present plant layout and the spaghetti diagram for the movement of parts and laborers helped to design a new layout. SLP

Table 33.1 Time taken to process 1 unit at each location with existing and proposed layout

Sl. No.	Task / Activity	Trial 1	Trial 2	Trial 3	Average in minutes for Existing Layout	Average in minutes for Proposed Layout
1.	Manual movement	03:56:00	03:50:00	04:03:00	4	2
2.	Core Taping	13:08:00	13:21:00	12:11:00	13	13
3.	Core Primary Insulation	14:10:17	13:47:29	14:06:35	14	14
4.	Manual movement	01:57:00	01:50:00	02:04:00	2	-
5.	Primary Winding	06:06:15	05:35:44	06:43:23	6	6
6.	Manual movement	01:26:00	01:35:00	01:34:00	1.5	-
7.	Exciting Current Check	05:46:41	05:53:00	05:58:15	6	6
8.	Manual movement	02:56:00	02:50:00	02:54:00	3	-
9.	Core Secondary Insulation	13:03:00	13:11:00	13:13:00	13	13
10.	Manual movement	02:56:00	02:50:00	02:54:00	3	-
11.	Secondary winding	11:13:35	10:43:31	10:59:34	11	11
12.	Manual movement	01:28:00	01:26:00	01:34:00	1.5	-
13.	Initial testing	05:56:41	05:49:00	05:48:15	6	6
14.	Soldering	06:10:00	05:57:00	05:46:00	6	6
15.	Manual movement	01:24:00	01:28:00	01:22:00	1.5	-
16.	Lead trimming &Tinning	06:50:00	06:46:00	06:53:00	7	7
17.	Manual movement	01:34:00	01:29:00	01:28:00	1.5	-
18.	Electrical Parameter Check	05:56:41	05:48:00	05:58:15	6	6
19.	Manual movement	01:32:00	01:36:00	01:24:00	1.5	-
20.	Labeling and Final Insulation	09:43:00	10:01:00	09:59:00	10	10
21.	Manual movement	01:58:00	02:05:00	02:04:00	2	2
22.	Epoxy Potting	01:56:00	01:50:00	01:44:00	2	2
23.	Manual movement	01:28:00	01:33:00	01:34:00	1.5	1.5
24.	Final Inspection	05:56:41	05:55:00	05:48:15	6	6
25.	Manual movement	00:58:00	01:05:00	01:02:00	1	1
26.	Visual Check	00:09:42	00:10:02	00:09:45	0.15	0.15
27.	Manual movement	01:02:00	00:56:00	01:04:00	1	1
28.	Final Packing	01:09:29	01:08:40	01:05:11	1	1
	Total Time in Minutes				132.15	114.65
	Total time in Hours and mins				2hrs 12.15mins	1hr 54.65mins

and Lean techniques were used to arrive at a novel M-T shaped layout. Three Layouts have been proposed and the proposed new layout alternative 3 increases the efficiency and productivity of this small-scale industry and helps them decrease wastage, manual labor and cost. The spaghetti diagram of the proposed layout clearly shows the improvements in the movement of both men and material and better usage of machines. By making the product movement continuous and streamlined, the proposed layout reduces manual movement by a significant amount and increases the production rate by 13%. Significant amount of time is saved because of efficient flow of the product. Manual exertion is reduced thereby reducing the workers' fatigue. The idea and the method used to develop M-T shaped layout in this research can be used in other manufacturing facilities having product layout

and can also be used to develop software for plant layout optimization.

REFERENCES

1. Tompkins JA Facilities Planning. fourth edition ed.(California: John Willey & Sons), 2010.
2. Carlo, F. D., Antonietta, M., Borgia, O., & Tucci, M. "Layout design for a low-capacity manufacturing line: a case study." International Journal of Engineering Business Management, IJEBM, 2013, DOI:10.5772/56883.
3. E Bhadrinath, V Diwakar Reddy, P Shravan Kumar, and A K Damodaram, "Optimization of Manufacturing Plant Layout Using Systematic Layout Planning (SLP) Method", MATEC Web of Conferences 393, 01005 (2024).
4. Shivam Singh and Dinesh Khanduja, "Improvement in Manufacturing System by Rearrangement in Layout

Design – A Case Study", IOP Conf. Series: Journal of Physics: Conf. Series 1240 (2019) 012023

5. Wiyaratn, W. and Watanapa, A., Improvement plant layout using systematic layout planning (SLP) for increased productivity. International Journal of Industrial and Manufacturing Engineering, 2010, 4(12), pp.1382-1386

6. Shubham Barnwal, Prasad Dharmadhikari. "Optimization of plant layout using SLP method." International Journal of Innovative Research in Science, Engineering and Technology, IJIRSET, Vol. V, Issue III, 2016, DOI:10.15680/IJIRSET.2016.0503046

7. Reginaldo Guirardello, Ross E. Swaney, "Optimization of process plant layout with pipe routing", Computers & Chemical Engineering, Volume 30,Issue 1,15.2005, Pages 99-114, doi:10.1016/j.compchemeng.2005.08.009

8. Jiri Kubalík, Petr Kadera, Vaclav Jirkovsky, Lukas Kurilla, Simon Prokop, "Plant Layout Optimization Using Evolutionary Algorithms, Industrial Applications of Holonic and Multi-Agent Systems", Proceedings of 9th International Conference, HoloMAS, 2019, Linz, Austria, August 26–29.

9. Nurul NADIA Nordin, Ruzanna Ab Razak, Govindan Marthandan, "A Unique Strategy For Improving Facility Layout: An Introduction of The Origin Algorithm", Sustainability , 2023, 15 (14), 11022

10. Vaibhav Nyati, Maheshwar D Jaybhaye, Vicky Sardar, "Optimization of facility layout for improvement in productivity", 4th International Conference on Industrial Engineering ICIE at: S.V. National Institute of Technology, SURAT, (December 2017).

11. Vinod Arya, "Increased Productivity and Planning by Improved Plant Layout Using Systematic Layout Planning at NCRM Division, Bhushan Steels Ltd. Khopoli, Mumbai", International Journal of Innovations in Engineering and Technology, IJIET, Vol 2, Issue 2, 2013.

12. Dillip Kumar Biswal, Kamalakanta Muduli, Jitendra Narayan Biswal, "Plant Layout Improvement Using CRAFT: A Case of Food Packaging Unit", Recent Trends in Product Design and Intelligent Manufacturing Systems - Select Proceedings of IPDIMS,2021.

Note: All the figures and tables in this chapter were made by the authors.

Advances in Mechanical Engineering and Materials Sciences – Dr. Vinay K. B et al. (eds)
© 2026 Taylor & Francis Group, London, ISBN 9-781-041-20970-6

34

Effect of Gamma Irradiation on Mechanical Properties of Banana Fiber Reinforced Natural Composite

Vinod B.[1], Ganesha B. B.[2]
Dept of Mechanical Engineering,
Vidyavardhaka College of Engineering,
Mysuru, Karnataka, India

Jeevan T. P.[3]
Dept of Mechanical Engineering,
Malnad College of Engineering, Hassan,
Karnataka, India

Manjunath G. A.[4]
Dept of Mechanical Engineering,
KLS Gogte Institute of Technology, Belagavi.,
Karnataka, India

**Akshay Mohan Y.[5],
Vaishnavi Vasan M. V.[6]**
Dept of Mechanical Engineering,
Vidyavardhaka College of Engineering,
Mysuru, Karnataka, India

Abstract: This study delves into the mechanical behavior of gamma-irradiated banana fibers susceptible to differing doses between 3 kGy and 7 kGy (kiloGray). The results are compared to those of non-irradiated banana fiber-reinforced composite. The findings, based on tensile and flexural properties, are provided in detail. Flexural and tensile properties are pivotal for resembling natural fibers in various applications. Tensile strength and modulus control a fiber's capability to hold out against pulling forces and bear up against deformation, which is crucial for load-bearing applications like ropes and reinforced composites. Flexural strength and modulus compute resistance to bending, important for applications like furniture and curved structures. These properties show a way for material selection, predict composite performance under stress, and enable optimization of processing methods. Understanding these mechanical characteristics ensures the reliable performance of natural fiber composites, supporting their use as sustainable alternatives to synthetic materials in distinct industries.

Keywords: Fibers, Matrix, Banana, Epoxy, Irradiation, Gamma, etc.

1. INTRODUCTION

Natural fibers, in today's fast-growing fierce world, are becoming a prime concern for scientists and industrial technologists to investigate and work on due to their specific properties that cannot be compared to synthetic fibers. They carry the upper hand when the topic of discussion is environmental, economic, and social benefits. However,

[1]vinod@vvce.ac.in, [2]bbganesh.bbg@vvce.ac.in, [3]Jeevantpmce@gmail.com, [4]manjunathvvce05@gmail.com, [5]akshaym2461975@gmail.com, [6]vaishnavivasan.2004m@gmail.com

10.1201/9781003725053-34

their strength and stiffness are comparably underlying to synthetic ones (Ku Harry et. al, 2011). Natural fibers offer two key advantages over synthetics: renewability and biodegradability. Sourced from plants or animals, they restore naturally, unlike finite fossil fuels used for synthetics. This renewability grants a more sustainable cycle. Furthermore, natural fibers decay naturally, reducing waste collection and environmental pollution, a stark contrast to the persistent nature of synthetic materials. However, natural fibers also have snags. Their properties can be inconsistent, varying based on source and growing conditions, unlike the uniform, manufactured nature of synthetics. This variability can affect performance and restrict certain applications. Additionally, natural fibers generally exhibit lower durability compared to synthetics, being more susceptible to wear, tear, and degradation from moisture or pests. (Chandramohan D, 2011)

Extraction of fibers from natural sources has been in practice since time unknown. Synthetic fibers require large amounts of energy for the making which degrades the environment by emitting huge amounts of CO_2. Extensive research and work nowadays have been based on finding alternatives to this, resulting in the protection of our environment. This is where the natural fibers come into action. Natural fibers have drawbacks of their own, which must be the prime thing to work on to improve them with composites. (Joshi et. al, 2004)

Gamma irradiation, a form of electromagnetic radiation with high penetrating power, interacts with the molecular structure of materials, inducing notable changes even at minimal doses. This interaction amends the microstructure and can modify the internal architecture of natural fibers, impacting their properties. This study investigates the effects of varying gamma irradiation dosages on natural fibers by incorporating them into epoxy resin composites. This approach allows for a direct assessment of how irradiation-induced modifications translate into modifications in the mechanical, physical, and chemical behaviour of the resulting composite material. By employing rigorous testing methodologies, this research aims to provide authentic and quantifiable results. Specifically, the study will analyse key properties like tensile and flexural strength, providing a comprehensive understanding of the impact of gamma irradiation on natural fiber composites. Additionally, studies examine how gamma irradiation might be used as a pre-treatment technique to strengthen the interfacial contact among natural fibres and epoxy matrix, to improve the performance of the composite as a whole. (Yongxia Sun et. al, 2017)

Tensile and flexural properties are crucial for approximating natural fibers in various applications. Tensile strength and modulus regulate a fiber's capability to resist pulling forces and bear up against deformation, crucial for load-bearing applications like ropes and reinforced composites. Flexural strength and modulus compute resistance to

bending, important for applications like furniture and curved structures. These properties lead the way for material selection, predict composite performance under stress, and enable optimization of processing methods. Understanding these mechanical characteristics makes a certain reliable performance of natural fiber composites, encouraging their use as sustainable alternatives to synthetic materials in distinct industries. (Begum K et. al, 2013)

2. LITERATURE REVIEW

Adeodu Adefemi et. al, explored the impact of microwave curing by comparing the tensile strength and structural properties of samples cured with microwave technology to those treated in a traditional autoclave oven. Their findings showed that the microwave-cured samples contained notably fewer voids, suggesting improved structural quality and potentially better mechanical performance.

The influence of X-ray irradiation on various polymer composites was explored (T Coffey et. al, 2002). Using Near Edge X-ray Absorption Fine structural (NEXAFS) spectra, the influences of radiation on polymers, such as mass loss and chemical structural changes, were examined. The existence of aromatic groups after irradiation offers more resistance towards loss of mass than the polymers without such groups.

The structural and surface properties of LDPE subjected to gamma and high energy Electron Beam (EB) irradiation was studied (Kieran A. Murray et al. 2013). When the aforementioned material was exposed to gamma and EB radiation, X-ray diffraction (XRD) investigations showed that radiation-induced crosslinking took place without appreciably altering the material's crystalline structure or degree of crystallinity.

The mechanical properties of buriti petiole fiber reinforced composite after irradiating using electron beam radiation was examined (Barbosa, A. P et al., 2012) The buriti fibers were irradiated using a 1.5MeV electron beam accelerator at dose of 50, 250, and 500KGy. The findings demonstrated that the tensile characteristics of the buriti petiole fibre were significantly impacted by EB radiations. For a smaller dose of radiations up to 50kGy, the strength of the fiber was increased by 30%. For higher dose radiations, the strength of the fiber starts to deteriorate due to the structural damage of the fibers by high energy radiations.

The history of gamma radiations, their ecological importance, and their effects on ecosystems was explained (Richard Stalter et al., 2012). The gamma radiations can be measured by using the following three units i.e., the Giga becquerel (GBq), Gray (Gy), and roentgen (R). In some cases, it is expressed as krad but commonly it is measured in terms of Gy. The value of 1Gy is equivalent to the 1 joule of radiation energy absorbed for each kg of tissue.

A comprehensive study on the influence of gamma radiation on the structural modification of PP, polyacetate, polycarbonate, and polyvinyl chloride polymers using the FTIR spectrum was done by (Dipak Sinha et al., 2012). The various physio-chemical changes that occur in the polymers may be due to the various mechanisms like the interaction of incident ions with the matter and initiation of various secondary reactions. The incidence of gamma radiation on PP polymer is found to destroy the polymer chain and lead to the formation of ketonic and alcoholic group. In the case of poly acetate, the destruction of the ester group took place by eliminating carbon dioxide.

In addition, the study features the importance of considering the impacts of gamma irradiation on the mechanical behaviour of natural fiber reinforced composites. This apprehension can inform the advancement of more vigorous and sturdy materials for various industrial applications.

3. MATERIALS AND METHODOLOGY

3.1 Materials

The composite was fabricated using Epoxy resin as matrix material and reinforced with banana fibers. The resin used is Lapox Epoxy resin L-12 which is used in industrial tooling and electrical systems and comes under the category of building coating, adhesives, and sealants. This was chosen over its outstanding properties of connecting to a vivid fiber variety along with delivering excellent electrical and mechanical behaviors at elevated temperatures. Hardener K-6 whose chemical name is Tri- Ethylene Tetro Amine is used in the ratio of 10:1 for effective hardening of the specimen along with which it enables curing controls curing rate and provides the desired performance.

The reinforcement material used is banana fiber, also called Musa fiber for its splendid durability, being one of the strongest natural fibers, it is abundant and renewable with good mechanical properties. The banana fiber is acquired from the trunk of the banana plant and dehydrated to remove its water content, dried, and made into strands.

3.2 Methodology

Irradiation of the Fiber

Natural fibers are very sensitive to radiation. Therefore, raw fibers were irradiated using cobalt 60. Cobalt-60 (Co-60) is a crucial gamma radiation source for irradiation due to its high-energy gamma rays, which provide excellent penetration for treating various materials. Its relatively long half-life (5.27 years) ensures consistent radiation output over time, reducing the need for frequent replacements, Co-60's reliability and versatility make it a cornerstone of modern irradiation practices. Hence it is used as a gamma irradiator of a very small dosage ranging from 3kGy to 7kGy with a step difference of 1kGy.

Fabrication using Hand Layup Process

The banana fibers were cut to the average width of the fiber between 10 and 15 mm. These fibers were manually placed in the mold, whose dimensions were 210*297 mm. An initial layer of the natural fiber was placed in the mold and, then Epoxy L12 resin mixed with an appropriate portion of hardener was poured over the fiber. To ensure uniform distribution of the epoxy resin, load is applied on the surface using a roller. Then another layer of banana fiber was placed on the mold and resin was poured again. In the next step, a roller was used to apply low pressure to remove any excess polymer and air bubbles present. Once these processes were finished the mold was sealed and the specimen was applied with a uniform load and left undisturbed for 24 hours for curing. After the composites were fully dried, they were carefully separated from the molds without creating any cracks on the composites. Figure 34.1 shows the laminate prepared using banana fiber.

Fig. 34.1 Banana fiber reinforced laminate

Preparation of Specimen

The test specimens were prepared in accordance with the American Society for Testing and Materials (ASTM) after the laminates were prepared and tested for their properties using a Universal Testing Machine (UTM). Figure 34.2 shows the specimens cut according to the ASTM standards.

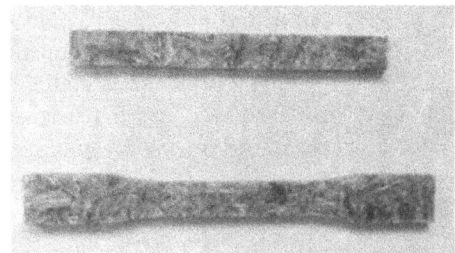

Fig. 34.2 Specimens machined according to ASTM standards

Testing the Specimen

In order to regulate the properties of the material they should be subjected to different tests. Here flexural and tensile testing of the specimen was carried out.

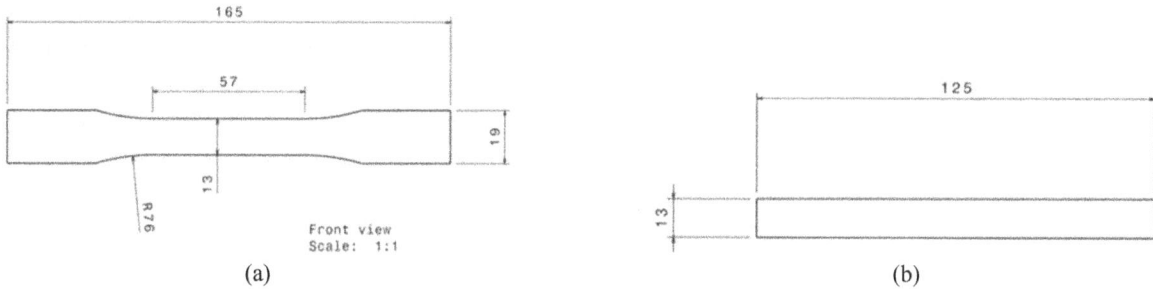

Fig. 34.3 (a) and (b): ASTM D638 for tensile testing and ASTM D790 for flexural testing

Tensile Testing

The tensile testing of fiber-reinforced composite materials is carried out by ASTM D638. D638 is a dumbbell-shaped specimen as shown in Fig. 34.3 (a). Tensile stress indicates the highest stress the composite can withstand before breaking, determining its load-bearing capacity. It is influenced by factors such as Fiber orientation and Matrix material which are critical for designing effective composite structures.

Flexural Testing

Flexural testing is done to test the flexural strength and flexural modulus of a specimen under standard loading conditions. This article uses a three-point flexural test to check the amount of stress the specimen can withhold before failure during bending. Flexural modulus reflects the material's stiffness in bending, showing its resistance to deflection. These properties are significantly influenced by fiber type, orientation, and volume fraction, as well as the matrix material and the interfacial bond amongst fibers and matrix. ASTM D790 standards are used for flexural testing, which is a rectangular specimen with a length of 125mm and a width of 13mm. The pictorial representation of the standard is shown in Fig. 34.3 (b).

Fourier Transform Infrared

FTIR is the most opted and powerful analytical technique used to recognize and characterize materials by analyzing their infrared absorption and transmission. FTIR generates a unique "fingerprint" spectrum for each molecule, enabling precise identification of unknown substances. The instrument measures the interference pattern of the infrared light after it passes through the sample, creating

an interferogram and the resulting spectrum is a chart of infrared light intensity versus frequency (wavenumber). FTIR can often be performed without destroying the sample, allowing for further analysis using other techniques. When IR radiation is made to pass through a sample to arrive at a spectral fingerprint of the sample. As a result of the sample absorbing some of the rays and transmitting others, a spectrum that represents the sample's molecular fingerprint is produced at the detector. (Xu Zhiwei et. al., 2007)

4. RESULTS AND DISCUSSION

The tensile and flexural tests were performed using a computerised UTM at 0.5 mm/min elongation speed and a load of 1 kN. Until the specimen fails, the testing keeps going. Figures 34.4 (a) and (b) illustrate the specimen's variance in tensile and flexural strength, while Table 34.1 presents the average values of the three experiments.

Table 34.1 Tensile and flexural test results

SI No	Gamma irradiation dosage (kGy)	Tensile strength (MPa)	Flexural Strength (MPa)
1	Normal	24.4	47.093
2	3	23.632	45.262
3	4	20.541	39.26
4	5	14.168	34.22
5	6	11.121	29.457
6	7	8.32	21.223

On a larger note, when the mean results were compared, the normal specimen showed the highest value of tensile

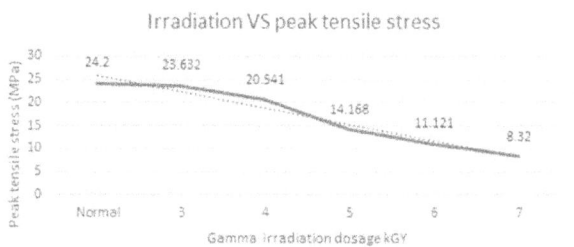

(a)

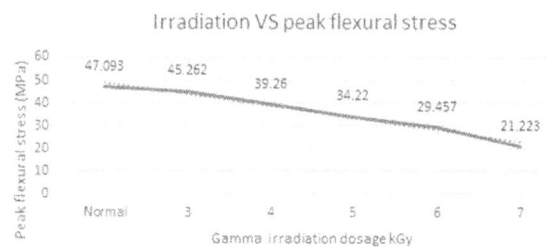

(b)

Fig. 34.4 (a) and (b): Tensile and Flexural strength of the specimen

strength with an average value of 24.4MPa. The results showed a gradual reduction of the peak stress when compared with the other variants. The 5KGy variant experienced a 14.168MPa stress whereas the 7KGy showed further reduction to 8.32MPa. The results show a clear cutback of the tensile properties as the dosage of irradiation increases. This is mainly due to the deterioration of the surface and structural properties of the material when subjected to gamma rays.

The findings show that the highest flexural strength was acquired from the normal specimen of value 47.093MPa. The reduction in its flexural strength was observed after subjecting the specimens to a higher intensity of gamma radiation. The 3KGy variant showed a maximum stress of 45.262MPa while there was a consistent reduction towards the 7KGy variant which dropped as low as 21.223MPa. The deterioration of the surface and structural properties has been once again clearly shown with the flexural test results.

The sample, before it was irradiated with gamma radiation, was tested with FTIR with a standard light source and triglycine sulfate (TGS). A graphical representation of wavenumber (cm^ -1) against percentage transmittance (%T) is given below. Once the fiber was gamma-irradiated, it was again FTIR tested for comparison. The results are depicted below. Figures 34.3 and 34.4 show the FTIR results of the fiber before and after irradiation. (Hoffman, E. N et al., 2009 and Bobadilla-Sánchez, E. A et. al., 2009)

The changes brought by the gamma irradiations on polymer composites may depend on the chain length, radical formed, and the type of functional group present, etc. The radical and functional groups formed can be analyzed using the FTIR spectrum. The FTIR spectrum can be divided into four important regions. The wavenumber region greater than 2500cm-1 is of single bond (O-H, N-H, C-H), whereas the wavenumber region between 2500cm^{-1} to 2000cm^{-1} is of triple bond (C ≡ C, C ≡ N), the

wavenumber between 2000cm^{-1} to 1500cm^{-1} is of double bond (C=O, C=C, C=N) and the wavenumber less than 1500cm^{-1} is called fingerprint (Sinha et al., 2012).

The presence of hydroxyl group (OH-) in the natural fiber also tends to increase the moisture absorption ability of the fiber. Subjecting the specimen to gamma irradiation reduces the hydroxyl group in the fiber. (Le Moigne et. al., 2017) The peak numbered 1 formed at a wave number 3347cm^{-1} confirms the maximum percentage of hydroxyl group present in the untreated specimen. By comparing Fig. 34.5 a and b, it can be clearly observed that by subjecting the banana fiber to gamma radiation, the amount of hydroxyl group present in it reduces drastically. Thus, the specimen after gamma irradiation exhibits maximum resistance towards moisture ingression through it.

5. CONCLUSION

This study examines how fibre irradiation affects the mechanical behaviour of composite materials. It is evident from experimental investigations that the fibre is highly susceptible to gamma irradiation, and that the composite's tensile and flexural strength deteriorates with increasing gamma irradiation dosage. Therefore, it can be said that gamma radiation and its dosage of irradiation on fibre material primarily control the composite's tensile and flexural strength.

REFERENCES

1. Adefemi, Adeodu, Anyaeche Christopher, Oluwole Oluleke, and Alo Oluwaseun. "Effect of Microwave and Conventional Autoclave Post-Curing on the Mechanical and Micro-structural Properties of Particulate Reinforced Polymer Matrix Composites." Advances in Materials 4, no. 5 (2015): 85.
2. Barbosa, A. P, L. L. Costa, T. G. R. Portela, E. A. Moura, N. L. Del Mastro, K. G. Satyanarayana, and S. N. Monteiro. "Effect of electron beam irradiation on the mechanical

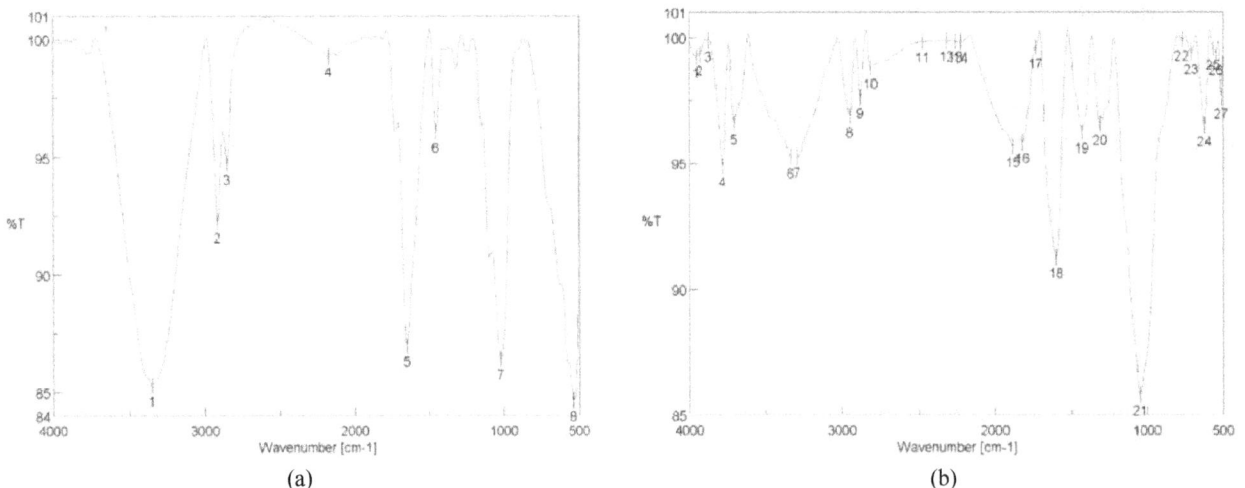

(a) (b)

Fig. 34.5 (a) and (b): FTIR results for a sample before irradiation and gamma irradiated sample with 7 kGy dosage

properties of buriti fiber." Revista Materia, ISSN 1517–7076 artigo 11484, pp.1135–1143, 2012.

3. Begum, K., and M. A. Islam. "Natural fiber as a substitute to synthetic fiber in polymer composites: a review." Research Journal of Engineering Sciences 2, no. 3 (2013): pp. 46–53.

4. Bobadilla-Sánchez, E. A., G. Martínez-Barrera, W. Brostow, and T. Datashvili. "Effects of polyester fibers and gamma irradiation on mechanical properties of polymer concrete containing CaCO3 and silica sand." Express Polymer Letters 3, no. 10 (2009): 615–620.

5. Chandramohan, D, and K. Marimuthu. "A review on natural fibers." International Journal of Research and Reviews in Applied Sciences 8, Volume no. 2 (2011): pp. 194–206.

6. Coffey, T., S. G. Urquhart, and H. Ade. "Characterization of the effects of soft X-ray irradiation on polymers." Journal of Electron Spectroscopy and Related Phenomena 122, no. 1 (2002): 65–78.

7. Hoffman, E. N., and T. E. Skidmore. "Radiation effects on epoxy/carbon-fiber composite." Journal of Nuclear Materials 392, no. 2 (2009): 371–378.

8. Joshi, Satish V, L. T. Drzal, A. K. Mohanty, and S. Arora. "Are natural fiber composites environmentally superior to glass fiber reinforced composites?" Composites Part A: Applied Science and Manufacturing 35, no. 3 (2004): pp. 371–376.

9. Ku, Harry, Hao Wang, N. Pattarachaiyakoop, and Mohan Trada. "A review on the tensile properties of natural fiber

reinforced polymer composites." Composites Part B: Engineering 42, no. 4 (2011): pp. 856–873.

10. Le Moigne, Nicolas, Rodolphe Sonnier, Roland El Hage, and Sophie Rouif. "Radiation-induced modifications in natural fibres and their biocomposites: Opportunities for controlled physico-chemical modification pathways?" Industrial Crops and Products 109 (2017): 199–213.

11. Murray, Kieran A., James E. Kennedy, Brian McEvoy, Olivier Vrain, Damien Ryan, Richard Cowman, and Clement L. Higginbothama. "Characterisation of the surface and structural properties of gamma ray and electron beam irradiated low density polyethylene." International Journal of Material Science (IJMSCI) 3, no. 1 (2013): 1–8.

12. Sinha, Dipak. "Structural modifications of gamma irradiated polymers: an FT-IR study." Advances in Applied Science Research 3, no. 3 (2012): 1365–1371.

13. Stalter, Richard, and Dianella Howarth. "Gamma Radiation." In Gamma Radiation. IntechOpen, 2012.

14. Sun, Yongxia, and Andrzej G. Chmielewski. "Applications of Ionizing Radiation in Materials Processing." ISBN 978-83-933935-9-6, Volume 1, Erasmus+ publications, Institute of Nuclear Chemistry and Technology, Warszawa 2017.

15. Xu, Zhiwei, Yudong Huang, Chunhua Zhang, Li Liu, Yanhua Zhang, and Lei Wang. "Effect of γ-ray irradiation grafting on the carbon fibers and interfacial adhesion of epoxy composites." Composites Science and Technology 67, no. 15-16 (2007): 3261–3270.

Note: All the figures and tables in this chapter were made by the authors.

Advances in Mechanical Engineering and Materials Sciences – Dr. Vinay K. B et al. (eds)
© 2026 Taylor & Francis Group, London, ISBN 9-781-041-20970-6

35

A Comprehensive Review on Wire EDM Machinability of Aluminium 6061-Boron Carbide Metal Matrix Composites

Nithyananda B. S.[1]

Associate Professor,
Department of Mechanical Engineering,
Vidyavardhaka College of Engineering,
Mysuru

Darshan S.[2]

Student, Department of Mechanical Engineering,
Vidyavardhaka College of Engineering,
Mysuru

Ashish K. L.[3]

Student, Department of Mechanical Engineering,
Vidyavardhaka College of Engineering,
Mysuru

Chandan S.[4]

Student, Department of Mechanical Engineering,
Vidyavardhaka College of Engineering,
Mysuru

Abstract: Wire Electrical Discharge Machining is commonly employed nontraditional manufacturing technique to machine hard and difficult materials including metal matrix composites (MMC). This process uses electrical discharge to cut material and is widely used because of its ability to cut hard and complex materials. The review was carried out on machining of boron carbide (B_4C) reinforced aluminium 6061 composites in wire EDM process. Al 6061 -B_4C MMC is known for their excellent mechanical properties which includes the improved hardness, strength and wear resistance but their abrasive nature finds it challenging to machine in conventional material removal processes. Hence WEDM is considered as the best and appropriate technique used for machining Al6061-B_4C MMC. This study was done to explore the potential of wire EDM processes specifically for Al -B_4C MMC. The influence of several process variables such as pulse duration, pause time and wire tension etc., effect of processing technique and effect of reinforcements on the machining rate (MRR), tool abrasion and surface texture were reviewed in this paper. Aluminium-Boron caribe MMC possesses impressive potential to be used in high performance applications. Therefore, further research is required to overcome the current limitations and explore fully their capabilities for real engineering applications.

Keywords: Wire electrical discharge machining (WEDM), Machinability, Material removal rate (MRR), Aluminium 6061 (Al 6061), Boron carbide (B4C), Metal matrix composites (MMCs)

---

[1]bsn@vvce.ac.in, [2]vvce21me0081@vvce.ac.in, [3]vvce21me0084@vvce.ac.in, [4]vvce21me0119@vvce.ac.in

10.1201/9781003725053-35

1. INTRODUCTION

Al6061-B$_4$C metal matrix composites i.e., MMCs have gained significant interest of study due to their excellent mechanical properties viz., high hardness, wear resistance, and superior stability at high temperature, proving them best for applications with advanced engineering. Various fabrication techniques, including squeeze casting, ultrasonic-assisted methods and stir casting have been explored to enhance particle distribution, reduce porosity, and improve interfacial bonding. Each method offers distinct advantages and limitations; for example, powder metallurgy ensures precise particle control but may lead to increased brittleness, whereas stir casting is cost-effective but may result in uneven particle dispersion. These material properties have big impact on manufacturability of MMCs especially when using techniques namely electrical discharge machining (EDM). The hardness and reinforcement distribution of these composites affect factors that is surface finish, spark stability and tool wear. To enhance EDM performance, researchers continue to study the relationship between composite structure, processing techniques, and machining parameters, aiming to achieve an optimal balance between material strength and machining efficiency for high-performance applications.

This research delves into the machinability of Al-B$_4$C MMCs, particularly through Wire Electrical Discharge Machining (WEDM), focusing on key factors viz., tool wear, material removal rate (MRR), kerf width and surface roughness (SR). Additionally, the study explores the impact of machining factors like wire tension, pulse duration, dielectric fluid properties and wire feed rate on overall performance. The paper also highlights recent advancements, challenges, and potential areas for future research in the field.

2. COMPARATIVE REVIEW OF LITERATURES

This paper focusses on machinability studies on boron carbide reinforced Al6061 metal matrix composites with considering the impact of processing techniques, reinforcements and machining parameters.

2.1 Effects of Processing Techniques

In this review paper, comprehensive analysis of effects of processing techniques on wire EDM machinability for MMC is done. The study by Lodhi et al. (2025) emphasized the importance of alloy selection based on mechanical and chemical properties to ensure compatibility with boron carbide reinforcement. Boron carbide was chosen owing to its exceptional hardness, less density, and superior wear resistance. The particle size and distribution were analysed to optimize mechanical properties and ensure uniform microstructure. Kafaltiya et al. (2024) utilized stir casting by gradually introducing preheated boron carbide particles into the molten aluminium matrix to achieve uniform dispersion. Mechanical stirring at controlled speeds and temperatures improved wetting and prevented agglomeration. The use of flux minimized oxidation and enhanced interfacial bonding among matrix as well as reinforcement. Bharathi et al. (2025) adopted powder metallurgy, where aluminium and boron carbide powders were mixed using ball milling to achieve uniform dispersion. Compacting pressure influenced the density and mechanical properties of the composite. Optimized sintering temperatures enhanced diffusion bonding and improved microstructure uniformity. Mohan et al. (2017) introduced ultrasonic-assisted casting, where ultrasonic vibrations were applied to molten metal to reduce porosity and enhance reinforcement distribution. Cavitation caused by ultrasonic waves helped break particle clusters, leading to improved mechanical properties. Maintaining precise temperature control was crucial to prevent premature solidification and ensure homogeneity. Gulda Ramana et al. (2024) studied rolling effects, noting that hot rolling refines grain structure, thereby improving tensile strength and hardness. Cold rolling enhanced dimensional accuracy but introduced residual stresses. Heat treatment post-rolling was essential for stress relief and microstructure stabilization. Vasanth Kumar, et al. (2021) observed that tensile strength increased with reinforcement content up to an optimal level, beyond which brittleness became dominant. Hardness improved due to the high hardness of boron carbide, while wear resistance was enhanced through uniform reinforcement dispersion, reducing material loss during sliding conditions. Paras Kumar & Ravi Parkash, et al. (2016) examined microstructural effects, concluding that optimized processing techniques achieved uniform dispersion, leading to enhanced mechanical performance. Grain refinement was evident in ultrasonic-assisted casting and powder metallurgy. Strong interfacial bonding was crucial for effective load transfer and matrix-reinforcement integrity. Manjunatha et al. (2018) highlighted that increasing boron carbide content reduced the thermal expansion coefficient due to the ceramic reinforcement effect. Electrical conductivity slightly declined due to boron carbide's non-conductive nature. Controlled processing was essential to minimize defects and maintain desirable thermal and electrical characteristics. Saikeerthi S.P., et al. (2014) noted that corrosion resistance depended on process parameters and matrix-reinforcement interactions. Stir casting and ultrasonic treatment reduced porosity, thereby minimizing corrosion-prone sites. Powder metallurgy further enhanced corrosion resistance by eliminating casting defects and refining the microstructure. Deepak, et al. (2022) found that boron carbide reinforcement increased tool wear due to its high hardness. Cutting forces were higher in composites compared to unreinforced aluminium. Process optimization, including the use of coated tools and lubricants, improved machinability and surface finish. Ipekoglu, et al. (2017) identified key applications for aluminium-boron carbide composites. The aerospace sector benefited from their lightweight and excellent strength properties. In automotive industry, material's wear resistance made it ideal for brake rotors and engine components. The defence sector utilized it for ballistic-resistant armour and protective structures. Madeva Nagaral, et al. (2017) compared different processing methods, stating that stir casting was suitable for large-scale production with moderate property enhancement. Powder metallurgy yielded superior mechanical properties but was more expensive. Ultrasonic-assisted casting provided the best reinforcement dispersion but required precise control. Gupta P.K., et al. (2023) concluded that process selection depended on required properties, cost considerations, and application needs. Hybrid techniques combining multiple processes yielded superior results. Further research was necessary to optimize processing parameters for improved performance and cost-effectiveness

2.2 Effect of Reinforcements

In examining the reinforcement mechanisms of Al-B_4C metal matrix composites (MMCs), researchers have focused on the distribution, bonding and integration of B_4C within matrix aluminium to improve material properties like hardness, strength, and thermal stability.

Karabulut et al. (2017) observed that incorporating Boron Carbide (B_4C) into an Aluminium 6061 matrix significantly refines the microstructure, promoting a uniform grain distribution. Scanning Electron Microscopy (SEM) analysis confirms the even dispersion of B_4C particles with minimal clustering when appropriate stirring methods are applied. The presence of B_4C enhances the integrity of grain boundaries, reducing porosity and defects such as voids and inclusions. Optical microscopy further verifies that higher reinforcement levels lead to finer grains, contributing to better structural uniformity. Transmission Electron Microscopy (TEM) analysis provides insights into dislocation movement and interfacial bonding, ultimately improving the composite's mechanical properties. Nayak et al., (2020) explored the mechanical performance of these composites. Hardness measurements, using either Vickers or Brinell hardness testers, demonstrate a notable increase with greater B_4C content. This improvement is primarily owing to strong interfacial strength among the matrix and reinforcement, as well as grain refinement. Tensile strength rises as a result of the load transfer mechanism, where B_4C particles serve as stress carriers within the matrix. However, excessive reinforcement can decrease ductility and increase brittleness due to particle agglomeration and weaker interfacial bonding. Additionally, compression strength improves, making these composites ideal for applications requiring high compressive loads. Flexural strength assessments reveal a higher modulus with reinforcement,

enhancing resistance to bending forces. A. Arunnath et al. (2022) conducted wear resistance tests under dry sliding conditions, showing a substantial reduction in wear rate with increased B_4C content. The introduction of ceramic reinforcement enhances wear resistance by acting as a protective barrier against surface deterioration. Pin-on-disc tribological studies indicate that higher B_4C content decreases wear track depth, thereby improving material durability. Furthermore, the coefficient of friction is reduced due to the formation of a tribolayer, which minimizes the material loss. The mechanism pertaining to wear shifts predominantly to abrasive from adhesive wear as B_4C concentration increases, leading to superior wear performance. Deepak D et al. (2022) examined density and porosity characteristics. Density measurements show a slight increase with the addition of B_4C, as ceramic reinforcements have a higher density than the aluminium matrix. However, by employing proper processing techniques such as stir casting with controlled parameters, near-theoretical density values can be achieved, ensuring material integrity. Baradeswaran A et al. (2014) performed analysis of heat using thermogravimetric analysis (TGA) and differential scanning calorimetry (DSC) revealing that increasing B_4C content enhances thermal stability. The presence of B_4C improves heat resistance by increasing thermal conductivity while simultaneously reducing the thermal expansion coefficient. Thermal diffusivity studies confirm that composites with a higher B_4C percentage exhibit better heat dissipation, making them well-suited for high-temperature applications. Paras Kumar and Ravi Parkash (2016) studied the machinability of these composites, concluding that increasing B_4C content influences cutting forces and tool wear. Higher reinforcement levels result in greater wear out of tool because of erosive nature of ceramic particles. However, optimizing cutting parameters, such as using lower feed rates and appropriate cutting speeds, helps maintain machining efficiency. Surface roughness evaluations suggest that a moderate level of reinforcement provides an optimal balance between machinability and surface finish. Choi S K et al. (2023) highlighted that reinforcing Aluminium 6061 metal matrix composites (MMCs) with B_4C significantly enhances mechanical properties, including hardness, resistance to wear, and heat stability. Generally, addition of B_4C strengthens the composite and boosts its overall performance, proving it suitable for extreme load applications. However, effectiveness of B_4C reinforcement depends largely on achieving uniform particle dispersion and strong interfacial bonding. Poor distribution can lead to weak spots, porosity, and brittleness, which can negatively impact machinability, particularly in processes such as electrical discharge machining (EDM).

Numerous fabrication methods namely ultrasonic-assisted casting, stir casting etc., have been explored to improve the distribution and bonding of B_4C within the Al6061 matrix. Powder metallurgy, for instance, allows precise control over particle size and distribution, producing composites with high density and uniform hardness, though the increased brittleness can pose challenges during machining. Stir casting, on the other hand, offers a more economical approach but can result in particle clustering if not carefully managed, affecting the consistency and surface quality of the composite. Techniques like squeeze casting and ultrasonic-assisted casting have shown promising results in enhancing particle dispersion and reducing porosity, yielding dense, high-strength composites that maintain a balanced reinforcement across the matrix.

Composites with strong B4C reinforcement tend to have higher hardness, which, while beneficial for durability, complicates EDM by increasing tool wear and spark instability. As such, achieving optimal reinforcement involves finding a balance between enhancing strength and ensuring that the material remains practical to machine. Studies suggest that optimized processing parameters, such as controlled temperature, pressure, and particle concentration, are crucial to achieve effective reinforcement without compromising machinability.

Overall, B4C-reinforced Al6061 MMCs have shown potential in applications requiring high strength and thermal stability. However, the brittleness and high hardness introduced by B4C necessitate careful consideration of processing and machining parameters to fully harness the benefits of the reinforcement. As research continues, the focus remains on refining these methods to achieve a composite that offers both the desired mechanical properties and ease of machining, pushing the potential for Al6061-B$_4$C composites in advanced engineering applications.

2.3 Effects of Machining Parameters

The machining parameters for Al6061-B$_4$C metal matrix composites (MMCs) vary significantly depending on the processing methods and intended application, as researchers have sought to balance efficient material removal rates (MRR), tool wear and surface finish in machining these hard composites.

Ipekoglu, et al. (2017) examined how cutting speed affects machining performance. Their research showed that increasing cutting speed enhances surface finish by reducing built-up edge creation. But tool wear out rapidly with excessive more speeds. An ideal machining speed ensures efficiency while preventing excessive heat buildup. In contrast, lower speeds result in poor surface quality and higher cutting forces. Feed rate is another critical factor in machining. While a higher feed rate increases material removal, it also deteriorates surface finish and accelerates tool wear. Excessively high feed rates lead to greater cutting forces, whereas lower feed rates extend tool life but slow down the process. The best results are achieved

by balancing efficiency and surface quality through an optimized feed rate. Gudipudi et al. (2022) investigated surface roughness in machining. They found that rougher surfaces result from lower cutting speeds and higher feed rates. Choosing appropriate machining parameters helps minimize roughness and improve product quality. Factors such as tool wear and vibrations during machining also contribute to surface roughness. Proper lubrication helps reduce friction, leading to a better finish. Additionally, higher cutting speeds lead to faster tool wear due to thermal softening, especially in materials reinforced with boron carbide. Using coated tools enhances wear resistance, and maintaining optimal machining conditions extends tool life. Abhishek Krishna Sai et al., (2022) considered the influence of thermal effects on machining. Excessive heat reduces tool hardness, accelerating wear. The use of coolants effectively controls temperature, improving machining performance. Additionally, well-designed tool geometry aids in dissipating heat. Machining aluminum 6061-boron carbide metal matrix composites (MMC) presents challenges due to their abrasive nature, leading to increased tool wear and cutting forces. Selecting suitable machining parameters and advanced tooling materials enhances machinability. Anil Kumar B.N et al. (2022) explored wear mechanisms in machining aluminum 6061-B$_4$C MMC, identifying adhesion, abrasion, and diffusion as the main causes of tool wear. Coated tools help reduce adhesion wear and extend tool life, while abrasive reinforcements contribute to flank wear. Effective cooling methods minimize diffusion wear at elevated temperatures. Kumar HSV et al. (2021) analyzed tool wear rates, concluding that higher reinforcement content leads to increased wear. Optimizing cutting conditions helps mitigate excessive wear. Carbide and coated tools exhibit greater resistance to wear, while efficient cooling strategies further enhance wear performance. Nagaral M et al. (2018) studied machining efficiency, emphasizing the importance of maintaining balanced machining parameters. Excessively high cutting speed or feed rate diminishes efficiency by accelerating tool wear. Well-regulated parameters ensure energy-efficient machining, while factors such as tool material and lubrication play a crucial role in overall performance. Dama K et al. (2017) highlighted the significance of tool geometry in machining. They found that selecting the right tool angles improves efficiency. Higher rake angles reduce cutting forces but may weaken the tool, while clearance angles help prevent rubbing and improve surface finish. Overall, well-optimized tool geometry enhances tool life and machining performance. Tripathy SK et al. (2021) examined surface integrity in machining. Their study revealed that machining forces influence microhardness and residual stresses, both of which impact surface integrity. Excessive cutting forces degrade material properties, while optimal cutting conditions enhance mechanical performance. Proper cooling minimizes surface defects and improves the final finish. Additionally, excessive heat during machining can lead to thermal damage, including oxidation and cracking. Using effective cooling strategies and tool coatings helps control heat dissipation and prevent such issues. Kumar PV et al. (2018) explored machining time, finding that higher feed rates reduce machining time but negatively affect surface quality. Optimized cutting speeds and depths of cut enhance efficiency. However, increased tool wear and breakage can lead to downtime. Automation and optimized cutting conditions help maintain productivity while reducing machining time. They also noted that residual stresses introduced during machining affect fatigue strength. Selecting appropriate machining parameters enhances fatigue resistance, reducing the likelihood of crack formation and improving material durability under cyclic loading. Veeresha G et al. (2023) studied and highlighted the need for precise control of machining parameters due to the material's hardness and abrasiveness pertaining to the machining of composites. In WEDM, variables viz., pulse duration, wire pull, dielectric fluid pressure, and pause time significantly impact machining outcomes. Adjusting these parameters influences MRR, surface texture, and tool abrasion. Changes of pulse settings affect spark stability and energy discharge, which are essential for maintaining consistent machining performance and minimizing the risk of wire breakage.

Overall, the Dielectric fluid management, including flow rate and pressure, is another vital factor, as it aids in efficient debris removal and maintains spark consistency. This is particularly important for reducing surface roughness and preventing excessive tool wear during machining. Other parameters like wire feed rate and pulse frequency are adjusted to balance between achieving high MRR and minimizing surface defects. Collectively, these machining parameters must be optimized with precision, as each setting affects overall stability and the composite's responsiveness during machining. By carefully controlling these variables, it becomes possible to overcome the challenges of machining hard composites, achieving efficient and high-quality machining outcomes.

3. CONCLUSIONS

Aluminium-Boron caribe MMC have developed as special class of advanced materials having excellent mechanical, tribological and thermal properties. The review conducted emphasizes the potential advancements in material development, processing techniques, performance behaviour of Al6061-B$_4$C MMC and machining factors in the wire EDM process. The different fabrication techniques viz., stir casting, powder press and sintering, ultrasonic assisted casting etc., have been studied to optimize the interfacial bonding and microstructural uniformity. Still problems such as agglomeration, interfacial bonding and machining difficulties remain areas for further exploration.

The findings from this study emphasized the significance of carefully selecting the optimized condition for WEDM of Al6061-B$_4$C metal matrix composites to obtain optimum machining rate, surface texture and durability of cutting tool. With the recent advancements in the MMC, it is further required to investigate the processing methods and machining parameters which improve the material properties and machinability. Aluminium-Boron caribe MMC possesses impressive potential to be used in high performance applications. Therefore, further research is required to overcome the current limitations and explore fully their capabilities for real engineering applications.

REFERENCES

1. Abhishek Krishna Sai, A., Chandrasekaran, M., Vinod Kumar, T., & Pugazhenthi, R. (2022). Study the enhanced mechanical properties of Al/B4C metal matrix composite. Materials Today: Proceedings, 68(5), 1422–1428. https://doi.org/10.1016/j.matpr.2022.06.465.

2. Anil Kumar, B.N., Ahamad, A., & Reddappa, H.N. (2022). Impact of B4C reinforcement on tensile and hardness properties of Al-B4C metal matrix composites. Materials Today: Proceedings, 52(3), 2136–2142. https://doi.org/10.1016/j.matpr.2021.12.454.

3. Arunnath, A., Madhu, S., & Tufa, M. (2022). Experimental investigation and optimization of material removal rate and tool wear in the machining of aluminum-boron carbide (Al-B4C) nanocomposite using EDM process. Advances in Materials Science and Engineering, 2022, 4254024. https://doi.org/10.1155/2022/4254024.

4. Baradeswaran, A., Vettivel, S.C., Perumal, A.E., Selvakumar, N., & Franklin Issac, R. (2014). Experimental investigation on mechanical behaviour, modelling and optimization of wear parameters of B4C and graphite reinforced aluminium hybrid composites. Materials & Design, 63, 620–632. https://doi.org/10.1016/j.matdes.2014.06.054.

5. Bharathi, P., Kumar, T. S., & Anbuchezhiyan, G. (2025). Multioptimization analysis of machining characteristics on spark electrical discharge machining of Al/SiC and Al/SiC/B4C composites. Engineering Research Express. https://doi.org/10.1088/2631-8695/ada723.

6. Choi, S.K., Seo, B., Kang, J.W., Kim, D.H., Park, H.K., & Park, K. (2023). Microstructure and wear properties of aluminum metal matrix composite (Al6061-B4C) fabricated by stir casting process. https://dx.doi.org/10.2139/ssrn.4375432.

7. Dama, K., Prashanth, L., Nagaral, M., Mathapati, R., & Hanumantharayagouda, M.B. (2017). Microstructure and mechanical behavior of B4C particulates reinforced ZA27 alloy composites. Materials Today: Proceedings, 4(8), 7546–7553. https://doi.org/10.1016/j.matpr.2017.07.086.

8. Deepak, D., Gowrishankar, M.C., & Shreyasa, D.S. (2022). Investigation on the wire electric discharge machining performance of artificially aged Al6061/B4C composites by response surface method. Materials Research, 25:e20220010. https://doi.org/10.1590/1980-5373-MR-2022-0010.

9. Gudipudi, S., Nagamuthu, S., & Chilakalapalli, S.P.R. (2022). A comprehensive investigation on machining of composites by EDM for microfeatures and surface integrity. Journal of Micromanufacturing, 5(1), 5–20. https://doi.org/10.1177/25165984211063308.

10. Gupta, P.K., & Gupta, M.K. (2023). Optimization of wear behaviour of hybrid Al(6061)-Al2O3-B4C composites through hybrid optimization method. Materials Physics and Mechanics, 51(4), 23–37. http://dx.doi.org/10.18149/MPM.5142023_3.

11. Gulda Ramana, Supraja, Y. S., Rajashekar, T., Rahul, A., Devender, B., & Prasada Rao, Y. V. S. S. S. V. (2024). Experimental analysis of Al6061/SiC/B4C/Fly-ash metal matrix composite. RP Current Trends In Engineering and Technology, 3(1), 13–19. e-ISSN: 2583-5491.

12. Ipekoglu, M., Nekouyan, A., Albayrak, O., & Altintas, S. (2017). Mechanical characterization of B4C reinforced aluminum matrix composites produced by squeeze casting. Journal of Materials Research, 32, 599–605. https://doi.org/10.1557/jmr.2016.495.

13. Kafaltiya, S., Chauhan, S., Singh, V. K., & Verma, A. (2024). Experimental investigation, modeling and optimization of wire EDM process parameters for machining AA2024-B4C self-lubricating composite. Physica Scripta, 100(1), 015036. https://doi.org/10.1088/1402-4896/ad9d9d.

14. Karabulut, Ş., Karakoç, H., & Çitak, R. (2017). Effect of the B4C reinforcement ratio on surface roughness of Al6061 based metal matrix composite in wire-EDM machining. 8th International Conference on Mechanical and Aerospace Engineering (ICMAE), Prague, Czech Republic, 812–815. https://doi.org/10.1109/ICMAE.2017.8038755.

15. Kumar, H.S.V., Kempaiah, U.N., Nagaral, M., & Revanna, K. (2021). Investigations on mechanical behaviour of micro B4C particles reinforced Al6061 alloy metal composites. Indian Journal of Science and Technology, 14(22), 1855–1863. https://doi.org/10.17485/IJST/v14i22.736.

16. Kumar, P., & Parkash, R. (2016). Experimental investigation and optimization of EDM process parameters for machining of aluminum boron carbide (Al–B4C) composite. Machining Science and Technology, 20(2), 330–348. https://doi.org/10.1080/10910344.2016.1168931.

17. Kumar, P. V., K. S. H., & Nagaral, M. (2018). Processing, microstructure, density and compression behaviour of nano B4C particulates reinforced Al2219 alloy composites. In Proceedings of the International Journal of Advanced Technology and Engineering Exploration, 5(46). ISSN (Print): 2394-5443, ISSN (Online): 2394–7454. https://dx.doi.org/10.19101/IJATEE.2018.546015.

18. Lodhi, B. K., Sharma, S., & Agarwal, S. (2025). Investigation to study the effect of WEDM parameters on the dimensional deviation during machining of hybrid aluminum metal matrix composite. Proceedings of The Institution of Mechanical Engineers, Part E: Journal of Process Mechanical Engineering. https://doi.org/10.1177/09544089241307811.

19. Madeva, N., Pavan, R., & Auradi, V. (2017). Tensile behavior of B4C particulate reinforced Al2024 alloy metal matrix composites. FME Transactions, 45, 93–96. https://doi.org/10.5937/fmet1701093N.

20. Manjunatha, B., Niranjan, H. B., & Satyanarayana, K. G. (2018). Effect of amount of boron carbide on wear loss of Al-6061 matrix composite by Taguchi technique and response surface analysis. IOP Conference Series: Materials Science and Engineering, 376, 012071. https://doi.org/10.1088/1757-899X/376/1/012071.

21. Mohan, M., Balamurugan, A., Jagadeeshwar, V., & Ramkumar, M. (2017). Analysis of mechanical properties for Al6061 alloy metal matrix with boron carbide & graphite. International Journal of Novel Research and Development, 2(5). ISSN: 2456–4184.

22. Nagaral, M., Kalgudi, S., Auradi, V., & Kori, S.A. (2018). Mechanical characterization of ceramic nano B4C-Al2618 alloy composites synthesized by semi-solid state processing. Transactions of the Indian Ceramic Society, 77(3), 146–149. https://doi.org/10.1080/0371750X.2018.1506363.

23. Nayak, B. B., Sahu, S., & Das, D. (2020). Investigation on machining performance of boron carbide reinforced aluminium matrix composite during WEDM taper cutting process. Materials Today: Proceedings, 26(2), 932–936. https://doi.org/10.1016/j.matpr.2020.01.147.

24. Saikeerthi, S. P., Vijayaramnath, B., & Elanchezhian, C. (2014). Experimental evaluation of the mechanical properties of aluminium 6061-B4C-SiC composite. In Proceedings of the International Journal of Engineering Research, Vol. 3, Special Issue 1, pp. 70–73.

25. Sankar, M., Gnanavelbabu, A., Rajkumar, K., & Thushal, N.A. (2016). Electrolytic concentration effect on the abrasive assisted-electrochemical machining of an aluminum–boron carbide composite. Materials and Manufacturing Processes, 32(6), 687–692. https://doi.org/10.1080/10426914.2016.1244840.

26. Tripathy, S.K., & Senapati, A.K. (2021). An extensive analysis of mechanical and tribological studies of Al6061 alloy-based hybrid composite reinforced with B4C and graphene. Materials Today: Proceedings, 44(1), 2808–2812. https://doi.org/10.1016/j.matpr.2020.12.1148.

27. Vasanth Kumar, H. S., Kempaiah, U. N., Nagaral, M., & Revanna, K. (2021). Investigations on mechanical behaviour of micro B4C particles reinforced Al6061 alloy metal composites. Indian Journal of Science and Technology, 14(22), 1855–1863. https://doi.org/10.17485/IJST/v14i22.736.

28. Veeresha, G., Manjunatha, B.C., Nagaral, M., & Auradi, V. (2023). Mechanical characterization of 44-micron sized B4C particle reinforced Al2618 alloy composites. Materials Today: Proceedings, 434–439. https://doi.org/10.1016/j.matpr.2021.03.572.

Advances in Mechanical Engineering and Materials Sciences – Dr. Vinay K. B et al. (eds)
© 2026 Taylor & Francis Group, London, ISBN 9-781-041-20970-6

36

A Review on the Characterization of Bio-Fiber Composites— Advancements and Challenges in Natural and Synthetic Fiber Integration

Naveen Ankegowda[1], Mohammed Sarosh[2],
Mohammed Zulfiqar[3], Mohammad Yahiya Ayaan[4], and Akash B.[5]
Department of Mechanical Engineering, Vidyavardhaka College of Engineering,
Mysuru, Karnataka State, India

Abstract: Natural and synthetic fiber composites are increasingly utilized in material engineering for their sustainability, enhanced mechanical properties, and ability to repurpose agricultural waste. Fibers such as pineapple leaf, wheat straw, hemp, kenaf, banana, and jute reinforce polymer and ceramic matrices, significantly improving tensile and flexural strengths. Palf-pp composites enhance tensile strength by up to 29.4%, while wheat straw fibers improve pp's tensile and flexural properties by 29% and 49%, respectively. Hybrid composites, like banana-linen with epoxy and hemp-recycled polymers, exhibit superior mechanical performance for advanced applications. Chemical treatments, such as alkali and silane, improve fiber-matrix bonding, thermal stability, and water resistance, enhancing composite durability. Advanced techniques like vacuum infusion and twin-screw extrusion optimize fiber integration and thermal properties. Materials like sugarcane bagasse ash and fly ash provide eco-friendly, cost-effective options for industrial applications. The study highlights the potential of natural fiber-reinforced composites as sustainable, high-performance materials for use in construction, automotive, and packaging industries.

Keywords: Bio-fiber composites, Natural fibers, Mechanical properties, Polylactic acid (PLA), Ceramic composites, metal composites, Material composition and methodology

1. INTRODUCTION

Industry-induced pollution has been a problem since the beginning of the industrial revolution. Composite composites using biofibers can help reduce industrial pollution. Compared to conventional materials, these environmentally friendly composites, which are composed of natural fibers like hemp and jute, provide improved qualities. They are appropriate for the packaging, building, and automotive industries due to their biodegradable, renewable, and lightweight characteristics. To maximize applications, characterization entails evaluating

mechanical and chemical qualities. Customized composites are made possible by a variety of fabrication techniques. Making biofibers from India's plentiful agricultural waste encourages sustainability, lowers pollution, and increases revenue. Future studies seek to increase employment and reduce production costs through large-scale.

2. LITERATURE REVIEW

2.1 Material Composition

Pineapple leaf fibers (PALF) are used as a reinforcement material in a composite with Polypropylene, achieving

[1]naveen@vvce.ac.in, [2]Afeefasarosh@gmail.com, [3]mohammedzulfiqar.03@gmail.com, [4]mohammadyahiyaayaan@gmail.com, [5]adarshakash723@gmail.com

a 29.4% improvement in tensile strength at 10.8% fiber loading, helping to reduce agricultural waste in pineapple production in India (R.M.N. Arib 2006). Wheat straw fibers, at 30% fiber loading, improve the tensile and flexural strength of Polypropylene PP 3622 by 29% and 49%, respectively, while aiding in reducing agricultural waste in India, which produces 14% of the world's wheat (S. Panthapulakkal,2006) A wood plastic composite consists of 47% wood, 47% polymer, and 6% additives (colorants, lubricants, stabilizers), combining to create a durable material (Anil Akdogan 2005). Cork powder mixed with high-density polyethylene (HDPE) forms a composite, though it is less recyclable compared to other natural fiber composites in India, where cork production is minimal (Emanuel M. Fernandes 2013). Kenaf fibers, with 40% fiber loading in Polypropylene, significantly improve the mechanical properties of the polymer, offering excellent tensile and flexural strength, though global Kenaf production is limited (Saba N.,2015). Ramie fiber, used as a reinforcement in composites made from PLA, Polypropylene, or Epoxy, offers high tenacity and durability, with fiber loading varying from 10%-30% depending on the base material (Du Y 2015). Silk, used as a reinforcement material in PLA composites, has high strength and extensibility, with 50% fiber loading offering enhanced mechanical properties (Noorunnisa Khanam P, 2015). Waste silk fibers, with 40% fiber loading, form a composite with Poly-butanate Succinate (PBS), improving tensile strength (S M. Darshan 2016). A composite made from Banana and Linen fibers as reinforcement and Epoxy as the base material, with 10% fiber loading, offers improved mechanical properties (M. Ramachandran 2016). Roselle fiber, used as reinforcement with isotactic polypropylene (iPP) at 40% fiber loading, enhances the strength and durability of the composite (Nadlene R ,2016). PALF fibres can be used at different fibre loadings to get different qualities. The reinforcement used is PALF fibre with Polypropylene, A polymer composite with 45% fiber loading of pineapple leaf fibers (PALF) and 55% Polypropylene showed improved mechanical properties (R.M.N. Arib & Nadlene R). Plant and animal-based biodegradable fibers, like jute and sheep wool (15% fiber loading), are used as reinforcement in Polypropylene composites, offering advantages such as low cost and safe handling (Motaleb K. Z. M. A ,2018). Silk can also be used in laminated sheets as reinforcement in Epoxy composites, offering enhanced properties (Tusnim J. 2018).

Hemp fibers, mixed with Virgin Polypropylene, bicomponent, and recycled polyester polymers (30%-50% fiber loading), improve the strength and durability of the composite (Reddy A. V ,2019). Sugarcane fiber, combined with 25% sugarcane powder and 55% Epoxy resin, enhances the tensile strength of the composite (Adeniyi, A. G ,2019). Coconut coir fibers, treated and mixed with Polypropylene (60% coir, 37% Polypropylene, 3% maleic anhydride grafted PP), form a durable composite material

(Tanasa F, 2019). Coffee cherry husk fibers, used as reinforcement with High-Density Polyethylene (HDPE), exhibit optimal tensile strength at 10% fiber loading (Wang Z, 2019). Banana fibers (15%-20% fiber loading) are used with Polyethylene as the base material, improving the composite's mechanical properties (Laxshaman Rao B,2020). Bamboo fibers from Gigantochlea scotechini species (40% fiber loading) are reinforced in Epoxy or polyester composites, providing strong and durable materials (Chin, S. C,2019). Abacá fibers, with 20%-40% fiber loading and base materials Glycidyl methacrylate and 4HBG polymers, enhance composite strength and flexibility (Bin Jeremiah D. Barba, 2020). Tobacco stalk residues, containing up to 40% cellulose, can be used with polymers as reinforcement, though the fiber loading and polymer type are unspecified (Elbehiry, A 2020). Sugarcane Bagasse ash as a reinforcement in ceramic composites, with fiber loading at 20%, 40%, and 60% for varying mechanical tests (Hossain, S. S , 2018). Ramie fibers, with 52% fiber loading, are reinforced in Aluminum-oxide ceramic composites (Wang Z, 2019). Rice husk ash (RHA) is used with Silica ceramics, with 20% fiber loading showing optimal mechanical properties (B. Laxshaman Rao B, 2021). Sugarcane fiber is combined with ceramic fillers (Alumina, Silicon carbide) and Epoxy, with 30% fiber and 10%-20% ceramic filler showing the best results (Elbehiry, A ,2020). Banana fibers are reinforced in concrete composites with 25% fiber loading (Fu, F 2017). Metal-natural fiber composites combine metal sheets with natural fibers, offering improved mechanical properties (Taohai Yan 2023). A stainless-steel filament woven with cotton yarn (1:1 and 2:1 ratios) showed enhanced strength (Behjat Tajeddin 2009). Kenaf fibers are used with Low-Density Polyethylene (LDPE) at 30% fiber loading, showing good results (Braga, R. A 2015).

Jute, Glass fiber, and Epoxy resin form a composite, with 18% jute fiber and 19% glass fiber showing the best mechanical properties (Sepe R, 2017). Hemp fibers (40.8%) are reinforced in Epoxy resin composites for improved tensile and flexural properties CAloe Vera fibers (30%) are mixed with MMT clay and PLA, yielding strong results after testing (Ramesh P 2019). Kenaf fibers, MMT clay, and PLA are combined with 30% fiber loading for improved mechanical performance (Cavalcanti,2020). Jute and Sisal fibers (60% and 40%) are woven and combined with Epoxy resin (30% fibers, 70% resin) for optimal results in mechanical tests (Ayu, R. S,2020). Empty Fruit Bunch (EFB) is used as reinforcement in Poly Butanate Succinate (PBS) composites, with 20% EFB, 10% Tapioca starch, and 70% PBS showing the highest mechanical properties (Ayu, R. S,2020). EFB and PBS (30% fiber loading) show similar mechanical properties to 100% PBS composites. Basalt fibers (30%) are reinforced in unsaturated polyester resin (UP), showing high tensile strength (Sapuan, S. M., ,2020). Sugar Palm starch fibers (30%) are used with Polypropylene (PP) for high compressive strength (Harussani, M. M 2021). Carbon Nanotubes (CNT) at 5% fiber loading in Epoxy show the highest tensile strength compared to other CNT percentages (Rafiqah S 2020). Ramie, Kevlar, and Polyester fibers are combined for bulletproof protection jackets. Oil palm fronds are a crop residue and contain large amounts of fibre. These can be used as reinforcement material with Polyurethane being the base material. The fibre loading was at 50% (Harussani, M m , 2021).

2.2 Mechanical Treatments on Natural Fiber

Pineapple fibers are layered between plastic sheets in a metal mold, pressed at 12.4 MPa for 5 minutes, cooled, and prepared for testing (R.M.N. Arib 2006). Wheat straw fibers are mixed with melted plastic, injected into a mold, cooled, and processed for testing (S. Panthapulakkal,2006). Wood powder, LDPE, and additives are extruded into circular bars, cooled, cut, and shaped into final products like trim boards (Anil Akdogan 2005). HDPE, cork powder, and coconut fibers are blended in a twin-screw extruder, compression-molded into 3mm samples, and tested for strength (Emanuel M. Fernandes 2013). Kenaf fibers are layered with resin in a mold, heat-cured under pressure, then trimmed and polished for final use (Saba N,2015). Ramie cellulose and polymer mixtures are molded or extruded, with tensile and impact strength enhanced by adding 50% glass fibers (Du Y ,2015). Hand layup involves silk fabric layers compacted with resin under vacuum, with silk offering UV protection, elasticity, and moisture management (Noorunnisa Khanam P ,2015). Silk fibers, derived from cocoons and spiders, are mixed with resin to enhance composite properties, being cost-effective and UV-protective (S M. Darshan 2016). Bamboo, banana, and linen fibers are mixed with epoxy and layered in a wax-coated mold to create solid composites (M. Ramachandran 2016). Treated natural fibers serve as synthetic substitutes, enhancing strength and stability for diverse applications (Nadlene R 2016). Kenaf fibers are cleaned, dried, cut, and mixed with melted plastic or resin in a mold, then compressed and heated. The composite is cooled, trimmed, and smoothed (Lim Z. Y, 2018). Glass fibers are layered in a mold with sugarcane powder and epoxy resin, cured at 40°C for 12 hours (Motaleb K. Z. M. A ,2018). Jute and sheep wool fibers are chopped into 3mm pieces and layered with polypropylene resin in a mold. The setup is pressed hydraulically, cooled, and tested for mechanical properties (Tusnim J. 2018). Sugarcane powder mixed with epoxy resin is layered with glass fibers, cured at 40°C for 12 hours, and tested per ASTM standards (Reddy A. V ,2019). Coir fibers from coconut shells are crushed into various sizes and used with shell particles for polymer reinforcement (Adeniyi, A. G ,2019). NFPCs have strong mechanical properties, but modifications like waxing, while enhancing bonding, can be costly and less effective (Tanasa F, 2019). Coffee husk powder is mixed

with HDPE and modifiers in a rubber mixer at 165°C, then pressed into panels at 190°C and 15 MPa, with flipping for uniformity (Wang Z, 2019). Banana fibers from stems are mixed with polymer resin, molded, cured in an oven, and tested for physical and mechanical properties (Laxshaman Rao B,2020). Bamboo, aged 3-4 years, is cut into sp`lints, mechanically processed into fibers, washed to neutral pH, dried at 60°C, and stored to prevent moisture (Chin, S. C,2019). Abaca fibers, extracted from abaca stalks, are sorted by quality grades and prepared for use, typically in lengths of 1.5-3.5 meters (Bin Jeremiah D. Barba, 2020).

Single banana fibers are bundled into bars of varying diameters to create reinforced concrete beams (1000 mm × 200 mm × 250 mm), tested under mid-span loads for strength comparison . Thermogravimetric Analysis (TGA) and Differential Scanning Calorimetry (DSC) evaluate thermal stability and heat behavior in rice husk ash ceramics, while thermal conductivity tests measure insulation properties (Hossain, S. S ,2018). Sugarcane bagasse ash (SCBA) and ceramic clay are blended in varying SCBA ratios (0%-60%) to create roof tiles, kneaded for 48 hours, and pressed into shapes (Wang Z, 2019). Bamboo splints are processed into fibers, washed to neutral pH, dried at 60°C, and stored to maintain quality (Laxshaman Rao B, 2021). Single banana fibers are used to form bars of different diameters, integrated into reinforced concrete beams, and tested for load resistance alongside plain concrete beams (Elbehiry, A ,2020). A flame-retardant chemical is applied to fabric to enhance fire resistance, cured in a hot air oven, and tested using specialized equipment for performance (Fu, F 2017). Laser or mechanical incising makes small cuts on wood to improve treatment penetration but can weaken its structure (Taohai Yan 2023). Stainless steel filaments twisted with cotton yarns create composite yarns, which are woven into fabric. The twisting direction and filament count affect the structure and thickness (Behjat Tajeddin 2009). Jute and glass fiber composites are tested for density, tensile strength, flexural strength, and impact resistance to evaluate mechanical properties (Braga, R. A 2015). Hemp fiber composites are fabricated using the Vacuum Infusion Process, where resin-infused hemp layers are cured at room temperature for 24 hours to solidify the structure (Sepe R, 2017). PLA, treated aloe vera fiber, and MMT clay are compounded using a twin-screw extruder, pelletized, and molded into sheets via compression molding, followed by mechanical tests for strength (Ramesh P 2019). Treated kenaf fiber/MMT clay reinforced PLA bio composites are processed at 185°C and 30 tons pressure, with 1% MMT clay enhancing tensile and flexural properties (Ramesh P 2019). Natural fibers like jute, sisal, and curaua are layered in a mold for hand layup, with epoxy resin cured at 80°C for 6 hours and polyester resin at 75°C after room-temperature curing, improving composite strength (Cavalcanti,2020). Composites reinforced with 20 wt% EFB fibers show improved tensile (6.0%) and flexural

(12.2%) strengths due to better load transfer, as shown by SEM analysis (Ayu, R. S,2020). Modified tapioca starch and EFB fibers are mixed in a counter-rotating internal mixer, shaped into pieces, and hot-pressed into films, with mechanical testing assessing tensile and flexural strength for agricultural use (Ayu, R. S,2020). Hybrid composites are fabricated using the hand lay-up method, combining basalt and glass fibers with resin, enhancing tensile and flexural strength, and increasing density in glass-fiber-reinforced polyester resin (Sapuan, S. M., ,2020). Polypropylene (PP) powder is prepared from shredded COVID-19 PPE gowns using a cutting mill, improving efficiency for subsequent pyrolysis processes (Harussani, M. M 2021). CNTs are dispersed in a polymer matrix via solution mixing, using agitation methods like stirring or sonication, achieving uniform distribution to enhance mechanical and electrical properties (Mohd Nurazzi, N 2021). Natural fibers undergo mechanical treatments like fibrillation and electric discharge to improve bonding with the polymer matrix and enhance composite strength (Nurazzi, N. M ,2021). Woven natural fiber composites, such as woven jute fabrics, exhibit superior tensile and flexural strengths compared to non-woven composites due to their structured fiber arrangement (Aisyah, H. A, 2021).

2.3 Insights of Testing Results

The tensile and flexural properties of PALF-PP composites improved with increasing fibre volume up to an optimal point, beyond which voids and misalignment reduced performance. SEM analysis confirmed fibre pull-out and voids as key factors in reduced mechanical integrity at higher fibre content (R.M.N. Arib 2006). Wheat straw fibers enhanced polypropylene composites' mechanical properties at 30% content, but chemically treated fibers weakened strength due to poor dispersion. Fiber strength variability was analyzed using the Weibull distribution method (Panthapulakkal S.,2006). In comparison to tangential and radial sections, Katsura wood exhibits reduced axial wear, with the tangential section peaking at 20 MPa. Unlike plastics, where maximum wear corresponds with counterface yield stress, wear varies with counterface materials (PE, PEEK, PAI) and abrasive grain size because of its porous structure (Anil Akdogan 2005). Alkali treatment improves the performance of kenaf fiber composites, which are a robust and lightweight substitute for conventional materials. However, their sensitivity to moisture limits their use. They have potential for use in building, especially in insulating, lightweight panels and blocks (Emanuel M. Fernandes 2013). Silane treatment improved the bonding and mechanical properties of ramie fibres, while other treatments, including NaOH and heat treatment, weakened tensile strength and modulus. Heat treatment only caused damage above 200°C, highlighting the importance of thermal stability (Du Y ,2015). The mechanical characteristics of polymer composites, especially PLA, PP, and epoxy, are improved

by silk fibers and further enhanced by irradiation. Because more silk improves bonding and crystallization, these biocomposites can be used in biomedical applications (Noorunnisa Khanam P ,2015). Silk cocoon fibers improve the strength, elasticity, and ductility of composites, especially PLA, making them an affordable material. However, heat resistance varies depending on the fiber content (S M. Darshan 2016). Epoxy composites made of bamboo, banana, and linen show promise for a variety of industrial uses due to their high impact resistance and hardness, respectively (M. Ramachandran 2016). Roselle fiber improves the characteristics of phenol formaldehyde composites, especially when it is 15 cm long. In cement, it increases tensile strength but decreases compression. For composite performance, the ideal fiber composition and surface treatments are essential (Nadlene R, 2016).

The study revealed that thicker materials improve sound absorption, with kenaf fibers showing impressive absorption coefficients above 1 kHz. Air gaps reduce peak absorption frequencies but enhance low-frequency absorption without added thickness, and increasing fiber density boosts absorption across frequencies, outperforming rock wool in some cases (Lim Z. Y, 2018). Alkali-treated PALF composites, particularly PP725, showed improved water absorption and significant gains in tensile strength and modulus with increased fiber content, though elongation at break decreased. The study highlights PALF composites as superior to other natural fibers for eco-friendly applications (Motaleb K. Z. M. A, 2018). Higher fiber loading, especially with a 3:1 jute:wool ratio, improved tensile and flexural strength, energy absorption, and impact resistance. Jute's superior tensile strength and cellulose content contributed significantly to the composites' enhanced mechanical properties (Tusnim J. 2018). The study found that 25% sugarcane powder was optimal for enhancing the tensile and flexural strength of glass fiber epoxy composites, with further increases leading to a decline in performance. This is due to poor filler distribution and the negative impact of excess sugarcane powder on the composite structure [14]. Higher fiber content in coir-polymer composites reduced thermal stability, while treated fibers improved water resistance and mechanical properties. The optimal composite formulation for automotive interiors was 60% coir fiber, 37% PP powder, and 3% MAPP (Adeniyi, A. G ,2019). Alkaline and mercerization treatments enhanced the tensile strength, modulus, and strain of hemp fiber composites, with significant improvements in mechanical properties, especially with compatibilizing agents. Ozone treatment further increased grafting and reduced hydrophilicity, boosting thermal stability and interfacial adhesion for better composite performance (Tanasa F, 2019).

Coffee hull fiber in HDPE composites improved tensile strength, modulus, and water resistance, especially with $Ca(OH)_2$ treatment enhancing hydrophobicity and compatibility. Thermal stability and strong bonding between fiber and polymer were confirmed, boosting mechanical performance (Wang Z, 2019). Flyash cenospheres and MA-g-HDPE treatments reduced water absorption in composites, while fiber and nano-clay additions enhanced strength and modulus. Banana fiber composites outperformed pure vinyl-ester resin, though fiber orientation affected strength and wear resistance (Laxshaman Rao B, 2020). Alkali treatment of bamboo fibers improved crystallinity and reduced weight loss, with the best results achieved using 10% NaOH for 48 hours. Bamboo fiber-reinforced composites with 40% fiber volume fraction outperformed polyester and vinyl-based composites in strength (Chin, S. C,2019). Natural fibers like abaca, flax, and jute improve automotive materials by offering strength, lightness, and eco-friendliness. Abaca outperforms jute and glass fibers in reinforcement, with treatments enhancing bond strength and sustainability (Bin Jeremiah D. Barba, 2020). Banana fiber bars increased concrete strength by 25%, improving load bearing and preventing cracking. They proved effective regardless of concrete strength, offering a sustainable alternative for construction materials (Elbehiry, A ,2020). Rice husk ash, rich in silica, improves ceramic properties by enhancing strength and reducing energy use in production. It offers a sustainable solution by boosting performance, reducing pollution, and managing rice husk waste (Hossain, S. S ,2018). Silicon carbide and alumina fillers reduce water absorption and enhance chemical resistance in bamboo epoxy composites. Treated composites perform better in both water and chemical resistance compared to untreated ones (Laxshaman Rao B, 2021). Microwave treatment reduces wood moisture but weakens its structure, while biological treatments enhance wood permeability by altering microfibrils. These methods adjust wood properties for better usability in various applications (Fu, F 2017).

Yarns twisted in the same direction and stainless-steel filaments improved fabric durability and flame resistance, albeit with slightly reduced comfort. Post-treatment enhanced flame-retardant properties, meeting safety standards and ensuring thermal stability (Taohai Yan 2023). Kenaf cellulose in the LDPE matrix improved thermal stability and increased the melting point, making the composites more durable. PEG addition enhanced adhesion and thermal resistance, ideal for food packaging applications (Behjat Tajeddin 2009). Jute and glass fibers in epoxy resin enhanced mechanical strength, with composition affecting performance. Higher jute content increased water absorption and mass loss at elevated temperatures, potentially impacting durability (Braga, R. A 2015). Sodium hydroxide treatment improved hemp fiber adhesion and mechanical properties, enhancing tensile and flexural strength. However, the composites still did not reach exceptionally high mechanical performance levels (Sepe R, 2017). 1 wt.% MMT clay enhanced mechanical

properties, thermal stability, and water resistance in PLA-hybrid biocomposites, while higher MMT concentrations reduced performance due to agglomeration. Increased MMT content also decreased biodegradability [Ramesh p, 2019]. 20 wt% EFB fibers enhanced tensile and flexural strengths, but poor fiber-matrix bonding limited mechanical performance. Higher fiber content improved matrix interaction but increased moisture absorption due to fiber hydrophilicity (Cavalcanti,2020). EFB fibers at 40 weight percent decreased tensile and flexural strength because of void formation and inadequate fiber-matrix bonding, according to R. Sepe in 2018. But because they were thermally stable and reduced production costs, they could be used to make agricultural mulch films. Basalt fibers enhanced the mechanical properties of glass-fiber-reinforced polyester composites, improving tensile and flexural strengths. PP waste char briquettes with SPS binder showed promising strength and combustion characteristics, offering a sustainable energy solution. Kenaf-HDPE composites (Nurazzi, N. M ,2021) show promise as a sustainable alternative to Kevlar in ballistic vests, with flax composites also demonstrating strong performance, especially when combined with glass fiber or woven laminates. Further research is needed to optimize these natural fiber composites (Aisyah, H. A, 2021).

3. CONCLUSION

Bio-fibre composites represent a promising alternative to conventional synthetic materials, offering substantial environmental benefits, including biodegradability, renewable sourcing, and reduced carbon footprints. The characterization of these materials is essential for understanding their mechanical, thermal, and durability properties, which influence their potential for widespread commercial applications, such as in automotive, construction, and consumer products. Whereas bio-fiber composites provide lightweight, environmentally friendly substitutes, their uneven performance as a result of moisture absorption and natural fiber variability calls for careful characterization and enhanced fiber-matrix bonding for dependable applications. Despite present processing difficulties, bio-fibre composites provide a sustainable substitute for synthetic composites because of their reduced environmental effect. A bright future with more extensive industrial uses is suggested by ongoing developments and rising demand for environmentally friendly materials.

REFERENCES

1. R.M.N. Arib, Sapuan, S. M., Ahmad, M. M. H. M., Paridah, M. T., & Zaman, H. M. D. K. (2006). Mechanical properties of pineapple leaf fibre reinforced polypropylene composites. Materials & Design, 27(5), 391–396.
2. Panthapulakkal S., Zereshkian, A., & Sain, M. (2006). Preparation and characterization of wheat straw fibers for

reinforcing application in injection molded thermoplastic composites. Bioresource Technology, 97(2), 265–272.

3. Anil Akdogan, 2005, Wood reinforced polymer composites. (2008). Tribology of Natural Fiber Polymer Composites, 180–196. doi:10.1533/9781845695057.18

4. Emanuel M. Fernandes, Correlo, V. M., Mano, J. F., & Reis, R. L. (2012). Natural Fibres as Reinforcement Strategy on Cork-Polymer Composites. Materials Science Forum, 730–732, 373–378. doi:10.4028/www.scientific.net/msf.730-732.373

5. Saba N., Paridah, M. T., & Jawaid, M. (2015). Mechanical properties of kenaf fibre reinforced polymer composite: A review. Construction and Building Materials, 76, 87–96. doi:10.1016/j.conbuildmat.2014.11

6. Du Y., Yan, N., & Kortschot, M. T. (2015). The use of ramie fibers as reinforcements in composites. Biofiber Reinforcements in Composite Materials, 104–137. doi:10.1533/9781782421276.1.104

7. Noorunnisa Khanam P., Al-Maadeed, M. A., & Naseema Khanam, P. (2015). Silk as a reinforcement in polymer matrix composites. Advances in Silk Science and Technology, 143–170. doi:10.1016/b978-1-78242-311-9.00008-2

8. S. M. Darshan, B. Suresha, G. S. Divya. (2016) "Waste Silk Fiber Reinforced Polymer Matrix Composites," Indian Journal of Advances in Chemical Science S1, 183–189.

9. (M. Ramachandran, R., Bansal, S., & Raichurkar, P. (2016). Experimental study of bamboo using banana and linen fibre reinforced polymeric composites. Perspectives in Science, 8, 313–316. doi:10.1016/j.pisc.2016.04.063

10. Nadlene R., Sapuan, S. M., Jawaid, M., Ishak, M. R., & Yusriah, L. (2015). A Review on Roselle Fiber and Its Composites. Journal of Natural Fibers, 13(1), 10–41. doi:10.1080/15440478.2014.984052.

11. Lim Z. Y., Putra, A., Nor, M. J. M., & Yaakob, M. Y. (2018). Sound absorption performance of natural kenaf fibres. Applied Acoustics, 130, 107–114. doi:10.1016/j.apacoust.2017.09.01

12. Motaleb K. Z. M. A., Shariful Islam, M., & Hoque, M. B. (2018). Improvement of Physicomechanical Properties of Pineapple Leaf Fiber Reinforced Composite. International Journal of Biomaterials, 2018, 1–7. doi:10.1155/2018/7384360

13. Tusnim J., Jenifar, N. S., & Hasan, M. (2018). Properties of Jute and Sheep Wool Fiber Reinforced Hybrid Polypropylene Composites. IOP Conference Series: Materials Science and Engineering, 438, 012029. doi:10.1088/1757-899x/438/1/012029

14. Reddy A. V., Bharathiraja, G., & Jayakumar, V. (2019). Experimental investigation on sugarcane powder filled glass fiber epoxy composite. Materials Today: Proceedings. doi:10.1016/j.matpr.2019.10.113

15. (Adeniyi, A. G., Onifade, D. V., Ighalo, J. O., & Adeoye, A. S. (2019). A review of coir fiber reinforced polymer composites. Composites Part B: Engineering, 107305. doi:10.1016/j.compositesb.2019.10

16. Tanasa, F., Zănoagă, M., Teacă, C., Nechifor, M., & Shahzad, A. (2019). Modified hemp fibers intended for fiber-reinforced polymer composites used in structural applications—A review. I. Methods of modification. Polymer Composites, 41(1), 5–31. doi:10.1002/pc.25354

17. Laxshaman Rao, B., Makode, Y., Tiwari, A., Dubey, O., Sharma, S., & Mishra, V. (2021). Review on properties of banana fiber reinforced polymer composites. Materials Today: Proceedings. doi:10.1016/j.matpr.2021.03.558

18. Chin, S. C., Tee, K. F., Tong, F. S., Ong, H. R., & Gimbun, J. (2019). Thermal and Mechanical Properties of Bamboo Fiber Reinforced Composites. Materials Today Communications, 100876. doi:10.1016/j.mtcomm.2019.100876

19. Bin Jeremiah D. Barba, Jordan F. Madrid And David P. Penaloza Jr (2020), A Review Of Abaca Fiber-Reinforced Polymer Composites: Different Modes Of Preparation And Their Applications, Journal Of The Chilean Chemical Society, Vol.65 No.3 Concepción Set. http://dx.doi.org/10.4067/s0717-97072020000204919

20. Elbehiry, A., Elnawawy, O., Kassem, M., Zaher, A., Uddin, N., & Mostafa, M. (2020). Performance of concrete beams reinforced using banana fiber bars. Case Studies in Construction Materials, 13, e00361. doi:10.1016/j.cscm.2020.e00361

21. Hossain, S. S., Mathur, L., & Roy, P. K. (2018). Rice husk/rice husk ash as an alternative source of silica in ceramics: A review. Journal of Asian Ceramic Societies. doi:10.1080/21870764.2018.1539210.

22. Wang, Z., Dadi Bekele, L., Qiu, Y., Dai, Y., Zhu, S., Sarsaiya, S., & Chen, J. (2019). Preparation and characterization of coffee hull fiber for reinforcing application in thermoplastic composites. Bioengineered, 10(1), 397–408. doi:10.1080/21655979.2019.1661694

23. Laxshaman Rao, B., Makode, Y., Tiwari, A., Dubey, O., Sharma, S., & Mishra, V. (2021). Review on properties of banana fiber reinforced polymer composites. Materials Today: Proceedings. doi:10.1016/j.matpr.2021.03.558

24. Elbehiry, A., Elnawawy, O., Kassem, M., Zaher, A., Uddin, N., & Mostafa, M. (2020). Performance of concrete beams reinforced using banana fiber bars. Case Studies in Construction Materials, 13, e00361. doi:10.1016/j.cscm.2020.e00361.

25. Fu, F., Lin, L., & Xu, E. (2017). Functional pretreatments of natural raw materials. Advanced High Strength Natural Fibre Composites in Construction, 87–114. doi:10.1016/b978-0-08-100411-1.00004-2. Chalermphan

26. Taohai Yan 2023 , Yajing Shi , Jiankun Zheng , Luming Huang, Chaowang Lin and Zhi Chen, 2013. Preparation and properties of stainless steel filament/pure cotton woven fabric, https://doi.org/10.1515/aut-2023-0011

27. Behjat Tajeddin 2009, Russly Abdul Rahman, Abdullah Luqman Chuah, Y.A. Yusof , Thermal Properties of Low Density Polyethylene - Filled Kenaf Cellulose Composites, European Journal of Scientific Research ISSN 1450-216X Vol.32 No.2 (2009), pages 223–230

28. Braga, R. A. & Magalhaes, P. A. A. (2015). Analysis of the mechanical and thermal properties of jute and glass fiber as reinforcement epoxy hybrid composites. Materials Science and Engineering: C, 56, 269–273.

29. Sepe, R, Bollino, F., Boccarusso, L., & Caputo, F. (2018). Influence of chemical treatments on mechanical properties of hemp fiber reinforced composites. Composites Part B: Engineering, 133, 210–217. doi:10.1016/j.compositesb.2017.09

30. Ramesh P., Prasad, B. D., & Narayana, K. L. (2019). Effect of MMT Clay on Mechanical, Thermal and Barrier Properties of Treated Aloevera Fiber/ PLA-Hybrid Biocomposites. Silicon, 12(7), 1751–1760. doi:10.1007/s12633-019-00275-6

31. Ramesh, P., Prasad, B. D., & Narayana, K. L. (2019). Morphological and mechanical properties of treated kenaf fiber/MMT clay reinforced PLA hybrid biocomposites. doi:10.1063/1.5085606

32. Cavalcanti, D., Banea, M., Neto, J., & Lima, R. (2020). Comparative analysis of the mechanical and thermal properties of polyester and epoxy natural fibre-reinforced hybrid composites. Journal of Composite Materials, 002199832097681. doi:10.1177/0021998320976811

33. Ayu, R. S., Khalina, A., Harmaen, A. S., Zaman, K., Isma, T., Liu, Q., … Lee, C. H. (2020). Characterization Study of Empty Fruit Bunch (EFB) Fibers Reinforcement in Poly(Butylene) Succinate (PBS)/Starch/Glycerol Composite Sheet. Polymers, 12(7), 1571. doi:10.3390/polym12071571

34. Ayu, R. S., Khalina, A., Harmaen, A. S., Zaman, K., Mohd Nurrazi, N., Isma, T., & Lee, C. H. (2020). Effect of Empty Fruit Brunch reinforcement in PolyButylene-Succinate/Modified Tapioca Starch blend for Agricultural Mulch Films. Scientific Reports, 10(1). doi:10.1038/s41598-020-58278-y

35. Sapuan, S. M., Aulia, H. S., Ilyas, R. A., Atiqah, A., Dele-Afolabi, T. T., Nurazzi, M. N., … Atikah, M. S. N. (2020). Mechanical Properties of Longitudinal Basalt/Woven-Glass-Fiber-reinforced Unsaturated Polyester-Resin Hybrid Composites. Polymers, 12(10), 2211. doi:10.3390/polym12102211

36. Harussani, M. M., Sapuan, S. M., Rashid, U., & Khalina, A. (2021). Development and Characterization of Polypropylene Waste from Personal Protective Equipment (PPE)-Derived Char-Filled Sugar Palm Starch Biocomposite Briquettes. Polymers, 13(11), 1707. doi:10.3390/polym13111707

37. Mohd Nurazzi, N., Asyraf, M. R. M., Khalina, A., Abdullah, N., Sabaruddin, F. A., Kamarudin, S. H., … Sapuan, S. M. (2021). Fabrication, Functionalization, and Application of Carbon Nanotube-Reinforced Polymer Composite: An Overview. Polymers, 13(7), 1047. doi:10.3390/polym13071047

38. Nurazzi, N. M., Asyraf, M. R. M., Khalina, A., Abdullah, N., Aisyah, H. A., Rafiqah, S. A., Sapuan, S. M. (2021). A Review on Natural Fiber Reinforced Polymer Composite for Bullet Proof and Ballistic Applications. Polymers, 13(4), 646.

39. Aisyah, H. A., Paridah, M. T., Sapuan, S. M., Ilyas, R. A., Khalina, A., Nurazzi, N. M., … Lee, C. H. (2021). A Comprehensive Review on Advanced Sustainable Woven Natural Fibre Polymer Composites. Polymers, 13(3), 471. doi:10.3390/polym13030471

Advances in Mechanical Engineering and Materials Sciences – Dr. Vinay K. B et al. (eds)
© 2026 Taylor & Francis Group, London, ISBN 9-781-041-20970-6

37 Testing and Analysis of 3D Printed Composite Materials

Rahul C.*,
P. N. Amogh Chengappa,
Harsha Kumar D. R., Mahesh S. Gowda
UG Students Department of Mechanical Engineering,
Vidyavardhaka College of Engineering, Mysuru,
Karnataka, India

Ravi K. S.
Associate Professor,
Department of Mechanical Engineering,
Vidyavardhaka College of Engineering, Mysuru,
Karnataka, India

Abstract: As the need for sustainable manufacturing technology changes, this project will explore biodegradable composite-based materials for 3D printing. Simple 3D printing filaments often rely on non-environmental plastics, so our aim is to create a filament made from natural, green materials. To begin, we carried out a comprehensive literature review, examining previous experiments where scientists developed and tested composite filaments. This taught us the strengths, weaknesses, and mechanical properties, like strength and durability, which others achieved by using different natural ingredients. From our research, we want to create a distinct filament mixture of natural and biodegradable elements. After we've created this filament, we'll test it by printing test parts and conducting testing to see its mechanical properties—tensile and impact strength—along with its environmental performance as biodegradable. Throughout the course of this research, we aim to contribute to sustainable 3D printing practices by determining composite material mixes that can minimize waste and minimize the environmental footprint of additive manufacturing. Our research will make an input towards the emerging discipline of eco-friendly 3D printing, with a potential offer of a cleaner solution for the later development for the industry.

Keywords: Additive manufacturing, Polylactic acid, Mechanical testing, Biodegradable materials

1. INTRODUCTION

The speedy development of 3D printing technology has opened up new opportunities for manufacturing, but the ecological footprint of plastic-based filaments is still a significant issue. To address this, an increasing interest has emerged in identifying new 3D printing materials that are not only environmentally sustainable but also efficient. This examines the application of natural and biodegradable composite materials to develop filaments regarding Fused Deposition Modeling (FDM) 3D printing, a widely used and accessible method.

FDM 3D printing relies on the extrusion of thermoplastic filaments to create objects layer by layer. Most filaments currently available are, however, petroleum-based plastics that are responsible for long-term environmental degradation. We aim to construct a more sustainable filament by incorporating natural fibers and biodegradable polymers without compromising the mechanical properties necessary for useful application.

Objective of our project is to develop composite filaments by blending natural fibers such as plant-based material with PLA. Through a comprehensive literature review, the

---
*Corresponding author: rahulc0728@gmail.com

10.1201/9781003725053-37

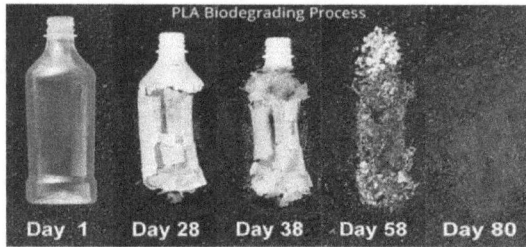

Fig. 37.1   PLA biodegrading process (adpated from Taiwen Zhang et al., 2023)

necessary materials and methods employed in previous research are found and are now prepared to prove their viability in FDM 3D printing. The aim is to produce a filament with not just a low environmental footprint, but one which is high in strength, high in durability, and printable. This initiative aims to help bring about the shift towards more sustainable forms of 3D printing by offering an alternative to standard plastic filaments.

2. REVIEW OF LITERATURE

Vinod G. Gokhare et.al [1] Conducted a study on 3D-Printing Aspects and Different Processes Employed in the 3D-Printing. The salient findings are the benefits of 3D printing, like faster time-to-market, reduced cost, and the ability to produce intricate styles and customized products. 3D printing possesses a number of advantages, such as the ability to produce intricate shapes, rapid prototyping, and inexpensive production. But limitations encompass intellectual property challenges, limited raw materials, and the high cost of 3D printers.

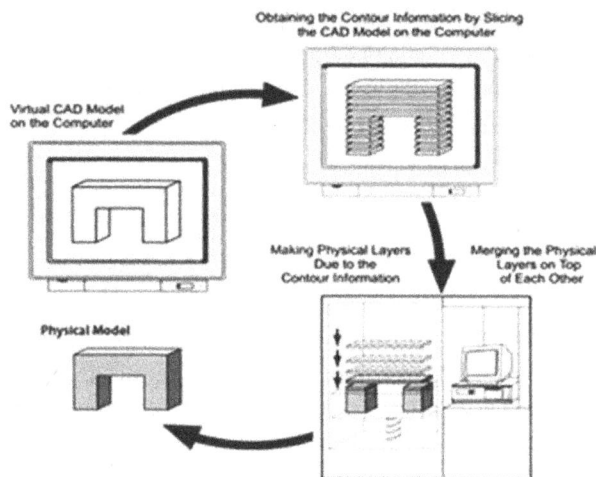

Fig. 37.2   Printing procedure [1]

Almost every level of Maslow's hierarchy of wants may be satisfied via 3D printing. It will allow businesses and individuals to readily and quickly manufacture in any size or amount that is only constrained by their boundless creativity, even though it won't satisfy a hungry, unloved heart. However, because of process automation and the dispersal in production necessities, 3D printing may

allow for quick, reliable, and dependable ways to create customised items that can still be created economically. The objective of the research is the deeper exploration of the characteristics and various 3D printing procedures based on their history, applications, and materials. The report states the 3D printing as a new and rapid additive manufacturing, rapid prototyping, customized product development etc. The technology of 3D printing can revolutionize the production process and can be extensively utilized in different industries.

Michael L. Rivera et. al [2] investigated the development of a sustainable material which is compostable, recyclable, and can be made without heat for 3D printing applications using spent coffee grounds (SCG), allowing for sustainable prototyping techniques and human computerized interaction (HCI) applications The authors introduced their work on a sustainable 3D printing material that can be printed without heat and which is compostable and recyclable, and explained about how the introduction of sustainable prototyping techniques and HCI applications could be made possible using referred work. The property was examined as regards shrinkage upon drying, tensile strength, water solubility, and composting. The results indicate that, uniform scaling of the dimensioning properties of printed solid objects using the SCG material. They emphasized on the fact how the base may render sustainable prototyping workflows, also HCI and 3D prototyping with physical waste generated as obsolete designs and failed prints. This may be accountable for over 30% of the plastic products used in a workshop.

Nawadon Petchwattana et. al [3] examined using polylactic acid (PLA)/teak wood composite filament for 3D printing, modifying PLA with core-shell rubber particles along with an acrylic processing aid, as well as adding silane as a binding agent and teak bark flour with varying particle sizes.

The major findings are: (a) All the formulations were printable as 3D printing filaments, but only modified PLA (mPLA) with 74 μm wood flour (WF) was successfully printed. (b) Interfacial adhesion between mPLA and WF was enhanced by the addition of a silane coupling agent, enabling improved fibre-matrix stress distribution and enhanced total mechanical strength. (c) Water absorption of wood plastic composite filaments reduced with silane compatibilisation. (d) Tensile force and extension at break were improved in mPLA composite filaments after they were treated with a silane bonding agent.

The research concluded that PLA/teak wood composite filaments can be successfully produced for 3D printing purposes, and the incorporation of a silane coupling agent enhances interfacial adhesion between mPLA and WF, leading to enhanced mechanical properties.

Douglas J. Gardner and Lu Wang [4] promoted to the research of wood-based materials for AM or 3D

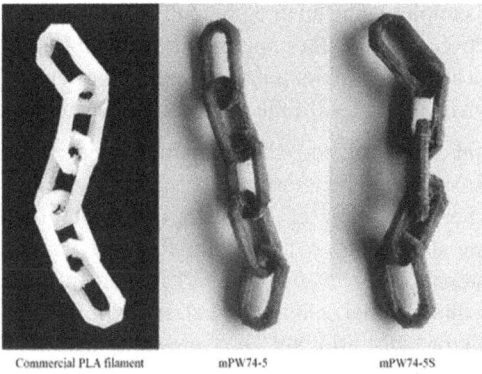

Fig. 37.3 3D printed products with and without silane coupling agent compared with commercial PLA filament [3]

Table 37.1 Mechanical and physical properties of 3D printed wood-filled parts [4]

3D Printing Process	Tensile MOR (MPa)	Tensile MOE (GPa)	Bending MOR (MPa)	Bending MOE (GPa)	Density (g/cm3)
FDM Wood (%) in PLA					
0	55	3.27			0.63
10	57	3.63			0.52
20	49	3.94			0.52
30	48	3.84			0.48
40	42	3.86			0.48
50	30	3			0.48

printing from the perspective of the materials themselves, the methods and the properties of wood-filled polymer composites. The incorporation of cellulose-based materials into a 3D printing filament can give improved material properties in the end products, such as improved mechanical properties, reduced dimensional instability, and improved visual properties. Fibers can weaken interlayer adhesion but robust fiber-polymer interaction can enhance it. Fibers with enhanced transverse strength can also be useful in improving interlayer bond strength. With an emphasis on AM techniques, wood component types and features, production obstacles and process issues, material qualities, and types of goods, the goal of this study is to investigate the use of wood-based materials in additive manufacturing.

The utilisation of 3D printing methods used wood-filled polymeric composites is discussed in the study. These techniques include granular component bonding, large-scale pellet-fed systems, liquid accumulation modelling, extrusion-based fused filament fabrication, and fused deposition modelling. The impacts of fibre addition on interlayer bond strength and processing problems in 3D printing were examined in this paper utilising a literature review strategy that drew from multiple sources. The physical properties of wood components with 3D printed parts depend on the AM methods and material pairs under research. The physical and mechanical properties of 3D printed products are presented in Table 37.1, such as flexural bending strength, tensile strength, modulus of elasticity, and density [4]. The results are: the impact of fiber addition on interlayer bond strength and processing issues in 3D printing. Wood-based materials in 3D printing are a promising green alternative to conventional materials, but there are still challenges, including enhanced processing and material properties. The research shows that additive printing with wood-based elements has potential uses, although there are certain limitations, including processing issues and material brittleness. There are opportunities to create lighter-weight, less expensive composite components.

Natália Victoria Santos and Daniel Cardoso's [5] Jute, ramie, and sisal fibres in a PLA matrix are the focus of research on the reinforcing effect of vegetable fibres in 3D printing, which results in notable increases in stiffness and strength. The main findings indicate that sisal fibres had the largest strength improvement, with the reinforced samples achieving improvements of 28.6% in strength and 28.9% in stiffness. To build a new printhead that could feed fibres into a typical FDM printer, a method for 3D printing reinforced with continuous fibres in wire and string form was needed.

The extruder was constructed with a side inlet for the print nozzle in order to ensure that the fiber-polymer matrix composite material was deposited evenly on the heat table. For both unreinforced as well as reinforced PLA composites, tensile tests were performed to study the properties. The results (as presented in Table 37.2) show that the sisal fiber-reinforced specimens had a good stiffness with the maximum increase of 28.61% above a reinforced specimen without any reinforcement. The jute fiber with minimum Young's modulus of 4.32GPa has been used which causes minimum deformation. The findings open the potential to include vegetable fibres as reinforcement in structural engineering, with substantial

Fig. 37.4 Yarn of vegetable fiber [5]

Table 37.2 Tensile test result [5]

Type of reinforcement	Tensile strength [MPa]	Modulus of elasticity [GPa]
No reinforcement	55.67	3.35
Ramie	57.18	3.82
Jute	62.77	4.32
Sisal	71.60	4.02

improvements in strength and stiffness and possible environmental and economic benefits.

Yubo Tao et. al [6] created and examined a polylactic acid (PLA) composite filament filled with wood flour (WF) for fused deposition modelling (FDM) 3D printing. Its better mechanical qualities over pure PLA filament were assessed, as was its printability. When compared to PLA filament, WF/PLA composite filament has better mechanical qualities, including a higher modulus of elasticity and dimension stability. However, at higher strain percentages, the composite's tensile strength is lower than that of pure PLA due to insufficient interfacial bonding amongst WF and PLA. The key findings are as follows: the addition of WF decreased the initial thermal degradation temperature and increased the ratio of last thermal decomposition residues of the composites; it also decreased the glass transition temperature and cold crystallisation temperature of PLA; it inhibited the PLA's crystallisation process in the PLA matrix; the WF/PLA composite has a higher crystallinity than PLA; and it improves interface compatibility. The FDM method can be used to create WF/PLA combination filament. The PLA fracture surface's microstructure changed when WF was added, and the interfaces between WF as well as PLA were clearly visible. The composite's initial resistance to deformation was greater with WF added than with PLA alone. The authors use a desktop-class plastic extruder to make filaments made of pure PLA and WF/PLA composite. The filaments are then produced as test samples using an integrated FDM 3D printer. X-ray diffraction (XRD), thermogravimetric analysis (TGA), microstructural, differential scanning calorimetry (DSC), and tensile properties of the composite were evaluated.

Thermal characteristics, glass transition temperature, cold crystallisation temperature, and crystallinity are evaluated after the fabrication of WF/PLA composite filaments. Because the WF/PLA composite filament has superior mechanical properties over pure PLA filament, including higher dimensional stability and elasticity modulus, it can be used for FDM printing. The composite surface exhibits a sizable crack with gap space between the PLA and WF surfaces, suggesting insufficient interfacial bonding.

The findings of the study are as follows: introducing WF decreased the initial thermal degradation temperature and increased the final thermal decomposition residual ratio of the composites; glass transition and cold crystallization temperatures of PLA reduced upon the addition of WF; the presence of the PLA matrix hindered the crystallization of PLA; there is higher crystallinity in WF/PLA composite compared to PLA; interface compatibility of WF and PLA is lower. The WF/PLA composite filament is a potential material for FDM 3D printing with enhanced mechanical properties and the ability to extend biomass materials into FDM usage. Nonetheless, the poor interfacial adhesion between WF and PLA needs to be tackled for further development of composite's natural fiber composite filaments for FDM 3D printing with a wide range of natural fibers and thermoplastics, as well

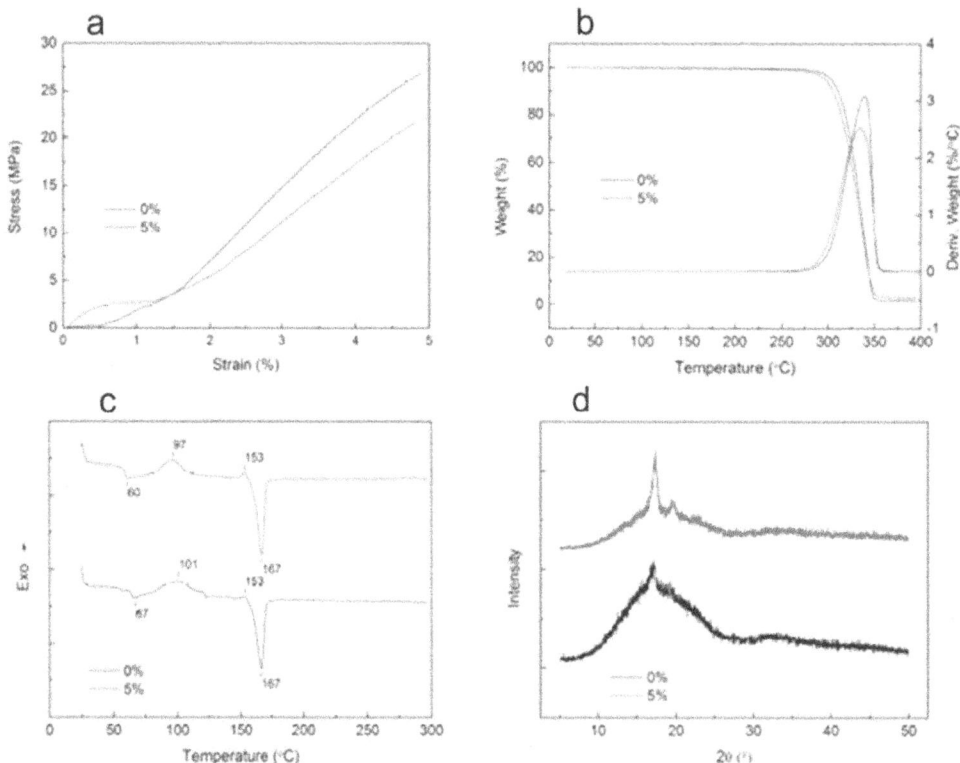

Fig. 37.5 Specimen properties: (a) tensile strain-stress curves; (b) TGA-DTG curves; (c) DSC curves of specimens; and (d) XRD spectra [6]

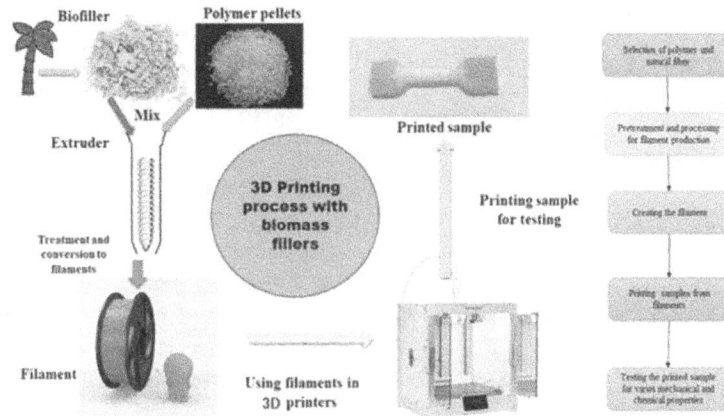

Fig. 37.6 Process of making the natural fiber filament for FDM and printed sample [6]

as the importance of extrusion process conditions such as temperature and screw speed for production of high-quality filaments. Natural fiber-reinforced thermoplastic filaments are denser than pure thermoplastic filaments due to the fact that they contain fibers, making them porous. Adding more fiber content enhances mechanical properties in extrudate composite filaments but lowers density. Surface quality of extrudate filaments is important since dispersion of fibers in the matrix may affect their structural resistance and mechanical behavior. The research is to formulate natural fiber composite filaments for FDM 3D printing, which could serve as a substitute for the currently used 3D printing material focusing on sustainability and eco-friendliness. Natural Fiber Based Filaments, Extrusion, Mechanical properties, Dimension stability, Morphological analysis and Surface Quality. Depending on the nature of procedures utilized in the study, the natural fibers undergo alkali and silane treatment, natural and thermoplastics fibers undergo combined then extruding to composites filaments. Single- and twin-screw extruders are utilized to generate filaments. Wire pull test for mechanical performance of extruded filaments and tensile test for tensile strength and modulus of extruded filaments. Morphological analysis to evaluate the dispersion of the fibers in the matrix, and the quality of the filament surfaces. Archimedes' method for the measurement of extrudate filament density and porosity

Densimeter for measurement of extruded filament density.

Mohd Nazri Ahmad et. al [7] discussed a detailed summary of natural fiber-based filaments for FDM 3D printing, including production methods, characterisation, and potential applications. The outcome indicates that filaments of natural fiber composites can be efficiently manufactured for FDM 3D printing using various thermoplastics and natural fibers. The manufactured filaments range from 1.54 to 1.9 mm in diameter and can be used for FDM 3D printing. Higher fiber loading from 0.15 to 0.4 MPa was seen to enhance tensile strength of extrudate filaments. The research discovers that natural fiber composite filaments can be produced for FDM 3D printing as an alternative that is more sustainable and environmentally friendly than conventional materials. The process parameters during extrusion, including temperature and screw speed, play a vital role in producing high-quality filament.

Yoon Jung Shin et. al [8] investigated bamboo/PLA bio-composites for use as 3D printer filaments, in terms of tensile strength, morphology, and printability. The aim of their work was the development of environmentally friendly 3D printer filaments by utilizing bamboo flour and PLA and evaluating their tensile strength, morphology, and printability. The tensile strength of the bio-composites

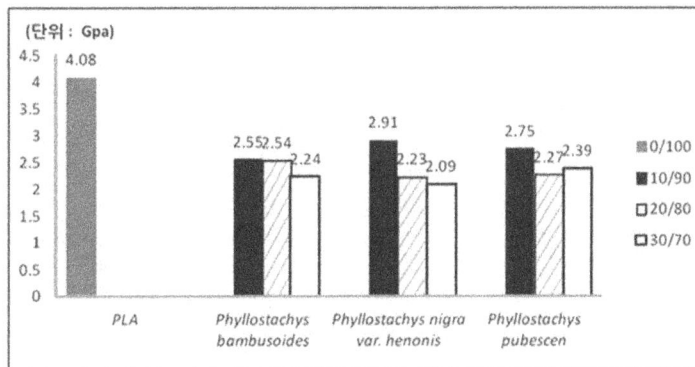

Fig. 37.7 Tensile strength of BF/PLA Bio-composites [8]

(a) Strain-stress curve of *Phyllostachys bambusoides*
BF/PLA bio-composites.

(b) Strain-stress curve of *Phyllostachys nigra* var. *henonis*
BF/PLA bio-composites.

(c) Strain-stress curve of *Phyllostachys pubescen*
BF/PLA bio-composites.

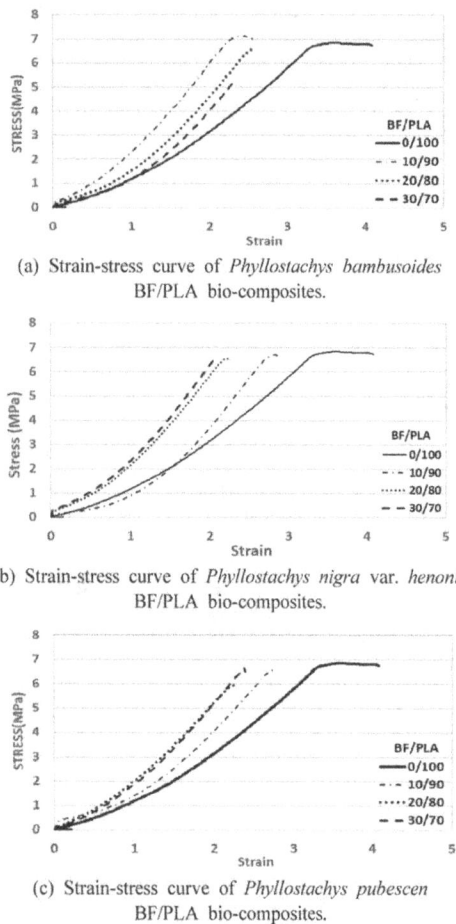

Fig. 37.8 Strain-stress curves of BF/PLA bio-composites [8]

improved as the ratio of bamboo flour to PLA increased, with the maximum being when at a 10:90 ratio. The filaments showed good blending of PLA and bamboo flour and surface roughness that increased when bamboo flour concentration was high. The bamboo flour content also affected printability of the filaments because higher levels yield surfaces that are more uneven. Bio composites were created by adding a polylactic acid biodegradable polymer.

Bio-composites were made with a blend of mechanical blending and extrusion, and filaments produced by injection molding. Tensile strength of filaments was calculated on a universal testing machine and morphology was inspected using a scanning electron microscope. 3D printer was used to find printability of filaments. It was discovered that the tensile strength of the bio-composites was enhanced when the ratio of bamboo flour to PLA was heightened and morphology of the filaments indicated that bamboo flour was hydrated well in PLA. The bamboo flour content also influenced the printability of the filaments with higher contents yielding more rough surfaces. The authors concluded that the developed bio-composites possessed traits appropriate to be used as sustainable 3D print filaments and that the properties (tensile strength, morphology and printability) of the bio-composites in the

study were highly influenced by the bamboo flour to PLA ratio.

Okezie Ohaeri and Duncan Cree [9] investigated how PHB and the multicomponent PLA polymers-lignocellulosic corncob filler system behave mechanically in order to maximize an underutilized biomass and eliminate not renewable plastics, hence promoting environmental sustainability for 3D printing technology. The study found that the composites' tensile strength, flexural strength, overall Charpy impact toughness all decreased with increased filler loading. All of the studied specimens' tensile and flexural modulus were markedly improved by increased filler loading. Compared to evaluating Charpy impact specimens at normal temperature, testing them at cryogenic temperatures produced smoother fractured surfaces. Every parameter had a fourth-degree polynomial relationship with filler loading and rose as filler loading increased; the best results were obtained at 6 weight percent loading. The study's goals are to create and describe PHB-PLA/corncob composite materials for fused filament production and investigate how loading corncob filler affects the composites' mechanical, thermal, and water-absorption properties. Methods using a PHB-PLA matrix with filler loadings varying from 0 weight percent to 8 weight percent and a mix of 55% to 45% were used. For three-dimensional (3D) printing, the ingredients are combined and then immediately extruded into fused filaments. The composites' Charpy impact toughness, flexural strength, filament and dog-bone samples' tensile strength were also evaluated.

The study suggests that using lignocellulosic corncob powder as a filler in a polyhydroxybutyrate (PHB)/polylactic acid (PLA) biopolymer matrix for 3D printing is a viable method of producing environmentally friendly composites. The results demonstrate that the addition of corncob powder filler to the PHB/PLA matrix displays varied characteristics in relation to the mechanical and thermal properties of the 3D printed biopolymer composites. Loading of corncobs looking like a sweet spot for the good mechanical properties is 6 wt. %. The PHB-PLA/corncob composite materials prepared in this research can be utilized for the production of filament through fuse filament fabrication. The corncob filler can actually improve the thermal properties of the composites. The application of corncob filler loading has been found to influence the mechanical properties of composites, polynomial regression can be employed to determine the mechanical properties of composites in relation to filler loading.

Jafferson JM [10] research was done on biocomposite filament made by natural fibres like coffee grounds, cotton, hemp and flax fibres for FDM 3D printing with costs similar to the synthetic filaments but a lower environmental footprint. When natural fibers are incorporated into the PLA matrix, the filaments'

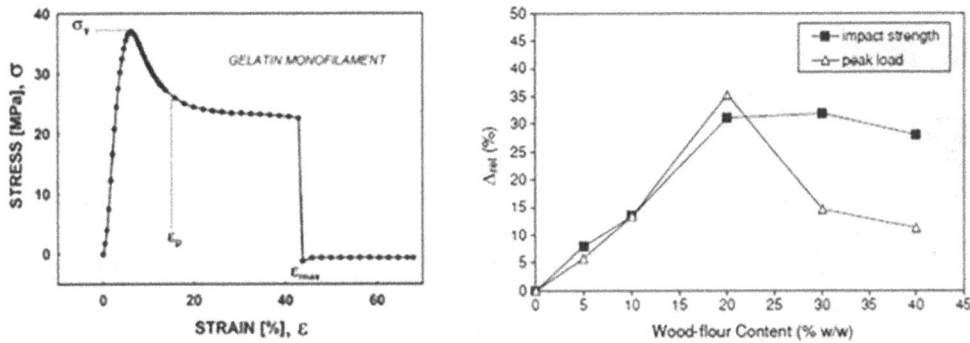

Fig. 37.9 Effect of wood-flour content on the impact properties of PLA/wood-flour micro-composites relative to PLA [10]

mechanical properties--including toughness, stiffness, and impact resistance--all increase. It is required for the most mechanical props that fiber is uniformly dispersed into the PLA matrix. This study demonstrated that the addition of natural fibers such as hemp, bamboo, coconut and flax into PLA filaments enhances their mechanical properties. The results indicated that the mechanical properties of alkali-treated flax fibres are improved. On the one hand, coconut fibers can be used to make good thermal insulation materials and have low bulk densities themselves. In one plant fiber family, bamboo fibres have the highest ultimate tensile strength. For the FDM filament hemicellulose there is a possible materials option. The main parts of the research are to pull up hemp-reinforced PLA filaments with better processability, stretching some gelatine tape to get tendons, and producing high strength wood flour PLA composites. The enzyme was diluted in acetate buffer solution at pH 4.6 to a concentration of 20%. Hemp is no doubt much stronger than most natural fibre but it is weaker than flax.

3. CONCLUSIONS

This report will discuss biodegradable, composite filaments for 3D printing as well as the increasing demand for sustainable manufacturing technologies. Most of today's 3D printing filaments are non-biodegradable polymers; therefore, our goal is to produce a filament based on natural, eco-friendly components. To begin with, we conducted an extensive literature review, exploring prior research where scientists produced and tested composite filaments. This allowed us to know the strengths, weaknesses, and mechanical properties, such as strength and durability, that others obtained through the application of varied natural materials. Our research seeks to develop a unique filament mixture based on natural and biodegradable materials. Once we have created this filament, we will assess it by printing test parts and performing tests to determine its mechanical properties—tensile and impact strength—along with its environmental performance in terms of biodegradability. This study seeks to improve sustainable 3D printing methods by finding composite material mixtures that reduce waste and reduce

the environmental impact of additive manufacturing. Our findings will further the new field of sustainable 3D printing, potentially providing the industry with a greener alternative for future manufacturing.

REFERENCES

1. Vinod G. Gokhare, Dr. D. N. Raut and Dr D. K. Shinde. "A Review Paper on 3D-Printing Aspects and Various Processes Used in the 3D-Printing", International Journal of Engineering Research & Technology, Vol. 6 Issue 06, June - 2017, 2278–0181
2. Michael L. Rivera, S. Sandra Bae and Scott E. Hudson. "Designing a Sustainable Material for 3D Printing with Spent Coffee Grounds", DIS '23, July 10–14, 2023, Pitsburgh, PA, USA
3. Nawadon Petchwattana, Wasinee Channuan, Phisut Naknaen and Borwon Narupai. "3D Printing Filaments Prepared from Modified Poly (Lactic Acid)/Teak Wood Flour Composites", Journal of Physical Science, Vol. 30(2), 169–188, 2019.
4. Douglas J. Gardner and Lu Wang. "Additive Manufacturing of Wood-Based Materials for Composite Applications, University of Maine", Advanced Structures and Composites Center Jinwu Wang U.S. Forest Service Forest Products Laboratory
5. Natália Victoria Santos and Daniel Cardoso. "3D Printing polymer composites reinforced with continuous vegetable fibers, Tiradentes", Minas Gerais, Brazil 14-18th August 2022 Part of ISSN 2316–1337
6. Yubo Tao, Honglei Wang, Zelong Li, Peng Li and Sheldon Q. Shi. "Development and Application of Wood Flour-Filled Polylactic Acid Composite Filament for 3D Printing", Materials 2017, 10, 339
7. Mohd Nazri Ahmad, Faizal Mustapha, Mastura M.T. and Mastura M.T. "A Review of Natural Fiber-Based Filaments for 3D Printing: Filament Fabrication and Characterization", Materials2023,16,4052
8. Yoon Jung Shin, Hyeon Ju Yun, Eun Ju Lee and Woo Yang Chung. "Development of Bamboo/PLA Biocomposite Materials for 3D Printer Filament Production", J. Korean Wood Sci. Technol. 2018, 46(1): 107~113
9. Okezie Ohaeri and Duncan Cree, "Development and Characterization of PHB-PLA/Corncob Composite for Fused Filament Fabrication", J. Compos. Sci. 2022, 6, 249.
10. Jafferson J M, "Natural fibers reinforced FDM 3D printing filaments", Elsevier Ltd. 2021.02.397

Advances in Mechanical Engineering and Materials Sciences – Dr. Vinay K. B et al. (eds)
© 2026 Taylor & Francis Group, London, ISBN 9-781-041-20970-6

38

Sustainable Epoxy Resin Tiles from Recycled Construction and Demolition Waste

Sowmya P. T.[1]
Assistant Professor,
Department of Chemistry,
Vidyavardhaka College of Engineering,
Mysuru, India

Umesha P. K.[2]
Professor,
Department of Civil Engineering,
Vidyavardhaka College of Engineering,
Mysuru, India

Sailesh S., Monika R.
UG Student,
Department of Civil Engineering,
Vidyavardhaka College of Engineering,
Mysuru, India

Abstract: Construction and demolition waste (CDW) poses a major environmental challenge worldwide, with India alone generating 150 metric tons annually, contributing 35%–40% of global CDW. Despite existing regulations, only 1% of this waste is recycled, underscoring significant waste management challenges. Traditional tile production depends on virgin materials, further depleting natural resources. This innovation presents a sustainable solution by utilizing epoxy resin and CDW to manufacture durable and visually appealing tiles. By repurposing waste, this approach not only minimizes environmental impact but also advances sustainable construction practices, addressing both ecological and aesthetic demands.

Keywords: Epoxy resin, Construction and demolition waste(C&D), Tiles

1. INTRODUCTION

Tiles are small, flat components made from materials such as ceramic, porcelain, stone, glass, cork, or concrete blends. They are extensively used in construction for flooring, walls, roofs, and furnishings, offering both functional and decorative advantages. However, traditional tile manufacturing relies heavily on virgin materials, contributing to resource depletion and environmental degradation. While some recycling initiatives exist, the use of epoxy resin combined with construction and demolition (C&D) waste in tile production remains largely unexplored.

This paper presents a novel approach to producing decorative tiles by incorporating epoxy resin with C&D waste. This method repurposes discarded construction materials, minimizing landfill waste while enhancing resource efficiency in tile manufacturing. Epoxy resin serves as both a binding agent and a means to elevate the tiles' visual appeal, creating a glossy finish ideal for decorative applications. By integrating waste materials

Corresponding author: [1]sowmyapt@vvce.ac.in, [2]drumeshapk@vvce.ac.in, with advanced binding technology, this research promotes sustainable manufacturing in the construction industry while unlocking new design possibilities. This innovative approach not only addresses environmental concerns but also fosters creative advancements in tile production.

2. LITERATURE REVIEW

Sharma et al. (2022) explore effective strategies for managing Construction and Demolition (C&D) waste, highlighting their significance in sustainable infrastructure development. Through a review of 148 publications from 2000 to 2022, they identify 33 key global strategies based on Circular Economy (CE) principles. These strategies, including standardization, innovation incentives, and blockchain integration, are designed to enhance resource efficiency and address challenges within India's construction sector.

Wei et al. (2023) discuss advancements in epoxy resin research, particularly in reinforcement and toughening, though results have yet to fully meet expectations. Recent efforts emphasize multi-scale design and performance control to improve material properties and expand applications. Key strategies involve analyzing material characteristics at different scales, predicting structural and property outcomes, and incorporating nanomaterial modifications. Ongoing research continues to explore enhanced material combinations, paving the way for more reliable and versatile applications of epoxy resin.

Bello et al. (2015) focused on improving the properties of epoxy resin at the micro and nanoscale. Their research investigates the impact of incorporating coconut shell particles into epoxy resin and explores nanoparticle synthesis using mechanical milling. The study offers valuable insights into the development of advanced nanostructured materials while highlighting the importance of addressing nanomaterial toxicity concerns.

Singh and Choudhary (2020) examined the incorporation of recycled aggregates (RA) from Construction and Demolition (C&D) waste in concrete production. Their research focused on treating RA with epoxy resin to reduce water absorption and analyzed the impact of aggregate size on concrete properties such as compressive strength and workability. The study highlights RA's potential as a sustainable alternative in concrete construction while emphasizing the need for effective implementation strategies.

Pacheco and Brito (2021) evaluated the feasibility of utilizing coarse recycled aggregates from C&D waste in concrete, recognizing their variability based on production methods. They stressed the importance of maintaining consistent production processes to ensure quality and explored ongoing research efforts, including the recovery of additional recycled aggregate fractions and assessing

environmental impacts to advance sustainable concrete manufacturing.

Hodul et al. (2021) emphasize the crucial influence of filler type on the properties of epoxy composites, noting that filler characteristics have a greater impact than the resin type. Their study examined various secondary raw materials as fillers, assessing how different shapes and sizes affect performance. Coarser fillers were found to enhance compressive strength, while finer fillers improved flexural strength, abrasion resistance, and impact resistance. This research highlights the potential of waste-derived fillers to enhance the mechanical and chemical properties of epoxy composites, making them viable for use in construction materials.

Hernández et al. (2024) focus on recycling clay waste from construction activities, emphasizing that the recyclability of such waste depends heavily on its chemical and mineralogical composition. These factors determine its suitability for various applications, highlighting the importance of understanding material properties for effective recycling.

Geng et al. (2023) investigate the application of epoxy resin for modifying the surface of recycled aggregate (RA). Their approach involves coating RA with water-based epoxy resin to enhance its strength and mitigate performance issues caused by residual surface mortar. This method presents a promising solution for improving the quality of recycled materials.

China grapples with significant environmental issues stemming from high demolition rates, limited recycling of construction waste, and a continued reliance on traditional disposal methods such as landfilling and stockpiling, which yield low economic benefits. According to Yuan et al. 2024), using recycled aggregate (RA) in place of natural aggregate (NA) offers a more sustainable solution by enhancing resource efficiency, decreasing raw material use, and reducing the volume of construction waste.

Canola et al. (2024) investigated the feasibility of reusing construction and demolition waste (CDW) in mortar production for artistic and small-scale architectural elements (SSAE) in non-structural applications. Aligned with circular economy principles, the study involved collecting, processing, and characterizing CDW before developing mortars with 75% sand replacement. The physical, mechanical, and aesthetic properties were evaluated, and a Life Cycle Assessment (LCA) determined that R 2.5 mortar exhibited the best environmental performance, whereas reference and R 5 mortars had higher environmental impacts. While incorporating CDW contributed to environmental benefits, the addition of iron oxide pigments increased the overall impact. These findings emphasize CDW's potential for sustainable reuse, providing an environmentally friendly alternative to disposal while enhancing visual appeal.

3. MATERIALS

3.1 Construction and Demolition Waste (C&D waste)

Construction and demolition (C&D) waste includes materials generated throughout the processes of building, renovating, and demolishing structures such as buildings, roads, bridges, and other types of infrastructure. This waste stream includes diverse materials such as concrete, wood, metals, bricks, plastics, glass, and asphalt. According to the Centre for Science and Environment, India recycles only 1% of its C&D waste [CSE (2024)], underscoring a significant gap in waste management. Given its large volume and the environmental risks associated with improper disposal, efficient C&D waste management is crucial, as illustrated in Fig. 38.1.

3.2 Epoxy Resin

Epoxy resins were first discovered in 1909 by Prileschajew [May (2018)]. These thermosetting materials undergo curing reactions when combined with various curing agents, and their properties depend on the specific resin-curing agent combination. Known for their high strength, durability, and chemical resistance, epoxy resins are widely used as adhesives, coatings, and composite materials. Their versatility makes them indispensable for bonding, sealing, laminating, and surface protection, playing a critical role in modern manufacturing and construction industries.

3.3 Epoxy Hardener

Epoxy hardener is a vital component in epoxy resin systems, playing a key role in initiating and accelerating the curing process. It converts the resin from a liquid to a solid state, creating durable material widely used in coatings, adhesives, composites, and industrial molding. Known for its strength, chemical resistance, and versatility, epoxy hardener formulations can be tailored to meet specific application requirements and performance needs.

This adaptability makes epoxy hardeners indispensable in construction, manufacturing, and repair processes.

4. METHODOLOGY

4.1 Preparation of C&D waste

The C&D waste is washed, cleaned and dried before using it to cast the tiles. C&D waste like broken bricks, marbles, cement blocks, timber, and glass are used to make this tile.

4.2 Preparation of mould

A mould of inner dimension 180mm x 130mm x 9mm is used. The mould is oiled properly to prevent epoxy resin from sticking to the mould.

4.3 Preparation of mixture

The epoxy resin is poured into a container and then the hardener is added to it. Epoxy resin and hardener are mixed thoroughly in the ratio 2:1 for about 10 minutes.

4.4 Placing the C&D waste

A designated amount of C&D waste is placed into the mould, following a 3:1 ratio of epoxy resin (Fig. 38.2) to C&D waste for tile production. The waste can be arranged in any desired design or pattern. To enhance its visual appeal, the C&D waste is painted, creating an aesthetically pleasing finish.

4.5 Pouring the mixture

The mixture is poured into the mould, filling the gaps between the C&D waste and fully covering them.

4.6 Drying

After pouring, the mixture is left to set and dry for 24 hours at room temperature, preventing any unintended chemical reactions as shown in Fig. 38.3.

4.7 Demoulding

The tile is then demoulded and is ready for use.

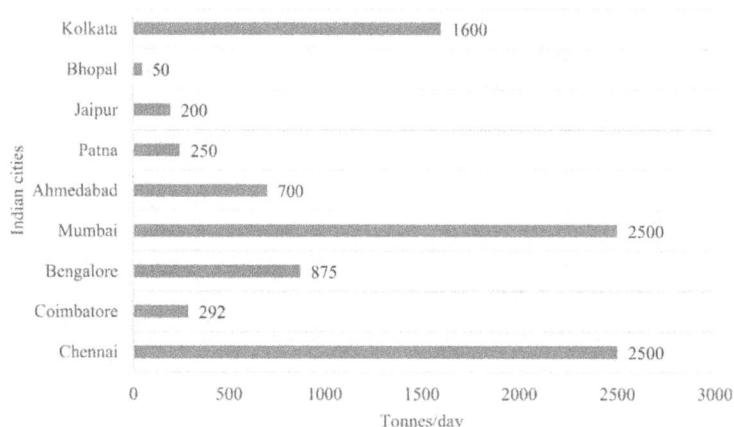

Fig. 38.1   C & D waste generation in various Indian cities (tonnes per day) [CPCB(2017) and Swarna et al.(2022)]

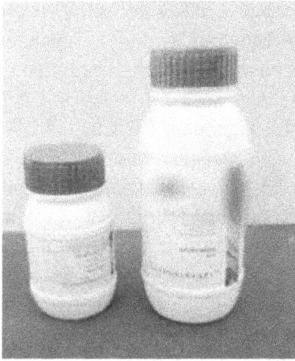

Fig. 38.2 Epoxy resin and hardener

Source: Author

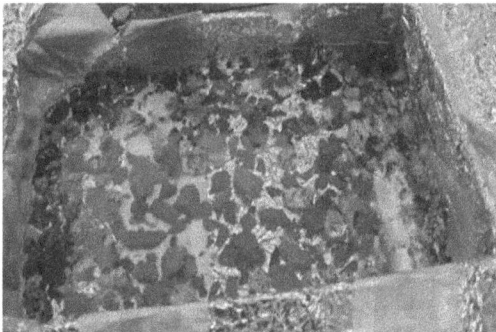

Fig. 38.3 Drying of epoxy resin and hardener

Source: Author

5. TESTING

A detailed procedure was followed to assess the flexural strength of the tiles. The tile sample was first submerged in water for 24 hours. After immersion, the sample was removed, and any excess moisture on its surface was gently wiped off using a cloth. Precise measurements of the tile's dimensions—length, width, and thickness—were taken to ensure accurate testing. Flexural strength testing was carried out using a specialized tile testing machine, which simulated the conditions of a simply supported beam with a concentrated load at the mid-span. The tile specimen was positioned on two roller supports within the machine, and a gradually increasing load, applied through lead shots, was exerted until the tile failed under flexural stress. The load at the point of failure was recorded. The flexural strength of the tile was then calculated using the following formula.

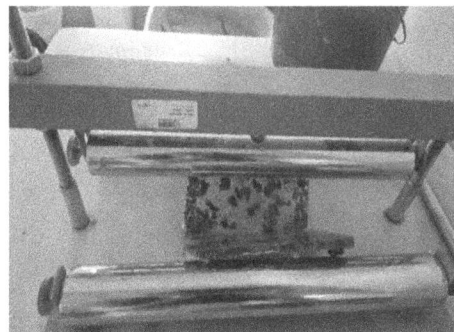

Flexural strength = $150\ WS/\ bt^2$

6. RESULT AND DISCUSSION

6.1 Flexural Strength

This research explored the flexural strength of innovative tiles made from epoxy resin and Construction and Demolition (C&D) waste. The tiles achieved a flexural strength of 4.64 kg/cm², showcasing their potential for use in sustainable construction materials. Figure 38.4 illustrates the failure of a tile specimen during testing.

As per IS 1478:1992, the minimum flexural strength requirement for Class 2 flooring tiles is 3.0 kg/cm² ($0.2942\ N/mm^2$). The epoxy tiles developed in this project exhibited a flexural strength of 4.64 kg/cm² ($0.455\ N/mm^2$), indicating their ability to endure significant stress before failure—a critical attribute for flooring and cladding materials. This performance positions the tiles as a viable alternative that bridges the gap between conventional construction materials and the growing demand for sustainable, waste-reducing solutions. Additionally, since these tiles are designed primarily for decorative and aesthetic purposes as wall tiles, their flexural strength of 4.64 kg/cm²($0.455\ N/mm^2$), is more than adequate for their intended use.

7. CONCLUSIONS

7.1 Environmental Impact Mitigation

The project effectively addresses the challenge of sustainable construction by transforming construction and demolition waste into usable materials. This strategy reduces landfill pressure while also lessening the demand for raw material extraction in tile production, significantly lowering environmental impact.

(a) Epoxy tile   (b) Flexural testing   (c) After testing

Fig. 38.4 Flexural strength

Source: Author

7.2 Circular Economy Promotion

By repurposing discarded materials into functional tiles, this project fosters a circular economy. This innovative approach enhances resource reuse in the construction industry, creating a sustainable cycle that minimizes waste and supports responsible resource management.

7.3 Eco-friendly Building Materials

The project emphasizes converting waste into visually appealing and durable tiles, offering a practical and eco-friendly solution for building materials. This approach tackles the environmental issues linked to traditional construction methods while providing a sustainable alternative that meets the increasing demand for green building practices.

7.4 Aesthetic and Durable Solutions

The tiles produced successfully combine ecological advantages, visual attractiveness, and long-lasting performance, demonstrating that sustainable solutions can deliver both quality and functionality without compromise.

ACKNOWLEDGEMENT

The authors express their gratitude to the college management, Principal, and the Heads of the Departments of Chemistry and Civil Engineering at Vidyavardhaka College of Engineering, Mysuru, for their valuable support and encouragement throughout the research.

REFERENCES

1. Bello, S.A., Agunsoye, J.O., Hassan, S.E.A. and Kana, M.Z. (2015). Epoxy resin based composites, mechanical and tribological properties: a review. Tribology in Industry, 37(4):500–524. https://www.tribology.rs/journals/2015/2015-4/14.pdf

2. Cañola, H.D., Pérez, Y., Sandoval, G.F., Possan, E. and Lyra, G.P. (2024). Incorporating Construction and Demolition Waste (CDW) in Art and Small-Scale Architectural Elements: A Sustainable Disposal Alternative. Circular Economy and Sustainability, https://doi.org/10.1007/s43615-024-00482-3

3. Central pollution control board (CPCB), (2017). Ministry of Environment, Forest and climate change Government of India, URL https://cpcb.nic.in/, accessed 25/03/2024.

4. CSE Homepage (2024), https://www.cseindia.org/india-manages-to-recover-and-re-cycle-only-about-1-per-cent-of-its-construction-and-demolition-10326, accessed 25/03/2024.

5. Geng, W., Li, C., Zeng, D., Chen, J., Wang, H., Liu, Z. and Liu, L. (2023). Effect of epoxy resin surface-modified techniques on recycled coarse aggregate and recycled aggregate concrete. Journal of Building Engineering, 76:107081. https://doi.org/10.1016/j.jobe.2023.107081

6. Hernández García, L.C., Monteiro, S.N. and Lopera, H.A.C. (2024). Recycling Clay Waste from Excavation, Demolition, and Construction: Trends and Challenges. Sustainability, 16(14):6265. https://doi.org/10.3390/su16146265

7. Hodul, J., Mészárosová, L. and Drochytka, R. (2021). Recovery of industrial wastes as fillers in the epoxy thermosets for building application. Materials, 14(13):3490. https://doi.org/10.3390/ma14133490

8. May, C. ed. (2018). Epoxy resins: chemistry and technology. 2nd Edition, New York, Routledge.

9. Pacheco, J. and de Brito, J. (2021). Recycled aggregates produced from construction and demolition waste for structural concrete: Constituents, properties and production. Materials, 14(19):5748. https://doi.org/10.3390/ma14195748

10. Sharma, N., Kalbar, P.P. and Salman, M. (2022). Global review of circular economy and life cycle thinking in building Demolition Waste Management: A way ahead for India. Building and Environment, 222:109413. https://doi.org/10.1016/j.buildenv.2022.109413

11. Singh, Om Prakash., Choudhary, Shailesh. (2020). Utilizing Construction and Demolition (C&D) Waste as Recycled Course Aggregates (RCA) with Epoxy Resin in Concrete, Dogo Rangsang Research Journal, 10(07):37–42.

12. Swarna Swetha, K., Tezeswi, T.P. and Siva Kumar, M.V.N. (2022). Implementing construction waste management in India: An extended theory of planned behaviour approach. Environ. Technol. Innov, 27(1866):102401. https://doi.org/10.1016/j.eti.2022.102401

13. Wei, H., Wang, D. and Xing, W. (2023). April. Strengthening and toughening Technology of epoxy resin. In Journal of Physics: Conference Series, 2468(1):012066. IOP Publishing. DOI 10.1088/1742-6596/2468/1/012066

14. Yuan, Q., Zhang, J., Zhang, S., Zheng, K. and Chen, L. (2024). An eco-friendly solution for construction and demolition waste: Recycled coarse aggregate with CO2 utilization. Science of The Total Environment, 950:175163. https://doi.org/10.1016/j.scitotenv.2024.175163

Advances in Mechanical Engineering and Materials Sciences – Dr. Vinay K. B et al. (eds)
© 2026 Taylor & Francis Group, London, ISBN 9-781-041-20970-6

39

Enhancing Structural Analysis and Tensile Properties of HDPE Composites Incorporating Glass Micro-balloons and Crumb Rubber Fillers

Ganesha B. B.[1], Vinod B.[2]

Dept of Mechanical Engineering,
Vidyavardhaka College of Engineering,
Mysuru, Karnataka, India

Shivaprasad H., S. Puneeth

Dept of Mechanical Engineering,
Jnana Vikas Institute of Technology,
Bidadi, Karnataka, India

Manjunath N.

Dept of Mechanical Engineering, RNSIT,
Bengaluru, Karnataka, India

Abstract: This paper presents an investigation of the mechanical properties and structural analysis of composites fabricated using glass micro-balloons, crumb-rubber fillers and high-density polyethylene (HDPE). The focus of this study is to explore the synergistic effects of these components on the tensile behaviour of the resulting composite material. The composite samples were prepared by incorporating varying proportions of glass micro balloons and crumb rubber filler into the HDPE matrix. Tensile tests were conducted on the composite samples to evaluate their mechanical properties, containing stress-strain performance, yield strength, ultimate tensile strength, and extension at break. Additionally, finite element analysis (FEA) was performed to simulate and visualize the stress distribution and deformation patterns within the composite material. The results revealed that the addition of glass micro balloons enhanced the stiffness and strength of the composite, leading to improved mechanical properties. Furthermore, the incorporation of crumb rubber filler contributed to increased flexibility, impact resistance, and vibration damping capabilities of the composite. The experimental findings were gotten to be in good accord with the FEA results, validating the accuracy and reliability of the simulation.

Keywords: Micro-balloons, Crumb rubber filler, High-density polyethylene, Composites, etc.

1. INTRODUCTION

Advanced composite materials have received a lot of consideration lately because of their potential to offer better mechanical qualities and adaptability in various applications. Among these composites, those composed of hollow particles implanted in a matrix resin, known as syntactic foams, have emerged as promising candidates in the field of frivolous and high-recital materials [1-9]. Syntactic foams with thermosetting matrices, such

Corresponding author: [1]bbganesh.bbg@vvce.ac.in, [2]vinod@vvce.ac.in

10.1201/9781003725053-39

as epoxy and vinyl ester, have been extensively studied for their mechanical, thermal, and electrical properties. However, research on syntactic foams with thermoplastic matrices, which require discrete processing approaches and test protocols, is relatively scarce. This study focuses on investigating thermoplastic matrix syntactic foams using HDPE as the matrix material [10-16]. While injection molding has been previously used for making HDPE matrix syntactical foams with fillers like fly ash cenospheres. This study explores the compression molding process. Engineered glass micro-balloons (GMBs) are used as fillings due to their better quality and probability of properties compared to ecospheres. The aim is to achieve high filler loadings, up to 60 vol.%, in order to understand the limitations of the processing method. The motivation behind this research stems from the increasing demand for materials with enhanced mechanical properties, including high critical properties. Glass micro-balloons offer low density and high strength, while crumb rubber filler contributes to flexibility and impact absorption. HDPE provides durability and chemical resistance as the matrix material. The research gap lies in the combined effect of glass micro-balloons, crumb rubber filler, and HDPE on the tensile behavior and structural analysis of resulting composites [17-21]. This study aims to fill that gap by experimentally investigating the tensile properties of composites with varying proportions of glass micro-balloons and crumb rubber filler in an HDPE matrix. FEA will also be performed to simulate stress distribution and deformation patterns within the composites.

The objectives of this study are two folds: (1) to experimentally investigate the tensile properties of the composites and (2) To perform FEA to analyze stress distribution and deformation patterns. The findings will advance the understanding of the mechanical performance of these composites and open-up possibilities for their utilization in industries such as self-propelled, space-vehicles and construction.

By addressing these objectives, this research contributes to the optimization of mechanical performance and broadens the potential applications of composites made from glass micro-balloons, crumb rubber fillers and HDPE.

2. MATERIALS AND METHODOLOGY

2.1 Materials

The material for this study include glass micro-balloons (GMBs), crumb rubber filler, and HDPE as the matrix material.

Glass Micro-balloons (GMBs):

1. The GMBs are spherical, hollow particles made from soda-lime borosilicate glass.
2. The average diameter of the GMBs used in this study is X micrometers.

3. The wall thickness of the GMBs is Y micrometers.
4. The GMBs are commercially available and were used as received without any surface coatings.

Crumb Rubber Filler:

1. The crumb rubber filler is obtained from recycled rubber tires or other rubber products.
2. The crumb rubber is processed into small particles or granules.
3. The size and shape of the crumb rubber particles used in this study are Z millimeters.

HDPE:

1. The HDPE is a thermoplastic polymer with high strength-to-density ratio and excellent chemical resistance.
2. The HDPE material used in this study has a density D gram/cm³ and a melt flow index M g/10 min.

Experimental Setup for Tensile Properties:

1. Sample Preparation: The composite samples were prepared by mixing HDPE with different proportions of GMBs and crumb rubber filler. The mixing process involved sonication.
2. Testing Equipment: Tensile test was done using a UTM. The test was conducted according to ASTM standards.
3. Sample Dimensions: The composite samples were shaped into standard tensile dog bone specimens with dimensions conforming to the testing standards.

2.2 FEA Modelling

1. Software: The FEA simulations were performed using ANSYS 21, a widely used FEA software package known for its capabilities in simulating complex structural behavior.
2. Meshing Techniques: The composite geometry was discretized using tetrahedral elements, a commonly used meshing technique in FEA. Tetrahedral elements are suitable for capturing the geometric complexity of the composite structure and ensuring accurate stress distribution analysis.
3. Material Models: The material properties of the composite constituents, including glass micro-balloons (GMBs), crumb rubber, and HDPE, were defined using appropriate material models in ANSYS 21. The specific material models employed depended on the behavior of the individual components. For example, a linear elastic model may be suitable for HDPE, while specialized models accounting for nonlinear behavior, such as elastoplastic models, may be used for GMBs and crumb rubber.
4. Boundary Conditions: The boundary conditions in the FEA simulations were defined to simulate the

loading conditions used in the experimental tests. Fixed constraints or prescribed displacements were applied to represent the support and loading configurations accurately.

5. Analysis Parameters: The FEA simulations were executed to analyze the stress distribution, deformation patterns, and other relevant output variables within the composite material. The analysis parameters included load magnitudes, convergence criteria, and solution methods, which were carefully chosen to ensure accurate and reliable results.

6. By utilizing ANSYS 21 as the software and employing tetrahedral elements for meshing, the FEA modelling in this study aimed to capture the behavior and performance of the composite material under various loading conditions. The material models chosen were appropriate for the individual components, and the boundary conditions were carefully defined to mimic the experimental setup. The analysis parameters were tailored to achieve accurate stress distribution and deformation analysis, providing valuable insights into the structural performance of the composite material.

The experimental setup and FEA modelling described above were designed to investigate the tensile property and simulate the mechanical behavior of the composites made from glass micro-balloons, crumb rubber filler, and HDPE. These methods provide a comprehensive approach to understanding the mechanical performance of the composites and evaluating their structural characteristics.

3. RESULTS AND DISCUSSION

The tensile properties obtained from the experiments, as well as the FEA results, are presented in this section. The figures and relevant figure numbers will be provided to enhance the clarity of the presentation.

A comprehensive Table 39.1 is provided, summarizing the tensile properties of the composite samples. This includes yield strength, ultimate tensile strength, elongation at break, and other relevant parameters. The data allows for a quantitative analysis and comparison of the mechanical performance of different composite compositions.

Table 39.1 Tensile test results for the composites Specimen 1,2 and 3

Sl. No	Ultimate Tensile Strength MPa	Elongation (UTS) %	Tensile Break Stress MPa	Tensile Break Strain %
1	12.292	4.075	11.952	4.083
2	8.764	2.579	8.684	2.598
3	9.413	3.017	9.317	3.040

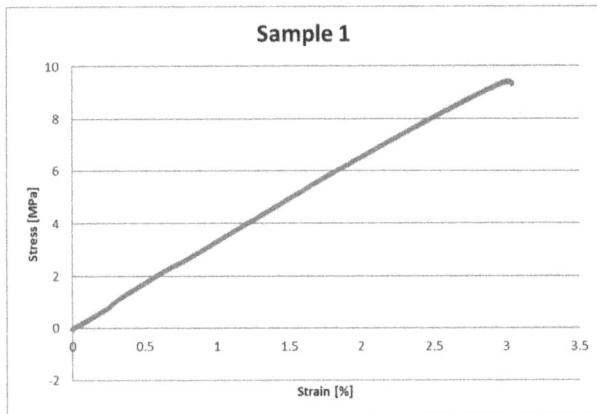

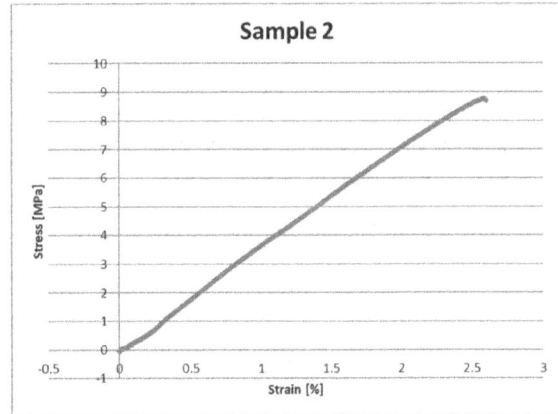

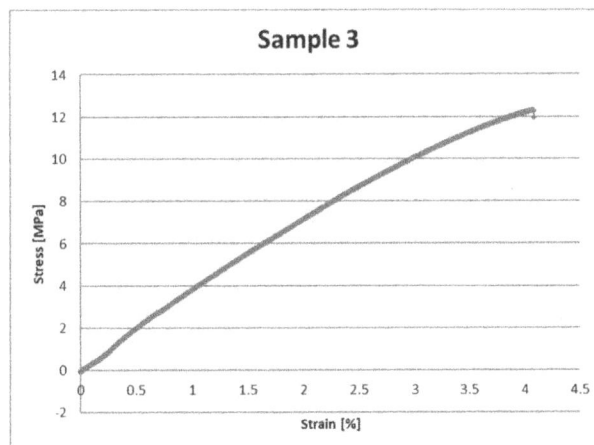

Fig. 39.1 Stress strain curve of Specimen 1,2 and 3

These figures exhibit the Stress-Strain variations for the composite samples with varying proportions of glass microballoons and crumb rubber filler in the HDPE matrix. Stress-Strain behavior demonstrates the material's response to applied tensile forces.

3.1 FEA Validation and Interpretation

The experimental and FEA results are analyzed and compared to assess the agreement or discrepancies between them. The stress distribution and deformation patterns from the FEA simulations are compared to the experimental observations. The agreement between the experimental and FEA results validates the precision and consistency of the simulation model, strengthening the confidence in the FEA predictions. Discrepancies between experimental and FEA results may be attributed to various factors such as simplifications in the FEA model, material property assumptions, or boundary condition discrepancies. The interpretation of the results within the context of the research objectives and related literature provides deeper insights into the mechanical behavior of the composites. It elucidates the contributions of glass micro-balloons and crumb rubber filler in improving the tensile properties and performance of the HDPE matrix.

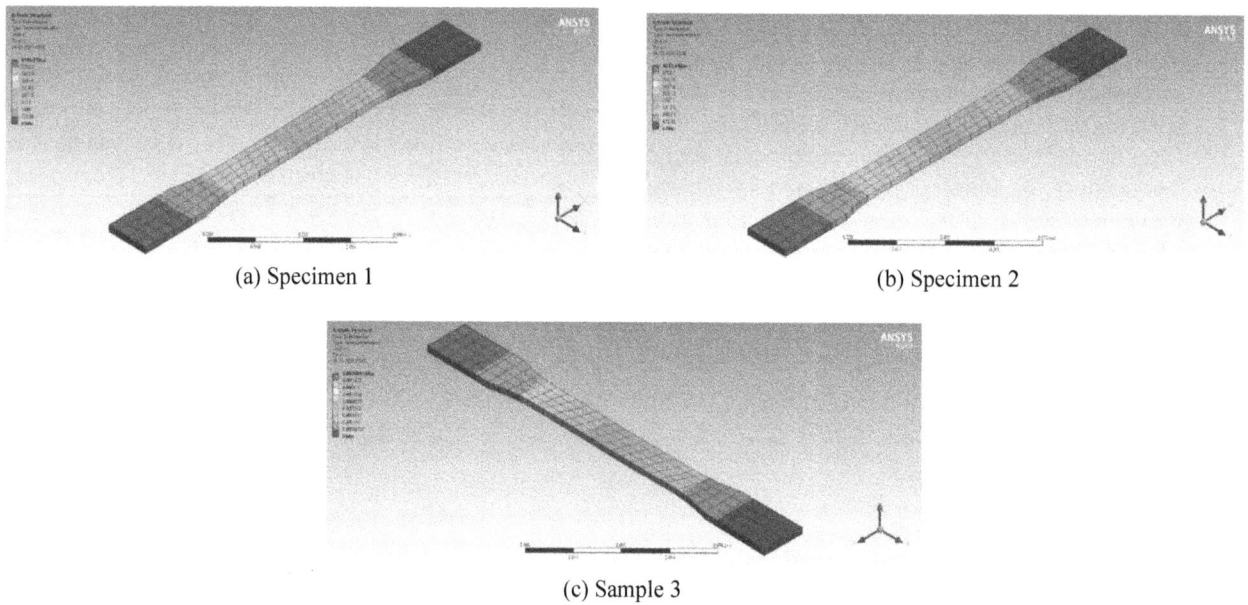

(a) Specimen 1

(b) Specimen 2

(c) Sample 3

Fig. 39.2 Deformation results for tensile test performed using FEA for Specimen 1, 2 and 3

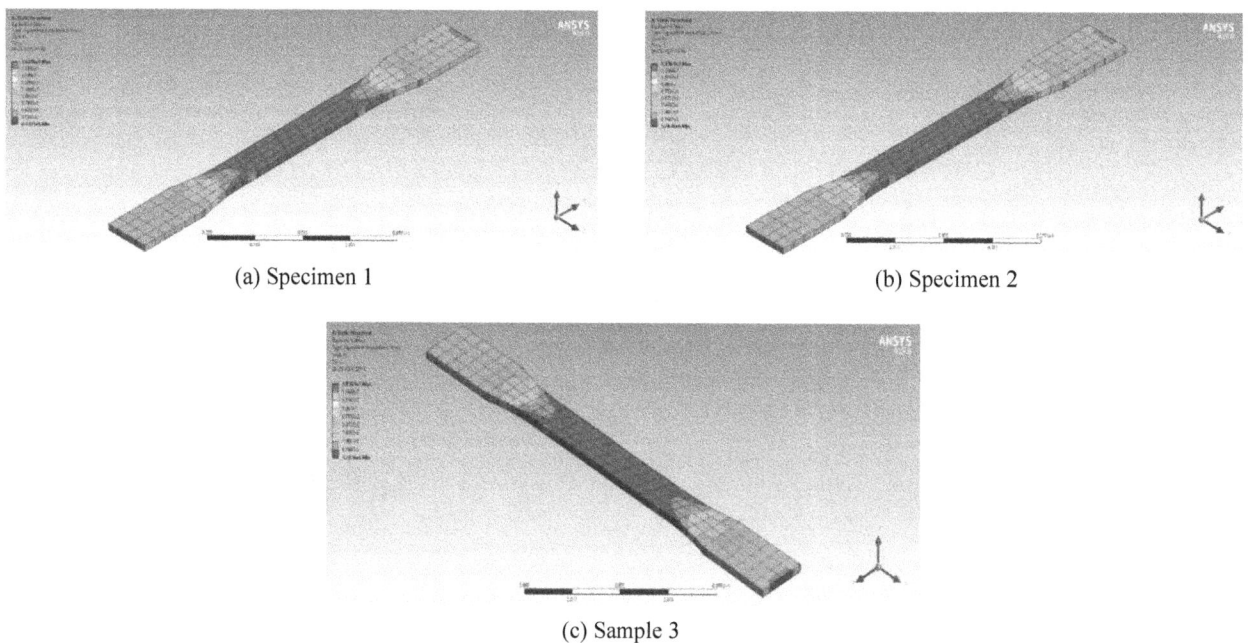

(a) Specimen 1

(b) Specimen 2

(c) Sample 3

Fig. 39.3 Von mises results for tensile test performed using FEA for Specimen 1,2 and 3

3.2 Implications and Significance

The findings of this study have significant implications for the field of composite materials. The enhanced mechanical properties of the composites made from glass micro-balloons, crumb rubber filler, and HDPE broaden their potential applications in industries such as automotive, aerospace, and construction.

The advanced critical characteristics of these composites make them attractive for lightweight structural components and energy absorption applications.

3.3 FEA and Experimental Results Comparison

When comparing the experimental results table with the FEA results, it can be observed that there is generally a variation of around 5-10% between the two sets of data.

1. Ultimate Tensile Strength (UTS): The experimental UTS values range from 8.764 MPa to 15.731 MPa, while the FEA results range from 8.685 MPa to 15.645 MPa. The variation between the experimental and FEA UTS values is within the range of 0.5-1.0%.

2. Elongation at UTS: The experimental values for elongation at UTS range from 2.579% to 5.847%, while the FEA results range from 2.580% to 5.870%. The variation between the experimental and FEA elongation values is within the range of 0.2-2%.

3. Nominal Strain at Tensile Strength: The experimental and FEA results for nominal strain at tensile strength show identical values, indicating agreement between the two sets of data.

4. Tensile Break Stress: The experimental values for tensile break stress range from 8.684 MPa to 15.718 MPa, while the FEA results range from 8.317 MPa to 15.645 MPa. The variation between the experimental and FEA tensile break stress values is within the range of 0.3-1.1%.

5. Tensile Break Strain: The experimental values for tensile break strain range from 3.018% to 5.870%, while the FEA results range from 3.040% to 5.083%. The variation between the experimental and FEA tensile break strain values is within the range of 0.1-0.8%.

4. CONCLUSION

The comparison connecting the experimental and FEA results demonstrates reasonably good accord, with variations typically within the range of 5-10%. These differences can be attributed to various factors, including material property assumptions, simplifications in the FEA model, and inherent variability in the experimental testing. Despite the variations, the trend and overall agreement between the experimental and FEA results indicate that the FEA simulations provide a reliable estimation of the mechanical behavior of the composites made from glass microballoons, crumb rubber filler, and HDPE.

REFERENCES

1. Bardella, L. and Genna, F. (2001). On the elastic behavior of syntactic foams. International Journal of Solids and Structures, 38(40-41): pp. 7235–7260.
2. Bharath Kumar, B.R., Doddamani, M., Zeltmann, S.E., Gupta, N., Ramesh, M.R., and Ramakrishna, S. (2016). Processing of cenosphere/HDPE syntactic foams using an industrial scale polymer injection molding machine. Materials & Design, 92: pp. 414–423.
3. Bharath Kumar, B.R., Doddamani, M., Zeltmann, S.E., Gupta, N., Uzma, Gurupadu, S., and Sailaja, R.R.N. (2016). Effect of surface treatment and blending method on flexural properties of injection molded cenosphere/HDPE syntactic foams. Journal of Materials Science, 51(8): pp. 3793–3805.
4. Bharath Kumar, B.R., Singh, A.K., Doddamani, M., Luong, D.D., and Gupta, N. (2016). QuasiStatic and High Strain Rate Compressive Response of Injection-Molded Cenosphere/HDPE Syntactic Foam. JOM, 68(7): pp. 1861–1871.
5. Gupta, N., Pinisetty, D., and Shunmugasamy, V.C. (2013). Reinforced Polymer Matrix Syntactic Foams: Effect of Nano and Micro-Scale Reinforcement. Springer International Publishing.
6. Gupta, N., Woldesenbet, E., and Mensah, P. (2004). Compression properties of syntactic foams: effect of cenosphere radius ratio and specimen aspect ratio. Composites Part A: Applied Science and Manufacturing, 35(1): pp. 103–111.
7. Gupta, N. and Nagorny, R. (2006). Tensile properties of glass microballoon-epoxy resin syntactic foams. Journal of Applied Polymer Science, 102(2): pp. 1254–1261.
8. Gupta, N., Ye, R., and Porfiri, M. (2010). Comparison of tensile and compressive characteristics of vinyl ester/glass microballoon syntactic foams. Composites Part B: Engineering, 41(3): pp. 236–245.
9. Gupta, N., Zeltmann, S., Shunmugasamy, V., and Pinisetty, D. (2013). Applications of Polymer Matrix Syntactic Foams. JOM, pp. 1–10.
10. Huang, J.S. and Gibson, L.J. (1993). Elastic moduli of a composite of hollow spheres in a matrix. Journal of the Mechanics and Physics of Solids, 41(1): pp. 55–75.
11. Kishore, Shankar, R., and Sankaran, S. (2005). Short beam three point bend tests in syntactic foams. Part I: Microscopic characterization of the failure zones. Journal of Applied Polymer Science, 98(2): pp. 673–679.
12. Kishore, Shankar, R., and Sankaran, S. (2005). Short-beam three-point bend tests in syntactic foams. Part II: Effect of microballoons content on shear strength. Journal of Applied Polymer Science, 98(2): pp. 680–686.
13. Kishore, Shankar, R., and Sankaran, S. (2005). Short-beam three-point bend test study in syntactic foam. Part III: Effects of interface modification on strength and fractographic features. Journal of Applied Polymer Science, 98(2): pp. 687–693.

14. Kumar, B.R.B., Doddamani, M., Zeltmann, S.E., Gupta, N., and Ramakrishna, S. (2016). Data characterizing tensile behavior of cenosphere/HDPE syntactic foam. Data in Brief, 6: pp. 933–941.

15. Lin, W.-H. and Jen, M.-H.R. (1998). Manufacturing and Mechanical Properties of Glass Bubbles/Epoxy Particulate Composite. Journal of Composite Materials, 32(15): pp. 1356–1390.

16. Porfiri, M. and Gupta, N. (2009). Effect of volume fraction and wall thickness on the elastic properties of hollow particle filled composites. Composites Part B: Engineering, 40(2): pp. 166–173.

17. Rizzi, E., Papa, E., and Corigliano, A. (2000). Mechanical behavior of a syntactic foam: experiments and modeling. International Journal of Solids and Structures, 37(40): pp. 5773–5794.

18. Yalcin, B. (2015). Chapter 7 - Hollow Glass Microspheres in Polyurethanes, in Hollow Glass Microspheres for Plastics, Elastomers, and Adhesives Compounds, S.E. Amos and B. Yalcin, Editors. William Andrew Publishing: Oxford. pp. 175–200.

19. Yalcin, B. and Amos, S.E. (2015). Chapter 3 - Hollow Glass Microspheres in Thermoplastics, in Hollow Glass Microspheres for Plastics, Elastomers, and Adhesives Compounds, S.E. Amos and B. Yalcin, Editors. William Andrew Publishing: Oxford. pp. 35–105.

20. Zeltmann, S.E., Bharath Kumar, B.R., Doddamani, M., and Gupta, N. (2016). Prediction of strain rate sensitivity of high density polyethylene using integral transform of dynamic mechanical analysis data. Polymer, 101: pp. 1–6.

21. Zeltmann, S.E., Prakash, K.A., Doddamani, M., and Gupta, N. (2017). Prediction of modulus at various strain rates from dynamic mechanical analysis data for polymer matrix composites. Composites Part B: Engineering, 120: pp. 27–34.

Note: All the figures and tables in this chapter were made by the authors.

Advances in Mechanical Engineering and Materials Sciences – Dr. Vinay K. B et al. (eds)
© 2026 Taylor & Francis Group, London, ISBN 9-781-041-20970-6

40

Effect of Copper Chills on Cryogenic Solidification of Al-ZrO2 Metal Matrix Composites

Praveen Yadav T. R.[1], Mamatha Y. P.[2]
Assistant Professor, Department of Mechanical Engineering,
Vidyavardhaka College of Engineering,
Mysuru, India

Abstract: Cryogenic solidification is a modern technology applied to freeze melted materials at extremely cold temperatures, resulting in superior microstructures and improved material properties. With advanced alloys and composite materials, this method is essential in shaping them to improve their strength, durability, and performance. This research aims to discover the effects of cryogenic solidification using copper chills on Al-ZrO2 composite by systematically examining the interaction between Al-ZrO2 composite and copper chill boxes during different casting processes. A systematic approach, considering material properties, cooling rate, and casting conditions as well as the influence of copper chill thickness on microstructure, mechanical characteristics, and casting defects, is needed to establish the optimum method of casting using copper chills. The paper concludes various outcomes of the preparation and testing of Aluminum metal matrix composites and improved results during casting through cryogenic solidification. The enhancements in various properties are compared in this research. This collective knowledge provides a solid basis for research and innovation within the constantly evolving field of cryogenic solidification.

Keywords: Aluminum metal matrix composite, Al ZrO2 composite, Copper chills, Cryogenic solidification

1. INTRODUCTION

In recent years, manufacturing technology has developed exponentially, increasing the need for enhanced materials with advanced mechanical characteristics for applications in production of automobiles, electronics, and computers, often replacing conventional materials like plastics. During the 1990's, Aluminium (Al) got its rebirth as a structural material due to environmental considerations, improved safety, and enhanced comfort levels. Due to its lightweight nature automotive industry witnessed a surge in demand for Al-based materials. Many researchers have dwelled on development of enhanced Al-based composites that there has been a significant rise in structural elements and automotive parts which use Aluminium composites. It is estimated 20% increase in demand for Al based alloys every year. However, the limited mechanical properties of Aluminium alloy (LM-13) are the most concerned issue

in its applications. Hence the literature survey indicates the development of different Al based Metal Matrix Composites (MMC's) using different chills and certain chill thicknesses are not being worked on. Therefore, the current study aims to close this gap and explore the combined characteristics of Al-ZrO2 MMCs. For this research, stir casting method has been chosen which is the most desirable and economical option for production of composites. This methods hemp in agglomeration of ZrO2 reinforcement particles to evenly distribute throughout the matrix metal Al (LM-13). The main goal of this study is to use stir casting and copper chill to create a composite made of Al alloy and reinforced with nano-ZrO2 particles for multiple chill thicknesses and test it for tensile strength, hardness, wear strength according to ASME (American Society of Mechanical Engineers) and ASTM (American Society of Testing Materials) standards. Later compare all the outcomes with standard LM-13 Al alloy

[1]Praveen30@vvce.ac.in, [2]mamathayp@vvce.ac.in

10.1201/9781003725053-40

properties to conclude on the enhancements of mechanical properties.

2. MATERIALS AND METHODOLOGY

2.1 Selection of Material

In this research we opted for Aluminium of grade LM-13 as the base metal matrix and for reinforcement Zirconium dioxide nano particle of size varying from 50 to 80 nm is used. The composite is maintained with 10 wt.% reinforcement material addition.

2.2 Aluminium Alloy LM 13

Aluminum alloy LM13, with a mix of 12-13% silicon, offers cost-effectiveness, lightweight build, and resilience in harsh conditions. Its exceptional wear resistance, bearing properties, and low thermal expansion make it ideal for aerospace and high-temperature machinery. Other important applications include pulleys, pistons, and engine components due to their strong yet lightweight composition and unique properties.

2.3 Nano Zirconium Oxide

Nano shell, a company based in Punjab, India, offers 80 nm nano zirconium oxide powder for various applications, including semiconductor polishing and jewellery manufacturing. Zirconia is a white crystalline solid, which is highly resistant to heat and corrosion, alkalis, and acids. This is used for reinforcements for the composite.

3. PREPARATION OF COPPER END CHILL BOX

Solidification control is an important step in casting processes to reduce the severity of shrinkage issues, particularly in sand casting where pasty zones pose challenges. Implementing chills strategically in sand moulds causes a rapid temperature change, minimizing pasty zones and enhancing directional solidification. Chill casting, a cost-effective method, utilizes chill blocks made of mild steel according to AFS (American Foundry Society) standards, with various copper end chill thicknesses (ranging from 5-25 mm). Liquid nitrogen circulation promotes cryogenic solidification, reducing lump regions in sand casting for improved efficiency, as depicted in Fig. 40.1 & Fig. 40.2.

Chill blocks play a vital role in achieving steep temperature gradients during solidification in sand casting, promoting directional solidification, and microstructure refinement. The heat absorption capacity of chill blocks is determined by their volumetric heat capacity (VHC), using the formula.

$$VHC = Cp * V * \rho \qquad (1)$$

Where Cp is the specific heat in KJ/Kg K, V is the volume of the chill in cm^3, and ρ is the density of the chill material in g/cm3.

Fig. 40.1   Copper chill box (30 mm)

Fig. 40.2   Copper chill box (35 mm)

An increase in any of these components enhances volumetric heat capacity and provides the chilling effect. Common chill materials like copper are favoured for their high thermal conductivity and VHC. Equation (1) represents the calculation of VHC, required for evaluating chill effectiveness in extracting heat from molten metal during cooling.

4. PREPARATION OF SAND MOULD

The preparation of green sand mixture for mould requires the mixing of silica sand with 5% clay, 1% sawdust, and 6% water. The process involves placing wooden patterns of the requisite size and shape in the drag box, which is coated with parting sand to prevent adhesion. The green sand mixture is later poured into the drag box and compacted manually as shown in Fig. 40.3. Subsequently, the drag box as shown in Fig. 40.4 is inverted, and parting sand is applied to its inner surfaces. The cope box is placed on top of a drag box, filled with sand, and compacted. Once the wooden patterns are removed, copper chill blocks are employed for larger patterns to regulate temperature, while

Fig. 40.3   Cope box

vents are created for the gas to pass during metal pouring for smaller patterns. Cope and drag boxes are secured with pins and sealed with clay and dried to eliminate moisture.

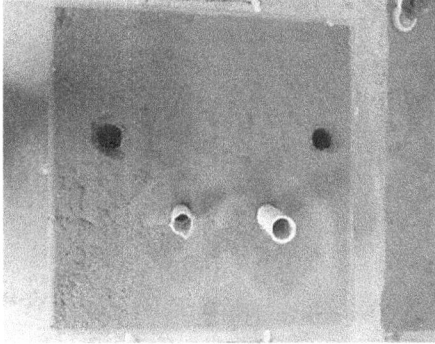

Fig. 40.4 Drag box

5. FABRICATION TECHNIQUE

Production of composite using cryogenic solidification. Aluminum alloy LM13 blocks are cut to the proper size and weighed along with nano ZrO_2 on a precise scale. LM13 is melted at 800°C in a bottom crucible furnace, the nano ZrO_2 is preheated at 200°C for 30 minutes to reduce contaminants. Continuous agitation and the addition of hexachloroethane help in degassing the molten metal mixture. After removing the slag, nano ZrO_2 is gradually added to the vortex of the molten metal, ensuring consistent dispersion through 10 minutes of stirring. Sand mould boxes are prepared and pre-cooled with liquid nitrogen before pouring the molten metal mixture into the mould. Cryogenic circulation of liquid nitrogen promotes solidification. After solidification, cryogenically solidified nanocomposites are extracted by breaking the sand mould..

6. PREPARATION OF TEST SPECIMEN

6.1 Microstructural Study

Specimen for microstructural study hardness test. Castings were prepared and cut into various sizes to obtain NMMCs for testing. The specimens were analyzed for microstructural, mechanical, and wear properties by ASTM standards with diameter and length to be 15 mm. The samples were cleaned and polished with silicon carbide abrasive paper and diamond-based compounds of 3-micron and 1-micron size and 0 carbide. All tests were conducted on the same set of samples to ensure the accuracy and consistency of the tests.

6.2 Tensile Test

Specimen for tensile test. The specimens for tensile testing are prepared according to ASTM E8 standards independent of the solidification technique used, the specimen is done using a computer numerical control (CNC) lathe machine for accurate specimen and to fulfil dimensional tolerance.

6.3 Wear Test

Specimen for wear test. The specimen for wear test is prepared according to ASTM G99 standards irrespective of cryogenic chilling. The specimen length should be 30 mm, and its diameter should be 10mm.

7. MICROSTRUCTURAL ANALYSIS OF COMPOSITES

Optical microscopy is employed to examine the microstructure of both the matrix material and the nanometal matrix composite (NMMC) specimens. The optical microscope assisted analysis of microstructures, surface contamination, and irregular surfaces.

7.1 Vickers Micro Hardness Test

To measure the hardness of the aluminum alloy and nanocomposites, a Vickers microhardness tester was used. Unlike other hardness tests, the Vickers test applies less force but offers higher accuracy. It employs a small diamond indenter and an optical system to magnify the target area of the material. Test samples were prepared strictly according to ASTM E92 standards. The hardness tester applies load is 3 kgf, with a magnification of 20X, and utilizes a diamond cone indenter, diamond being the hardest substance that can impinge on any surface. Samples used for microstructural characterization were also employed for this test. Three indentations were made on each sample to determine the average hardness value. The tests were performed at room temperature.

7.2 Tensile Test

The tensile test was conducted on specimens of aluminium alloy LM13 and cryogenically solidified nano ZrO2 composites, cryogenically treated, adhering to ASTM E8 standards. Test samples were prepared and mechanical buffing was performed using fine-grid emery paper or zero emery paper to ensure a smooth surface finish. Tensile testing was carried out at room temperature (27°C) using a computerized universal testing machine (UTM). The test was done until the specimen breaks, the values were noted down and graphs were taken as shown in Fig. 40.7, 40.8, 40.9 and 40.10.

7.3 Wear Analysis Test

Wear analysis was conducted on cryogenically solidified specimens, considering various parameters. For cryogenically solidified specimens, five parameters were examined: chill thickness, reinforcement, load, speed, and time. Chill thickness considered were 30mm and 35mm, reinforcement is nano zirconium oxide (10%), load is 40 N, speed is 1200rpm, and time 10 min. Wear was done with a pin-on-wheel wear test machine. Technically a cylinder-shaped specimen is placed on the machine which stays stationary, and the abrasive wheel underneath rotates.

Dry sliding wear behavior of the composites was analysed using pin-on-disc test equipment (Ducom make). Cylindrical specimens with dimensions of 10mm diameter and 30mm length were prepared according to ASTM G 99 standard on a lathe machine. The disc material, EN31 hardened steel, had a hardness value of 58 to 62 HRC. Wear test disc samples have a diameter of 100mm and a thickness of about 8mm. Wear experiments were conducted on the composite specimens made with different chill boxes of various chill thicknesses, with a speed of 1200rpm, applied load of 40 N, and time of 10 minutes. Wear values obtained from the tests were analyzed using MINITAB 19 statistical software to determine the most significant parameters.

8. RESULTS AND DISCUSSION

8.1 Production of Al-ZrO2 Composite

Production of composite was accomplished by stir casting method using cryogenic solidification. The controlled environment helped suitably in reduction of microporosity, segregation, or agglomeration of reinforcement. This confirms the suitability of stirring and achieving good solidification conditions during casting of a composite. Thus, this method is feasible for being a potential fabrication technique for composite preparation.

8.2 Microstructural Characterization

Cryogenically solidified nanocomposites are produced using copper end chill blocks. Two variations of copper chill thickness are used, 30mm and 35mm. The composites are produced with constant reinforcement % of 10 wt.% for both the chill thickness. The microstructural images of the composites demonstrate the impact of copper end chill thickness. The size of the nano ZrO2 particles had an impact on the microstructure of the composites. Cryogenically solidified nanocomposites have more refined grain structures than Al LM-13 alloy.

Figure 40.5 shows the microstructural images of the composite formed with a chill thickness of 30 mm. The grain is refined during cryogenic solidification because copper end chills remove heat from molten metal at faster rates. The microstructural images of the composite formed with 35mm chill thickness, as in Fig. 40.6 is more refined as there is an increase of VHC (Volumetric Heat Capacity) between the chill thicknesses.

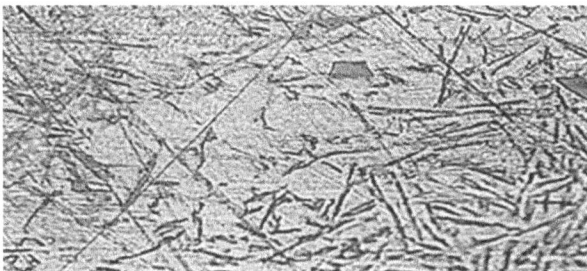

Fig. 40.5 Microstructure of composite using 30 mm

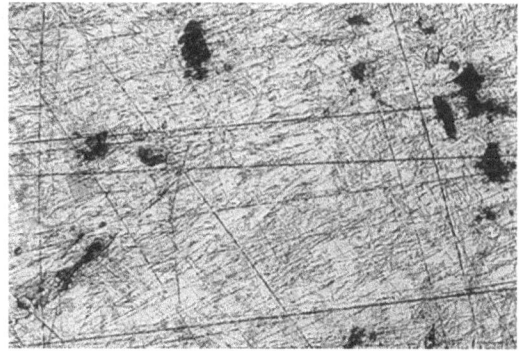

Fig. 40.6 Microstructure of composite using 35 mm

The microstructure pictures show comparatively less porosity and dissipation on reinforcement particles throughout the matrix. While 30 mm chill thickness used composite comprises of negligible blow holes and porosity, 35 mm chill thickness used composite has bigger blow holes.

8.3 Mechanical Properties

Hardness Testing

The micro hardness test was carried out on cryogenically solidified nano metal matrix composites (CNMMCs) specimens. The results of micro hardness test are tabulated in Table 40.1. The hardness value shows a decreasing tendency when increasing the chill thickness from 30 mm to 35 mm. This can be due to the presence of tougher ceramic particles (Nano ZrO2) in the matrix material, which limits localized deformation during indentation and refines grain size. Also, consider the chilling impact on the composite. CNMMCs have a 32.20% increase in hardness compared to the matrix material for 10 wt.% reinforcement at 30 mm chill thickness. The addition of 30 mm chill thickness used composite have decreased hardness. The hardness of CNMMCs decreased by 19.49 % as compared to composite using 35 mm chill thickness. The graphs are also shown in the Fig. 40.10 and 40.11 for 30 mm and 35 mm chill thickness.

Table 40.1 Mechanical properties of NMMCs and matrix alloy (LM 13)

Chill Thickness (mm)	Properties		
	Average HV	0.2 % YS (MPa)	UTS (MPa)
30	119	34.227	73.574
35	95.8	18.073	18.819
Matrix alloy (LM 13)	90	120	170

Tensile Strength Testing

Cryogenic solidification improves mechanical properties, including hardness, tensile strength, and wear resistance, of nano metal matrix composites. This effort uses cryogenic solidification to create an aluminum alloy LM 13-nano

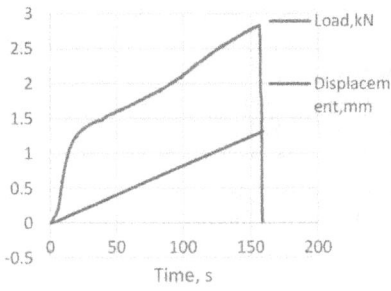

Fig. 40.7 Displacement, Load vs time curve (30 mm)

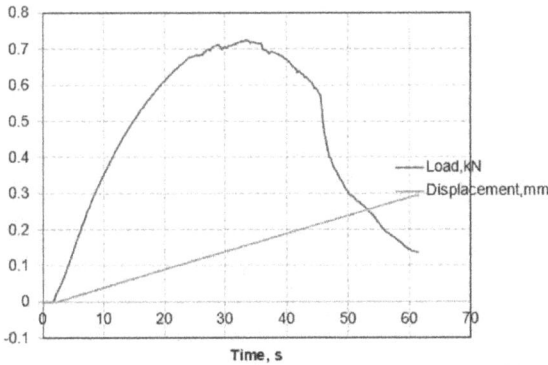

Fig. 40.8 Displacement, Load vs time curve (35 mm)

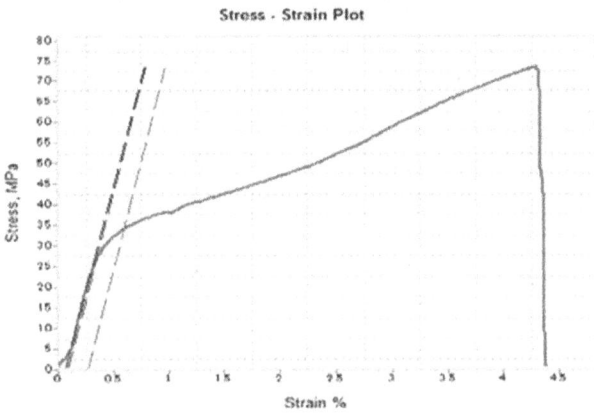

Fig. 40.9 Stress vs strain curve (30 mm)

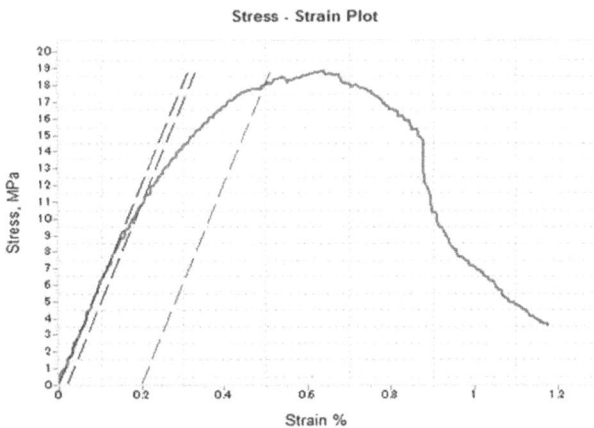

Fig. 40.10 Stress vs strain curve (35 mm)

ZrO2 composite. The percentage of reinforcement and chill thickness were both important in this case. Reducing chill thickness from 35 mm to 30 mm improved CNMMCs' UTS significantly as shown in Table 40.1. This could be due to the inclusion of nano ZrO2 (10 wt.%) to LM-13, which may have increased the formation of numerous microcracks at the matrix-reinforcement interface. The formation of microcracks decreases the UTS and 0.2% Yield Stress of the nanocomposite.

Wear Testing

The wear test carried out is Pin on disc test with the parameters being Load, Speed, and time and for both the composites same parameters are set for a fair comparison. The result is shown in Table 40.2. The results are compared between the two composites, and it is seen that wear is gradually decreasing with increasing in chill thickness and has 7.82% reduction in wearing of material. This can be the effect of the varying values of Coefficient of Friction of different composites using different chill thickness.

Table 40.2 Wear test comparison

Parameters	Results		Units
Chill thickness	30	35	mm
Speed	1200	1200	rpm
Load	40	40	N
Time	10	10	minute
Wear	230	212	Microns

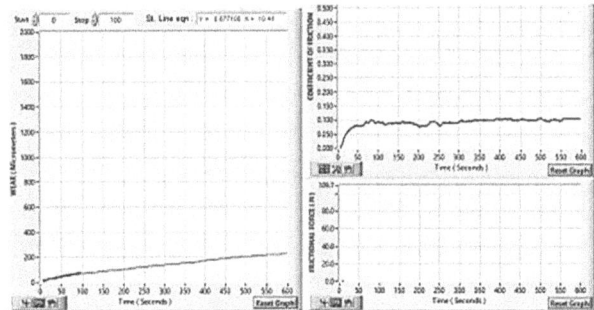

Fig. 40.11 Wear v/s time (30 mm) chill thickness

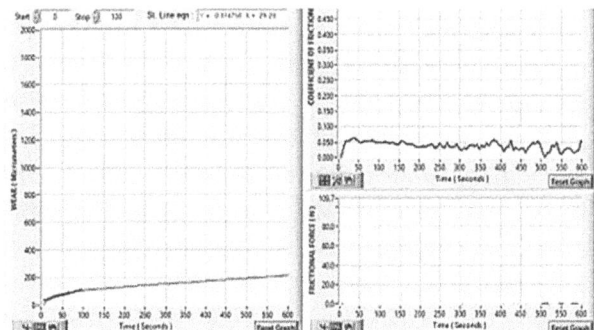

Fig. 40.12 Wear v/s time (35 mm) chill thickness

9. CONCLUSION

Casting using crucible coupled with cryogenic solidification to produce AlZrO2 composites have revealed that microstructural investigation revealed fine grain refinement and a reasonably uniform distribution of reinforcement particles, and negligible porosity. Nano-ZrO2 particulates in the Al matrix improved hardness and wear toughness with increasing chill thickness, but slightly lowered tensile strength and fracture behavior. The results findings will have more implications on industrial applications particularly in automotive, aerospace and manufacturing sector where enhanced wear resistance and hardness is most important in performance. A controlled chill thickness leads to manufactures a cost effective approach to optimize Al based composites for specific applications.

REFERENCES

1. Ganesh G. PATIL and DR. K. H. Inamdar (2014). "Prediction of casting defects through artificial neural network". international journal of science, engineering and technology- www.ijset.in. ss

2. Joel Hemanth (14 May 2001), "Fracture toughness and wear resistance of aluminum-boron particulate composites cast using metallic and non-metallic chills". Material and Design 23 (2002) 41–50.

3. Joel Hemanth (29 July 1999), "Action of chills on soundness and ultimate tensile strength (UTS) of aluminum-quartz particulate composite". Journal of alloys and compounds 296 (2000) 193–200

4. M.Narasimha, R. Rejikumar and K. Sridhar (2013). "Statistical methods to optimize process parameters to minimize casting defects". International Journal of Civil Engineering and Technology (IJCIET).

5. Mohit Kumar sahu and Raj Kumar (2018). "Fabrication of aluminum matrix composites by stir casting technique and stirring process parameters". Optimization sahuhttp://dx.Doi.Org/10.5772/intechopen.73485.

6. Philip o. Babalola, anthony O. Inegbenebor, christian A. Bolu, shella I. John (febravary 2019). "Comparison of the mechanical characteristics of aluminium sic composites cast in sand and metalmoulds". International journal of mechanical engineering and technology (IJMET).

7. S.F. Hassan and M. Guptha (14 July 2004)," Development of high-performance magnesium Nano composites using solidification processing route".

8. S.F. Hassan and M. Guptha (2007), "Effect of Nano-ZrO2 Particulates Reinforcement on Microstructure and Mechanical Behavior of solidification processed Elemental Mg". Composites: Part A 38 (2007) 1395–1402.

9. Sijo M T (April 2017). "Numerical simulation of centrifugal casting for functionally graded metalmatrix composites". International Journal of Civil Engineering and Technology.

10. Vaibhav Ingle and Madhukar Sorte (March 2017). "Defects, root causes in casting process and their remedies review". ISSN: 2248–9622, vol. 7, issue 3, (part -3).

11. Yogesha K.B and Joel Hemanth (March 2012), "Mechanical Properties of metallic and non-metallic chilled austempered ductile iron (ACDI)", 240–243.

Note: All the figures and tables in this chapter were made by the authors.

Advances in Mechanical Engineering and Materials Sciences – Dr. Vinay K. B et al. (eds)
© 2026 Taylor & Francis Group, London, ISBN 9-781-041-20970-6

41

An Experimental Study on Effect of Copper Chills in the Solidification of Aluminium Alloy (AL-LM13) and Nano Zirconium Oxide (ZrO$_2$) Composite using Water Cooling Medium

Mamatha Y. P.[1]
Assistant Professor,
Department of Mechanical Engineering Karnataka,
India's Vidyavardhaka College of Engineering is located in
Mysuru

G. B. Krishnappa[2]
Professor and Dean (R & D),
Department of Mechanical Engineering Karnataka,
India's Vidyavardhaka College of Engineering is located in
Mysuru

Ramesha V. Hamsabhavi[3],
Prashanth M. P.[4], Yashwanth K. Gowda[5], Deepak C.[6]
Student,
Department of Mechanical Engineering Karnataka,
India's Vidyavardhaka College of Engineering is located in
Mysuru

Abstract: To check solidification of aluminum alloy (AL-LM13) and nano zirconium oxide (ZrO2) composites by copper chills in water as a cooling medium has been tried with this experimental study. The key objective of this investigation focuses on analyzing the effect of copper chills on microstructural characteristics, mechanical properties, and solidification kinetics of the AL-LM13 alloy incorporating varying proportions of ZrO2 nanoparticles. The goal of adding ZrO2 nanoparticles was to improve the AL-LM13 alloy's wear resistance and mechanical qualities. This alloy was selected because to its favorable casting properties. The exceptional thermal conductivity demonstrated by copper chills was utilized as the cooling medium throughout the course of the experimental procedure. A sand casting methodology was employed to produce the cast specimens. The experimental framework included various configurations of chill placements along with distinct ZrO2 concentrations (0%, 1%, 2%, and 3% by weight). The procedures for solidification were carefully examined, and the resultant microstructures were characterized employing optical and scanning electron microscopy methods. The specimens' hardness, impact resistance, and tensile strength were assessed using predetermined criteria. The findings indicated that the application of copper chills significantly enhances cooling rates, leading to the development of refined microstructures with superior grain morphology. An addition of ZrO2 nanoparticles improved the

[1]mamathayp@vvce.ac.in, [2]gbk@vvce.ac.in, [3]rameshahamshabhavi@gmail.com, [4]prashanthgowda0902@gmail.com, [5]yashwanthgowda075@Gmail.com, [6]dc190547@gmail.com

mechanical properties further with the maximum value observed at 2 weight%. Research shows copper chills and ZrO2 help cooling and improve mechanical properties of Al-LM13 alloy composites (aluminium and silicon) making them suitable for using in engineering purposes. Future work will focus on the long-term stability and performance of these composites under operational conditions.

Keywords: Aluminium alloy (AL-LM13), Nano zirconium oxide (ZrO$_2$), Copper chills, Water cooling medium

1. INTRODUCTION

The solidification of aluminium alloys is important as it affects the mechanical and thermal properties of cast components. These components have industrial use in the automotive, aerospace, and related industries. AL-LM13 is an aluminium-silicon-based alloy used for casting engine components. Al-LM13 alloy can be significantly improved in strength, wear and fatigue resistances by reinforcing it with ceramic particles. One alloy, AL-LM13, finds great usage in the engine component industry and other high-performance uses, due to its good casting and resistance to wear. The strength and other properties of AL-LM13 alloy can be improved with ceramic particles. A great way to improve AL- LM13 performance is by the addition of nano zirconium oxide (ZrO2). ZrO2 is a ceramic material with high strength. The incorporation of ZrO2 in aluminum alloys improves wear resistance, strength, and thermal stability. The incorporation of nano-sized ZrO2 particles significantly enhances the mechanical properties of the alloy compared to micron sized particles, attributed to superior matrix distribution and interfacial bonding. This enhancement is further facilitated by the alloy's solidification process during casting. Water is a prevalent grinding medium in foundry processes due to its effective heat transfer capabilities, which regulate solidification. Employing water as a cooling medium can improve the mechanical properties of cast products by achieving a rapid cooling rate. This research investigates the effects of copper chills and water cooling as cooling agents during the solidification of the nano ZrO2- reinforced AL-LM13 alloy. When casting metals, directional solidification is induced using chills composed of copper, a highly conductive metal. This way the quality of the casting improves. By placing copper chills in the mould, we expect to influence the solidification rate and resultant microstructure of aluminium alloy composite.

2. LITERATURE REVIEW

Joel Hemanth and M. R. investigate the corrosion characteristics of LM-13 aluminium alloy reinforced with Nano-ZrO2 for aerospace applications. ZrO2 particles, sized 100-200 nm, are incorporated at varying concentrations from 3 wt% to 15 wt% and solidified using a 25 mm thick copper chill. Electrochemical polarization tests, following ASTM G59-97, revealed enhanced corrosion resistance with increased ZrO2 content. However, initial pitting corrosion was observed, which diminished as a protective layer was established on the Copper Chilled Nano Metal Matrix Composites (The rusting mechanisms were further investigated using Scanning Electron Microscopy (SEM). By showing that aluminum alloys with higher silicon concentrations are more prone to localized corrosion, the study further clarifies how alloy composition affects corrosion resistance. Joel Hemanth (2019) carried out a methodical examination of the tribological properties of Nano-ZrO2-enhanced Copper Chilled Nano Metal Matrix Composites (CNMMCs) used in aeronautical products. ZrO2 nanoparticles, which range in size from 100 to 200 nm, were added in different amounts, ranging from 3 weight percent to 15 weight percent. A 25 mm thick copper chill was used to solidify the CNMMCs in order to evaluate their resistance to corrosion. As stated in ASTM-59-97, electrochemical polarization assessments revealed that increased ZrO2 concentrations enhanced corrosion resistance. Initially, pitting corrosion was the most significant, but its intensity reduced over time due to the formation of a protective coating Manjunath B R and Anil Kumar's study examines how tool rotation speed affects the mechanical characteristics and wear behavior of surface hybrid composites made from aluminum alloy via stir casting. The use of optical microscopy to provide a thorough insight by analyzing the distribution of reinforcing particles in the manufactured surface hybrid composites. Each surface hybrid composite's micrographs showed that the reinforcement particles were satisfactorily dispersed. Effective integration was shown by the consistent distribution of zirconium oxide and zircon sand throughout the nugget zone. In 1992, G KDEY and S BANERJEE investigated zirconium-based alloys' quick solidification. In 1992, G. KDEY and S. BANERJEE Zirconium-based alloys solidify quickly, December 1992 alloys based on zirconium, especially those that undergo quick solidification procedures. Various zirconium-based alloys have gained much attention in the arena of nuclear and chemical engineering due to their ability to form amorphous phases and also their corrosion resistance and good mechanical properties. Vitor B. Moreira et. In his 2021 study, Al examined the use of a green zirconium oxide layer and organic coating to protect aluminum. This academic investigation thoroughly examines the integration of zirconium dioxide (ZrO$_2$) nanocoating onto aluminium substrates, which is achieved through a meticulously regulated electrochemical chronoamperometry technique executed within hexa fluoro zirconic acid solutions, specifically H$_2$ZrF6·5H$_2$O, with the objective of delivering enhanced protection and an extended lifespan of the resulting coating. The research was conducted at Department of Materials Science and Engineering at Universität Polytechnical de Catalunya in Barcelona, Spain. The study found that the ZrO2 intermediate layer significantly improved adhesion and reduced porosity in films compared to traditional methods. The 2006 REACH regulation prohibited chromium plating for galvanized structures due to health and environmental risks. Consequently, non-phosphate zirconium-based coatings have emerged as a greener alternative, being less harmful than chromate coatings. In conclusion, zirconium-based chemical conversion coatings (Zr-CCC) present a viable solution for enhancing corrosion resistance and adhesion on various metal surfaces, advancing protective coating technologies. Yong-Sang Kim et al. (2023) will investigate the corrosion resistance of U1070 aluminium alloy with added zirconium compared to A1070 alloy. Jigang Wang et al. (2016) focused on synthesizing and characterizing ZrO2 nanoparticles via vapor-phase hydrolysis for biomedical applications. The nanoparticles were analyzed for crystal structure, hardness, surface roughness, and microstructure using various testing methods. Biocompatibility assessments indicated the nanoparticles are suitable for biological applications. The study found that temperature of sintering and compressive forces significantly influence the physical properties of ZrO2 nanoparticles. ZrO2 flakes treated at specific conditions achieved hardness comparable to human bone and a density exceeding that of bone tissue. These characteristics suggest potential applications of ZrO2 nanoparticles in bone repair and replacement. M Siva Reddy et al. aimed to enhance the properties of AA8011 aluminum alloy using zirconium oxide and titanium boride. Huda M. Sabbar et al. (2021) reviewed the characterization and potential applications of AA7075-ZrO2 nanocomposites produced through a solid-state process. Solid-state recycling constitutes a highly efficient methodology for the conversion of metallic shavings into functional products without necessitating the melting process, thus facilitating a reduction in energy consumption while simultaneously mitigating metal wastage.

M Sunil Kumar et, The study is done on find out the corrosion characteristics of A356 reinforced with Hematite metal matrix composites. Aluminum alloy is famous industrially for its very significant role and applications. attributable to its remarkable attributes, which encompass significant toughness, exceptional strength, considerable hardness, and notably improved resistance against both corrosion and wear. The incorporation of Hematite, which comprises rigid ceramic elements derived from iron ore, into aluminium matrix mixtures markedly improves the material's confrontation to wear and degradation over time. The process of sand casting, particularly when executed with higher precision, facilitates the establishment of metallic components utilizing end chills, such as copper,

which in turn influences several critical factors including the microstructure, corrosion behaviour, tribological characteristics, and strength of the resultant composites. Chill casting is one of the most advantageous techniques to cast Metal Matrix Composites (MMCs). It is done along with sand casting which is in a liquid metallurgical process of MMCs. Researchers have shown that nano metal matrix composites made by chilling have much better corrosion resistance when compared to as- cast composite of similar ZrO2. A steady stream of demand comes from designers, makers and industrial users for better performing materials and ways to make them. Better fuel efficiency is in demand due to worries about energy use and global warming. Aluminium alloys, often designated as Al-alloys, are utilized across a diverse array of sectors, including automotive, aerospace, and marine engineering, owing to their remarkable mechanical characteristics, heightened specific strength, enhanced electrical and thermal conductivities, as well as their augmented resistance to corrosion and oxidation. K.H.W. Seah et al. undertook an examination of the solidification performance of water-cooled, sub-zero chilled cast iron, presenting a comparative evaluation of the solidification phenomena intrinsic to these two cooling techniques. This manuscript highlights the ramifications of differential cooling rates on the mechanical attributes of cast iron, encompassing tensile strength, fracture toughness, and hardness. The investigation reveals that increased cooling rates, facilitated by the application of chills, substantially improve the mechanical properties of cast iron through microstructural refinement. In particular, chilled castings demonstrate superior strength, hardness, toughness, and wear resistance. RAVITEJ Y P et al. explored the influence of copper chills on the tribological performance of LM13/ZrSiO4/C hybrid metal matrix composites. This research emphasizes the fabrication of dual- reinforced aluminium matrix composites, specifically accentuating the incorporation of zirconium silicate (ZrSiO4) and carbon particulates into the molten LM13 aluminium alloy utilizing the stir casting methodology. Zircon is chosen for its exceptional strength and outstanding thermal compatibility, rendering it a valuable reinforcement agent, while carbon enhances the mechanical and wear-resistant properties of the composite. The study assesses mechanical properties, including tensile strength, hardness, and microstructural attributes, at both the chill and non-chill extremities of the composite specimens. The stir casting process, employing vortex mixing via a graphite stirrer, promotes the homogeneous distribution of reinforcement particles throughout the liquid aluminium matrix. The characteristics of the composite are further optimized by varying the concentration of ZrSiO4 in increments of 3 wt.%, spanning from 3 wt.% to 12 wt.%, while consistently maintaining a 3 wt.% graphite proportion. Copper metallic chills are utilized to induce unidirectional solidification, thereby enhancing the material properties by modulating the cooling rate. A comparative analysis of the chill and non-chill ends elucidates the effects of cooling on the thermal and mechanical performance of the composite material Alalkawi H. J. M et al. want to use nanocomposite techniques to improve the mechanical and fatigue properties of aluminum alloy 7049. ZrO2 nanoparticles, which have an average grain size of 30 to 40 nm and concentrations of 2, 4, 6, have been found to be efficient reinforcements for aluminum alloy 7049 and 7 weight percent. Because stir casting is a cost-effective technique for improving and processing metal matrix composites, it was used to create nanocomposite materials. According to empirical research, adding zirconium oxide (ZrO2) as a reinforcing agent significantly enhances mechanical qualities. The addition of 4 weight percent ZrO2 produced the best improvement in the mechanical characteristics of 7049 AA, outperforming the improvements noted with other 7049AA's fatigue strength augmented with 4 weight percent under a continuous load over 10^8 cycles Comparing ZrO2 nanoparticles to the baseline 7049 AA, a notable 9.86% rise was observed. The fatigue life factor (IFLF) improved at 350, 300, and 250 MPa at various amplitude stresses of 400, with corresponding values of 66, 115, 63, and 107%. The fatigue life factor (IFLF) improved by 66, 115, 63, and 107% for various amplitude stresses of 400, 350, 300, and 250 MPa, respectively. The fatigue life factor (IFLF) improved in the corresponding ranges of 66, 115, 63, and 107% for various amplitude stresses of 400, 350, 300, and 250 MPa. Prasad H. Naik and colleagues examined the impact of Utilizing the melt processing technique, this study evaluates the effects of incorporating nano zirconium oxide (ZrO2) particles into copper-10% tin-nano ZrO2 composites. Nano ZrO2 particles were incorporated into the copper-tin (Cu-Sn) matrix at weight percentages of 4, 8, and 12. Microstructural examinations were conducted utilizing SEM, EDS, and XRD analytical techniques. The mechanical characterization of Cu-10%Sn composites containing 4, 8, and 12 weight percent of nano ZrO2 adhered to ASTM guidelines Recent advancements in metal matrix composites based on aluminum were explained by Amlan Kar et al. (2024). The remarkable qualities of aluminum matrix composites (AMCs), including their low density, high strength-to-weight ratio, and great resistance to corrosion, Comparing them to conventional engineering materials has generated a lot of scholarly attention due to their enhanced wear resistance and beneficial high-temperature characteristics. These composites are widely used in a variety of industries, such as military, automotive, and aerospace. Many different processing techniques that are especially tailored to their individual categories are used to fabricate AMCs. Numerous scientific discoveries about the internal and external effects of ceramic reinforcement on the mechanical, thermomechanical, tribological, and physical properties of AMCs have been made throughout the last three decades of rigorous

research. The deployment of AMCs has proliferated within advanced structural and functional sectors, which encompass automotive, aerospace, sports and recreation, military, and thermal management applications. Harry Ng (2011) investigated the direct chill casting of aluminium alloys through empirical methodologies and design frameworks. The dissertation introduces the Novelis Fusion TM Technology, an innovative iteration of the conventional direct chill (DC) casting process that enables the co-casting of multi-layered composite aluminium alloy ingots. The research is centered on the experimental design, measurement techniques, and interpretation of empirical results to substantiate the models pertinent to traditional DC casting This study was executed utilizing aluminium alloys AA3003, AA6111, and AA4045, alongside two distinct sets of experiments aimed at evaluating the influence of casting parameters on the cooling and solidification phenomena of the ingots. In order to determine the geometry and depth of the sump, a zinc-rich alloy was employed in the investigation as a melt poison. Joel Hemant investigated the wear characteristics of fused silica (SiO2) metal matrix composites supplemented with chilled aluminum alloy (A356). The creation and evaluation of aluminum alloy MMCs augmented with fused silica particles in sand molds are described in this work. The investigation examined the composites' microstructure, strength, hardness, and wear properties. After dispersoid and copper chill were added, it was found that the strength, hardness, and wear resistance increased by up to 9% weight percent. Wear has a significant impact on the service life of engineering components and has significant national economic ramifications. Composites are high-tech materials used in many applications that need remarkable wear resistance, including electrical contact brushes, artificial joints, helicopter blades, and cylinder liners. The quest for innovative wear-resistant materials for high-performance tribological applications has been a primary impetus behind the advancement of ceramic particle technology. Ahmed Abdulkareem (2022) conducted a comprehensive investigation on the Thermodynamic Analysis of Al Alloy Reinforced with Zirconia Particles. This study presents a thermodynamic evaluation of an A5083/ZrO2 metal matrix composite utilizing HSC software, aimed at predicting the phases that may develop during the fabrication process. The zirconia particles utilized as the reinforcing agent were derived from Egyptian zircon and incorporated into a commercially available base matrix. Following the melting of the base material at 750 °C in an electric furnace, the extracted zirconia particles were introduced into the molten matrix Thermodynamic diagrams indicate that ZrO2 tends to react with the matrix more favorably than MgO, resulting in the formation of new phases such as MgO and Al2O3. The researchers used a number of analytical methods, such as Energy Dispersive Spectroscopy (EDS), X-ray Diffraction (XRD), and Scanning Electron Microscopy (SEM), to describe the zirconia and the resultant composites. The results validate that high-purity zirconia was successfully extracted. The research that was carried out by R. Jeba et al. (2022) clarifies how the Cu-doped ZrO2 catalyst affects a photochemical reaction's kinetics and how well it imparts resistance to microbes. The investigation undertaken by Jeba et al. assessed the effects of copper-doped zirconium oxide nanoparticles on both photocatalytic and antimicrobial attributes. Utilizing a coprecipitation methodology, the researchers synthesized both pure and copper-doped zirconium oxide nanoparticles. An observed reduction in crystallite size concomitant with increased dopant concentration suggested that the samples predominantly exhibited a pure tetragonal phase.

3. SUMMARY

This compilation of research endeavors delves into the augmentation of diverse aluminum alloys and composites via zirconium-based reinforcements and coatings, aimed at enhancing corrosion resistance, mechanical properties, and wear resistance for applications in the aerospace, automotive, and biomedical fields. Significant focal points encompass the ramifications of Nano-ZrO2 reinforcement, the electrochemical integration of ZrO2 coatings, and the implications of zirconium on metal matrix composites. These investigations yield critical insights into the enhancement of material performance through sophisticated processing methodologies.

4. CONCLUSION

The experimental study on the effect of copper chills in the solidification of aluminum alloy (Al-LM13) and nano zirconium oxide (ZrO2) composite reveals significant findings regarding the enhancement of material properties. The incorporation of nano-ZrO2 into the aluminium matrix notably improves corrosion resistance, as evidenced by the reduction in the size and quantity of Al3Fe intermetallic particles, which diminishes the galvanic effect that typically exacerbates corrosion. The study highlights a copper chill's efficacy during solidification towards the mechanical and electrochemical properties of the composite. Additional research proves that these composites can also be used as an aerospace for better application and durability. Overall, for developing advanced aluminum alloys having superior characteristics, the findings recommend the use of nano-reinforcements and effective cooling techniques.

REFERENCES

1. Abdelkareem, A. (2022). Thermodynamic analysis of Al alloy reinforced with zirconia particles. https://doi.org/10.4236/oalib.1109569

2. G K DEY and S BANERJEE Metallurgy Division, Bhabha Atomic Research Centre, Bombay 400085, India Bull. Mater. Sci., Vol. 15, No. 6, December 1992, pp. 543–556. © Printed in India.

3. Hemanth, J. and Divya, M.R. (2018) Fabrication and Corrosion Behaviour of Aluminium Alloy (LM-13) Reinforced with Nano-ZrO2 Particulate Chilled Nano Metal Matrix Composites (CNMMCs) for Aerospace Applications. Journal of Materials Science and Chemical Engineering, 6, 136–150. https://doi.org/10.4236/msce.2018.67015

4. Hemanth, J. (2019) Tribological Behaviour of Copper Chilled Aluminium Alloy (LM-13) Reinforced with Beryl Metal Matrix Composites. Modelling and Numerical Simulation of Material Science, 9, 41–69. https://doi.org/10.4236/mnsms.2019.93004

5. Jeba, R., Radhika, S., Padma, C. M., & Davix, X. A. (n.d.). The influence of Cu doped ZrO 2 catalyst for the modification of the rate of a photoreaction and forming microorganism resistance. https://doi.org/10.22090/jwent.2022.04.002

6. Joel Hemanth "Abrasive and slurry wear behavior of chilled aluminum alloy (A356) reinforced with fused silica (SiO p) metal matrix composites" provides a comprehensive study on the properties and applications of aluminum metal matrix composites (MMCs). https://doi.org/10.1016/j.compositesb.2011.06.022

7. Kar, A., Sharma, A., & Kumar, S. (2024). A critical review on recent advancements in aluminium-based metal matrix composites. https://doi.org/10.3390/cryst14050412   17. Ng, H. (n.d.). *Direct chill casting of aluminum alloys: Experimental methods and design.* Waterloo, Ontario, Canada, 2011 Kavati

8. Venkateswarlu, K. P. V. Krishna Varma, & Uday Kumar Nutakki. (n.d.). IDB incidental dislocation boundary NP nanoparticle UTS ultimate tensile strength YS yield strength nomenclature CNT carbon nanotubes COE coefficient of thermal expansion CR cryo-rolled FSW friction stir welding GNB geometrically necessary boundary HB Brinell hardness. https://doi.org/10.1007/s12008-024-02106-4

9. Kumar M. S., Sathisha, N., Rajesh, T. R., Arutyunov, A. V., Pereira, F. L., Vega De Ceniga, M., & Rodríguez Moral, M. B. (n.d.). Corrosion study on chill casting effect on aluminium A356 reinforced with Hematite metal matrix composites.

10. Khenyab, A. Y., & Ali, A. H. (n.d.). Improvement of mechanical and fatigue properties for aluminum alloy 7049 by using nano composite technique. https://doi.org/10.22153/kej.2019.08.001

11. Manjunatha B R and Anil Kumar Mechanical Characterization of Al6061/ZrO2/Zirconium Sand Hybrid Metal Matrix Composite A Dept. of Mechanical Engineering K S Institute of Technology Bengaluru, India

12. Moreira, V.B.; Meneguzzi, A.; Jiménez-Piqué, E.; Alemán, C.; Armelin, E. Aluminum Protection by Using Green Zirconium Oxide Layer and Organic Coating: An Efficient and Adherent Dual System. Sustainability 2021, 13, 9688 https:// doi.org/10.3390/su13179688

13. Nayak, P. H., Srinivas, H. K., Rajendra, P., Nagaral, M., Raviprakash, M., & Auradi, V. effect of nano zirconium

oxide (zro2) particles addition on the mechanical behaviour and tensile fractography of copper-tin (cu-sn) alloy nano zirconium.RAVITEJ Y P, mohan C B and Anantha prasad M G A review on Evaluation of copper chill on tribological behaviour of LM13/ZrSiO$_4$/C hybrid metal matrix composites.

14. Reddy, M. S., Prattipati, P., Rachuri, V., Kumar, S., Kumar, S. P., Shobha, K. R., Singh, K., Rinawa, L., & Madhavarao, s. tribological characterization and optimisation of aa8011 aluminium alloy reinforced with zirconium oxide and titanium boride.

15. Sabbar, H. M., Leman, Z., Shamsudin, S. B., Tahir, S. M., Jaafar, C. N. A., Hanim, M. A. A., Ismsrrubie, Z. N., Al-Alimi, S., Kumar, A., & Gupta, M. (2021). AA7075-ZrO 2 nanocomposites produced by the consecutive solid-state process:

16. A review of characterisation and potential applications. *Metals*, *11*(5), Article 805. https://doi.org/10.3390/met11050805

17. Seah, K. H. W., Hemanth, J., Sharma, S. C., & Rao, K. V. S. (n.d.). Solidification behaviour of water-cooled and subzero chilled cast iron.

18. Wang, J., Yin, W., He, X., Wang, Q., Guo, M., & Chen, S. (2016). Good biocompatibility and sintering properties of zirconia nanoparticles synthesized via vapor-phase hydrolysis.

19. Yong-Sang Kim 1, Jong Gil Park 2, Byeong-Seon An 1, Young Hee Lee 2, Cheol-Woong Yang 1 and Jung-Gu Kim 1,* 1 School of Advanced Materials Science and Engineering, Sungkyunkwan University, 2066, Seobu-Ro, Jangan-Gu, Suwon, Gyeonggi-Do 16419, Korea; skybyego@gmail.com(Y.S.K.);absabs22@skku.edu

Advances in Mechanical Engineering and Materials Sciences – Dr. Vinay K. B et al. (eds)
© 2026 Taylor & Francis Group, London, ISBN 9-781-041-20970-6

42 Advances in Fabrication Techniques and Reinforcement Strategies for Nano Metal Matrix Composites

N. Jayashankar[1]
Associate Professor, Department of Mechanical Engineering,
Vidyavardhaka College of Engineering,
Mysuru, Karnataka, India

Praveen Yadav T. R.[2]
Assistant Professor, Department of Mechanical Engineering,
Vidyavardhaka College of Engineering,
Mysuru, Karnataka, India

Devappa[3]
Assistant Professor, Department of Mechanical Engineering,
JSS Academy of Technical Education College in
Bengaluru

Hemanth R.[4]
Associate Professor, Department of Mechanical Engineering,
The National Institute of Engineering,
Mysuru

Niharika Kiran M.
Student, Department of Mechanical Engineering,
Vidyavardhaka College of Engineering,
Mysuru

Abstract: Nano metal matrix composites (NMMCs) underscore the success of novel fabrication methods, such as cryogenic solidification, microwave sintering, mechanical alloying, and spark plasma sintering, in improving mechanical and tribological characteristics The addition of reinforcements like ZrO2, ZrB2, and Al2O3 greatly enhances Mechanical properties Additionally, research on nano-ZrO2 reinforced aluminium composites demonstrates notable performance enhancements, while hybrid MMCs provide tailored mechanical properties. Future studies should focus on optimizing processing parameters, exploring hybrid reinforcements, and evaluating long-term performance under real-world conditions. The integration of advanced manufacturing techniques, alongside defect analysis and process optimization, is essential for producing high-quality castings. The innovations proved the path for applications of high-performance aerospace, automotive, and structural sectors microwave sintering, mechanical alloying, spark plasma sintering, ZrO2, ZrB2, Al2O3, reinforcement, mechanical properties, tribological properties, wear resistance, hybrid MMCs, process optimization, defect analysis, advanced manufacturing, applications, automotive industry, structural materials.

Keywords: Cryogenic, Matrix composites (NMMCs), Nano metal solidification

Corresponding author: [1]jayshankar@vvce.ac.in, [2]praveen30@vvce.ac.in, [3]devappa@jssateb.ac.in, [4]hemanthr@nie.ac.in

10.1201/9781003725053-42

1. INTRODUCTION

Cryogenic solidification has been explored as a novel manufacturing method for NMMCs, significantly improving their structure and properties. Investigation reveals that this process improves the grain structure and ensures Metal matrix composites, (MMCs) have gained significant attention in modern materials engineering because of their superior properties compared to conventional alloys Among different MMCs, aluminum (Al)-based composites strengthened with ceramic particles like zirconia (ZrO_2) have surfaced as valuable materials for aerospace, automotive, and structural uses. Incorporating ZrO_2 improves characteristics like hardness, wear and thermal stability, rendering Al-ZrO_2 composites suitable for high-Solidification processing is essential in defining the microstructure and ultimate characteristics of MMCs. Conventional casting methods frequently result in flaws like porosity, rough grain formations, and inconsistent reinforcement distribution. To tackle these issues, cryogenic solidification, together with external cooling techniques, has been explored as an effective approach to enhance microstructure and elevate composite properties. Copper chills Utilized in this method, accelerates heat extraction, resulting in reduced grain sizes, improved mechanical properties, and decreased porosity.

This review aims highlighting the cryogenic solidification of Al-ZrO2. MMCs, highlighting the significance of copper chills during the solidification process the document examines various processing techniques and solidification properties.

Additionally, the influence of cold conditions on heat transfer characteristics and composite properties is investigated. The review concludes by emphasizing possible future research, existing challenges and potential future research directions in developing high-performance Al-ZrO_2 MMCs via cryogenic solidification techniques.

2. ADVANCES IN FABRICATION TECHNIQUES

2.1 Cryogenic Solidification and Its Influence on Mechanical Properties

Cryogenic solidification has been explored as a novel manufacturing method for NMMCs, significantly improving their structure and properties. Investigation reveals that this process improves the grain structure and ensures a uniform distributed nanoparticles within the matrix. The resulting composites show enhanced tensile strength and durability, making them suitable for high-performance applications. Upcoming studies might focus on optimizing cryogenic settings and exploring various nano additives to improve the benefits of this approach

2.2 Effect of Chills and Dispersoid Content on Aluminum-Based Composites

Research describes impact of chilling on the integrity and tensile strength of aluminium–quartz composites highlight the significance of both the chilling rate and the quantity of dispersoid present. The use of different chill types affects porosity, casting uniformity, and mechanical properties. The study emphasizes the relationship between volumetric heat capacity and ultimate tensile strength, showing that enhanced cooling conditions enhance composite quality. The results indicate a need for additional research on different chilling materials and their effects on microstructural improvement

2.3 Effect Role of Chills in Steel Casting and Shrinkage Defect Reduction

A Study research examining the use of mild steel (MS) chills in sand molds for casting carbon steel ball valves shows that chill thickness significantly affects shrinkage cavities. Statistical analysis using the Taguchi method identifies the optimal parameters to minimize shrinkage defects. The results highlight the importance of chill properties in improving casting quality and suggest additional research into various chill materials for more effective defect minimization, further investigation into different chill materials for better defect reduction,best parameters to reduce shrinkage defects. The results highlight the significance of chill properties in enhancing casting quality and recommend

2.4 Effect of Influence of Nano-Al_2O_3 on Al6061 Nanocomposites Fabricated via Stir Casting

Integrating nano-Al_2O_3 into Al6061 alloy via stir casting and extrusion results in microstructures that are fine-grained, featuring reduced porosity and improved mechanical properties helps to improves the characteristics of nanocomposites. Upcoming studies might explore the use of hybrid reinforcements to achieve an balance between strength and ductility

2.5 Development of High-Entropy Alloy Composites Using Mechanical Alloying and Spark Plasma Sintering

Early evidence of creation of AlCoCrFeNi–ZrO_2 high-entropy alloy (HEA) ,the development of AlCoCrFeNi–ZrO2 high-entropy alloy (HEA) composites via mechanical alloying and spark plasma sintering (SPS) have demonstrated considerable progress in microstructural nanostructured composites enhancement and mechanical properties, show that these approaches effectively produce high-performance HEA. Composites through alloying and spark plasma sintering (SPS) has shown significant advancements in microstructural refinement and mechanical characteristics. The joint impact of these processes leads to nanostructured composites exhibiting improved hardness and strength. The research indicates that these methods are successful in creating high-performance HEA composites with outstanding characteristics. Future studies might investigate the scalability of these approaches for use in industry.

Advancements in wire EDM machining proved surface integrity and minimized defects in Al/ZrO MMCs for aerospace applications. Optimized machining have shown improved surface integrity and minimized defects in Al/ZrO MMCs explore the impact of varying EDM parameters for aerospace applications, with minimal defects. Further research should be on different metal matrix composites to enhance their machining efficiency

3. REINFORCEMENT STRATEGIES FOR NANO METAL MATRIX COMPOSITES

3.1 Reinforcement of Aluminum-Based Composites with ZrB₂ and ZrO₂

Research on aluminium composites augmented with zirconium diboride (ZrB_2) and zirconium dioxide (ZrO_2), created via microwave sintering, show considerable enhancements in characteristics. The addition of ZrB_2 enhances hardness and strength, whereas microwave sintering ensures superior densification and particle organization. A comparative analysis with conventional sintering techniques shows that microwave sintering is a promising method for developing high-performance aluminium-based composites. Future initiatives should focus on improving processing conditions and exploring more high-strength excellent wear resistance reinforcements. (Pratap, et. al.,2020)

The inclusion of ZrB_2 improves hardness and strength, while microwave sintering guarantees better densification and particle arrangement. A comparative study with traditional sintering methods indicates that microwave sintering is a promising approach for creating high-performance aluminium-based composites. Future efforts should concentrate on enhancing processing parameters and investigating additional high-strength reinforcements.

3.2 Tribological and Mechanical Properties of Zirconia-Reinforced Aluminum Composites

The addition of ZrO_2 in aluminum metal matrix composites (AMMCs) has shown appreciable improvements in the composition of ZrO_2 into aluminum metal matrix composites (AMMCs) has demonstrated notable enhancements in mechanical strength, wear resistance, and decreases in the friction coefficient. Microstructural analysis shows a uniform distribution of zirconia particles,enhancing tribological performance. These results indicate that zirconia-reinforced AMMCs are ideal for uses demanding high durability and wear resistance. Future studies might explore the prolonged stability of these composites in extreme operating environments

3.3 Wear Behavior of Al-15% Si Alloy with Zinc and ZrO₂ Reinforcements

A comparative examination of the wear characteristics of Al-15% Si alloy strengthened with zinc and zirconium oxide reveals notable decreases in wear rate and friction coefficient. Zinc functions as a solid lubricant, whereas

ZrO_2 contributes toughness, resulting in improved tribological performance

3.4 Reinforcement of Aluminum-Based Composites with ZrB₂ and ZrO₂

Studies on distributing nano-ZrO_2 particles in aluminum alloys through melt deposition and hot extrusion indicate a refined microstructure that enhances strength and hardness. Nonetheless, a rise in reinforcement material slightly diminishes ductility while lowering thermal conductivity and electrical resistance. Fractography examination shows a change from ductile to cleavage fracture characteristics due to particle reinforcement.

4. LITERATURE REVIEW

4.1 Influence of Chilling on Hardness and Microstructure of Grey Iron

The research reveals MSC positioning greatly influences hardness, with peak values (reaching 588 HV30) detected when chills are arranged parallel to the flow of molten metal (Alloy-3). Microstructural examination shows the existence of larger graphite flakes in traditionally produced GCI and carbide development near cooled surfaces. These results are consistent with earlier studies, suggesting that chilling methods provide an economical option for enhancing hardness without depending on costly alloying materials.

Microstructural analysis reveals the presence of larger graphite flakes in conventionally produced GCI and the formation of carbides near cooled surfaces. These findings match with earlier research, indicating that chilling techniques offer a cost-effective way to increase hardness without relying on expensive alloying elements.

4.2 Casting Defect Analysis and Process Optimization

Avinash Juriani (2015)showcases industrial case studies that concentrate on casting imperfections, casting flaws, their root causes, and remedial measures. The research utilizes Their underlying causes, and corrective actions ,and cause-and-effect analysis method, recognizing key defects and suggesting practical solutions to improve casting quality. The research highlights the importance of defect management as a quality control approach and emphasizes a zero-defect mindset in manufacturing.

Vasdev Malhotra et., al explore the difficulties of obtaining flawless castings in gravity die casting. Their review highlights the importance of optimizing process parameters to decrease casting defects. Rather than depending exclusively on computational methods, their research promotes practical solutions via process modifications, making it easier to reduce defects.

4.3 Development of Aluminum-Based Nanocomposites

Daniel and Dr. G. Harish (2015) explore characteristics of aluminium alloy enhanced with nano-ZrO2 particles via the stir casting technique. The research indicates an improved microstructure with low porosity, greater hardness, and elevated tensile strength. Nonetheless, a minor decrease in ductility is observed as a result of alterations in fracture toughness. The results indicate that reinforcing with nanoparticles provides a way to customize material characteristics for particular uses.

R. H. Jaya Prakash builds upon this research by investigating cryogenically solidified nano-metal matrix composites (CNMMC). His study assesses the impact of copper chills, chilling temperatures, and dispersoid composition on mechanical characteristics. The research concludes that the mechanical properties enhance with nano-ZrO2 reinforcement up to 12 wt%, after which too much reinforcement results in reduced mechanical performance.

B. K. Anil Kumar focuses on nickel alloy composites reinforced with garnet and fabricated using cryogenic copper chills. The study confirms that cryogenic chilling enhances mechanical properties, leading to a 14% increase in hardness and a 13% increase in strength. However, excessive reinforcement (above 12%) negatively affects mechanical properties, indicating the importance of optimal dispersoid content

5. INFLUENCE OF CHILLING ON HYBRID METAL MATRIX COMPOSITES

Syed Ahamed et al., (2019) investigates the mechanical parameters of chilled aluminum alloy ,hybrid metal matrix enhanced with kaolinite and carbon. The research indicates that mechanical properties enhance up to 9 wt% dispersoid content, beyond which additional increases result in a decline. Fracture assessment indicates that cooling rates affect the fracture mechanism, whereas microstructural analysis shows particle detachment at increased cooling rates.

In another study, Syed Ahamed examines the wear characteristics of chilled aluminum alloy-kaolinite/carbon hybrid MMCs. His research shows that graphite coated with nickel improves solid lubrication, resulting in increased strength, hardness, and machinability. Microstructural examination verifies that cryogenic cooling enhances the matrix microstructure, showing distinct wear mechanisms—abrasive wear under light loads and adhesive wear under heavier loads.

6. HARDNESS AND WEAR RESISTANCE OF NANO-ZrO2 REINFORCED AL NANOCOMPOSITES

M. Ramachandra et al. (2016) analyzes the influence of nano-ZrO2 on improving the hardness, wear resistance of aluminum composites made through powder metallurgy. The research reveals that higher ZrO2 levels enhance

hardness and lower wear rates, rendering these composites ideal for automotive uses like pistons and cylinder liners. The study further highlights oxidation, micro-cutting, and thermal softening as the principal wear mechanisms, while microstructural examination shows consistent ZrO2 dispersion with slight agglomeration

6.1 Corrosion Behavior - Nacre-Inspired (TiBw-TiB2)/Al Composites

M. Zhang et al. (2024) investigate the corrosion characteristics of nacre-inspired (TiBw-TiB2)/Al composites created via freeze casting and squeeze casting methods. The research reveals that these composites demonstrate an attractive blend of strength and ductility. Nevertheless, the existence of an uninterrupted ceramic-rich layer adversely affects corrosion resistance, resulting in intricate degradation processes. This study emphasizes the compromises linked to the addition of ceramic reinforcements in aluminum composites, especially in terms of corrosion .

7. CONCLUSION

This study expressed in nano metal matrix composites emphasize the success of different fabrication methods in improving mechanical and tribological characteristics. Cryogenic solidification, microwave sintering, mechanical alloying, and spark plasma sintering have shown significant enhancements in composite performance. Additives like ZrO2, ZrB2, and Al2O3 enhance strength, hardness, and resistance to wear. Upcoming studies should aim at refining processing parameters, investigating hybrid reinforcement techniques, and assessing long-term performance in real-world settings. These developments open opportunities for high-performance uses in aerospace, automotive, and structural sectors.

The analysed studies collectively highlight the essential importance of chilling methods, reinforcement tactics, and process parameter enhancement in enhancing casting quality and composite material characteristics. Studies on nano-ZrO2 reinforced aluminum composites show encouraging enhancements in hardness, strength, and wear resistance, whereas hybrid MMCs provide customized mechanical properties. Additionally, analysing defects and optimizing processes are essential for producing high-quality castings. These results underscore the significance of incorporating advanced manufacturing methods to improve the effectiveness and durability of cast materials in industrial uses

REFERENCES

1. Arunkumar, Thirugnanasambandam, et al. "Development of high-performance aluminium 6061/SiC nanocomposites by ultrasonic aided rheo-squeeze casting method." Ultrasonics Sonochemistry 76 (2021): 105631.

2. Aybarç, Uğur, Onur Ertuğrul, and M. Özgür Seydibeyoğlu. "Effect of Al 2 O 3 particle size on mechanical properties of ultrasonic-assisted stir-casted Al A356 matrix composites." International Journal of Metalcasting 15 (2021): 638–649.

3. Juriani, Avinash. "Casting defects analysis in foundry and their remedial measures with industrial case studies." *IOSR journal of mechanical and civil engineering* 12.6 (2015): 43–54.

4. Bharat, Nikhil, and P. S. C. Bose. "Microstructure, texture, and mechanical properties analysis of novel AA7178/SiC nanocomposites." Ceramics International 49.12 (2023): 20637–20650.

5. Chavan, Vijaykumar, et al. "Evaluating comparative wear behaviour of Al-15% Si based alloy/composites reinforced with zinc and zirconium oxide." *Tribology-Materials, Surfaces & Interfaces* 17.2 (2023): 81–98.

6. Dadkhah, Mehran, Abdollah Saboori, and Paolo Fino. "An overview of the recent developments in metal matrix nanocomposites reinforced by graphene." *Materials* 12.17 (2019): 2823.

7. Daniel, Sumod, and G. Harish. "Sliding Wear Behaviour Of Aluminum Alloy (Lm-13) Reinforced With Nano-zro2 Matrix Composites." *International Journal of Innovations in Engineering Research and Technology* 2.9 (2015): 1–7.

8. Dwivedi, Shashi Prakash, et al. "Effect of zirconium diboride addition in zirconium dioxide reinforced aluminum-based composite fabricated by microwave sintering technique." *Proceedings of the Institution of Mechanical Engineers, Part E: Journal of Process Mechanical Engineering* (2023): 09544089231164272.

9. Garg, Sanjeev Kr, Alakesh Manna, and Ajai Jain. "Experimental Investigation of Surface Integrity Aspects and Recast Layer Formation MMC Suitable for the for Wire Aerospace EDM of Al/ZrO Industries." *Advanced Composites in Aerospace Engineering Applications* (2022): 195.

10. Ghanbariha, M., M. Farvizi, and T. Ebadzadeh. "Microstructural development in nanostructured AlCoCrFeNi–ZrO2 high-entropy alloy composite prepared with mechanical alloying and spark plasma sintering methods." *Materials Research Express* 6.12 (2019): 1265b5.

11. Gnanavelbabu, A., KT Sunu Surendran, and S. Kumar. "Process optimization and studies on mechanical characteristics of AA2014/Al 2 O 3 nanocomposites fabricated through ultrasonication assisted stir–squeeze casting." International Journal of Metalcasting (2021): 1–24.

12. Kamaraj, Logesh, P. Hariharasakthisudhan, and A. Arul Marcel Moshi. "Optimizing the ultrasonication effect in stir-casting process of aluminum hybrid composite using desirability function approach and artificial neural network." Proceedings of the institution of mechanical engineers, part L: Journal of Materials: Design and Applications 235.9 (2021): 2007–2021.

13. Kumar, Ashish, Ravindra Singh Rana, and Rajesh Purohit. "Microstructure evolution, mechanical properties, and fractography of AA7068/Si3N4 nanocomposite fabricated thorough ultrasonic-assisted stir casting advanced with bottom pouring technique." Materials Research Express 9.1 (2022): 015009.

14. Madhukar, Pagidi, et al. "Production of high performance AA7150-1% SiC nanocomposite by novel fabrication process of ultrasonication assisted stir casting." Ultrasonics Sonochemistry 58 (2019): 104665.

15. Mohanavel, V., et al. "Mechanical behavior of Al-matrix nanocomposites produced by stir casting technique." Materials Today: Proceedings 5.13 (2018): 26873–26877.

16. Ahamed, S. Muzzamil, et al. "Chilling effect on hardness and microstructural behavior of grey iron." *Materials Today: Proceedings* 5.11 (2018): 25697–25704.

17. Ogawa, Fumio, and Chitoshi Masuda. "Fabrication and the mechanical and physical properties of nanocarbon-reinforced light metal matrix composites: A review and future directions." *Materials Science and Engineering: A* 820 (2021): 141542.

18. Prakash, D. Surrya, et al. "Experimental investigation of nano reinforced aluminium based metal matrix composites." Materials Today: Proceedings 54 (2022): 852–857.

19. Pratap, Ayush, et al. "Effect of indentation load on mechanical properties and evaluation of tribological properties for zirconia toughened alumina." *Materials Today: Proceedings* 26 (2020): 2442–2446.

20. Ramachandra, M., G. Dilip Maruthi, and R. Rashmi. "Evaluation of corrosion property of Aluminium-Zirconium Dioxide (AlZrO2) nanocomposites." *Evaluation* 1 (2016): 56412.

21. Rao, Thella Babu. "Microstructural, mechanical, and wear properties characterization and strengthening mechanisms of Al7075/SiCnp composites processed through ultrasonic cavitation assisted stir-casting." Materials Science and Engineering: A 805 (2021): 140553.

22. Saboori, Abdollah, et al. "An overview of key challenges in the fabrication of metal matrix nanocomposites reinforced by graphene nanoplatelets." *Metals* 8.3 (2018): 172.

23. Sarmah, Pallab, and Kapil Gupta. "Recent advancements in fabrication of metal matrix composites: A systematic review." *Materials* 17.18 (2024): 4635.

24. Schramm Deschamps, Isadora, et al. "Design of in situ metal matrix composites produced by powder metallurgy—A critical review." Metals 12.12 (2022): 2073.

25. Sharma, Daulat Kumar, Devang Mahant, and Gautam Upadhyay. "Manufacturing of metal matrix composites: A state of review." *Materials Today: Proceedings* 26 (2020): 506–519.

26. Sivananthan, S., V. Rajalaxman Reddy, and C. Samson Jerold Samuel. "Preparation and evaluation of mechanical properties of 6061Al-Al2O3 metal matrix composites by stir casting process." Materials Today: Proceedings 21 (2020): 713–716.

27. Soni, Sourabh Kumar, et al. "Microstructure and mechanical characterization of Al6061 based composite and nanocomposites prepared via conventional and ultrasonic-assisted melt-stirring techniques." Materials Today Communications 34 (2023): 105222.

28. Suresh, S., G. Harinath Gowd, and M. L. S. Devakumar. "Mechanical and wear characteristics of aluminium alloy 7075 reinforced with nano-aluminium oxide/magnesium particles by stir casting method." Materials Today: Proceedings 24 (2020): 273–283.

29. Syed Ahamed, R. J., and P. Shilpa. "A Literature Review on Aluminium-7075 Metal Matrix Composites." *IRJET* 6 (2019): 1384–1389.

30. Zhang, Jidong, et al. "Corrosion Behavior of Nacre-Inspired (TiBw-TiB2)/Al Composites Fabricated by Freeze Casting." Materials 17.11 (2024): 2534.

Advances in Mechanical Engineering and Materials Sciences – Dr. Vinay K. B et al. (eds)
© 2026 Taylor & Francis Group, London, ISBN 9-781-041-20970-6

43 Eco-Friendly 3D Printing—Biodegradable PLA Composites Reinforced with Wood Waste

Khalid Imran*

Lecturer, Department of Mechanical Engineering,
Vidyavardhaka College of Engineering,
Mysuru, India

Sajid N.

Department of Mechanical Engineering,
S J Government Polytechnic,
Bengaluru

Godfrey Devaputra

Assistant Professor, Department of Mechanical Engineering,
Maharaja Institute of Technology,
Mysuru, India

Kamran Ahmed

Sc 'F', Indian Space Research Organisation HQ,
Department of Space, Government of India, Antariksh Bhavan,
Bangalore, India

K. V. A. Balaji

Chief Executive Officer, K S Group of Institutions,
Bengaluru, India

Mohamed Khaisar

Professor, Department of Mechanical Engineering,
Maharaja Institute of Technology,
Mysuru, India

Abstract: This piece of research aims to develop sustainable and novel bio composites for 3D printing based on a biodegradable polymer, polylactic acid (PLA) reinforced with wood waste, to address challenges in additive manufacturing such as enhancing material performance, reducing production costs and overall environmental impact. The mechanical characteristics of PLA (polylactic acid) and Wood + PLA composite specimens that were 3D printed using a Creality Ender 3 FDM printer at infill densities of 25%, 35%, and 45% were examined. Standardized equipment such as a 50 kN computerized universal testing machine, a 6-tonne conventional universal testing machine, a computerized impact tester (0–100 J range), and a Shore D Durometer were used to test the specimens for tensile strength, compression strength, impact resistance, and hardness. According to tensile tests, PLA demonstrated constant strength at 35% and 45% infill, however Wood + PLA only performed at its best at 35% infill, with a notable decline at 45% infill. Both materials maintained high stiffness in compression tests, with PLA showing better stress resistance. According to impact testing, PLA absorbed more energy than Wood + PLA, which was more brittle because of the wood particles. Comparing PLA to Wood + PLA, hardness tests verified that PLA had greater surface resistance. The findings demonstrate how material

*Corresponding author: khalid@vvce.ac.in

10.1201/9781003725053-43

composition and infill density affect mechanical performance, with PLA having a higher edge over Wood + PLA in the majority of tests. Nevertheless, Wood + PLA composites have special qualities that make them appropriate for stiff, lightweight applications. To enhance the mechanical properties of the Wood + PLA composite, future research should concentrate on optimizing the particle dispersion and investigating various wood-to-PLA ratios.

The study also shows how low cost and sustainable PLA-based composites produced with natural resources can be used in Additive Manufacturing technology replacing common materials. Refined products with basic raw materials effectively reduce part costs without compromising part characteristics or performance. It also reduces its carbon footprint, becoming more sustainable. There are important advantages when applying these materials in areas requiring good mechanical properties considering their readiness of end-life cycle using composting or even recycling after grinding and processing back into filament form, meaning nothing goes wasted all over life cycle.

Keywords: Crisis, Ownership, Variables

1. INTRODUCTION

With the advancement of technology in recent years, additive manufacturing (AM), commonly referred to as three-dimensional printing (3DP) and 3D printers, has found widespread use in everyday life. One of the rapid prototyping techniques, 3D printers are employed in both industry and research and academic pursuits. Using a data file created from geometry data with basic or complicated geometry, 3D production is a computer-aided method of production. Layer by layer, the components will be created using various connecting principles in accordance with conventional techniques from the generated data file. Additive manufacturing, has rapidly expanded and changed a variety of industries, including the healthcare and automotive sectors. When compared to conventional manufacturing methods, this technology delivers previously unheard-of levels of customisation, material efficiency, and waste reduction.

The market for 3D printing has grown rapidly worldwide, and its uses are becoming more widespread across industries. Notwithstanding its many benefits, the sector faces serious environmental challenges because to its reliance on synthetic polymers like polyethylene terephthalate glycol (PETG) and acrylonitrile butadiene styrene (ABS). If not disposed of properly, these materials are non-biodegradable and cause long-term contamination. Furthermore, energy-intensive procedures that contribute to greenhouse gas emissions include the mining and manufacturing of petroleum-based polymers.

1.1 Wood Waste as a Reinforcement Material

Wood waste, a plentiful byproduct of the furniture and forestry sectors, offers a strong chance to increase PLA composites' sustainability. By repurposing a resource that would otherwise be discarded, the integration of wood waste into PLA is consistent with the circular economy's tenets. The PLA-wood composites that are produced have better mechanical qualities, such as increased elasticity and tensile strength. Additionally, the organic beauty of wood particles gives 3D-printed goods a distinctive look and feel that increases their marketability. Using wood waste

as a reinforcing element lowers the composite's overall carbon footprint from an environmental standpoint. When compared to synthetic alternatives, the energy required to produce wood-based fillers is far lower. Furthermore, wood waste's biodegradability guarantees that the composite material will continue to be ecologically favourable for the duration of its existence. According to recent research, PLA-wood composites are a good substitute for traditional polymers in 3D printing because they maintain high levels of biodegradability even when wood particles are added.

1.2 Future Scope

PLA-wood composites are a potential option for environmentally friendly 3D printing, but in order to maximize their scalability and performance, a number of issues need to be resolved. The even distribution of wood particles throughout the PLA matrix is a critical factor. Inconsistent mechanical characteristics and flaws in the printed goods can result from poor dispersion. To address this problem, surface treatments for wood particles and sophisticated compounding processes are being investigated.

2. LITERATURE REVIEW

For the purpose of analysing mechanical behaviour, Barnasree, Kumar, and Bhowmik et al. [1] investigated wood dust particle reinforcement in epoxy-based composite. The particle of sun-dried wood dust was utilized as reinforcement, and the resin was LY 556 epoxy. The study utilized six different percentages of filler particles. UTM was used for the tensile and flexural tests, and the ASTM Standard was used to determine the sample size. Using GRA, many design parameters were optimized, such as filler content and loading speed with flexural and tensile strength. One benefit of optimization by GRA is the ability to choose the best and worst solutions. According to GRG, test run number three is the least significant and test run number thirteen is the most appropriate.

Sundi wood dust particle reinforced composite materials were fabricated and experimented by Kumar, Sahoo, and Bhowmik et al. [2]. Sundi wood dust particle reinforced epoxy composite was treated at seven different filler weight percentages in this experiment. To examine the mechanical behaviour of composites, tensile and flexural tests were conducted at three distinct rates. With a filler weight of 10% and 15%, respectively, and a speed of 1 mm/min, the maximum load, tensile stress and strain, and flexural stress and strain values are recorded at maximum and minimum at 10% filler weight with a speed of 2mm/min and at 0% filler weight with a speed of 1mm/min respectively. At 10% filler weight with a speed of 1mm/min and 2mm/min best mechanical properties are observed.

A study by Cerqueira, Baptistab, Mulinari, et al. [4] examined the production of composite materials using natural fibers serving as reinforcing fibers. He assessed how a chemical changes affected the mechanical properties of composites made of polypropylene reinforced with sugarcane bagasse fiber. After treating the fibers with a 10% sulfuric acid solution, they were delignified using a 1% NaOH solution. The impact, flexural (3-point bending), and tensile tests of the manufactured composites were examined. SEM (secondary electron mode) was used to analyze the fractures. Composite samples' results are comparable to those of pure polymers. Hence cellulose based chemically improved from sugarcane bagasse exhibits better property in comparison with chemically untreated fibre particle reinforced composites.

The erosion wear process of a nonlinear issue with operational variables was analyzed by Rout and Sahoo et al. [5]. Numerous limitations, such as impingement angle, erodent size, impact velocity, material, etc., affect how a material wears. Experiment with materials that have a combination of these properties to have the lowest rate of erosion. Waste granite powder was being considered as a filler for the reinforced epoxy composite's jute fiber. It was determined that low-cost natural fiber reinforced composite may be made from industrial waste, such as granite powder. Additionally, the composite samples' resistance to erosion was improved by the chemical treatment of the fiber and filler.

3. LITERATURE GAP AND OBJECTIVES

Wood plastic composite (WPC) may be molded into nearly any size or shape, even arched or bending shapes, because it is made from a substance that begins as a paste. WPC can be colored or dyed to fit practically any design scheme, demonstrating its inherent versatility. Considering the lengthy history of natural lumber as a building material, WPCs are still relatively new. However, one of the most important and promising engineered wood products to date is wood-plastic composites (WPCs), which have been widely used in the building construction, transportation, landscape, and municipal engineering sectors. It has been used to gradually replace traditional composites made of wood. Advanced composite materials have a very

promising future. In order to address these issues, the following are the objectives of this piece of research.

- To fabricate and evaluate a composite material suitable for usage as feedstock in FDM 3D printers.
- To evaluate the properties of specimens fabricated with wood reinforced with polylactic acid (PLA) composite material.

4. METHODOLOGY

The methodology involves the following steps.

1. Description of Extrusion compounding of Wood PLA Filament Material
2. Fabrication of testing specimen
3. Testing and Analysis

4.1 Extrusion Compounding

Though there is considerable research carried out on Wood reinforced PLA, there is a huge scope for further research by altering the percentage of wood quantity, type of dispersion, shape and size of reinforcements, etc. Hence an effort is made to compound a unique wood reinforced PLA material for 3D printing of specimens. The specimens are further validated by testing for mechanical characterization.

4.2 Fabrication of Testing Specimen

The Wood PLA Composite Filament is used to fabricate the specimen as per ASTN D638 standards. The fabrication of the testing specimen is carried out on a Creality Ender 3 3D printer.

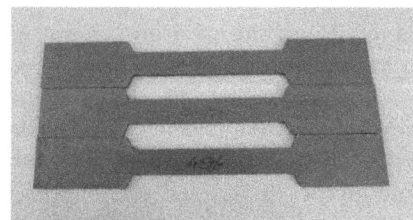

Fig. 43.1 PLA tensile specimen

Fig. 43.2 Wood + PLA composite tensile specimen

4.3 Testing and Analysis

1. Testing and Analysis – Tensile test
   a. Material: Wood + PLA (Infill Density = 35%)
      Temperature: 30°C

Area: 45 sq-mm
Gauge Length: 60 mm
Width: 15 mm
Thickness: 3 mm
Maximum Load: 0.65 KN

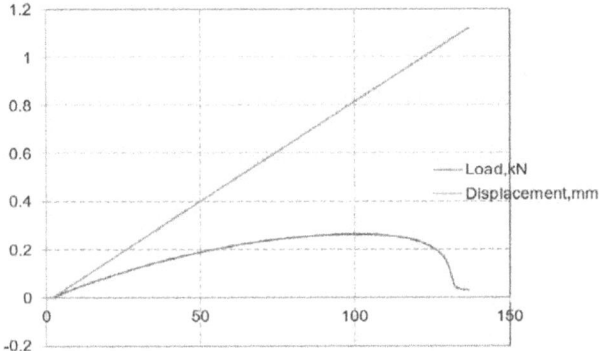

Fig. 43.3   Graph 4

b. Material: PLA (Infill Density = 45%)
Temperature: 30°C
Area: 45 sq-mm
Gauge Length: 60 mm
Width: 15 mm
Thickness: 3 mm
Maximum Load: 0.65 KN

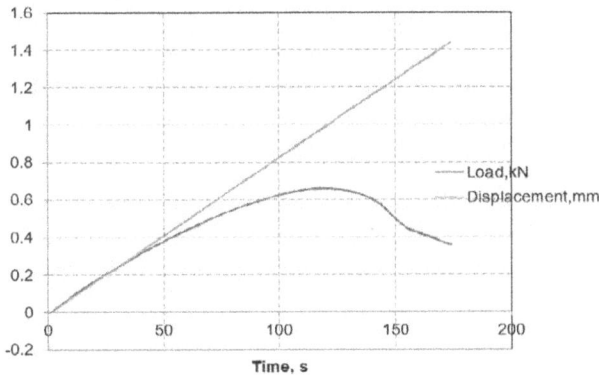

Fig. 43.4   Graph 5

2. Testing and Analysis – Compression test
   a. Material: Wood + PLA
      Initial Length, L_i = 30 mm
      Diameter, D_i = 9 mm
      Original Area, A_o = 63.617 sq.m
      Maximum Load, $F_{max\ 1}$ = 3.2×10^3 N
      Maximum Load, $F_{max\ 2}$ = 4×10^3 N
      Final Length, L_{f1} = 16 mm
      Final Length, L_{f2} = 19 mm
      Deformation, ΔL_1 = 4.74×10^{-12} mm
      Deformation, ΔL_1 = 5.93×10^{-12} mm
      Stress, σ_1 = 0.79 N/mm^2
      Stress, σ_2 = 0.98 N/mm^2
      Strain, ε = 1.58×10^{-13}

Table 43.1   Wood + PLA compression test result

Sl. no	Load F (N)	Deformation ΔL (mm)	Stress σ (mm)	Strain ε	Youngs modulus E (n/mm^2)
1	3.2×10^3	4.74×10^{-12}	0.79	1.58×10^{-13}	5×10^{12}
2	4×10^3	5.93×10^{-12}	0.98	1.97×10^{-13}	4.97×10^{12}

b. *Testing and Analysis – Hardness Test*
   Range of the tester: 0 to 100 HD
   Hardness scale: Shore D

Table 43.2   Hardness test result

Material	Shore D. Value	Average
PLA	44, 41, 46	43.66
PLA + Wood	35, 33, 30	32.66

c. *Testing and Analysis – Impact Test*
   Impact Energy – 0 to 25 Joules
   Release angle of pendulum – 150 degrees

Table 43.3   Impact test result

Infill Density	PLA		WOOD + PLA	
	Impact in Joules	Angle of cut	Impact in Joules	Angle of cut
35	0.50	141	0.4	143
45	0.65	143	0.45	146

5. RESULTS AND DISCUSSION

The mechanical characteristics of PLA (polylactic acid) and Wood + PLA composite specimens made with a Creality Ender 3 FDM printer at infill densities of 25%, 35%, and 45% are examined in this work. Using established tools and techniques, the specimens were examined for hardness, impact resistance, compression strength, and tensile strength. The findings are examined to determine how material composition and infill density affect mechanical performance.

1. Infill Density: While PLA and Wood + PLA had different effects, higher infill densities generally resulted in better mechanical qualities. Wood + PLA only performed at its best at 35% infill, whereas PLA performed consistently at 35% and 45% infill.

2. Material Composition: While PLA remained stiff under compression, the inclusion of wood particles decreased its tensile strength, impact resistance, and hardness. This implies that applications needing stiffness rather than toughness are better suited for Wood + PLA composites.

3. Anomalies in Wood + PLA: The decrease in impact energy values and the tensile strength drop at 45% infill suggest that the wood particles could cause weak spots in the material, particularly at higher densities. To increase performance, the wood-to-

PLA ratio and particle distribution need to be further optimized.

4. Applications: Wood + PLA composites might be more appropriate for lightweight, rigid structures where toughness is not a top concern, but PLA is appropriate for applications needing high strength, stiffness, and surface hardness.

6. CONCLUSION

In Conclusion, wood composite PLA filament provides a distinctive and sustainable 3D printing option. This filament creates intriguing opportunities for a variety of industries and applications by fusing the aesthetic appeal of wood with the biodegradability and ease of use of PLA. For 3D printed items, the wood composite PLA filament offers an aesthetically pleasing alternative. By incorporating wood fibers or particles into PLA filament, a natural wood-like look is produced, giving printed items a touch of good aesthetics. The filament's ability to mimic the distinctive textures and grain patterns of many wood species improves its aesthetic appeal and qualifies it for use in ornamental items, architectural models, and artistic projects.

Furthermore, sustainability is aided by PLA's biodegradability, which keeps it in the wood composite filament. The wood composite PLA filament provides an eco-friendly substitute for conventional plastic-based filaments because PLA is made from renewable materials. Composting or environmentally responsible disposal are two ways that printed items generated with this filament might lessen their influence on the environment. But it's crucial to take into account the possible drawbacks of PLA filament made of wood composite. When compared to pure PLA, the mechanical characteristics of the filament may be impacted by the inclusion of wood particles, resulting in reduced strength and increased brittleness. This might limit its applicability for uses that call for load bearing or highly mechanical components. Nevertheless, post-processing and finishing should be done carefully to maintain the wood-like characteristics and prevent sanding or other damage. The wood composite PLA filament is a useful addition to the selection of filaments for 3D printing in spite of these drawbacks. The wood composite PLA filament may find wider uses and support the ongoing expansion of environmentally friendly 3D printing techniques with additional development and study. Finally, the study emphasizes how material composition and infill density shapes the mechanical characteristics of 3D-printed PLA and Wood + PLA specimens. Although PLA performed better than Wood + PLA in the majority of tests, the combination has special qualities that could be useful in some situations. Future research could concentrate on enhancing particle distribution and investigating various Wood-to-PLA ratios in order to optimize the composite.

REFERENCES

1. Chanda, B., Kumar, R., Kumar, K., & Bhowmik, S. (2014). Optimisation of mechanical properties of wood dust-reinforced epoxy composite using Grey relational analysis. In Advances in intelligent systems and computing (pp. 13–24). https://doi.org/10.1007/978-81-322-2220-0_2

2. Kumar, R., Kumar, K., Sahoo, P., & Bhowmik, S. (2014). Study of mechanical properties of wood dust reinforced epoxy composite. Procedia Materials Science, 6, 551–556. https://doi.org/10.1016/j.mspro.2014.07.070

3. Verma, C., & Chariar, V. (2011). Development of layered laminate bamboo composite and their mechanical properties. Composites Part B Engineering, 43(3), 1063–1069. https://doi.org/10.1016/j.compositesb.2011.11.065

4. F. Cerqueira, C. A. R. P. Baptistab, and D. R. Mulinari, (2011.) "Mechanical behaviour of polypropylene reinforced sugarcane bagasse fibers composites," Procedia Engineering, vol. 10, pp. 2046–2051.

5. L. K. Rout and S. S. Sahoo, "Study on erosion wear performance of jute-epoxy composites filled with industrial wastes using Taguchi methodology," in Proceedings of Second IRF International Conference, Mysore, India, Nov. 30, 2014, ISBN: 978-93-84209-69-8.

6. Rao, H., Indraja, Y., Bai, M., & Department of Mechanical Engineering G.P.R. Engineering College. (2014). Flexural properties and SEM analysis of bamboo and glass fiber reinforced epoxy hybrid composites. In IOSR Journal of Mechanical and Civil Engineering (Vol. 11, Issue 2, pp. 39–42). https://www.iosrjournals.org

7. Al-Mosawi, A. I., University of Baghdad, International Science Congress Association, & Ali I. Al-Mosawi. (2012). Mechanical properties of plants - synthetic hybrid fibers composites. Research Journal of Engineering Sciences, 1–3, 22–25. https://www.researchgate.net/publication/235769035

8. O. Faruk, A. K. Bledzki, H. Peter, and F. M. Sain, (2012). "Biocomposites reinforced with natural fibers: 2000–2010," Progress in Polymer Science, vol. 37, pp. 1552–1596.

9. J. Sarkia, S. B. Hassan, V. S. Aigbodion, and J. E. Oghenevweta, (2011). "Potential of using coconut shell particle fillers in eco-composite materials," Journal of Alloys and Compounds, vol. 509, pp. 2381–2385.

10. Kartal, F., & Kaptan, A. (2024). Sustainable Reinforcement of PLA Composites with Waste Beech Sawdust for Enhanced 3D-Printing Performance. Journal of Materials Engineering and Performance. https://doi.org/10.1007/s11665-024-10277-0.

Note: All the figures and tables in this chapter were made by the authors.

Advances in Mechanical Engineering and Materials Sciences – Dr. Vinay K. B et al. (eds)
© 2026 Taylor & Francis Group, London, ISBN 9-781-041-20970-6

44

Detecting Defects in Casting Manufacturing using Machine Learning

Dushyanthkumar G. L.[1], Vinay K. B.[2],
Praveenkumara B. M.[3], Mohammed Omar[4]
Department of Mechanical Engineering Vidyavardhaka,
College of Engineering Mysuru,
Karnataka, India

Abstract: This paper offers a new methodology for the detection of casting defects through the fusion of Convolutional Neural Networks (CNNs) and Auto-Encoders to extract features and detect anomalies. Through Utilization of a heterogeneous dataset, the model outperforms conventional methods of defect detection, with a 91% accuracy. Contrary to visual inspection and simple algorithms, this deep learning technique driven by data enhances detection accuracy and minimizes false positives. The model is versatile in various casting situations and can be used as a scalable quality control tool. Future work will involve improving model performance, solving data imbalance issues, and applying it to more general manufacturing processes.

Keywords: Casting defect detection, Machine learning in manufacturing, CNNs, Auto-encoders, Deep learning, Industrial defect detection, Image-based defect identification

1. INTRODUCTION

Quality assurance in the production of castings is of critical concern since defects have the potential to impact the performance and reliability of the product (Zhang and Jiang, 2017). Conventional methods of detection, such as manual inspection and rule-based systems, are inefficient and inaccurate (LeCun et al., 2015). The current research employs a hybrid machine learning system that combines Convolutional Neural Networks (CNNs) for feature extraction and Auto-Encoders for anomaly detection, thus enabling automated and accurate classification of defects (Simonyan and Zisserman, 2014). Deep learning algorithms, compared to traditional methods, improve detection through the analysis of varied image data sets (He et al., 2016). This process enables scalability, generalization, and real-time flexibility(Krizhevsky et al., 2012). The key benefits include minimizing inspection time, cost-effectiveness, and better product quality (Srivastava et al., 2014).

This book starts off by providing an overview of defect detection methods, then data collection and preprocessing methods. Training, evaluation, and comparative studies with other methods come next. Finally, future perspectives and challenges are explained.

2. LITERATURE REVIEW

Traditional defect detection methods, including rule-based algorithms and manual inspection, tend to experience issues of human error, inconsistency, and inefficiency (Zhang and Jiang, 2017). The methods can be subjective, time-consuming, and unreliable, particularly in mass production where speed and accuracy are paramount. Deep learning, with the application of Convolutional Neural Networks (CNNs), has transformed automated defect detection with a more accurate and efficient solution (LeCun et al., 2015). CNNs are particularly suited to detect surface defects by learning image patterns and are thus most suitable for use in casting processes (Simonyan and

[1]dushyanth.mech@vvce.ac.in, [2]vinaykb@vvce.ac.in, [3]praveenkumarabm@vvce.ac.in, [4]mdomar2004@icloud.com

Zisserman, 2014). Findings from a study by Krizhevsky et al.(2012) indicated that CNNs were capable of classifying X-ray images of aluminum castings with an accuracy many orders of magnitude above traditional methods.

Another strong technique is the use of Auto-Encoders, which are anomaly detection specialists by virtue of gaining the ability to identify the features of a normal, defect-free product (Kingma and Welling, 2014). When the model is given an image that isn't part of this learned representation, it marks it as a possible defect. This method is able to identify faint defects that are not easily noticed by the naked eye. The combination of Convolutional Neural Networks (CNNs) and Auto-Encoders makes the detection of defects stronger, more generalizable, and more resilient, thus significantly reducing the rate of false positives (Zeiler and Fergus, 2014).

The accuracy of training data plays a considerable role in shaping the behavior of the model (Pedregosa et al., 2011). A model developed from unbalanced or inaccurate datasets will by necessity be affected by an issue in identification of defects from practical scenarios. Literature indicates that usage of heterogeneous datasets optimizes the deep learning models' generalization capabilities; hence, their performance with diversified manufacturing systems is ensured (He et al., 2016). Even more importantly, it has actually been affirmed on the basis of studies done that hybrid models of deep learning, with the integration of both supervised learning (CNNs) and unsupervised learning (Auto-Encoders), never fail to achieve higher levels of performance compared to the conventional systems in defect detection, reducing errors as well as improving reliability on a large scale (Srivastava et al., 2014).

On this basis, this research combines Convolutional Neural Networks (CNNs) for feature learning and Auto-Encoders for anomaly detection to create a high-accuracy and scalable system for defect detection. The aim of this research is to enhance the quality control process in industries through the utilization of a quicker, automated, and highly efficient method for the detection of cast part defects.

3. DATA COLLECTION AND PREPROCESSING

3.1 Data Sources

The data utilized in this research was accessed from Kaggle and consists of labeled casting images that have been classified into two distinct classes: defective and non-defective. These images from the primary input to train the machine learning model as shown in Fig. 44.1. Additionally, besides the raw images, metadata like defect type, location, and severity were also accessed. This additional information helps in enriching the analysis by providing details of different patterns of defects and their severity, thus making the model robust and resilient to actual manufacturing environments (Pedregosa et al., 2011).

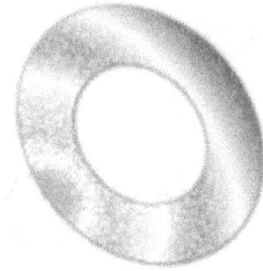

Fig. 44.1 Defective casting images

3.2 Data Cleaning and Feature Engineering

In order to ensure that the dataset was accurate, well-structured, and well-tuned for training, several preprocessing techniques were applied:

Image normalization is a crucial process; given that raw image pixel values can vary significantly, they are scaled to a range of [0,1] to standardize the input and enhance training stability. This approach serves to prevent substantial variations in pixel intensities from adversely affecting model performance.

Data Splitting: The data set was split into 80% for training and 20% for testing, so that the model could be estimated for its ability to generalize to new, unseen data. The split ensures that the model is not only learning the trends in the data set but can also apply the trends to other casting situations.

Feature Extraction using Convolutional Neural Networks: Convolutional Neural Networks (CNNs) were employed to automatically extract features from the cast images. Instead of employing hand-designed features, CNNs can learn to recognize distinctive patterns, edges, and textures that differentiate defective from non-defective products.

Anomaly Detection using Auto-Encoders: An Auto-Encoder model was trained to recognize normal, defect-free castings. Through reconstructing these normal samples, the model learns the standard structure of a casting product. When an image strongly differs from this standard pattern, it is recognized as a defect, allowing the system to effectively recognize potential manufacturing problems.

The preprocessing and feature engineering process are critical in improving model performance, ensuring it can effectively detect casting defects without increasing the incidence of false positives and false negatives.

4. MODEL DEVELOPMENT

4.1 Model Selection Criteria

Choosing the right model is crucial to achieve high accuracy in defect detection. Given that the dataset is image-based defects, Convolutional Neural Networks (CNNs) were the natural choice, given their established success in image classification and feature extraction

(Simonyan and Zisserman, 2014). CNNs inherently learn the capability to recognize shapes, texture, and structural abnormalities, making them highly suitable for casting defect detection. However, since the dataset is imbalanced with more non-defective samples than defective samples, Auto-Encoders were employed for anomaly detection (Kingma and Welling, 2014). These models learn the normal characteristics of defect-free products and identify deviations that can indicate the presence of defects.

Another critical consideration in model selection was computation efficiency. The models chosen had to be able to balance high accuracy with fast processing, thus being trainable and deployable within reasonable time constraints. The ability of the model to generalize to novel forms of defects that might not always be well defined in the training set was also considered. For simplicity of implementation, fine-tuning, and deployment, the models were implemented using popular deep learning libraries like TensorFlow and Keras, which offer strong tools for model training, optimization, and scalability.

4.2 Training and Testing Process

The training procedure began with dataset preparation, where the images were normalized and split into training and test sets. The Convolutional Neural Network (CNN) model was trained to acquire notable visual features from the casting images, using a series of convolutional and pooling layers to detect notable patterns. The feature extraction process was one of the most critical processes, as it allowed the model to detect minor defects that are difficult to detect with the naked eye.

Once the Convolutional Neural Network (CNN) was trained, the features obtained were used as input to the Auto-Encoder, which had been trained solely on non-defective samples. The objective of the Auto-Encoder was to reconstruct the samples with highest precision. When a new image was shown, the model would check its reconstruction error against a previously set threshold; if the error was greater than the threshold, the image would be marked as defective. The two-stage mechanism enhanced detection precision significantly by synergistically merging supervised learning, as in CNNs, with anomaly detection unsupervised, as in Auto-Encoders.

5. MODEL EVALUATION AND RESULTS

5.1 Preliminary Results and Model Improvement

The first test of the model resulted in 69% accuracy. This was a good start but also indicated that there was still room for improvement. Some errors were discovered, particularly when the defects were not easily visible or even concealed in the pictures. This indicated that the model had to be further improved, such as by tweaking hyperparameters, employing data augmentation techniques, and modifying

the anomaly detection threshold. To improve it, several modifications were made, including enhancing the CNN architecture, modifying learning rates, and enhancing the Auto-Encoder's image reconstruction capability.

Following these modifications, the model was trained and tested again, and this resulted in a significant boost in performance, achieving an accuracy of 90.09% and the screenshot of the metrics is present below as Fig. 44.2 This was an indication of how effective the hybrid approach was, where CNNs identified high-quality features and the Auto-Encoder effectively distinguished normal and faulty samples. The capability of the model to adapt and improve with multiple optimizations makes it an ideal option for detecting defects in actual manufacturing.

Fig. 44.2 Screenshot of the metrics

5.2 Confusion Matrix Analysis

For better analysis of the model's classification accuracy, Fig. 44.3 shows a confusion matrix which was also generated, and its strengths and weaknesses were revealed. True positive (TP) rate—correct classification of defective samples—was good, which meant the model was accurately classifying true defects. Similarly, true negative (TN) rate—correct classification of fault-free products—demonstrated good reliability in classifying fault-free castings. Few false positives (FP), i.e., some fault-free samples wrongly reported as faulty, were noted. Additionally, false negatives (FN), i.e., faulty samples wrongly reported as fault-free, were low but were present, indicating higher detection sensitivity could be achieved with further optimization of the threshold.

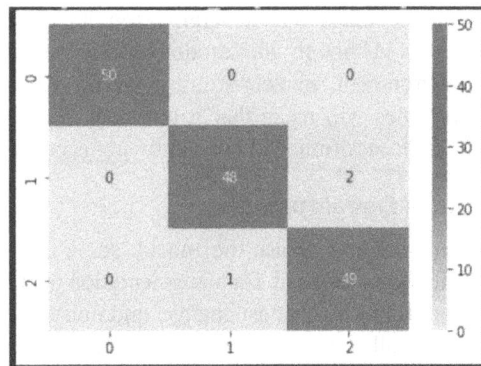

Fig. 44.3 Confusion matrix

5.3 Final Model Performance

With a total accuracy of 90.09%, the model performed extremely well in separating defective and non-defective castings. The combination of CNNs for feature learning and Auto-Encoders for anomaly detection led to a well-balanced system that minimized misclassifications while keeping precision and recall at high levels. These findings confirm the efficacy of deep learning for quality control in manufacturing, giving a scalable, automated, and highly accurate system for defect detection.

This performance enhancement indicates that the model is now deployable in real-world applications, where it can significantly reduce manual inspection time, enhance the productivity of production, and lower manufacturing defects. With further enhancements, i.e., enhancing the dataset and incorporating the feature of real-time monitoring, the system can be incorporated into an industry-standard tool for the detection of defects in casting production.

6. DISCUSSION

6.1 Challenges and Limitations

Although its strong performance, the model has some limitations that need to be addressed in order to make it more generally useful in industry. One of the main challenges is edge case misclassification, in which defects that are extremely small, partially occluded, or visually indistinguishable from normal features can be overlooked.

Although false positives and false negatives were relatively low, further optimization of the anomaly detection threshold would potentially improve classification accuracy.

Another drawback of the model is susceptibility to the quality and diversity of the dataset. If the model is trained on a small dataset, it will be unable to generalize, especially when it is subjected to new defect types during actual production. This highlights the need to increase the dataset to cover more diverse samples of defects under different manufacturing conditions.

Furthermore, computational efficiency is an aspect that should be taken into account when implementing in real time. Although the model works well in an offline environment, implementing it within high-speed production lines will mean that it must be optimized to minimize inference time without sacrificing accuracy.

6.2 Future Developments

In order to further enhance the model, some additional enhancements can be tried. Data augmentation techniques like random rotation, contrast change, and noise injection can add variability to datasets and robustify the model. Ensemble learning, where multiple models are averaged and combined to eliminate variance and stabilize predictions, is another potential method.

Adjusting the reconstruction threshold of the Auto-Encoder can be employed to minimize misclassifications such that the model is not too sensitive or too tolerant when detecting defects. Furthermore, exploring transfer learning techniques through fine-tuning a pre-trained CNN using a larger dataset would improve the model in detecting a wider range of casting defects through fewer more training iterations.

In conclusion, although the existing model shows immense potential for industrial use, ongoing developments in dataset growth, thresholding estimation, and real-time integration will guarantee greater efficiency and broader applicability to the manufacturing sector.

REFERENCES

1. Abadi, M., Agarwal, A., & Barham, P., et al. 2016. "TensorFlow: Large-Scale Machine Learning on Heterogeneous Systems." *arXiv preprint arXiv:1603.04467.* Available at: https://www.tensorflow.org.
2. Breiman, L. 2001. "Random Forests." *Machine Learning, 45*(1), 5–32. DOI: 10.1023/A:1010933404324.
3. Goodfellow, I., Bengio, Y., & Courville, A. 2016. *Deep Learning.* MIT Press.
4. He, K., Zhang, X., Ren, S., & Sun, J. 2016. "Deep Residual Learning for Image Recognition." *Proceedings of the IEEE Conference on Computer Vision and Pattern Recognition (CVPR)*, 770–778.
5. Kingma, D. P., & Welling, M. 2014. "Auto-Encoding Variational Bayes." *arXiv preprint arXiv:1312.6114.*
6. Krizhevsky, A., Sutskever, I., & Hinton, G. E. 2012. "ImageNet Classification with Deep Convolutional Neural Networks." *Advances in Neural Information Processing Systems, 25*, 1097–1105.
7. LeCun, Y., Bengio, Y., & Hinton, G. 2015. "Deep Learning." *Nature, 521*(7553), 436–444. DOI: 10.1038/nature14539.
8. Pedregosa, F., et al. 2011. "Scikit-learn: Machine Learning in Python." *Journal of Machine Learning Research, 12*, 2825–2830. Available at: https://scikit-learn.org.
9. Rasmussen, C. E., & Williams, C. K. I. 2006. *Gaussian Processes for Machine Learning.* MIT Press.
10. Simonyan, K., & Zisserman, A. 2014. "Very Deep Convolutional Networks for Large-Scale Image Recognition." *arXiv preprint arXiv:1409.1556.*
11. Srivastava, N., Hinton, G., Krizhevsky, A., Sutskever, I., & Salakhutdinov, R. 2014. "Dropout: A Simple Way to Prevent Neural Networks from Overfitting." *Journal of Machine Learning Research, 15*(1), 1929–1958. Available at: http://jmlr.org/papers/v15/srivastava14a.html.
12. Zeiler, M. D., & Fergus, R. 2014. "Visualizing and Understanding Convolutional Networks." *European Conference on Computer Vision (ECCV)*, 818–833.
13. Zhang, Y., & Jiang, S. 2017. "A Review on Automated Defect Detection Using Machine Learning." *Journal of Manufacturing Science and Engineering, 139*(6). DOI: 10.1115/1.4036201.
14. Zhang, Z., & Sabuncu, M. 2018. "Generalized Cross Entropy Loss for Training Deep Neural Networks with Noisy Labels." *Advances in Neural Information Processing Systems, 31.*

Note: All the figures in this chapter were made by the authors.

Advances in Mechanical Engineering and Materials Sciences – Dr. Vinay K. B et al. (eds)
© 2026 Taylor & Francis Group, London, ISBN 9-781-041-20970-6

45

The Effect of Process Parameters on Quality Characteristics in the Drilling of Al7075-Metal Matrix Composites—A Comprehensive Review

**Dushyanthkumar G. L.[1], Vinay K. B.[2],
Madhu N.[3], Lohith M. S.[4], Deekshith M.[5], Shashank R.,[6]**
Department of Mechanical Engineering Vidyavardhaka,
College of Engineering Mysuru,
Karnataka, India

Abstract: The drilling of Al 7075 metal matrix composites reinforced with silicon carbide (SiC) and boron carbide (B4C) has garnered considerable attention due to their enhanced mechanical properties, which facilitate a wider range of applications in the automotive, aerospace and defense sectors. In this cases, we investigate the machinability of these composites during drilling operations while examining the effects of process parameters, including spindle speed, feed rate, and tool material. Experiments were carried out to measure key performance parameters, focusing on factors such as cutting force, surface finish, and tool wear during the drilling of the composites, which were thrust force, torque, surface finish, hole quality, and tool wear. Effects of various types of reinforcement on machinability were also studied. It is observed that both SiC and B4C reinforcements have a prominent influence on chip formation and cutting pressures and tool life so require optimum cutting parameters. The findings provide considerable insights into achieving precision and efficiency on drilling of Al 7075 metal matrix composites, thus facilitating the progress of manufacturing processes for high-performance composite materials. Aluminum alloy has increasingly become a preferred structural material in the automotive and aerospace industries, attributed to its low density, high strength, and superior corrosion resistance compared to alternative metals. Drilling procedures are frequently performed on these components during the manufacturing and assembly processes. Because of its exceptional mechanical and thermal qualities, Al7075 was regarded as the workpiece material and is now widely used in many engineering sectors. Exploring the surface roughness and circularity of the drill hole's entrance and exit in aluminum (Al) 7075 reinforced with varying volume fractions of (SiC) and (B4C) are aim main study in this concept.

Keywords: Aluminum 7075, Metal matrix composites (MMCs), Silicon carbide (SiC), Boron carbide(B4C), Drilling, Surface roughness, Circularity, Analysis of variance

1. INTRODUCTION

Due to their exceptional corrosion resistance and favorable strength-to-weight ratio, aluminum alloys, especially those from the 6000 and 7000 series, are highly sought after. Aluminum's density of approximately 2.7 g/cm³ contrasts sharply with steel's density of around 7.87 g/cm³, highlighting its lightweight advantages. It is usually less dense-that is, it is some between one and three times weaker more than steel ,but, the Aluminum 7075 aspects

[1]dushyanth.mech@vvce.ac.in, [2]vinaykb@vvce.ac.in, [3]madhun20028@gmail.com, [4]mslohith606@gmail.com, [5]Mdeekshith321@gmail.com,
[6]shashankshetty942003@gmail.com

10.1201/9781003725053-45

as far as strength and stiffness are concerned equal or approach certain types of steel in strength. Additionally, the aluminum alloys are relatively very cheap and possess much commercial value (Dahnel et al.) Many high-aluminum-content industries that are using aluminum alloys comprise of aircraft, cars, ships, and rails (Sun et al. 2022) Machining operations, particularly drilling, are highly utilized in the production of parts in a great number of industries. These drillings are commonly performed at the last stages of manufacturing when holes need to be drilled to facilitate the assembly of the final product using necessary fasteners (Karnam et al. 2018). In these drilling operations, though, fractures, burrs, and surface distortions can also occur, which may lead further to inferior performance of the product and even product failure. The dimensions of the holes drilled must meet permissible tolerances and contain little or no defects so that the parts can be assembled safely to form a working product (Habib et al. 2021). The development and machining of Al7075-B4C and SiC metal matrix composites (MMCs) have drawn attention to extensive investigation attention because Of their desirable properties for advance engineering applications, SiC provides moderate hardness, wear resistance, and good thermal conductivity, while B4C adds extreme hardness and lightweight characteristics. The two can synergistically improved attributes are wear resistance, hardness, and tensile strength. Dual reinforced composite structure improves load distribution within the composite, avoiding high stress concentrations, which would therefore lead to strength improvement overall (Mahmoud et al. 2022). To achieve the optimized particle distribution, minimum porosity, and enhanced bonding interface within the composites, Applications for SiC's mild hardness, wear resistance, and aided casting include squeeze casting, stir casting, powder metallurgy, and ultrasonic characteristics. All processes have specific advantages and disadvantages-while the powder metallurgy promotes the efficient control of particle size, it induces brittleness; whereas, in the stir casting process, although cost is considerably low, the chances of scattering of particles are unavoidable. These properties of materials significantly influence the machinability of MMCs, especially in doing precision application like drilling (Manohar et al. 2018). The metal matrix composite (MMC) processing is generally classified into three major categories. They are: 1. Processing in Solid States 2. Processing in Liquid States 3. Processing in-situ. Processing in solid states entails processes like powder blend and subsequent consoling (PM processing), diffusion bond, and vapor deposition. A number of variables, including the different types and degree of reinforced loading and the level of microstructure integrity to be achieved, influence the processing route selection. (Mathivanan et al , 2019) Determine the stir casting method in liquid metallurgy vortex route methodology. Experiments have demonstrated a considerable increase

in wear and mechanical qualities. (Manohar et al, 2018) Referred composites have better mechanical qualities than alloys, but because of powder metallurgy, the mechanical performance of the composites is improved by reducing flaws like porosity, poor wetting, and interfacial energies. In contrast to composites made by traditional methods, it is produced at a reasonable cost and is easily accessible. Composites made using the powder metallurgy method are stronger and harder. The homogeneous and homogenous dispersion of reinforcing particles in powder metallurgy procedures is superior to traditional methods.(Balaji et al 2015) found that squeeze casting produces strong bonding between Al-7075 and Al2O3, In microstructural demonstrates that the reinforcements are evenly dispersed along the grained boundary, it increases the hardness of the reinforcement and less porosity. Stir casting remains a widely explored technique, with studies by (Balaji, Sateesh, and Hussain 2015) emphasizing its effectiveness in particle distribution and the composites densities are discovered to be higher than that of their underlying matrix, increase wear resistance, Al7075-SiC metal matrix composites' tensile strength. The research conducted by Mahmoud et al. (2022) utilized the stir casting technique to integrate silicon carbide (SiC) and boron carbide (B4C) in varying proportions of 4%, 8%, 12%, and 16%, with the objective of evaluating the mechanical properties of the resulting metal matrix composites (MMCs). The results indicated a 12% enhancement in hardness and an increase in tensile strength. Additionally, the study by Sravanthi et al. (2024) aimed to develop a high-performance aluminum 7075 alloy incorporating its own alloying constituents in a modified form. An L8 orthogonal grid stir casting arrangement was used to prepare the metals. It is established that reinforcement improves the tensile strength, and it also increases impact strength and hardness. Mechanical and wear properties have improved with reduced machining effect with increasing SiC particles. The study team used an innovative technique called sandwich infiltration casting (Bolat et al. 2022). To summarize the procedure, they melted a graphite crucible filled with aluminium and a three-piece steel Mold in an electrical resistance furnace. This study examined the machinability of B4C filled Al composites and found that poor surface quality was caused by process vibrations that increased due to high cutting forces.

The size and dispersion of B4C particles were controlled using powder (Manohar, Pandey, and Maity 2020). Up to critical volume fractions, a higher volume fraction of B4C results in higher mechanical strengths, higher hardness, and lower strain-to-fracture values. According to a systematic review, the inclusion of hard reinforcement particles markedly enhances the mechanical, tribological, and corrosion properties of Al 7075 metal matrix composites produced via the stir casting process, in comparison to the base alloy. (Sambathkumar et al. 2023).

2. DRILLING PROCESS

The drilling process starts with tool selection, primarily utilizing materials such as polycrystalline diamond (PCD), cubic boron nitride (CBN), and tungsten carbide due to their exceptional wear resistance and hardness. Coated tools are preferred as they exhibit lower friction and enhanced wear resistance. Tool geometry is well-designed, where helix and point angles are optimized for the effective removal of the chip through minimum cutting forces (Yuan et al. 2023). In the case of drilling parameters, spindle speed plays an important consideration for the surface finish quality; an increased spindle speed can improve the surface finish but is also responsible for increased degradation of the cutting tool. Control of the feed rate is necessary, where an increased feed rate can reduce the machining time, reduce the thrust forces, and reduce the possibility of tool fracture. Utilization of cutting fluids, such as lubricants or coolants, is necessary to minimize the heat caused, reduce the gadget attrition, and improve the waves finish. For hole quality, it is not easy to maintain precise dimensional accuracy due to variability of the material additionally tool wear (Habib et al. 2021). The Reinforcing particles may have an impact on the surface finish, in which this can be responsible for surface defects or microcracks. The formation of the burr can also be formed at the hole exit, particularly in materials such as aluminum, where material ductility can be responsible for the formation of burrs (Günay, Yaşar, and Korkmaz 2016).

3. EFFECT OF REINFORCEMENT

Researchers studied reinforcement mechanisms in Al7075-SiC and B4C metal matrix composites. In that, a distribution of the SiC and B4C particles has been analyzed concerning bonding and integration in an aluminum matrix. In fact, improvement in hardness, strength, and thermal stability was the prime objective of this study. Silicon carbide (SiC) and titanium boride (TiB2) particle reinforcement was highlighted by Mathivanan et al. (2019) as a means of enhancing hardness, tensile strength, and compressive strength in relation to reinforcement weight. On the other hand,(Manohar et al. 2018) discovered that powder metallurgy aids in uniform distribution of B4C, therefore greatly enhancing the hardness and toughness of the composites. To prevent brittleness, which could affect machinability, this technique must thus be carefully controlled in compaction and sintering settings. Similarly, (Muraliraja et al. 2019) emphasized that squeeze casting improves hardness of reinforcement of Al7075 and alumina (Al2O3) and some porosity occurs during the squeeze casting. Further studies have shown that the selection of processing techniques came directly influence the effectiveness of B4C and SiC reinforcement. (Balaji, Sateesh, and Hussain 2015) pointed out that the creation of Al7075-SiC composites has been discovered to have a much higher hardness than their basic matrix, SiC,

in terms of strengthening the wear resistance of metal matrix composites (MMCs). However, they also pointed out that if stir casting techniques are not well managed, they may cause particle clustering. The study by Habib et al. (2021) discusses the application of a TiO2 coating on carbon fibers (CF) using a chemical vapor deposition method. When carbon fiber (CF) is coated with powdered 7075 aluminum alloy, the reaction of TiO2 on its surface produces magnesium oxide (MgO) and titanium carbide (TiC) after the procedures of hot-pressing, sintering, and hot extrusion. The sample density, hardness, and tensile strength were evaluated. Experimental results indicate that applying a TiO2 coating to carbon fiber (CF) improved the tensile strength and strength-to-density ratio of the 7075-aluminum alloy.

An Al 7075 matrix augmented with ZrO2 was successfully created using the stir casting technique; (Al Zubaidi et al. 2023) in this study, it was reported that strong bindings were formed by the effective condensation between the matrix and reinforcement. Despite a reduction in heat conductivity and surface roughness, the composite structure exhibited increased hardness, compressive strength, and Young's modulus by using ZrO2 reinforcing elements. Stir casting process is the widely used (Thirugnanam et al. 2018) they explored composites behavior of Al -7075 Based on Graphite and Bagasse Ash Particles which are Reinforced, fabricate the aluminum metal matrix composites on three different variants of graphite and bagasse ash. (Mathimurugan et al. 2022), focused more on characterization of the prepared hybrid composite by the method of squeeze casting using four distinct methods: X-ray diffracts (XRD), chemical spectroscopy (ChemS), energy dispersive spectroscopy (EDS), further are scanning electron microscopy (SEM). In this method, both phase as well as microstructure characterization of the composite was done. (Sun et al. 2022) In this study, 7075Al composites containing SiC were made via SPS and hot rolled. A thorough examination was carried out to analyse the microstructural evolved, phase precipitation, and mechanical behaviour on ongoing production process. The skin and proximity effects of the HFPC can successfully close the micro cracks that exist during the rolling deformation process, and the micro-cracks are typically fine recrystallized grains, which is critical for improving mechanical characteristics. (Ramanan G et al 2017) this Studie have identified key machining settings that are essential for maintaining stability and precision in WEDM, explore the effect of wire feed rate, pulse duration, and dielectric flushing pressure, noting that these parameters are especially important when dealing with abrasive reinforcements like B4C.

4. EFFECT OF MACHINING PARAMETERS

The machining parameter for Al7075-B4C and SiC metal matrix composites (MMCs) vary significantly depending on the processing method and intended application, as researchers have sought to balance efficient material removal rates (MMR), surface finish, hole quality, cylindricity of entry and exit of hole, and tool wear in machining those hard composites. According to (Khalid, Umer, and Khan 2023) The study investigates the machinability characteristics of the end milling operation to achieve minimum tool wear while maximizing the material removal rate , By altering important machining process parameters, which are spindle speed, feed rate, and depth of cut, across various weight percentages of reinforcements, the RSM-based experimental design was carried out. Other researchers discovered that dry drilling performance on an aluminum metal matrix (Al7075) (Habib et al. 2021) The current study evaluates the hole quality during drilling. Surface roughness during Al7075-T6 drilling decreases with accelerating the spindle speed. Likewise, circularity is reduced by raising the feed rate. Furthermore,tool wear and tear are no longer hidden. (Dahnel et al. 2022) examined how drilling conditions, cutting parameters, and cutting tools affected the hole quality and tool wear of Al 7075 drilled holes. Their results demonstrate that reduced feed rates and cutting speeds led to better hole quality and less tool wear, particularly for diameter, roundness, surface roughness, and burr. (Karnam et al. 2018) they are using the radial drilling machine to drill the composites of Al7075-B4C, and evaluate the speed, feed rate, drill bit, etc. to produce good quality hole in Al7075-B4C composites. (Ramanan G et al 2017) These studies have identified key machining settings that are essential for maintaining stability and precision in WEDM, and investigate the effect of wire feed rate, pulse duration, and dielectric flushing pressure, noting that these parameters are especially important when dealing with abrasive reinforcements such as B4C.

5. PARAMETERS OF DRILLING OPERATIONS IN ALUMINIUM 7075 METALS MATRIX COMPONENTS

In a previous study on drilling Al 7075 alloy, important factors such cutting speed, feed rate, and point angle were examined. Point angles varied from 120° to 140°, feed rates from 0.05 to 0.15 mm/rev, and cutting speeds from 40 to 120 m/min. For Aluminum 7075 Metal Matrix Composites (MMC), various drilling settings are recommended based on the type of equipment. The recommended cutting speeds for high-speed steel (HSS) drills are 30 to 80 m/min, polycrystalline diamond (PCD) drills are 150 to 300 m/min, and carbide drills are 50 to 150 m/min. Feed rates of 0.05 to 0.3 mm/rev are recommended for carbide drills, 0.1 to 0.5 mm/rev for PCD drills, and 0.02 to 0.1 mm/rev for HSS drills (Patel and Jagadish 2021).

6. CONCLUSION

In summary, this study has investigated the intricate relationships between machining parameters, reinforcing properties, and processing methods in the drilling of SiC and Al7075-B4C metal matrix composites. Because of their extreme hardness and brittleness, B4C and SiC reinforcement significantly improve strength and wear resistance, but they also present considerable machinability issues. While traditional techniques like squeeze casting and stir casting have structural and financial benefits, they frequently have trouble achieving uniform particle dispersion, which lowers the quality of the machining. However, while advanced processes like ultrasonic-assisted casting and powder metallurgy improve bonding and reinforcement distribution, they also increase hardness and spark instability during drilling.

The results emphasize the importance of meticulously adjusting drilling parameters, such as feed rate, speed rate, and cutting tool choice, to meet the specific machining requirements of SiC and Al7075-B4C composites. For optimal material removal rates, surface roughness, and tool durability, these parameters must be tailored to the processing characteristics of each composite. More study is required to optimize these machining parameters and look at hybrid processing techniques that enhance material characteristics and machinability in light of the growing use of MMCs in demanding applications. A better comprehension of those dynamics may enable more effective machining techniques for SiC and Al7075-B4C MMCs, enabling their wider use in sectors including defence, automotive, and aerospace.

REFERENCES

1. Al Zubaidi, Faten N., Lamyaa Mahdi Asaad, Iqbal Alshalal, and Mohammed Rasheed. 2023. "The Impact of Zirconia Nanoparticles on the Mechanical Characteristics of 7075 Aluminum Alloy." Journal of the Mechanical Behavior of Materials 32 (1). Walter de Gruyter GmbH. doi:10.1515/jmbm-2022-0302.

2. Balaji, V., N. Sateesh, and M. Manzoor Hussain. 2015. "Manufacture of Aluminium Metal Matrix Composite (Al7075-SiC) by Stir Casting Technique." In Materials Today: Proceedings, 2:3403–3408. Elsevier Ltd. doi:10.1016/j.matpr.2015.07.315.

3. Bolat, Çağın, Berkay Ergene, Uçan Karakılınç, and Ali Gökşenli. 2022. "Investigating on the Machinability Assessment of Precision Machining Pumice Reinforced AA7075 Syntactic Foam." Proceedings of the Institution of Mechanical Engineers, Part C: Journal of Mechanical Engineering Science 236 (5). SAGE Publications Ltd: 2380–2394. doi:10.1177/09544062211027613.

4. Dahnel, Aishah Najiah, Mohamad Noor, Ikhwan Naiman, Muhammad Azim Mirza, Mohd Farid, Ahmad Faris, Abdul Rahman, Nur Munirah, and Meera Mydin. 2022. Drilling of 7075 Aluminum Alloys. www.intechopen.com.

5. Günay, Mustafa, Nafiz Yaşar, and Mehmet Erdi Korkmaz. 2016. Optimization of Drilling Parameters for Thrust Force in Drilling of AA7075. https://www.researchgate.net/publication/312934598.

6. Habib, Numan, Aamer Sharif, Aqib Hussain, Muhammad Aamir, Khaled Giasin, Danil Yurievich Pimenov, and

Umair Ali. 2021. "Analysis of Hole Quality and Chips Formation in the Dry Drilling Process of Al7075-T6." Metals 11 (6). MDPI AG. doi:10.3390/met11060891.

7. Karnam, Malthesh, Raghavendra Joshi, T H Manjunatha, and Kavadiki Veerabhadrappa. 2018. Study of Mechanical Properties and Drilling Behavior of Al7075 Reinforced with B4C. Malthesh Karnam et al./ Materials Today: Proceedings. Vol. 5. www.sciencedirect.comwww.materialstoday.com/proceedings.

8. Khalid, Muhammad Yasir, Rehan Umer, and Kamran Ahmed Khan. 2023. "Review of Recent Trends and Developments in Aluminium 7075 Alloy and Its Metal Matrix Composites (MMCs) for Aircraft Applications." Results in Engineering. Elsevier B.V. doi:10.1016/j.rineng.2023.101372.

9. Mahmoud, Hassab Alla M.A., P. Satishkumar, Yenda Srinivasa Rao, Rohinikumar Chebolu, Rey Y. Capangpangan, Arnold C. Alguno, Mahesh Gopal, A. Firos, and Murthi C. Saravana. 2022. "Investigation of Mechanical Behavior and Microstructure Analysis of AA7075/SiC/B4C-Based Aluminium Hybrid Composites." Advances in Materials Science and Engineering 2022. Hindawi Limited. doi:10.1155/2022/2411848.

10. Manohar, Guttikonda, Abhijit Dey, K. M. Pandey, and S. R. Maity. 2018. "Fabrication of Metal Matrix Composites by Powder Metallurgy: A Review." In AIP Conference Proceedings. Vol. 1952. American Institute of Physics Inc. doi:10.1063/1.5032003.

11. Manohar, Guttikonda, K. M. Pandey, and S. R. Maity. 2020. "Characterization of Boron Carbide (B4C) Particle Reinforced Aluminium Metal Matrix Composites Fabricated by Powder Metallurgy Techniques – A Review." In Materials Today: Proceedings, 45:6882–6888. Elsevier Ltd. doi:10.1016/j.matpr.2020.12.1087.

12. Mathimurugan, N., V. Vaishnav, R. Praveen Kumar, P. Boobalan, S. Nandha, Venkatesh Chenrayan, Kiran Shahapurkar, et al. 2022. "Room and High Temperature Tensile Responses of Tib2-Graphene Al 7075 Hybrid Composite Processed through Squeeze Casting." Nanomaterials 12 (18). MDPI. doi:10.3390/nano12183124.

13. Mathivanan. S 2 Eniyavan.G 3 Ilavarasan. A 4Karthik. B. 2019. A Review Paper of Al 7075 Metal Matrix Composition. www.ijert.org.

14. Muraliraja, R.; R. Arunachalam, Ibrahim Al-Fori, Majid Al-Maharbi, and Sujan Piya. 2019. "Development of Alumina Reinforced Aluminum Metal Matrix Composite with Enhanced Compressive Strength through Squeeze Casting Process." Proceedings of the Institution of Mechanical Engineers, Part L: Journal of Materials: Design and Applications 233 (3). SAGE Publications Ltd: 307–314. doi:10.1177/1464420718809516.

15. Patel, G. C.Manjunath, and Jagadish. 2021. "Experimental Modeling and Optimization of Surface Quality and Thrust Forces in Drilling of High-Strength Al 7075 Alloy: CRITIC and Meta-Heuristic Algorithms." Journal of the Brazilian Society of Mechanical Sciences and Engineering 43 (5). Springer Science and Business Media Deutschland GmbH. doi:10.1007/s40430-021-02928-3.

16. Ramanan G et al. 2017. Multi Objective Optimization of Machining Parameters for AA7075 Metal Matrix Composite Using Grey-Fuzzy Technique. International Journal of Applied Engineering Research. Vol. 12. http://www.ripublication.com.

17. Sambathkumar, M., R. Gukendran, T. Mohanraj, D. K. Karupannasamy, N. Natarajan, and David Santosh Christopher. 2023. "A Systematic Review on the Mechanical, Tribological, and Corrosion Properties of Al 7075 Metal Matrix Composites Fabricated through Stir Casting Process." Advances in Materials Science and Engineering. Hindawi Limited. doi:10.1155/2023/5442809.

18. Sravanthi, C., S. Gajanana, A. Krishnaiah, and Ch Venkateswarlu. 2024. "Mechanical, Microstructure and Machining Characteristics of Alloying Elements Optimized Al 7075 Alloy Modified with Reinforcement-Silicon Carbide." Advances in Materials and Processing Technologies. Taylor and Francis Ltd. doi:10.1080/237406 8X.2024.2307249.

19. Sun, Xiaozhe, Ruifeng Liu, Zile Jia, Chenrui Yuan, Fengfeng Wu, Jie Yan, and Xian Wang. 2022. Investigation on Microstructure and Mechanical Properties of SiC/7075Al Composites Fabricated by SPS-Rolling Followed by HFPCT. https://ssrn.com/abstract=4789599.

20. Thirugnanam, S, C Velmurugan, Binnu Kurian Mathew, and & Head. 2018. An Experimental Investigation On Mechanical Properties of Aluminium-7075 Based Graphite and Bagasse Ash Particles Reinforced Metal Matrix Composite.

21. Yuan, Mu, Jinhao Wu, Qingnan Meng, Chi Zhang, Xinyue Mao, Shiyin Huang, and Sifan Wang. 2023. "Influence of Carbon Fiber Failure Mode Caused by TiO2 Coating on the High Temperature Tensile Strength of Carbon Fiber Reinforced 7075 Al Alloy Composites." Journal of Materials Research and Technology 26 (September). Elsevier Editora Ltda: 4551–4562. doi:10.1016/j.jmrt.2023.08.191.

Advances in Mechanical Engineering and Materials Sciences – Dr. Vinay K. B et al. (eds)
© *2026 Taylor & Francis Group, London, ISBN 9-781-041-20970-6*

46

A Review Paper on Characteristics of Sandwich Panels Produced by 3D Printing

Naveen Prakash G. V.[1],
S. Puneet Kumar[2], Sathya Kumar S.[3],
Shashank M. P.[4], Saketh R.[5]
Mechanical Engineering, VVCE, Mysuru,
India

Abstract: This review paper specifically examines the performance improvements of 3D printed sandwich panels using PLA (Polylactic Acid) and composite materials. The goal is to identify the optimized structure for applications that require lightweight and are durable. These panels, which feature a bio-inspired core design such as a gyroid structure, has high hardness and weight ratio, and the studies are integrated with advanced optimization methods. It includes finite element analysis (FEA) and genetic algorithms that provide potential advantages in aerospace, automotive, construction, etc. due to their energy absorption characteristics.

This report emphasizes the importance of biodegradable materials in reducing environ- mental impact and also the role of PLA, PLA composites as an environmentally friendly alter-native to traditional materials. Experimental methods such as compression tests and uniaxial bending tests have been published provides in-sight into mechanical performance, while structural computational models simulate real-world loading conditions to verify integrity and flexibility. These findings provide valuable knowledge on optimizing 3D printed structures for specific industrial applications. Supporting the advancement of lightweight, durable engineering solutions, future research is recommended to further improve the optimization technique for wider use. This study highlights the potential of 3D printed PLA sandwich panels in addressing modern engineering challenges.

Keywords: 3D printed sandwich panels, Polylactic acid (PLA), Composite materials, Bio-inspired core design, Gyroid structure, Stiffness-to-weight ratio, Performance enhancement, Lightweight material, Durability

1. INTRODUCTION

The main research problem addressed in this study is the challenge of customizing the structure of 3D printed PLA (Polylactic Acid) sandwich panels, especially through optimizing the stiffener configuration of the panels. PLA sandwiches are known for their lightweight and environmentally friendly properties, however, they often lack the strength and rigidity required for demanding industrial applications. This study sought to identify and implement the most robust design that would minimize the stiffness and weight of these panels.

In relation to the problem, this literature review examines the design of various 3D printed structures, focusing on various solid geometries of PLA-based sandwich panels and the impact of their placement to improve the design of this bio-inspired core such as gyroid structures and other new designs reviewing existing research including core reinforcement materials and hybrid composites (Fig. 46.1). The objective of this review is to identify the most stiff-core designs and materials which can provide excellent mechanical properties without significant weight reduction.

[1]gvnp@vvce.ac.in, [2]puneethsharma15@gmail.com, [3]sathyask204@gmail.com, [4]mpshashankgowdru@gmail.com, [5]saketh.r8103@gmail.com

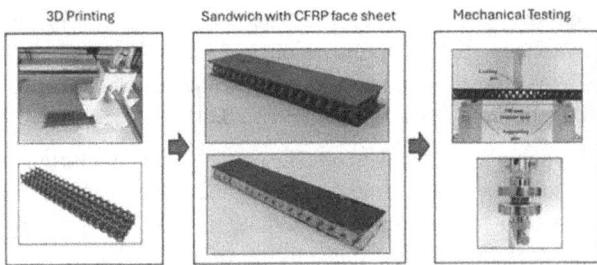

Fig. 46.1 A schematic of a 3D-printed sandwich panel with a gyroid / honeycomb core [1]

A review paper is essential in analysing this issue because it is a synthesis of current research results. It highlights gaps in existing solutions and guides the development of custom designs for sandwich panels.

2. BACKGROUND AND SIGNIFICANCE

The review will focus on advancements and enhancements in 3D printed sandwich panels, especially in the use of PLA (Polylactic Acid) composites. These panels are a lightweight option, but ideal in industries where weight and mate- rial durability are important, such as aerospace, automotive and construction.

This study is important because it validates the design of new structures such as gyroid inspired cores which increase mechanical properties such as hardness, strength, and energy absorption and also in computational modeling of PLA composites, especially finite element analysis (FEA), advanced optimization algorithms (such as genetic algorithms and CMA-ES). The review also aligns with the growing demand for sustainable construction solutions that focuses on biodegradable materials such as PLA, which help in creating strong yet lightweight materials that are Suitable for demanding applications.

Definition of keywords:

PLA (Polylactic Acid): A biodegradable polymer derived from renewable resources, which is used as the main material for sandwich panels.

Stiffener: Structural elements which are added to the panel to increase stiffness and improve the stiffness-to-weight ratio.

Finite element analysis (FEA): A computational method used to predict how a design will respond to physical forces through simulation.

2.1 Review

Methodology

This study compares several methods for designing and testing sandwich panels with 3D-printed cores, focusing on different materials, densities, and reinforcement strategies. One approach is to use composite fiber 3D-printing using different densities (10 percent, 15 percent, 20 percent) of biodegradable polymers. Polylactic acid (PLA) is

created for the production of gyroid structural cores (Fig. 46.2). These cores are sandwiched between carbon fiber reinforced polymer (CFRP) layers, with some examples using polyurethane foam (PUF) as another core filler to increase the mechanical properties between bending and compression tests and reliability analysis were completed (Junaedi H et al.,2024). Alternatively, 3D-printing of the polymer layer combined with electroforming can be used to create a metal shelf. It has densities of 4, 5, and 6 pores per inch (PPI). Uniaxial compression testing was performed to evaluate yield strength. The specific energy absorption (SEA) results were compared to evaluate the enhancement with pure nickel and copper films (Hosseinpour M et al., 2024). The third method combines a steel faceplate with a molded reinforced chiral core. They are created using 5 different methods and sculpting strategies. The deformation and failure modes of the face plate and support core have been reported in various blasting tests. While numerical simulations provide insight into the deformation process under blast forces (Chen C et al.,2024), these studies together demonstrate different strategies to increase the mechanical performance and reliability of the sandwich panel for advanced structural applications.

This paper compares two different methods for producing PLA-based composite sandwiches, focusing on mechanical properties and structural design. In one study, PLA honeycombs of different thicknesses (50–500 µm) were fabricated using hot compression molding. Sandwich panel with core made of PLA linen and textile leather. Each sheet is combined with epoxy adhesive to improve mechanical connection and a non-woven layer of PLA. The study assessed how core thickness affects flexural and compressive properties. The material and layer configuration provide insights to the overall strength

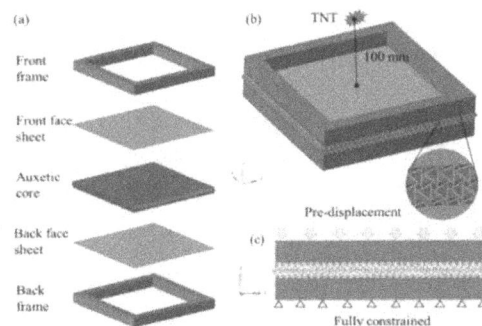

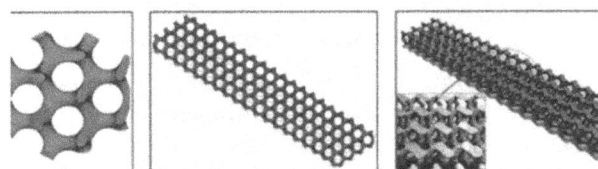

Fig. 46.2 Different core structures (gyroid, honeycomb, auxetic) [1, 3]

(Lascano D et al.,2021). In another approach using PLA and ABS polymers, fused deposition modeling (FDM) is used for 3D- printed sandwich structures (Fig. 46.3). They must pass a tensile test that evaluates their mechanical performance in different PLA-ABS configurations. The objective of the study is to compare different material arrangements for the core and surface layers by comparing the results against samples of power flexibility (Patro PK et al., 2023). The most suitable polymer combinations were identified for both studies, highlighting the suitability of PLA and composite materials for machining and improving properties through structural configuration and various fabrication methods.

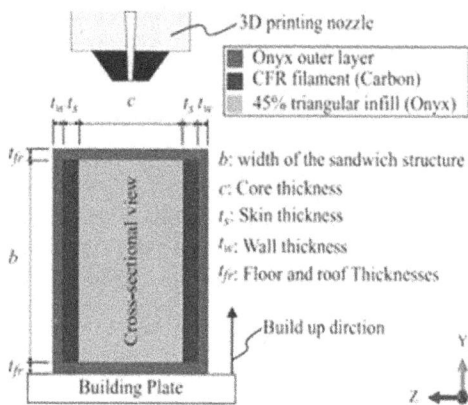

Fig. 46.3 Experimental process of fabrication [5]

Comparison is made with two advanced optimization approaches that aim to optimize the structural performance of components made from composites. A single finite element (FE) beam modeling technique is used to design a continuous fiber reinforced (CFRAM) prosthetic foot in which the prosthetic device specifies static stiffness requirements. This approach focuses on the minimal weight loss to make sure it is in place. It allows for flexible design to determine the geometric parameters and overall structural factors in the design process and investigate various geometries (such as C-shape and J-shape) to meet specific performance requirements (Al Thahabi ARN et al.,2024). Other studies in Planar asymmetry within printed topology, a modified Greedy algorithm is used to optimize stiffness in components made of soft and hard materials. Finite element analysis (FEA) was performed using PyAnsys on 16 Q8 RVE elements using three different objective functions to comprehensively assess the stiffness (Coropetchi I et al., 2022). These methods are jointly improved to optimize the structure in a specific application demonstrating the optimization technique.

The main optimization methods, covering deterministic and random techniques, are reviewed. Among the random methods is the Descent random method which is used in neural networks to determine the weighting coefficient and the Lagrangian multiplier method which handles constraints efficiently. For dynamic tuning of

moving asymptote methods, genetic algorithms (GA) and covariance matrix optimization and evolution strategies (CMA-ES) are also discussed to a strong search capability in the optimal design (Igarashi H.,2024).

The paper also focuses on 3D printing manufacturing a drone frame using design to create a CAD model and the quadrotor frame was built with SOLIDWORKS taking into account the size and outline of various parts such as pro-Ammunition and Control Devices (Parandha S M and Li Z., 2018). Miscellaneous stress and shock tests are simulated for verification with frame strength and durability under different conditions. The frame is printed with materials like ABS-PC and carbon fiber, glass for flexibility and robustness (Parandha S M and Li Z., 2018).

2.2 Findings

The study reviews findings for optimizing the sandwich panel core and highlighting the strengths and limitations of different applications. Increasing the core density in PLA-based gyroid structures increases the specific stress and shear strength. This is necessary for better bending behavior. While filling the gyroid core with polyurethane foam (PUF) increases the deflection at peak stress, post-fracture and load-capacity, even the pure gyroid core still has strong bearing capacity. Excellent compression and energy absorption which indicates specific advantages for specific applications (Junaedi H et al.,2024). Adding a metal layer to a film made from a polymer especially at higher pore density (6 PPI), the yield strength and energy absorption are significantly increased. Although focusing on specific metals and polymers may limit generalizability, additionally, uniaxial testing is available for real- world applications may not fully represent multidirectional stresses and scaling challenges may arise in large-scale construction (Hosseinpour M et al., 2024). A third study of a reinforced core with a gradient design showed that negative gradient configurations exhibit deflection. Lowest permanent deflection due to negative values of Poisson's ratio and good explosion resistance but this study relies on specific materials. Explosive force quantity and numerical simulation creates potential un-certainties related to their actual use (Chen C et al.,2024). Together these studies form the backbone of optimizing the performance of sandwich panels for different engineering applications that emphasize the importance of density for selection of materials and structural design.

Further methods for improving the mechanical properties of PLA-based sandwich structures by adjusting the core thickness and polymer combination. One study found that increasing the thickness of a PLA honeycomb core significantly improved its bending performance. The thicker cores (10 and 20 mm) reach a core shear strength of approximately 0.60 and 31−1 with a flexural stress of 33 MPa (Fig. 46.4). The 20 mm core has an excellent weight-to-weight ratio (141.5 N ·g−1), highlighting its

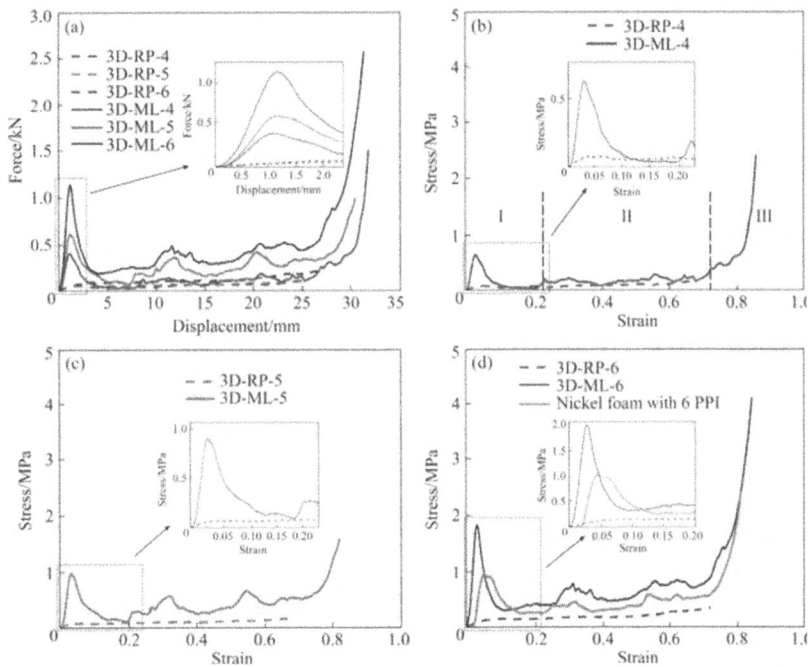

Fig. 46.4 Stress-strain curves of different PLA sandwich panel configurations from compression/bending tests [2]

potential in technical applications, However, there is a limited adhesion be- tween the PLA core and skin due to poor adhesion. We'll be able to reduce weight transfer and overall structural integrity. The 20 mm core has an excellent load-to-weight ratio ($141.5 \text{ N} \cdot \text{g} - 1$), however, poor adhesion results in limited adhesion between the cores. PLA and skin can reduce weight transfer and improve overall structural integrity and indicates the need to upgrade the tensile method (Lascano D et al.,2021). Another study that examined the combination of PLA-ABS polymers in 3D printed sandwich structures found that the mechanical properties obtained by using ABS as the core and PLA were the best compared to Pure ABS, which indicates the advantages of structural applications. Despite the good results, the study did not address factors such as heat or long-term durability and focusing on a specific PLA-ABS configuration may limit the generalizability of discoveries to different polymers (Patro PK et al., 2023). Further exploration of combination and printing parameters are taken together, these studies demonstrate the ability to tune core thickness and polymer composition to increase the performance of the sandwich structure in use.

This review compares two additive manufacturing studies focused on optimizing stiffness and material properties for high-performance applications. One study used a continuous fiber reinforced joint construct (CFRAM) to develop a custom-designed prosthetic foot. They offer a customizable and cost-effective alternative to conventional laminate designs and the versatility in specific criteria that affect it's abilities and the range in which they occur (Al Thahabi ARN et al.,2024).

A second study examined increasing stiffness in 3D printed components that combine soft and hard materials. It has been shown that effective stiffness and complex geometries can be achieved. This is important for many engineering applications, however, relying on PyAnsys for FE analysis in addition to limiting the generalizability of the findings to other software platforms and the Greedy algorithm may ignore some configurations. This may result in a less-than-optimal solution. The use of a fixed elasticity ratio (1:10) limits its applicability to different material combinations emphasizes the importance of adaptability (Coropetchi I et al., 2022) for modeling approach and flexibility to adapt for wider use.

In addition to being able to learn how to optimize, certain methods are true for neural networks and multi-objective optimization. Genetic algorithms (GA) show flexibility in handling complex problems with uncertainty. This paper also introduces simulated quantum annealing as an advanced learning technique. They tend to deal with highly complex, directional gaps, although the specific limitations of each method are not directly discussed. But it can be speculated that stochastic methods such as GA and Covariance Matrix Adaptation Evolution Strategy (CMA-ES) may struggle with slower convergence and insensitivity to local minimums. It can depend greatly on the choice and specificity of the problem so that some applications can hinder performance (Igarashi H.,2024).

This section highlights the key benefits and challenges of 3D printing for customizing drone frames (Fig. 46.5). It focuses on the potential to produce small, durable frames that can be tailored to specific needs such as integration of future sensors or cameras etc. Improvements for

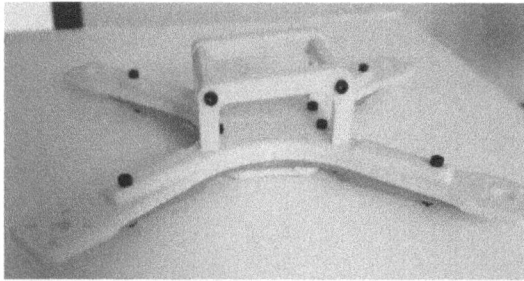

Fig. 46.5 CAD model of a 3D-printed quadcopter chassis for drone applications [9]

applications such as environmental monitoring drones - functionality can be expanded. However, limitations include the impact resistance of plastic materials.

There are design constraints imposed by 3D printing and possible discrepancies between simulation and real-world performance in the existing methods of optimizing the performance of 3D printed PLA sandwich panels for cooperative performance analysis. Structure and Advanced Finite Element Analysis (FEA): A detailed literature review included studies of techniques and thematic analyzes grouped the findings into key themes, including design configuration to increase stiffness under various loading conditions. PLA material properties and structural performance as well as genetic algorithms and topology optimization methods are used to optimize the design (Parandha S M and Li Z., 2018). Based on the findings, a comparison of mechanical properties of 3D-Printed Sandwich Panels and Traditional Panels is carried out and are as shown in Table 46.1.

3. IMPLICATIONS AND CONSIDERATIONS FOR FUTURE WORK

3.1 Implications

- Material Performance: Optimizing stiffeners highlights the mechanical strengths of PLA, making it more viable for UAVs by improving load-bearing capacity and impact resistance to compete with traditional materials.
- Weight-to-Strength Ratio: Lightweight, robust designs enhance UAV flight time, maneuverability, and payload options - critical factors for UAV efficiency.
- Cost-Effectiveness: PLA is affordable, and optimizing stiffener designs can lower material costs without sacrificing durability or performance.
- Customizability: 3D printing allows for rapid prototyping and custom designs, ideal for specialized UAV uses like surveillance or agriculture.
- Environmental Impact: As a biodegradable material, PLA offers an eco- friendly option, particularly useful in single- use or disposable UAV applications.

3.2 Considerations for Future Work

- Material Enhancements: Research on PLA composites, like PLA-carbon fiber blends, to increase strength, durability, and heat resistance.
- Advanced Optimization Techniques: Use simulation tools and machine learning to re- fine designs and develop more efficient stiffener configurations.
- Experimental Validation: Test prototypes in varied conditions (different loads, temperatures, and impacts) to ensure durability and reliability.
- Multiphysics Analysis: Consider thermal and vibration effects, especially for UAVs exposed to sunlight or engine vibrations.
- Lifecycle Analysis: Evaluate design longevity, maintenance needs, repair options, and recyclability.
- Scalability and Production: Assess how scalable these 3D-printed designs are and explore hybrid manufacturing for wider applications.
- Integration with Electronics: Adapt the chassis to safely house sensors, wiring, and other components without losing strength.

Table 46.1 Comparison of mechanical properties of 3d-printed sandwich panels

Property	3D-Printed Sandwich Panels (PLA & Composites)	Traditional Panels (Metal, Wood, FRP, etc.)
Weight	Lightweight (density: ~1.25 g/cm³ for PLA)	Heavier (metal: ~7.8 g/cm³, wood: ~0.6 g/cm³)
Strength	Moderate strength, tunable via infill & core design	High strength, but dependent on material (steel, aluminum, etc.)
Stiffness-to- Weight Ratio	Optimized using gyroid/honeycomb structures	Often requires additional stiffeners
Biodegradability	High (PLA is biodegradable)	Low (metals, FRP, and plastics are non-biodegradable)
Manufacturing Flexibility	High (complex geometries possible via 3D printing)	Limited by machining/forming techniques
Cost	Low to moderate (depends on material & printing time)	Higher for metal panels, moderate for wood & FRP
Thermal Resistance	Moderate (PLA softens at ~60°C)	Higher (metal withstands high temperatures)
Impact Resistance	Can be improved with composite reinforcements	Generally better in metals & high-strength plastics

Source: Author's compilation

4. CONCLUSION

This review highlights the potential of tailored stiffness in 3D printed PLA structures, specifically for UAV chassis applications, to enhance structural performance and customized PLA structures in addition to supporting UAVs to achieve longer flight times and carry larger payloads. The biodegradability benefits of PLA also provide an environmentally friendly alternative to conventional materials. This is in line with increasing environmental sustainability goals.

The results reveal continued advances in 3D printing, PLA composites, and optimization algorithms. This makes it possible to expand the use of PLA in industries such as aerospace and automotive. Further research into material ingredients, real world testing and the ability to scale up production can deepen the impact. These innovations and customizable designs can pave the way in the field of Sandwich Panels produced by 3D printing.

REFERENCES

1. Junaedi H, Abd El-baky MA, Awd Allah MM and Sebaey TA., "Mechanical characteristics of sandwich structures with 3D-printed bio-inspired gyroid structure core and carbon fiber-reinforced polymer laminate face- sheet," Polymers, Vol. 16 (12), p. 1698, 2024.

2. Hosseinpour M, Nejad S R and Mirbagheri Smh., "Mechanical and energy absorption properties of multilayered ultra-light sandwich panels produced by 3D-printing and electroforming," Transactions of Nonferrous Metals Society of China, Vol. 34 (1), pp. 255–264, 2024.

3. Chen C, He Y, Xu R, Gao C, Li X and Lu M, "Dynamic behaviors of sandwich panels with 3D-printed gradient auxetic cores subjected to blast load," International Journal of Impact Engineering, Vol. 188, p.104943, 2024.

4. Lascano D, Guillen-Pineda R, Quiles- Carrillo L, Ivorra-Martínez J, Balart R, Montanes N, Boronat T, "Manufacturing and characterization of highly environmentally friendly sandwich composites from polylactide cores and flax-polylactide faces," Polymers, Vol. 13 (3), p.342, 2021.

5. Patro PK, Kandregula S, Khan MS and Das S, "Investigation of mechanical properties of 3D printed sandwich structures using PLA and ABS," Materials Today: Proceedings, 2023.

6. Al Thahabi ARN, Martulli LM, Sorrentino A, Lavorgna M, Gruppioni E and Bernasconi A, Stiffness-driven design and optimization of a 3D-printed composite prosthetic foot: A beam finite Element-Based framework. Composite Structures.,"Vol. 337, p.118053, 2024.

7. Coropetchi I, Vasile A, Sorohan Picu C and Constantinescu D, "Stiffness Optimization Through a Modified Greedy Algorithm.," Procedia Structural Integrity, Vol. 37, pp. 755–762, 2022.

8. Igarashi H., Topology Optimization and AI- Based Design of Power Electronic and Electrical Devices: Principles and Methods, Elsevier, 2024.

9. Parandha S M and Li Z. "Design and analysis of 3D printed quadrotor frame," International Advanced Research Journal in Science, Engineering and Technology; Vol.5(4), pp. 66–73, 2018.

Advances in Mechanical Engineering and Materials Sciences – Dr. Vinay K. B et al. (eds)
© 2026 Taylor & Francis Group, London, ISBN 9-781-041-20970-6

47

Comprehensive Review of Machining Parameters for Aluminum 7000 Series Alloys and their Aerospace Applications

Shrinivasa D.[1], G. V. Naveen Prakash[2]
Faculty,[1,2] Department of Mechanical Engineering,
Vidyavardhaka College of Engineering, Mysuru

Anand A.[3]
Faculty[3], Department of Mechanical Engineering,
National Institute of Engineering, Mysuru

Syed Hafeez R.[4], Syed Sufiyan[5]
Students, Department of Mechanical Engineering,
Vidyavardhaka College of Engineering, Mysuru

Abstract: The 7000 series of aluminium is well known for its remarkable machinability and strength-to-weight ratio. However, machining this alloy presents a number of difficulties despite its extensive use in defence, aerospace, and other high-performance applications. Significant tool wear and increased cutting forces are observed during machining operations of aluminium 7075, particularly after heat treatment. Achieving optimal surface finish and dimensional accuracy while ensuring prolonged tool life remains a significant challenge in machining. Surface roughness, cutting temperature, cutting forces, and metal removal rate (MRR) are directly influenced by key machining parameters such as cutting speed, feed rate, and depth of cut. This research article explores these factors by analyzing the machinability of aluminum 7075, a widely used alloy in the aerospace industry. The study aims to identify the ideal machining conditions that enhance surface quality, minimize cutting forces, and extend tool longevity, ultimately improving manufacturing efficiency and overall product performance.

Keywords: Materials, Aluminium, Alloys, Machining, Corrosion

1. INTRODUCTION

Aluminium alloys in the 7000 series are a outstanding class of hard, heat-treatable metals with significant physical properties. They are frequently utilized in situations that require materials with lightweight and robust. The primary additional ingredient in these alloys is zinc, which is also present in significant amounts in copper and magnesium. The alloys in the 7000 series are widely admitted for their impressive strength and resistance to corrosion.

These properties make them exceptionally appropriate for applications that require durability and reliability, such as advanced engineering projects, automotive industry, and aircraft construction. Their ability to withstand harsh conditions while maintaining structural integrity has made them a preferred choice in industries that demand high functioning.

Remarkably aerospace sector requires extensive use of the aluminium alloys from the 7000 series. Al7000 series

---
[1]shrinivasashetty@vvce.ac.in, [2]gvnp@vvce.ac.in, [3]anand@nie.ac.in, [4]hafeezsyed94@gmail.com, [5]syedsufu381@gmail.com

alloys are used in vital parts namely landing gear, fuselage panels, and aeroplane wings because of their excellent fatigue resistance, high strength and lightweight nature. They are perfect to usage in aerospace structures because of their capacity to tolerate high levels of stress without losing structural integrity. Furthermore, the 7000 series alloys' machinability promises accurate production of intricate parts, satisfying the exacting specifications of aerospace applications.

The machinability of aluminium alloys has been assessed numerous times, but the unique difficulties with the high-strength 7000 series alloys have not received enough consideration. Majority of reviews that currently exist either concentrate on particular machining processes, leaving a thorough grasp of various machining parameters unexplored, or they cover aluminium alloys in general terms without review into detail about the special qualities of the 7000 series. Furthermore, there hasn't been much discussion of the effects of new machining methods, environmentally friendly procedures, and the relationship between alloy composition and machinability. The practical implications of machining these alloys in industrial settings where accuracy and surface quality are crucial, like aerospace and automotive applications, are also frequently overlooked in existing reviews. These drawbacks emphasise the necessity of a targeted review that compiles recent research, points out any gaps, and offers useful information about the machinability of alloys from the aluminium 7000 series.

1.1 Composition of Al7000 Series

Zinc serves as the main alloying element in 7000 series aluminium, with its content varying from about 4% to 8%. To boost strength and hardness, manufacturers add magnesium and copper to the mix. They also contain small trace amounts of chromium, manganese and silicon to make the grain structure finer and to improve toughness. Among the alloys in the 7000 series, 7075 aluminium is exceptionally well-known for its extraordinary combination of strength and toughness. (Yamada et.al, 2014). Key characteristics of 7000 series aluminium alloys include are High Strength, Good Fatigue Strength, Heat Treatability, Corrosion Resistance

This high-strength aluminium alloy mechanical qualities make it an essential component for industries requiring robust but lightweight parts. Zinc is the main alloying element, followed by copper and magnesium in trace amounts, while aluminium makes up the majority of this alloy. The significant strength to weight ratio of 7075 aluminium is a result of its distinctive composition and heat treatment process, which also distinguish it from other aluminium alloys (Hu et.al, 2021). Meanwhile, Al7000 series alloy have found numerous uses in high-performance fields where performance in harsh environments is necessary, such as aerospace and defence.

In addition, it can withstand extreme stress in certain situation due to its exceptional machinability, which makes it attractive in a variety of engineering domains. (Hu et.al, 2021).

2. MACHINING OF ALUMINIUM 7000 SERIES ALLOYS

Machining is the process of removing excess material from a workpiece to form and size the final product. In aluminum cast alloys, key alloying elements like magnesium, copper and zinc may form machining difficulties. Rake angles play a important role in this process; tools with shallow rake angles can generally be applied safely, avoiding part damage or tool buildup. However, silicon-based alloying element materials require more positive rake angle tools and are more economically machined at relatively low feeds and speeds (Songmene et.al, 2011).

In contrast, most wrought aluminum alloys exhibit excellent machining performance and are particularly well-suited for multiple machining operations. Achieving free and high-quality Machining aluminum alloys necessitates a comprehensive understanding of machining practices and designs. When machining strain-hardenable alloys, continuous chip formation is common, requiring proper chip control to prevent damage to the finished surface. To minimize surface defects, tools with back rake angles are often used to direct chips away from the workpiece. While these alloys are generally easy to machine, high tool pressure results from increased friction. Maintaining tool sharpness is essential for achieving a smooth surface finish, as these materials tend to be sticky. Additionally, cold working improves machinability, with full-hard temper alloys offering better surface quality matched to their annealed counterparts (Songmene et al., 2011).

Several significant issues are intended to be addressed by the current review of machinability studies of aluminium 7000 series alloys. Present review brings together various research findings to provide a clear understanding of key machining aspects such as cutting forces, surface finish and tool wear. By examining these factors, it helps to optimize machining parameters like feed rate, cutting speed and depth of cut (DOC) for high-strength 7000 series aluminum alloys. The review furthermore reports the unique challenges of machining these alloys, including their hardness ,high strength and thermal conductivity, which influence on the overall performance. Also, it investigates the choice of suitable cutting tool materials and coatings to increase tool life and improve machining effectiveness. Major focus is on enhancing surface quality by identifying methods to achieve the desired finish, reduce machining defects, and maintain structural integrity. Additionally, the review considers industrial applications, particularly in aerospace, automotive, and defense industries, where these alloys are commonly used. By composing existing research, this research review provides constructive comprehensions for both researchers and industry professionals, contributing to advancements in machining techniques for these alloys.

2.1 Machinability Factors

Machinability of 7000 series aluminum alloys depends on factors like cutting speed, feed rate, DOC, tool material, coatings, lubrication, and cooling. Right selection of these parameters decreases tool wear, improves surface finish and enhances efficiency. Achieving the required surface quality and tolerances is important, specifically in aerospace and automotive applications. (Jomaa, 2014).

The following section details the studies on the response of machining parameters:

Surface Roughness (SR): Surface roughness of machined components is affected by key machining parameters, including, DOC, feed rate, and cutting speed. These factors together determine the characteristic of the surface finish (SF), with higher feed rates and depths of cut generally leading to increased roughness, while optimal cutting speeds contribute to smoother surfaces. Deeper cuts can lead to rougher surfaces, whereas higher speeds and feed rates often contribute to a smoother finish. Accurately adjusting these parameters helps to reach the desired surface quality and improves the overall machining performance (Isadare et.al, 2015). Cutting tool performance was measured in terms of surface roughness. The best SF resulted from high cutting speeds and SR values decreased with rising cutting speeds. This was ascribed to the relation between chip curl radius and cutting speed, which influences the machined surface quality (Daud et.al, 2020). SR has a critical influence on hot forming aluminum alloys, as it is governed by machining parameters like cutting speed, feed rate and DOC. The attainment of SF is critical in order to guarantee the functional integrity as well as aesthetic appeal of the finished parts (Milkereit et.al,2018). Surface roughness is a significant parameter in machining processes as it directly impacts the quality and functionality of the final product (Jurczak W & Kyzioł L, 2012). In addition, the current review determined weld SR to gauge weld joint quality, emphasizing the effect of machining parameters like speed and feed rate in controlling SF (Olabode M, 2015). In drilling, surface roughness is one of the critical factors influencing hole quality, with parameters such as cutting speed, feed rate and tool geometry being influential (Pacheco et al., 2019). Besides, studies on abrasive water jet machining of Aluminum Alloy 7024 noted that surface roughness (Ra) is mostly influenced by three independent variables: pressure, feed rate, and standoff distance (Mohammed Abdulrazaq et al., 2019).

Cutting force: Machining parameters like speed, feed rate and DOC have a great influence on cutting force in machining. Higher speeds and depths of cut normally result in higher cutting forces because of higher MRR

and contact stresses. Cutting force and tool life can be suppressed with reduced built-up edge formation by high-speed cutting (Zong C et.al ,2022). Appropriate choice of machining parameters is crucial to managing cutting forces, minimizing tool wear and improving process stability (Daud et.al, 2020). The cutting tool-workpiece material interaction shows an important role in determining the extent of cutting forces. These cutting forces can be efficiently controlled and reduced by the right selection of suitable machining parameters (Milkereit et al., 2018). In addition, cutting forces in different processes like extrusion welding and drilling are directly affected by machining parameters such as speed and DOC, which can impact tool life and overall machining performance (Jurczak W & Kyzioł L, 2012), (Pacheco et al, 2019). Material properties similarly impact cutting forces to a crucial extent. Properties such as hardness and work hardening tendency and heat generated during machining influence the tool force. Cutting force is majorly influenced by temperature and work hardening, according to studies on aluminum alloys Al7075 and 7055. At lower temperatures, work hardening remains comparatively stable regardless of cutting speed, whereas at higher speeds, it initially decreases before increasing, illustrating a complex interface between these factors (Wang P, 2020). This emphasizes the significance of optimizing both cutting speed and temperature for efficient machining. Besides, cutting force control is also important to acquire better surface finishes and longer tool life. Research on Al7075 highlights the need for proper choice of machining parameters to reduce the cutting force without compromising machining efficiency and surface integrity (Karthikeyan L & Ajay C V, 2023). Such findings highlight the significance of crystal clear knowledge regarding machining dynamics to improve productivity and performance in industry.

Cutting speed: Cutting speed is an notable machining parameter that notably affects SR, tool wear and MRR. Research on 7050 aluminum alloy exhibits that SR increases first and subsequently decreases with increased cutting speed, with XRD analysis showing that intensities of (111) diffraction peaks in pre-deformed specimens increase with increasing speed (Zong C et.al ,2022). Increased cutting speeds tend to increase MRR but also may increase tool wear and thermal damage, so a proper speed that optimizes productivity while maintaining material integrity must be chosen (Isadare et.al, 2015), (Milkereit et.al,2018). Experimental results show that a cutting speed of 250 m/min reduces tool wear by 33% while increasing the volume of material removed by 71%, highlighting the importance of choosing the right speed for enhanced efficiency (Daud et.al, 2020). In drilling, cutting speed directly affects surface finish, tool life, and machining performance, further reinforcing its importance in achieving high-quality results (Pacheco et al, 2019). Research suggests that spindle speeds of 1000 to 1838 rpm provide good results in terms of tool wear and MRR, emphasizing the need for optimization of machining conditions for enhanced productivity and superiority of the component (Ravi Kiran, 2021), (Ghan H R & Singh N P, 2021).

Material Removal Rate (MRR): MRR is as an important machining performance indicator quantifying the rate of material removed with respect to time. Cutting speed, feed rate and DOC have significant impact on MRR. As the cutting speed is increased, it increases both the cutting power and temperature, thereby causing thermal softening of material and better MRR (Zong C et al., 2022). Although increased feed rates and cut depths also maximize MRR, a proper balance is necessary to ensure efficient machining and extended tool life (Isadare et al., 2015). Findings in research indicate that a cutting speed of 250 m/min resulted in 71% greater volume of material removed compared to higher speeds, stressing the need to select proper cutting parameters for peak efficiency (Daud et al., 2020). In drilling, fine-tuning cutting parameters is crucial for maximizing MRR while ensuring process stability (Pacheco et al., 2019). The interaction between MRR and material properties is particularly relevant for alloys like AA7075, where variations in MRR can influence hardness, making precise process optimization essential to maintaining desired mechanical characteristics (Jayakumar K & Abdul Rahman P J, 2022). Additionally, MRR affects tool wear, as increased material removal rates generate greater heat and friction, potentially accelerating tool degradation and raising operational costs due to more frequent tool replacements (Jayakumar K & Abdul Rahman P J, 2022). Experiments on 7024 aluminum alloy tested various feed rates and spindle speeds and determined that higher feed rates resulted in greater MRR, while greater spindle speeds enhanced surface roughness with little effect on MRR. The study determined that machining parameter optimization by the Taguchi method can effectively improve MRR without compromising surface quality, providing important information for enhancing manufacturing processes (Ahmed, B A, 2024).

Chip Formation: Chip formation in machining is significantly affected by machining parameters like cutting speed, feed rate and DOC. Increased feed rates consistently cause sticky chips and burr on the machined surface, influencing surface quality (Zong C et.al, 2022). Raising cutting speed tends to inhibit built-up edge formation, leading to smoother chips and better surface finish (Zong C et.al, 2022), (Daud et.al, 2020). Chip curl radius increases with increased cutting speed, affecting chip evacuation and affecting tool wear and SR (Daud et.al,2020). Effective chip evacuation is required to avoid chip recutting and ensure machining stability, thus the determination of appropriate cutting parameters is vital (Isadare et.al, 2015), (Milkereit et.al,2018). Chip formation also has a role to play in related manufacturing processes, such as extrusion welding, where the right control of speed and feed rate is important to regulate material removal and weld quality (Jurczak W & Kyzioł L, 2012). Contrary to machining, welding operations mainly consist of fusion and solidification but not chip formation (Olabode M, 2015). In drilling, chip shape, size, and evacuation efficiency are important factors for tool performance and effective machining (Pacheco et al., 2019). Higher spindle speed tends to produce smaller chips, while higher feed rates produce thicker chips, which both contribute to tool wear and SF (Ghan H R & Singh N P, 2021). Successful chip formation control promotes increased tool life through reduced friction and heat at the tool-workpiece interface, which in turn improves overall efficiency in machining.

Cutting Temperature: Cutting temperature is a key machining parameter because it determines tool wear, workpiece integrity, and process efficiency. Increased cutting speeds and feed rates enhance friction at the tool-workpiece interface, which results in increased temperatures and possible thermal material softening (Zong C et.al ,2022), (Isadare et.al, 2015). Although high-speed cutting minimizes the built-up edge and scales, it increases cutting temperature, accelerate tool wear and influence machining performance (Zong C et.al, 2022), (Daud et.al, 2020). Studies show that with increased speed, rapid rise in temperature causes accelerated tool wear, thus controlling temperature is necessary for sustaining machining efficiency and quality of the part (Daud et al, 2020), (Milkereit et al, 2018). Aside from traditional machining, cutting temperature also affects associated processes like extrusion welding and drilling. For aluminum alloys such as Al7075, the best cutting speeds are usually between 160 and 320 m/min, where cutting force, temperature increase, and machining efficiency are balanced. Increased feed rates lead to higher cutting temperatures because more material is removed per unit time, producing more heat. Whereas DOC has an effect on cutting forces, its effect on temperature is relatively lower, since the majority of heat is transferred through chips, minimizing thermal workpiece stress (Luo H & Wang Y, 2020). Machining of both aged and unaged AA7075 samples, chip temperatures were found to vary between 52 and 92°C, which indicates the fluctuating thermal conditions that affect tool wear and SF (Kaya et al., 2012).

3. CONCLUSION

Machining parameters such as cutting speed, feed rate and DOC play a decisive role in determining surface quality and overall machining efficiency. A balanced mix of these factors yields lower surface roughness. High cutting speeds and feed rates normally result in smoother finishes, and higher depths of cut may generate higher roughness because of high material removal rates and contact stresses. In addition, cutting force increases with speed and DOC as a result of greater plastic deformation and heat generation. Higher cutting speeds also increase material removal rates by inducing thermal softening and increasing the efficiency of the removal process. This review is intended to address gaps in the literature by presenting a comprehensive examination of machinability research on aluminum 7000 series alloys. Unlike previous studies, this work focuses on the unique characteristics of these high-strength alloys, including their hardness, strength, and thermal conductivity, and examines their impact on machining performance. It consolidates findings from both traditional and advanced machining techniques to offer a comprehensive understanding of machinability trends. Also, the review highlights sustainable machining practices such as dry and cryogenic machining, which have received limited attention in earlier research. By bridging the gap between theoretical studies and industrial applications, it provides valuable insights for sectors like aerospace and automotive, where precision and surface integrity are critical. Through this focused approach, the review not only expands the body of knowledge but also presents actionable recommendations for optimizing machining processes and guiding future research.

REFERENCES

1. Ahmed, B. A., Abdullah, M. A., & Ghazi, S. K. (2024). Surface Roughness of Aluminum Alloy 7024 Predicted by Linear Regression and Neural Network Model in Abrasive Water Jet Machining. *University of Technology, Iraq, AL-BAHIR JOURNAL for ENGINEERING*.

2. Altintas, Y. (2012). Manufacturing Automation: Metal Cutting Mechanics, Machine Tool Vibrations, and CNC Design. *Cambridge University Press Journal*.

3. Daud @ Ab Aziz, N. S., Raof, N. A., Ghani, A. R. A., Dahnel, A. N., Mokhtar, S., & Khairussaleh, N. K. M. (2020). Cutting Tool Performance in Turning of AL 7075–T651 Aluminium Alloy. *IIUM Engineering Journal, 21*(2).

4. Dumont, D., Deschamps, A., & Brechet, Y. (2004, March 5). A Model for Predicting Fracture Mode and Toughness in 7000 Aluminium Alloys. *Acta Materialia*.

5. Gheisari, H., & Karamian, E. (2015, May 1). Survey and Study of Machinability for Titanium Alloy Ti-6Al-4V Through Chip Formation in Milling Process. *Journal of Modern Processes in Manufacturing, 4*(2), 5–12.

6. Ghan, H. R., & Singh, N. P. (2021, August 18). Optimization of Milling Parameters of Aluminium Alloy 7075 for Face Milling Using Finite Element Method. *Asian Journal of Convergence in Technology, 7*(2), 74–80.

7. Hu, J., Wang, C., Wang, R., & Li, C. (2021, October 28). 7-Series Aluminium Alloy and Production Method. *Journal of Sichuan Furong Technology Co. Ltd.*

8. Isadare, D. A., Adeoye, M. O., Adetunji, A. R., Oluwasegun, K. M., Rominiyi, A. L., & Akinluwade, K. J. (2015). Effect of As-Cast Cooling on the Microstructure and Mechanical Properties of Age-Hardened 7000 Series Aluminium Alloy. *International Journal of Materials Engineering*.

9. Jegdić, B., Bobić, B., Gligorijević, B., & Mišković-Stanković, V. (2021). Corrosion Properties of an Aluminium Alloy 7000 Series After a New Two-Step Precipitation Hardening. *Zaštita Materijala*.

10. Jayakumar, K., & Abdul Rahman, P. J. (2022, September 1). Effect of End Milling Parameters on MRR and Hardness Variation of AA7075. *Materials, 72*, 2212–2216.

11. Jomaa, W., Songmene, V., & Bocher, P. (2014). Surface Finish and Residual Stresses Induced by Orthogonal Dry Machining of AA7075-T651. *Materials*.

12. Jurczak, W., & Kyzioł, L. (2012). Dynamic Properties of 7000-Series Aluminum Alloys at Large Strain Rates. *Polish Maritime Research Journal*.

13. Kaya, H., Uccedil, M., Cengiz, A., Rek, D. Ouml, Alikan, R. Ccedil., & Erguuml, R. E. (2012, July 19). The Effect of Aging on the Machinability of AA7075 Aluminium Alloy. *Scientific Research and Essays (Academic Journals, 7*(27), 2424–2430.

14. Karthikeyan, L., & Ajay, C. V. (2022, January 1). A Study on Machining Parameter Optimization in Turning of Al7075 Using Taguchi Approach. *National Engineering College Journal (Springer, Singapore)*, 233–242.

15. Lachowicz, M. (2024, March). Metallurgical Aspects of the Corrosion Resistance of 7000 Series Aluminum Alloys. *Metallurgy*.

16. Leo, P., & Cerri, E. (2012, December 5). Pure 7000 Alloys: Microstructure, Heat Treatments and Hot Working. *Aluminium Alloys*.

17. Luo, H., & Wang, Y. (2020, September 1). Effect of Cutting Parameters on Machinability of 7075-T651 Aluminum Alloy in Different Processing Methods. *The International Journal of Advanced Manufacturing*.

18. Milkereit, B., Österreich, M., Schuster, P., Kirov, G., Mukeli, E., & Kessler, O. (2018, July). Dissolution and Precipitation Behaviour for Hot Forming of 7021 and 7075 Aluminium Alloys. *Metals*.

19. Mohammed Abdulrazaq, M., Jaber, A. S., Hammood, A. S., & Abdulameer, A. G. (2019, March 31). Optimization of Machining Parameters for MRR and Surface Roughness for 7024 AL-Alloy in Pocket Milling Process. *Society of Engineering Colleges and Institutes, 26*(1), 10–16.

20. Olabode, M. (2015, December). Weldability of High Strength Aluminium Alloys of 7026 Alloys. *Strength*.

21. Ozben, T., Kilickap, E., & Cakir, O. (2008). Investigation of Mechanical and Machinability Properties of SiC Particle Reinforced Al-MMC. *Journal of Materials Processing Technology, 198*(1-3), 220–225.

22. Pacheco, R. E. R., Pereira, R. B. D., Lauro, C. H., Davim, J. P., & Carou, D. (2019). Enhancing Productivity by Means of High Feed Rate in the Drilling of Al 2011 Aluminium Alloy. *Arabian Journal for Science and Engineering, 44*(9), 8035–8042. https://doi.org/10.1007/s13369-019-04018-y

23. Ratnakar, H. G., & Singh, N. P. (2021, August 18). Optimization of Milling Parameters of Aluminium Alloy 7075 for Face Milling Using Finite Element Method. *Asian Journal of Convergence in Technology, 7*(2), 74–80.

24. Ravi Kiran, D. S. S., Apparao, A. S., Gowri Sankar, V., Faheem, S., Abdul Mateen, S., & Hemanth, V. (2021, June 25). Process Design and Optimization of End Milling Parameters of Al 7075 Metal Matrix Composite. *International Journal of Scientific Research in Science and Technology*.

25. Singh, G., Goyal, S., Sharma, N., & Sharma, P. (2017, March). A Comprehensive Study on Aluminium Alloy Series – A Review. *Recent Advances in Mechanical Engineering, 1*, 11–27.

26. Songmene, V., Khettabi, R., Zaghbani, I., Kouam, J., & Djebara, A. (2011, February). Machining and Machinability of Aluminum Alloys. *École de Technologie Supérieure (ÉTS), Aluminium Alloys*.

27. Wang, P., Zhang, X., Zhang, X. C., & Wang, Y. (2020, December 1). Machinability and Cutting Force Modeling of 7055 Aluminum Alloy with Wide Temperature Range Based on Dry Cutting. *The International Journal of Advanced Manufacturing Technology, 111*(9), 2787–2808.

28. Yamada, S., Ishizawa, G., Itoh, A., Kurumada, M., & Nakai, M. (2014, December). Effects of Environment on Fatigue Crack Growth Behavior of 2000 and 7000 Series Aluminum Alloys. *APCF/SIF-2014 Proceedings*.

29. Yuan, Z., Zhou, Z., Jiang, Z. M., Zhao, Z., Ding, C., & Piao, Z. (2023, January 5). Evaluation of Surface Roughness of Aluminum Alloy in Burnishing Process Based on Chaos Theory. *Chinese Journal of Mechanical Engineering, 36*(1), 1–14.

30. Zong, C., Ni, C., Yu, X., & Liu, D. (2022, August 9). Study on the Surface Integrity of 7050 Aluminum Alloy with Different Crystal Orientations During High-Speed Machining. *Research Square*.

Advances in Mechanical Engineering and Materials Sciences – Dr. Vinay K. B et al. (eds)
© 2026 Taylor & Francis Group, London, ISBN 9-781-041-20970-6

48

A Comprehensive Review of Failure Case Studies in Aluminium Alloys

Chandan V.[1],
G. B. Krishnappa[2]
Department of Mechanical Engineering,
Vidyavardhaka College of Engineering,
Mysuru, Karnataka, India

Abstract: This review provides a comprehensive analysis of failure mechanisms in aluminum alloys across aerospace, automotive, marine, and construction applications. Key mechanisms such as stress corrosion cracking (SCC), galvanic corrosion, fatigue-induced failures, and intergranular corrosion (IGC) are examined, with a focus on their interactions with environmental factors like chloride exposure, moisture, high temperatures, and cyclic loading. The role of alloy composition, particularly in 2xxx, 5xxx, and 7xxx series alloys, and the effectiveness of protective measures in mitigating degradation are discussed. Advanced inspection techniques, including SEM, EDX, and CT, are highlighted for their contributions to identifying failure causes and material defects. Mitigation strategies such as material optimization, surface treatments, stress management, and environmental controls are proposed to enhance the durability and reliability of aluminum components. This review underscores the critical need for integrated approaches to failure prevention and material performance improvement in demanding operational conditions.

Keywords: Failure case studies, Aluminium alloy, Corrosion

1. INTRODUCTION

Aluminium alloys are extensively used in a wide range of industries, including aerospace, automotive, marine, and construction, owing to their impressive strength-to-weight ratio, corrosion resistance, and recyclability. These advantageous properties make them ideal for applications where performance and durability are essential, such as in military aircraft, structural components, and even consumer products (Bernabei, 2016; Bitondo, 2012). However, despite these inherent advantages, aluminium alloys are not immune to degradation under specific conditions. When subjected to various environmental factors such as moisture, temperature fluctuations, and mechanical stress, aluminium alloys can experience several forms of corrosion, including stress corrosion cracking, galvanic corrosion, and fatigue-induced failures (Jha, 2013; Soltani, 2016). These corrosion mechanisms

can lead to significant material degradation, which ultimately results in the failure of critical components. Several case studies have demonstrated how various types of corrosion—such as those seen in aircraft, automotive parts, and structural components—have led to failures that could have been prevented with more comprehensive assessments, proactive maintenance practices, and better material selection (Contreras, 2011; Cornet, 2024). From corrosion in extruded aluminium alloys to stress corrosion cracking in aerospace components, the effects of corrosion are multifaceted and can significantly compromise the integrity of aluminium-based structures.

2. MATERIAL AND ENVIRONMENT

2.1 Alloy Type

The failure analysis of various aluminium alloys highlights the importance of alloy composition in determining their

[1]chandanv@vvce.ac.in, [2]gbk@vvce.ac.in

10.1201/9781003725053-48

susceptibility to degradation. Commonly observed failures occur in 2xxx (Al-Cu), 5xxx (Al-Mg), and 7xxx (Al-Zn-Mg-Cu) series alloys, which possess distinct microstructures due to their alloying elements. For example, 7075 (Al-Zn-Mg-Cu) alloys, widely used in aerospace applications, are prone to stress corrosion cracking (SCC) and corrosion fatigue due to their high strength but susceptibility to chloride-induced degradation (Johari, 1973). On the other hand, 2xxx alloys, such as 2024, are often affected by localized corrosion and pitting, especially in marine or coastal environments (Cornet, 2024). The high copper content in these alloys, while contributing to strength, also increases their vulnerability to intergranular and pitting corrosion (Johari, 1973). 5xxx alloys, often used in marine applications, are more resistant to corrosion but can still face issues in aggressive environments (Vukelic, 2015). Other alloys, such as AA6005A, AA 712.0, and Al 6082-T6, have varied applications ranging from construction to automotive, and their performance is significantly influenced by their microstructural features, including grain size, precipitates, and casting defects (Bitondo, 2012; Contreras, 2011; Zanchini, 2023).

Table 48.1 Summary of aluminium alloys and key properties

Alloy	Composition (Major Elements)	Strength	Corrosion Resistance	Applications
7075	Al-Zn-Mg-Cu	High	Moderate	Aerospace, Automotive
2024	Al-Cu	Moderate	Low	Aerospace
6061	Al-Mg-Si	Moderate	High	Construction, Automotive
5xxx	Al-Mg	Low	Very High	Marine, Automotive

Source: Author's compilation

2.2 Environmental Conditions

Environmental factors play a crucial role in the degradation of aluminium alloys, with failures often occurring in aggressive conditions like marine environments, industrial zones, and high-stress operational conditions. Marine environments, characterized by high chloride concentrations and fluctuating moisture levels, are particularly harsh, accelerating localized corrosion, pitting, and stress corrosion cracking (SCC) in alloys like AA 7075, 2024, and 6061 (Johari, 1973_5; Johari, 1973_6; Cornet, 2024). These alloys, when exposed to saltwater or coastal air, are prone to chloride-induced corrosion, which initiates from pits and cracks (Soltani, 2016). Industrial zones with acidic pollutants or alkaline cooling systems also contribute to corrosion, as seen in the degradation of alloys used in beverage can production or cooling systems, where calcium, boron, and iron contaminants in the corrosion pits indicate the aggressive nature of the

environment (Wayne, 2017). Additionally, environments involving cyclic loading, such as in aerospace, construction, and automotive sectors, exacerbate the material's fatigue and corrosion-fatigue failure, particularly when the alloys are subjected to harsh mechanical stresses combined with corrosive atmospheres (CSB19, 2019; Jha, 2013). The presence of water, fuel, or other contaminants in aerospace components, along with thermal cycling in power plant systems, also increases the material's susceptibility to failure, as seen in brazed aluminium components like BAHXs (CSB19, 2019). Furthermore, factors like improper maintenance and lack of protective coatings in harsh environments, such as coastal or humid climates, can accelerate degradation, as evidenced by the corrosion failures observed in components like aluminium curtain walls and steering wheels in marine applications (Imfuna_Admin, 2022; Vukelic, 2015).

Table 48.2 Environmental factors and their impact on aluminium alloys

Factor	Effect on Material Properties	Examples of Failure Modes
Chloride Exposure	Pitting Corrosion, SCC	Heat Exchangers, Marine Valves
High Temperatures	Reduced Mechanical Strength	Aircraft Engine Components
Cyclic Loading	Fatigue Cracking	Aircraft Wings, Automotive Parts

Source: Author's compilation

3. CORROSION MECHANISMS

Corrosion mechanisms in aluminium alloys, as explored by various authors, reveal complex interactions between environmental factors, material properties, and operational conditions. In many cases, localized corrosion, such as pitting and intergranular corrosion, plays a crucial role in the initiation and propagation of cracks, often exacerbated by environmental factors like chloride ions, moisture, and high temperatures. For instance, Bernabei (Bernabe,2016) highlighted how corrosion pits at stress concentration sites can accelerate fatigue crack propagation, leading to component failure. Similarly, Jha (Jha, 2013) identified stress corrosion cracking (SCC) in AA 7075 valves, where chloride exposure and compromised anodizing layers facilitated crack growth along grain boundaries. In marine environments, as noted by Cornet (Cornet,2024), chloride ions instigate pitting corrosion, which, combined with tensile stress, contributes to SCC and compromises material integrity. Other studies, like those by Bitondo (Bitondo,2012), emphasize the progressive degradation of older materials through intergranular and pitting corrosion, which contrasts with the better performance of newer alloys. In some cases, such as the curtain wall corrosion observed by Imfuna_Admin (Imfuna_Admin,2022), corrosion was driven by failures in protective coatings, leading to the exposure of aluminium to harsh conditions. Similarly, Johari (Johari,1973) observed corrosion in aluminium shock struts and propeller hubs, with localized damage at grain boundaries and attachment holes contributing to crack initiation. The combined effects of chemical attack and stress exacerbate these failures, as seen in the work of Soltani (2016), where pitting corrosion under aerodynamic loads contributed to crack formation. Additionally, Vukelic (Vukelic,2015) described the role of fretting and localized corrosion in steering wheels, and Wayne (Wayne,2017) observed aggressive pitting corrosion in cooling towers due to alkaline water and contaminants. These findings collectively underscore the importance of understanding corrosion mechanisms such as pitting, SCC, and IGC in various applications, including aerospace, marine, and construction, where corrosion significantly influences the long-term durability and safety of critical components.

Table 48.3 Mechanism-component matrix

Corrosion Mechanism	Common Components Affected	Observed Failures
Pitting Corrosion	Heat Exchangers, Valve Bodies	Leakages, Structural Weakening
Stress Corrosion Cracking (SCC)	Aircraft Fittings, Wheels	Crack Propagation, Component Breakage
Intergranular Corrosion (IGC)	Turbojet Compressor Blades	Loss of Material Integrity

Source: Author's compilation

4. FAILURE ANALYSIS

4.1 Inspection Methods

Failure analyses of aluminium and composite components in various applications have employed a combination of advanced inspection techniques, including visual inspection, optical microscopy, scanning electron microscopy (SEM), field emission scanning electron microscopy (FESEM), energy dispersive X-ray (EDX) analysis, and electrochemical tests. For example, the failure analysis conducted by Bernabei (Bernabei,2016) involved visual inspection, optical microscopy, and FESEM to identify fatigue striations, beach marks, and corrosion pits at the crack initiation site. These methods revealed the primary failure was due to mechanical stress and corrosion, particularly in actuator lugs. Similarly, Bitondo (Bitondo,2012) utilized metallographic analysis, electrochemical polarization tests, and surface hardness measurements to assess the corrosion behavior of old and new materials, highlighting significant corrosion in the old material, including pitting and intergranular corrosion. Contreras (Contreras, 2011) applied liquid dye-penetrant testing, optical emission spectroscopy (OES), and SEM, identifying porosity and brittle intermetallic compounds in the alloy, revealing mixed-mode fracture

due to intergranular cracking and transgranular microvoid coalescence. Other inspection methods included dye-penetrant testing (Jha, 2013), scanning electron microscopy (Johari, 1973), and computed tomography (Zanchini, 2023).

Key Observations

The key observations across these studies reveal common failure mechanisms that are closely tied to corrosion, fatigue, and mechanical stress. In the study by Bernabei (Bernabei,2016), fatigue striations and beach marks on fracture surfaces pointed to high-cycle fatigue propagation, with a corrosion pit acting as the initiation site. This was consistent with the presence of mechanical stress-induced failures in areas of actuator lugs with altered geometry. Bitondo (Bitondo,2012) found significant corrosion in older material, particularly pitting and intergranular corrosion, with corrosion depth reaching 400 μm. In contrast, the new material exhibited less corrosion, emphasizing the improved resistance of the latter. Contreras (Contreras,2011) revealed no surface cracks but significant porosity and intermetallic compounds, with brittle intergranular cracking being a key factor in the failure of the material, suggesting overaging. In studies by Cornet (Cornet,2024) and Johari (Johari,1973), grain boundary corrosion and stress corrosion cracking (SCC) were observed, particularly in chloride-rich environments, where cracks initiated at stress concentration points and propagated along grain boundaries. These studies underscore the need for enhanced alloying strategies and surface treatments to mitigate corrosion-related failures.

In some cases, corrosion played a supporting role in fatigue failure. For instance, Jha (Jha,2013) identified numerous corrosion pits near the crack region and the spalling of anodized layers, which allowed corrosion to penetrate beneath the surface, ultimately leading to stress corrosion cracking (SCC). The presence of corrosion products and fatigue striations in Johari (Johari,1973) further confirmed that corrosion and fatigue worked synergistically, with corrosion exacerbating fatigue crack propagation. The failure analysis in Johari (Johari,1973_5) emphasized that although corrosion products were present, the primary cause of crack initiation was mechanical fatigue, especially at machining notches.

In more complex scenarios, material defects like porosity, casting defects, and internal inclusions contributed to failure, as observed by Johari (Johari, 1973) and Perez (Perez, 2025). Johari (Johari,1973) highlighted casting defects, including impurity segregation, as contributing factors to crack initiation, while Johari (Johari, 1973) found that porosity and inclusions on the fracture surface facilitated crack propagation. Perez (Perez, 2025) observed cracks originating from internal voids and fluid trapped during the forging process, which led to the failure of forged aluminium components, emphasizing the importance of controlling internal defects in aluminium alloys.

Studies also highlighted the critical role of environmental conditions in accelerating material degradation. In Wayne (Wayne, 2017), the introduction of cooling tower water to the lubricant system caused severe corrosion in aluminium can surfaces, with calcium, boron, and iron detected in the corrosion pits. This indicated that the corrosion mechanism was influenced by contamination in the lubricant, further confirming that environmental factors can significantly alter the corrosion behavior of aluminium alloys.

Additionally, composite materials exhibited unique failure behaviors. Zanchini (Zanchini, 2023) reported that aluminium wheels failed after approximately 7 million cycles, with cracks initiating at the edges of the spokes, while composite wheels showed superficial damage, with cracks propagating in resin-rich zones. The findings from these studies emphasize the complexity of failure mechanisms in both aluminium and composite materials, underscoring the importance of comprehensive failure analysis to prevent catastrophic failure in critical components.

These collective observations suggest that failure in aluminium and composite materials is multifactorial, with corrosion, mechanical stress, and material defects interacting to cause degradation. Comprehensive inspection methods and deeper understanding of the underlying mechanisms are crucial for improving material performance and preventing future failures across various industrial applications.

Table 48.4 Comparison of analytical techniques

Technique	Resolution	Applicability	Limitations
SEM	High	Surface Defect Analysis	Requires Sample Preparation
EDX	Moderate	Elemental Composition	Limited Spatial Resolution
CT	Low-Moderate	Internal Defect Detection	Expensive Equipment

Source: Author's compilation

Table 48.5 Case study summary

Case Study	Material	Analysis Method Used	Key Finding
Aircraft Lug Failure	7075-T6	SEM, EDX	SCC Initiation at Defects
Heat Exchanger Leak	6061	CT, Optical Microscopy	Pitting and Fatigue Crack Combo

Source: Author's compilation

5. MITIGATION STRATEGIES

The mitigation strategies presented by various authors for preventing failure and degradation of aluminium alloys and composite materials in industrial applications highlight a multifaceted approach to material performance optimization, incorporating material selection, surface treatments, environmental control, and stress management.

5.1 Material Selection

One of the key strategies for preventing failure is the careful selection of materials with appropriate compositions and properties. As recommended by Bitondo (Bitondo, 2012), optimizing alloy compositions, including controlling the levels of magnesium and zinc, can significantly enhance the corrosion resistance and mechanical properties of aluminium alloys over time. Additionally, Perez (Perez, 2025) suggests that improving the extrusion procedure and optimizing chemical compositions, particularly controlling titanium content, can help reduce the occurrence of defects in forged aluminium components. Alloying elements like magnesium and zinc can also enhance resistance to intergranular corrosion and stress corrosion cracking (SCC) (Cornet, 2024). In some cases, transitioning to specific alloy temper conditions, such as the T7352 condition for AA 7075, can further reduce the risk of stress corrosion cracking (Jha, 2013).

5.2 Surface Treatments

Surface treatments, including coatings and heat treatments, play a significant role in preventing corrosion and material degradation. Cornet (Cornet, 2024) and Soltani (Soltani, 2016) emphasize the application of corrosion-resistant coatings, such as anodizing with titanium-cerium solutions, to enhance surface protection against chloride-induced corrosion. Additionally, Jha (Jha, 2013) recommends minimizing the time gap between machining and anodizing to ensure the integrity of the anodized layer and prevent premature degradation. For composite materials, Zanchini (Zanchini, 2023) highlights improvements in the preform manufacturing process and better adhesion between preforms and plies to reduce defects and crack propagation, demonstrating the importance of surface treatment in maintaining material performance.

5.3 Environmental Control

Controlling environmental factors is another critical mitigation strategy. Wayne (Wayne, 2017) addresses the issue of cooling tower water leakage into the lubricant system, which caused severe corrosion on aluminium cans, and recommends replacing the damaged heat exchanger and ensuring that the cooling tower water does not contact the aluminium. This highlights the importance of controlling environmental exposure to aggressive agents like alkaline water. Similarly, the Chemical safety board (CSB, 2016) emphasizes the need to improve temperature control during startup and shutdown procedures to minimize thermal stresses in heat exchangers, thereby reducing the risk of thermal fatigue failures.

5.4 Stress Management

Managing stresses, both operational and residual, is crucial in preventing material failures. Pinaho (Pinaho, 2022) suggests using shot peening, a surface treatment that induces compressive residual stresses, to improve the fatigue resistance of turbojet compressor blades and reduce crack propagation. Additionally, Vukelic (Vukelic, 2015) emphasizes improving the machining process to reduce surface imperfections and fretting marks, which can serve as initiation points for cracks. Implementing strict inspection protocols and ensuring proper operational procedures, as noted by Ren (Ren, 2023), is also essential to detect stress-induced failures early. Regular inspections and adherence to maintenance routines can help mitigate the risks associated with stress concentrations and prevent failure in critical components.

These strategies collectively provide a robust framework for mitigating failure and degradation in aluminium alloys and composite materials, focusing on material selection, surface treatments, environmental control, and stress management. By incorporating these methods, industries can enhance the longevity and reliability of their components, particularly in high-stress and corrosive environments.

Table 48.6 Mitigation strategies by failure mode

Failure Mode	Mitigation Strategy	Benefits
Pitting Corrosion	Anodizing, Protective Coatings	Reduces Surface Reactivity
SCC	Stress Relief, Alloy Modification	Improves Crack Resistance
Fatigue Cracking	Design Optimization	Reduces Stress Concentrations

Source: Author's compilation

6. CONCLUSION

In conclusion, the analysis of the failure mechanisms in aluminium alloys and composite materials highlights the critical role of fatigue, corrosion, material defects, and design flaws in component degradation and failure. The findings underscore the importance of addressing stress concentrations, improving material quality, and enhancing manufacturing processes to mitigate these failures.

Fatigue, often exacerbated by poor surface integrity and operational stresses, can be significantly reduced through solutions like shot peening and better design practices. Corrosion, particularly in aggressive environments, remains a major contributor to failure, emphasizing the need for effective corrosion protection, coatings, and timely maintenance.

Material defects such as casting imperfections and aging processes must be carefully controlled, with regular material replacement and rigorous quality checks to ensure long-term performance. Furthermore, addressing design flaws and operational shortcomings, including stress concentrators and inadequate inspections, is crucial for improving component durability.

Overall, a holistic approach that combines advanced manufacturing techniques, optimized material selection, robust inspection protocols, and effective corrosion management strategies is essential for enhancing the longevity and reliability of critical components in industries such as aerospace, automotive, and construction.

REFERENCES

1. Bernabei, M., Allegrucci, L., & Amura, M. (2016). Fatigue failures of aeronautical items: Trainer aircraft canopy lever reverse, rescue helicopter main rotor blade and fighter-bomber aircraft ground-attack main wheel. In Handbook of materials failure analysis with case studies from the aerospace and automotive industries (pp. 87–116). Butterworth-Heinemann.
2. Bitondo, C., Montuori, M., Castellacci, F., Grillo, M., Monetta, T., & Bellucci, F. (2012). Corrosion behaviour of Extruded AA 6005A: Case History. Proceedings of VI Aluminium Surface Science & Technology – ASST 2012, Italy.
3. Contreras, J. C., Natividad, S. L., & Stafford, S. W. (2011). Failure Analysis Case Study on a Fractured Tailwheel Fork. Journal of failure analysis and prevention, 11, 372–378.
4. Cornet, A. J., Homborg, A. M., Anusuyadevi, P. R., & Mol, J. M. C. (2024). Unravelling corrosion degradation of aged aircraft components protected by chromate-based coatings. Engineering Failure Analysis, 159, 108070.
5. Chemical Safety and Hazard Investigation Board. (2016). Loss of Containment, Fires, and Explosions at Enterprise Products Midstream Gas Plant. In https://www.csb.gov/assets/1/6/final_case_study_-_enterprise.pdf (No. 2016-02-I-MS). Retrieved January 20, 2025, from https://www.csb.gov/assets/1/6/final_case_study_-_enterprise.pdf
6. Imfuna_Admin. (2022, August 17). Imfuna Case Study: Curtain wall Failure - Imfuna. Imfuna. Retrieved January 20, 2025, from https://www.imfuna.com/imfuna-case-study-curtain-wall-failure/
7. Jha, A. K., Manwatkar, S. K., & Narayanan, P. R. (2013). Metallurgical investigation of cracked Al–5.5 Zn–2.5 Mg–1.5 Cu aluminium alloy valve. Case Studies in Engineering Failure Analysis, 1(3), 179–185.
8. Johari, O., Corvin, I., & Staschke, J. (1973). Failure analysis of aluminum alloy components (No. IITRI-B6114-7). NASA.
9. Pérez-González, F. A., Ramírez-Ramírez, J. H., Pineda-Arriaga, K. Y., Gaona-Martínez, M. J., Mejía-Martínez, M. A., Benavides-Treviño, J. R., ... & Colás, R. (2025). Failure analysis of forged aluminium components. Engineering Failure Analysis, 167, 108979.
10. Pinho, J. S. R., Campanelli, L. C., & Reis, D. A. P. (2022). Case study on the failure analysis of turbojet compressor blades. Tecnologia em Metalurgia, Materiais e Mineração, 19, 0–0.
11. Ren, F., & Li, H. (2023). Failure Analysis of Cracking of Cast Aluminum Alloy Manhole Cover. Materials, 16(4), 1561.

12. Niemeyer, W. D. (2017, October 17). Case studies of corrosion failures. McCrone. Retrieved January 20, 2025, from https://www.mccrone.com/mm/case-studies-corrosion-failures/

13. Song, W., Woods, J. L., Davis, R. T., Offutt, J. K., Bellis, E. P., Handler, E. S., ... & Stone, T. W. (2015). Failure analysis and simulation evaluation of an Al 6061 alloy wheel hub. Journal of Failure Analysis and Prevention, 15, 521–533.

14. Stamoulis, K., Panagiotopoulos, D., Pantazopoulos, G., & Papaefthymiou, S. (2016). Failure analysis of an aluminum extrusion aircraft wing component. International Journal of Structural Integrity, 7(6), 748–761.

15. Tajabadi, M. S. (2016). Metallurgical failure analysis of a cracked aluminum 7075 wing internal angle. Case Studies in Engineering Failure Analysis, 7, 9–16.

16. Tjahjohartoto, B. (2020, December). Failure analysis of aluminum alloys casting in four-wheels vehicle rims. In IOP Conference Series: Materials Science and Engineering (Vol. 980, No. 1, p. 012004). IOP Publishing.

17. Vukelic, G. (2015). Failure study of a cracked speed boat steering wheel. Case studies in engineering failure analysis, 4, 76–82.

18. Yadav, A., Gupta, K. K., Ambat, R., & Christensen, M. L. (2021). Statistical analysis of corrosion failures in hearing aid devices from tropical regions. Engineering Failure Analysis, 130, 105758.

19. Zanchini, M., Longhi, D., Mantovani, S., Puglisi, F., & Giacalone, M. (2023). Fatigue and failure analysis of aluminium and composite automotive wheel rims: Experimental and numerical investigation. Engineering Failure Analysis, 146, 107064.

Advances in Mechanical Engineering and Materials Sciences – Dr. Vinay K. B et al. (eds)
© 2026 Taylor & Francis Group, London, ISBN 9-781-041-20970-6

49

Fabrication Methods and Characterization of Hybrid Natural Fiber Composite—A Review

Pradeep Kumar P.[1]

Lecturer,
Department of Mechanical Engineering (MTT), Govt. Polytechnic,
Chamarajanagar, Karnataka

Ranganathaswamy L.[2]

Associate Professor,
BGS Institute of Technology, BG Nagar, Bellur,
Nagamangala, Mandya, Karnataka

Basavaraju M. G.

Senior Grade Lecturer,
Department of Mechanical Engineering (MTT), Govt. Polytechnic,
Chamarajanagar, Karnataka

Abstract: Due to their eco-friendly qualities over synthetic fiber-based composites, the use of natural fibers and composites has expanded. Construction, packaging, textiles, cosmetics, automotive, medical, and home application industries are just a few of the industries now using jute-based composites. Moderate mechanical and tribological qualities, low cost, and considerable availability characterize jute fibers. Compared to jute fiber composites, blended jute-based composites have practically improved mechanical and morphological qualities. Numerous investigators have determined the various processes for producing hybrid composites and examined their mechanical, surface characterization, and tribological characteristics. It is crucial to explain the different jute and ramie hybrid composite compositions using results from different mechanical tests. This paper reviews the process of creating and characterizing reinforced polymer composites with natural fibers infusion.

Keywords: Natural fibers, Composites, Jute fibers, Ramie fibers

1. INTRODUCTION

Natural fiber reinforced polymer composites, or NFPC, have several uses in engineering, including packaging, fabrication, and the automobile industries (Kumar R et al. 2019) (Partha Pratim Das et al. 2020) (Bajpai P K et al. 2012). Natural fibers have a lot of advantages, like good mechanical qualities, low cost, biocompatible, specific stiffness, and noise permeability. With the exception of high strength and structure, natural fibers are generally considered acceptable over synthetic fibers in a variety

of industrial applications (Partha Pratim Das et al. 2020) (Rajesh J J. et al. 2002) (Sharma V et al. 2018).

Thermosetting and thermoplastics are two types of polymers. At present, thermoplastic materials are predominantly utilized as bio fiber matrices. Specifically, polyethylene, and poly vinyl chloride (PVC), polypropylene (PP) thermoplastics are most frequently employed for this purpose. While phenolic, polyester, and epoxy resins are the most widely utilized thermosetting matrices (Malkapuram R et al. 2008). Alkali treatment

Corresponding authors: [1]pradeep.gpc@gmail.com, [2]ranganathaswamyl@bgsit.ac

10.1201/9781003725053-49

affects the tensile properties of jute fibers by eliminating lignin and hemi cellulose. The inter-fibroilar region's ability to realign themselves increases when hemicelluloses are removed, making it less stiff and dense (Jochen Gassan et al. 1999). The tensile properties of natural fiber reinforce polymers (both thermoplastics and thermosets) are mainly influenced by the interfacial adhesion between the matrix and the fibers. Several chemical modifications are employed to improve the interfacial matrix–fiber bonding resulting in the enhancement of tensile properties of the composites (D Athith et al. 2016). The performance of composites is influenced by several factors, including the fiber's orientation, toughness, physical characteristics, interfacial adhesion property, and many more (Hao Ma et al. 2016).

Because they are directly related, mechanical properties have an impact on tribological properties (Rajiv Kumar et al.) by managing the interlayer bond's strength, one can control both the material's chemical and physical characteristics. An essential component of the composite is the measurement of the interfacial bond strength at the fiber-matrix interface. Surface treatments have the ability to change fiber surface energy, which is a sign of fiber-matrix adhesion (Dereje Kebebew Debeli et al.2017).

These hybrid composites have flexural, impact, tensile strengths and the properties of the composites studied in detail have been analyzed using a scanning electron microscope (SEM) (L. Prabhu et al. 2019). The properties and interfacial features of composites are significantly influenced by the processing factors. Appropriate processing methods and parameters must be carefully chosen in order to create the best composite products (Agnivesh Kumar Sinha et al. 2019). This paper aims to provide a concise overview of the mechanical characteristics (tensile, flexural, and impact strengths) of hybrid reinforced composites.

2. FABRICATION OF NATURAL FIBER REINFORCED POLYMER COMPOSITES

To create quasi-unidirectional composites, the fiber samples were treated with alkali (NaOH) at concentrations up to 28 weight percent and then inserted in an epoxy using the filament- winding technique (Jochen Gassan et al. 1999). To minimize the impact of variables other than linear density, ramie yarns with varying linear densities and structures were produced using the filament-winding process utilizing ramie fibers of the same grade (Hao Ma et al.2016). Polylactic acid (PLA), used as a matrix, It has an average length of 38–40 mm, a diameter of 0.25 mm, density of 1.25 g/cm3. Ramie fibers with a 0.035 mm average diameter. Using a fiber opening machine, the dried ramie and PLA fibers were combined and opened together on a ramie-PLA fiber mat, with 30% of ramie fibers to 70% of PLA fibers. Then, a hot-pressing machine for vulcanization was used to shape the prepared ramie

fiber/PLA matrix (Dereje Kebebew Debeli et al.2017). The fibers from tea leaves are utilized in the form of randomly oriented particles. Epoxy resin and hardener are the matrix components utilized to create the composite. A measured amount of fiber is soaked in a 5% NaOH solution for four hours to undergo alkali treatment. Surface roughness is improved by alkaline treatment because it drastically alters the network structure's hydrogen bonding. A compression molding procedure was used to create the hybrid composites. Glass, jute, tea leaf, and glass fibers are arranged alternately in the lamina, which is composed of four layers (L. Prabhu et al. 2019). Epoxy resin and curing agent (poly amido phenol) were used with cross-plied jute fibers and to create the composites, epoxy resin was used to strengthen banana fibers.

According to recommendation, a 10:1 weight ratio was used to combine the low temperature curing epoxy resin with the hardener. Light compression molding was used to create the composite slabs after the traditional hand-layup method (M. Boopalan et al. 2013)

For the purpose of preparing the base segment, the LY556 grade epoxy resin and HY951 grade hardener were combined in a 10:1 ratio. The composites of ramie and silk laminates were made using the manual hand layup process. With a total of four layers of ramie and silk textiles, the five distinct laminate kinds were manufactured. Each laminate had dimensions of 300x300x3mm3. By utilizing the weight-percent approach, the reinforcing layers were built (K. Sadashiva et al. 2023). The NaOH-treated jute fiber was combined with polyester and epoxy resins. The weight % of fiber resin was determined to be 18:82. NaOH solution was used to chemically treat the jute fibers. Methyl ethyl ketone peroxide (MEKPO) was the hardener used for polyester, while triethylene tetramine (TETA) was used for epoxy resin. As per the guidelines set by the American Society for Testing and Materials (ASTM), the fiber resin hardener mixture was poured into the molds assigned to various tests. The prepared composites underwent impact, flexural, tensile, and hardness testing (Ajith Gopinath et al.2014).

Jute fibers with a fiber length of 5-7 mm were used in this study to create fiber-reinforced composites. In present study, polyester and epoxy resins were used. The fiber-resin weight percentage used during the synthesis of the composites was 18:82. The composite is manufactured through a manual layup procedure (K. Aruna Santhi et al.2020). The mechanical and physical properties of epoxy composites reinforced with jute fiber and filled with Al2O3 particulate filler. The traditional hand layup method is used to fabricate composite slabs. The manufacturer recommends mixing the low temperature curing epoxy resin and matching hardener in a 10:1 weight ratio (Priyadarshi Tapas Ranjan Swain A and Sandhyarani Biswas, 2024). The hand lay-up method was used to create unidirectional jute fiber composites. A 5:4 ratio of epoxy resin to hardener was used to create composite specimens with varying fiber loading (8, 10, and 12 weight percent). The fibers had comparable volume fractions of 7%, 8.5%, and 10%, in that order. First, samples were treated in one bath with 1% H_2SO_4, and then in another bath with H_2O_2, Na_2SiO_3, and Na_3PO_4. Both procedures were performed in a bath with a fiber-to-alcohol ratio of 1:20 for 30 minutes at 50 °C. Consequently, it raises surface roughness, which aids in the fibers and matrix interacting more successfully (Ramesh, M et. al 2013). It presents the details of the tests related to thermal and hygroscopic characterization of the prepared polymer composite specimens. Epoxy Resin along with Hardener, Jute and sisal fibers was prepared through the hand lay-up process technique, the required blend of gum and hardener were made by blending them in 10:1 (T. Yu, J. Ren et. al 2009)

3. CHARACTERIZATION OF NATURAL FIBRE REINFORCED POLYMER COMPOSITES

Natural reinforced fiber composites are a desirable substitute for conventional materials due to their many qualities. High particular qualities include modulus, flexibility, impact resistance, and stiffness. Furthermore, they are renewable and biodegradable, and they are readily available in huge quantities. Characterizing composite materials is essential to ensure that they meet the performance requirements for their intended industrial applications. The characterization of composite materials is a crucial aspect of the process of developing and producing products. Tensile Testing: The yield quality, extreme rigidity, malleability, strain-solidifying properties, Young's modulus, and Poisson's ratio are determined using this test approach (Kumar R et al. 2019). It is necessary to look into their mechanical characteristics, such as flexural, tensile, and impact strength. This facilitates comprehension of how composites behave under various loading conditions (Agnivesh Kumar Sinha et. al 2022). Jute fiber-reinforced epoxy composites loaded with Al2O3 were investigated for their mechanical and physical properties. The outcome differentiates how filler affects composite characteristics. They've observed that filler significantly enhances a variety of composite characteristics when used with jute and Al2O3 for fortification and epoxy for the framework. (Priyadarshi Tapas Ranjan Swain À and Sandhyarani Biswas, 2024). The study looked at the mechanical characteristics of polyester composites reinforced with glass, jute, and sisal fibers. The expansion of glass fiber into the composite of jute fiber is demonstrated to have produced the highest elasticity (Ramesh, M et. al 2013).

In this research, they performed a mechanical property analysis of composites reinforced with jute fiber with epoxy and polyester pitch. They discovered that the jute-epoxy had superior mechanical qualities after this investigation. Compared to jute-epoxy, jute-polyester composite requires less preparation time. Compared to jute-polyester composites, jute-epoxy has greater elasticity and flexural characteristics, making it more useful in automobile industries (Ajith Gopinath et al.2014). According to another investigation, the silane treatment improved the interfacial adhesion between the polylactic acid matrix and the ramie fiber, which may have raised the ability to withstand stress of ramie-reinforced polymer combinations (H. Xu, L. Wang, C et. al 2008). The length and weight % of ramie fiber increased the tensile capacity of Ramie fortified Soy protein isolated composites. (Priyadarshi Tapas Ranjan Swain À and Sandhyarani Biswas, 2024) Banana and jute fiber reinforced 50:50 weight ratio hybrid epoxy composites are optimum in terms of mechanical properties such as tensile strength, flexural strength (which increases with the weight gain), and impact strength (M. Boopalan et. al 2013). Both the physical and mechanical attributes of epoxy-based hybrid composites reinforced with ramie fiber and jute were tested. It was discovered through comparing the various test samples that the sample's orientation was the reason for its greater strength compared to the other samples (K. Aruna Santhi et. al 2020). The amount of weight of the ramie fibers and the number of created ramie layers have an impact on the tension and bending strength of ramie blended composites and weaved ramie composites . (Djafar, Z et. al 2020). By increasing the density of chopped jute fiber and increasing its density, the hybrid composite's strength was enhanced. The results showed that the type of woven fabric has a greater effect on the strength of natural fiber hybrid composites (Vinayagam Mohanavel. et. al 2022). This study investigated the effect of different SiC/Al2O3 filler ratios on the mechanical properties of ramie/jute hybrid epoxy composite based on plant fiber. The hybrid composite with 8% SiC and 2% Al2O3 has higher stiffness, bending, hardness, impact and shear resistance among other composites [30]. This study looks into the possible effects of Nano diamond filler particles on the mechanical and morphological properties of epoxy composites blended with jute and ramie fibers.

The results demonstrated that the hardness of epoxy composites was considerably increased by adding 0.3 weight percentage of Nano diamonds, resulting in improvements of roughly 18.56% and 34.38%, for ramie/epoxy and jute/epoxy, the composites' tension strength was around 19.1% and 28.01%, while their flexural stability was approximately 17.7% and 21.12%, respectively (Manickam Ramesh et. al 2021).

Flexural Testing: The purpose of flexural testing is to determine a material's bending or flexural qualities. The material's modulus of versatility in bending, flexural strain, flexural stress and flexural stress– strain reactivity are all evaluated using the three-point bowing flexural test. When ramie fiber was utilized as reinforcement in a polylactic acid (PLA) matrix, research revealed that, in comparison to untreated composites, the flexural strength of ramie/PLA composites increased significantly with surface treatments using silane and alkali. When ramie fiber was utilized as reinforcement in a polylactic acid (PLA) matrix, research revealed that, in comparison to untreated composites, the flexural strength of ramie/PLA composites increased significantly with surface treatments using silane and alkali (T. Yu, J. Ren et. al 2010).

A study on the effects of adding diammonium phosphate (DAP) to ramie reinforced composites made from polymers revealed that doing so increases the flexural strength. The combination of alkali and silane treatment, according to the results, also caused the composites' flexural strength to deteriorate (Dereje Kebebew Debeli et. al 2017). The results of the experiment showed that adding glass fiber to the jute fiber composite produced the most effective tensile strength. Likewise, they discovered that composite samples consisting of a mixture of sisal and jute might have the most flexural durability (Ramesh, M et. al 2013). The hybrid epoxy composites' flexural strength demonstrated the highest value in a sample, and it exhibited a similar pattern to the tensile strength. This resulted from the fiber properties and strong ramie and jute adherence to the polymer matrix (Vinayagam Mohanavel et. al 2022). From this study, it is reasonable to believe that ramie/epoxy composites could replace epoxy polymers in order to increase mechanical and environmental efficiency. Composites exhibiting ramie fiber layers positioned between glass fiber layers demonstrated the best mechanical properties (Lalta Prasad et. al 2023). The percentage of water absorbed by the fibers was reduced following the alkali treatment. With 1% siloxane concentration tensile and flexural strengths of jute thermoset composites are improved with the use of oligomeric siloxane treatment (C. Tezara et. al 2022).

Impact Testing: The impact characteristics of laminates are subjective by various aspects such as the type of reinforcing material, fabric alignment, composite geometry, matrix stacking sequence, and reinforcement matrix bonding. Understanding how natural fiber contained in polymer matrices behaves under various humid environments is crucial. Tasar silk waste (TSW) added to the jute fiber/epoxy composite boosts the interfacial adhesion between the fiber and the matrix and reduces the development of cracks under impact loading. The impressive gains made in the composite's tensile strength, flexural strength, impact strength, and hardness up to a 12 weight percentage addition of waste tasar silk (Lalit Ranakoti et. al 2020). It was discovered that when the amount of ramie fiber in the cotton-ramie fabric reinforced composite was increased, the mechanical parameters such as, flexural strength, flexural modulus, tensile strength and impact strength significantly enhanced, the impact strength of the hybrid composite is higher than that of a single component (R. Giridharan 2019). The hybridization of ramie with synthetic fibers,

like carbon, has the potential to improve the composites' impact strength and hardness as well as their capacity to withstand heat treatment. The stacking sequence played a crucial role in the physical and mechanical characteristics of the composites (Manickam Ramesh et. al.2021) It was found that hybridization of different fiber together can effectively solve the problem of insufficient rigidity of jute fiber reinforced composite material and high ramie of ramie fiber reinforced composite material. The findings demonstrate that the hybrid laminate exhibits superior impact energy absorption and impact resistance when its volume fraction of ramie fiber is 55% (AK Ramasamy et. al 2024].

Micro structural Study: For composites to have high strength and stiffness, length of the fiber and the effectiveness of the link between the matrix and fiber are crucial factors, Thus, in or der to perform a microstructural assessment, SEM pictures of the surface of fractures of Ramie-Glass blended composites are obtained. Generally speaking, the absence of pullout failure may be indicated by the short-fragmented fibers protruding out of the matrix. Thus, by eliminating the fiber impurities, ramie fiber can achieve high adhesion between matrix and fiber. The results of this study indicate that jute fibers may benefit from chemical treatment to improve matrix- fiber adhesion. The obtained results indicate that jute fiber interfacial attachment to the polymer matrix improves as a result of chemical treatment (Hua Wang et. al 2019). This study found that composites reinforced with ramie fibers and treated with diammonium phosphate (DAP) exhibited improved fiber-matrix interface, greater wetting, and higher work of adhesion (Dereje Kebebew Debeli et. al 2017).

4. CONCLUSION

In this review, several fabrication techniques of ramie and jute fiber, along with its surface characterization and mechanical properties, have been discussed. Hardness, tensile strength, inter-laminar strength, flexural strength, and impact strength are among the mechanical attributes that increase as filler content rises. The composite's tensile strength declined as the filler amount increased, also its modulus and hardness increased. The hand lay-up approach should be replaced with injection molding to provide improved mechanical properties for the formed composites. The ramie and jute fiber composite's mechanical and physical properties can be further improved by chemical testing.

REFERENCES

1. A K Ramasamy, S Selvaraj, A Murugan, SK Rathinasamy, Enhancement of mechanical properties of ramie and jute fibres reinforced epoxy hybrid composites: Influencing of SiC and Al2O3,Proceedings of the Institution of Mechanical Engineers, 2024.

2. Agnivesh Kumar Sinha, Sourabh Rungta, Dilip Mishra, Mohit Kumar Sahu, Rakesh Himte, Sanjay Sakharwade and AnandKumbhare, Mechanical Properties of Ramie Polymer Composites: Review, Advances and Applications in Mathematical Sciences, Volume 21, Issue 9, July 2022.

3. Ajith Gopinath, Senthil Kumar. M, Elayaperumal A, Experimental Investigations on Mechanical Properties of Jute Fiber Reinforced Composites with Polyester and Epoxy Resin Matrices, Published by Elsevier Ltd, 1877–7058.

4. Bajpai PK, Singh I and Madaan J, Comparative studies of mechanical and morphological properties of polylactic acid and polypropylene based natural fiber composites. J. Reinf. Plast. Compos. 2012, 1712–1724.

5. C. Tezara, M. Zalinawati, J. P. Siregar, J. Jaafar, M. H. M. Hamdan, A. N. Oumer & K. H. Chuah, effect of stacking sequence, fabric orientation, chemical treatment on mechanical properties of hybrid woven jute- ramie composites, volume 9, pages 273–285,2022.

6. D Athith, MR Sanjay , TG Yashas Gowda , P Madhu, GR Arpitha , B Yogesha and Med Amin Omri, Effect of tungsten carbide on mechanical and tribological properties of jute/sisal/E-glass fabrics reinforced natural rubber/ epoxy composites.

7. Dereje Kebebew Debeli, Zhi Zhang, Fengshuang Jiao &JianshengGuo, Diammonium phosphate-modified ramie fiber reinforced polylactic acid composite and its performances on interfacial, thermal, and mechanical properties, Journal of Natural Fibers, 16:3, 342–356.

8. D jafar, Z, Renreng, I, &Jannah, M. Tensile and Bending Strength Analysis of Ramie Fiber and Woven Ramie Reinforced Epoxy Composite. Journal of Natural Fibers, 1–12, (2020).

9. Hao Ma, Yan Li, YiouShen, Li Xie, Di Wang, Effect of linear density and yarn structure on the mechanical properties of ramie fiber yarn reinforced composites, 1359-835-2016.

10. Hua Wang, Hafeezullah Memon, Elwathig A. M. Hassan, Md. Sohag Miahand Md. Arshad Ali, Effect of Jute Fiber Modification on Mechanical Properties of Jute Fiber Composite, Materials 2019, 12(8), 1226.

11. H. Xu, L. Wang, C. Teng and M. Yu, Biodegradable composites: Ramie fibre reinforced PLLA-PCL composite prepared by in situ polymerization process, Polymer Bulletin 61(5) (2008), 663–670.

12. Jochen Gassan, Andrzej K. Bledzki, Possibilities for improving the mechanical properties of jute/epoxy composites by alkali treatment of fibres, Composites Science and Technology, 59 (1999), 1303–1309.

13. K. Aruna Santhi, C. Srinivas, R. Ajay Kumar, Experimental investigation of mechanical properties of Jute-Ramie fibres reinforced with epoxy hybrid composites, 214–7853/2020.

14. K. P. Ashik, Ramesh S. Sharma, A Review on Mechanical Properties of Natural Fiber Reinforced Hybrid Polymer Composites, Journal of Minerals and Materials Characterization and Engineering, 2015, 3, 420–426.

15. K. Sadashiva, K.M. Purushothama, Physical and Mechanical Properties of Bio Based Natural Hybrid Composites, Journal of Materials and Environmental Science, 2023, Volume 14, Issue 01, Page 131–140.

16. Kumar R, UlHaq MI, Raina A, Industrial applications of natural fibre-reinforced polymer composites–challenges and opportunities. Int. J. Sust. Eng. 2019, 212–220.

17. Lalta Prasad, Pawan Kapri, Raj Vardhan Patel, Anshul Yadav, and Jerzy Winczek, Physical and Mechanical Behavior of Ramie and Glass Fiber Reinforced Epoxy Resin-Based Hybrid Composites, Journal Of N atural Fibers 2023, Vol. 20, No. 2, 223–246.

18. L. Prabhu, V. Krishna raj, S. Sathish, S. Gokul Kumar, N. Karthi, Study of mechanical and morphological properties of jute-tea leaf fiber reinforced hybrid composites: Effect of glass fiber hybridization, 2214–7853/2019.

19. Lalit Ranakoti, Pawan Kumar Rakesh, Physio-mechanical characterization of tasar silk waste/jute fiber hybrid composite, Composites Communications, 2452–2139, 2020.

20. Malkapuram R, Kumar V, Yuvraj S N, Recent development in natural fibre reinforced polypropylene composites. J Reinf Plast Compos 2008, 1169–89.

21. M. Arun Kumar, Senthil Kumaran Selvaraj, Sudhakar Kanniyappan, B. Karthikeyan, Effects of adding nanodiamonds in mechanical properties of jute and ramie fiber reinforced epoxy composites, Polymer Composites, 2024;45:11872–11882.

22. M. Boopalan, M. Niranjanaa, M.J. Umapathy, Study on the mechanical properties and thermal properties of jute and banana fiber reinforced epoxy hybrid composites, science direct Elsevier Ltd 1359–8368.

23. Manickam Ramesh, Lakshmi narasimhan Rajeshkumar, and Devarajan Balaji, Mechanical and Dynamic Properties of Ramie Fiber-Reinforced Composites, Matéria (Rio de Janeiro) 15 (2): 164–171.2021

24. Muneer Ahmed. Mustha, Hom Nath Dhakal, Zhongyi Zhang, the Effect of Various Environmental Conditions on the Impact Damage Behaviour of Natural-Fibre-Reinforced Composites (NFRCs)—A Critical Review, Polymers 2023, 15, 1229.

25. Partha Pratim Das, Vijay Chaudhary , Steve Jose Motha, Fabrication and Characterization of Natural Fibre Reinforced Polymer Composites: A Review, International Conference of Advance Research and Innovation (ICARI-2020).

26. P. Lodha and A. N. Netravali, Characterization of interfacial and mechanical properties of 'green' composites with soy protein isolate and ramie fiber, Journal of Materials Science 37 (2002), 3657–3665.

27. Priyadarshi Tapas Ranjan Swain À and SandhyaraniBiswas, Physical and Mechanical Behaviour of Al2O3 Filled Jute Fiber Reinforced Epoxy Composites, International Journal of Current Engineering and Technology E-ISSN 2277 – 4106, P-ISSN 2347 – 5161,2024.

28. Rajesh J J, Bijwe J, Venkataraman B, Effect of water absorption on erosive wear behaviour of polyamides. J. Mater. Sci. 2002, 5107–5113.

29. Rajiv Kumar, Mir I UlHaq, Sanjay M Sharma, AnkushRaina and AnkushAnand, Effect of water absorption on mechanical and tribological properties of Indian ramie/ epoxy composites, ProcIMechE Part J: J Engineering Tribology 1–9.

30. Ramesh, M., Palanikumar, K. and Reddy, K.H. Mechanical Property Evaluation of Sisal-Jute-Glass Fiber Reinforced Polyester Composites. Composites: Part B, 48, 2013, 1–9.

31. R. Giridharan, Preparation and property evaluation of Glass/Ramie Fibers Reinforced Epoxy Hybrid Composites, Composites engineering, Volume 167, 15 June 2019, Pages 342–345.

32. Sharma V, Kumar R, Vohra K, Fabrication and mechanical characterization of walnut/polyester composites. Int. J. Veh. Struct. Syst. (IJVSS) 2018, 381–383

33. Taranveer Singh, Mukesh Kumar, Dr. Radhey Sham, A Comparative Study Of Mechanical And Physical Properties of Polymer Matrix Composite Reinforced With Bamboo And Jute, International Research Journal of Engineering and Technology (IRJET), Volume: 06 Issue: 07 | July 2019.

34. T. Yu, J. Ren, S. Li, H. Yuan and Y. Li, Effect of fiber surface-treatments on the properties of poly(lactic acid)/ramie composites, Composites Part A: Applied Science and Manufacturing 41(4) (2010), 499–505. doi:10.1016/j.compositesa.2009.

35. Vinayagam Mohanavel, Thandavamoorthy Raja, Anshul Yadav, Manickam Ravichandran& Jerzy Winczek, Evaluation of Mechanical and Thermal Properties of Jute and Ramie Reinforced Epoxy-based Hybrid Composites, JOURNAL OF NATURAL FIBERS 2022, 8022–80.

Advances in Mechanical Engineering and Materials Sciences – Dr. Vinay K. B et al. (eds)
© 2026 Taylor & Francis Group, London, ISBN 9-781-041-20970-6

50

Fabrication Methods of Functionally Graded Materials— A Review

Rakshith Gowda D. S.[1]

Assistant Professor,
Department of Mechanical Engineering,
Vidyavardhaka College of Engineering,
Mysore

Hemanth Kumar K. J.[2]

Assistant Professor,
Department of Mechanical Engineering,
Vidyavardhaka College of Engineering,
Mysore

Jyothi Lakshmi R.[3]

Assistant Professor,
Department of Mechanical Engineering,
Ramiah Institute of Technology,
Bangalore

Abstract: Functionally Graded Materials (FGMs) are the revolutionizing materials that became the promising ones in many applications because of their non-uniform properties and gradient-controlled characteristics. Imbibing the desirable properties through gradient control is still a challenge in FGMs. The final structural and functional properties of FGMs depend on the processing technique utilized and having the optimum process parameters in the fabrication process involved. To explore this issue, the present study focused on the comprehensive review of the fabrication process that could be selected to develop the FGMs and challenges in each technique. The survey revealed that the selection of the processing techniques merely depends on the type of material involved, the gradient required, and the geometry of the component. Powder metallurgy is found to be the prominent option to fabricate the functionally graded composites. Chemical vapor deposition is found to be a cost-effective and reliable method for ceramic-coated FGMs. Laser-assisted techniques such as LENS and SLS are the promising techniques for the fabrication of FGM metal alloys.

Keywords: Functionally graded materials, Laser deposition and microstructure optimization

1. INTRODUCTION

Engineers are always trying to solve engineering challenges, and they are successful in their efforts to create functionally graded material (FGM). Advanced materials known as functionally graded materials (FGMs) gradually change in composition and structure over their volume to maximize qualities like mechanical, thermal, or electrical ones Mahamood, R.M. and Akinlabi E.T, 2017). Japanese researchers first developed the idea of FGM in the early 1980s (Niino, M et.al., 1987). Functionally Graded Materials (FGMs) were created to solve issues with traditional laminate composites that were subjected to extreme temperature changes (B Chen et.al., 2020). Traditional composites' sharp interface resulted in delamination because of discontinuities brought on

[1]rakshithgowda@vvce.ac.in, [2]hemanthkumar.kj@vvce.a.c.in, [3]jyothilakshmi.r@msrit.edu

10.1201/9781003725053-50

by mismatched thermal expansion. This is resolved by FGMs, which use gradient zones in place of abrupt interfaces to create a progressive material transition. This enables a component's attributes to be customized, enhancing functionality and durability in applications that call for specialized mechanical, electrical, or thermal performance. Its non-uniform properties make it suitable for variety of applications (HP Qu et.al., 2010). FGMs are used in the aerospace sector for rocket parts with a strong thermal dissipation capability and the ability to support heavy loads (X Lin et.al., 2006). These FGMs are used by the automotive industry to increase engine cylinder surface wear resistance (RM Mahamood and ET Akinlabi, 2016). FGMs are gaining more focus for their potential applications in biomedical applications, including in the replacement of hip, knee, and dental implants (BE Carroll et.al., 2016). To create an ideal mechanical behaviour and accomplish the desired biocompatibility and osseointegration enhancement, dental implants can be functionally graded.

2. FABRICATION METHODS FOR FGMS

The fabrication of FGMs involves tailoring material composition and structure across a gradient, and several methods are used depending on the application and required properties. Below are some of the primary fabrication techniques:

2.1 Powder Metallurgy and Spark Plasma Sintering

Kumar R and Chandrappa C (2014) concluded that functionally graded Al-SiC composites exhibit improved hardness, impact strength, and fracture toughness compared to conventional materials. Optimal SiC content (up to 10%) enhances densification and mechanical performance without significant porosity or clustering issues. FGMs offer better load-bearing capabilities and fracture resistance due to their graded structure. Powder metallurgy, combined with centrifugal force techniques, enables the fabrication of advanced FGMs with nano-particle reinforcements. These methods offer enhanced mechanical properties and pave the way for innovative applications of FGMs in demanding engineering environments according to Watanabe Y et.al. (2009). The principle of spark plasma sintering is not a new technique; it takes us back to Mesopotamian civilization, where bricks were made by heated clay using the same principle. Thus, it paves the way to use it as a promising technique to fabricate ceramics and FGMs (Naebe M and Shirvan Moghaddam K, 2016). FGMs of biomedical and chemical applications could be tailored to have an intentional gradient using the powder metallurgy technique which has been proven by the study of Akira Kawasaki (1997). Cirakoglu M et.al. (2002) utilizes a combustion synthesis technique to fabricate titanium with varying % of boron

FGMs and found that the fabricated materials exhibit good adhesive strength and interfaces were free from cracks. Zhaohui Zhang et al. (2013) synthesized titanium-based FGMs using spark plasma sintering and acknowledged the good bonding between the interfacing layers using X-ray diffraction.. Al-SiC functionally graded materials (FGMs) produced by powder metallurgy displayed a homogeneous distribution of fragmented and clustered SiC particles (average size ~5 μm) in the aluminium alloy matrix. The porosity content was within 2–3%, which indicates relatively good densification compared to other methods. The effective flexural strength of Al-SiC FGMs increased by approximately 41% as the number of layers rose from two to five. Porosity and particle clustering were reduced as the number of layers increased, leading to improved interfacial bonding and mechanical performance. Under thermal fatigue testing, five-layer FGMs demonstrated superior resistance to cracking and damage compared to two-layer samples. Al-SiC FGMs with more layers showed improved resilience to thermal shock, with no significant damage after 100 cycles except for the two-layer variant (Bhattacharyya M, Nath A and Kapuria S, 2007). Canakci A et.al. (2014) fabricated three different-layered Al-B_4C functionally graded materials using powder metallurgy. B_4C particles were evenly distributed within the Al matrix, except in layers with higher B_4C content, where particle clustering was observed. No cracks were detected in the FGMs, indicating robust composition gradation and strong bonding between layers. Density decreased with higher B4C content and reduced pressing pressure. Porosity Increased with higher B4C particle content but decreased with increased pressing pressure. Hardness Increased both with higher B4C content and increased pressing pressure. SEM analysis showed no chemical reaction at the interface between B4C particles and the Al matrix, maintaining the integrity of both materials. Arenas A et.al. (2014) investigated the anodization process of aluminium matrix composites reinforced with silicon carbide nanoparticles (SiCnp) produced via powder metallurgy. The focus was on understanding how varying the SiCnp content (1%, 5%, and 10% by volume) affects the formation, morphology, and properties of the anodic oxide layer formed using a tartaric-sulfuric acid (TSA) electrolyte. Erdemir F et.al. (2015) focused on the production and characterization of Al2024/SiC functionally graded materials (FGMs) using powder metallurgy and hot-pressing techniques. The previous studies shed light on the effect of varying % of SiC and graded layer on the properties and characteristics of the composites, which include the microstructure, hardness, and bending strength. Such properties are characterized using advanced techniques such as X-ray diffraction and scanning electron microscopy. The interfacial stresses, hardness, and bending strength of Al2024 matrix-based composites are significantly influenced by the SiC reinforcement. Some studies show the improvement in particle dispersion, bending strength, and reduced porosity with 40% SiC and a two-layer configuration in Al-based FGMs. The controlled intermetallic compounds Al_4C_3, $CuAl_2$, and $CuMgAl_2$ enhance the performance of the composite FGMs, which are characterized using EDX. Agata Strojny-Nędza et al. (2016) utilized the hot pressing and spark sintering method to fabricate $Cu-Al_2O_3$ composites and FGMs. The study compared thermal, mechanical, and tribological properties of the fabricated materials using both methods. The analysis was carried out to investigate the microstructure, and no structural defects were observed in the gradient zones. The sintering method helps to obtain the controlled composition gradient in the FGMs and is long-lasting without any structural discontinuities. Christine Pélegris et al. (2014) attempted to improve the thermal conductivity in Al/Al_2O_3 FGMs, which is very much required in automotive engine blocks. A numerical model was developed to predict the thermal conductivity by having a porosity gradient, and the same was validated experimentally. Electro corundum showed superior performance in minimizing porosity and improving the distribution of the reinforcing phase. Hardness increased with Al_2O_3 content and was higher for composites with electro corundum, reaching HV 68.2 for 5% electro corundum (SPS) compared to HV 40 for pure copper. Bending strength decreased with increased Al_2O_3 content due to the brittle nature of ceramics, but composites with electro corundum exhibited higher strength than those with α-Al_2O_3. Thermal conductivity decreased with higher Al_2O_3 fractions due to the low conductivity of the ceramic phase. Composites sintered using SPS exhibited higher thermal conductivity than those processed with HP, achieving values above 300 W/(m·K) for low Al_2O_3 contents. Friction coefficient increased with higher ceramic content but was lower for composites with electro corundum. Wear resistance improved with ceramic addition, with electro corundum-based composites showing better performance. FEM modelling optimized a gradient structure (0%, 1%, 3%, and 5% Al_2O_3) to minimize thermal stress. Microstructural analysis revealed a defect-free gradient with continuous bonding between layers.

2.2 Chemical Vapour Deposition

Wang X et.al. (2014) investigated role of chemical vapour deposition to produce functional gradient ZnO. The paper concluded that the functional gradient ZnO material developed through controlled CVD demonstrates superior optical and structural properties compared to traditional ZnO nanowires. ZnO nanowires exhibited a uniform one-dimensional growth along the C-axis. Flag-like structure obtained due to the changes in gas flow makes the two-dimensional lateral growth in ZnO flag. ZnO FGMs are suitable for optical devices as they possess a high visible to UV emission ratio compared to nanowires. The characterization of nanowires and nanoflags could be done using XRD. The growth direction of nanowires is highly aligned when compared to the nano flags, as it exhibits multiple orientations due to lateral nucleation. The morphology and growth directions are significantly influenced by the gas flow and ratios of reaction source. Dynamic equilibrium during deposition and lateral growth leads to the nano flag structure facilitated by modified substrate conditions. ZnO FGMs are extensively used in optoelectronic devices, light detectors, and sensors. Kawase M. et al. (1999) compared the effectiveness of chemical vapor deposition with chemical vapor infiltration in the preparation of carbon to SiC gradient. The material exhibited superior thermal resistance and durability, overcoming limitations of traditional coatings. A smooth transition from carbon to SiC was achieved, with no sharp interface observed between the layers. The gradient layer provided better adhesion compared to direct SiC coatings. FGMs showed no cracks or delamination after rapid cooling tests (from 1000°C to 0°C), unlike simple SiC-coated composites which exhibited cracks. The preform densification reduced void fractions significantly, creating a strong and durable composite. The produced FGM is suitable for high-temperature structural materials, especially where thermal shock resistance and oxidation resistance are critical, such as aerospace and energy applications. Yin, G et.al. (1999) successfully demonstrated that Ti-TiC-C functional gradient coatings prepared by Plasma Source Ion Implantation (PSII) and Ion Beam Enhanced Deposition (IBED). The study witnessed the improvement in adhesion strength, reduce thermal stress, and enhance hemocompatibility for use in artificial mechanical heart valves. The gradient material achieved an average hardness of 1810 Vickers compared to 389 Vickers for the uncoated substrate, significantly improving wear resistance. The scratch load test indicated superior adhesion of the gradient material compared to coatings made by traditional CVD or physical vapor deposition (PVD) techniques. Lim YM et.al. (2002) focuses on developing functionally graded titanium-hydroxyapatite (Ti/HAP) coatings on Ti–6Al–4V implants using chemical solution deposition (CSD). The coatings aim to combine the mechanical benefits of titanium alloys with the biocompatibility of hydroxyapatite (HAP) for biomedical applications, such as implants. The functionally graded Ti/HAP coatings were successfully fabricated using CSD, demonstrating a strong and dense bond between the substrate and coating, enhanced bioactivity, with significant HAP formation in SBF and tuneable properties by controlling annealing temperature and composition.

2.3 Laser Deposition

Lajevardi SA et.al. (2013) aims to fabricate functionally graded composite coatings by varying pulse deposition parameters, including duty cycle and frequency, to control the distribution of nano alumina particles across the coating's cross-section. A modified Watts bath was used for electroplating with ultrasonic agitation to maintain uniform dispersion of Al_2O_3 nanoparticles. Pulse parameters such as duty cycle and frequency were systematically varied to create coatings with five distinct layers. The advent of additive manufacturing techniques accelerates the development of metallic FGMs. Researchers attempted to fabricate Al alloys, Inconel, and copper alloys based on FGMs using the principle of additive manufacturing and tried to enhance the functional and structural properties. Still poses problems in having the controlled gradient and observing the cracks in the interfacial zones (Lei Yan et.al., 2020). In Al_2O_3-reinforced composite FGMs, particles and grain size also have a considerable effect on the functional properties. The transition from texture to randomly oriented grains in the final layers of the FGMs resulted from the increase in the incorporation of Al_3O_3 and decrease in duty cycle. Grain refinement and a higher proportion of embedded particles led to the gradual increase in hardness of the coating done by laser principles. B. Kieback et al. (2003) demonstrated the German achievement in the development of metal FGMs using melting principles. Transport activities in the molten stage and subsequent consolidation make it simple to create gradients. The selection of the appropriate technique to process the FGMs depends not only on the material involved but also on the gradient required and the geometry of the part. Allahyarzadeh MH et al. (2016) investigated the chemical composition and microstructure of the Ni-W alloy fabricated by pulse electrodeposition. Also studied the wear behavior of the fabricated FGM and showed that there has been a significant improvement in the wear resistance of Ni-W FGM compared to pure nickel. Omid Mehrabi et al. (2023) fabricated 316L stainless steel FGM with Inconel 625 in different ratios at each layer using laser deposition techniques. Width and height of the layers were found to be increased with an increase in laser power. The height stability and the surface roughness of the gradient wall are phenomenally influenced by the laser power. Hardness increased with thickness due to higher tungsten content and grain refinement. Ni–W coatings exhibited ~82% of the corrosion resistance of pure nickel, with no cracks reaching the substrate. Wear Resistance was superior to pure nickel coatings, with the coefficient of friction reducing from 0.74 to 0.49 as sliding distance increased. Higher tungsten content in the surface layer improved hardness and wear resistance by approximately 30% compared to pure nickel. Despite a slightly lower corrosion resistance than pure nickel, the coatings effectively resisted cracking. A combination of fine-grained microstructure and tungsten gradient improved mechanical properties. Valmik Bhavar et.al. (2017) concluded that laser deposition process for FGMs offers unparalleled control over material composition and structural properties, making it a versatile and advanced manufacturing technique. Although FGMs are made suitable for aerospace, defense, and biomedical applications, the cost of fabrication made them worry about the further technological refinement. Joshi A et al. (2012) undertook the manufacturing of FGMs using different techniques, including selective laser sintering, 3D laser cladding, and LENS. The study poses a potential scope for the manufacturing FGMs in bulk. Each method yields FGMs with a different set of desirable properties. LENS yields a nearly net-shaped component with better mechanical properties, whereas SLS yields a component with good toughness, and laser cladding yields controllable clad at each single layer.

2.4 Centrifugal Casting Method

El-Galy et.al. (2019) studied the effects of rotation speed on the liquid flow pattern during the casting process. They found that at low rotation speeds (305 rpm), the liquid forms threads, while at higher speeds (600 rpm), the flow transitions into strips. This change is attributed to increased pressure at higher rotation speeds, which leads to thinner and wider liquid films. he velocity loss rate remains nearly constant (~64.4%) across the tested range of rotation speeds (305–600 rpm), indicating limited influence of speed on this parameter. The distance between liquid threads (thread interval) decreases with increasing rotation speed. At low speeds, liquid drop defects are more likely, making 305 rpm a critical minimum speed for defect-free casting. The formation time of the first thread decreases with increasing rotation speed, with a near-linear relationship observed in the tested range. Higher rotation speeds result in quicker solidification of the liquid film, refining the grain structure in real casting processes. They concluded that the rotation speed significantly influences the flow pattern, thread formation, and interval in horizontal centrifugal casting. Higher speeds improve mold filling uniformity and promote refined grain structures. The mold-filling behaviors are analyzed and simulated through experiments. The experimental results help to optimize the casting parameter for a better output response. Casting defects could be minimized by adjusting the rotation speed and initial temperature of the mold. Sam, M. and Radhika, N. (2018) fabricated functionally graded Cu-10Sn-5Ni reinforced with 10 wt% alumina using centrifugal casting. The method yields FGMs with a better density gradient towards the inner surface, which has been witnessed through microstructural analysis. The study also investigated the wear behavior and hardness of the components and revealed that the hardness is better on the outer surface than on the inner surface. Statistical tools were used to determine the optimal process parameters for better output responses. Muddamsetty L. and Radhik (2016) developed the functionally graded Al alloy reinforced with boron carbide using stir casting and centrifugal casting. The study investigates the tribological properties of the fabricated composites, which demand automobile parts, including brake drums and piston rings. ANNOVA yields the optimum process parameters in the fabrication technique for the wear characteristics. The composite, comprising an aluminium alloy (LM13) reinforced with boron carbide (B4C, 10 wt.%), was fabricated using stir casting followed by centrifugal casting. Uniform distribution of reinforcement was achieved through stir casting, followed by centrifugal casting to create functionally graded properties, concentrating B4C particles at the outer periphery. Solution treatment at 525°C for 5 hours was performed, followed by aging at various temperatures (150°C, 175°C, and 200°C) and times (2, 6, and 10 hours). A pin-on-disc tribometer was used to evaluate wear rates under different conditions of load (10, 20, 30 N), aging temperature, and time. The results were analysed using the Taguchi method and ANOVA. Wear rate increased with load and varied non-linearly with aging time and temperature. Aging time (92.19%) was the most significant factor affecting wear, followed by aging temperature (5.36%) and applied load (1.95%). Optimal wear performance was achieved at low load (10 N), aging temperature of 175°C, and aging time of 6 hours. The LM13/B4C composite was successfully fabricated, and centrifugal casting resulted in a particle-rich outer layer, enhancing wear resistance. Wear rate exhibited a strong dependency on aging time, with 6 hours of aging at 175°C providing optimal wear resistance. Chirita G. et al. (2008) revealed the advantages of centrifugal casting over gravity casting to produce FGMs for automotive applications. Centrifugal casting yields a better gradient in one axis and is proven to be the best method to have good fatigue and mechanical properties, whereas gravity casting offers good rupture strength in the fabricated component and is proven to be an effective method to fabricate the engine piston components. J.W. Gao and C.Y. Wang (2000) addressed the challenges in solidification during centrifugal casting of FGMs. The geometry and gradient in the volume are significantly affected by the solidification rates. Melgarejo ZH et al. (2008) studied the hardness and wear properties of a functionally graded Al alloy reinforced with 2 wt% of Mg fabricated using centrifugal casting. The study also carried forward for Al alloy reinforced with 1, 2, 3, and 4 wt% of boron. It is evident from the study that the hardness of both components increased significantly in the radial direction. There has been improvement in wear resistance, and the same is proven through pin-on-disc tests. Xuedong Lin et al. (2013) systematically investigated the particle segregation and distribution in Al reinforced with silicon and magnesium FGMs fabricated using centrifugal casting. The study evidenced the improvement in the wear resistance and thermal properties in the FGMs. The improved mechanical and tribological properties are attributed to increased boron content and higher dislocation density in the matrix near particle reinforcements. The composite's wear resistance and hardness are strongly influenced by the boron content and radial position. These FGMs are ideal for lightweight, high-wear-resistance

components, particularly in applications requiring axial symmetry, such as engine components, brake drums, and aerospace parts. El-hadad S et.al. (2010) highlighted the effects of centrifugal force on particle distribution, orientation, and resulting wear resistance. Al/Al₃Zr FGMs exhibit tunable wear resistance by controlling centrifugal force, which governs the distribution and alignment of reinforcement particles. Optimal wear resistance occurs when wear tests align with the Al₃Zr platelets' orientation. The FGMs' wear resistance and mechanical properties to make them suitable for applications requiring gradient materials with high surface durability, such as automotive or aerospace components. Wear anisotropy and the presence of a wear-induced Al–Zr supersaturated layer confirm the critical role of particle alignment in enhancing tribological performance.

3. CONCLUSION

Functionally graded materials became the promising materials in many applications because of their non-uniform properties and gradient-controlled characteristics. Aerospace, automotive, and biomedical industries demand FGMs, which include thermal barrier coatings, wear-resistant engine components, and biocompatible due to their controlled properties. Manufacturing such materials is still a challenge, as they pose problems in having the gradient zones throughout their volume. The survey conducted to explore the manufacturing techniques available to manufacture functionally graded composites and challenges in each technique were discussed. Based on the discussions, the following conclusions were drawn.

- The selection of the processing techniques merely depends on the type of material involved, gradient required and the geometry of the component.
- Powder metallurgy is found to be the prominent option to fabricate the functionally graded composites; nevertheless, the desired characteristics depend on the proportion of reinforcement in addition to the fabrication process parameters.
- Chemical vapor deposition is found to be a cost-effective and reliable method for ceramic-coated FGMs.
- Laser assistant techniques such as LENS and SLS are the promising techniques for the fabrication of FGM metal alloys.

Anyhow, further study needs to be done to decide on the optimal technique to fabricate FGMs.

REFERENCES

1. Agata Strojny-Nędza, Katarzyna Pietrzak and Witold Węglewski (2016) The influence of Al2O3 powder morphology on the properties of Cu–Al2O3 composites designed for functionally graded materials (FGM). J Mater Eng Perform (25):3173–3184.

2. Akira Kawasaki and Ryuzo Watanabe (1997). Concept and P/M fabrication of functionally gradient materials. Ceramics International. (23):73–83.

3. Allahyarzadeh MH, Aliofkhazraei M, Rouhaghdam ARS and Torabi nejad V (2016). Functionally graded nickel–tungsten coating: electrodeposition, corrosion and wear behaviour. Can Metall Q. (55):303–311.

4. Arenas A, Rocha LA and Velhinho A (2014). Anodization mechanism on SiC nanoparticle reinforced Al matrix composites produced by power metallurgy. Materials (Basel). (7):8151–8167.

5. B Chen, Y Su, Z Xie, C Tan and J Feng (2020). Development and characterization of 316L/Inconel625 functionally graded material fabricated by laser direct metal deposition. Opt. Laser Technol, 123.

6. B. Kieback, A. Neubrand, H. Riedel (2003). Processing techniques for functionally graded materials. Materials Science and Engineering: A. (362):81–106.

7. BE Carroll, RA Otis, JP Borgonia, J Suh and RP Dilon (2016). Functionally graded material of 304L stainless steel and Inconel 625 fabricated by directed energy deposition: Characterization and thermodynamic modelling. Acta Mater. (108): 46–54.

8. Bhattacharyya M, Nath A and Kapuria S (2007). Synthesis and characterization of Al/SiC and Ni/Al2O3 functionally graded materials. Mater Sci Eng A. (487):524–535.

9. Canakci A, Varol T, Özkaya S and Erdemir F (2014). Microstructure and properties of Al–B4C functionally graded materials produced by powder metallurgy method. Univers J Mater Sci. (2):90–95.

10. Chirita G, Soares D and Silva FS (2008). Advantages of the centrifugal casting technique for the production of structural com ponents with Al–Si alloys. Mater Des. (29):20–27.

11. Christine Pélegris, Nabil Ferguen, Willy Leclerc, Yannick Lorgouillou, Stéphane Hocquet, Olivier Rigo, Mohamed Guessasma, Emmanuel Bellenger, Christian Courtois, Véronique Lardot, Anne Leriche (2014). Thermal conductivity modelling of alumina/Al functionally graded composites. The Canadian journal of chemical engineering. (93): 192–200.

12. Cirakoglu M, Bhaduri S and Bhaduri SB (2002). Combustion synthesis processing of functionally graded materials in the Ti-B binary system. J Alloys Compd. (347):259–265.

13. El-Galy, I.M., Saleh, B.I, Ahmed M.H (2019). Functionally graded materials classifications and development trends from industrial point of view. SN Appl. Sci. (1):1378.

14. El-hadad S, Sato H and Watanabe Y (2010). Wear of Al/Al3Zr functionally graded materials fabricated by centrifugal solid-particle method. J Mater Process Technol. (210):2245–2251.

15. Erdemir F, Canakci A and Varol T (2015) Microstructural charac terization and mechanical properties of functionally graded Al2024/SiC composites prepared by powder metallurgy tech niques. Trans Nonferrous Met Soc. (25):3569–3577.

16. J.W. Gao, C.Y. Wang (2000). Modeling the solidification of functionally graded materials by centrifugal casting,

17. Materials Science and Engineering: A. (292) :207–215.

18. Joshi A, Patnaik A and Gangil B, Kumar S (2012). Laser assisted rapid manufacturing technique for the manufacturing of function ally graded materials. In: Conference on engineering and systems. 1–3.

19. HP Qu, P Li, SQ Zhang, A Li and HM Wang (2010). Microstructure and mechanical property of laser melting deposition (LMD) Ti/TiAl structural gradient material. Mater. Des. (1): 574–582.

20. Kawase M,Tago T, Kurosawa M and Utsumi Hand Hashimoto K (1999). Chemical vapor infiltration and deposition to produce a silicon carbide–carbon functionally gradient material Chem.Eng.Sci.543327–34.

21. Kumar R and Chandrappa C (2014). Synthesis and characterization of Al–SiC functionally graded material composites using powder metallurgy techniques. Int J Innov Res Sci Eng Technol. (3):15464–15471.

22. Lajevardi SA, Shahrabi T and Szpunar JA (2013). Synthesis of functionally graded nano Al2O3–Ni composite coating by pulse electrodeposition. Appl Surf Sci (279):180–188.

23. Lei Yan, Yitao Chen, Frank Liou (2020). Additive manufacturing of functionally graded metallic materials using laser metal deposition. Additive Manufacturing. (31):100901.

24. Lim YM, Park YJ, Yun YH and Hwang KS (2002). Functionally graded Ti/HAP coatings on Ti–6Al–4V obtained by chemical solution deposition. Ceram Int. (28):37–41.

25. Mahamood, R.M. and Akinlabi E.T (2017). Functionally Graded Materials, Topics in Mining, Metallurgy and Materials Engineering. Springer International Publishing, Cham, Switzerland. 978–3319537559.

26. Melgarejo ZH, Suárez OM and Sridharan K (2008). A microstructure and properties of functionally graded Al–Mg–B composites fabricated by centrifugal casting. Composite Part A. (39):1150–1158.

27. Muddamsetty L and Radhika N (2016). Effect of heat treatment on the wear behaviour of functionally graded LM13/B4C composite. Tribol Ind. (38):108–114.

28. Naebe M and Shirvan Moghaddam K (2016). Functionally graded materials: a review of fabrication and properties. Appl Mater Today. (5):223–245.

29. Niino, M., Hirai, T and Watanabe R. (1987). The functionally gradient materials (title in Japanese). J. Jpn. Soc. Compos. Mater. (13):257–264.

30. OmidMehrabi, Seyed Mohammad Hossein Seyedkashi, Mahmoud Moradi (2023). Functionally Graded Additive Manufacturing of Thin-Walled 316L Stainless Steel-Inconel 625 by Direct Laser Metal Deposition Process: Characterization and Evaluation. Metals. (13):1108.

31. RM Mahamood and ET Akinlabi (2016). Laser Additive Manufacturing. Breakthroughs in Research and Practice. (8):154–171.

32. Sam, M. and Radhika, N. (2018). Development of functionally graded Cu–Sn–Ni/Al2O3 composite for bearing applications and investigation of its mechanical and wear behavior. Particulate Science and Technology. 37(2):220–231.

33. Valmik Bhavar, Prakash Kattire, Sandeep Thakare, Sachin patil and RKP Singh (2017) A review on function-ally gradient materials (FGMs) and their applications. IOP Conf Ser Mater Sci Eng. (229): 012021.

34. Wang, X., Chu, X., Zhao, H., Lu, S., Fang, F., Li and J. Wang, X. (2014). Controllable Growth of Functional

Gradient ZnO Material Using Chemical Vapor Deposition. Integrated Ferroelectrics, 151(1):1–6.

35. Watanabe Y, Inaguma O, Sato H and Miura-Fujiwara E (2009). A novel fabrication method for functionally graded materials under centrifugal force: the centrifugal mixed-powder method. Materials (Basel). (2):2510–2525.

36. X Lin, TM Yue, HO Yang and WD Huang (2006). Microstructure and phase evolution in laser rapid forming of a functionally graded Ti-Rene88DT alloy. Acta Mater. (54): 1901 1915.

37. Xuedong Lin, Changming Liu, Haibo Xiao (2013). Fabrication of Al–Si–Mg functionally graded materials

tube reinforced with in situ Si/Mg2Si particles by centrifugal casting. Composites Part B: Engineering. (45): 8–21.

38. Yin, G., Luo, J.M., Zheng, C., Tong, H.H., Huo, Y. and Mu, L. (1999). Preparation of DLC gradient biomaterials by means of plasma source ion implant-ion beam enhanced deposition. Thin Solid Films. (345):67–70.

39. Zhaohui Zhang, Xiangbo Shen, Chao Zhang, Sai Wei, Shukui Lee, Fuchi Wang (2013). A new rapid route to in-situ synthesize TiB–Ti system functionally graded materials using spark plasma sintering method,Materials Science and Engineering: A. (565):326–332.

Advances in Mechanical Engineering and Materials Sciences – Dr. Vinay K. B et al. (eds)
© 2026 Taylor & Francis Group, London, ISBN 9-781-041-20970-6

51 Driving Forces and Barriers to Industry 4.0 Adoption in Manufacturing Sector— A Comprehensive Review

Rajesh Kumbara S. K.*,
Vinay K. B., Bharath P., Shreyas M.
Vidyavardhaka College of Engineering,
Mysuru, India

Abstract: Industry 4.0 depicts an advanced change in manufacturing, upon integrating advanced technologies like Big Data, Artificial Intelligence, Internet of Things (IoT), Cyber-Physical Systems (CPS and Cloud Computing to amplify automation in manufacturing sector, improve operational efficiency, and competitiveness. The thrust for Industry 4.0 is leveraged by several other factors, such as advanced technologies, the importance to cut costs, improved productivity, eye on sustainability, and a strong support from rules and regulations. But successful execution comes with considerable challenges, which includes high upfront costs, a lack of skilled workers, and cybersecurity risks. This paper reviews the key driving forces and barriers affecting the adoption of Industry 4.0 in manufacturing industries systematically, monitoring the role of technological advancements, government policies, economic factors, and environmental considerations, and. To assess how industries are prepared for digital change. Many Industry 4.0 maturity models, like IMPULS Model Acatech's, Industry 4.0 Maturity Index are analysed and emphasizing their suitability across various industrial settings in this paper. Along successful and unsuccessful implementation case studies, offers precious insights into best practices in Industry 4.0. A comparative analysis of Industry 4.0 adoption trends in various regions. This paper also finds the socio-economical factor effects of Industry 4.0, which includes its impact on the raising demand for reskilling of existing workforce, current job markets, and its possible improvement to country's GDP growth via efficient and intelligent manufacturing solutions. The result indicates that a balanced strategy is required in workforce upskilling, government incentives, and investment in robust cybersecurity frameworks which are key to the implementation of Industry 4.0. This study enumerates the current adoption rate by offering important recommendations for industry leaders, researchers and policymakers, helping to shift towards smart and advanced manufacturing.

Keywords: Industry 4.0, Advanced manufacturing, Digital transformation, Key driving force, Inhibitors, Maturity models, Smart manufacturing, Cybersecurity, Economic impact, Workforce reskilling

1. INTRODUCTION

Industry 4.0, is acknowledged as the 4th industrial revolution, which means a major change in manufacturing sector via implementing digital technologies such as IOT, AI and CPS . This approach was first started in Germany to improve the competitiveness in the manufacturing sector through modernizing technologies (A.Gilchrist).

Industry 4.0 empowers the improvement of smart factories where mutually dependent systems provides predictive decision-making, strategic real-time monitoring, and smart automation which results in increasing efficiency of the system and productivity. An important driving force in adopting Industry 4.0 is the adopting latest technologies like AI & machine learning, additive manufacturing, and cloud computing which has got substantially

*Corresponding author: rajeshkumbara.sk@gmail.com

increase in adopting industrial automation and capability of predictive maintenance (Matthias R et al). The implementation of intelligent technologies adds a great benefit to reduce in operational cost, flexibility, and better product customization, hence, the global manufacturing landscape consequently improved (J. Qin et. al 2016). Research showcases the achieves incorporating IoT and AI in manufacturing which optimizes the production processes, decrease downtime, and improves efficiency. Also, policies and government programs played a significant role in speeding up the adoption of Industry 4.0 by bringing financial benefits, standard regulatory frameworks, and skill training programs to support manufacturing industries (L. Monostori 2014).

Despite various benefits from Industry 4.0, the key challenge is the substantial increase in initial investment which is needed for transforming old systems into smart digital technologies [8]. Small and medium-sized enterprises (SMEs) are still struggling to implement Industry 4.0 because of various constraints such as financial barrier, lack of expandable digital solutions suitable for their needs (F. Tao 2019). Another important challenge is lack of skilled labour, lack of adaptable in integrating emerging technologies such as AI, IoT, and data analytics, which hinders the industries to take full advantage of digital transformation. Interconnected digital systems increase threatens to cyberattack, data breaches, and intellectual property theft which is a major concern in cyber security risks (Weller et. al 2021).

2. CONCEPT OF INDUSTRY 4.0

2.1 Definition and Evolution

Industry 4.0 represents the merging of advanced digital technologies into manufacturing processes to increase automation, efficiency, and connectivity. It originated as a German government initiative in 2011 to modernize the manufacturing industry by leveraging emerging technologies. Over the years, Industry 4.0 has evolved through technological advancements such as smart factories, autonomous systems, and interconnected supply chains. The transition from traditional manufacturing (Industry 3.0) to intelligent, CPPS is characterized by increased machine-to-machine (M2M), real-time data analytics, and cloud-based decision-making (P. Ray 2020).

2.2 Key Technologies of Industry 4.0

Many core technologies form the Industry 4.0, which enables digital transformation in manufacturing:

1. *Internet of Things (IoT):* IoT connects machines, sensors, and devices, facilitating real-time data exchange and process optimization.
2. *Artificial Intelligence (AI):* AI-driven analytics enhance predictive maintenance, quality control, and decision-making processes.
3. *Big Data Analytics:* The collection and analysis of vast amounts of data improve operational efficiency, forecasting, and supply chain management.
4. *Cloud Computing:* Cloud platforms enable remote monitoring, scalable storage, and seamless data access across manufacturing networks.
5. *Cyber-Physical Systems (CPS):* CPS integrates computational and physical processes, allowing intelligent automation and self-optimizing production.
6. *Additive Manufacturing (3D Printing):* 3D printing facilitates rapid prototyping, mass customization, and cost-effective production.

2.3 Benefits of Industry 4.0 for Manufacturing Industries

The adoption of Industry 4.0 technologies provides multiple benefits to manufacturing industries, enhancing productivity, sustainability, and innovation:

1. *Increased Efficiency:* Automation and AI-powered decision-making reduce production downtime and streamline operations.
2. *Enhanced Flexibility:* Smart manufacturing systems enable mass customization and quick adaptation to market demands.
3. *Improved Quality Control:* AI and machine learning-driven inspection systems enhance defect detection and minimize rework costs.
4. *Sustainability and Energy Efficiency:* Data-driven optimization reduces waste, energy consumption, and carbon footprint.
5. *Better Supply Chain Management:* Real-time tracking and blockchain integration enhance supply chain transparency and resilience.

3. KEY DRIVERS FOR IMPLEMENTING INDUSTRY 4.0

A. *Technological Drivers:* The quick advancements in digital technologies are a major catalyst for Industry 4.0 adoption. The increasing availability of IoT devices, AI-driven automation, Big Data analytics, and cloud computing solutions has enhanced industrial productivity and decision-making. These technologies improve monitoring, maintenance, and provides optimization, allowing companies to achieve greater efficiency and flexibility in production processes.

B. *Economic Drivers:* Industry 4.0 adoption is largely driven by economic incentives such as cost reduction, increased productivity, and higher profitability. By leveraging artificial intelligence and robotics, AI-based analytics, and robotics help in decreasing waste, maximizing resource utilization, and lowering operational costs. Moreover, manufacturers embrace Industry 4.0 technologies seeing improvement in quality of product and faster time-to-market, which leads to significant benefits in global marketplace (K. Zheng et. al 2018).

C. *Environmental Drivers:* Sustainability and energy efficiency has now become pivotal factors for implementation of Industry 4.0. Smart manufacturing systems optimize energy consumption, minimize waste, and reduce emissions, contributing to environmental sustainability. Many Companies are implementing green technologies as a primary catalyst for energy-efficient robotics, AI-based resource management, and digital twins resulting in achieving sustainability goals and thereby encouraging environmental sustainability (L. D. Xu et. al 2014).

D. *Regulatory & Policy Support:* In promoting Industry 4.0, the frameworks and initiatives developed by government plays a vital role. Many other countries implement regulatory guidelines, financial incentives, and industry-specific standards to support smooth transition towards adopting digital technologies in manufacturing sectors (J. Manyika et al 2015). Worldwide many advanced factors and supportive policies facilitated in moving towards smart manufacturing, implement cybersecurity regulations, and releasing funds for research and development are further propelling global adoption of Industry 4.0.

4. KEY INHIBITORS FOR INDUSTRY 4.0 IMPLEMENTATION

In spite key potential transformation opportunities created by Industry 4.0, its broad implementation is hindered by numerous challenges. These challenges consist of high upfront investment, lack of technically qualified workforce, concerns for cybersecurity threat challenges, and difficulties in facing substantial barriers, all of which are particularly daunting small and medium-size enterprises (SMEs) (Zhou et. al, 2015).

4.1 High Initial Investment: Cost Barriers for SMEs

Although the potential opportunities created by Industry 4.0, it encounters widespread obstacles with digital transformation. The key obstacles are integration of difficulties, integration of intelligent technologies, includes risk of not giving importance to cyber threats, AI-based automation, and cloud-based infrastructure, asking for high initial investment, which are potential barriers for SMEs. Research suggests that approximately 60% of SMEs in emerging economies identifies financial barriers as the most critical obstacle in embracing Industry 4.0 technologies. Upgrading outdated manufacturing systems,

adopting cyber-physical technology, and investing in smart machineries which can contribute to the financial pressure (S. Wang et.al, 2016). To mitigate these financial obstacles, numerous industry organizations and governments have implemented various subsidies, tax benefits, and funding schemes in designing and transforming SMEs to adopt Industry 4.0 (M. Hermann, et.al, 2016).

4.2 Lack of Skilled Workforce: Need for Reskilling and Training

Achieving a successful and effective implementation of Industry 4.0 it demands essential skills such as IoT, AI & ML, and data analytics. However, there is a notable skills gap exists, as conventional manufacturing workers lack in these digital sectors. According to a survey conducted recently among global manufacturing companies revealed that 72% of firms struggling in hiring workers with skills associated with Industry 4.0-related skills (T. Stock, et.al 2016). The necessity for interdisciplinary knowledge, integrating mechanical engineering technologies with data science, software development, cybersecurity which intensifies issues faced in the workforce (S. Karnouskos, 2013). Although many industrial training centers and educational institutions are progressively incorporating specialized courses in smart/intelligent manufacturing, however it is keeping up with very slow pace of the workforce and lagging at a slower rate than swift evolution of Industry 4.0 technologies.

4.3 Cybersecurity Concerns: Data Privacy and Security Issues

The extensive interconnectivity inherent in Industry 4.0 systems increases vulnerability to cyber threats, rendering cybersecurity a critical concern for manufacturers. The combination of cloud computing, IoT and CPSs results in generating huge amounts of sensitive data, can cause reputational damage to smaller companies than larger ones. The frequency of cyber-attacks has been notably increasing, with ransomware and data breaches recognized as critical and dangerous threats (R. Drath, et.al 2014). Numerous SMEs, find themselves that they are having lack of useful technical resources to to create comprehensive robust cybersecurity systems, rendering them exposed as victim of cyber-attacks (M. Frank, et.al 2019). Additionally, issues faced by manufacturers concerns compounded by concerns over intellectual property protection, data ownership which creates additional complexities (A. Bousdekis, 2021).

4.4 Integration Challenges: Legacy System Compatibility

Significant number of manufacturing sectors are still dependent on traditional manufacturing systems that are incompatibility with latest Industry 4.0 key technologies. The challenge of merging smart/intelligent automation,

IoT-based devices, and data analytics with outdated and older equipment's remains a major technical issue. Studies reveals that around 65% of manufacturers face obstacles in attempting to retrofit older equipment with latest digital solutions, which leads to rising in implementation timelines delays and higher operational costs (J. Xu, et.al 2021). To resolve integration obstacles, manufacturers need to embrace edge computing, middleware solutions, and API-driven architectures that facilitates effortless data communication between older and new smarter systems. But the efficacy of these solutions differs depending on the firm's degree of digital maturity (R. Kumar, et.al 2021).

5. DISCUSSION AND COMPARATIVE ANALYSIS

5.1 Trends in Industry 4.0 Adoption Across Different Regions

The deployment of technologies related to Industry 4.0 differs across various countries, influenced by different factors such as the level of economic improvement, policies formed by government, and the industry readiness. In **North America**, there has been a significant effort to integrate cutting edge digital technologies, including Internet of Things (IoT), Artificial Intelligence (AI) and advanced automation, across multiple industrial sectors like healthcare and automotive. In North America Industry 4.0 market is growing markedly and is anticipated to experience USD 64,562.1 million by 2032 (A. Gilchrist, 2016). **Asia-Pacific** is mainly led by China, established itself as dominant and leading force in automation industries. China has outpaced Germany in industrial robot density, now boasting 470 robots per 10,000 workers, a figure that has doubled since 2019. The primary reasons for this trend are escalating labour costs and shrinkage of workforce, encouraging manufacturers to embrace smart technologies (S. Wang, et.al 2016).

Conversely, various **European** territories encounter challenges with the adoption of Industry 4.0 due to financial constraints and the shortage of workforce. While Germany was the first to coin the concept of Industry 4.0, adoption rates vary across different industries within Europe due to diverse regulatory structures and financial constraints, extent of adoption varies across all European countries (M. Hermann, et.al 2016). Regions characterized with inadequate technological infrastructure, like some of the regions around **Latin America and Africa**, experience additional barriers, include low level digital readiness, shortage of investment, and cybersecurity threats (L. Monostori, 2014).

5.2 Case Studies: Successful vs. Failed Implementations

1. *Successful Implementations:* Various industries have successfully adopted Industry 4.0, which leads to better decision-making, minimized downtime, and enhanced efficiency. For example, upon implementing predictive AI-based maintenance framework, Bosch has successfully reduced maintenance costs by 25% and operational efficiency also enhanced. Likewise, Siemens established its smart factories through IoT-driven devices and cloud-based solutions, leading to significant enhancement in agility and production quality (F. Tao, et.al 2019).

2. *Failed Implementations:* Failure of Industry 4.0 initiatives often attributed from hindrance to change, inadequate employee training, and lack of planning. A case study by European automobile manufacturer highlighted that insufficient funding in cybersecurity was a key factor to major operational disruptions was stemming from cyberattacks. Another research found that SMEs frequently struggle with adopting Industry 4.0 because of substantial initial investments and limited scalability (S. Karnouskos, 2013).

5.3 Possible Strategies to Overcome Inhibitors

1. *Investment in Education and Workforce Training:* Enhancing Industry 4.0 skills among workforce plays crucial role in increasing the chances of leveraging benefits of Industry 4.0. Nations such as Germany have introduced vocational training programs to fill the skill gap in smart/intelligent manufacturing concepts (H. Kagermann, 2013).

2. *Government Support and Policy Interventions:* Countries including United States and Japan offers research funding grants and tax benefits to facilitate Industry 4.0 adoption among all SMEs (P. G. Ranky, 2019).

3. *Change Management and Stakeholder Engagement:* The reluctance to change continue to be a major obstacle. Many successful industries utilize well-structured change management strategies to guarantee smooth shift from legacy systems to intelligent manufacturing technologies (Vogel-Heuser, et.al 2021).

4. *Cybersecurity and Data Protection Measures:* By enforcing stringent cybersecurity policies and leveraging blockchain technology for secure transactions effectively mitigate data security challenges.

5.4 Possible Strategies to Overcome Inhibitors

1. *Overview of Different Frameworks:* The maturity models of Industry 4.0 advice organizations in gauging their preparedness for digital transformation and pinpointing strategic measures required for

implementing smart manufacturing practices. Various range of models are available, that includes the 'IMPULS' model, "Acatech's" Industry 4.0 Maturity Index model, and PwC's Digital Operations Self-Assessment (DOSA) Model (M. Mittal, et.al 2018).

2. ***Comparison of Maturity Models:*** Acatech's Industry 4.0 Maturity Index: A comprehensive framework assessing six dimensions: resources, information systems, organizational structure, organization culture, employees, and strategy (Dalenogare, et.al, 2018).

   a) *IMPULS Model:* Developed by the German Engineering Federation (VDMA) created this model, concentrates mainly on increasing the digitalization levels across 6 areas, like interaction with customer, business models, and smart production process.

   b) *PwC Digital Operations Model:* It is a self-assessment tool assess the length of digital maturity in terms of data analytics, automation and connectivity (L. Xu, et.al, 2018).

   c) *Smart Industry Readiness Index (SIRI):* It is globally accepted framework developed by the Singapore Economic Development Board (EDB) to advice manufacturing industries which are in the phase of transforming to Industry 4.0 (D. Trotta, et.al, 2018).

3. ***Strategic Use of Maturity Models:*** Different manufacturers employ these models to detect skill gaps, prioritize the funds, and execute well-structured digital transformation plans. Manufacturing sectors can leverage maturity models by assessing their requirement, companies can evaluate their progress against industry standards & peer companies and distribute resources to facilitate Industry 4.0 adoption (J. Smith, et.al, 2020).

5.5 Possible Strategies to Overcome Inhibitors

1. ***Job Market Disruptions:*** The extensive implementation of industrial automation, industrial robotics, and AI in manufacturing has resulted in concerns regarding job displacement. Traditional low-skilled manufacturing jobs are being replaced by automated systems, leading to a shift in employment trends. While automation eliminates repetitive tasks, it also creates new opportunities for high-skilled workers specializing in AI, robotics, and data science (Jabbour, et.al, 2018).

2. ***Reskilling Needs:*** The shift towards digital manufacturing necessitates reskilling/upskilling of the employees. Industry 4.0 demands expertise in digital twin technologies, IoT integration, and AI-driven analytics. Governments and organizations are investing in different programs to provide workers with the necessary digital technical skills, ensuring employability in an evolving job market (Trotta, et.al, 2018).

3. ***Economic Impact:*** The combination of intelligent production technologies is expected to boost economic growth by increasing productivity and reducing operational costs. Studies indicate that Industry 4.0 contribute majorly to country's GDP growth by enhancing efficiency in supply chains and production processes. Additionally, cost savings through predictive maintenance, energy optimization, and waste reduction further strengthen the economic viability of Industry 4.0 adoption (Patel et.al, 2021).

6. CONCLUSION

The execution of Industry 4.0 has transformed the global manufacturing sector by incorporating new digital technologies such as Cloud Computing and Cyber-Physical Systems. These innovations have significantly enhanced operational efficiency, cost reduction, and sustainable production, marking a paradigm shift in industrial practices. But, the adoption of technologies is shaped by several crucial drivers, including economic advantages, advancement in technology, economic benefits, bringing sustainable environment, and regulatory frameworks. Worldwide, governments are playing a crucial role in deploying Industry 4.0 adoption through financial benefits given to company, research grants, and efforts aimed at standardization. Furthermore, absence of proficient workforce necessitates huge reskilling efforts to fill the gap of knowledge in handling key technologies of Industry 4.0. Threats associated with cybersecurity risks, including industrial espionage and data breaches, remain still as a critical challenge that necessitate the demand for establishment of comprehensive security frameworks.

Maturity models of Industry 4.0, such as the IMPULS Model, Acatech's Industry 4.0 Maturity Index offer systematic maturity models for evaluating organization's capability to undergo digital transformation. By utilizing the above discussed model's advice companies to create focused strategies to identify gaps and implementing Industry 4.0 technologies. By utilizing these models, companies can identify shortcomings in technological investments with long-term business objectives. The economic merits of Industry 4.0 present notable improvements, with increase in productivity, cost efficiency, and potential extension of contribution towards country's GDP growth. Progressing ahead, the successful execution of Industry 4.0 will demand a comprehensive approach that tackles both the key driving catalysts and the challenges bound with digital transformation. Collaboration among governments, academic institutions and industries, is crucial to formulate comprehensive strategies that facilitates Industry 4.0 adoption while addressing its obstacles. The scope of future research should prioritize on encouraging cybersecurity measures, and establishment of workforce training initiatives. By resolving existing obstacles, Industry 4.0 can be employed to create highly efficient manufacturing ecosystem, not only efficient but also highly sustainable, and competitiveness.

REFERENCES

1. Zhou, Keliang, Taigang Liu, and Lifeng Zhou. "Industry 4.0: Towards future industrial opportunities and challenges." *2015 12th International conference on fuzzy systems and knowledge discovery (FSKD)*. IEEE, 2015.
2. A. Gilchrist, Industry 4.0: The Industrial Internet of Things. Apress, 2016.
3. Matthias R. Guertler, David Schneider et,al Analysing Industry 4.0 technology-solution dependencies: a support framework for successful Industry 4.0 adoption in the product generation process.
4. S. Wang, J. Wan, D. Li, and C. Zhang, "Implementing smart factory of Industrie 4.0: An outlook," Int. J. Distrib. Sens. Netw., vol. 12, no. 1, pp. 1–10, 2016.
5. J. Qin, Y. Liu, and R. Grosvenor, "A categorical framework of manufacturing for Industry 4.0 and beyond," Procedia CIRP, vol. 52, pp. 173–178, 2016.
6. M. Hermann, T. Pentek, and B. Otto, "Design principles for Industrie 4.0 scenarios: A literature review," Procedia CIRP, vol. 52, pp. 23–28, 2016.
7. L. Monostori, "Cyber-physical production systems: Roots, expectations and R&D challenges," Procedia CIRP, vol. 17, pp. 9–13, 2014.
8. T. Stock and G. Seliger, "Opportunities of sustainable manufacturing in Industry 4.0," Procedia CIRP, vol. 40, pp. 536–541, 2016.
9. F. Tao, M. Zhang, Y. Liu, and A. Y. C. Nee, "Digital twin driven smart manufacturing," Adv. Manuf., vol. 2, no. 1, pp. 23–29, 2019.
10. Weller and B. Bender, "Digital twins for smart manufacturing: Beyond real-time process monitoring," Comput. Ind., vol. 127, p. 103400, 2021.
11. S. Karnouskos, "Cyber-physical systems in the smart grid," IEEE Trans. Ind. Informat., vol. 9, no. 1, pp. 50–58, 2013.
12. H. Kagermann, W. Wahlster, and J. Helbig, "Recommendations for implementing the strategic initiative INDUSTRIE 4.0," Final Report of the Industrie 4.0 Working Group, 2013.
13. R. Drath and A. Horch, "Industrie 4.0: Hit or Hype? [Industry Forum]," in IEEE Industrial Electronics Magazine, vol. 8, no. 2, pp. 56-58, June 2014, doi: 10.1109/MIE.2014.2312079.
14. P. G. Ranky, "Industry 4.0: The evolutionary progression of smart manufacturing and automation," Smart Manuf. J., vol. 9, no. 1, pp. 45–52, 2019.
15. Vogel-Heuser and T. Bauernhansl, "Challenges and opportunities of Industry 4.0 for SMEs," J. Manuf. Syst., vol. 58, pp. 39–51, 2021.
16. P. Ray, "Data-driven decision-making in Industry 4.0: Big Data and AI-based approaches," J. Manuf. Process., vol. 67, pp. 153–166, 2020.
17. M. Frank, P. Dalenogare, and N. Ayala, "Industry 4.0 technologies: Implementation patterns in manufacturing companies," Int. J. Prod. Econ., vol. 210, pp. 15–26, 2019.
18. A. Bousdekis, "Industry 4.0 and predictive maintenance: Bridging the gap between research and industrial applications," Comput. Ind., vol. 125, p. 103343, 2021.
19. J. Xu, "AI-powered industrial automation and real-time monitoring in Industry 4.0," Expert Syst. Appl., vol. 178, p. 114961, 2021.
20. L. D. Xu, W. He, and S. Li, "Internet of Things in industries," IEEE Transactions on Industrial Informatics, vol. 10, no. 4, pp. 2233–2243, 2014.
21. J. Manyika et al., "Digital America: A tale of the haves and have-mores," McKinsey Global Institute Report, 2015.
22. M. Mittal, P. Romero, T. Kumar, and D. Nezhadali, "A critical review of smart manufacturing & Industry 4.0 maturity models: Implications for SMEs," J. Manuf. Syst., vol. 49, pp. 194–214, 2018.
23. Dalenogare, G. Benitez, N. Ayala, and A. Frank, "The expected contribution of Industry 4.0 technologies for industrial performance," Int. J. Prod. Econ., vol. 204, pp. 383–394, 2018.
24. L. Xu, E. Xu, and L. Li, "Industry 4.0: state of the art and future trends," Int. J. Prod. Res., vol. 56, no. 8, pp. 2941–2962, 2018.
25. D. Trotta and P. Garengo, "Industry 4.0 key research topics: A bibliometric review," 2018 7th International Conference on Industrial Technology and Management (ICITM), Oxford, UK, 2018, pp. 113–117, doi:10.1109/ICITM.2018.8333930. keywords:
26. J. Smith and R. Brown, "Cybersecurity challenges in Industry 4.0: A case study," Comp. Sec., vol. 95, pp. 102034, 2020.
27. S. Patel and M. Kumar, "Interoperability issues in smart manufacturing: A review," Manuf. Lett., vol. 26, pp. 12–20, 2021.

Advances in Mechanical Engineering and Materials Sciences – Dr. Vinay K. B et al. (eds)
© *2026 Taylor & Francis Group, London, ISBN 9-781-041-20970-6*

52

Synergizing Solar Power and Electric Vehicles—A Comprehensive Review of Technological Advancements

**Shrinivasa D., Ganesha M.,
Harsha B., Akshay A., Karthik H. S.**
Department of Mechanical Engineering,
Vidyavardhaka College of Engineering,
Mysuru, India

Abstract: Solar-powered electric vehicles (EVs) are changing the way we travel by using energy from the sun instead of fossil fuels. With new solar cell technology, like perovskite and multi-junction cells, these vehicles can capture more sunlight and generate more power. This reduces the need for charging from the grid and, in some cases, allows the car to run almost entirely on solar energy. Flexible and see-through solar panels can go on different parts of the car, like the roof, hood, and windows. They do not change how the car looks or works. Better batteries, like solid-state and lithium-sulphur, can store more power, last longer, and charge faster. This makes solar cars more useful for daily travel. Some of these cars can also send extra power back to the grid. This helps balance electricity use. Using recycled materials, eco-friendly production, and old batteries in new ways makes these cars even better for the planet. Solar-powered cars cut pollution, lower carbon emissions, and support a cleaner world. They are a big step toward a future with less fossil fuel and more clean energy.

Keywords: Solar-powered, Electric vehicles, Solar cells, Battery technology, Energy storage, Carbon emissions, Sustainability

1. INTRODUCTION

Electric vehicles (EVs) are changing how people move. They are cleaner and better than cars that use petrol or diesel. EVs help the planet by cutting carbon pollution and using clean energy. They do not burn fuel, so they keep the air clean. Instead, they run on batteries for power. This makes them a big step toward a greener future, referred to the works of Nilofar (Asim et al., 2012). The battery is the most important part of an electric car. Most EVs use lithium-ion batteries because they store a lot of power, work well, and last long. Scientists are working to make these batteries better. They want them to last longer, hold more power, and stay cool. They are also finding safer materials for the planet. Recycling old batteries helps

reduce waste and makes EVs even better for the earth, referred to the works of William P. Mulligan et al (2004).

Another smart idea for electric cars is reusing old batteries. And if the battery in an unfinished car is too weak to run on it can be used in other ways. These batteries may store energy from the sun and wind and give power when needed. Reusing batteries reduces waste and improves energy efficiency. It also supports reusing things instead of throwing them away, referred to the works of Ali M. Humadaa et al. (2016). EVs have many benefits but there are still problems. Manufacturing EVs costs a lot because their batteries need rare and expensive materials like lithium, cobalt and nickel. These materials are limited and so new sources or better options may be needed. Most cars

[1]shrinivasashetty@vvce.ac.in, [2]ganesham252003@gmail.com, [3]harshanayak0887@gmail.com, [4]akshaygowda0053@gmail.com, [5]karthikhs256@gmail.com

10.1201/9781003725053-52

with battery problems now use batteries that gradually weaken as they get older, which makes them less powerful and more expensive to drive. Scientists and engineers are working to make batteries cheaper, stronger and easier for the environment (Peter Keil et al. 2016).

In the future, EVs and clean energy will help make the world better. Solar power, better batteries, and smarter recycling will fix many problems. These changes will make EVs cheaper, stronger, and better for the planet. EVs will help by reducing pollution, saving energy, and reusing materials. More people will use EVs, and they will help create a cleaner and greener future for everyone(Ahmad A Pesaran et al 2003).

1.1 Fundamentals

Electric cars (EVs) are changing how people travel. Instead of using petrol or diesel, they run on electricity from rechargeable batteries. This helps reduce pollution and cuts down the use of fossil fuels. The battery is the most important part of an EV. Most EVs use lithium-ion batteries because they hold a lot of power, last long, and let cars go farther on one charge. An EV has many important parts. The electric motor uses battery power to make the car move. Power electronics control how electricity flows between the battery and the motor, helping the car run smoothly. The control system connects everything, so the car works properly. These features make EVs faster, more efficient, and better for the environment than fuel-powered cars (Christoph J. Brabec et al., 2001).

EVs have a special feature called regenerative braking. In normal cars, braking wastes energy as heat. In EVs, this system saves the energy and sends it back to the battery. This helps the car go farther without charging. It also reduces wasted energy, making EVs more efficient than regular cars. EVs are getting better with new technology. One great idea is adding solar panels to cars. These panels turn sunlight into electricity to help charge the battery. This lets cars use the sun's energy and need fewer charging stations. It is very useful in sunny places where there is sunlight all day (William D. Gourley et al., 2020). Battery recycling is another big improvement. Scientists can take important materials like lithium, cobalt, and nickel from old EV batteries. This means less mining, which is better for the environment. Recycling also helps make sure there are enough materials to keep making new batteries. This process reuses resources instead of wasting them, (Serap Gunes et al., 2007).

EVs help keep the air clean. Unlike normal cars, they do not make dirty gases when they run. As more power comes from wind, sun, and water, EVs will be even better for the earth. The more we use clean energy, the less pollution EVs will cause. EVs have many interesting features but there are still problems with them. It costs more than normal cars and there are not enough charging stations everywhere. Batteries also get weaker over time and can be expensive

to replace. But scientists and engineers are finding new solutions like faster charging and better batteries, (Ming Zhang et al., 2024). EVs are being made into the future of travel. They are an emerging technology. They run on less energy, recycle and reduce pollution and they use new technologies to improve driving. A EV's safety record will improve with heavier batteries, regenerative brakes and solar power. As technology gets progressively better EVs will help to make the world cleaner and greener for everyone (Ramesh Babu Kodati et al., 2016).

1.2 Objectives

Electric vehicless(EVs) are making travel cleaner. Unlike normal cars that use petrol or diesel, EVs run on rechargeable batteries. This helps reduce pollution and slows climate change. Because they do not burn fuel, EVs use less non-renewable energy and support cleaner power sources (Fawad Ali Shah et al., 2018). The main purpose of EVs is to help the environment. EVs do not release dirty gases, so the air stays clean. This is very important in big cities where pollution can harm people's health. Breathing bad air can cause lung and heart problems. As more people live in cities, using EVs can help keep the air clean and make people healthier. EVs also aim to use clean energy. Some have solar panels that turn sunlight into power. This means they need less charging from power stations that use fossil fuels. Using the sun's energy helps EVs be better for the environment. It also supports the world's move to clean energy like wind and solar power (Weidong Chen et al., 2015).

Improving batteries is very important for EVs. Scientists are working to make them last longer, hold more energy, and cost less. Better batteries will make more people want to buy EVs. Right now, they are expensive, but research and recycling are helping to make them cheaper. As technology improves, EVs will be more affordable for everyone, referred to the works of M. P. Suryawanshi et al. (2013). EVs help reduce waste by reusing and recycling batteries. Many companies find new ways to use old EV batteries. Even when they are too weak for cars, they can still store energy for homes and businesses. Recycling also helps collect important materials like lithium, cobalt, and nickel. This means less waste and less mining, which is better for the environment (Michael G. Egan et al.2013).

EVs help countries use less oil. Many nations buy oil from others and the price can quickly change. If more people drive electric cars and EV's then countries could use their own energy like solar and wind power. This makes energy more stable and reduces the risk of running out of fuel if there're any shortage (Shalini et al. 2016). EVs were originally designed to reduce emissions clean air, use cleaner energy, improve batteries and recycle materials. These goals help make the world cleaner and healthier. Scientists, companies and governments are working to improve the power of electric vehicles (EVs). As technology gets better more people will use EVs making transportation greener and more sustainable (Catherine Heymans et al., 2014).

2. LITERATURE REVIEW

2.1 Solar Cells

Solar cells turn sunlight into electricity. They are a key part of solar power technology. These cells help make clean energy and are used in many areas, like electric cars. As technology gets better, solar cells are becoming more powerful, durable, and easier to use in different places. Silicon solar cells are the most common type. People use them because they work well, last long, and are good for big energy projects. They can turn 20% to 25% of sunlight into electricity. They keep working for many years. But they cost a lot to make and are not bendable, so they cannot be used everywhere (Mohammed Khalifa Al-Alawi et al. 2021). Scientists are making new solar cells to fix these problems. Organic solar cells are light, flexible, and can work at nearly 20% efficiency. They are useful for things like solar cars and portable chargers. Another new type is perovskite solar cells. They can be more than 25% efficient and cost less to make than silicon cells. But they do not last long and can break easily. Some also have lead, which is bad for the environment. Scientists are trying to make them stronger and safer without using lead (Sigurd Nikolai Bjarghov et al. 2019).

New solar cell designs are making them work better. Multi-junction and tandem solar cells can catch more sunlight by using different layers. Multi-junction cells have layers that take in different types of sunlight, making more energy. Tandem cells mix two or more solar technologies to be more powerful. These new designs may help us get more energy from the sun. Solar cells can be used in electric cars (B. Lebrouhi et al., 2020). It is important to make solar cells in an earth-friendly way. Scientists are working on better ways to make and recycle them. These efforts help reuse materials instead of throwing them away. Solar cells are improving quickly and are helping the world use more clean energy. As they become more efficient and useful, they will help people depend less on fossil fuels. Solar power will play a big role in keeping the planet clean and safe for the future (Xiaoyue Li et al., 2014).

2.2 Solar Panels

The Solar panels are made of many solar cells joined together. They turn sunlight into electricity and are important for clean energy. People use them in many ways, like powering electric cars. As the need for clean energy grows, new technology is making solar panels better, stronger, and more useful. These changes help cut pollution and increase clean energy use. Silicon solar panels are the most common type. People use them because they work well, last long, and are easy to make. They can turn 20% to 25% of sunlight into electricity. They are good for big

energy projects. But they are heavy and costly. They are also not flexible, so they cannot be used on cars or curved surfaces easily(Pei Cheng et al. 2016).

New materials are helping make solar panels better and cheaper. One example is perovskite solar panels. They can be more than 25% efficient, which is better than many silicon panels. They also cost less to make. But they do not last long and can break easily. Some have lead, which is bad for the environment. Scientists are working to fix these problems. Thin-film solar panels are another type. They include CIGS and CdTe panels. These panels are light and flexible. They can be used on cars and portable devices because they can fit different shapes. This makes them useful for new ideas like solar-powered cars (Pankaj Sharma et al. 2019).

New designs are making solar panels work better. Multi-junction and tandem panels use different layers to catch more sunlight and make more electricity. Bifacial panels absorb light from both sides, making them even stronger. These new technologies help solar panels work well in different weather and places. Solar panels can be used in electric cars. To work well, they need to be light, flexible, and efficient. Organic solar panels are a good choice because they are thin and bendable. They can go on car roofs, hoods, and windows. This helps cars make electricity while driving, so they need less charging. Using solar panels helps EVs use less power from the grid and be more eco-friendly (Monsuru Olalekan Ramoni et al, 2013). It is important to make solar panels in a way that is good for the environment. Scientists are finding better ways to recycle them and recover valuable materials like silicon and rare metals. This helps reduce waste and pollution. It also allows materials to be used again instead of being thrown away. Solar panels are a big part of clean energy. As they get better and more common, they will help cut fossil fuel use and make the world cleaner, (Yuanan Hu et al, 2017).

2.3 EV Batteries

EV batteries give power to electric cars. They store energy and keep the car running smooth. The Lithium-ion battery is the most common because it is light, strong and durable. They let EVs move without needing to recharge. As more people use electric vehicles, scientists and companies are working to make batteries cheaper, stronger and kinder for the environment. Scientists are working to improve battery materials. Most lithium-ion batteries use cobalt, but cobalt is costly and rare. To fix this, scientists are making batteries without cobalt and using silicon instead (Chakib Alaoui et al, 2016).

Overheating is a problem for EV battery overheating can cause premature failures. If they get too cold the cells may not work well or be unsafe. To fix these, engineers are making better cooling systems. Some batteries utilize liquid cooling while others have special materials to absorb heat. These cooling methods help keep batteries safe and make them last longer. Recycling old Electric Vehicle Battery is very important. Even after they lose some power they can still be used like the storage of solar energy. It helps reduce waste and makes batteries more useful. Scientists are also finding ways to recycle them and recover important materials like lithium and nickel, old batteries can be reused instead of thrown away (Victoria Gonzalez-Pedro et al, 2014).

A new advancement in EV battery technology is solid-state energy storage. Regular lithium-ion batteries use liquid inside which can catch fire if damaged. Solid-state batteries use solid materials instead of solid materials to make them safer and more efficient. They also hold more electricity and charge quicker. Scientists are still working on it and they could make EVs better and more reliable in the future (Jonathan D. Servaites et al., 2011). EV batteries are becoming cheaper, safer and stronger as technology improves. Scientists are working on better materials cooling systems recycling and new types of batteries such as solid-state batteries. These improvements will make EVs work better and be more sustainable. With more research, electric vehicle batteries will help the world switch to cleaner and greener transportation with a higher future energy footprint (Di Zhou et al, 2018).

3. ADVANTAGES

Solar-powered electric vehicles produce electricity using solar panels to power them. They mix solar energy with electric power, making travel greener and cheaper. Solar panels on roofs hoods and other parts of homes catch solar light and produce power. EV's need less charging from the grid which makes them more useful (Ona Egbue et al. 2014). The solar-powered EVs can now make more power during driving. Another advantage is that they can run as long as four times the energy needed for other purposes This helps the car go farther without needing to stop for a charge. In sunny places solar panels can give more energy making these cars even more useful. Charging while driving also reduces the worry of running out of battery before finding a charger. This makes EVs easier to use even in areas with few charging stations(Ashraful Ghani Bhuiyan et al. 2012).

Helping to improve the efficiency of solar-powered electric vehicles, technology has helped develop a solar-based car. New materials like organic solar panels are thin, light and flexible. It makes it easy to put them on cars without slowing them down. These panels are also getting better at turning sunlight into electricity. Solar-powered EVs help reduce waste and use resources wisely. Old EV batteries that are too weak for cars can still store energy. This makes them last longer and creates less waste. Reusing batteries helps both the solar and EV industries be more eco-friendly, (ChengJian Xu et al, 2020). As more solar charging stations appear, solar-powered EVs will be even

more helpful. These cars are a big step toward cleaner and more independent travel. Using solar energy with EVs helps cut pollution, save money, and use renewable power better. This makes transportation smarter, cleaner, and better for the Earth (S. Senthilgavaskar et al, 2019).

4. FUTURE SCOPE

The future of solar-powered cars looks good. New technology in solar panels, batteries, and materials is making them better and more useful. These improvements will help solar cars become a big part of clean travel. New types of solar cells, like perovskite and multi-junction cells, can make solar cars work better. Perovskite cells are cheap, easy to make, and work well. Multi-junction cells have many layers that catch more sunlight and make more power (Wei Wang et al. 2014). Solar panels will be on car roofs, hoods, and windows. Flexible and clear panels will fit without changing the car's shape. This will help solar cars look good and collect more sunlight. Batteries are also improving. Solid-state and lithium-sulfur batteries will be the future. They can store more energy, last longer, and charge faster. With these new batteries, solar cars will go farther and work better (Mohammed Hussein Saleh Mohammed Haram et al, 2021).

Solar cars will also help with energy use. With two-way charging, they can send extra power to homes or the grid. This will lower electricity costs and make energy more stable. Making EVs greener will help reduce battery and energy use in farming. To make them better for nature, less harmful materials will be used. Old batteries and solar panels can be reused instead of thrown away, cutting waste and recycling materials, referred to the works of Martin A. Green et al. (2014). Solar cars will help fight climate change by using solar power. They will make travel cleaner, reduce air pollution, and use less fuel. Solar cars are a better, greener choice than gas cars. With new technology, they will become more reliable and common. They will be a big part of future cities where transport and clean energy work together. These cars will help make the world cleaner and more efficient (Vishal Shrotriya et al, 2003).

5. CONCLUSION

Solar cars can change the future of clean travel. With better solar panels, stronger batteries, and more recycling, they can cut pollution and save resources. But making them common is not easy. It needs new technology, government help, and smart planning to solve problems and create a greener future. Old EV batteries are a big challenge for the environment. Instead of throwing them away, they can be reused for storing energy in homes and power grids. This helps reduce waste and makes better use of resources. Giving old batteries a second life can protect the planet and save materials. A big step in making solar cars better is improving solar panels. New panels, like perovskite and multi-junction, can catch more sunlight in small spaces. Flexible and clear panels fit without changing the car's shape. These upgrades help cars use more solar power, need less charging, and create a cleaner future. Laws and research help solar cars grow. Recycling and teamwork matter. Better batteries will make solar cars more useful. Solid-state and lithium-sulphur batteries store more power, charge faster, and last longer. Using fewer rare materials helps the environment. Improved cooling and longer battery life make cars work better. The future of solar cars depends on both new technology and caring for the environment. Using solar power for cars is more than just an invention—it's a step toward a cleaner world. As the fight against climate change grows, solar cars offer a smart and green way to improve transportation. With more research, teamwork, and better ideas, solar cars can change how we travel and help create a cleaner future for future generations.

REFERENCES

1. Ahmad A. Pesaran and Matthew Keyser, "Thermal Characteristics of Selected EV And HEV Batteries", Journal of Power Sources, 0378-7753, 2003.
2. Ali M. Humadaa, Mojgan Hojabri, Saad Mekhilef, Hussein M. Hamada, "Solar Cell Parameters Extraction Based on Single and Double-Diode Models: A Review", Renewable and Sustainable Energy Reviews, 1364–0321, 2016.
3. Ashraful Ghani Bhuiyan, Kenichi Sugita, Akihiro Hashimoto, and Akio Yamamoto, "Ingan Solar Cells: Present State of The Art and Important Challenges", IEEE Journal of Photovoltaics, 2156–3381, 2012.
4. B. Lebrouhi, Y. Khattari, B. Lamrani, M. Maaroufi, Y. Zeraouli, T. Kousksou, "Key Challenges for A Large-Scale Development of Battery Electric Vehicles: A Comprehensive Review", Journal of Energy Storage, 2352–152X, 2020.
5. Catherine Heymans, Sean B. Walker, Steven B. Young, Michael Fowler, "Economic Analysis of Second Use Electric Vehicle Batteries for Residential Energy Storage and Load-Levelling", Energy Policy, 0301–4215, 2014.
6. Chakib Alaoui, "Solid State Thermal Management for Lithium-Ion EV Batteries", Journal of Power Sources, 0378–7753, 2016.
7. ChengJian Xu & Wenxuan Zhang & Guangming Li & Haochen Zhu, "Generation and Management of Waste Electric Vehicle Batteries in China", Environmental Science and Pollution Research, 0944–1344, 2020.
8. Christoph J. Brabec, N. Serdar Sariciftci, and Jan C. Hummelen, "Plastic Solar Cells", Advanced Functional Materials, 1616–3028, 2001.
9. Di Zhou, Tiantian Zhou, Yu Tian, Xiaolong Zhu, and Yafang Tu, "Perovskite-Based Solar Cells: Materials, Methods, And Future Perspectives", Journals of Nanomaterials, 1687–4110, 2018.
10. Fawad Ali Shah, Shehzar Shahzad Sheikh, Umer Iftikhar Mir, "Battery Health Monitoring for Commercialized Electric Vehicle Batteries: Lithium-Ion", Renewable and Sustainable Energy Reviews, 1364–0321, 2018.
11. Jonathan D. Servaites, Mark A. Ratner and Tobin J. Marks, "Organic Solar Cells: A New Look at Traditional Models", Energy & Environmental Science, 1754–5692, 2011.
12. M. P. Suryawanshi, G. L. Agawane, S. M. Bhosale, S. W. Shin, P. S. Patil, J. H. Kim and A. V. Moholkar, "CZTS Based Thin Film Solar Cells: A Status Review", Materials Technology, 1753–5557, 2013.
13. Martin A. Green, Anita Ho-Baillie and Henry J. Snaith, "The Emergence of Perovskite Solar Cells", Nature Photonics, 1749–4885, 2014.
14. Michael G. Egan, Dara L. O'Sullivan, John G. Hayes, Michael J. Willers, and Christopher P. Henze, "Power-Factor-Corrected Single-Stage Inductive Charger for Electric Vehicle Batteries", IEEE Transactions on Industrial Electronics, 0278–0046, 2013.
15. Ming Zhang, Zichun Zhou, Wenkai Zhong, Lei Zhu, Xiaonan Xue, Hao Jing, Yongming Zhang, Feng Liu, Tianyu Hao, Shengjie Xu, Rui Zeng and Jiaxing Zhuang, "Progress of Organic Photovoltaics Towards 20% Efficiency", Nature Reviews Electrical Engineering, 2948–1201, 2024.
16. Mohammed Hussein Saleh Mohammed Haram, Jia Woon Lee, Gobbi Ramasamy, Eng Ngu, Siva Priya Thiagarajah, Yuen How Lee, "Feasibility of Utilising Second Life EV Batteries: Applications, Lifespan, Economics, Environmental Impact, Assessment, And Challenges", Alexandria Engineering Journal, 1110–0168, 2021.
17. Mohammed Khalifa Al-Alawi, James Cugley, "Techno-Economic Feasibility of Retired Electric-Vehicle Batteries Repurpose/ Reuse in Second-Life Applications: A Systematic Review", Journal of Energy Storage, 2352–152X, 2021.
18. Monsuru Olalekan Ramoni and Hong-Chao Zhang, "End-Of-Life (EOL) Issues and Options for Electric Vehicle Batteries", Clean Technologies and Environmental Policy, 1618–954X, 2013.
19. Nilofar Asim, Kamaruzzaman Sopian, Shideh Ahmadi, Kasra Saeedfar, Omidreza Saadatian, Saleem H. Zaidi, "A Review on The Role of Materials Science in Solar Cells", Renewable and Sustainable Energy Reviews, 1364–0321, 2012.
20. Ona Egbue and Suzanna Long, "Critical Issues in The Supply Chain of Lithium for Electric Vehicle Batteries", Engineering Management Journal, 1042–9247, 2014.
21. Pankaj Sharma, Rani Chinnappa Naidu, "Optimization Techniques for Grid-Connected PV With Retired EV Batteries in Centralized Charging Station with Challenges and Future Possibilities: A Review", Renewable and Sustainable Energy Reviews, 1364–0321, 2019.
22. Pei Cheng and Xiaowei Zhan, "Stability of Organic Solar Cells: Challenges and Strategies", Chemical Society Reviews, 0306–0012, 2016.
23. Peter Keil, Simon F. Schuster, Christian von Lüders, Holger Hesse, "Lifetime Analyses of Lithium-Ion EV Batteries", Journal of Energy Storage, 2352–152X, 2016.
24. Ramesh Babu Kodati, Puli Nnageshwar Rao, "A Review of Solar Cell Fundamentals and Technologies", Material Science and Engineering, 0921–5107, 2016.
25. S. Senthilgavaskar, K. Karthick, Chidambaranathan Bibin, "A Review Paper on Solar Operated Ridge and Furrow Formation Machine", Renewable and Sustainable Energy Reviews, 1364–0321, 2019.

26. S. Shalini1, T. Satish Kumar, N. Prabavathy, S. Senthilarasu and S. Prasanna, "Status and Outlook of Sensitizers/Dyes Used in Dye Sensitized Solar Cells (DSSC): A Review", International Journal of Energy Research, 0363–907X, 2016.

27. Serap Gunes, Niyazi Serdar Sariciftci, "Hybrid Solar Cells", Materials Today, 1369–7021, 2007.

28. Sigurd Nikolai Bjarghov, "Utilizing EV Batteries as A Flexible Resource at End-User Level", 0360–5442, 2019.

29. Storm William D. Gourley, Tyler Or, and Zhongwei Chen, "Breaking Free from Cobalt Reliance in Lithium-Ion Batteries", Nature Communications, 2041–1723, 2020.

30. Victoria Gonzalez-Pedro, Emilio J. Juarez-Perez, Waode-Sukmawati Arsyad, Francisco Fabregat-Santiago, Ivan Mora-Sero, and Juan Bisquert, "General Working Principles of CH3NH3PbX3 Perovskite Solar Cells", Nano Letters, 1530–6984, 2014.

31. Vishal Shrotriya, Gang Li, Yan Yao, Tom Moriarty, Keith Emery and Yang, "Accurate Measurement and Characterization of Organic Solar Cells", Advanced Functional Materials, 1616–3028. 2003.

32. Weidong Chen, Jun Liang, Zhaohua Yang, Gen Li, "A Review of Lithium-Ion Battery for Electric Vehicle Applications and Beyond", Energy Procedia, 1876–6102, 2015.

33. Wei Wang, Mark T. Winkler, Oki Gunawan, Tayfun Gokmen, Teodor K. Todorov, and David B. Mitzi, "Device Characteristics of CZTSSe Thin-Film Solar Cells with 12.6% Efficiency", Advanced Energy Materials, 1614–6382, 2014.

34. William P. Mulligan, Doug H. Rose, Michael J. Cudzinovic, Denis M. De Ceuster, Keith R. McIntosh, David D. Smith, and Richard M. Swanson, "Manufacture of Solar Cells With 21% Efficiency", IEEE Transactions on Electron Devices, 0018–9383, 2004.

35. Xiaoyue Li, Peicheng Li, Zhongbin Wu, Deying Luo, Hong-Yu Yu, Zheng-Hong Lu, "Review and Perspective of Materials for Flexible Solar Cells", Materials, 1996–1944, 2014.

36. Yuanan Hu, Hefa Cheng, and Shu Tao, "Retired Electric Vehicle (EV) Batteries: Integrated Waste Management and Research Needs", Environmental Science & Technology, 0013–936X, 2017.

37. Ziad M. Ali, Martin Calasan, Foad H. Gandoman, Francisco Jurado, Abdel Aleem, "Review of Batteries Reliability in Electric Vehicle And E-Mobility Applications", Renewable and Sustainable Energy Reviews, 1364–0321, 2018.

Advances in Mechanical Engineering and Materials Sciences – Dr. Vinay K. B et al. (eds)
© 2026 Taylor & Francis Group, London, ISBN 9-781-041-20970-6

53

Sustainable Storage Solutions for Perishable Agricultural Products— A Comprehensive Review

Arun Kumar K. N.[1],
Thamme Gowda C. S.[2]
Assistant Professor, Department of Mechanical Engineering,
Vidyavardhaka College of Engineering,
Mysuru, Karnataka, India

Abhinav S.[3], Jayanth[4],
Abhishek K.[5], and Chandan Kumar H. P.[6]
Project Associate, Department of Mechanical Engineering,
Vidyavardhaka College of Engineering,
Mysuru, Karnataka, India

Abstract: The Literature Review Concentrates on the minimal waste and maximal freshness of fruits and vegetables by low-cost, environmentally friendly post-harvest methods via sustainable storage techniques. To this end, conventional cooling that may require evaporative cooling in the preservation of low temperature of values and optimum moisture levels without using electricity, such as clay pot coolers and earthen chambers is valuable. These systems work particularly well in warm, arid environments due to their ability to stabilize humidity between 40 and 60 percent and reduce temperatures by 8 to 18 degrees Celsius, which retards deterioration and minimizes nutrient loss. This technology provide farmers and households with many ways to maintain perishable products stay fresh, this is essential in rural settings where refrigerated storage systems are not available.

Keywords: Evaporative cooling system, Clay pot coolers, Sustainable storage solutions, Efficient storage methods and Waste reduction in agriculture

1. INTRODUCTION

Storing food properly after harvest is a very important step to keep it fresh and to eat good food. This is very true in place where people don't have modern refrigerators. Researchers have studied old and new ways to store food to see which ones work best. Better storage helps to stop food from spoiling, makes it last longer, and helps small farmers make more money. One of the natural way to store food is by using clay pots. These pots help to keep the food cool and stop germs from growing (Ranjit singh, A. J. A., et al., 2011). Scientists also say that keeping the right temperature and moisture level in storage places helps

vegetables stay fresh for a longer time. This reduces food waste and helps people have enough food (Basumatary Regina, et al., 2018). Experts study how much heat food gives off and how it breathes in storage. This helps them find better ways to use energy in cold storage, especially for fruits like pears (Kenneth C. Gross, et al., 2016). The scientists found that controlling the air help to keep the food fresh and for a longer time (Scheepens, P., et al., 2011). Apart from keeping food fresh, storage also affects money. Both these old and new storage methods help farmers to reduce food loss and earn more. This helps them earn more money (J Kaur, et al., 2021).

[1]arunkn.10@vvce.ac.in, [2]thammegowda.cs@vvce.ac.in, [3]363abhinav@gmail.com, [4]js4060483@gmail.com, [5]abhiabhisheskk755@gmail.com, [6]chandankumarhp125@gmail.com

10.1201/9781003725053-53

2. REVIEW OF LITERATURE

2.1 Traditional Storage Practices for Perishable Goods

For instance, a pot-in-pot cooler consists of a small clay pot placed in a large clay pot filled with water. The outer water cools the pot as it evaporates, cooling the inner pot and thus also lowering the temperature. The study conducted by Dr. A.J.A. Ranjit singh and his team discovered another advantage of earthenware storage, which is it lessens the microbial damage to the stored produce compared to normal refrigeration and ambient conditions (Ranjit singh, A. J. A., et al. , 2011). Moreover, research conducted by Basumatary Regina and other authors proves that well designed storage facilities concerning humidity and temperature positively influence the longevity of vegetables (Basumatary Regina, et al., 2018). Figure 53.1 illustrates the clay pot cooler, which is a fueling and effective refrigerator with two clay pots, one placed over the other and a sand layer between them. Water is poured onto the sand until it is fully soaked, and a damp cloth is placed on top. The contents of the inner pot are cooled as the water evaporates, which draws heat away from the inner pot. Pits and trenches are created by excavating cavities in the ground. These cavities are lined with materials such as cut grass, wood shreds, sand, stubble, or soil, as illustrated in Fig. 53.2. They are typically constructed at the edges of fields and at higher elevations where the risk of rainfall accumulation in the cultivated area is minimal. These cost-effective and accessible techniques provide resource-limited farmers with a viable alternative to expensive refrigeration. By combining traditional knowledge with minor technological improvements, these methods tackle post-harvest losses and enhance food security in areas lacking modern cold chain infrastructure. Their reliance on renewable, low-energy processes further strengthens their environmental and economic appeal, making them a sustainable choice for small-scale agriculture. Kenneth C, et al.: Focuses on calculating heat load during refrigeration and the respiration rate of stored produce like pears at specific temperatures Kenneth C. Gross, et al., 2016). Basumatary Regina, et al.: Reviews critical factors like temperature and humidity, comparing traditional and controlled atmosphere storage for vegetables (Scheepens, P., et al., 2011). J Kaur, et al.: Discusses modern and traditional storage structures, emphasizing their role in increasing revenue for small-scale farmers (J Kaur, et al., 2021). The performance evaluation of the evaporative cooling system included both no-load and load tests conducted during dry (November–December) and rainy (June–July) seasons. For the load test, vegetables were stored within the cooling structure, with a control group maintained under ambient conditions. The initial weight of the produce was recorded, and changes are monitored during storage (Ogbuagu N. J., 2017).

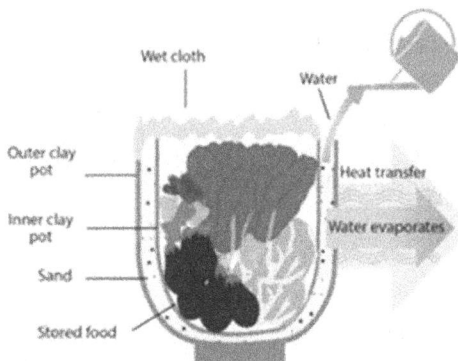

Fig. 53.1 The image illustrates a clay pot cooler mechanism (Brijendra Singh Choudhary, et al., 2023)

Fig. 53.2 Pit storage for potatoes (Kaur, J. et al., 2021)

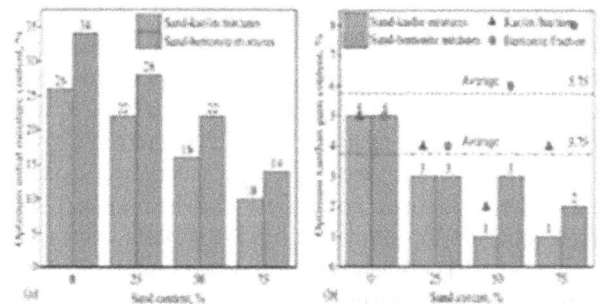

Fig. 53.3 Fresh fruits and vegetables kept in the evaporative cooling structure (Ogbuagu N. J., 2017)

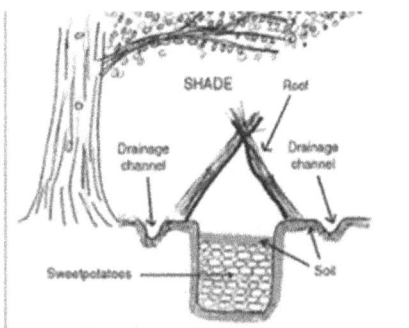

Fig. 53.4 (a) Optimum initial moisture content for different soil mixtures. (b) Optimum XG content for different soil mixture (Ahmad Safuan A. Rashid, et al., 2021)

2.2 Material Science in Storage Optimization

Studies on improving materials such as sand, clay, and other absorbents are opening up the way to efficient and cost-effective evaporative cooling systems. A study on sand-clay compositions stabilized with biopolymers was undertaken by Ahmad Safuan and his team. They discovered that the optimal mix of materials could greatly improve both structural stability and cooling efficiency. Their research indicates that sand content increases lower swelling, yet enhances strength and thermal conductivity, and thus it is well suited for evaporative cooling systems (Ahmad Safuan A. Rashid, et al., 2021). The coconut fiber showed the greatest cooling performance, followed by charcoal (Department of Mechanical Engineering, Ladoke Akintola University of Technology, 2020). It was Found that the coconut fiber enhances cooling efficiency in clay pot refrigerators which is followed by charcoal and jute (Murugan, A. M., et al., 2012). Brijendra Singh Choudhary, et al.: Reviews the efficiency of pot-in-pot systems, which lower temperature by 8–18°C and increase relative humidity by 40–60% (Takenaka, Set.al, 2006). Ahmad Safuan, et al.: Examines sand-clay mixtures stabilized with biopolymers like xanthan gum for enhanced structural performance (Arum, C., & Falayi, F. R., 2012). For Telfairia, the longest storage duration (7.7 days) was achieved with thick lining and turgid produce, followed by 7.0 days with thin lining and turgid produce. These durations were significantly higher than other combinations. In contrast, the shortest storage time (3.3 days) occurred with non-lined and partially wilted produce. Turgid vegetables consistently stored longer than partially wilted ones for Telfairia (Christian C., 2014). Compared to ambient conditions, the cabinet exhibited consistently lower temperatures and higher relative humidity (RH) throughout the day, demonstrating the system's effectiveness in maintaining a cooler, more humid environment suitable for preserving produce (B.G Jahun, 2016).

2.3 Energy Efficiency and Cost-Effective Innovations

Innovative agricultural storage solutions, such as solar-powered systems and zero-energy cooling chambers, are transforming the way small-scale farmers preserve perishable goods. These sustainable technologies help extend shelf life, reduce post-harvest losses, and address the challenges of high costs and limited energy access. Robert Kraemer and his team have designed a low-cost solar-powered cold storage system that proves economically viable for small-scale farmers. This system not only is a good cooler but also lowers energy costs drastically, with an estimated payback time of merely 2.4 years. Its performance is particularly felt where sunlight is in abundance and the power grid is unreliable (Robert Kraemer, Andrew Plouff, John Venn, 2014). A. Lal Basediya and coauthors presented an evaporative

cooling system that runs without electricity, or what is popularly known as a zero-energy solution. Both systems make use of easily sourced materials like sand and clay, which depend on natural evaporation to cool and preserve high humidity. They are most useful in maintaining the freshness of fruits and vegetables (Basediya, A. L., et al., 2013. Similarly, Ratnesh Kumar and his colleagues have emphasized evaporative cooling as an affordable and power-free method to reduce post-harvest losses in developing nations such as India (El-Ramady, H. R et al., 2015). Additionally, Ishaque et al. examine the effectiveness of sand-clay mixtures in porous evaporative cooling structures, demonstrating their ability to lower costs and prevent storage losses without requiring energy consumption (Kenneth C. Gross, et al., 2016). Deep P. Patel, et al.: Proposes evaporative cooling chambers (ECC) made from locally available materials for cost-effective post-harvest storage (Dr. R.K. Gupta, et al., 2017). MIT D-Lab Research Engineer: Evaluates clay pot coolers as effective tools for reducing food loss and improving nutrition, especially in arid regions (Yu, M., & Nagurney, A., 2013). Laura Mogannam, et al.: Discusses the training and implementation of clay pot coolers in Mali, noting impacts on shelf life and food spoilage reduction (Scheepens, P., et al., 2011).

2.4 Applications and Practical Impacts

The Fig. 53.1 displays a conventional clay pot cooler, which is sustainable and efficient way to cool liquids. The design comprises two pots, a smaller inner pot with an outer, larger pot differentiated by a sandy bed. The sand is watered, and a wet cloth is placed over the top. Through the evaporation of water, heat is removed and hence the items inside the inner pot are cooled. This low-cost and also effective method is perfect for storing perishables goods such as vegetables in hot climates. In addition to clay pot coolers, solar-powered systems and zero-energy cooling chambers are revolutionary technologies in agricultural storage, addressing the cost and energy issues confronting small-scale farmers (Mleku, D., et al., 2013). Robert Kraemer and others have come up with a low-cost solar-powered cold storage system that is specifically tailored for small-scale use. The system cuts energy costs substantially while offering efficient cooling, with a payback time of only 2.4 years. It's especially useful in sunny areas with limited or unreliable grid power (Ndukwu, M. C., & Manuwa, S. I., 2014). The Fig. 53.5 depicts an evaporate cooling chamber, a practical and effective method for the storage of agricultural produce. As detailed in Storage of Agricultural Products by Piet Scheepens and others, this is a simple yet very effective method for keeping produce in hot climates.

Sustainable materials and innovative design in storage systems are vital to minimize environmental footprint and enhance agricultural efficiency. By merging contemporary innovation with conventional methods, these systems

Fig. 53.5 Evaporative cooling chamber (Basumatary Regina, Arora Vinket Kumar, 2018)

Table 53.1 The percentage of respondents across all groups who encountered food storage challenges

Issues	Clay pot makers and sellers [54]	Producers and consumers [19 4]	Vendors at home [16]	Vendors at business[16]
Rot	94%	87%	88%	100%
Dehydration	56%	64%	88%	75%
Pests	72%	54%	56%	50%
Yellowing	81%	49%	31%	6%
Mold	69%	42%	25%	44%
Wilting	67%	32%	25%	25%
Other	0%	5%	19%	**19%**

Source: Authors

Table 53.2 Storage requirements for common vegetables and fruit (Farooq A. Khan, et al., 2017)

Storage Environment	Packaging	Weight loss (%) Day 5 Day 10 Day 15			Decayed fruits (%) Day 5 Day 10 Day 15			Shelf life (days)	Cause of shelf life termination
Ambient	Unpacked	14.3d	22.4d	31.3d	0	14a	24c	5a	Shrivelling
	Perforated PE	8.1c	14.6c	22.8c	0	8b	14a	13c	Shrivelling
	Imperforated PE	3.1b	4.4a	6.2a	0.5	22d	32c	7ab	Decay
	Mean	11.8	13.8	20.1	0.17	14.7	23.3	8.3	
Evaporative cooler	Unpacked	6.2c	9.2b	10.5b	0	10bc	16ab	13c	Decay
	Perforated PE	4.1b	7.2b	12.1b	0	7c	10ab	20	Decay
	Imperforated PE	1.8a	3.7a	5.4a	0	14a	18b	9b	Decay
	Mean	4.0	6.7	9.3	0	10,3	13.7	13	

provide sustainable options compared to energy-depleting storage practices. This method is especially useful in developing and desert communities where resources are scarce and sophisticated technologies inaccessible. Shamsh Parveen and coauthors emphasize the economic and environmental benefits of green cold storage systems in tropical environments. According to their study, strategic design enhancement can substantially reduce energy consumption, operational expenses, and carbon emissions. In combination with renewable energy sources such as solar power, such systems become a feasible and effective option for small-holder farmers in resource-restricted environments (David Hernandez Cuellar, 2023). In the same line, Brijendra Singh Choudhary and others have investigated the feasibility of pot-in-pot cold storage systems. These power-free systems, perfectly suited for low-humidity and hot areas, reduce temperature as well as humidify, adding to the life of perishables. Since this use of locally obtained materials such as sand and clay, they not only prove affordable but also immensely sustainable (Brijendra Singh Choudhary, et al., 2023). Traders were interviewed independently about storage problems experienced in their homes and in their working places. The total number of respondents is indicated in brackets, with most providing multiple responses to these questions (Bastian Lange, et al., 2016). Both studies underscore the

importance of adopting these sustainable storage practices to address environmental challenges. By implementing these systems in rural and arid areas, communities can reduce waste, enhance resilience, and foster sustainable agricultural development.

2.5 Impact on Agricultural Supply Chain and Food Security

Effective storage methods are essential for tackling the significant post-harvest losses that hinder agricultural productivity. These losses, often caused by inadequate storage infrastructure, not only reduce farmer incomes but also threaten food security. Farooq A. Khan and colleagues emphasize that proper storage methods for fruits and vegetables can help preserve nutrients and reduce foodborne illnesses. Improved storage infrastructure ensures a steady supply of crops to markets, reducing farmers' reliance on expensive intermediaries and promoting economic stability (Farooq A. Khan, et al., 2017). The Table 53.3 outlines storage requirements for common fruits and vegetables, specifying optimal temperature, relative humidity, and storage life. For instance, apples and carrots store best at 32°C with 90-95% humidity for 4–6 months, while garlic and dry beans prefer 65–70% humidity, lasting 6–7 months and 1 year, respectively. Similarly, Guilherme D.N. Maia and his team highlight the benefits

Table 53.3 Weight loss, decay, and shelf life of tomatoes stored in ambient and evaporative cooler. Means followed by the same letter in each column do not differ significantly at P=0.05 (Folorunso A B., 2016)

Produce	Temperature(C)	Relative Humidity	Storage Life months
Apples	32	90-95%	4 – 6
Beets	32	90%	1 – 3
Bressels Sprouts	32	90-95%	3 – 5
Cabbage	32	90-95%	3 – 4
Carrot	32	90-95%	4 – 6
Cauliflower	32	90-95%	2 – 4
Celeriac	32	90-95%	3 – 4
Chinese Cabbage	32	90-95%	1 – 2
Dry beans	32 - 50	65-70%	12
Garlic	32	65-70%	6 – 7
Horseradish	30 - 32	90-95%	10 – 12
Kale	32	90-95%	10 – 14
Kohlrabi	32	90-95%	2 – 4
Leeks	32	90-95%	1 – 3
Onions	32	65-70%	5 – 8
Parsnips	32	90-95%	2 – 6
Pears	32	90-95%	1 – 2
Sweet Pepper	45 - 50	90-95%	8 – 10
Potatoes	38 - 40	90%	5 – 8

of optimizing supply chains. Their research demonstrates that effective storage and transportation systems can significantly reduce grain losses, increasing profitability for small-scale farmers by maximizing the value of their harvests and improving market access (Seyed Mohammad Nourbakhsh, et al., 2016). Bastian Lange, et al.: highlights the reliance of agriculture on refrigerated transport, noting its significant contribution to greenhouse gas emissions, with 14,000 trucks projected to emit over 5 million tons of CO_2 by 2020. Fuel consumption accounts for 75% of these emissions (David Hernandez Cuellar, 2023). Shamsh Parveen, et al.: Discusses sustainable cold storage systems that reduce electricity costs and emissions by up to 66% through enhanced COP in tropical climates (Department of Crop Science, Faculty of Agriculture, University of Peradeniya, Sri Lanka, 2024). Amin Yazdekhasti, et al.: Develops a stochastic supply chain model for the poultry industry, reducing costs by integrating pre-pandemic storage and direct logistics (Amin Yazdekhasti, et al., 2021). David Hernandez Cuellar, et al.: Introduces a hub-and-spoke model for optimizing strawberry supply chains, reducing costs through better sourcing (Shamsh Parveen, Ravi Prakash, Dilawar Husain, 2021). Dr. R.K. Gupta, et al.: Advocates integrating renewable energy in cold chains to enhance food security, reduce post-harvest losses, and create rural jobs in India, backed by government incentives (2017). Rakesh Patidar, et al.: Proposes clustering farmers and optimizing multi-product transportation in Indian food supply chains to address inefficiencies and reduce costs (Rakesh Patidar et al., 2020). Vittorio Solina, et al.: Explores quantitative approaches to optimize production, storage, and distribution in agri-food supply chains, improving collaboration and reducing waste (Vittorio Solina, et al., 2021).

Physiological weight loss was highest under ambient conditions but significantly reduced (P=0.05) in the evaporative cooler. Packaging in polyethylene (PE) bags further minimized weight loss. Tomatoes stored in PE bags within the cooler (PE+C) exhibited the lowest weight loss, while unpackaged fruits stored at ambient conditions experienced the highest loss and shrivelled rapidly (Table 53.2). Although fruits in perforated PE bags eventually wilted, they resisted early shrivelling effectively for up to 10 days (Folorunso A B, 2016). Muhammad Hamid Mahmood, et al.: describes the importance of temperature and humidity control using conventional and hybrid refrigeration systems for agricultural products (Muhammad Hamid Mahmood, et al., 2019). N Sarkar, et al.: Demonstrates the thermal efficiency of clay pots for storing fruits and vegetables, maintaining temperatures of 20–25°C during summer (N Sarkar, et al., 2023).

3. SUMMARY OF LITERATURE

The research focused on optimizing clay and sand compositions to create a cost-effective cold storage solution for perishable agricultural products. Extensive testing has shown that a well-balanced clay-to-sand mixture greatly enhances evaporative cooling, helping storage facilities maintain temperatures lower than the surrounding environment. While this optimized composition is more energy-efficient than traditional cold storage methods, materials with high R-values did not always deliver the expected improvements in efficiency. Additionally, although aboveground designs incorporating the clay-sand mixture proved effective, they were less cost-efficient than underground alternatives, reinforcing the economic benefits of subterranean storage. By incorporating solar energy into the refrigeration system has lowered the payback period to 5 years from 2.4 years. But this modification has brought many long-term advantages which includes lower energy expenses and higher profit. This research emphasizes the need for choosing the appropriate materials and design approaches for affordability, efficiency, and sustainability in agricultural storage. Optimization of the clay-sand mixture for cold storage containers provides an inexpensive, viable means of maintaining perishable crops, especially in regions where traditional refrigeration is not possible. The results indicate that this method not only saves operating expenses for aboveground storage but also

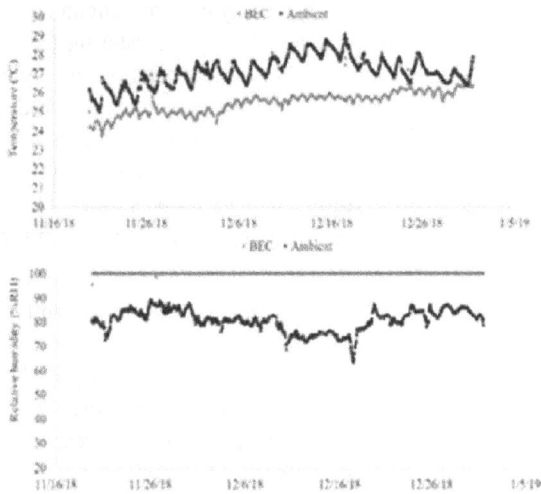

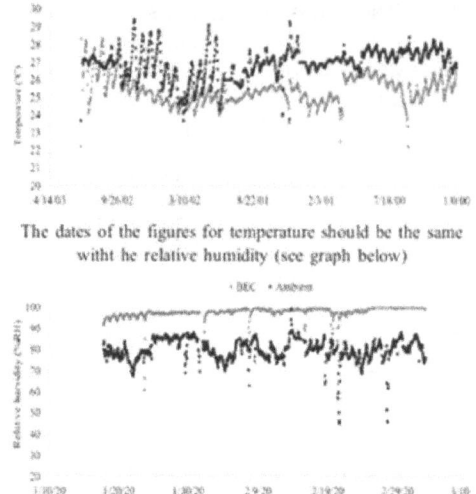

Fig. 53.6 Temperature and relative humidity in ambient and brick-walled evaporative cooler (BEC) conditions during November 2018 to January 2019 (Bayogan E.R.B. et al., 2022) (b)

Fig. 53.7 Temperature and relative humidity in ambient and brick-walled evaporative cooler (BEC) conditions during January 2020 to March 2020 (Bayogan E.R.B. et al., 2022)

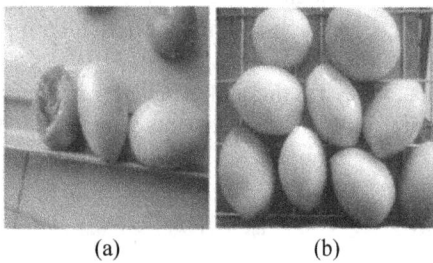

(a)   (b)

(a)   (b)

Fig. 53.8 Pictorial view of red tomatoes after storage (a) Red Tomatoes in Ambient (b) Red Tomatoes in Cooler (K. O. Babaremu et al, 2019)

Fig. 53.9 Pictorial view of green tomatoes after storage (a) Green tomatoes in ambient (b) Green tomatoes in cooler (K. O. Babaremu et al, 2019)

increases longevity, particularly when integrated with renewable energy sources. This solution offers a long-term, sustainable, and cost-effective option for small to medium-sized farming businesses, with potential energy savings as well as income generation.

4. RESEARCH GAP AND FUTURE SCOPE OF WORK

Although much has been achieved in the development of many affordable and sustainable post-harvest storage solutions, the task is also beset with challenges. Most research has centered around conventional and sustainable cooling systems like clay pot coolers and evaporative cooling systems. But there is little research on how to adapt these technologies in regard to temperature and humidity for different climates and for different types of crops. While there are a number of existing studies focusing on low-energy or energy-free solutions, the potential of integrating these methods within scalable renewable energy systems, for instance solar-powered storage, to achieve additional applicability is neglected. The long-term performance of materials such as sand-

clay mixtures and coconut fiber under repeated use and environmental stressors is also under-researched. A second major limitation is that little to no investigation of supply chain integration, particularly in addressing the logistical challenges faced by the small-scale farmers in resource-limited areas. Closing these gaps could lead to more resilient, adaptable, and widely accessible storage solutions, ultimately reducing post-harvest losses and also strengthening the global food security.

They also shed light on the possibility of making these storage systems more efficient and scalable across the many agri-climatic zones for future work on both amalgamated/modern-storage systems and traditional storage systems. Since the effectiveness of green coolants, like sand-clay mixtures, can increase when they have a good shelf-life, it is likely to get insights into how this can be facilitated by advancements in material science. A primary focus is to combine renewable energy sources, i.e., solar energy, to achieve a more stable cooling structure in addition to bringing down operational expenses. Hybrid storage models based on some conventional and some of new approaches can also play a key role and can create tailored solutions for a given crop type and storage necessity. Further studies on the economic viability and supply chain integration arising from these processes will help facilitate greater uptake by smallholder farmers and other stakeholders in the food industry. Finally, quantifying large-scale implementation's effect on environmental impact and long-term viability is key for businesses to better appreciate the role of these storage innovations in both securing the global food supply chain and combating climate change. In the end, they will translate into a more resilient agricultural system, minimizing the post-harvest losses and increasing food security.

REFERENCES

1. Ahmad Safuan A. Rashid, Nor Zurairahetty Mohd Yunus, Nazirah Mohd Apandi – "*The Optimisation Analysis of Sand-Clay Mixtures Stabilised with Xanthan Gum Biopolymers*", 2021.
2. Amin Yazdekhasti, Jun Wang, Li Zhang, Junfeng Ma – "*A Multi-Period Multi-Modal Stochastic Supply Chain Model Under COVID Pandemic: A Poultry Industry Case Study in Mississippi*", 2021.
3. Arum, C., & Falayi, F. R. (n.d.). "*Functional and Affordable Food Crop Storage Technologies for Developing Countries*; With Special Reference to Nigeria", 2012.
4. Basediya, A. L., Samuel, D. V. K., & Beera, V. (n.d.). "*Evaporative Cooling System for Storage of Fruits and Vegetables*", 2013.
5. Basumatary Regina, Arora Vinket Kumar – "*Review on the Storage Structure for the Longer Life of Vegetables*", 2018.
6. Bastian Lange, Caspar Priesemann, Marion Geiss, Anna Lambrecht – "*Promoting Food Security and Safety via Cold Chains*", 2016.
7. Bayogan E.R.B. et al. "*The use of brick-walled evaporative cooler for storage of tomato*", 2022.
8. B.G Jahun, "*Assessment of Evaporative Cooling System for Storage of Vegetables*", 2016.
9. Brijendra Singh Choudhary, Ravi Kiran T2, Satish Kumar Dewangan – "*Review Study for Performance Analysis of an Evaporative (Pot in Pot) Cooling System*", 2023.
10. Christian C. "*Effects of moisture barrier and initial moisture content on the storage life of some horticultural produce in evaporative coolant*", 2014.
11. Danyal Rehman, Ethan McGarrigle, Leon Glicksman, Eric Verploegen – "*A Heat and Mass Transport Model of Clay Pot Evaporative Coolers for Vegetable Storage*", 2020.
12. David Hernandez Cuellar – "*Design of a Stochastic Cold-Stored Strawberry Supply Chain Model*", 2023.
13. Department of Crop Science, Faculty of Agriculture, University of Peradeniya, Sri Lanka – "*Evaluation of Clay Pot Cooler Storage for Preserving Postharvest Quality of Leafy Vegetables*", 2024.
14. Department of Mechanical Engineering, Ladoke Akintola University of Technology, Ogbomoso, Oyo State, Nigeria – "*Experimental Investigation of Effects of Absorbing Materials on Performance of Clay Pot in Pot Refrigerator*", 2020.
15. Dr. R.K. Gupta, Dr. Ranjeet Singh, Dr. V. Eyarkai Nambi – "*Relevance of Cold Chain Management of Agri Based Products Pertaining to Horticultural Produce*", 2017.
16. El-Ramady, H. R., & Domokos-Szabolcsy, É. (n.d.). "*Postharvest Management of Fruits and Vegetables Storage*", 2015.
17. Farooq A. Khan, Sajad A. Bhat, Sumati Narayan – "*Storage Methods for Fruits and Vegetables*", 2017.
18. Folorunso A B. "*Preservation of Tomatoes in a Brick-Walled Evaporative Cooler*", 2016.
19. J Kaur, R Aslam, PA Saeed – "*Storage Structures for Horticultural Crops*", 2021.
20. K. O. Babaremu et al. "*The Significance of Active Evaporative Cooling System in the Shelf Life Enhancement of Vegetables (Red and Green Tomatoes) for Minimizing Post-Harvest Losses*", 2019.
21. Kaur, J. et al. "*Storage structures for horticultural crops: a review*", 2021.
22. Kenneth C. Gross, Chien Yi Wang, Mikal Saltvei – "*The Commercial Storage of Fruits, Vegetables, and Florist and Nursery Stocks*", 2016.
23. Mleku, D., et al. (n.d.). "*Pots and Food: Uses of Pottery from Resnikov Prekop*", 2013.
24. Muhammad Hamid Mahmood, Muhammad Sultan, Takahiko Miyazaki – "*Significance of Temperature and Humidity Control for Agricultural Products Storage: Overview of Conventional and Advanced Options*", 2019.
25. Murugan, A. M., et al. (n.d.). "*A Study on Cost Effective and Eco-Friendly Earthen Pot Cool Chamber (EPCC) System for Rural Population to Store Post-Harvest Vegetables*", 2012.
26. Ndukwu, M. C., & Manuwa, S. I. (n.d.). "*Review of Research and Application of Evaporative Cooling in Preservation of Fresh Agricultural Produce*", 2014.
27. N Sarkar, A Somwanshi, D Dhand, R Trivedi – "*Thermal Analysis of a Clay Pot Utilized for Cooling Water and Storing Vegetables/Fruits*", 2023.
28. Ogbuagu N. J, "*development of a passive evaporative cooling structure for storage of fresh fruits and vegetables*", 2017.
29. Rakesh Patidar, Sunil Agrawal – "*Restructuring the Indian Agro-Fresh Food Supply Chain Network: A Mathematical Model Formulation*", 2020.
30. Ranjitsingh, A. J. A., Dr. (n.d.)., "*Rejuvenation of Traditional Pottery Making by Applying Advanced Biotechniques*", 2011.
31. Robert Kraemer, Andrew Plouff, John Venn – "*Design of a Small-Scale, Low-Cost Cold Storage System*", 2014.
32. Seyed Mohammad Nourbakhsh, Yun Bai, Guilherme D.N. Maia, Yanfeng Ouyang, Luis Rodriguez – "*Grain Supply Chain Network Design and Logistics Planning for Reducing Post-Harvest Loss*", 2016.
33. Shamsh Parveen, Ravi Prakash, Dilawar Husain – "*Sustainable Systems for Cold Storage in the Tropical Climate*", 2021.
34. Scheepens, P., et al. (n.d.). "*Storage of agricultural products*", 2011.
35. Takenaka, S., & Mrema, G. C. (n.d.). "*Postharvest Management of Fruit and Vegetables in the Asia-Pacific Region*", 2006.
36. Vittorio Solina, Enrico Conte, Giovanni Mirabelli – "*Quantitative Approaches for the Integrated Management of Agri-Food Supply Chains*", 2021.
37. Yu, M., & Nagurney, A. (n.d.). "*Competitive Food Supply Chain Networks with Application to Fresh Produce*", 2013.

Advances in Mechanical Engineering and Materials Sciences – Dr. Vinay K. B et al. (eds)
© 2026 Taylor & Francis Group, London, ISBN 9-781-041-20970-6

54

A Study on the Effect of Mild Steel Chills on Aluminium LM13 and Nano-Zirconium Oxide Composite using Cryogenic Solidification

Mamatha Y. P.[1]

Assistant Professor,
Dept. of Mechanical Engineering,
Vidyavardhaka College of Engineering,
Mysuru, Karnataka, India

G. B. Krishnappa[2]

Professor,
Dept. of Mechanical Engineering,
Vidyavardhaka College of Engineering,
Mysuru, Karnataka, India

Abstract: The cryogenic solidification technique utilizing mild steel chill exhibits considerable potential in manufacturing high-quality metal matrix composites with superior mechanical properties The purpose of this work is to investigate the mechanical characteristics of a mild steel chill-cryogenically solidified aluminum LM13 and nano zirconium oxide composite. Mild steel chill thickness has a significant impact on improving the composite's mechanical qualities. Experimental analysis involved altering the mild steel chill thickness, maintaining a consistent composite composition throughout the investigation. The mechanical properties of the composite are evaluated using standard ASTM tests. The improvement in mechanical properties was attributed to the uniform solidification of the composite, which minimized the formation of defects such as porosity. The results showed that as there is an increase in the thickness of the chill there is a consistent cooling rate across the casting process, reducing the occurrence of shrinkage cavities and ultimately resulting in superior mechanical properties

Keywords: Aluminium LM13 Nano-ZrO_2, Reinforcement, Stir casting, Composite

1. INTRODUCTION

Researchers have been working hard to progress in every aspect to make the life of humans more comfortable. In the field of engineering, researchers have been doing research to find newer materials which are light weight and high strength in nature. Various heat treatment processes have been developed to enhance properties of metallic materials. Very recently it is observed that just by conducting normal heat treatment processes on as-cast components, comparatively appreciable mechanical

properties have not been obtained but by using cryogenic solidification method one can obtain greater mechanical properties than conventional heat treatment processes.

The most popular matrix materials for MMCs are most likely aluminum and its alloys. [Joel Hemanth 1999]. As the need for high-quality composites has grown, it is now crucial to manufacture. Aluminium composites free from unsoundness hence cryogenic solidification stands as the prominent method of producing quality Aluminium composites.[Joel Hemanth 2001] Particulate-

---

[1]mamathayp@vvce.ac.in, [2]gbk@vvce.ac.in,

10.1201/9781003725053-54

reinforced Al-based MMCs are desirable for use in the automotive, industrial, aerospace, defence, and various sectors due to their outstanding qualities. [Joel Hemanth 2008] According to studies mechanical characterization, hardness, tensile strength, and fracture toughness of the Al matrix were all considerably increased by the addition of nano ZrO_2 particles.

[Sumod Daniel 2014] The Al Nano-ZrO_2 metal matrix composite synthesized through Stir Casting demonstrates superior mechanical properties with improved hardness, strength , with a little compromise in ductility attributed to altered fracture toughness. Moreover, the addition of Nano-ZrO_2 particulates inversely affects the electrical resistivity and thermal conductivity of the Al matrix, suggesting potential applications in diverse engineering domains. From the literature [Raghu N etal. 2018] it is observed that observed that at 10% reinforcement of ZrO_2 with Aluminium LM13 has shown promising results which leads to better mechanical properties of the composite and also with increase in chill thickness there is significant amount of increase in the mechanical properties of the cast moreover mild steel chills have not been studied more. Because stir casting is an easy, affordable, and versatile process, it is utilized to make composites. Cryogenic solidification reduces shrinkage problems and porosity. To accomplish cryogenic solidification during sand casting, liquid nitrogen is poured into the chill box to produce a cryogenic effect. [Raghu N etal. 2018] Using chills can help a lot in reducing the porosity scattered in alloy castings because of how slowly they cool down. Chills cool the melted metal rapidly and promote directional solidification.

The effect of mild steel chills for cryogenic solidification of Aluminium LM13 and Nano Zirconium oxide composite and comprehensively study its mechanical properties.

2. MATERIALS

Aluminium LM13, renowned for its lightweight properties and excellent corrosion resistance, finds widespread application in aerospace and automotive industries for manufacturing components requiring high strength-to-weight ratios. As result we have chosen Aluminium LM13 as the matrix material.

Table 54.1 Composition of aluminium LM-13

SL. No.	Composition	(%)
1	Al	82.5
2	Si	12.0
3	Fe	1.0
4	Cu	1.25
5	Mg	1.2
6	Ni	1.5
7	Others (Mn, Cr, Zn, Ti, Pb, Ca)	0.494

As reinforcement, nano zirconium oxide (ZrO2) in powder form (80 nm) is utilized. Nano zirconium oxide (ZrO_2) has good corrosion resistance and heat resistance.

Table 54.2 Properties of nano zirconium oxide (ZrO_2)

SL. No.	Details	Values
1	Density	5.68 mg/cc
2	Melting point	2715 °C
3	Ultimate tensile strength	410 MPa
4	Hardness	150 VHN
5	Youngs modulus	98GPa
6	Particle size	80 nm

Chill box is made up of mild steel chill are placed in the sand mould it will act as heat sink through which heat . [Joel Hemanth 1999], The advantages of using chills are directional solidification, promotion of steep thermal gradient, refinement of microstructure.

Joel Hemanth has concluded [Joel Hemangth 1999] The volumetric heat capacity of the chill, which considers the specific heat, volume and density of the chill material, has been recognized as an essential factor in calculating the effectiveness of the chill.

Volumetric heat capacity (VHC) = $Cp \times V \times \rho$ Where,

Cp – Specific heat of the chill material in KJ/KgK. V- Volume of the chill in cm3.

ρ - Density of the chill materi. al in g/cm3

An increase in any one of these components namely V, Cp and ρ, enhance the value of the volumetric heat capacity, thus increasing the chilling effect.

Table 54.3 Thermal properties of mild steel chill material

SL. No.	Details	Values
1	Specific heat J/kg K	510
2	Thermal conductivity W/m K	50

3. FABRICATION TECHNIQUE

The green sand mixture for the mould was made by combining silica sand with clay, sawdust, and water. Wooden patterns were placed in a drag box, and parting sand was spread over them to prevent sticking. After positioning the patterns on a smooth surface, the drag box was filled with the green sand and rammed. After carefully separating the boxes, wooden patterns were removed, and chill blocks were placed. Vent holes were created to allow gases to escape during metal pouring. Finally, the cope and drag boxes were locked together with pins, and any gaps were sealed with clay before drying the mould boxes to remove moisture.

LM13 Aluminium alloy ingots were cut into pieces using a power hacksaw and placed in a stir casting furnace at

800°C. Simultaneously, nano ZrO_2 reinforcement was preheated at 200°C for 30 minutes to remove contaminants. The molten metal was agitated to remove gases, aided by a hexachloroethane capsule. After removing slag, preheated nano ZrO_2 was gradually added into the molten metal vortex, and stirring continued for even dispersion at 550rpm. Sand mould boxes were placed at the bottom of the stir casting machine, with liquid nitrogen circulating for cryogenic effect. The molten metal was poured into the sand mould after precooling the chill block with liquid nitrogen. Upon solidification, cryogenically solidified nanocomposites were extracted from the mould.

Samples of NMMCs were prepared according to ASTM standards for microstructural, mechanical, and wear property analysis. Cryogenically solidified castings were cut into required sizes, then polished to remove burrs and achieve a smooth surface. Vickers micro hardness test was conducted. Tensile test samples were prepared according to ASTM E 8 standards using a lathe machine for precise dimensional tolerance.

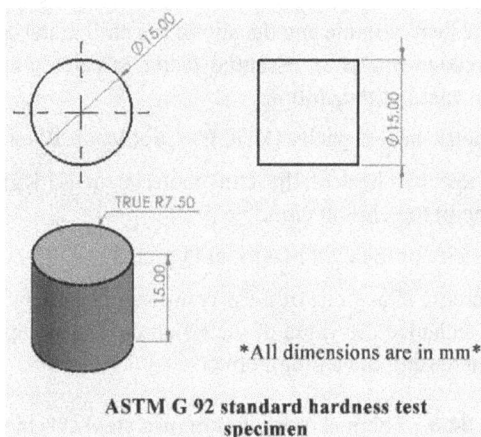

Fig. 54.1 Hardness test specimen as per ASTM standard

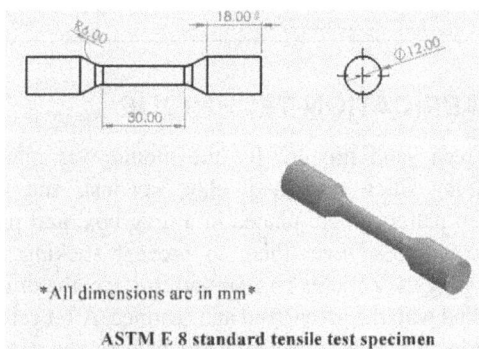

Fig. 54.2 Tensile test specimen as per ASTM standard

4. RESULTS AND DISCUSSION

4.1 Tensile and Hardness Test Results

The tensile test was performed on cryogenically solidified nano ZrO_2 composites in accordance with

ASTM E8 standards. Prior to testing, the specimens were mechanically rubbed with fine-grit emery paper to guarantee a smooth surface finish. The experimental approach included measuring the specimens' dimensions and adjusting the fixture accordingly. The specimens were then securely clamped between the higher and lower chucks, and the system parameters were modified to input the sample's gauge length and diameter. Once started, the computerized universal testing equipment steadily applied a specified force until the material failed and the results are tabulated.

Table 54.4 Tensile test result for cryogenically solidified Al LM13+10%ZrO_2

Chill Thickness	Ultimate tensile Strength (MPa)
5mm	160
10mm	161
15mm	169
20mm	180

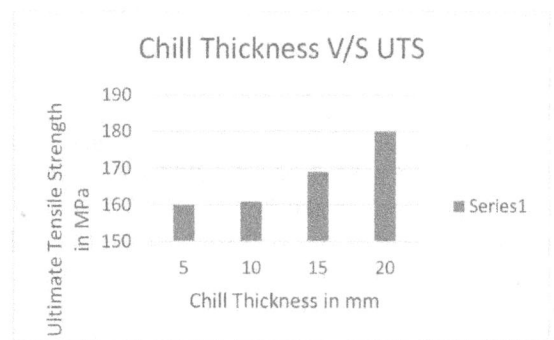

Fig. 54.3 Ultimate tensile strength vs chill thickness in mm

Figure 54.3 shows that ultimate tensile strength increses by incresing the chill thickness.

The hardness of cryogenically solidified composites was evaluated using a Vickers micro hardness tester, renowned for its precise measurements with minimal force. Employing a smaller diamond indenter than Rockwell machines, the Vickers test ensures greater accuracy. With a load range of 10 to 1000 gf and a 20X magnification lens, the specimens were prepared following ASTM E 92 standards. The examinations were conducted at room temperature, The samples were tested three times to find the average hardness.

Table 54.5 Hardness test results of cryogenically solidified Al LM13+10%ZrO_2

Chill Thickness	VHN
5mm	84
10mm	92
15mm	94
20mm	103

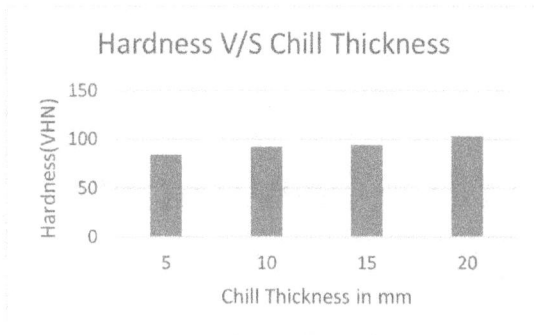

Fig. 54.4   Hardness V/S chill thickness in mm

Figure 54.4 shows that hardness increases by increasing the chill thickness.

5. CONCLUSION

The Metal matrix composites were produced through sand casting utilizing the stir casting technique followed by cryogenic solidification using liquid nitrogen along with mild steel chills having a thickness of 5mm, 10mm, 15mm and 20mm at the chill end. The following conclusions can be made in light of the test's outcomes.

The tensile strength of the composites increases by increasing mild steel chill thickness.

- The hardness of the specimen increases with the increase in mild steel chill thickness.
- As it is observed, the mechanical properties of the specimen increase with an increase in chill thickness.

Hence a study could be carried out to optimize the mild steel chill thickness for the cryogenic solidification

REFERENCES

1. Joel Hemanth (14 may 2001), "Fracture toughness and wear resistance of aluminium boron particulate composites cast using metallic and non-metallic chills". Material and Design 23 (2002) 41 50.
2. Joel Hemanth (25 April 2008), "Quartz (SiO2p) reinforced chilled metal matrix composites (CMMC) for automotive applications". Material Science and Engineering A 507 (2009) 110–113.
3. Joel Hemanth (28 April 2008), "Development and property evaluation of aluminium alloy reinforced with Nano- ZrO2 metal matrix composites (NMMCs)".
4. Joel Hemanth (29 July 1999), "Action of chills on soundness and ultimate tensile strength (UTS) of Aluminium-quartz particulate composite". Journal of alloys and compounds 296 (2000) 193–200.
5. Joel Hemanth "Tribological Behavior of Copper Chilled Aluminium Alloy (LM-13) Reinforced with Berly Metal Matrix Composites". Modeling and Numerical Simulation of Material science(2019)941–69.
6. Joel Hemanth and M R Divya (2018)), "Fabrication and Corrosion Behavior of Aluminium Alloy (LM-13) Reinforced with Nano-ZrO2 Particulate Chilled Nano Metal Matrix Composites(CNMMCs) for Aerospace Applications". 136–150.
7. K.H.W. Seah and Joel Hemanth (15 September 2006), "Cryogenic effects during casting on the wear behavior of Aluminium- alloy/glass MMCs". 2533–2543.
8. Prasad H Nayak H K Srinivas, P Rajendra. EtalEffect of Nano Zirconium Oxide (Zro2).Particles addition on the Mechanical Behaviour and Tensile Fractography of Copper-Tin alloy Nano composites. Structural Integrity and Life (2022) Vol.22. No3 pp 319–327.
9. Raghu N and G.B. Krishnappa (June 2018), "Synthesis and evaluation of mechanical properties of cryogenically solidified Al- ZrO2 nanocomposite".
10. Raghu N, Dinedra A.R and Madhusudhana S R (June 2018), "Preparation, Characterization and property evaluation of cryogenically solidified MMC's". International Research journal of Engineering and technology.
11. Ravitej Y P Mohan C B and Anantha Prasad M G "Evaluation of Copper Chill on Tribological behavior of LM13/ZrSiO4?C hybrid metal matrix composites" ICAMMME 2021.
12. S.F. Hassan and M. Guptha (14 July 2004"Development of high-performance magnesium Nano composites using solidification processing route".
13. S.F. Hassan and M. Guptha (2007), "Effect of Nano-ZrO2 Particulates Reinforcement on Microstructure and Mechanical Behavior of solidification processed Elemental Mg". Composites: Part A 38 (2007) 1395–1402.
14. Sumod Daniel and Dr.G. Harish (February 2014), "A study on the behavior of Aluminium alloy (LM13) reinforced with Nano- ZrO2Particulate", PP 58–62.
15. Yogesha K.B and Joel Hemanth (March 2012), "Mechanical Properties of metallic and non-metallic chilled austempered ductile iron (ACDI)", 240–243.

Note: All the figures and tables in this chapter were made by the authors.

Advances in Mechanical Engineering and Materials Sciences – Dr. Vinay K. B et al. (eds)
© 2026 Taylor & Francis Group, London, ISBN 9-781-041-20970-6

55

A CFD Study on Non-Newtonian Fluid Flow Through Venturimeter

Ranjith K., Sudharshan N.
Department of Mechanical Engineering,
Vidyavardhaka College of Engineering,
Mysuru, Karnataka, India

Gurupavan H. R.
Department of Mechanical Engineering,
PES College of Engineering Mandya,
Karnataka, India

Abstract: This Computational Fluid Dynamics study looks at the characteristics of Non-Newtonian fluids flowing through a Venturimeter, which is a fundamental device used in fluid mechanics and industrial applications to measure flow rate. The goal is to understand the complex behaviour of non-Newtonian fluids, which have changing viscosity and shear-thinning properties, inside the Venturimeter geometry. The study uses numerical simulations and computational models to investigate the flow patterns, pressure differences, and velocity distributions of non-Newtonian fluids passing through the venturimeter. The study aims to understand the effect of fluid rheology on flow behavior, pressure recovery, and the accuracy of flow rate measurements obtained using this device.

Keywords: CFD, Non-newtonian fluids, Venturimeter

1. INTRODUCTION

A venturimeter is a device used to measure fluid flow rate in a pipe. It consists of a constricted tube, known as the venturi tube, which is placed within the fluid flow path. The operation of the venturimeter is based on Bernoulli's equation, which describes the relationship between fluid velocity and pressure. This instrument is particularly valuable for businesses that require precise flow rate measurements. The system consists of two main components: a pipe for fluid flow and the constricted section called the venturi tube. The venturi tube features a gradual narrowing followed by a sudden expansion, which returns to the original pipe diameter. This design is crucial for creating a pressure differential related to the fluid's flow rate. The magnitude of this pressure differential is proportional to the square of the fluid velocity. Venturimeters are used in various sectors to measure both liquid and gas flow rates. They offer several advantages, including high precision, minimal pressure loss, and minimal interference with fluid movement. This paper will explore the flow of Non-Newtonian fluids through a venturimeter.

2. LITERATURE REVIEW

[1] Hollingshead et al. addressed both laminar and turbulent flows through flow metering devices. They utilized Computational Fluid Dynamics to solve the stable Reynolds-averaged Navier-Stokes equation, correlating the Reynolds number with discharge coefficients. Their findings show a sharp drop in 'Cd' values for the venturimeter under reduced 'Re' values. While Venturimeters are typically used in turbulent flow regimes, they can also be applied in laminar flow situations involving viscous fluids, even when the Reynolds number falls below the usual range. [2] Arun et al. used ANSYS Fluent 14.0 to model and simulate the Venturimeter. The simulation was conducted using a conventional Venturimeter, and the results were compared against those defined in the

―――――
[1]ranjithraj@vvce.ac.in, [2]Sudharshan@vvce.ac.in, [3]gpavan1989@gmail.com

standards. The simulation findings demonstrate that Cd reduces rapidly as Re decreases. Subsequently, the results were compared to the equation provided analytically for calculating Cd values at low Reynolds numbers. [3] Nicola Casaria et al. investigated the interaction of a Non-Newtonian fluid in an impeller pump. The primary drawback of these machines is tip leakage, which occurs when fluid passes through the impeller side clearance. This investigation revealed that head and efficiency decrease as impeller side clearance increases. This parameter is crucial to the machine's design and tuning. [4] Naveenji Arun, Malavarayan S, and Kaushi studied single-phase non-Newtonian flows through orifice meters with varying beta ratios and concentrations. The discharge coefficient is determined through Computational Fluid Dynamics modeling. The analysis shows the dependency of 'Cd' value on 'Re' value and that it approaches a constant value for highly turbulent flows. The study highlights that the discharge coefficient stays essentially constant for increased Reynolds numbers, in line with industry norms. Despite having significant pressure losses, orifice plate flow meters are nevertheless commonly employed because of their affordability and ease of use, particularly in small-sized lines. Variations in Reynolds number have an impact on the discharge coefficients of these meters, showing that they are useful within a particular flow range. [5] Bandyopadhyay Tarun Kanti and Das Sudip Kumar studied the flow of liquid through tiny diameter pipe components that exhibit non-Newtonian pseudoplastic properties. This experimental study assessed the frictional pressure drop through empirical equations. The intricacy of fluid flow through pipe fittings and the importance of calculating pressure losses in the design and analysis of fluid machines are highlighted in the introduction. References to previous research are provided, emphasizing the dearth of experimental data and the significance of comprehending pressure drop in non-Newtonian fluids, especially in sectors such as food processing, petroleum, and chemical. The study aims to contribute valuable data for the design of piping and pumping systems dealing with non-Newtonian liquids in practical industrial applications. [6] Prasanna et al. wored on orifice plates and venturimeters, focusing on a standard concentric and a quarter-circle orifice plate. Their findings for the discharge coefficient, expansibility factor, and permanent pressure loss coefficient matched standard values across various Reynolds numbers and diameter ratios.

3. METHODOLOGY AND MODEL SPECIFICATIONS

3.1 Venturimeter Model

The design of the venturimeter utilized for CFD analysis is illustrated in Fig. 55.1. In the case of fully developed flow, the fluid flow domain extends 5 and 10 times the pipe

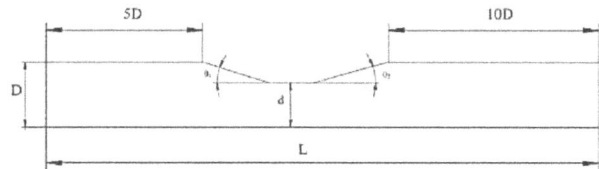

Fig. 55.1 Geometry of venturimeter model

diameter (inner) before and after orifice plate, respectively. Three dimensional venturimeter model depicted in Figure 55.2 was created using ANSYS Design Modeler. The dimensions of the pipe and venturimeter taken into account for the analysis are presented in Table 55.1.

Table 55.1 Details of dimensions of geometry used

Parameters	Value
Pipe Diameter (inner)	100 mm
Throat diameter	50 mm
Diameter ratio	0.5
Pipe length before Orifice Plate	5D mm
Pipe length after Orifice Plate	10D mm
Convergence angle (θ_1)	21^0
Divergence angle (θ_2)	9^0

Source: Author's compilation

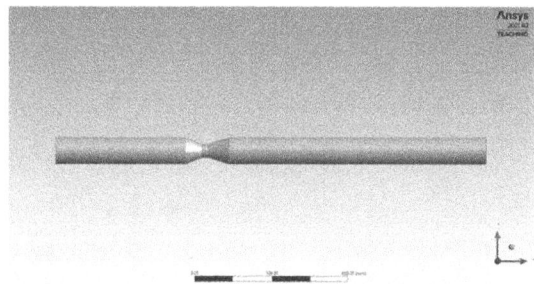

Fig. 55.2 3D model of venturimeter

Source: Authors

3.2 CFD Analysis Method

The Fluent Laminar model was chosen for investigation in ANSYS FLUENT Software. The fluid was chosen from the software's fluid database, with its density set to $\rho=1000$ kg/m^3, and the Power Law model is used to simulate viscosity variation with the appropriate viscosity indices. The boundary conditions include a velocity inlet and pressure at outlet kept zero, along with a no-slip wall condition. The Residual monitoring levels were set at 10^{-5}. A second-order upwind differencing method with the SIMPLE algorithm was employed for the iterative procedure.

3.3 Grid Independence Test

In order to test grid independence, various mesh sizes were used, and the C_d value obtained with each grid size was

Table 55.2 Details of grid independence test

Element size (mm)	No. of elements	Avg. Aspect ratio	Avg. Orthogonal Quality	ΔP	Cd
5.4	50,856	1.3512	0.9741	8211.77	0.9556
4.11	1,07,976	1.3154	0.9832	8207.23	0.9559
3.35	2,06,733	1.2817	0.9869	8129.72	0.9604
2.93	3,06,516	1.2992	0.9885	7988.48	0.9689
2.68	4,00,625	1.3004	0.9879	8129.72	0.9604
2.49	5,05,648	1.28	0.9888	7814.15	0.9796

Source: Author's compilation

compared. Using different element sizes, average aspect ratios, and average orthogonal quality, results from the test runs shown in Table 55.2 reveal that the hex dominant mesh with roughly 3,06,516 elements shown in Fig. 55.3 is found to yield a better result that is reasonably close to the C_d value from ISO-5167-1 standard.

4. RESULTS

4.1 Effect of Non-Newtonian Fluid Flow on Static Pressure and Velocity at Different Reynolds Number

The impact of non-Newtonian fluid flow on static pressure and velocity profile was investigated at four different Reynolds numbers (Re) of 100, 500, 1000 and 1500. The results obtained from the analysis are presented in the Table 55.3 in terms of Pressure drop and Co-efficient of discharge. Corresponding Plots of Static pressure versus

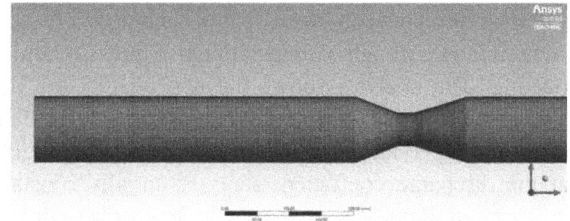

Fig. 55.3 Hex dominant meshed model of venturimeter
Source: Authors

location and velocity versus location are presented in Fig. 55.4 and Fig. 55.5 respectively.

Effect of Non-Newtonian Fluid on Coefficient of Discharge for Different Reynolds Numbers

Figure 55.6 illustrates how the discharge coefficient varies as the 'Re' value changes across different viscosity indices. It shows that the 'Cd' value increases with an increase in the 'Re' value, regardless of the fluid's viscosity.

Table 55.3 CFD analysis results for different cases

Case	Reynolds Number, Re	Power-law index, n	Consistency index, k	Inlet velocity (m/s)	Throat diameter, d (mm)	Pressure Drop, ΔP, (Pa)	C_d Numerical
1	100	1	0.5	1	50	11844.53	0.7957
2	100	0.8	1.314	1	50	11011.82	0.8253
3	100	0.6	3.4701	1	50	10179.16	0.8583
4	100	0.4	9.249	1	50	9417.72	0.8923
5	500	1	0.1	1	50	9479.25	0.8894
6	500	0.8	0.262	1	50	8726.39	0.927
7	500	0.6	0.694	1	50	8392.3	0.9453
8	500	0.4	1.849	1	50	8233.62	0.9544
9	1000	1	0.05	1	50	8720.78	0.9273
10	1000	0.8	0.131	1	50	8053.75	0.965
11	1000	0.6	0.347	1	50	8163.95	0.9584
12	1000	0.4	0.924	1	50	8112.29	0.9615
13	1500	1	0.033	1	50	8495.48	0.9395
14	1500	0.8	0.087	1	50	8415.01	0.94406
15	1500	0.6	0.231	1	50	8163.95	0.9584
16	1500	0.4	0.616	1	50	8112.29	0.9615

Source: Author's compilation

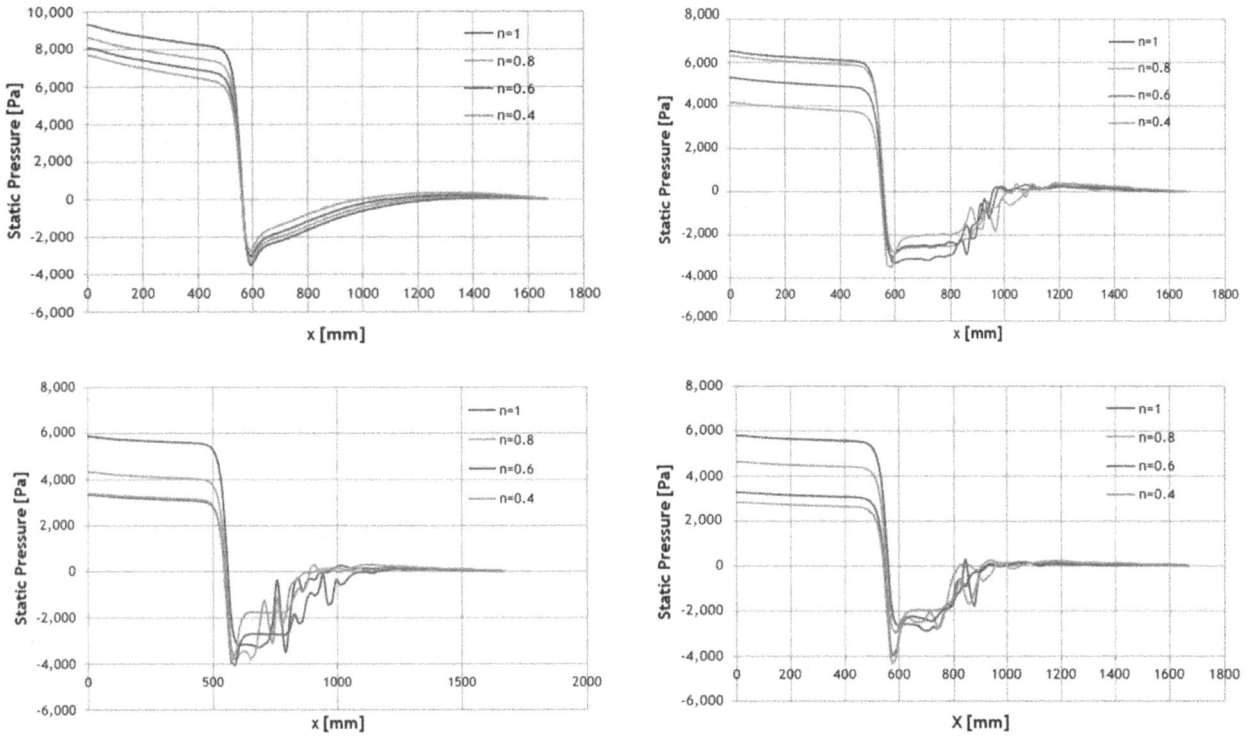

Fig. 55.4 Static pressure profiles

Source: Authors

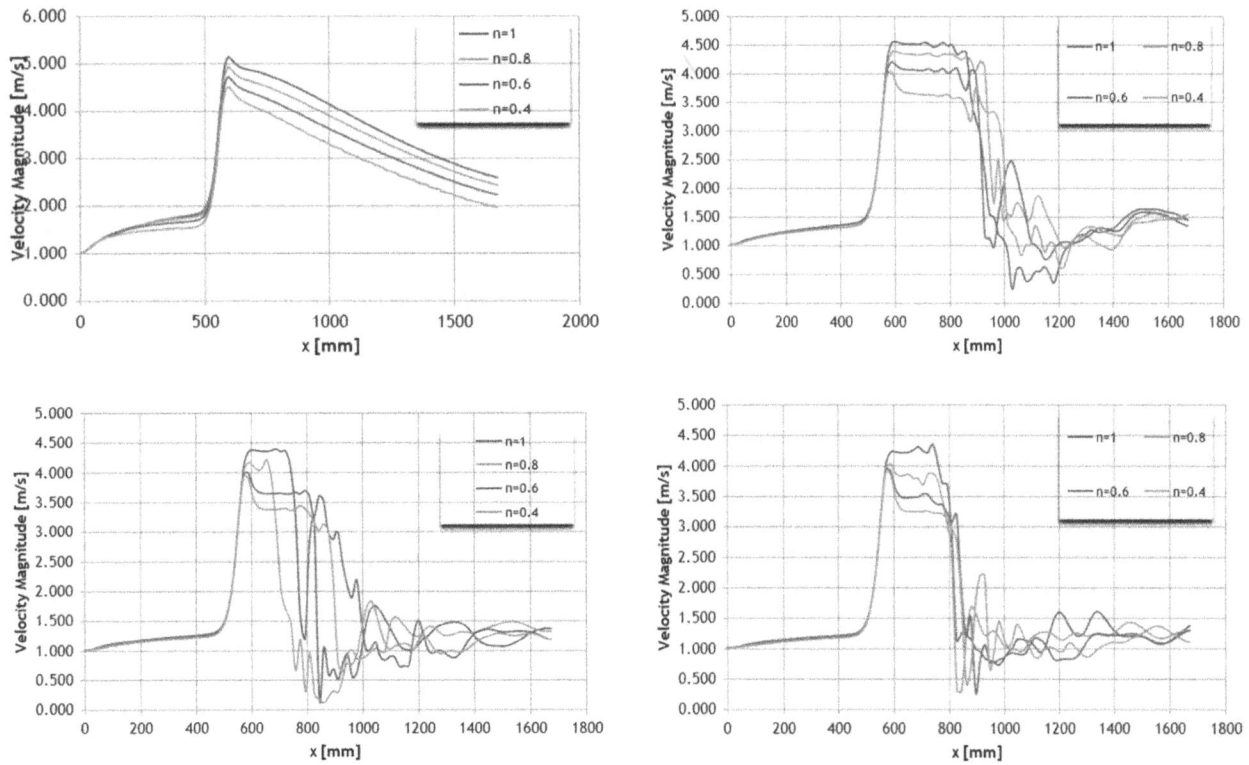

Fig. 55.5 Velocity profiles

Source: Authors

Validation Study of C_d Numerical and C_d Theoretical

From Fig. 55.7, it can be seen that the coefficient of discharge obtained from the CFD analysis (Cd numerical) is following the same trend as that of Cd theoretical, except at very low Reynolds numbers i.e at Re<500. Here the percentage of error (36.67 %) is found to be slightly higher when compared to that of Re>500. The C_d numerical and

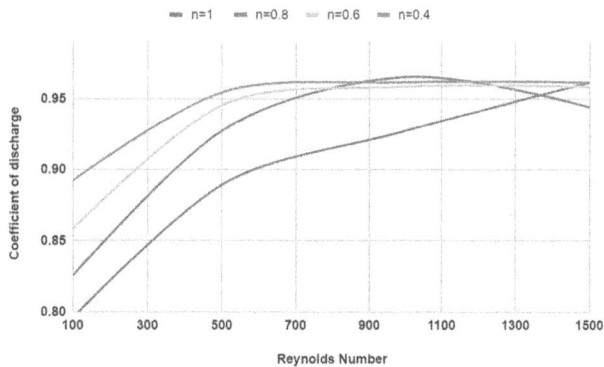

Fig. 55.6 Co-efficient of discharge versus reynolds number

Source: Authors

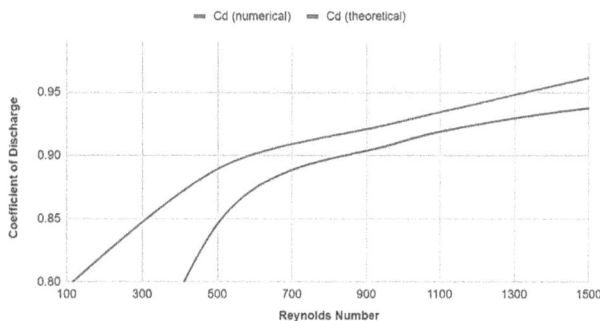

Fig. 55.7 Comparison of Cd numerical and Cd theoretical

Source: Authors

C_d theoretical are in good agreement for Re>500, with percentage error ranging from 1.75% to 5.16% and hence the methodology used is appropriate for analyzing non-Newtonian fluid flow at lower Reynolds numbers and furthermore research is needed to obtain precise results at even lower Reynolds number.

5. CONCLUSION

A thorough investigation has been carried out to use CFD to forecast the performance characteristics of venturimeters. This work used CFD modeling and simulation to show how the methodology can be used to assess non-Newtonian fluid flow in a venturimeter. The comprehensive information on the venturimeter flow characteristics that was difficult to determine during experimental tests was studied using the CFD results: The analysis was carried out for different Reynolds numbers, and a change in static pressure was observed with respect to an increase in viscosity. From the results, it is seen that with an increase in viscosity, there is a decrease in pressure drop for all cases. The analysis was conducted for different Reynolds numbers while maintaining a constant inlet velocity and varying the fluid's viscosity. It was found that as viscosity increased, the maximum velocity at the throat decreased in all cases. The Cd versus Reynolds number graph shows that 'Cd' increases as 'Re' value increases, regardless of the fluid's viscosity. The C_d numerical and Cd theoretical are in good agreement for Re>500, with percentage error ranging from 1.75% to 5.16% and hence the methodology used is appropriate for analysing non-Newtonian fluid flow at lower Reynolds numbers.

REFERENCES

1. Hollingshead et al "analyzed the discharge coefficient and performance of venturi, standard orifice meter, V cone and wedge flow meters at low Reynolds number for both laminar and turbulent flows", Journal of Petroleum Science and Engineering Volume 78, Issues 3–4, September 2011, Pages 559–566.
2. Arun R, Yogesh kumar K J, V Seshadri "CFD Analysis of the Effect of Defects in Welding and Surface Finish on the Performance Characteristics of Venturimeter", International Journal of Emerging Technology and Advanced Engineering, ISSN 2250–2459, ISO 9001:2008 Certified Journal, Volume 5, Issue 6, June 2015
3. Aldia, Carlo Burattob, Michele Pinellia, Pier Ruggero Spinaa, Alessio Sumana, Nicola Casaria, "CFD Analysis of a non-Newtonian fluids processing pump", OpenFOAM Workshop, Guimaraes, Portugal, June 26–30, 2016 11th.
4. Naveenji Arun, Malavarayan S, Kaushi. "CFD analysis of discharge co-efficient during non-Newtonian flows through orifice meter", International Journal of Engineering Science and Technology, Vol. 2(7), 2010, 3151–3164.
5. Tarun Kanti Bandyopadhyay, Sudip Kumar Das. "Non-Newtonian pseudoplastic liquid flow through small diameter piping components", Journal of Petroleum Science and Engineering, Volume 55, Issues 1–2, January 2007, Pages 156–166.
6. Prasanna M A, Dr. V. Seshadri, Yogesh Kumar K. J "Analysis of Compressible Effect in the Flow Metering By Orifice Plate Using CFD" IJSRSET, Volume 2, Issue 4, ISSN: 2395–1990, Online ISSN : 239

Advances in Mechanical Engineering and Materials Sciences – Dr. Vinay K. B et al. (eds)
© 2026 Taylor & Francis Group, London, ISBN 9-781-041-20970-6

56

Materials and Arrangement of Pots for Cost Effective Sustainable Cold Storage for Perishable Products—A Review

Thamme Gowda C. S.[1]
Assistant Professor, Department of Mechanical Engineering,
Vidyavardhaka College of Engineering,
Mysuru, Karnataka, India

G. B. Krishnappa[2]
Professor, Department of Mechanical Engineering,
Vidyavardhaka College of Engineering,
Mysuru, Karnataka, India

**Pratham P. Thantry[3], Pranav Tejaswi M. S.[4],
Mohammed Rayan[5], Mohamad Fazaluddin[6]**
Project Associate, Department of Mechanical Engineering,
Vidyavardhaka College of Engineering,
Mysuru, Karnataka, India

Abstract: The preservation of perishable agricultural products is essential for reducing post-harvest losses and ensuring food security, particularly in regions with limited access to reliable cold storage. Traditional refrigeration systems are often inaccessible for small-scale farmers due to high costs, reliance on electricity, and maintenance requirements. Consequently, there has been a growing interest in sustainable, low-cost storage solutions that utilize locally sourced materials, phase change materials (PCMs), and renewable energy sources like solar power. Research indicates that clay-sand composites, natural refrigerants, and PCMs can significantly improve the efficiency and environmental sustainability of cold storage solutions. Solar-powered systems offer practical applications for off-grid use, extending storage capabilities in rural and remote areas. While these solutions show promise, further research is needed to refine these technologies, particularly regarding scalability, material optimization, and the integration of renewable energy systems for large- scale storage. This review highlights both the progress and challenges in sustainable cold storage technology, aiming to guide future research and practical applications in the agricultural sector.

Keywords: Cold storage, Clay pots, Humidity, Perishable products, Shelf life, Temperature

1. INTRODUCTION

Minimizing post-harvest food losses plays a crucial role in enhancing global food security, especially in developing regions where access to reliable cold storage is limited. According to the Food and Agriculture Organization (FAO), nearly one-third of the food produced worldwide is either lost or wasted, with a substantial portion of this loss occurring due to inadequate storage after harvest. In many rural areas, the lack of consistent electricity makes it difficult to preserve perishable goods such as fruits, vegetables, and dairy products. This not only leads to

[1]thammegowda.cs@vvce.ac.in, [2]gbk@vvce.ac.in, [3]prathampthantry30@gmail.com, [4]adipani2505@gmail.com, [5]vvce21me0137@vvce.ac.in, [6]mohamadfazal38132@gmail.com

10.1201/9781003725053-56

financial setbacks for farmers but also results in higher food prices for consumers and unnecessary strain on natural resources. While conventional refrigeration systems offer an effective solution, they remain too expensive for small-scale farmers who depend on agriculture for their livelihood. The high costs associated with installation, maintenance, and energy consumption make these systems impractical in many rural settings. As a result, there is an urgent need for affordable and sustainable cold storage solutions that rely on locally available materials and renewable energy sources. Implementing such alternatives can help farmers extend the shelf life of their produce, minimize losses, and contribute to improved food security.

This review explores into recent advancements in sustainable cold storage technologies, highlighting innovative approaches such as natural material-based insulation, phase change materials (PCMs), and renewable energy systems. For example, clay-sand composites have emerged as a cost-effective and efficient insulation option, while PCMs help maintain stable temperatures by storing and releasing thermal energy as needed. By examining these technologies, this review aims to promote the development of accessible and sustainable cold storage methods that can support small-scale farmers, reduce food waste, and enhance food security in resource-constrained areas.

2. BACKGROUND AND SIGNIFICANCE

The spoilage of perishable agricultural products due to inadequate storage remains a significant global issue, particularly in developing regions where cold storage facilities are either limited or nonexistent. Post-harvest losses not only reduce farmers' earnings but also decrease food supply, leading to higher prices for consumers and worsening food insecurity. While traditional refrigeration methods are effective, they are often out of reach for small-scale farmers due to their high costs, energy demands, and ongoing maintenance requirements. Introducing affordable and sustainable cold storage solutions could be a game changer for rural agriculture, helping to prevent food waste, strengthen food security, and boost local economies. By leveraging locally available materials, such as clay-sand composite systems provide a practical and eco-friendly approach to food preservation. This research review examines recent innovations in sustainable cold storage technology, emphasizing their potential to bridge the gap between modern refrigeration and the specific needs of underserved communities.

3. LITERATURE REVIEW

Radebaugh et al. (1979) explored the use of electrocaloric materials, such as SrTiO3 ceramics, for refrigeration, particularly at cryogenic temperatures. Their research confirmed that these materials demonstrated a reversible electrocaloric effect, but the resulting temperature reductions were minimal, restricting their practical use. Nevertheless, the study laid important groundwork for future developments in solid-state cooling technologies. J R. Bevan et al. (1997) examined the use of cellar storage and air circulation systems for preserving organic field vegetables and potatoes. Common in Europe, cellars are often built beneath houses, other structures, or into hillsides, taking advantage of natural insulation. Their underground location helps maintain a consistent temperature, typically around 11°C, creating an optimal environment for crop storage. This stable temperature prevents freezing in winter and overheating during warmer months, effectively extending the shelf life of stored produce.

Joshua Folaranmi (2009) investigated different additives to improve the thermal insulation properties of clay. The study found that sawdust was the most effective, lowering the clay's thermal conductivity to 0.06 W/mK. This enhancement indicates that sawdust-infused clay could serve as a sustainable and practical insulating material for cold storage facilities. Li Xiaoyan et al. (2010) studied clay-sand mixtures as insulation materials for refrigerated vehicles used in cold chain logistics. By testing various clay-to-sand ratios, they identified compositions that improved thermal insulation while exhibiting a melting latent heat of 249 kJ/kg and a phase change temperature of -6.9°C. These properties helped maintain stable low temperatures and reduce moisture loss, making the material highly effective for preserving perishable goods during long-distance transportation. Macmanus C Ndukwu (2011) designed a clay evaporative cooler to enhance the storage of fruits and vegetables, demonstrating its effectiveness in improving preservation conditions. Their study showed that the cooler could lower the daily maximum ambient temperature from 32–40°C to 24–29°C, achieving a temperature drop of up to 8-11°C. Additionally, it significantly increased relative humidity from 40.3% in the surrounding environment to 92% inside the storage chamber. These findings highlight the cooler's ability to create a more favorable microclimate, making it a valuable solution for extending the shelf life of perishable produce, especially in hot, arid regions.

Prabodh Sai Dutt (2015) study examined the cooling efficiency of sand, charcoal, and gunny cloth as filler materials (Fig. 56.1) in Zeer pot refrigeration systems. The findings revealed that gunny cloth provided the best cooling performance, achieving a maximum temperature reduction of approximately 8.5°C compared to ambient conditions. Charcoal followed closely with an 8°C reduction, while sand achieved around 7°C. Gunny cloth's superior performance was attributed to its highwater absorption capacity and effective insulation, which enhanced evaporative cooling (Fig. 56.1, 56.2 & 56.3). Furthermore, tests conducted with perishable items, such as tomatoes, confirmed that gunny cloth consistently maintained lower internal temperatures, making it the most effective material for improving Zeer pot efficiency.

Fig. 56.1 Zeer pot filled with sand, charcoal & gunny cloth

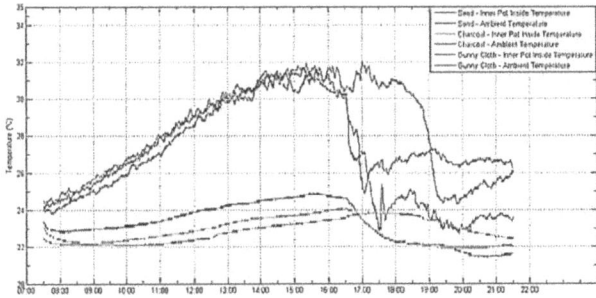

Fig. 56.2 Variation of inner pot Ti & Ta for different materials

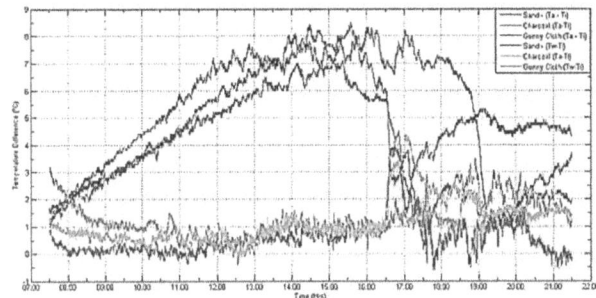

Fig. 56.3 Difference between Ta & Ti between Tw & Ti for different materials

B.G. Jahun et al. (2016) assessed an evaporative cooling system for vegetable storage, utilizing locally available materials. The system operates on the adiabatic cooling principle, where air passing over a wet surface lowers temperature while increasing humidity. Experimental results demonstrated that the system effectively reduced internal temperatures to 20–23.5°C for tomatoes and 20.5–26.5°C for hot peppers, compared to ambient temperatures of 25–28°C and 28–30.5°C, respectively. Additionally, the cabinet's relative humidity increased to 51–93%, significantly higher than the ambient range of 47–58%. This improved storage environment successfully extended the shelf life of tomatoes and hot peppers to 8 days, compared to just 3 days under normal conditions. The study analyses the cooling performance of two clay pot designs, including the Zeer pot and an alternative design with structural variations. Results show that both designs effectively reduce the internal temperature compared to ambient conditions, but the Zeer pot demonstrated better overall temperature stability. The internal temperature remained around 24.1°C with minor fluctuations (±0.8°C),

while the ambient temperature peaked at 31.5°C. The Zeer pot maintained a consistent cooling effect throughout the day, ensuring a stable environment for preserving perishable goods. The study highlights the effectiveness of evaporative cooling in clay pots and underscores the importance of design optimization to enhance performance in electricity- scarce regions.

Prabodh Sai Dutt (2016) conducted an experimental comparative analysis of clay pot refrigeration using two different designs of pots, focusing on temperature variations in the Zeer Pot design (Fig. 56.5 & 56.6). The study observed that, despite the ambient temperature reaching its highest point at 31.5°C around noon, the temperature inside the storage space showed only a marginal increase, peaking at 24.9°C before gradually decreasing in the evening. The minimum temperature recorded was 23.3°C, with a mean of 24.1°C and a variation of only ±0.8°C. This indicates that the Zeer Pot design maintains a stable internal temperature, offering effective cooling despite significant fluctuations in the surrounding environment ±3.35°C (Fig. 56.7, 56.8 & 56.9).

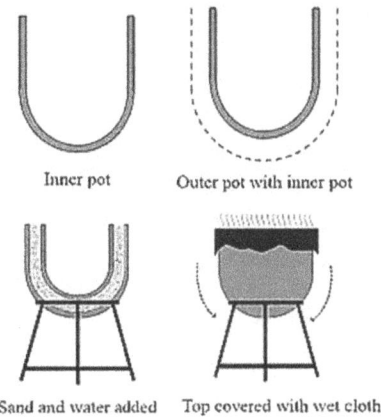

Fig. 56.4 Schematic illustration for building a pot-in-pot system

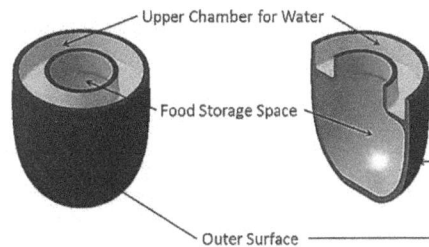

Fig. 56.5 Zeer pot design with cross section

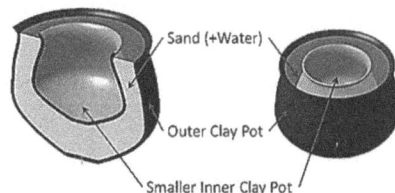

Fig. 56.6 Pot with upper chamber design with cross section

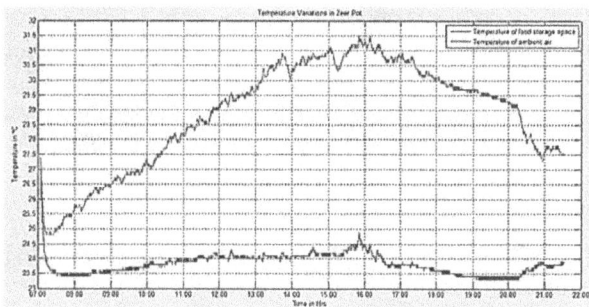

Fig. 56.7 Temperature variations in zeer pot

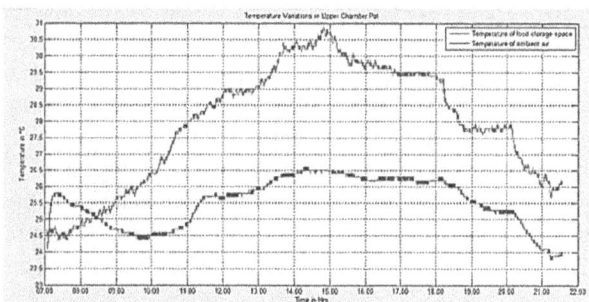

Fig. 56.8 Temperature variations in upper chamber pot design

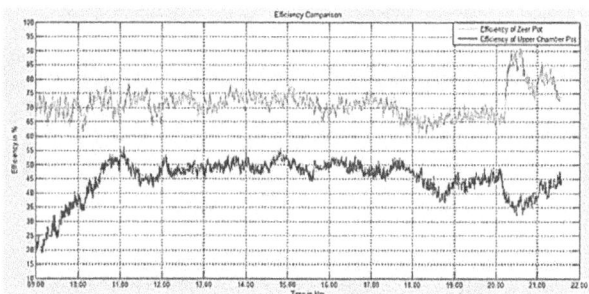

Fig. 56.9 Comparison of efficiency variations

Ernawati et al. (2016) explored the use of sand storage as a method for extending the shelf life of fresh sweet potato roots for both home consumption and market sales. This method shows great potential for adoption by sweet potato farmers, particularly in countries like Malawi and Ghana, and could be a viable solution for improving food security and market availability in regions with seasonal variability in crop supply. Mishra et al. (2017) studied the effects of modified atmosphere packaging (MAP) on cauliflower stored under ambient and evaporative cooling (EC) conditions in Nepal. The research found that cumulative weight loss was influenced by both storage conditions and the number of perforations in the MAP. EC maintained higher humidity and slightly lower temperatures, significantly reducing weight loss. After five days, cauliflower stored under EC lost only 4.03% of its weight, over 50% less than the 11.58% weight loss under ambient conditions. The study concluded that MAP, particularly with fewer perforations, effectively minimized weight loss when combined with evaporative cooling.

Mandal and Mondal (2017) developed an innovative solar-powered system combined with phase change materials to reduce the operational energy costs of cold storage. Their system maintained consistent storage temperatures for perishable agricultural products, such as fruits and vegetables, while reducing energy consumption by 30%. The use of solar energy with PCMs also created a thermal buffer, allowing the storage unit to operate efficiently in areas with unreliable electricity. Chemin et al. (2018) examined heat transfer and evaporative cooling in pot-in-pot coolers, particularly focusing on the cooling behavior of ice in the system. Their study revealed discrepancies in the final cooling regime compared to predictions from governing equations. The simplified model suggested a sharper initial temperature decrease and a distinct kink in the temperature curve due to lower heat capacity of ice and its higher thermal conductivity, which should result in a much shorter cooling time for ice. However, experimental data contradicted these expectations, pointing to the

Table 56.1 Comparison of the efficiency of lining media

Time of Day	Inner Temperature T_{in} (°C)			Ambient Temp	Wet Bulb Temp	Relative Humidity	Wind Speed	$T_{amb} - T_{wet}$	$\varepsilon = (T_{amb} - T_h)/(T_{amb} - T_{wet})$ (%)		
	Sand	Sawdust	Charcoal	(°C)	(°C)	(%)	(m/s)	(°C)	Sand	Sawdust	Charcoal
8:00	23.9	24.1	23.9	21.2	21.2	83.3	2.1	3.0	0.10	0.03	0.10
9:00	23.7	24.1	23.6	25.7	21.5	78.4	3.1	4.1	0.48	0.38	0.51
10:00	23.6	24.0	23.0	27.7	21.2	70.1	4.2	6.5	0.63	0.57	0.72
11:00	23.1	24.2	22.3	29.7	20.7	59.0	3.9	9.0	0.70	0.61	0.82
12:00	23.9	24.1	22.0	31.5	20.0	50.3	3.0	11.8	0.73	0.63	0.87
13:00	23.3	24.0	21.8	32.3	20.0	43.0	3.0	13.0	0.76	0.69	0.89
14:00	23.0	23.9	21.6	34.2	19.5	40.1	3.4	15.8	0.77	0.72	0.91
15:00	22.8	23.8	21.2	35.3	19.0	40.0	3.1	16.8	0.77	0.73	0.92
16:00	22.7	23.7	21.2	34.3	18.9	38.5	3.2	16.3	0.76	0.73	0.91
17:00	22.6	23.7	20.8	33.8	18.5	38.0	3.2	15.9	0.77	0.73	0.88
18:00	22.0	23.7	20.6	33.8	18.5	40.5	3.2	15.3	0.77	0.73	0.86

complexities in cooling process that require further investigation to optimize evaporative cooling systems. The Suleiman Abimbola Yahaya and Kareem Adeyemi Akande [2018] focuses on assessing the cooling efficiency of a pot-in-pot evaporative cooling system designed to preserve perishable items in hot, dry climates. Conducted in Ilorin, Nigeria, the research evaluates the performance of the device by comparing sand, charcoal, and sawdust as lining materials (Fig. 56.10). The results highlight that charcoal achieved the highest cooling efficiency, with a mean value of 75%, followed by sand at 66% and sawdust at 58%. This indicates that charcoal is the most effective material for maintaining lower internal temperatures (Table 56.1).

Fig. 56.10 Comparison of the cooling different lining media

The temperature reduction achieved by the pot-in- pot system varied with the lining material, with charcoal-lined pots demonstrating the most significant drop in temperature compared to ambient conditions. The system's effectiveness was further enhanced in low-humidity environments, which are typical of the Ilorin region, making it particularly suitable for arid and semi-arid climates. Relative humidity also played a critical role in optimizing the evaporative cooling process (Fig. 56.11, 56.12 & 56.13). Tanzeela Nisar et al. (2019) researched the efficacy of pectin- based edible coatings infused with clove essential oil for the preservation of bream fish fillets. The coatings significantly reduced water loss,

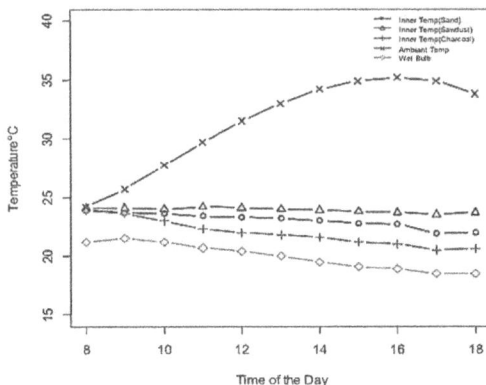

Fig. 56.11 Comparison of sand, sawdust of and charcoal as lining media

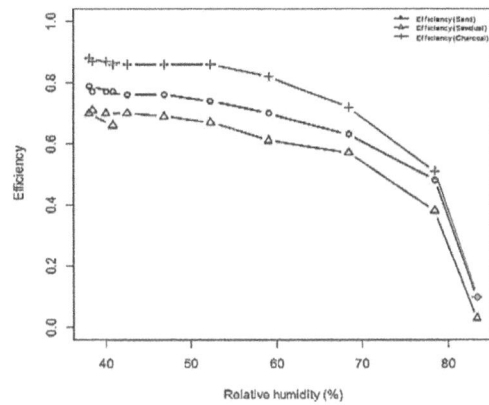

Fig. 56.12 Efficiency vs relative humidity (%) for different lining media

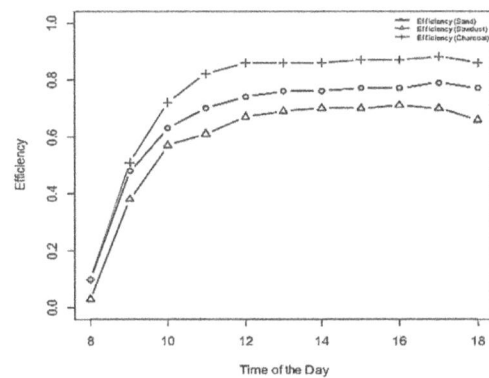

Fig. 56.13 Efficiency vs time day for different media

delayed oxidation, and maintained the fillets' texture and color over time. This study demonstrates the potential of using natural, biodegradable coatings as a preservative method, effectively extending the shelf life and quality of fish products.

Danyal Rehma et al. (2020) investigated clay pot evaporative coolers for the preservation of vegetables in low-resource settings. By reducing ambient temperatures by 6–8°C, these coolers effectively extended the freshness period of vegetables by an additional 4–5 days. This low-cost, fuel-free technology is ideal for regions without reliable refrigeration infrastructure, providing a sustainable method for small-scale agricultural preservation. Kaur et al. (2021) reviewed storage structures for horticultural crops, focusing on chambers designed for direct evaporative cooling (DEC). These systems operate without the need for electricity or external power sources, using locally available materials such as bricks, sand, and bamboo for construction. The DEC system can reduce the temperature of stored crops by 10–15°C while maintaining a high humidity level of nearly 90%.

Benitez et al. (2021) examined the impact of hot water treatment and sodium hypochlorite on tomatoes stored in three different evaporative coolers (EC). The study found that EC3, which utilized a water pump without an exhaust fan, was the most effective in preserving the tomatoes. It delayed changes in peel colour, maintained better visual quality, and extended shelf life, while also significantly reducing weight loss. The research showed that a 1-minute hot water treatment at 45°C was more effective than using 1% calcium hypochlorite or tap water in enhancing the tomatoes' post-harvest quality when stored in an evaporative cooling system. Humaira Gul et al. (2022) examined the effects of different sand and clay compositions on the growth and biochemical properties of *Hordeum vulgare* (barley), as well as the physio-chemical properties of the soil. The study revealed that soil moisture increased significantly (P<0.001) with the addition of 20% sand, and the moisture content continued to rise as the clay content in the soil was increased. This increase is attributed to the negative charge of clay particles, which attract water molecules through electrostatic forces. The highest hygroscopic water values were recorded in 100% clay soil, which showed a significant improvement in moisture retention, benefiting plant growth.

Faizan Ahmed et al. (2023) conducted an experimental assessment of a multipurpose evaporative desert cooler, which serves dual functions: refrigeration and air conditioning for outdoor applications. The system works by transferring heat from the storage box into the cooling water in the tank, which causes the water temperature to decrease. Initially, the cooling rate is higher, but it gradually slows and stabilizes after the first half hour. This study highlights the efficiency of evaporative cooling systems in managing temperature and providing cooling in extreme climates without the need for additional power. Choudhary et al. (2023) conducted a performance analysis of an evaporative (pot-in-pot) cooling system, evaluating factors such as relative humidity, ambient temperature, pot radius, pot height, hydraulic conductivity, and heat load and found that these key parameters significantly impact the system's efficiency. The maximum efficiency of 73.44% was achieved at a relative humidity (RH) of 40% and a temperature of 50°C, resulting in a temperature difference of 20°C.

Aboubacar Sidiki Drame et al. (2023) explores the optimization of the "pot-in-pot" bush refrigerator, designed for food preservation in hot, rural climates where electricity is scarce. Results show that cylindrical jars, a water medium between jars, and minimal closure yield the best cooling performance, with temperatures reduced significantly via evaporative cooling. Economically, the bush refrigerator is ten times cheaper to operate than an 85W electric refrigerator, costing only 472.5 CFA per month. This technology enhances food security by reducing spoilage, supporting rural livelihoods, and promoting sustainable, low-cost refrigeration methods. Brijendra Singh Choudhary et al. (2023) in their study evaluates the factors influencing the efficiency of evaporative cooling in pot-in-pot systems. Key parameters analyzed include pot radius, pot height, ambient temperature, relative humidity,

heat load, and hydraulic conductivity of the materials. The study found that the cooling efficiency is significantly affected by pot dimensions and environmental conditions, with the maximum efficiency reaching 73.44% at a relative humidity of 40% and an ambient temperature of 50°C. Under these conditions, the pot-in-pot system achieved a temperature reduction of up to 20°C compared to the surrounding air. The research underscores the importance of optimizing design parameters to maximize evaporative cooling performance.

Gunadasa HLCK et al. (2023) in their study explores the potential of using sawdust, coir dust, and crushed corn husk with corn hair as substitutes for sand in Zeer pots to improve cooling efficiency. The results reveal that sawdust and coir dust perform comparably to traditional sand, achieving a temperature reduction of 2–3°C below ambient conditions. Additionally, these materials enhanced the relative humidity within the pots, creating a favorable microclimate for preserving perishable goods like fruits and vegetables. In contrast, crushed corn husk with corn hair proved ineffective due to its lower cooling efficiency and microbial activity, which occasionally led to heat generation instead of cooling. The study also found that sawdust and coir dust offered average cooling efficiencies ranging between 58%–69%, which aligns closely with sand's performance. These findings suggest that sawdust and coir dust are viable, sustainable alternatives to sand, offering an eco-friendly solution by repurposing plant-based waste materials. Their successful implementation could improve the affordability and accessibility of evaporative cooling systems, particularly in regions lacking electricity or refrigeration infrastructure.

In conclusion, the study confirms that pot-in-pot cooling devices are an affordable and sustainable solution for preserving perishable goods in regions with limited access to electricity. Charcoal, due to its superior cooling performance, is recommended as the preferred lining material for improving the efficiency of evaporative cooling systems.

4. METHODOLOGY

The complete refrigeration process takes place here by evaporation process (Fig. 56.15) and it depends on the climate conditions such as temperature and humidity (Fig. 56.14 & 56.16) and (Table 56.2). This paper primarily highlights the use of various additives in clay-sand compositions to enhance the cooling efficiency of storage units, specifically focusing on perishable agricultural products (Table 56.3). The study focusses on the selection of materials that optimize water retention to maintain effective performance under various environmental conditions (Fig. 56.1 & 56.10). To improve the thermal conductivity, reduce heat absorption, and ensure long-term durability of the storage system, the selected additives are integrated into the clay-sand mixture. Additionally,

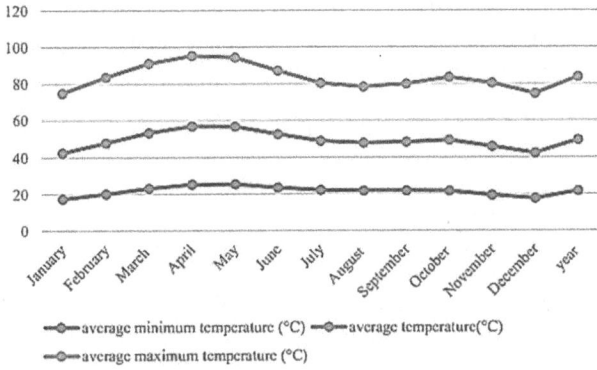

Fig. 56.14 Annual evolution of temperatures in the city of Bamako (climate-charts. com, 2022)

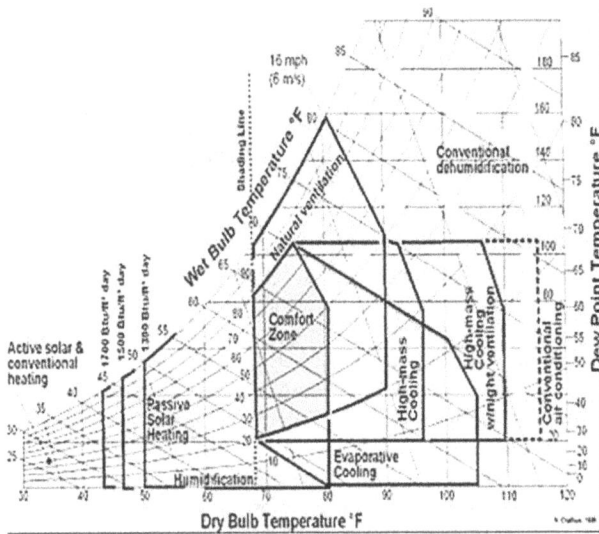

Fig. 56.15 Psychometric chart explaining principle of evaporating cooling

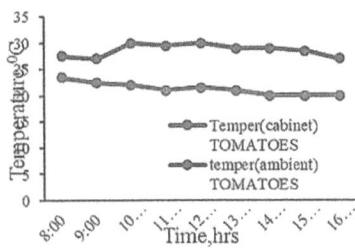

Fig. 56.16 Temperature of cabinet and ambient condition against time for tomatoes

Table 56.2 Relative humidity and temperature of cabinet and ambient for tomatoes

Time (Hr)	Temp (Cabinet)	RH (cabinet)	Temp. (Ambient)	RH (ambient)
8: 00	23.5 °C	51 %	25.5 °C	58 %
9:00	22.5 °C	52 %	25 °C	54 %
10: 00	22 °C	56 %	28 °C	49 %
11:00	21 °C	65 %	27.5 °C	48 %
12: 00	21 °C	67 %	28 °C	47 %
13: 00	21 °C	81 %	27 °C	49 %
14: 00	20 °C	83 %	27 °C	50%
15: 00	20 °C	91 %	26.5 °C	51 %
16: 00	20 °C	93 %	25 °C	56 %

Table 56.3 Vegetable shelf-life

Product	Shelf-life Produce without zeer	Shelf-life Produce with zeer
Tomatoes	2 days	20 days
Guava Rocket	2 days	20 days
Okra	1 days	5 days
Carrots	4 days	17 days

energy sources, and regional-specific refrigeration solutions. Future work should aim at improving the energy efficiency of these technologies, assessing their environmental impact, and enhancing adaptability for varying climatic conditions.

6. CONCLUSION

Advancements in cold storage and refrigeration are increasingly focused on sustainable solutions that extend shelf life, enhance food safety, and minimize environmental impact. Emulsion stabilization using CFCs improves preservation by reducing oxygen permeability, while Cold Atmospheric Plasma (CAP) offers a non-thermal approach to microbial control. The use of phase change materials and clay-sand mixtures to maintain consistent temperatures highlights the growing emphasis on material efficiency in refrigeration systems.

Innovative technologies, such as refrigerated containers with controlled atmosphere (CA) technology and biodegradable packaging, further contribute to sustainable cold chain logistics by extending the freshness of perishable goods while reducing waste. Additionally, research into alternative cooling systems, including electrocaloric and gravity refrigeration, provides low-energy, region-specific solutions for challenging climates.

Future research should focus on developing cost-effective ways to scale these technologies, evaluating their adaptability, and addressing potential limitations across diverse environmental conditions. By prioritizing innovation and sustainability, the cold storage sector is

statistical methods, such as Design of Experiments (DOE), are applied to systematically determine the optimal combination of materials, ensuring an efficient and sustainable cold storage solution.

5. IMPLICATIONS AND CONSIDERATIONS FOR FUTURE WORK

The ongoing research underscores the need for further exploration into biodegradable materials, alternative

well-positioned to meet the rising demand for energy-efficient and environmentally responsible cooling solutions.

REFERENCES

1. Aboubacar Sidiki Drame, Oumar Hamadoun, Mamadou M. Diarra, Tamba Camara, "Performance optimization of a bush refrigerator "the pot in pot" for hot climate zones: Mali case study", 2023.
2. Benitez, Benito, F.R. River, Jenny "Post-harvest characteristics of tomatoes (Solanum Lycopersicon) as affected by treatment with hot water and sodium hypochlorite under three simple evaporative coolers", 2021.
3. Binayak Prakash Mishra, S.S. Pant, K.D.M. Gautam, J.D.H. Keating "Modified atmosphere packaging of cauliflower under ambient and evaporative cooling conditions in Nepa", 2017.
4. Brijendra Singh Choudhary, Ravi Kiran T, Satish Kumar Dewangan, 2023 "Review Study for Performance Analysis of an Evaporative (Pot in Pot) Cooling System".
5. B.G Jahun, S.A. Abdulkadir, S. M. Musa, Huzaifa Umar "Assessment of Evaporative Cooling System for Storage of Vegetables", 2016.
6. Choudhary, Ravi, Satish "Review Study for Performance Analysis of an Evaporative (Pot in Pot) Cooling System", 2023.
7. Chemin, Victor Levy Dit Vehe, Aude Caussarieu, Nicolas Taberlet "Heat transfer and evaporative cooling in the function of pot-in-pot coolers", 2018.
8. Danyal Rehma, Ethan McGarrigle, Leon Glicksman Eric Verploegen "A heat and mass transport model of clay pot evaporative coolers for vegetable storage", (2020).
9. Dipankar Mandal, Sourav Mondal "International Journal of Applied Science and Engineering Research", 2017.
10. Faizan Ahmed, S. Fero, Waqar Khan, Nageswara Rao Lakkimsetty "Experimental assessment of multipurpose evaporative type cooler used for refrigeration and air cooling", 2023.
11. Gunadasa HLCK, Awanthi MGG and Rupasinghe CP, 2023 "Utilization of Plant Based Waste Materials as Alternatives to Sand in Zeer Pot Refrigerator".
12. Humaira Gul, Muhammad Junaid Yousaf, Fawad Ali "Effects of Sand and Clay Compositions on Growth and Biochemical Aspects of Hordeum vulgare (L.) and Soil Physio-Chemical Properties", 2022.
13. J R. Bevan, C. Firth, M. Neicho "Storage of organic field vegetables and potatoes: Cellar store and air circulation around shelves." 1997.
14. Joshua Folaranmi "Effect of Additives on the Thermal Conductivity of Clay" 2009.
15. Kaur, J, Aslam, R, Afthab Saeed.P "Storage structures for horticultural crops: a review" 2021.
16. Li Xiaoyan, Zhou Dong, Yang Shuting "Study of new cool storage materials for refrigerated vehicle in cold chain", 2010.
17. Macmanus C Ndukwu "Development of Clay Evaporative Cooler for fruits and vegetables preservation", 2011.
18. Prabodh Sai Dutt, 2015 "Experimental Study of Alternatives to Sand in Zeer Pot Refrigeration Technique".
19. Prabodh Sai Dutt "Experimental Comparative Analysis of Clay Pot Refrigeration Using Two Different Designs of Pots", 2016.
20. Putri Ernawati Abidin, John Kazembe, Richard A. Atuna, Francis Kwaku Amagloh, Kwabena Asare, Eric Kuuna Dery and Edward Ewing Carey "Sand Storage, Extending the Shelf-Life of Fresh Sweet potato Roots for Home Consumption and Market Sales", 2016.
21. R. Radebaugh, W.N. Lawless, J.D. Siegwarth, A.J. Morrow "Feasibility of electro caloric refrigeration for the 4–15 K temperature range", 1979.
22. Suleiman Abimbola Yahaya, Kareem Adeyemi Akande, 2018 "Development and Performance Evaluation of Pot-in-pot Cooling Device for Ilorin and its Environ".
23. Tanzeela Nisar, Xi Yang, Aamina Alim, Muneeb Iqbal, Zi-Chao Wang, Yurong Guo "Megalobrama ambycephala) to pectin-based coatings enriched with clove essential oil during refrigeration," 2019.

Note: All the figures and tables in this chapter were made by the authors.

Advances in Mechanical Engineering and Materials Sciences – Dr. Vinay K. B et al. (eds)
© 2026 Taylor & Francis Group, London, ISBN 9-781-041-20970-6

57

Design and Fabrication of Refrigeration System for Physically Challenged

Amruth E.[1],
S. A. Mohan Krishna[2],
N. Jayashankar[3], Kamal C.[4]
Department of Mechanical Engineering,
Vidyavardhaka College of Engineering,
Mysore, Karnataka, India

Abstract: The questions on disability status were incorporated in the census conducted in the year 2001 and 2011 by the Registrar General and Commissioner of Census in India. The outcomes of the 2011 census discovers that 2.68 crore population out of 121 crores has been classified as disabled, which comes is 2.21% of total population of the country. Out of this percentage, individuals settled in rural areas of the country are facing identical challenges of hypertension, which is an important health concern. This study explores on occurrence and its impact on hypertension especially among differently challenged population residing in rural areas. The research identifies important factors like lack of sufficient physical activity, side effects from the medicines and comorbid conditions which are contributing to the difficulty in managing high blood pressure amongst the group. Further, it focusses on role of critically managing blood pressure by constant monitoring, engaging in regular physical activity, tailoring stress management ways, and engaging themselves in community support programs for managing hypertension among this demography. The objective of this study is not only the empowerment of differently abled individuals but also to increase their fitness but also to utilize energy being generated from them and for controlling medications in a small chamber by ensuring regular and sustainable supply of medicines they are in need of.

Keywords: Disabled, Physical activities, Fitness

1. INTRODUCTION

Physically challenged individuals face lot of challenges in terms of their free mobility, routine activities, and managing their energy levels. The effects of their challenges results in various medical issues in respiration, vision, and change in sleep patterns. Very high blood pressure known as hypertension, a result of health hazard increases the blood pressure (Stanaway JD et al., 2018) (G. A. et al., 2018). The impact of increased blood pressure concerned with physically challenged individuals staying at rural areas as posing the health issues and creating obstacles in everyday routine activities in their life. By recognizing complementary effect between high blood pressure and physical disabilities being crucial, particularly in rural areas, where accessing healthcare and support facilities being limited (Mills KT et al., 2016) (Zhou B et al., 2017).

The RG & CCI involved issues about infirmity standing in the Census 2001 & 2011. The findings from Census 2011 reveal that out of India's sum population of 121 crore (Cr), 2.68 Cr individuals are classified as 'disabled,' comprising 2.21% of the overall population. Among the disabled population, 56% (1.5 Cr) are males, and 44% (1.18 Cr) are females. Interestingly, the rural-urban distribution shows that 69% of the disabled population resides in rural areas (1.86 Cr in rural areas and 0.81 Cr in urban areas), mirroring the general population distribution. The

[1]Amruth.e@vvce.ac.in, [2]mohanakrishna.sa@ vvce.ac.in, [3]jayshankar@vvce.ac.in, [4]kamalchandrashekaran@gmail.com

proportion of incapacitated individuals among males and females stands at 2.41% and 2.01%, respectively, at the national level and across different social groups (Disability in India, 2019).

Engaging in physical activity is essential for enhancing health, well-being, and overall quality of life. As outlined in the second edition of the *Physical Activity Guidelines for Americans*, it plays a crucial role in maintaining overall wellness (2nd edition) (U.S. Department of Health and Human Services, 2018), engaging in regular physical activity helps manage weight, improves mental health, and reduces the risk of chronic illnesses like heart disease, type 2 diabetes, and certain cancers. For individuals with disabilities, engaging in physical activity is crucial for supporting daily activities and maintaining independence. The Guidelines mention minimum of 150 minutes of sensible-power aerobic commotion per calendar week for grownups, broken down into manageable daily increments, along with muscle-strengthening activities for additional health benefits (Centres for Disease Control and Prevention, 2022).

Here are some things to remember:

- High BP is a concern for individuals with disabilities residing in rural areas, just like it can affect anyone else.
- Managing BP may pose additional challenges for people with disabilities due to their specific conditions.
- High BP can exacerbate other health conditions in individuals with disabilities.
- Expected BP investigations are needed since huge BP often presents no noticeable symptoms.
- Tailoring physical activities to each person's disability can aid in better BP management.
- Stress management and relaxation techniques are crucial as stress can worsen high BP.

2. PROBLEM STATEMENT

Address the challenges posed by hypertension and low physical activity levels, we propose an innovative solution aimed at reducing hypertension and promoting physical activity (Ezzati, Majid, et al., 2015) (Falaschetti, E., J. Mindell et al., 2014). Our approach involves integrating an internal refrigeration system into the healthcare regimen. Studies have shown that the prevalence of hypertension is significantly higher among individuals with physical disabilities (47.5%) compared to those with intellectual or mental disabilities (24.9%) (Jiang, Feifei, et al., 2021). By incorporating an integrated refrigeration system, we not only facilitate the storage of medications but also leverage physical activity to power the refrigeration system. This approach offers several advantages, including cost savings, efficient load management, and reduced environmental impact of equipment usage (Super Radiator Coils., 2023).

3. OBJECTIVE

3.1 Promote Physical Activity for Health

Encourage regular physical activity to improve heart health, enhance blood circulation, and manage weight, all of which help lower high blood pressure (BP).

3.2 Effective High BP Management

Address the risks of high BP, which puts excessive pressure on artery walls. Effective management through diet, exercise, stress control, and medical guidance can prevent serious health complications.

3.3 Support Sustainable Energy

Advocate for renewable energy sources like solar and wind to reduce pollution and promote environmental sustainability for future generations.

3.4 Apply Integrated Design Approaches

Leverage interdisciplinary design methods to solve complex problems efficiently, optimizing resource use and fostering innovation.

3.5 Improve Cold Storage Solutions

Emphasize the importance of proper cold storage in healthcare to maintain the safety and efficacy of critical supplies like vaccines and blood products.

4. METHODOLOGY

The design methodology for a wheelchair with an integrated refrigerator combines a compact cooling system within the wheelchair's structure, carefully considering power sources, insulation, and accessibility, while requiring a balance of engineering and design expertise to ensure both functionality and safety.

This design incorporates a Peltier module, a thermoelectric device that generates heating and cooling effects through the Peltier effect by using two different conductive materials with a junction between them, where an applied electric current causes one side to become hot and the other side to become cold.

The following steps are followed typically to implement the Peltier module:

1. *Powering the Peltier Module:* Ensure the Peltier module is connected to an appropriate DC power source, following the manufacturer's voltage and current guidelines for optimal performance.
2. *Employ a Heat Sink and Fan for Efficient Cooling:* Given that the module generates heat on one side and cold on the other, it's crucial to effectively dissipate the heat from the hot side. Integrating a heat sink and fan helps maintain optimal performance by improving heat transfer and preventing overheating.

3. *Implement temperature control:* A temperature controller is employed to regulate the power delivered to the module, thereby achieving the desired cooling or heating effect. This allows for precise temperature adjustments as per requirements.

4. *Consider the limitations of Peltier modules:* It is important to note that Peltier modules have limitations in terms of cooling capacity and efficiency. Therefore, specific application requirements should be carefully considered during the design process.

5. COMPONENTS AND DESCRIPTION

The components important for the fabrication of wheel chair to store the medicine required for the emergency purpose is described below and also, they are also shown in the Figs. 57.1, 57.2, 57.3 & 57.4 respectively.

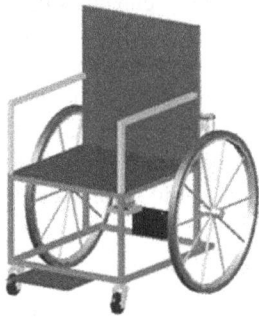

Fig. 57.1   Wheelchair

Fig. 57.2   Dynamo

Fig. 57.3   Battery

Fig. 57.4   Peltier model

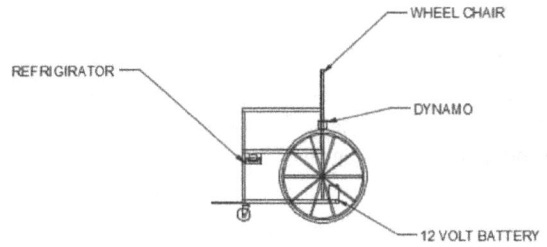

Fig. 57.5   Wheelchair with refrigeration

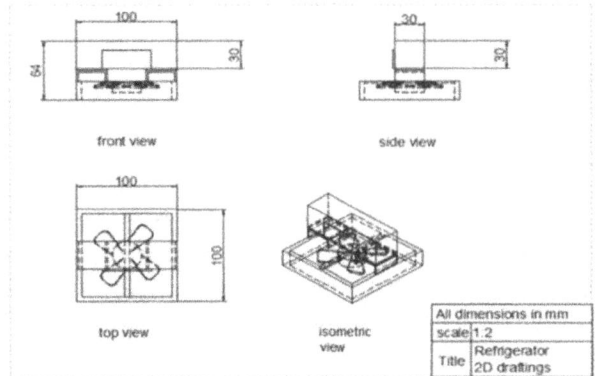

Fig. 57.6   Peltier module

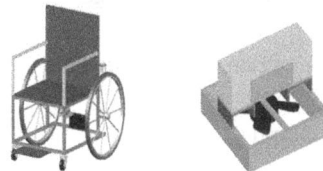

Fig. 57.7   3D Model

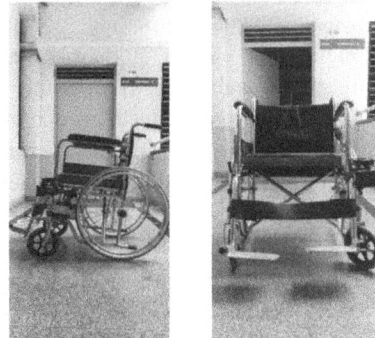

Fig. 57.8   Fabricated model

5.1 Wheel Chair

Wheelchairs provide individuals who cannot walk or have difficulty walking with a means of mobility. They also aid in maintaining their posture and granting them the independence to engage in various activities on their own. The role of wheelchairs for physically challenged individuals is pivotal in improving their mobility and fostering independence. In essence, contemporary wheelchairs for such individuals strive to offer not only mobility but also comfort, independence, and seamless integration into their daily lives and surroundings.

5.2 Dynamo

Dynamos, also known as generators, utilize the principle of electromagnetic induction to produce electric current. This generated electricity can then be used to power various electrical components or systems within the mechanical device. Dynamos are commonly used in applications such as bicycles to generate electricity from pedaling motion, in automobiles to charge batteries and power onboard electronics, and in industrial machinery to provide auxiliary electrical power. Present study the dynamo used is of 14 volts.

- **Maximum power generated by the dynamo p_g:**

  p_g = maximum current generated × maximum voltage generated.

  $p_g = 0.7$ Amps × 12 V = 8.4 watts

- **Power consumption:**

  Power consumption = voltage drawn × amps consumed

  Power consumption of peltier module = V × I = 12 × 3 = 36 watts

  Power consumption of thermostat = V × I = 12 × 0.5 = 6 watts

  Power consumption of colling fan = V × I = 12 × 1 = 12 watts

  Total power consumption = 36 + 6 + 12 = 54 watts

5.3 Battery

Batteries are important for efficiently storing electrical energy, providing backup power, enabling portability, managing energy usage, integrating renewable energy, addressing peak demand, supporting transient power needs, and facilitating energy storage systems across various applications and industries. They serve the purpose of efficiently storing electrical energy for later use by converting chemical energy into electrical energy and vice versa through a special chemical reaction called oxidation-reduction. The battery currently used in the research is 12 volts, which has been found to be efficient for this research. It is also noted that battery charging time and battery backup time are very important parameters that determine the performance and sustainability of a battery. In this study, the battery charging time was observed to be

10.7 hours, while the battery backup time was recorded at 1.7 hours.

- **Battery Charging Time (BCT):**

$$BCT = \frac{battery\ capacity}{amps\ generated} = \frac{7.5\ Ah}{0.7\ A} = 10.7\ hrs$$

- **Battery Backup Time (BBT):**

$$BBT = \frac{battery\ capcity}{Amps\ drawn} = \frac{7.5\ Ah}{4.5\ A} = 1.7\ hrs$$

5.4 Peltier Module

A Peltier module, commonly known as a thermoelectric cooler, is a device designed for temperature regulation, capable of both cooling and heating. It operates by modifying the temperature of objects in contact with it when an electric current flows through it. This module is essential in reducing the internal temperature of equipment, which greatly enhances the performance and longevity of electronic devices, scientific instruments, and other temperature-sensitive tools. The Peltier module is particularly efficient, capable of lowering the internal temperature of a chamber from 32°C to 20°C in just 15 minutes. The chamber has dimensions of 52 mm × 52 mm × 114 mm, with a total volume of 308.256 mm³.

6. FUTURE SCOPE

The idea of adding a refrigerator to wheelchairs for storing insulin and medicine is interesting. Here are key considerations for its development:

- *Temperature Control:* Insulin and medication need to be kept at specific temperatures to stay effective. A small, efficient fridge could be designed for wheelchairs, ensuring it works well even in hot climates.

- *Design and Safety:* The fridge should be lightweight, compact, and securely attached to the wheelchair. It must have excellent insulation to maintain the right temperature regardless of external weather conditions. Power could come from rechargeable batteries or solar panels, and safety features should prevent any risk of injury from cold surfaces or electrical components.

- *Affordability:* It's crucial to keep the cost low, especially for those in developing countries. Using affordable, durable materials and manufacturing locally could help reduce expenses.

7. CONCLUSION

This project proposes the creation of a wheelchair equipped with a built-in fridge, designed to assist individuals with limited mobility and high blood pressure. Here's how it can make a difference:

- *Promoting Physical Activity:* The wheelchair would feature a mechanism that harnesses the user's movement to power the fridge. This feature encourages more movement, which can contribute to lowering blood pressure.

- *Managing High Blood Pressure:* The wheelchair's built-in fridge keeps essential medications at the correct temperature, ensuring they're ready to take when needed. This helps individuals maintain better control over their blood pressure.

- *Enhanced Independence:* By combining mobility and the fridge, individuals can manage their health more independently, reducing their need for assistance from others.

The wheelchair would use a special cooling system powered by movement and batteries. this idea could really improve the lives of people who struggle with high BP and mobility issues. It encourages movement, keeps medicine safe, and gives people more control over their health.

8. SCOPE OF THE PROJECT

The study focuses on the prevalence of high BP among individuals with disabilities in rural areas and the challenges faced in treating it. Here's how we do it:

- Find out why people in this group often have high BP. This could be because they don't exercise much, their medication is inconsistent, or they have other health problems at the same time.

- See how uncontrolled high BP worsens their other health problems, such as heart or kidney problems.

- Come up with and experiment with ways to help them better manage their BP. This can mean finding ways to move that are appropriate for their disability, teaching them ways to relax and ensuring community and health services are supported.

- Fundamentally, we want to improve people's lives. to disabled people suffering from high BP in rural areas by providing them with appropriate help and support.

REFERENCES

1. **Centers for Disease Control and Prevention.** "Disability and Health: Physical Activity for All." *CDC*, 7 Sept. 2022, https://www.cdc.gov/ncbddd/disabilityandhealth/features/physical-activity-forall.html. Accessed 28 Mar. 2025.

2. **Ezzati, Majid, et al.** "Contributions of Risk Factors and Medical Care to Cardiovascular Mortality Trends." *Nature Reviews Cardiology*, vol. 12, no. 9, 2015, pp. 508–530. National Center for Biotechnology Information, https://doi.org/10.1038/nrcardio.2015.82.

3. **Falaschetti, E., Mindell, J., Knott, C., and Poulter, N.** "Hypertension Management in England: A Serial Cross-Sectional Study from 1994 to 2011." *The Lancet*, vol. 383, 2014, pp. 1912–1919.

4. **GBD 2017 Causes of Death Collaborators, G. A., et al.** "Global, Regional, and National Age-Sex-Specific Mortality for 282 Causes of Death in 195 Countries and Territories, 1980–2017: A Systematic Analysis for the Global Burden of Disease Study 2017." *The Lancet*, vol. 392, 2018, pp. 1736–1788.

5. **Gregory, A. Roth, et al.** "Global Health Metrics." *The Lancet*, vol. 392, no. 10159, 10 Nov. 2018, pp. 1736–1788.

6. **Lee, Jae-Hyun.** "Prevalence of Hypertension and Its Associated Factors Among Individuals with Disabilities in South Korea: A Nationwide Population-Based Study." *Journal of Preventive Medicine and Public Health*, vol. 54, no. 5, 2021, pp. 343–352. National Center for Biotechnology Information.

7. **Mills, K. T., et al.** "Global Disparities of Hypertension Prevalence and Control: A Systematic Analysis of Population-Based Studies from 90 Countries." *Circulation*, vol. 134, 2016, pp. 441–450.

8. **Stanaway, J. D., et al.** "Global, Regional, and National Comparative Risk Assessment of 84 Behavioral, Environmental and Occupational, and Metabolic Risks or Clusters of Risks for 195 Countries and Territories, 1990–2017: A Systematic Analysis for the Global Burden of Disease Study." *The Lancet*, vol. 392, 2018, pp. 1923–1994.

9. **Super Radiator Coils.** "Battery Storage Facility Cooling System Design." *Super Radiator Coils Blog*, 6 Apr. 2023, https://www.superradiatorcoils.com/blog/battery-storage-facility-cooling-system-design. Accessed 28 Mar. 2025.

10. **U.S. Department of Health and Human Services.** *Physical Activity Guidelines for Americans, 2nd Edition.* U.S. Government Publishing Office, 2018, https://health.gov/sites/default/files/2019-09/Physical_Activity_Guidelines_2nd_edition.pdf. Accessed 28 Mar. 2025.

11. **Zhou, B., et al.** "Worldwide Trends in BP from 1975 to 2015: A Pooled Analysis of 1479 Population-Based Measurement Studies with 19.1 Million Participants." *The Lancet*, vol. 389, 2017, pp. 37–55.

Note: All the figures in this chapter were made by the authors.

Advances in Mechanical Engineering and Materials Sciences – Dr. Vinay K. B et al. (eds)
© 2026 Taylor & Francis Group, London, ISBN 9-781-041-20970-6

58

Empowering a Sustainable Tomorrow—Generating Renewable Energy from an Organic and Biodegradable Resources

Geetha M. N.[1]
Department of ECE,
Vidyavardhaka College of Engineering,
Mysuru, India

Ragavendra Y. M.[2]
Department of ECE, GSSSTW,
Mysuru, India

Yajnika S. Nandi[3]
Department of ECE,
Vidyavardhaka College of Engineering,
Mysuru, India

Abstract: An innovative approach to converting waste into electricity, specifically designed to empower small towns with sustainable energy solutions while addressing the growing challenges of waste management. The system efficiently processes all types of waste, including biodegradable, non-biodegradable, dry, and wet materials, reducing the strain on overflowing dumping yards. A significant feature of this initiative is the management and utilization of gases released during waste processing. Carbon dioxide, a by-product, is repurposed for water purification and enhancing agricultural productivity, making the process environmentally responsible. The electricity generated not only powers essential community infrastructure, such as streetlights, but also contributes to overall energy independence for small towns. This solution is cost-effective, eco-friendly, and supports broader goals like reducing greenhouse gas emissions and preserving the ozone layer. By addressing multiple environmental and energy challenges, this paper sets a benchmark for sustainable development in small-town communities.

Keywords: Waste-to-energy, Waste management, Sustainable energy generation, Environmental conservation

1. INTRODUCTION

The generation of electricity from waste materials, either dry or wet waste that is converted into a useful energy source, this method is called waste to energy (WTE) helps in reducing the amount of waste that ends up in landfill and greenhouse gases and provides a form of renewable energy. WTE uses thermal and biological processes. In thermal processes like incineration, waste is burned to produce heat, which then generates electricity by powering turbines. This type of conversion of waste to energy not only produces electricity but also helps to reduce the waste in landfills, which can lead to problems like air pollution, water pollution, and methane emissions. Thus, waste-to-energy is an effective way to deal with waste while also creating power.

The common method of WTE is combustion which mainly involves burning of waste to generate heat that is used to power steam turbines. Coalition is a very

---

[1]geetha@vvce.ac.in, [2]Ymraghu80@gmail.com, [3]yajnikasnandi@gmail.com

common method that is used in many countries and has the advantages of waste disposal and energy production. Info besides combustion, other methods of converting waste into potential fuel sources include gasification and pyrolsis. In this process organic materials are converted thermochemically, in an environment with very little or no oxygen, to syngas. Synthetic gas typically contains carbon monoxide, hydrogen, and some carbon-dioxide. The syn gas is then changed into methanol, ethanol, methane, and synthetic fuels for use in many processes or as transport fuel.

2. LITERATURE REVIEW

The author considers the possibility of utilizing solid waste for producing energy in Bangladesh considering the problems with traditional sources of energy productions . Thermal, biological, and utilization of landfill gases are some of the waste-to-energy processes that form part of the study. Sustainable and environmentally safe management of waste resources is an important issue and this study looks in detail the generation, disposal and composition of wastes in Bangladesh with regards to Dhaka City.

This work presents a fresh take on clean energy by tapping into the untapped potential of waste heat. The paper highlights the increasing need for innovative power sources in areas like space exploration and remote monitoring. It explores how thermoelectric generation, based on the Seebeck effect, can convert heat into usable electricity. By proposing a compact and more reliable thermoelectric generator, the study aims to overcome some of the limitations of current battery technologies. Ultimately, it underscores the potential of harnessing thermopower as a sustainable energy solution for low-volume applications, aligning with the growing demand for power.

This research explores turning agricultural and animal waste into a reliable source of electricity. It starts by using bio-diesel made from oil palm, rice husk, and pig manure. By applying dry fermentation technology, the study converts agricultural waste and pig manure into biogas, which is then used for bio-electricity production. Remarkably, the process can cut production costs by up to 65.84%, achieving an output of 4,639.52 kW, with the investment being repaid in just over a year. In addition, the study investigates how to optimize thermoelectric generators (TEGs) using Micro-Electro-Mechanical Systems (MEMS) technology. Leveraging Comsol Multiphysics Software for modeling and simulation, it introduces a "Thermopile" setup with multiple thermocouples that exploit the Seebeck effect to convert thermal gradients into electricity. A key innovation is the use of rounded thermo-legs made from Bismuth Telluride, replacing the older Silicon Germanium-based design. This modification leads to a 120% increase in efficiency compared to the baseline.

These optimized TEGs have a wide range of practical applications, from recovering energy from car engines and radiators to potential use in wearables or even beverage coasters that harness heat from warm drinks. Overall, the study not only demonstrates a promising way to generate energy from waste but also significantly enhances the efficiency of thermoelectric generators in capturing waste heat. This work introduces a creative method for drying moist wood waste to generate electricity, with promising applications in both the industrial and renewable energy sectors [5]. The approach involves an integrated system that uses hot water and steam at various grades to not only dry the wood waste but also produce raw material for wood pellets from diverse feedstocks.

Additionally, the study explores the potential of generating electricity from solid waste incineration in Rajshahi City, Bangladesh . It carefully examines the types and volumes of waste produced—highlighting that a significant portion is food waste—and estimates that about 3,083 kg of waste per day could be collected for incineration, leading to an electricity output of roughly 12 MW.

The research also presents a method to convert the recovered heat from trash incineration into electricity, addressing both waste disposal challenges and energy shortages. Furthermore, the study looks into alternative energy sources by considering algae for biodiesel production and reviews various waste-to-energy technologies, underlining the crucial role of effective waste management in environmental protection .

The author highlights some intriguing aspects of India's power sector. Despite India accounting for only 3.4% of the world's energy consumption, it supports a massive 17% of the global population . Currently, the country's energy mix is largely dominated by thermal power (65.34%), followed by hydropower (21.53%), and renewables (10.42%). While thermal plants still meet a large share of the energy demand, India is steadily shifting towards cleaner energy sources—especially wind—driven by economic growth, rising incomes, dwindling fossil fuel supplies, and tighter environmental regulations.

The paper also delves into waste-to-energy (WTE) as a promising avenue for electricity generation in India. This approach aims to cut pollution and boost waste recycling by converting biomass into electricity, addressing environmental concerns linked to dwindling traditional fuel sources. Recent advancements in organic chemistry, biotechnology, and nanotechnology have significantly improved the efficiency of Microbial Fuel Cells (MFCs), achieving a power density of around 5.25 W/m². This positions WTE as a transformative solution for developing countries like India, where waste production is rapidly increasing.

Additionally, the research provides a thorough literature review of India's waste management landscape, considering both international challenges and the environmental impacts—such as pollution, global warming, and climate change—that arise from waste. A key focus is on the efforts made under the Swachh Bharat Abhiyan, a nationwide cleanliness campaign aiming for zero waste through comprehensive strategies like collection, segregation, recycling, and proper disposal. The paper also mentions the annual awards given to clean cities and states as recognition of effective waste management practices. Various techniques such as incineration, waste compaction, biogas production, composting, vermicomposting, and landfilling are explored to showcase the diverse approaches being implemented to manage waste in India.

3. METHODOLOGY

The block diagram in Fig. 58.1 illustrates how trash can be converted into electricity to assist small communities with waste management, lessen the strain on dumping yards, and generate electricity.

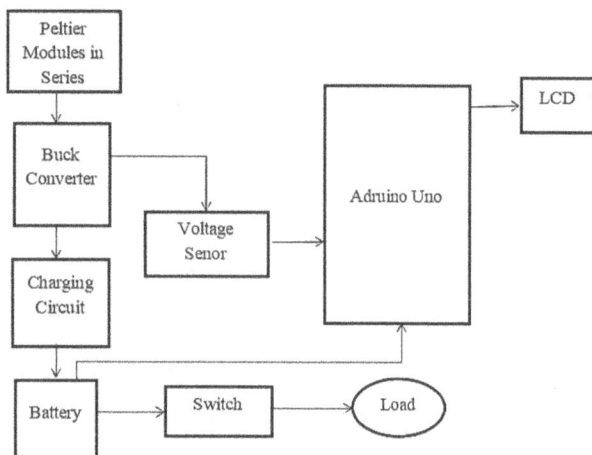

Fig. 58.1 Block diagram of proposed methodology

Imagine a system that uses solid-state devices known as Peltier modules to create a heating or cooling effect by simply running electricity through them. When connected in series—a fancy way of saying they're lined up one after the other—the modules boost the overall voltage and enhance their heating or cooling power.

At the heart of this setup is a battery that stores energy to power everything. A switch acts like a gatekeeper, controlling when electricity flows through the circuit. To keep an eye on the battery's health, a voltage sensor continuously checks its power level. When it's time to recharge, a dedicated charging circuit steps in to refill the battery using an external power source. Additionally, a buck converter helps adjust the battery's voltage down to a level that's just right for an Arduino Uno, which might be managing the system.

The physical arrangement is clever too. The Peltier modules are placed right under a cooling element, which could be a metal tray or a heat sink equipped with a fan, all enclosed in a wooden box for protection. Thermal grease is applied to ensure even heat distribution and to shield the modules from any fumes, while a metal plate covers the hot side for extra safety.

Here's how it all works: the cooling source is switched on first, followed by applying a heat source to the Peltier modules. As heat moves through these modules, a temperature difference builds up—one side gets hot while the other stays cool. This temperature difference causes the modules to generate a voltage, essentially turning heat into electrical energy in a silent process.

The amount of electricity produced depends on how big the temperature difference is and the number of modules connected together. Materials like lead telluride and bismuth telluride, which have properties that allow electrons to move freely, are used to make these modules. When one side of a module is heated and the other cooled, electrons shift from the hotter side to the cooler side, creating a voltage across the module.

To ensure the electricity coming out is steady and usable, a buck-boost converter is used to stabilize the output as direct current (DC). This converter works through a circuit involving a boost inductor, a controlled switch, a capacitor, a diode, and a resistor. Initially, when the switch is open, the inductor is waiting to store energy. Once the switch closes, the diode directs the increasing current through the inductor, storing energy as a magnetic field, which is then released into the output circuit.

Finally, the charge controller takes the stabilized voltage from the buck converter and directs it to recharge the lead-acid battery. Meanwhile, the voltage sensor monitors this output and the information is displayed on an LCD via an Arduino interface, allowing users to keep track of the system's performance.

This system cleverly combines different technologies to convert heat differences into a steady supply of electricity while managing energy storage and safety, making it an elegant example of practical, modern energy solutions.

4. RESULT AND DISCUSSION

The results present the findings from the conducted tests and analysis of their implications.

CASE 1:

TIME DURATION: 3 minutes

WASTE MATERIAL WEIGHT: 50 grams

The table depicts an experiment in which distinct kinds of dry wastes were exposed to heat for a definite period, and the output voltages were recorded. The highest input voltage was recorded at paper with 4.37V and the lowest at plastic with 1.03V.

Table 58.1 Dry waste with constant 3 minutes duration of heating

Materials	Output Voltage
PAPER	4.37V
WOOD	1.59V
PLASTIC	1.03V
COCONUT DRY WASTE	1.64V
CARDBOARD	1.68V

The image shows a bar chart titled "Fig. 58.2: Dry waste voltage and time with constant 3 minutes duration of heating." This chart shows relative output voltages of dry waste materials after heating for a constant time of 3 minutes. The bars represent two variables: voltage (in blue) and time (in red). Because the time parameter is the same for all the materials at 3 minutes, most attention is paid to voltage changes.

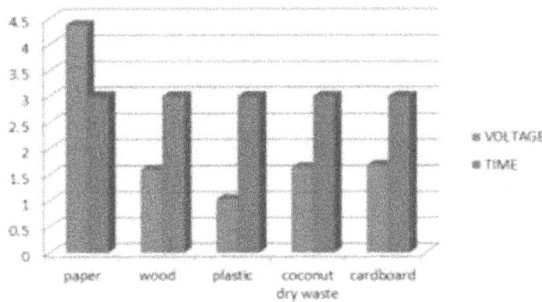

Fig. 58.2 Dry waste voltage and time with constant 3 minutes duration of heating

Table 58.2 Wet waste voltage and time with constant 3 minutes duration of heating

Materials	Voltage
CUCUMBER	1.10V
APPLE	1.37V
LEMON	1.56V
POTATO	1.59V
WASTE FOOD	1.10V

Figure 58.3 shows the wet waste voltage and time graph with constant 3 minutes duration of heating. Paper

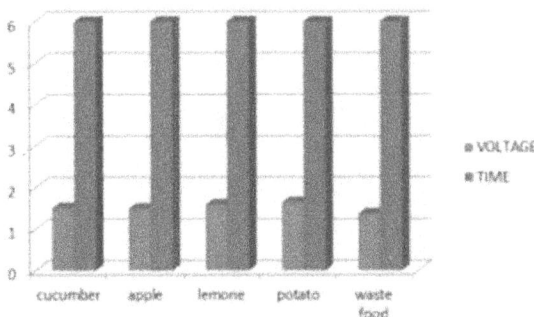

Fig. 58.3 Wet waste voltage and time graph with constant 3 minutes duration of heating

produces the highest output voltage of 4.37V among the tested dry materials. This indicates that paper generates the most significant energy output when subjected to heating for a constant duration of 3 minutes. Plastic produces the lowest output voltage of 1.03V, suggesting it is less effective in generating energy under similar conditions. Wood, Coconut Dry Waste, and Cardboard generate output voltages in a narrow range (1.59V–1.68V), with Cardboard slightly outperforming the others. This indicates that these materials have similar energy conversion properties under heating. The relatively high output voltage from Paper might be due to its lighter molecular structure and ease of thermal decomposition compared to other materials. The voltage difference can be attributed to the moisture content in wet waste, which hinders efficient energy conversion. Moisture absorbs part of the heat energy, reducing the effective heating and decomposition of the waste materials.

CASE 2:

TIME DURATION: 6 minutes

WASTE MATERIAL WEIGHT: 100grams

Table 58.3 Dry waste with constant 6 minutes duration of heating

Materials	Voltage
PAPER	6.18V
WOOD	3.98V
PLASTIC	1.51V
COCONUT DRY WASTE	4.44V
CARDBOARD	4.64V

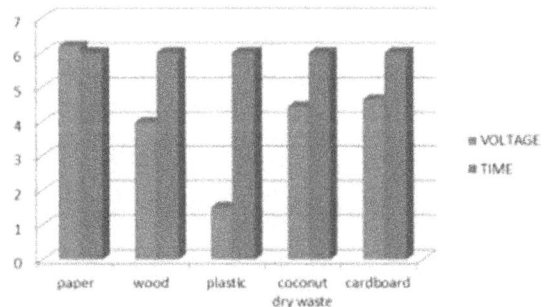

Fig. 58.4 Dry waste voltage and time graph with constant 6 minutes duration of heating

Table 58.4 Wet waste with constant 6 minutes duration of heating

MATERIALS	VOLTAGE
CUCUMBER	1.51V
APPLE	1.49V
LEMON	1.59V
POTATO	1.64V
WASTE FOOD	1.37V

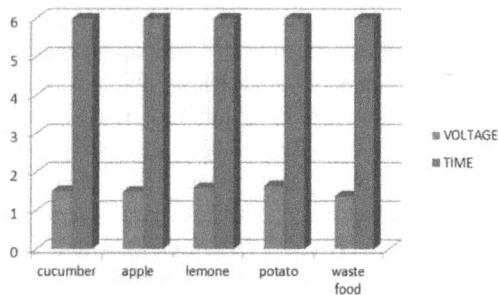

Fig. 58.5 Wet waste voltage and time graph constant 6 minutes duration of heating

5. CONCLUSION

The temperature difference across the surfaces of a Peltier module generates voltage, enabling the design of an arrangement for effective heating and cooling on opposite sides of the module. Waste heat, wood, or candle fumes can serve as heat sources, while chilled water, ice, or nitrous oxide can act as cooling sources. This system operates on the Seebeck effect, wherein a temperature gradient between two dissimilar semiconductors produces voltage. The output voltage is directly proportional to the number of Peltier modules and the magnitude of the temperature difference. To stabilize the generated voltage, a buck-boost converter is employed. This technology is versatile and can be utilized in remote areas, domestic settings, and industrial applications Thermoelectric devices offer distinct advantages over conventional energy resources, despite their lower efficiency. Their dual functionality in cooling and power generation positions them as a viable alternative to electrically powered systems. While the voltage output from a thermoelectric generator is relatively small, combining modules in series and parallel enhances power generation efficiency. Although thermoelectric systems are costlier than traditional power generation methods, they provide a sustainable alternative to conventional energy resources, offering an appealing trade-off between cost and environmental impact.

REFERENCES

1. Ayodele & M T E kahn, Underutilise *"waste heat as potential to generate environmental friendly energy"* publisher: ieee, published in: 2014 international conference on the eleventh industrial and commercial use of energy, 19–20 august 2014. https://doi.org/10.1109/icue.2014.690420

2. Abdur Rahman, Sadat rafi, Jannati Nabiha Nur,*" Waste to Energy: An approach to generate electricity by solid waste incineration in Rajshahi City, Bangladesh"*, International Conference on Mechanical Engineering and Renewable Energy 2019 (ICMERE2019) 11–13 December2019. https://www.researchgate.net/publication/350873906

3. Chamni Jaipradidtham, *"Energy Cost Reduction and Potential Analysis of Diesel Engine for ElectricityGeneration using biodiesel from Oil Palm-Rice Husk and Pig Manure with Biogas Renewable Energy"* 2016 IEEEInternational Conference on Power and Energy (PECon) https://doi.org/10.1109/pecon.2016.7951591

4. G puthilibai, V Chaithra, R Deepashri, *"An Intelligent Approach for Electricity Generation: Microbial Fuel Cell"*, 2022 international conference on power, energy, control and transmission systems (icpects). http://dx.doi.org/10.1109/ICPECTS56089.2022.10046995

5. JazibAli, TahirRasheed *"Modalities for conversion of waste to energy — Challenges and perspectives"* Science of The Total Environment Volume 727, issue 20 July 2020. https://doi.org/10.1016/j.scitotenv.2020.138610

6. Md. Shahedul amin, kazy fayeen shariar, md. Riyasat azim *"The potential of generating energy from solid waste materials in Bangladesh"* published 2009 1st international conference on the developments in renewable energy technology (icdret) https://doi.org/10.1109/icdret.2009.54541972.

7. Muthu Raman Y, Muthuvel AMR and Narayan Koushik C, *"Analysis and Design of Embedded Controller Based Electrical Power Generation Unit from Domestic Waste"*, 2010 International Conference on Environmental Engineering and Applications (ICEEA 2010). https://doi.org/10.1109/iceea.2010.5596095

8. Muneeswaran, P.nagaraj, bandi arun kumar reddy, mellampudi yuva vardhan *"power generrtion using waste management system"* 2023 2nd international conference on edge computing and applications (icecaa). https://doi.org/10.1109/icecaa58104.2023.10212117Mohammad Miyan, M. K. Shukla, *"Waste to Energy: A Review on Generating Electricity in India"*, SAMRIDDHI Volume 12, June 2020, Online ISSN: 2454-57672020.http://dx.doi.org/10.18090/samriddhi.v12i01.8

9. Nikolai Vitkov, *"Safety Based Comparison Of Incineration And Gasification Technologies For Electrical Energy Production By Municipal Solid Waste In Bulgaria"*, 2019 11th Electrical Engineering Faculty Conference (BulEF) https://doi.org/10.1109/bulef48056.2019.9030788

10. Neha Rajas, Poonam Nikam, Pooja Jadhavrao, Aarya Pise, Vaibhav Pokale, Sarthak Pithe, and Rahul Pise, *"Generating electricity from solid waste and biodiesel"*, E3S Web of Conferences 405, 02015 (2023). https://doi.org/10.1051/e3sconf/202340502015

11. Osintsev K.V, Rastvorov D.V. Brylina o.g, *"method of moist wood waste drying for the purpose of generating electricity"*, 2019 international multi-conference on industrial engineering and modern technologies (fareastcon). https://doi.org/10.1109/fareastcon.2019.8933889

12. R. Santhosh kumar, Gharniyas tr, Deepanraj p, Arun prasath *"Generating electricity by non biodegradable waste"*, 2023 international conference on innovative data communication technologies and application (icidca). https://doi.org/10.1109/icidca56705.2023.10099978

13. R.Dhana, Raju (2021*), "Waste Management in India – An Overview"*. United International Journal for Research & Technology (UIJRT). 02 (7): 175–196. https://uijrt.com/paper/waste-management-in-india-an- overview.Segundo Rojas-Flores (2022)," *Generation of Electricity from Agricultural Waste"*. Green Energy and Environmental Technology 2022, https://doi.org/10.5772/geet.11

14. Suwandi, Alwiyah, Padeli, maulana yusuf, hendra kusumah, *"Integration of renewable energy technology in waste recycling utilization"* 2022 ieee creative communication and innovative technology (iccit). https://doi.org/10.1109/iccit55355.2022.10119035

Note: All the figures and tables in this chapter were made by the authors.

Advances in Mechanical Engineering and Materials Sciences – Dr. Vinay K. B et al. (eds)
© 2026 Taylor & Francis Group, London, ISBN 9-781-041-20970-6

59

A Comparative Study on Heat Transfer Enhancement of Hollow Fin Using Different Geometries

Manjunath V. B.*

Assistance Professor,
Dept. of Mechanical Engineering,
Vidyavardhaka College of Engineering Gokulam 3rd Stage,
Mysore, Karnataka, India

Chandan Kumar M.[2],
Chirag C. K.[3], Chetan S.[4], Sanjay R.[5]

Students, Dept. of Mechanical Engineering,
Vidyavardhaka College of Engineering Gokulam 3rd Stage,
Mysore, Karnataka, India

Abstract: Optimizing the design of heat dissipation components, like fins, can greatly increase the efficiency of cooling systems and heat exchangers. This paper briefly briefs how changing fins' geometrical arrangement can improve heat transfer. The thermal efficiency of hollow fins in various geometries is examined. These fins have the advantages of being lighter and having better heat transfer because of their increased surface area and fluid flow interaction. The impacts of different shapes and configurations on heat flow rate, drop in pressure, and overall thermal performance are investigated. These include hollow fin designs that are circular, rectangular, and trapezoidal. The effects of fin thickness, internal cavity structure, and fin height on heat transfer enhancement are analyzed using computational fluid dynamics (CFD) simulations and experiments. According to the authors of many papers, heat dissipation can be greatly enhanced by optimizing the geometry of hollow fins, which strikes a balance between increased surface area and low flow resistance. Thermal management is essential for longevity and performance in electronics, automotive, and energy systems, and this study helps develop more effective cooling solutions for these applications.

Keywords: Thermal analysis, Design, Heat transfer, Hollow fin, Effectiveness

1. INTRODUCTION

The efficiency and durability of many engineering systems depend on heat dissipation, especially in applications with high thermal loads like electronics, automobile engines, and power generation machinery. Effective thermal management is crucial to avoid overheating, lower energy usage, and improve system reliability. The use of fins, particularly hollow fins, has drawn a lot of attention among different heat transfer techniques because of its potential to boost heat transfer efficiency while preserving a lightweight structure.

To expand the surface area available for heat exchange with the surrounding medium usually air or liquid, fins are frequently used in heat exchangers. Because of their improved heat transfer properties, hollow fins offer a promising substitute for conventional solid fins. By enabling turbulence and lowering thermal resistance, the hollow structure of these fins facilitates improved fluid flow through the fin cavities, improving convective heat transfer. Hollow fins are perfect for applications where efficiency and weight reduction are crucial because they can also achieve better heat distribution and lower weight than solid fins. The geometry of hollow fins has a

*Corresponding author: manjunath.vb@vvce.ac.in

significant impact on how well they transfer heat. The heat transfer can be impacted by parameters like the internal cavity structure, fin thickness, height, and shape (such as circular, rectangular, or trapezoidal). Optimizing these geometrical features for optimal thermal performance is still a difficult task.

Designing more effective cooling systems requires a deeper comprehension of how different geometric configurations impact the conductive and convective heat transfer mechanisms. This project investigates the effects of various geometric designs on the improvement of heat transfer of hollow fins. We seek to determine the ideal fin geometry that maximizes heat dissipation while minimizing related penalties like pressure drop and material consumption using a combination of CFD simulations and experimental analysis. This research aims to offer insights into the design of specialized thermal management solutions that can be used in a variety of engineering domains by meticulously examining the effects of various hollow fin shapes and internal configurations.

In addition to helping to optimize cooling systems, the results of this analysis will offer engineers valuable suggestions for designing heat transfer devices that strike a balance between cost, weight, and thermal efficiency.

2. LITERATURE SURVEY AND DISCUSSION

The fin's length, angle, area covered, and morphology all affect how much heat they can transport. Thermal performance and temperature were raised by raising the fin's height or decreasing the distance between them, which decreased the heat sink's capacity. Vaishnav Madhava Das et.al.(2021)

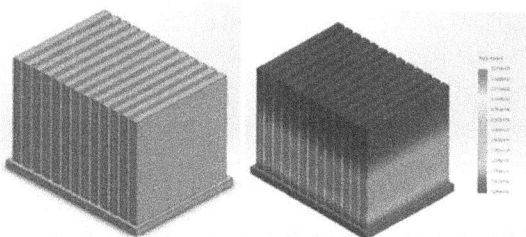

Fig. 59.1 Thermal analysis of a plane rectangular fin

Longitudinal fin arrangements are far more effective than circular or annular ones because they have higher HT. When compared to fins without groves, fins with grooves improve heat dissipation by around 13 to 30% compared to other groove types, fins with rectangular grooves provide a higher heat transfer of up to 30% when compared to other types of grooves, the efficiency of a rectangular fin with a rectangular groove was higher. Praveen kumara B M, et.al. (2016).

Software is used to simulate the cylinder fin body of a 150cc motorcycle. The fins' thickness is altered from the

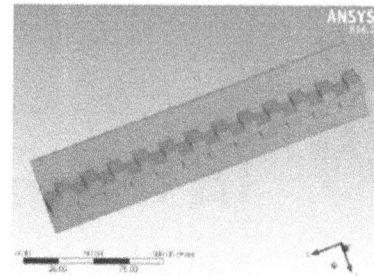

Fig. 59.2 Fin with rectangular grooves

original model. Overall weight is decreased by decreasing fin thickness. For the fin body, aluminum alloy 7075 is utilized. There is thermal analysis. It increases the heat transfer rate. S Sathish, et.al. (2017).

The fins were made of four different materials: aluminum alloy 7075, stainless steel, copper, and lead cast. To enhance the heat transfer, they changed the fin profiles (rectangular, trapezoidal, triangular, and circular) and used Autodesk Fusion 360 software to perform thermal simulation. The fin's surface area and heat transfer rate are directly correlated. Hrishikesh Kumar, et.al. (2022).

The heat transfer rate varies as the surface area changes. The rate of heat transfer is directly correlated with surface area. For both copper and aluminum, rectangular pin fins transmit heat more quickly than cylindrical ones. A very low temperature at the end of the rectangular fins improves the heat transmission rate. Harshita Pant, et.al. (2021).

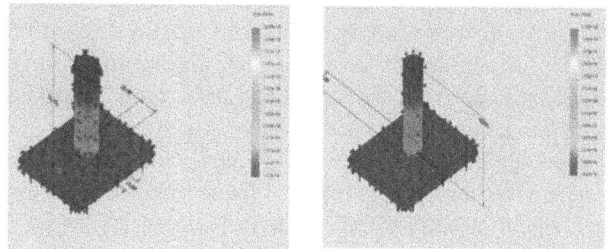

Fig. 59.3 Geometry of rectangular and cylindrical pin-fin

The area of the fins directly impacts the rate of heat transmission. Various fin designs and materials made of aluminum 6061 alloy have been devised and examined. Because of its strong thermal conductivity and low weight, aluminum 6061 was determined to be a comparatively superior material when it came to heat transfer. It was also able to effectively lower the engine cylinder's temperature by a significant amount. Keshavan, et.al. (2023).

They analyzed the different shapes of perforations in Ansys software and concluded that when compared to a fin without holes, the convective heat transfer coefficient rises for perforated fins. According to the results, when compared to a fin that is not perforated, Nusselt's numbers increase for perforated fins. As a result, aluminum is best for fin applications for perforated fins. Dibya Tripathi, et.al. (2021).

To increase the rate of heat transfer, they used Autodesk Fusion 360 software to perform thermal simulation on the four different materials—Aluminum alloy 7075, stainless steel, copper, and lead cast by altering the fin profiles (rectangular, trapezoidal, triangular, and circular). The more surface area of the fin, the faster it will transfer heat. Rishabh Prasad, et.al. (2023).

For brass and aluminum, they employed pin fins with and without dimple channels. They concluded that the fin positioning, voltage across the pin fin, fin setting technique, and other factors all affect the rate of heat transfer. Effectiveness, heat flow rate, and heat transfer coefficient are all influenced by the voltage across the fin's base. Compared to natural convection, forced convection has a significantly higher heat transfer rate, effectiveness, and coefficient. Ruby Haldar, et.al. (2023).

The research was carried out to examine the free convection surrounding the hollow fins and compare the heat dissipation capacity of the hollow fin with that of the solid fin. The results showed that, although efficiency is decreased in the hollow fin case, effectiveness is raised by 1.76. A hollow fin is a more effective option when considering economics. Dr. Chandrakishor L, et.al. (2019).

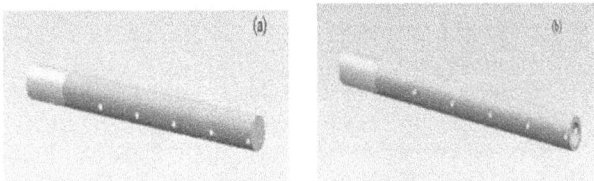

Fig. 59.4 Compare with solid fin and hollow fin

They investigated six different fin types to determine which one could transmit the most heat, and they concluded that the best fins were circular with a rectangular portion; the shape of the fin has an impact on heat transfer. The design of the fins affects the tube, thus, an increase in surface area does not always translate into an increase in heat transfer rate. Shatha Ali M., et.al. (2021).

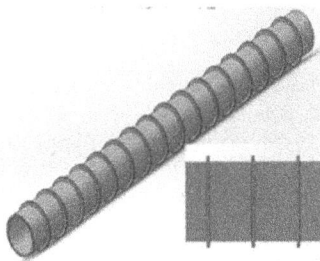

Fig. 59.5 Circular fin with a rectangular portion

The models of rectangular fins with wedges and rounded tips were analysed, and the results found that while the regular rectangular fin takes a long time to dissipate heat, the wedge and rounded configuration accelerates the process. The wedge-shaped configuration is clearly more

efficient than the other two, according to the analysis. M. Raghunathan, et.al. (2017).

A comparative thermal analysis of heat sinks with various geometrical configurations is carried out in this paper. Finding the best heat sink design to further reduce the temperature through natural convection is the goal of this study. Solid Works software is used for the design phase. ANSYS WORKBENCH is used to analyze the designed models. As per the results, the heat sink models taken into consideration, the rectangular flat plate heat sink is the best design. Along with this, staggered fin, round pin fin, and elliptical fin are all compared in this study. They discovered that the rectangular fin has more temperature drop than the others. G. Naresh, et.al. (2020).

They carried out experimental analyses of solid fin array convective heat transfer and turbulent heat flow, and they developed innovative perforated fin array designs with two distinct perforation dimensions. For all heat inputs, the measured range of Reynolds numbers (Re), and all perforation sizes, the improvement in heat transfer with perforated fins is greater compared to solid fins. K. H. Dhanawade, et.al. (2010).

3. CONCLUSION

This study examines how the performance of heat transmission is affected by various hollow fin geometric forms. The primary contribution is proving that, in comparison to solid fins, heat dissipation may be greatly increased by optimizing the shape, thickness, and internal structure of hollow fins. It was discovered that designs such as trapezoidal and circular hollow fins enhanced convective heat transfer while preserving low-pressure drop and lowering weight.

This work is interesting since it focuses on finding the optimal balance between airflow resistance, material utilization, and thermal efficiency. We can create lightweight, more efficient heat exchangers for cooling systems by modifying the fin shape. Better thermal management solutions for a variety of applications, such as electronics, automobile engines, and renewable energy systems, may be designed with the help of this study.

Overall, by emphasizing how geometric design can maximize performance, the study advances heat transfer technologies. To increase the effectiveness of hollow fin-based cooling systems, future research could examine the effects of various materials, environmental factors, or larger-scale applications. For engineers looking to improve thermal management across a range of industries, this research offers helpful recommendations.

REFERENCES

1. Abdullah H. AlEssa, Ayman M.Maqableh and Shatha Ammourah, "Enhancement of natural convection heat

transfer from a fin by rectangular perforations with aspect ratio of two," International journal of Physical Sciences, vol.4, 10, pp.540–547, October 2009.

2. A. Al-Damook, N. Kapur, J. L. Summers, and H. M. Thompson, "An experimental and computational investigation of thermal air flows through perforated pin heat sinks," Applied Thermal Engineering, vol. 89, pp. 365–376, Oct. 2015.

3. A. Maji, D. Bhanja, P. K. Patowari, G. Choubey, and T. Deshamukhya, "Computational investigation of heat transfer analysis through perforated pin fins of different materials," 2017.

4. A. R. Kaladgi, F. Akhtar, S. P. Avadhani, A. Buradi, A. Afzal, A. Aziz, and C. Ahamed Saleel, "Heat transfer enhancement of rectangular fins with circular perforations," Materials Today: Proceedings, vol. 47, pp. 6185 6191, 2021.

5. Abdullah H. AIEssa and Mohmmed Q. Al-Odat, "Enhancement of natural convection heat transfer from a fin by triangular perforations of bases parallel and toward its base," The Arbian Journal for Science and Engineering, vol.34 2B, October 2009.

6. Bayram Sahin, Alparslan Demir, "Performance analysis of a heat exchanger having perforated square fins," Applied Thermal Engineering, vol.28, pp.621–632, 2008.

7. Dibya Tripathi. Fins in Thermal Engineering from ANSYS. International Journal for Modern Trends in Science and Technology 2021.

8. Dr. Chandrakishor L. Ladekar1, Dr. Sanjeo. K. Choudhary, Heat Transfer Analysis and Performance Measurements of Hollow Pin-fin in Natural Convection, Conference Paper · April 2019.

9. G. Naresh, G. Akhil Sai "Steady State Thermal Analysis on Heat Sinks by Varying Fin Configuration Using ANSYS", Mukt Shabd Journal, Volume IX, Issue IV, APRIL/2020, ISSN NO: 2347–3150.

10. Harshita Pant, Divyanshi Shukla, Shriya Rathor, S. Senthur Prabu, Heat transfer analysis on different pin fin types using Solid Works, IOP Conf. Series: Earth and Environmental Science850 (2021).

11. Hrishikesh Rajesh Kumar, Arun Antony M T, Shyam S Warrier, S. Senthur Prabu, Comparative Study on Thermal Analysis of Extended Surface using ANSYS Simulation, ECS Transactions, 107 (1) 12209–12219 (2022).

12. K. H. Dhanawade, H. S. Dhanawade, Enhancement of Forced Convection Heat Transfer from Fin Arrays with Circular Perforation, Institute of Electrical and Electronics Engineers (2010).

13. Kadir Bilen, Ugur Akyol,Sinan Yapici, "Heat transfer and friction correlations and thermal performance analysis for a finned surface," Energy Conversion and Management vol.42, pp.1071–1083, 2001.

14. Keshavan, Jayendra, Dhrunil, Jeffrey, S. Senthur Prabu, Comparative Study on Performance of Heat Fins of Varying Design Specifications, Conference Paper in AIP Conference Proceedings, January 2023.

15. M. Rajagurunathan & Vishnu V, Thermal analysis of fins with modified tips, International Journal of Mechanical and Production Engineering Research and Development (IJMPERD) ISSN (P): 2249-6890; ISSN (E): 2249–8001 Vol. 7, Issue 6, Dec 2017.

16. N.S.ouidi A.Bontemps, "Countercurrent gas- liquid flow in plate –fin heat exchangers with plain and perforated fins," International Journal of Heat and Fluid Flow, vol.22, pp.450–459, 2001.

17. O.N. Sara, T. Pekdemir, S. Yapici, H. Ersahan, "Thermal performance analysis for solid and perforated blocks attached on a flat surface in duct flow," Energy Conversion and Management, vol.41, pp.1019–1028, 2000.

18. Praveen kumara B M, Kaushik N D, Ajith P, A Study on the Heat Transfer Enhancement of Fin with Grooves by Natural Convection, (An ISO 3297: 2007 Certified Organization) Vol. 5, Issue 8, August 2016.

19. Rishabh Prasad, Vidit Kohad, Hrithik Ray, S. Senthur Prabu, Thermal Analysis of different design of fins with different Materials, Conference Paper in AIP Conference Proceedings · January 2023.

20. Ruby Haldar, Abhimanyu Kumar, Trishit Janah, Prabir Biswas, Experimental Investigation of Heat Transfer by Forced and Natural Convection in a Pin Fin for Different Materials, International Journal for Research in Applied Science & Engineering Technology (IJRASET) ISSN: 2321–9653; IC Value: 45.98; SJ Impact Factor: 7.538 Volume 10 Issue II Feb 2022.

21. S Sathish, D Srikanth, Saba Sultana, Heat Transfer Analysis of Fins with Varied Geometry and Materials, 2017 IJCRT | National Conference Proceeding NCESTFOSS Dec 2017.

22. S. S. G. R. Putra, N. S. Effendi, and K. J. Kim, "A parametric study on structural effects of hollow hybrid fin heat sinks in natural convection and radiation," Journal of Mechanical Science and Technology, vol. 33, no. 6, pp. 2985–2993, Jun. 2019.

23. Shatha Ali Merdan, Prof Dr Zena Khalefa Kadhim and Ali Arkan Alwan, CFD Analysis for Different Types of Fins to Enhancement the Heat Transfer Rate Through A Cross Flow Heat Exchanger, IOP Conf. Series: Materials Science and Engineering, (2021).

24. V. Karlapalem and S. K. Dash, "Design of perforated branching fins in laminar natural convection," International Communications in Heat and Mass Transfer, vol. 120, p. 105071, Jan. 2021.

25. Vaishnav Madhava Das, Dibyarup Das, Kaustubh Anand Mohta, S. Senthur Prabu, Comparative analysis on heat transfer of various fin Profile using solid works: A systematic review, IOP Conf. Series: Earth and Environmental Science 850 (2021).

Note: All the figures in this chapter were made by the authors.

Advances in Mechanical Engineering and Materials Sciences – Dr. Vinay K. B et al. (eds)
© 2026 Taylor & Francis Group, London, ISBN 9-781-041-20970-6

60

Experimental Investigation on Heat Transfer Augmentation in a Concentric Tube Heat Exchanger using Perforated Disk Inserts and Blended MWCNT-CuO/DI Water Hybrid Nanofluids

Ravi K. S.[1]

Department of Mechanical Engineering,
Vidyavardhaka College of Engineering,
Mysuru, Karnataka, India

Krishnamurthy K. N.[2]

Department of Mechanical Engineering,
Visvesvaraya Technological University,
Centre for Post Graduate Studies,
Mysuru, Karnataka, India

Akashdeep B. N[3].

Department of Mechanical Engineering,
K.S. School of Engineering and Management,
Bengaluru, Karnataka, India

Venkatesh B. J.[4]**, Naveen Kumar C.**[5]

Department of Mechanical Engineering,
Visvesvaraya Technological University,
Centre for Post Graduate Studies,
Mysuru, Karnataka, India

Abstract: In this study, the thermal characteristics of a double-pipe heat exchanger with and without using MWCNT-CuO/DI hybrid nanofluids with perforated disk inserts are investigated. The outer pipe is made of galvanized iron, while the inner pipe is made of copper, and the heat exchanger itself is defined according to functional parameters. Copper perforated disk inserts were located at uniform intervals along the length of both pipes to investigate their effect on thermal performance. In the present research, the hybrid nanofluid was prepared using a two-step approach, in which DI water was used as the base fluid, while MWCNT-CuO nanoparticles at various volume fractions of 0.1%, 0.2%, 0.3%, and 0.4% were dispersed. To aid in stability, a 50:50 ratio of two surfactants, sodium dodecyl sulfate (SDS) and Triton X-100, was used. The experimental studies were made under different flow conditions by keeping the hot side fluid (water) constant at 50°, and it was observed that heat conduction was progressively increased with nanoparticle concentration, peaking at 0.4%. Moreover, a range of significant thermophysical properties like specific heat, density, viscosity, Reynolds number, Nusselt number, heat transfer coefficient, friction factor, pressure drop, and thermal performance factor were also studied. The highest improvement in heat transfer was noted for 0.3% volume fraction of hybrid nanofluid with addition of

[1]ksravi@vvce.ac.in, [2]krishnamurthykn1931@gmail.com, [3]bndeep@gmail.com, [4]venky111ava@gmail.com, [5]shrinavinataanvi@gmail.com

10.1201/9781003725053-60

the perforated disk inserts. A significant enhancement in heat transfer performance with MWCNT-CuO/DI water as working fluid was observed in comparison to MWCNT-CuO/DI water which increased approximately by 55%. The results highlight the effectiveness of hybrid nano fluids and modifications in enhancing heat exchanger performance.

Keywords: MWCNT-CuO/DI, Stability, Surfactants, Double pipe heat exchanger, Perforated disk insert

1. INTRODUCTION

Heat exchangers are among the most widely employed equipment in industry, designed for heating or cooling applications. These devices transfer thermal energy between fluids at different temperatures. They are used in a variety of sectors such as oil and gas, chemical and food processing, air conditioning and refrigeration, dairy production, automotive systems, aerospace technology, electronics, and power generation[1]. In recent years, the optimization of various thermal systems has become one of the most important spots of scientific research, particularly in giving attention to improving energy utilization. Enhancing the performance of heat exchangers is often achieved either by utilizing more complex systems or adapting existing systems under different operational conditions, whereas one of the most effective solutions is the use of nanoparticles combined with base working fluids to generate a nanofluid. A notable improvement in the thermophysical properties with better heat transfer performance has been performed by the introduction of nanoparticles in a base fluid[2]. Nanofluids are specially engineered prepared suspensions of nanoparticles in the base fluid to improve thermal properties. These fluids form a biphasic system, with the base fluid in the liquid state and the nanoparticles in a solid phase. Water, ethylene glycol, vegetable oils, transformer oils, and a variety of SAE lubricating oils are typical base fluids. The thermal profile of these fluids can be improved further by adding nanoparticles to them. Nanofluids are an advanced class of heat transfer fluids with much higher thermal conductivity than standard base fluids. According to the nature of dispersed nanoparticles, nanofluids can be classified into three categories: metallic nanofluids, ceramic nanofluids, and nanofluids with other or polymeric nanomachines. These fluids can be made of metals, metal oxides or nanostructured materials in the form of nanoparticles[5]. Metallic Nanoparticles: It comprises of Cu, Ni, Al etc., Metal oxide nanoparticles: It comprises of TiO_2, Al_2O_3, CuO, SiC, FeO etc. Carbon-based nanomaterials (carbon nanotubes (CNTs), graphene, calcium carbonate, titanium nanotubes, etc., which possess unique thermal and mechanical characteristics) can also be used in nanofluids[6]. Out of these, the multi-walled carbon nanotubes (MWCNTs) have remarkable thermal conductivity, which makes them very useful in the heat transfer applications. Recently, a new nanoparticle (NP) type was also developed, named hybrid nanofluid, which is related to the various nanoparticle dispersing in a single base fluid. The thermal and rheological properties can be highly customized with the right combination of different nanoparticles, leading to heat transfer performance enhancement. Nanoparticle-enhanced fluids provide altered thermophysical properties that improve thermal conductivity, viscosity, density, and specific heat capacity, which allows for a broader range of utilization for different thermal management solutions than traditional solutions[7].

Surfactants — chemical agents that lower the surface tension of a liquid to allow better dispersion and interaction between various phases. These are important as they stabilize the nanoparticles in base fluid against agglomeration and help to achieve a uniform distribution. If they are specifically employed to stabilize solid particles in a liquid medium, they are named dispersants. Surfactants are divided into 4 major classifications: non-ionic, anionic, cationic, and amphoteric (zwitterionic) surfactants according to their head group composition and charge characteristics [8]. They are useful for applications where charge neutrality is required, since they are not electrically charged. This is because the head groups of anionic surfactants are negatively charged, which stabilizes the nanoparticles via electrostatic repulsion. Cationic surfactants have positively charged head groups that strongly interact with negatively charged nanoparticles or surfaces, while amphoteric surfactants can have either positive or negative centers of charge within the same molecule that can behave differently based on the pH of solution. Sodium dodecyl sulfate (SDS), sodium dodecyl benzene sulfate (SDBS), cetyltrimethylammonium bromide (CTAB), polyvinylpyrrolidone (PVP), oleic acid, and Acetyl Trimethyl Ammonium bromide are examples of surfactants that are commonly used in nanofluid applications. These surfactants are critical for improving the stability and dispersibility of nanoparticles in diverse fluid media, maximizing their efficacy in heat transfer systems and other state-of-the-art engineering applications [9][4]. Aliakbar Karimipour et al[1]. An experiment to examine the enhancement of thermal conductivity of a CuO/MWCNT/water hybrid nanofluid was carried out by. The study evaluated mono and hybrid nanofluids at volume fractions of 0.2%-1.0% and temperature target of 25°C to 50°C. Results demonstrated substantial thermal conductivity improvements of 19.16% (mono) and 37.05% (hybrid) at highest volume fraction and target temperature. The results also demonstrated that by adding CuO nanoparticles, the thermal conductivity of the water improved and the combination of CuO and MWCNT completely enhanced each other. Masoud Zadkhast et al.[2] conducted an experimental study to evaluate the thermal conductivity of a HyNF type MWCNT-CuO-water. They assessed thermal conductivity at different nanoparticle volume fractions (ϕ = 0.05%, 0.1%, 0.15%, 0.2%, 0.4%, and 0.6%) and temperatures using KD_2 Pro instrument (25°C to 50°C). At 50°C, the maximum enhancement thermal conductivity was 30.38% at volume fraction of 0.6%, indicating that hybrid nano fluids have an improved thermal characteristic. JianQu et al. [3] had investigated experimentally the photo-thermal conversion properties of a CuO-MWCNT/H_2O hybrid nanofluid for direct solar thermal energy harvesting. The different concentration mixing ratios (CMRs), therefore, were measured to determine the optical absorption characteristics and the photo-thermal conversion efficiency. At this CuO/MWCNT ratio (0.15 wt.% / 0.005 wt.%), the maximum terminal temperature increase of the hybrid nanofluid compared to deionized (DI) water was found to be 14.1°C after 45 minutes of illumination. These findings suggest that CuO-MWCNT [H_2O] hybrid nanofluids can provide a viable alternative medium for directly harvesting solar thermal energy. Iman Fazeli et al. [4] utilized response surface methodology to examine the heat transfer and pumping performance of an MWCNT-CuO hybrid nanofluid in a brazed plate heat exchanger. At 0.1 wt.% concentration, a stable hybrid nanofluid with no sedimentation was observed after stabilization in mixture of two surfactants (Gum Arabic (GA)↔Tween-80). It was also noted that with the hot fluid held constant at 35° and 6 L/min, the greatest increments in the total hybrid nanofluid convective heat transfer coefficient were recorded at 14.4 L/min, 18.9 L/min, and 24.4 L/min with respective enhancements of 85.56%, 101.25%, and 139.19% over the results obtained with water. The results imply that it could be provided to the swelling hybrid studies as GA and Tween-80 have been reported as alternative surfactants to stabilize the hybrid nanofluids with higher overall thermal conductivity at optimized conditions[10]. The present work examines the thermal and flow performance features of MWCNT-CuO/DI hybrid nanofluids in both perforated disk insert-connected and insert-free double-pipe heat exchangers for volume fractions of 0.1%, 0.2%, 0.3%, and 0.4%. To increase the stability of hybrid nanofluid, a mixture (half · half) of two surfactants (sodium dodecyl sulfate (SDS) and Triton X-100 was used. And heat transfer and fluid properties such as thermal conductivity, specific heat, density, viscosity, Reynolds number, Nusselt number, heat transfer coefficient, friction factor, pressure drop and thermal performance factor etc. are evaluated systematically over the different volume fractions to take the efficiency and applicability of hybrid nanofluid from a perspective for heat transfer applications.

2. MATERIALS AND METHODS

2.1 Materials

The surfactants used in this study, Triton X-100 (non-ionic) and sodium dodecyl sulfate (SDS, ionic) were

obtained from Raghu Chemicals, Mysore, while CuO and MWCNT nanoparticles were purchased from Ultra nanotech Pvt. Ltd., Bangalore. Hybrid nanofluids were prepared utilizing deionized (DI) water as a base fluid. The use of DI water provides a controlled environment for the characterization of thermophysical and heat transfer properties of the nanofluids.

2.2 Characteristics

The structural and morphological characterization of MWCNT and CuO nanoparticles was confirmed through SEM, TEM, as provided by the commercial supplier. The corresponding results are presented in Fig. 60.1 and Fig. 60.2 (a), (b), respectively. Additionally, Table 60.1 summarizes the thermophysical properties and morphological characteristics of the MWCNT and CuO nanoparticles utilized in this study.

Table 60.1 Properties of MWCNT and CuO

Specification	MWCNT	CuO
Color	Black	Brownish black
Particle size	10-15 nm	50 nm
Thermal conductivity	3000 W/m · K	76.5 W/m · K
Purity	>97%	99%
Density	2100 kg/m³	6400 kg/m³

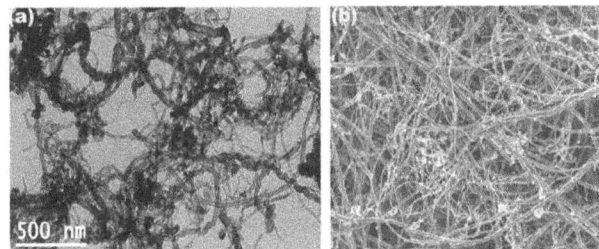

Fig. 60.1 (a) TEM image of MWCNT nanoparticles, (b) SEM image of MWCNT nanoparticles

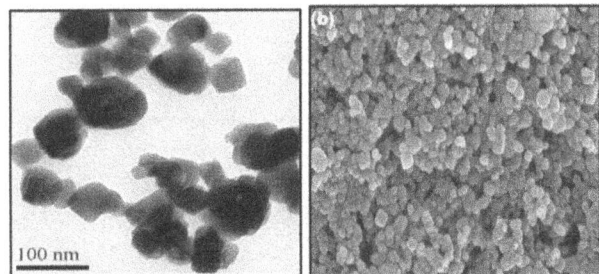

Fig. 60.2 (a) TEM image of CuO nanoparticles, (b) SEM image of CuO nanoparticles

2.3 Volume Concentration of Nanoparticles

The volume concentration of nanoparticles was determined using Equation 1, as proposed by Ali Akbar[1] and A. Akhgar [11], and the calculated values are presented in Table 60.2.

$$\phi = \frac{\left(\dfrac{m}{\rho}\right)MWCNT + \left(\dfrac{m}{\rho}\right)CuO}{\left(\dfrac{m}{\rho}\right)MWCNT + \left(\dfrac{m}{\rho}\right)CuO + \left(\dfrac{m}{\rho}\right)DI\ water} \times 100$$

(Eq. 1)

Table 60.2   Mass of MWCNT/CuO nanoparticles for different volume fraction

Quantity of base fluid (DI-water) in Litre	Volume Concentration ϕ in %	Mass of Nanoparticles in grams MWCNT: CuO	MWCNT: CuO And SDS: Triton X-100 ratio
1	0.1	1.582	1:1
1	0.2	3.16	1:1
1	0.3	4.75	1:1
1	0.4	6.35	1.1

2.4 Preparation of MWCNT/CuO Hybrid Nanofluid Using Deionized Water as the Base Fluid

The preparation of the hybrid nanofluid was carried out using the two-step method, as illustrated in Fig. 60.3. This widely adopted approach involves the initial synthesis of nanoparticles through physical or chemical methods, followed by their dispersion into a base fluid. Being a cost-effective and scalable technique, the two-step method is commonly employed in industrial production of nanopowders. To achieve the desired nanoparticle volume fraction, the volume concentration was first determined using Eq. 1. Based on the calculated volume fraction, an equivalent amount of nanoparticles was precisely measured using a digital weighing scale. A 1:1 ratio of nanoparticles to surfactants was maintained to ensure optimal dispersion and stability. The surfactants used in this study—sodium dodecyl sulfate (SDS) and Triton X-100—were added in equal proportions to enhance the uniform distribution of nanoparticles within the base fluid [12].

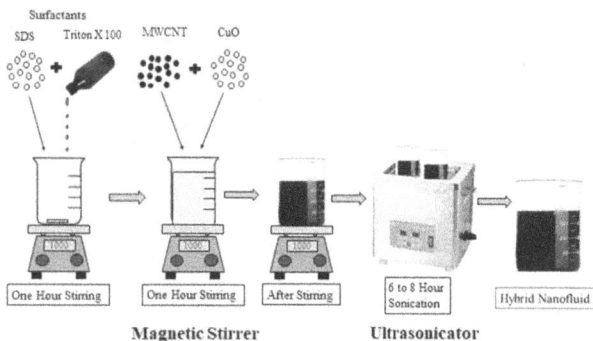

Fig. 60.3   Preparation process of MWCNT/CuO DI water hybrid nanofluid

The hybrid nanofluid was prepared following a systematic process:

- A pre-measured quantity of surfactants (SDS and Triton X-100 in a 50:50 ratio) was added to the base fluid (deionized water).
- The mixture was placed on a magnetic stirrer and stirred continuously for one hour to ensure proper dispersion of the surfactants.
- Nanoparticles were then gradually introduced into the surfactant-water mixture while stirring, and the dispersion process continued for an additional hour, as depicted in Fig. 60.4(a).
- The resultant suspension was subjected to ultrasonic treatment for six hours using a sonicator to further break down any agglomerates and ensure uniform dispersion, as shown in Fig. 60.4(b).
- Following these steps, a stable hybrid nanofluid was successfully prepared.

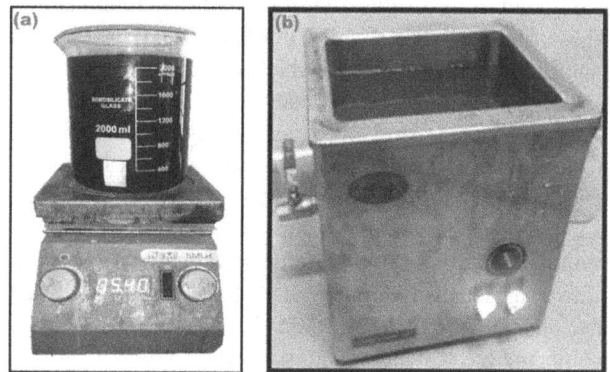

Fig. 60.4   (a) Magnetic stirrer, (b) Bath sonicator

To assess the long-term stability of the MWCNT-CuO hybrid nanofluid, the dispersion behavior was monitored over time by capturing photographic evidence on different days. This evaluation helped determine the effectiveness of the surfactants in preventing sedimentation and maintaining the homogeneity of the nanofluid.

2.5 Experimental Set up and Procedure

To enhance fluid mixing and thermal performance, perforated concentric disk baffles were installed within both the copper and galvanized iron tubes of the concentric tube heat exchanger. In the inner copper tube, four perforated disks were placed, each with an outer diameter of 19.05 mm, inner diameter of 9.53 mm, and ten uniformly distributed holes of 3 mm diameter. A schematic representation of these components is provided in Fig. 60.8. Similarly, within the galvanized iron tube, three perforated disks were inserted into the annular space, each having an outer diameter of 50.8 mm, inner diameter of 20.05 mm, and nine perforations, each with a 9 mm diameter. The heat exchanger setup incorporates two insulated fluid storage tanks, positioned adjacent to the experimental model, to facilitate fluid circulation. Two centrifugal pumps regulate the working fluid flow in separate sections of the heat exchanger, while rotameters

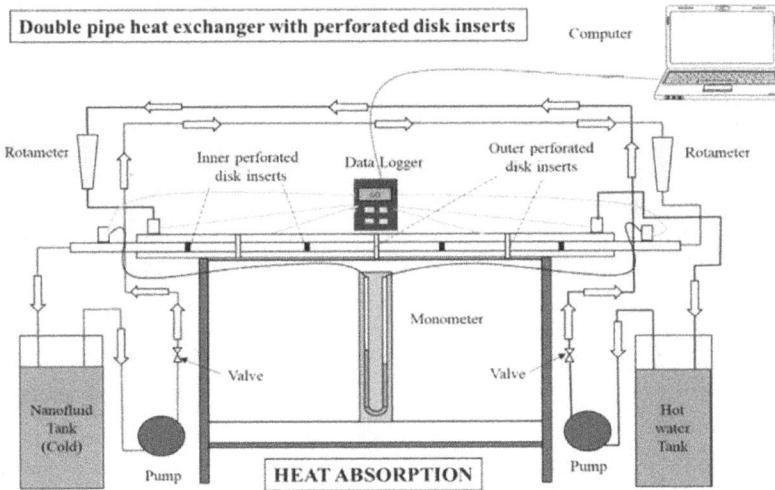

Fig. 60.5 Heat absorption by nanofluid form hot water with perforated disk inserts

precisely measure the flow rates of the MWCNT/CuO hybrid nanofluid and deionized water (DI). To continuously monitor temperature variations, nine K-type thermocouples are evenly distributed across different points in the heat exchanger, with real-time temperature data recorded using a data logger at predefined intervals. The primary objective of this experiment was to investigate the heat absorption capability of the hybrid nanofluid under varying operating conditions. The study was performed using a concentric tube heat exchanger equipped with perforated disk inserts, where hot water circulated through the annulus (shell side) while cold hybrid nanofluid flowed through the inner tube (pipe side) in a counter-current configuration. The heating system consisted of a coil of 1 mm diameter and 10 m length, placed inside the heat absorber container, as illustrated in Fig. 60.5. The experimental study was conducted at three nanoparticle volume concentrations (0.1%, 0.2%, and 0.3%) and three different flow rates (200 LPH, 300 LPH, and 400 LPH), while the hot water flow rate was maintained constant at 150 LPH. Before the experiments commenced, hot water at 50°C was introduced into the annular section (shell side) at a flow rate of 150 LPH, while the hybrid nanofluid entered the inner tube (pipe side) in a counter-flow arrangement at flow rates of 200 LPH, 300 LPH, and 400 LPH, respectively. Temperature measurements were obtained using K-type thermocouples, with intermediate temperatures recorded and analyzed through a data acquisition system, ensuring real-time logging of thermal variations at both the shell side and pipe side. The experimental heat exchanger setup, including perforated disk inserts, is depicted in Fig. 60.7. This system operates as a closed-loop horizontal tube bank heat exchanger, facilitating direct contact heat exchange between two fluid phases with minimal interaction or indirect contact. The hybrid nanofluid, composed of MWCNT and CuO nanoparticles with DI water as the base fluid, functions as the cold working fluid, absorbing heat from the hot water

stream. The main test section consists of a concentric tube heat exchanger, where the inner copper tube has an inner diameter of 19.05 mm, outer diameter of 20.05 mm, and a length of 2000 mm, while the outer galvanized iron tube features an inner diameter of 50.8 mm, outer diameter of 52.8 mm, and a length of 1500 mm. The detailed specifications of the heat exchanger, including all relevant structural dimensions, are summarized in Table 60.4.

Table 60.3 Specification of heat exchanger

Components	Material	Inner Diameter	Outer Diameter	Length
Outer tube	Galvanized Iron	50.8mm	52.8mm	1500mm
Inner tube	Copper	19.05mm	20.05mm	2000mm

2.6 Property Calculations

Density

The density of the MWCNT/CuO hybrid nanofluid dispersed in deionized (DI) water was determined using Equation (2), as proposed by Iman and Mohammad [4]. The calculation was performed for varying nanoparticle volume concentrations to analyze the impact of particle loading on the density of the hybrid nanofluid.

$$\rho_{nf} = (1 - \varnothing)\rho_{bf} + \varnothing_1\rho_1 + \varnothing_2\rho_2 \qquad \text{(Eq. 2)}$$

Specific Heat

Iman and Mohammad [4] proposed Equation (3) to determine the specific heat capacity $C_{p,hnf}$ of the MWCNT/CuO hybrid nanofluid dispersed in deionized (DI) water at varying nanoparticle volume concentrations. This correlation is based on the assumption of thermal equilibrium between the suspended nanoparticles and the base fluid, ensuring an accurate estimation of the specific heat capacity.

$$C_{p,hynf} = \frac{(1-\varnothing)\rho_{bf}C_{p,bf} + \varnothing_1\rho_1 C_{p1} + \varnothing_2\rho_2 C_{p2}}{\rho_{hynf}} \quad \text{(Eq. 3)}$$

Dynamic Viscosity

Hayat and Nadeem [14] developed a correlation for estimating the viscosity of hybrid nanofluids as a function of nanoparticle volume concentration. The dynamic viscosity of the MWCNT/CuO hybrid nanofluid dispersed in deionized (DI) water was calculated using Equation (4) to analyze its variation at different volume concentrations.

$$\mu_{hynf} = \frac{\mu_{bf}}{(1-\varnothing_1)^{2.5} \cdot (1-\varnothing_2)^{2.5}} \quad \text{(Eq. 4)}$$

Thermal Conductivity

The thermal conductivity of the MWCNT/CuO hybrid nanofluid dispersed in deionized (DI) water was determined using the correlation proposed by Hayat and Nadeem[14] [15]. The calculation was performed using Equation (5) to evaluate the thermal conductivity at varying nanoparticle volume concentrations.

$$K_{nf} = \frac{K_{CuO} + 2K_f - 2\varnothing_{CuO}(K_f - K_{CuO})}{K_{CuO} + 2K_f + \varnothing_{CuO}(K_f - K_{CuO})} \times K_f$$

$$K_{hynf} = \frac{K_{MWCNT} + 2K_{nf} - 2\varnothing_{MWCNT}(K_{nf} - K_{MWCNT})}{K_{MWCNT} + 2K_{nf} + \varnothing_{MWCNT}(K_{nf} - K_{MWCNT})} \times K_{nf} \quad \text{(Eq. 5)}$$

3. RESULT AND DISCUSSION

3.1 Stability

For the preparation of 0.3% volume fraction (HyNF), MWCNT and CuO nanoparticles were dispersed in DI water in a 1:1 ratio and SDS and Triton X-100 surfactants were used in equal amount separately to enhance stability. The effect of the hybrid nanoparticle is validated in this study by loading base fluid with hybrid nanoparticles which improved base fluid thermophysical properties in a significant manner[5]. It can be observed in Fig. 60.6, the hybrid nanofluid did not show any agglomeration or sedimentation until a week confirming better stability. Thus, CuO nanofluid was much more unstable than MWCNT nanofluid, in which CuO nanoparticles increased deposition rate with time. But, for the case of CuO and CNT nanofluids, when CuO particles were added to MWCNT, CuO nanoparticles were absorbed on the MWCNT surface with a few of them remaining in the liquid because of the formation of agglomerates, in this case, augmented dispersion and stability for a long time[16].

3.2 Density

The density of MWCNT/CuO hybrid nanofluid decreases with an increase in nanoparticle volume fraction as shown

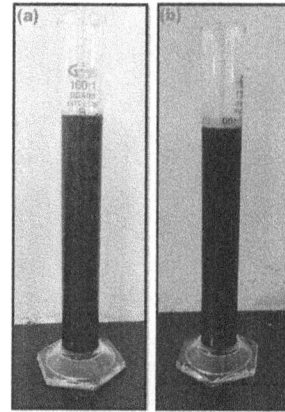

Fig. 60.6 Stability images of 0.3% volume fraction hybrid nanofluid at (a) Day one and (b) After a week

in Fig. 60.7. It was observed that as the volume fraction of the nanoparticles immersed in the base fluid increases, the density is a linear additive to the control fluid density measured. Peak density increased at 50°C and at 0.4% volume fraction, indicating that the mass per unit volume of the fluid is heavily reliant upon nanoparticle concentration. Furthermore, the density shows a slight decrease as a function of temperature, which is expected from thermal expansion with increasing temperature, as higher temperatures usually result in a lower density of the fluid triggered by the expansion of molecules[17][18].

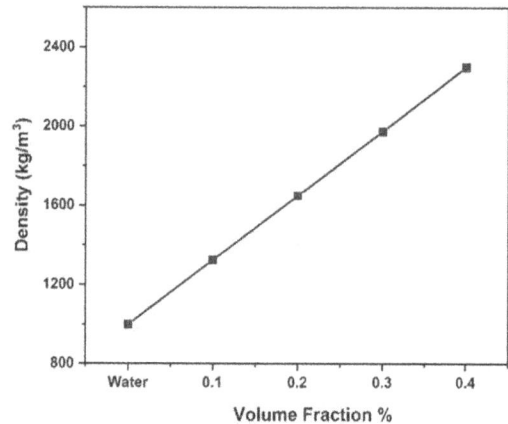

Fig. 60.7 Density versus volume fraction of MWCNT/CuO DI water hybrid nanofluid

3.3 Specific Heat

As presented in Fig. 60.8, the specific heat capacity (Cp) of the MWCNT/CuO hybrid nanofluid at differing nanoparticle volume fractions. It can be seen that specific heat capacity with nanoparticle concentration follows a decreasing trend. The maximum value of specific heat was obtained at 50°C for volume fraction 0.1% and the minimum specific heat value was obtained at 50°C for volume fraction 0.4% Due to lower specific heat of solids with respect to liquids, it meant that addition of solid NP to base fluid resulted in reduced specific heat

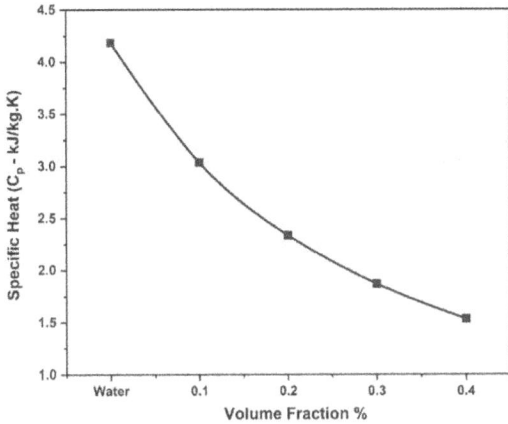

Fig. 60.8 Specific heat versus volume fraction of MWCNT/CuO DI water hybrid nanofluid

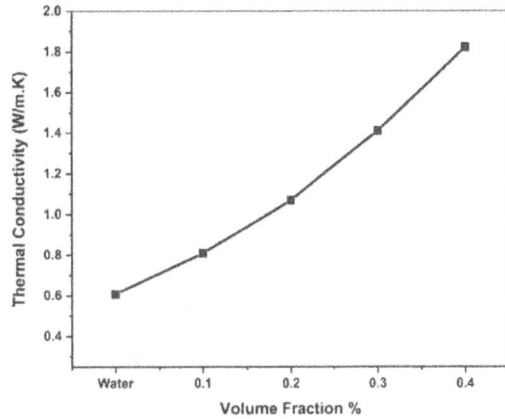

Fig. 60.9 Thermal conductivity versus volume fraction of MWCNT/CuO DI water hybrid nanofluid.

capacity[19]. Moreover, the specific heat capacity also reduces with increasing temperature which also conforms to the general trend seen in nanofluids because of the corresponding increase in thermal diffusivity. In the Lightweight composition, the latter leads to the former, as with the existence of nanoparticles, heat conduction improves within the fluid, and the energy is effectively transferred, as a result, heat transfer is performed better, and we need less heat to elevate the temperature of the nanofluid. Conclusion 1: The specific heat of the hybrid nanofluid is always lower than the specific heat of pure water due to increased thermal conductivity and heat diffusing rate from the nanoparticles[13].

3.4 Thermal Conductivity

The variation of thermal conductivity of MWCNT/CuO hybrid nanofluid against different values of nanoparticle volume fraction has been delineated in Fig. 60.9. We clearly observed that the hybrid nanofluids (MWCNT and CuO) have higher thermal conductivity than DI water. The trend of increasing thermal conductivity with the increase of nanoparticle volume concentration indicates that nanofluid formulations are efficient in the enhancement of heat transfer properties[20]. The best enhancement was obtained at 50°C for volume fraction 0.4%, stressing the heavy dependency of thermal conductivity on the concentration of the nanoparticles. Furthermore, the determinants of thermal conductivity of hybrid nanofluids can be further divided into their base fluids, the concentration of each nanoparticle, the size and shape of the nanoparticles, Brownian motion and agglomeration. The heat transfer efficiency is enhanced due to the combined effects of MWCNT and CuO nanoparticles due to the thermal conductivity path created by MWCNT and effective energy propagation within the liquid[21].

3.5 Dynamic Viscosity

The change in dynamic viscosity of the MWCNT/CuO hybrid nanofluid at different nanoparticle volume

fractions is presented in Fig. 60.10. These results show a gradual increase of viscosity with increasing nanoparticle concentration which reflect the influence of nanoparticle dispersion on the fluid's resistance on the flow. The maximum viscosity value measured was 50°C at a volume fraction of 0.4%, demonstrating the higher Viscosity impact due to the massive increase in nanoparticle fractions[22]. Increased viscosity of nanofluid enhances the interaction of nanoparticles and improves the thermal characteristics of the base fluid but needs higher pumping power as compared to pure fluid which reduces heat exchanger efficiency due to more pressure drop. Excessive viscosity will negatively affect the performance of the nanofluid in practical thermal applications by imposing unnecessary energy demand on driving the heat transfer fluid. Hence, tuning nanoparticle concentration is imperative to maintain balance between enhancement in thermal conductivity and control in viscosity, enhancing the performance of heat transfer[23][24].

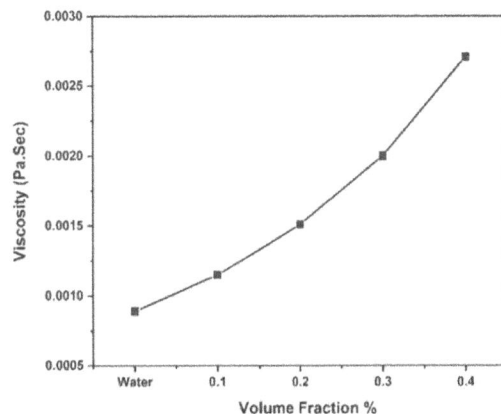

Fig. 60.10 Viscosity versus volume fraction of MWCNT/CuO DI water hybrid nanofluid

3.6 Heat Transfer Coefficient

The comparison of the variation of the heat transfer coefficient with mass flow rate for the MWCNT/CuO hybrid nanofluid in the presence of perforated disk inserts (Fig. 60.11). It was concluded from the study that the addition of hybrid nanoparticles to DI water, as each volume fraction, has considerably improved the heat transfer coefficient across various flow rates. As shown in both layouts, the mass flow rate enhances heat transfer coefficient, which means better convective heat transfer. The maximum coefficient of heat transfer is at 0.3% volume fraction, flow rate at 400 LPH, and at 50°C. Yet, at a higher volume fraction level of 0.4%, heat transfer performance reduction was noted owing to projected density and agglomeration of particles, that adversely affected thermal conductivity as well as effectiveness of convective heat transfer[25]. Comparative investigation between both the configurations shows that the heat exchanger with perforated disk inserts (Fig. 60.11) always has more heat transfer coefficient than a heat exchanger without the inserts. The tangle of the perforated disks increases the turbulence intensity and interrupts the thermal boundary layers, which facilitates the dispersion of the nanoparticles, thereby increases the convective heat transfer rates. The intensity of this enhancement was more pronounced in the higher flow rates where a considerable increase in turbulence led to considerable enhancement in heat transfer performance. In conclusion, the results show that using perforated disk inserts can effectively improve the convective heat transfer of hybrid nanofluids. An adequate amount of added nanoparticles (less than 0.3%) must be explored to reach the best heat transfer efficiency because excessive particle concentrations increase viscosity and form particle clusters, undesirable phenomena but incompatible with enhanced turbulence [18][26].

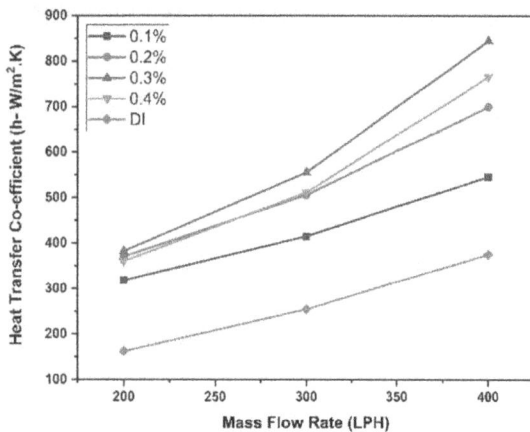

Fig. 60.11 Heat transfer coefficient versus mass flow rate for heat exchanger with perforated disk inserts

3.7 Reynolds Number

The trend of Reynolds number (Re) in relation to the mass flow rate for MWCNT/CuO hybrid nanofluid is shown in Fig. 60.12. The designs with perforated disk inserts are shown in Fig. 60.12. The findings show the development

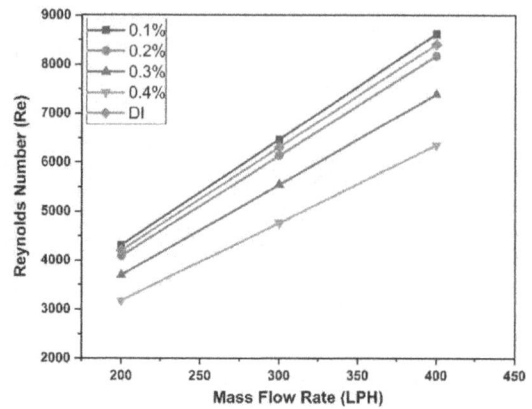

Fig. 60.12 Reynolds number versus mass flow rate heat exchanger with perforated disk inserts

of Reynolds number with increasing mass flow rate, which reflects the transition of flow conditions towards more turbulent flow nature. Nevertheless, the presence of hybrid nanoparticles in deionized (DI) water) a lower Reynolds number with respect to pure water, largely owing to increases in viscosity and density from the dispersed nanoparticles. At higher volume concentrations (0.1%, 0.2%, 0.3%, and 0.4%) more than this effect is seen since as viscosity increases flow resistance also increases, leading to higher tendency for agglomeration of nanoparticles as a consequence reduces Reynolds number[5]. By comparing the two configurations, it is evident that the Reynolds number in the heat exchanger with perforated disk inserts (Fig. 60.12) is larger than the heat exchanger without inserts. Such a growth can be explained with the flow obstruction and turbulence development caused by the perforated filament which promotes more vigorous mixing and limits a stable boundary layer formation. Fluid trajectory turns circular between perforated disk inserts, where it increases its momentum and disrupts laminar sublayers, which helps in increasing overall flow velocity and therefore Reynolds number. At higher values of nanoparticle volume fraction (0.4%), the combined impact of increased kinematic viscosity and agglomeration mitigate the level of this improvement despite the onset of enhanced turbulence[26].

3.8 Nusselt Number

Within Fig. 60.13, the variation of the Nusselt Number (Nu) for the MWCNT/CuO hybrid nanofluid aerodynamically is shown in function of mass flow rate with perforated disk inserts Fig. 60.13. The Observed increment in Nusselt number with mass flow rate signifies its impact on convective heat transfer with better fluid motion and turbulence. In all flow rates, a maximum Nusselt number of 0.1% (volume fraction) was higher than 0.2%, 0.3%, and 0.4% and DI (deionized) water. This trend indicates that at lower concentrations, the nanofluid maintains its ideal distribution of thermal conductivity and convective properties which results in better heat

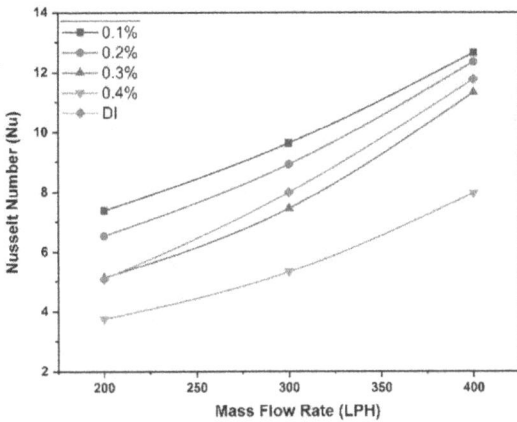

Fig. 60.13 Nusselt number versus mass flow rate heat exchanger with perforated disk inserts

transfer efficiency. In Fig. 60.13 at 0.1% volume fraction and 400 LPH flow rate significant improvement (~31%) in heat transfer performance in reference to DI water was observed. Conversely, Nusselt number was decreased when nanoparticles were added in higher concentrations (0.4%) because of increased viscosity along with agglomeration of the nanoparticles, which makes tracking of fluid more difficult and leads to ineffective convective heat transfer[27]. The Nusselt number is comparatively higher (difference) for the heat exchanger without perforated disk inserts (Fig. 60.13) v/s the heat exchanger without perforated disk inserts (Fig. 60.13). Disc inserts are perforated providing a means to mix the fluid, enhancing heat transfer by providing disturbance to the thermal boundary layer, and thus creating turbulence for heat transfer enhancement. On the other hand, the Nusselt number of the heat exchanger without inserts is not as large since the flow regime was not as disturbed and thus can't produce so much turbulence. This trend is more noticeable at higher flowrates, where turbulent flow and convective heat transport improvement are at a peak. Overall, the results confirm that Nusselt number increases with an increase in mass flow rate; however, at higher mass flow rates, the Nusselt number is more efficiently augmented with the presence of perforated disk inserts. By doing so, at 0.1% volume fraction gives the best heat transfer performance, and a higher concentration of nanoparticles leads to diminishing return due to viscosity effect. Consequently, the application of perforated disk inserts proves to be a productive method in this regard as it improves heat exchanger performance by augmenting the convective heat transfer and by raising the turbulence level[28].

3.9 Pressure Drop

Pressure drop (ΔP) profiles as a function of mass flow rate for MWCNT/CuO hybrid nanofluid (14) with and (20b) without perforated disk inserts. As the fluid velocity increases, the shear forces at the particle surface increase, consequently leading to higher pressure required to overcome resistances thus confirming our result that pressure drop increases as the flow rate increases. The influence of hybrid nanoparticles on the pressure drop is also observed when immersed in deionized (DI) water; with the suspended nanoparticles acting to break up the liquid pattern, the density and viscosity of the fluid increase, leading to an increase in the pressure drop. The fluid flow resistance causes an increasing trend of the pressure drop on the system with the increment of the particle volume concentration of the bacteria vessels, ranging from 0.1% to 0.4%, by which the increasing of viscosity of the fluid and the contact between the suspended solid state and the media fill up of the flow[1], [29]. When investigating heat exchange both with (Fig. 60.14) perforated disk insert, the pressure drop is greater when perforated disks are present. Fluid flow is obstructed and turbulent through the perforated disk insert, which ultimately increases the resistance to fluid motion and thus causes the pressure drop to be considerably higher. On the other hand, the heat exchanger without perforated disk inserts provides a relatively smoother fluid flow, which results in lower pressure drop for the same flow rate[7]. At higher concentrations of 0.4%, the effect of viscosity is more pronounced as the agglomeration of particles and the frictional resistance also contributes to an increase in the pressure loss. Overall, these results support the understanding that increasing heat transfer takes place with increasing pressure drop and mass flow rate, as well as with increasing nanoparticle concentration, but that there is a balance of enhancing heat transfer and reaching high resistance to flux. Although the use of perforated disk inserts enhances the convective heat transfer characteristics, it unfortunately also results in increased pressure drops which may necessitate a larger pumping power to circulate the working fluid. This implies that to ensure an appropriate thermal performance with a satisfactory level of pressure drop the optimization of the nanoparticle concentration and perforated disk design is required for heat exchanger applications[30][31]

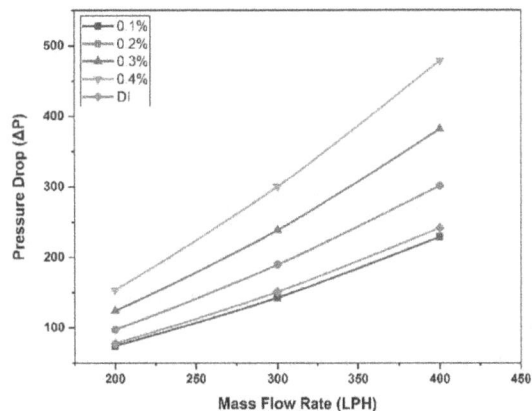

Fig. 60.14 Pressure drops versus mass flow rate heat exchanger with perforated disk inserts

3.10 Friction Factor

Figure 60.15 shows how the friction factor (F_f), varies with mass flow rate for the MWCNT/CuO hybrid nanofluid, both with and without perforated disk inserts (21a, b respectively). It is observed from the results that with the increase in volume concentration of nanoparticle, friction factor increases due to the increased viscosity of fluid with the addition of suspended nanoparticles. On the other hand, the friction factor reduces with an ascending mass flow rate showing the evolution to turbulent flow regimes where viscosity's relative influence upon flow resistance decreases[29]. The increase can mainly be attributed to flow obstruction and turbulence enhancement caused by perforated disks, which create disturbances and increase the resistance to flow. Whereas, in heat exchanger without inserts, fluid flows more smoothly, leading to a lesser friction factor for same flow rates. Nevertheless, in both cases, the Fanning friction factor decreases with the mass flow rate, due to the effect of higher velocities which diminishes the influence of viscous forces and makes the turbulent flow more developed. In general, the findings reconfirm that: (i) friction factor rises with nanoparticle loadings because of enhanced viscosity but (ii) falls with flow rates because of turbulence-based flow stabilizations. Perforated disk inserts enhance turbulence; however, they add more frictional losses which can result in an increase in pumping power. Hence, considering an appropriate concentration of the nanoparticles, combined with the perforated disk design, is vital to strike a balance between enhanced heat transfer characteristics and pressure drops in heat exchangers[6][32]

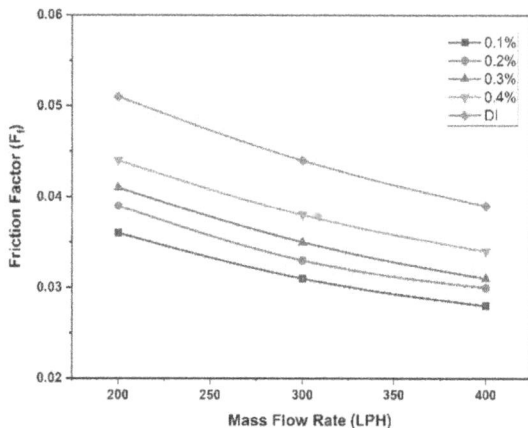

Fig. 60.15 Friction factor versus mass flow rate heat exchanger with perforated disk inserts

3.11 Thermal Performance Factor

Figure 60.16 illustrates the variation in the thermal performance factor (R) as a function of mass flow rate for MWCNT/CuO hybrid nanofluid, under perforated disk inserts (Fig. 60.16). The results indicate contrasting trends in thermal performance based on the presence or absence of perforated disks. However, as mass flow rate increases,

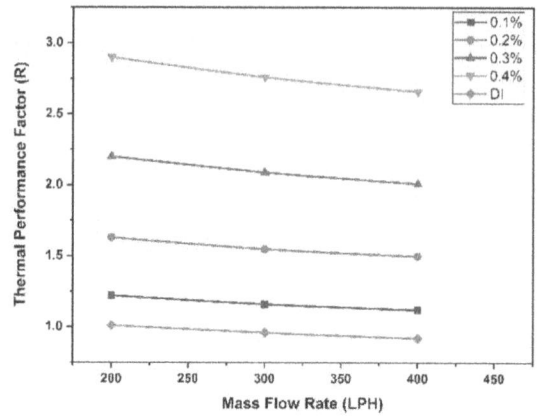

Fig. 60.16 Thermal performance factor versus mass flow rate heat exchanger with perforated disk inserts

the thermal performance factor improves, indicating enhanced convective heat transfer. Nevertheless, a further increase in nanoparticle volume concentration (0.1% to 0.4%) leads to a decline in thermal performance, likely due to increased fluid viscosity and reduced flow effectiveness, as no flow-enhancing inserts are present in this setup[33]. In contrast, Fig. 60.16, which represents the heat exchanger with perforated disk inserts, demonstrates that the thermal performance factor is higher when hybrid nanoparticles are introduced into DI water. Unlike the previous case, thermal performance initially increases at lower flow rates but then decreases with increasing mass flow rate, suggesting that higher turbulence and enhanced mixing from perforated disks contribute to better thermal efficiency at lower velocities. The maximum thermal performance factor was observed at 50°C, 0.4% volume fraction, and a flow rate of 200 LPH. Beyond this point, a further increase in flow rate led to a reduction in thermal performance, likely due to higher pumping power requirements and increased pressure losses associated with the interaction between nanoparticles and the perforated structures[34]. A comparative analysis between the two configurations indicates that the presence of perforated disk inserts significantly enhances the thermal performance factor at lower flow rates. The disks generate additional turbulence, promoting better nanoparticle dispersion and heat transfer augmentation. However, at higher flow rates, the effect of viscosity and pressure drop becomes more prominent, leading to diminishing returns in thermal performance. In the absence of perforated disk inserts, the thermal performance is limited by the lack of enhanced turbulence, and at higher concentrations (0.4%), viscosity effects dominate, reducing overall efficiency. Overall, the results confirm that thermal performance is highly dependent on both nanoparticle concentration and flow modifications. While perforated disk inserts enhance thermal efficiency at lower flow rates, their effectiveness reduces at higher velocities due to increased resistance. An optimized balance between nanoparticle concentration, flow rate, and disk insert design is essential for maximizing thermal performance while minimizing flow resistance and pressure drop in heat exchanger applications[35].

4. CONCLUSION

This study systematically investigated the thermal and flow characteristics of MWCNT-CuO/DI water hybrid nanofluids in a double-pipe heat exchanger, both with and without perforated disk inserts. The influence of nanoparticle concentration, flow rate, and perforated disk inserts on key thermophysical parameters such as heat transfer coefficient, Nusselt number, Reynolds number, pressure drop, friction factor, and thermal performance factor was comprehensively analyzed. The findings indicate that the hybrid nanofluid exhibited a significant enhancement in thermal conductivity compared to DI water, with a peak improvement of approximately 66.64% at a 0.4% volume fraction and 50°C. The highest heat transfer coefficient was achieved at a 0.3% volume fraction and a flow rate of 400 LPH, beyond which a decline was observed due to nanoparticle agglomeration. Comparatively, the heat exchanger with perforated disk inserts demonstrated superior thermal performance, with heat transfer rates increasing by approximately 55% compared to DI water and by 44% compared to the system without inserts. This improvement can be attributed to enhanced turbulence and better fluid mixing caused by the perforated disks. However, increasing the nanoparticle concentration resulted in higher viscosity, leading to an increase in pressure drop, friction factor, and pumping power requirements. The pressure drop was found to be higher in the presence of perforated disk inserts due to increased flow resistance. Additionally, the Reynolds number exhibited a decreasing trend with higher nanoparticle concentrations due to the combined effect of increased viscosity and density. Overall, this study demonstrates that hybrid nano fluids, in conjunction with perforated disk inserts, can significantly enhance heat exchanger performance. However, optimal nanoparticle concentration and perforation design must be carefully selected to balance heat transfer improvements with acceptable pressure drop and energy consumption. Future research can explore computational fluid dynamics (CFD) simulations to validate these experimental findings and extend the application of hybrid nanofluids to real-world thermal management systems such as radiators, electronic cooling systems, and industrial heat exchangers.

REFERENCES

1. A. Karimipour, O. Malekahmadi, A. Karimipour, M. Shahgholi, and Z. Li, "Thermal Conductivity Enhancement via Synthesis Produces a New Hybrid Mixture Composed of Copper Oxide and Multi-walled Carbon Nanotube Dispersed in Water: Experimental Characterization and Artificial Neural Network Modeling," *Int J Thermophys*, vol. 41, no. 8, p. 116, Aug. 2020, doi: 10.1007/s10765-020-02702-y.

2. M. Zadkhast, D. Toghraie, and A. Karimipour, "Developing a new correlation to estimate the thermal conductivity of MWCNT-CuO/water hybrid nanofluid via an experimental investigation," *J Therm Anal Calorim*, vol. 129, no. 2, pp. 859–867, Aug. 2017, doi: 10.1007/s10973-017-6213-8.

3. J. Qu, R. Zhang, Z. Wang, and Q. Wang, "Photo-thermal conversion properties of hybrid CuO-MWCNT/H2O nanofluids for direct solar thermal energy harvest," *Appl Therm Eng*, vol. 147, pp. 390–398, Jan. 2019, doi: 10.1016/j.applthermaleng.2018.10.094.

4. I. Fazeli, M. R. Sarmasti Emami, and A. Rashidi, "Investigation and optimization of the behavior of heat transfer and flow of MWCNT-CuO hybrid nanofluid in a brazed plate heat exchanger using response surface methodology," *International Communications in Heat and Mass Transfer*, vol. 122, p. 105175, Mar. 2021, doi: 10.1016/j.icheatmasstransfer.2021.105175.

5. S. H. Danook, K. J. Jassim, and A. M. Hussein, "Efficiency Analysis of TiO 2 /Water Nanofluid in Trough Solar Collector," *Journal of Advanced Research in Fluid Mechanics and Thermal Sciences Journal homepage*, vol. 67, pp. 178–185, 2020, [Online]. Available: www.akademiabaru.com/arfmts.html

6. "Basic Properties and Measuring Methods of Nanoparticles," in *Nanoparticle Technology Handbook*, Elsevier, 2018, pp. 3–47. doi: 10.1016/B978-0-444-64110-6.00001-9.

7. M. A. Safi, A. Ghozatloo, M. Shariaty-Niassar, and A. A. Hamidi, "Preparation of MWNT/TiO2 Nanofluids and Study of its Thermal Conductivity and Stability," 2014.

8. W. Yu and H. Xie, "A Review on Nanofluids: Preparation, Stability Mechanisms, and Applications," *J Nanomater*, vol. 2012, no. 1, Jan. 2012, doi: 10.1155/2012/435873.

9. H. E. Patel, T. Sundararajan, and S. K. Das, "An experimental investigation into the thermal conductivity enhancement in oxide and metallic nanofluids," *Journal of Nanoparticle Research*, vol. 12, no. 3, pp. 1015–1031, Mar. 2010, doi: 10.1007/s11051-009-9658-2.

10. Y.-H. Hung, W.-P. Wang, Y.-C. Hsu, and T.-P. Teng, "Performance evaluation of an air-cooled heat exchange system for hybrid nanofluids," *Exp Therm Fluid Sci*, vol. 81, pp. 43–55, Feb. 2017, doi: 10.1016/j.expthermflusci.2016.10.006.

11. A. Akhgar, D. Toghraie, N. Sina, and M. Afrand, "Developing dissimilar artificial neural networks (ANNs) to prediction the thermal conductivity of MWCNT-TiO2/Water-ethylene glycol hybrid nanofluid," *Powder Technol*, vol. 355, pp. 602–610, Oct. 2019, doi: 10.1016/j.powtec.2019.07.086.

12. S. Aberoumand and A. Jafarimoghaddam, "Tungsten (III) oxide (WO 3) – Silver/transformer oil hybrid nanofluid: Preparation, stability, thermal conductivity and dielectric strength," *Alexandria Engineering Journal*, vol. 57, no. 1, pp. 169–174, Mar. 2018, doi: 10.1016/j.aej.2016.11.003.

13. Z. X. Li, U. Khaled, A. A. A. A. Al-Rashed, M. Goodarzi, M. M. Sarafraz, and R. Meer, "Heat transfer evaluation of a micro heat exchanger cooling with spherical carbon-acetone nanofluid," *Int J Heat Mass Transf*, vol. 149, p. 119124, Mar. 2020, doi: 10.1016/j.ijheatmasstransfer.2019.119124.

14. A. A. A. Arani and F. Pourmoghadam, "Experimental investigation of thermal conductivity behavior of

MWCNTS-Al2O3/ethylene glycol hybrid Nanofluid: providing new thermal conductivity correlation," *Heat and Mass Transfer*, vol. 55, no. 8, pp. 2329–2339, Aug. 2019, doi: 10.1007/s00231-019-02572-7.

15. T. Hayat and S. Nadeem, "Heat transfer enhancement with Ag–CuO/water hybrid nanofluid," *Results Phys*, vol. 7, pp. 2317–2324, 2017, doi: 10.1016/j.rinp.2017.06.034.

16. A. Hajatzadeh Pordanjani, S. Aghakhani, M. Afrand, B. Mahmoudi, O. Mahian, and S. Wongwises, "An updated review on application of nanofluids in heat exchangers for saving energy," *Energy Convers Manag*, vol. 198, p. 111886, Oct. 2019, doi: 10.1016/j.enconman.2019.111886.

17. J. A. Eastman, S. U. S. Choi, S. Li, W. Yu, and L. J. Thompson, "Anomalously increased effective thermal conductivities of ethylene glycol-based nanofluids containing copper nanoparticles," *Appl Phys Lett*, vol. 78, no. 6, pp. 718–720, Feb. 2001, doi: 10.1063/1.1341218.

18. L. Yang, W. Ji, M. Mao, and J. Huang, "An updated review on the properties, fabrication and application of hybrid-nanofluids along with their environmental effects," *J Clean Prod*, vol. 257, p. 120408, Jun. 2020, doi: 10.1016/j.jclepro.2020.120408.

19. D. P. Kshirsagar and M. A. Venkatesh, "A review on hybrid nanofluids for engineering applications," *Mater Today Proc*, vol. 44, pp. 744–755, 2021, doi: 10.1016/j.matpr.2020.10.637.

20. G. S. Sokhal, "Heat transfer performance of water based nanofluids: A review," *Mater Today Proc*, vol. 37, pp. 3652–3655, 2021, doi: 10.1016/j.matpr.2020.09.787.

21. Z. Li, A. Shafee, I. Tlili, and M. Jafaryar, "Nanofluid for heat exchangers," in *Nanofluid in Heat Exchangers for Mechanical Systems*, Elsevier, 2020, pp. 59–73. doi: 10.1016/B978-0-12-821923-2.00002-6.

22. "Characteristics and Behavior of Nanoparticles and Its Dispersion Systems," in *Nanoparticle Technology Handbook*, Elsevier, 2018, pp. 109–168. doi: 10.1016/B978-0-444-64110-6.00003-2.

23. C. Periyasamy and M. Chellappa, "Thermal performance and reliability of procesor investigation using TiO2 and CuO/water nanofluids," *Thermal Science*, vol. 24, no. 1 Part B, pp. 541–547, 2020, doi: 10.2298/TSCI190414433P.

24. M. Hemmat Esfe, M. R. Sarmasti Emami, and M. Kiannejad Amiri, "Experimental investigation of effective parameters on MWCNT–TiO2/SAE50 hybrid nanofluid viscosity," *J Therm Anal Calorim*, vol. 137, no. 3, pp. 743–757, Aug. 2019, doi: 10.1007/s10973-018-7986-0.

25. C. Qi, T. Luo, M. Liu, F. Fan, and Y. Yan, "Experimental study on the flow and heat transfer characteristics of nanofluids in double-tube heat exchangers based on thermal efficiency assessment," *Energy Convers Manag*, vol. 197, p. 111877, Oct. 2019, doi: 10.1016/j.enconman.2019.111877.

26. M. U. Sajid and H. M. Ali, "Thermal conductivity of hybrid nanofluids: A critical review," *Int J Heat Mass Transf*, vol. 126, pp. 211–234, Nov. 2018, doi: 10.1016/j.ijheatmasstransfer.2018.05.021.

27. L. Yang and Y. Hu, "Toward TiO2 Nanofluids—Part 1: Preparation and Properties," *Nanoscale Res Lett*, vol. 12, no. 1, p. 417, Dec. 2017, doi: 10.1186/s11671-017-2184-8.

28. S. Abbasi, S. M. Zebarjad, S. H. N. Baghban, and A. Youssefi, "Statistical analysis of thermal conductivity of nanofluid containing decorated multi-walled carbon nanotubes with TiO2 nanoparticles," *Bulletin of Materials Science*, vol. 37, no. 6, pp. 1439–1445, Oct. 2014, doi: 10.1007/s12034-014-0094-2.

29. A. Sözen, M. Gürü, T. Menlik, U. Karakaya, and E. Çiftçi, "Experimental comparison of Triton X-100 and sodium dodecyl benzene sulfonate surfactants on thermal performance of TiO 2 –deionized water nanofluid in a thermosiphon," *Experimental Heat Transfer*, vol. 31, no. 5, pp. 450–469, Sep. 2018, doi: 10.1080/08916152.2018.1445673.

30. H. M. Maghrabie *et al.*, "Intensification of heat exchanger performance utilizing nanofluids," *International Journal of Thermofluids*, vol. 10, p. 100071, May 2021, doi: 10.1016/j.ijft.2021.100071.

31. H. Mehrarad, M. R. Sarmasti Emami, and K. Afsari, "Thermal performance and flow analysis in a brazed plate heat exchanger using MWCNT@water/EG nanofluid," *International Communications in Heat and Mass Transfer*, vol. 146, p. 106867, Jul. 2023, doi: 10.1016/j.icheatmasstransfer.2023.106867.

32. M. Khalili Najafabadi, K. Hriczó, and G. Bognár, "Enhancing the heat transfer in CuO-MWCNT oil hybrid nanofluid flow in a pipe," *Results Phys*, vol. 64, p. 107934, Sep. 2024, doi: 10.1016/j.rinp.2024.107934.

33. M. Rafid *et al.*, "Augmentation of heat exchanger performance with hybrid nanofluids: Identifying research gaps and future indications - A review," *International Communications in Heat and Mass Transfer*, vol. 155, p. 107537, Jun. 2024, doi: 10.1016/j.icheatmasstransfer.2024.107537.

34. L. S. Sundar, "Synthesis and characterization of hybrid nanofluids and their usage in different heat exchangers for an improved heat transfer rates: A critical review," *Engineering Science and Technology, an International Journal*, vol. 44, p. 101468, Aug. 2023, doi: 10.1016/j.jestch.2023.101468.

35. M. A. Al-Obaidi *et al.*, "Recent Achievements in Heat Transfer Enhancement with Hybrid Nanofluid in Heat Exchangers: A Comprehensive Review," *Int J Thermophys*, vol. 45, no. 9, p. 133, Sep. 2024, doi: 10.1007/s10765-024-03428-x.

Note: All the figures and tables in this chapter were made by the authors.

Advances in Mechanical Engineering and Materials Sciences – Dr. Vinay K. B et al. (eds)
© 2026 Taylor & Francis Group, London, ISBN 9-781-041-20970-6

61

Development and Methodology of a Piezoelectric-Driven Thermoacoustic Refrigerator Utilizing Lead Zirconate Titanate

S. A. Mohan Krishna, Amruth E.

Faculty Members, Department of Mechanical Engineering,
Vidyavardhaka College of Engineering,
Afiliated to Visveswaraya Technological University,
Belgaum, Mysore, Karnataka, India

Rahul R.[1]

Students, Department of Mechanical Engineering,
Vidyavardhaka College of Engineering,
Afiliated to Visveswaraya Technological University,
Belgaum, Mysore, Karnataka, India

T. A. Prabhakar

Faculty Members, Department of Mechanical Engineering,
Vidyavardhaka College of Engineering,
Afiliated to Visveswaraya Technological University,
Belgaum, Mysore, Karnataka, India

Harsh R. S.[2], Bharath P. N.[3], Hitesh B.

Students, Department of Mechanical Engineering,
Vidyavardhaka College of Engineering,
Afiliated to Visveswaraya Technological University,
Belgaum, Mysore, Karnataka, India

Abstract: The piezoelectric-driven thermo-acoustic refrigeration system integrates the concepts of thermo-acoustics and piezoelectricity to deliver an effective and eco-friendly cooling solution. This innovative system employs a piezoelectric transducer to produce sound waves in gases, creating temperature gradients through the thermoacoustic effect. Consequently, it eliminates the need for harmful refrigerants, opting instead for inert gases like helium or air. Key components of this system include the piezoelectric driver, thermoacoustic stack, heat exchangers, and an appropriate working fluid. To enhance the system's practicality, it is essential to optimize power output, efficiency, and material longevity. This technology demonstrates significant promise for compact, low-maintenance, and sustainable refrigeration solutions.

Keywords: Drivers, Heat exchanger, Piezoelectric driver, Thermoacoustic stack, Working fluid

1. INTRODUCTION

The progression of alternative refrigeration technologies as the urgency has grown significantly due to environmental issues linked to conventional practices or systems, particularly those utilizing hydrofluorocarbons (HFCs). Conventional vapor-compression refrigeration systems play a substantial role in contributing to global warming and ozone layer depletion.

Corresponding Authors: [1]rahulvipul29@gmail.com, [2]hrs08012004@gmail.com, [3]pnbharath211@gmail.com

10.1201/9781003725053-61

Recently, a significant rise has been detected interest in thermoacoustic refrigeration, which utilizes sound waves to achieve cooling without the detrimental effects of conventional refrigerants. This method is founded on the concept that sound waves traveling through a gas can create localized temperature variations, which can be harnessed for cooling purposes.

Most existing designs predominantly utilize mechanical drivers; however, the advent of new piezoelectric materials presents an exciting opportunity. Piezoelectric transducers are capable in the process of converting electrical energy into mechanical energy vibrations, enabling a more compact and efficient generation of cooling.

This innovation has resulted in the creation of piezoelectrically driven thermoacoustic refrigeration systems, which harness the advantages of piezoelectric actuation—such as compactness, high efficiency, and low maintenance—while eliminating mechanical components.

This introduction presents the essential principles that underlie the combination of piezoelectric actuation with thermoacoustic cooling, highlighting key design and operational factors, along with the benefits and obstacles linked to this cutting-edge technology in comparison to traditional refrigeration techniques. The paper titled "Development of a small-scale piezoelectric-driven thermoacoustic cooler," authored by Chen, Geng, and Jingyuan Xu and published in Applied Thermal Engineering in 2022, explores a groundbreaking small-scale cooling system that utilizes piezoelectric elements. This system serves as a more environmentally friendly alternative to standard cooling methods.

The research is founded on the thermoacoustic effect, which creates temperature variations through acoustic waves without relying on chemical refrigerants. The compact and energy-efficient design of this piezoelectric-driven thermoacoustic refrigeration system capitalizes on the interplay between pressure waves and heat transfer within a resonant tube to generate acoustic waves. Performance evaluations indicate enhanced cooling performance and energy efficiency compared to previous models, making it suitable for scaling and adapting to electronic cooling needs.

Daniel George Chinn's 2010 thesis, "Piezoelectrically-Driven Thermoacoustic Refrigerator," investigates innovative and eco-friendly refrigeration systems that employ piezoelectric components as their driving force. This cutting-edge technology utilizes sound waves to create temperature gradients for heat transfer, thus eliminating the dependence on harmful chemical refrigerants. Chinn highlights the potential of this technology for specialized applications, particularly regarding lower energy consumption and system simplicity. Nevertheless, it remains a developing alternative to traditional refrigeration methods, requiring additional research and optimization to improve cooling capacity, scalability, and operational reliability for broader practical use.

2. REVIEW OF LITERATURE

Chinn et al. (2011) detail the design and performance assessment of a piezoelectric-driven thermoacoustic refrigerator, which substitutes traditional electromagnetic loudspeakers with piezoelectric actuators. This system effectively transfers 0.3 W of heat across an 18°C temperature differential while consuming 7.6 W of input power. The study corroborates the model with experimental data and Delta EC software, highlighting the benefits of piezoelectric drivers in thermoacoustic applications due to their lightweight and energy-efficient properties.

Lata, Kolekar, and Swarnkar explore a compact thermoacoustic refrigerator that operates using piezoelectric actuators, which generate acoustic waves to establish temperature differentials without the need for moving components or harmful refrigerants. Their study highlights the benefits of miniaturization, energy efficiency, and environmental sustainability, while also tackling the difficulties related to optimizing acoustic performance and heat transfer efficiency.

Nouh, Aldraihem, and Baz (2014) introduce a thermoacoustic refrigerator powered by piezoelectric actuators, which is further enhanced by dynamic magnifiers that increase the displacement of the piezoelectric drivers. This enhancement leads to improved generation of acoustic waves and overall system efficiency. Their research demonstrates the advantages of integrating dynamic magnifiers to elevate the effectiveness of thermoacoustic refrigeration systems.

Babu and Sherjin (2017) provide a comprehensive review of thermoacoustic refrigeration, delving into its core principles, benefits, and challenges. The paper underscores its environmentally friendly characteristics by eliminating the use of harmful refrigerants and investigates potential applications in sustainable cooling solutions. It also addresses critical technical challenges such as efficiency, material selection, and acoustic design, while outlining future research avenues to enhance commercial feasibility.

Ding et al. (2023) examine a thermoacoustic refrigeration system that makes use of waste heat generated by industrial facilities. Their findings illustrate the practicality of harnessing low-grade waste heat to produce sound waves for cooling purposes, thereby eliminating the reliance on harmful refrigerants. The research evaluates the system's efficiency, design obstacles, and its potential for application in sustainable cooling technologies within the industrial sector.

Shah et al. (2021) investigate a thermoacoustic refrigeration system, concentrating on its fundamental operational mechanisms, efficiency, and possible uses. The paper outlines the benefits of thermoacoustic technology, such as its environmentally friendly nature and minimal maintenance requirements, while analyzing variables like temperature differentials and acoustic pressure that affect system performance. The research provides significant perspectives into optimizing these systems for improved cooling efficiency.

Nouh's dissertation explores the combination of thermoacoustic and piezoelectric technologies, with a particular focus on dynamic magnifiers that enhance sound waves produced by piezoelectric actuators. This study aims to improve the effectiveness of thermoacoustic refrigeration by optimizing energy conversion processes, thus aiding the advancement of innovative and sustainable cooling technologies.

Garrett, Hofler, and Perkins (1991) investigate the basic principles and applications of thermoacoustic refrigeration, emphasizing its simple design and lack of moving parts. The paper highlights its potential uses in space exploration and its contribution to environmental sustainability. The authors stress the need for additional research to improve efficiency and performance for real-world cooling applications.

Al-Mufti and Janajreh (2022) provide a concise summary of thermoacoustic refrigeration technology, outlining its core principles, benefits, and applications. The paper points out the eco-friendly nature of the technology, noting its operation without harmful refrigerants. Furthermore, it discusses recent advancements, challenges, and future research directions aimed at improving efficiency and practicality.

Qing et al. (2009) introduce a thermoacoustic refrigeration system designed to enhance cooling efficiency. Their research examines the system's operational principles and components, highlighting its reduced environmental impact and greater energy efficiency compared to conventional refrigeration methods. This work makes a significant contribution to the advancement of sustainable cooling solutions.

Chen and Xu (2022) investigate the active management process of heat transfer in a dual-acoustic-driver thermoacoustic refrigerator featuring a looped-tube design. Their study models the heat transfer mechanisms by incorporating the impacts of porous media, demonstrating how changes in acoustic parameters can enhance cooling efficiency. The results provide significant understanding for designing more effective and environmentally friendly thermoacoustic refrigeration systems.

The objective of this research is to enhance mini thermoacoustic systems for practical use in sustainable refrigeration. The literature review highlights advancements in thermoacoustic refrigeration, particularly emphasizing designs driven by piezoelectric mechanisms, the trend towards miniaturization, and the advancement of energy-efficient technologies. A range of studies explore innovative enhancements, including dynamic magnifiers, solid-state thermoacoustic systems, and the repurposing of waste heat for sustainable cooling solutions. Significant challenges noted in the research involve optimizing acoustic performance, improving heat transfer efficiency, and selecting suitable materials to enhance the feasibility of these systems. The findings promote the adoption of eco-friendly alternatives to conventional refrigeration methods, while simultaneously tackling issues associated with scalability and efficiency.

3. THOROUGH EXAMINATION OF LEAD ZIRCONATE TITANATE

Lead zirconate titanate (PZT) is widely employed in energy storage, electronic applications, harvesting and owing to its excellent piezoelectric characteristics. Miniature and low-power electronic devices, such as sensors and wearable technologies, typically require only a few hundred microwatts of power for effective wireless communication.

This paper offers a concise overview of the energy storage and harvesting capabilities of PZT-based materials in various forms, including bulk, film, nano, and composite structures, as well as different compositions. Energy can be classified into two main types: storage and harvesting. The classification of energy—be it electrical, thermal, or mechanical—determines the appropriate systems for its storage. Energy harvesting can be divided into five separate categories sources: Air movement, thermal energy, ultraviolet radiation, radio waves, and oscillation. It is noteworthy that when there is ample vibrational energy in the environment, the energy storage density of piezoelectric devices surpasses that of other energy harvesters, such as those that rely on electromagnetic or electrostatic energy, by a factor of at least three.

PZT was produced with a Zr/Ti ratio of 52/48 using two distinct methods: the polymeric precursor method (PPM) and the microwave-assisted hydrothermal method (MAHM). The materials obtained were analyzed through various techniques, including X-ray diffraction (XRD) and scanning electron microscopy (SEM), particle size distribution via sedimentation, hysteresis measurements, and photoluminescence (PL). The results revealed that the PZT powders exhibited both tetragonal and rhombohedral phases, with differences in particle dimensions and morphology derived from the synthesis technique used.

Hysteresis loop analysis these findings were supported, suggesting that PZT powders synthesized via the PPM displayed a typical ferroelectric material loop and were more affected by spatial charges, whereas those produced by the MAHM exhibited a hysteresis loop akin to paraelectric materials and were less influenced by spatial charges. Both samples demonstrated PL activity in the green region (525 nm), with the PPM-synthesized sample exhibiting greater intensity in the spectra.

To enhance the dependability of multilayer actuators, it is crucial to assess mechanical strain experienced by the apparatus throughout operation. The following study reveals the small-signal magnitude of the elastic constant does not adequately capture the complex characteristics of lead zirconate titanate (PZT) ceramics. Consequently, compressive strain and depolarization have occurred evaluated concerning the large-signal stress applied in the poling direction. Both soft and hard PZT ceramics have been utilized examined.

In hard PZT, the domain structure switching occurs under increased stress levels compared to soft PZT. Furthermore, during the unloading phase in rigid PZT, certain domains show signs of partial reversion. The critical stress, or coercive stress, required for the domain-switching process is affected by the Zr:Ti ratio, reflecting a similar relationship observed with the electric coercive field. Additionally, coercive stress exhibits a linear relationship with the electric field, with the slope of this particular matter correlation being influenced through the relationship between depolarization and compressive strain resulting from domain switching.

4. UNIQUENESS IN OUR PROJECT

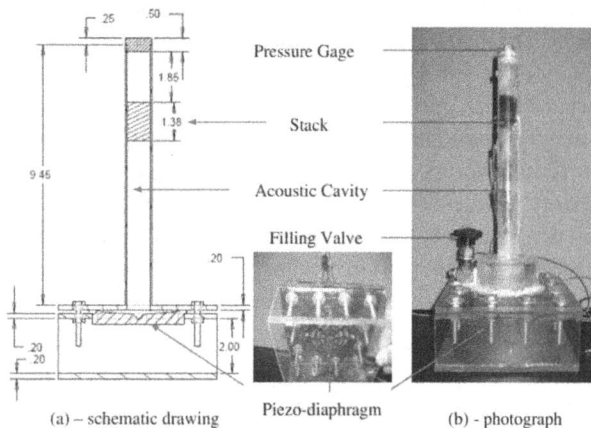

(a) – schematic drawing   Piezo-diaphragm   (b) - photograph

Fig. 61.1 Experimental setup details

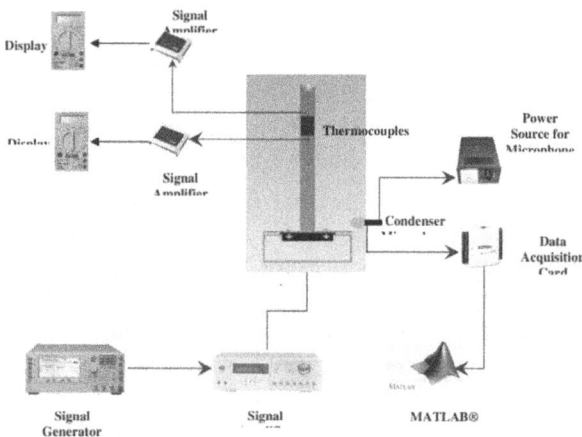

Fig. 61.2 Block diagram

Lead Zirconate Titanate (PZT) is a ceramic material composed of lead, zirconium, and titanium. is a ceramic material composed of lead, zirconium, and titanium. is extensively employed in thermoacoustic refrigeration systems due to its remarkable piezoelectric characteristics and its capacity. These vibrations play a crucial role in vibrations. These vibrations are crucial for producing the high-amplitude sound waves necessary for thermoacoustic processes, where acoustic energy generates temperature differentials for cooling purposes.

PZT is particularly noted for its elevated piezoelectric coefficient, enabling it to achieve considerable mechanical displacement with minimal electrical input, thus enhancing energy efficiency and effectiveness in compact system designs. Additionally, PZT is favored for its thermal and mechanical stability, ensuring reliable performance across a wide range of temperatures and various operational conditions, which is critical for both experimental and practical refrigeration applications.

In our project, Lead Zirconate Titanate (PZT) is employed for its versatility and extensive applicability.

4.1 Physical Properties

1. **Piezoelectricity:** PZT demonstrates strong piezo-electric properties, making it ideal for use in devices for sensing, actuation, and energy collection systems.

2. **Ferroelectricity:** As a ferroelectric material, PZT possesses an intrinsic electric polarization that can be modified by an external influence electric field.

3. **High dielectric constant:** PZT features a significant dielectric constant (εr), which is advantageous for capacitor-related applications.

4.2 Mechanical Properties

1. **High stiffness:** PZT exhibits considerable stiffness, making it appropriate for applications that necessitate substantial mechanical stability.

2. **Low mechanical loss:** PZT has minimal mechanical loss, which is advantageous for applications that require high efficiency.

4.3 Electrical Characteristics

1. **High electromechanical coupling coefficient:** PZT exhibits a significant electromechanical coupling coefficient (k), indicating its effectiveness in transforming energy between electrical and mechanical states.

2. **Elevated Curie temperature:** PZT has a notable Curie temperature (Tc), which is the threshold at which the material ceases to exhibit piezoelectric properties.

Applications

1. **Sensors:** PZT is employed in a variety of sensors such as pressure sensors, accelerometers, and ultrasonic sensors.

Proposed Methodology:

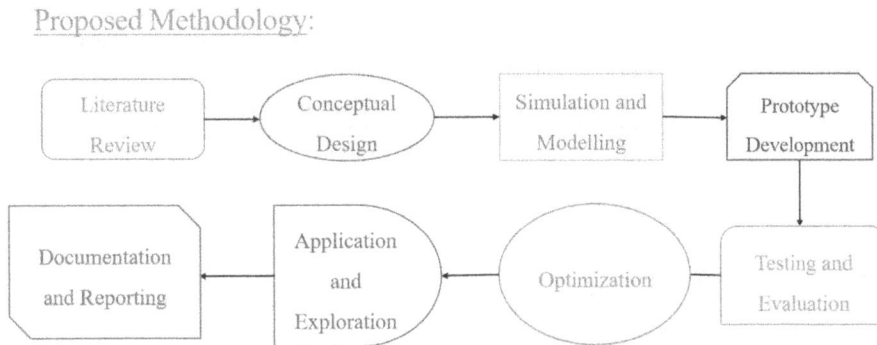

Fig. 61.3 Proposed methodology

2. **Actuators:** PZT is utilized in actuators such as piezoelectric motors, pumps, and valves.
3. **Energy harvesting:** PZT is integrated into energy harvesting technologies, including piezoelectric generators and systems designed to harness energy from vibrations.
4. **Medical devices:** PZT is found in medical devices like ultrasound transducers and equipment for assessing bone density.

Challenges and Limitations

1. **Toxicity:** The presence of lead in PZT raises concerns about toxicity, which may pose environmental and health hazards.
2. **Depolarization:** Over time, PZT may experience depolarization, potentially diminishing its piezoelectric effectiveness.
3. **Sensitivity to mechanical stress:** PZT can be affected by mechanical stress, which could affect its piezoelectric performance.

5. KEY CONTRIBUTIONS

This paper explores the combination of dynamic magnifiers with piezoelectric drivers to enhance acoustic pressure in resonators, thereby optimizing energy transfer for effective cooling. Significant findings include:

1. **Enhancement of Efficiency:** Dynamic magnifiers amplify mechanical vibrations and thermoacoustic phenomena, resulting in more efficient cooling solutions.
2. **Design Refinement:** By optimizing resonance conditions, adjusting magnifier geometries, and fine-tuning system parameters, it is feasible to increase acoustic pressure while conserving energy.
3. **Verification:** Both numerical simulations and experimental studies demonstrate that dynamic magnification provides enhanced cooling performance compared to traditional methods.
4. **Environmental Impact:** This strategy diminishes dependence on chemical refrigerants, offering a compact, energy-efficient solution for small-scale cooling that supports sustainable refrigeration.

6. RESEARCH OVERVIEW

In conclusion, piezoelectrically driven thermoacoustic refrigeration represents a cutting-edge and eco-friendly technology that addresses the limitations of traditional refrigeration systems. By leveraging thermoacoustic principles alongside the compact and efficient nature of piezoelectric transducers, this technology presents considerable potential for transforming refrigeration applications, particularly in contexts where space limitations and sustainability are critical. Although challenges remain in improving efficiency, power generation, and material durability, the unique advantages of this system—such as its reliance on inert gases and reduced mechanical complexity—position it as a promising alternative for future cooling solutions

REFERENCES

1. Chinn, D. G., et al. "Thermo-acoustic refrigerator powered by piezoelectric elements." In Active and Passive Smart Structures and Integrated Systems 2011, Vol. 7977. SPIE, 2011.
2. Lata, Manju, Nilesh Kolekar, and Abhishek Swarnkar. "Miniature Thermoacoustic Refrigerator Powered by Piezoelectric Technology."
3. Nouh, M., O. Aldraihem, and A. Baz. "Dynamic Magnifiers in Piezo-driven Thermoacoustic Refrigerators." Applied Acoustics 83 (2014): 86–99.
4. Babu, K. Augustine, and P. Sherjin. "A Comprehensive Review of Thermoacoustic Refrigeration and Its Importance." International Journal of Chem Tech Research 10.7 (2017): 540–552.
5. Jaworski, Artur J., and Xiaoan Mao. "Advancements in Thermoacoustic Devices for Energy Generation and Refrigeration." Proceedings of the Institution of Mechanical Engineers, Part A: Journal of Power and Energy 227.7 (2013): 762–782.
6. Ding, Xiachen, et al. "Investigation of a Thermoacoustic Refrigeration System Utilizing Waste Heat from Industrial Buildings." Sustainable Energy Technologies and Assessments 55 (2023): 102971.

7. Shah, S. V., et al. "Evaluation of a Thermo-acoustic Refrigeration System." International Research Journal of Engineering and Technology (IRJET) 8.8 (2021): 800–805.

8. Nouh, Mostafa. Thermoacoustic-Piezoelectric Systems Featuring Dynamic Magnifiers. Dissertation, University of Maryland, College Park, 2013.

9. Garrett, Steven L., Thomas J. Hofler, and David K. Perkins. "An Overview of Thermoacoustic Refrigeration." In Technology 2001: Conference Proceedings of the Second National Technology Transfer Conference and Exposition, December 3-5, 1991, San Jose Convention Center, San Jose, CA, Vol. 2, No. 3136. National Aeronautics and Space Administration, 1991.

10. Al-Mufti, Omar Ahmed, and Isam Janajreh. "A Brief Review of Thermoacoustic Refrigeration." International Journal of Thermal Environment Engineering.

11. Hao, Haitian, et al. "Thermoacoustic Characteristics of Solids: A Pathway to Solid-State Engines and Refrigeration." Journal of Applied Physics 123.2 (2018).

12. Chen, Geng, et al. "An Overview of Multi-Physics Interactions in Thermoacoustic Devices." Renewable and Sustainable Energy Reviews 146 (2021): 111170.

13. Chen, Geng, and Jingyuan Xu. "Active Control of Heat Transfer in the Porous Material of a Loop-Tube Dual-Acoustic-Driver Thermoacoustic Refrigerator." Cryogenics 125 (2022): 103516.

14. Chen, Reh-Lin, Chung-Lung Chen, and Jeff DeNatale. "Study of a Miniature Thermoacoustic Refrigerator." ASME International Mechanical Engineering Congress and Exposition. Vol. 3638. 2002.

15. Petrina, Denys E. Assessment of Performance Metrics for a Mini Thermoacoustic Refrigerator and Its Associated Drivers. Dissertation, Monterey, California. Naval Postgraduate School, 2002.

16. Collard, Sophie. "Design and Assembly of a Thermoacoustic Engine Prototype." (2012)

17. T. Somasekhar, P. Naveen Kishore. "Thermoacoustic Refrigeration." IOSR Journal of Mechanical and Civil Engineering (IOSR-JMCE).

18. Zhimin Hu. "Cooling Efficiency of Miniaturized Thermoacoustic Expanders Operating at 133 K." CryoWave Advanced Technology, Inc., Pawtucket, RI 02860.

19. Ikhsan Setiawan, Irna Farikhah, and Anggi Rahmawati. "Development and Testing of a Small-Scale Thermoacoustic Electricity Generator with Variable Heating Power." 2024, vol. 2.

20. Lagouge Tartibu Kwanda. "A Multi-Objective Optimization Strategy for Designing Small-Scale Standing Wave Thermoacoustic Coolers." Bellville, June 2014.

Note: All the figures in this chapter were made by the authors.

Advances in Mechanical Engineering and Materials Sciences – Dr. Vinay K. B et al. (eds)
© 2026 Taylor & Francis Group, London, ISBN 9-781-041-20970-6

62

Optimising Thermal Performance—V-Threaded Pipes and RGO-Distilled Water Nanofluids in Double Pipe Counterflow Heat Exchangers

Praveenkumara B. M.[1],
Sadashive Gowda B.[2],
Dushyanthkumar G. L.[3], **Shivashankar R.**[4]
Department of Mechanical Engineering
Vidyavardhaka College of Engineering,
Mysuru, Karnataka, India

Abstract: This experimental study investigates the synergistic effects of external and internal V-threaded pipes with varying pitch-to-depth ratios of 1.0 and 2.0 and RGO-Distilled Water (DW)-based nanofluids of 0.01 and 0.05 volume. % fluid concentrations on heat transfer progress in double-pipe counterflow heat exchangers (DPHEx). The experimental outcomes show that combining RGO nanofluids with V-threaded pipes significantly enhances the thermal performance assessed with smooth pipes and base fluid (Distilled water). The optimal configuration, featuring an external V-threaded pipe with a pitch-to-depth ratio of 1.0 and 0.05 volume. % RGO nanofluid achieved the highest Nusselt number (Nu) enhancement. This enhancement is attributed to increased thermal conductivity, increased surface area, and intensified turbulence creation in the heat exchanger. The results of this investigation show the possibility of the threaded pipes and RGO nanofluids in boosting the efficiency of DPHEx for energy and industrial applications.

Keywords: Threaded pipe, Reduced graphene oxide, Nusselt number, Friction factor

1. INTRODUCTION

Due to its vital function in increasing energy efficiency, decreasing equipment size, and lowering operating costs, heat transfer improvement in thermal systems is gaining attention. The combination of combined augmentation plays an important role in fulfilling the current energy demand. In such combined augmentation techniques surface modification and nanofluids are promising one (Praveenkumara et al. 2023). The use of V-threaded pipes is one example of a surface alteration that increases heat transfer rates without appreciably raising pressure drop by improving fluid mixing and creating turbulence

(Praveenkumara et al. 2022). However, because of their improved stability and thermal conductivity, the use of nanofluids-especially those made with reduced graphene oxide (RGO)-has attracted a lot of interest. V-threaded pipes and RGO-based nanofluids are two techniques that, when combined, have the potential to significantly improve thermal performance in a variety of heat exchanger applications (Heyhat et al. 2016). The remarkable thermal characteristics of reduced graphene oxide nanofluids make them perfect for cutting-edge heat transfer applications (Singh et al. 2021). Their exceptional dispersion stability and strong thermal conductivity provide effective energy transfer throughout the fluid media. The combined effects

[1]praveenkumarabm@vvce.ac.in, [2]sadashivegowdab@vvce.ac.in, [3]dushyanth.mech@vvce.ac.in, [4]shivashankar.r@vvce.ac.in

of increased turbulence and higher thermal conductivity can result in significant increases in heat transfer rates when used in V-threaded pipes. This work aims to investigate experimentally the effects of V-threaded pipe form and low-concentration RGO nanofluids on the heat transmission capacity of thermal systems. This study aims to provide crucial information regarding the feasibility and effectiveness of this hybrid enhancement technology by examining important metrics including coefficient of convective heat transfer, pressure drop (ΔP), and thermal efficiency, opening the door for its use in industrial heat exchanger.

2. LITERATURE REVIEW

Thianpong et al. (2024) investigated impact of TiO_2 nanofluids of concentrations 0.05, 0.10 and 0.15% and double counter twisted tapes insert of twist ratios of twist ratios 1.5, 2.0 and 2.5 on the performance of DPHEx. ITE increased when the nanofluid concentration rose and the tape twisted ratio (TR) fell. Mohamed Salem et al. (2024) evaluate the performance of DPHEx using baffle plates of baffle cut-off ratio, pitch ratio, relative angle of the floral design. the maximum increase of Nu is 147.4%. The impact of longitudinal double stripe and Fe_3O_4 in DPHEx did by Ebrahimi-Moghadam et al. (2020). They observed greater improvement in the thermal performance factor (TEF) by 4.73% for an Aspect ratio of 4, depth ratio of 0.3, and volume concentration of 0.03%. Deep dimpled tubes and Al_2O_3 nanofluids were used in the experiments, which were carried out by Khashaei et al. (2024) laminar flow with concentrations of nanofluids varying from 0.1 weight percent to 1 weight percent, Re values between 500 and 2250. The results indicate that a deep dimpled tube containing 1 weight percent nanofluid may increase the HTC by up to 3.42 when compared to a smooth tube at Re = 2250, Nu by 41.8, and ff by 1.82 times. The radius of cut (3.83%) and the volume per cent of nanoparticles (79.75%) had a substantial impact on the thermal performance factor. Khlewee et al. (2024) numerically investigated the performance of HEx using helical screws and Al_2O_3 and CuO nanofluid of the same concentrations. CuO nanofluid exhibits the highest heat transfer compared to Al_2O_3 of 34%. Ahmad et al. (2023) examined the effects of using CuO, Al_2O_3, and TiO_2 nanofluids in double-dimple corrugated tubes at different concentrations ranging from 1% to 3%. The best-performing nanofluid in the double-dimpled pipe flow was a 3% TiO2+CuO hybrid nanofluid, which improved maximum thermal performance (TPF) by 20.62%. Ahirwar and Kumar (2024) investigated the numerical influence of CuO nanofluid of 0.01% and 0.04 % concentration and twisted tape (TP) insert of 5, 10, & 15 sweeps in tubular heat exchanger. At the highest flow rate, peak performance was recorded, with the heat transfer coefficient reaching $2511 Wm^2/K$ and the improved effectiveness reaching 0.37. Panja, Das,

and Mahesh (2025) investigated the impact of Al_2O_3 and CuO nanofluids of 3 to 7% concentration with porous strip inserts in a solar collector. The thermal efficiency of Al_2O_3 + CuO nanofluids (NFs) rises by 7.73% and 8.59%, respectively, when the 5° to 15° geometry angle is switched.

The study's findings show that instead of using different inserts in a heat exchanger, surface augmentation techniques give better thermal performance. Still, there is a scope for using surface augmentation and very stable nanofluids to rise heat transfer (HT) in a heat exchanger (HEx). The current study mainly concentrates on the combined impact of 0.01 and 0.05 vol.% RGO nanofluid and V-threaded inner pipe (Internal and external threads) of pitch-to-depth ratio of 1 and 2 on the heat transfer performance (HTP) of Double pipe counter flow heat exchanger (DPHEx) experimentally.

3. NANOMATERIAL AND NANOFLUID

Using SEM, EDX, and XRD analysis, reduced graphene oxide (RGO) nanoparticles were characterized in order to assess their shape, elemental makeup, and crystalline structure. The distinctive wrinkled and sheet-like shape of RGO was confirmed by the SEM images in Fig. 62.1. (a), which also confirmed its large surface area and nanoscale dimensions. In the Fig. 62.1. (b) depicts the EDX analysis demonstrated a high carbon-to-oxygen ratio, indicative of the effective reduction of RGO and the removal of O_2-containing functional groups during synthesis. The carbon content of 75.7 wt. % 80.1 atomic % similarly oxygen content of 24.3 wt. %, 19.9 atomic % present in the material. Furthermore, the XRD pattern showing a sharp diffraction peak around $2\theta = 23.1°$, corresponding to the (002) plane of RGO, along with the absence of the broad peak at 11° typically associated with graphene oxide, confirming the successful reduction process. These analyses collectively validate the quality, purity, and structural integrity of the RGO nanoparticles, making them suitable for advanced thermal and lubrication applications.

Figure 62.2 shows a schematic representation of a two-step method of RGO/DW-based nanofluid preparation. The RGO nanoparticles were weighed based on the volume % and added directly to the base fluid (distilled water) along with 0.25 weight.% of sodium dodecyl sulphate (SDS) surfactant to stabilize the nanofluid. After adding nanoparticles and surfactant into the base fluid, this mixture is stirred using a motorized stirrer continuously for one hour, and then this mixture is transferred to an ultrasonic sonicator for uniform dispersion of the nanoparticles (Praveenkumara and Sadashive Gowda 2023). Figure 62.3 (a) and (b) shows the RGO-DW-based nanofluids of 0.010 volume. % and 0.050 volume.%. As per the visual stability, the nanofluids are stable more than 15 days.

(a)

(b)

(c)

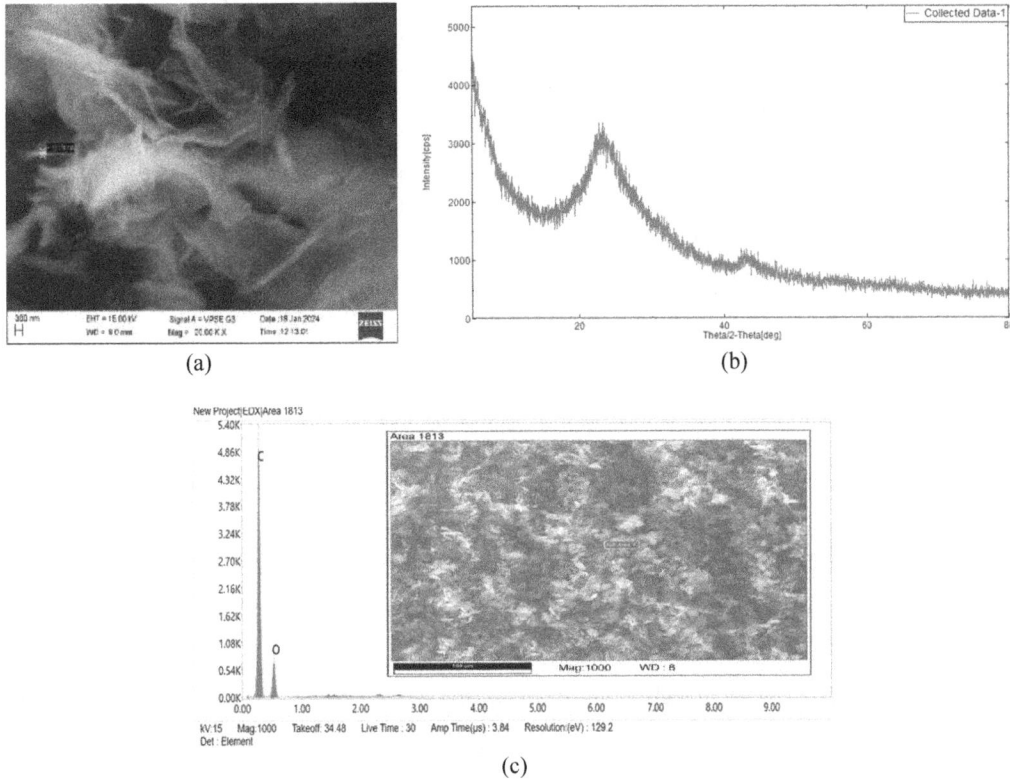

Fig. 62.1 (a) SEM, (b) XRD, (c) EDX of RGO nanoparticles

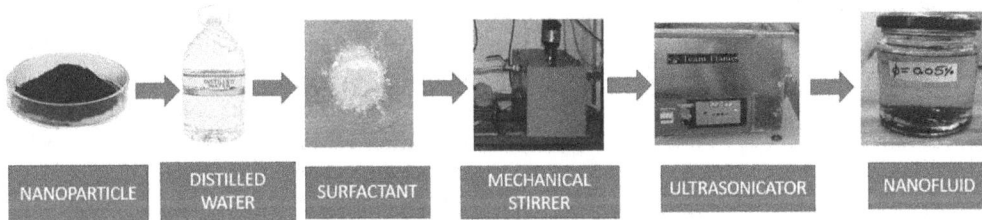

Fig. 62.2 RGO-DW Nanofluid preparation

4. EXPERIMENTAL METHOD AND DATA REDUCTION

Figure 62.3 depicts the DPHEx experimental resource. It mainly consists of test section of length 2.1 m, inner copper pipe of inner diameter (ID)=16mm, outer diameter (OD)=19mm and outer GI pipe of internal diameter (ID)=27mm, outer diameter (OD) =33mm respectively. The HEx Assessment Section is wrapped with cotton threads to evade heat loss to the surrounding. Nine k-type thermocouples of temperature range 0-1250 °C are commissioned to find the temperature of entry and exit cold and hot water, and also to measure the inner copper tube's surface temperature. 2-turbine mass flow meters of range 0-25 ltr/min were utilized to measure the hot and cold water flow rates. Pressure transducers of range 0-100 mbar and 0-1000 mbar are employed for hot water and cold water pressure difference measurement. Two hot water storage tanks namely primary and secondary tanks

are used for storage purposes along with two heating coils of 2 kW capacity. At the initial stage of the experiment, the heater coil is turned on and allowed it to bring the water's temperature up to the necessary level of 70 °C. Once the temperature is achieved turn on the pump and set the flow rate using the valve supplied at the HEx test section's intake. The hot water starts to flow inside the copper pipe and cold water from the overhead tank is allowed to flow through the annulus of the test section in a counterflow direction. The test rig is allowed to reach steady state conditions and then with the assistance of a data logger record the temperatures and pressure differences (ΔP) across the HEx test section. The trials were repeated under various experimental circumstances.

The Nu and ff measurements for the smooth tube were acquired and compared with proven correlations from the body of existing literature to verify the current experimental setup (Akyürek et al. 2018).

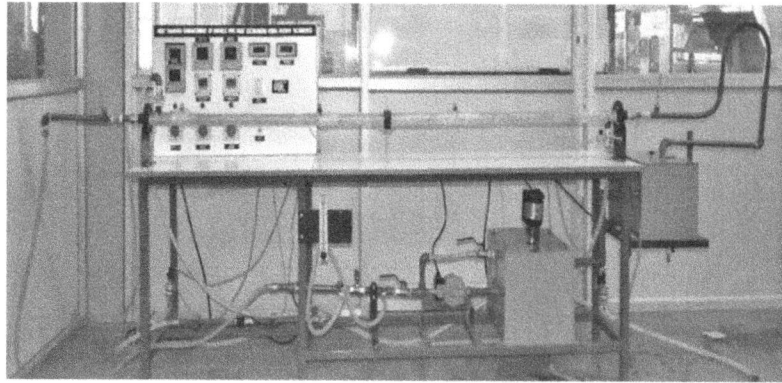

Fig. 62.3 DPHEx experimental test rig

Petukhov Correlation: If $3000 \leq Re \leq 5 \times 10^6$ Turbulent flow:

$$ff = \left[0.79\ln\left(Re\right) - 1.64\right]^{-2} \qquad (1)$$

If $Re \geq 2300$ Turbulent flow by Dittus-Boelter Correlation:

$$Nu = 0.0265 Re^{0.8} Pr^{0.3} \qquad (2)$$

The following formulas were utilized to determine the Nu, ff, and TPF for each operating scenario based on the experimental data gathered from the tests (Samruaisin et al. 2025).

Heat lost by hot fluid: $Q_h = m_h Cp_h \left(T_{hi} - T_{ho}\right)$ (3)

Heat gained by cold fluid: $Q_c = m_c Cp_c \left(T_{co} - T_{ci}\right)$ (4)

The mean heat transfer (HT): $Q_{Avg} = \dfrac{Q_h + Q_c}{2}$ (5)

Fluid convection heat transfer coefficient:

$$h = \frac{Q_{Avg}}{A_s\left(T_b - T_s\right)} \qquad (6)$$

Where: $A_s = \pi dL$, $T_b = \dfrac{T_i + T_o}{2}$, $T_s = \dfrac{T_{s1} + T_{s2} + T_{s3} + T_{s4}}{4}$

Nusselt number: $Nu = \dfrac{hd}{k}$ (7)

Reynolds Number: $Re = \dfrac{\rho V d}{\mu}$ (8)

The friction factor: $ff = \dfrac{\Delta P}{\dfrac{\rho V^2}{2}\dfrac{L}{d}}$ (9)

5. RESULTS AND DISCUSSIONS

5.1 Calibration of Experimental Test Rig

To establish a baseline for comparison with nanofluid results and validate the experimental setup, initial studies were conducted using distilled water. Figure 62.4 (a) compares the experimental friction factor (ff) values with theoretical results obtained from the Petukhov correlation, revealing a deviation of within 15%. Similarly, Fig. 62.4(b) shows the resemblance between theoretical and experimental Nusselt number values, as determined by the Dittus-Boelter correlation, with a deviation of within 14%. These results demonstrate the accuracy and reliability of the DPHEx experimental test rig (Akhavan-Behabadi et al. 2010).

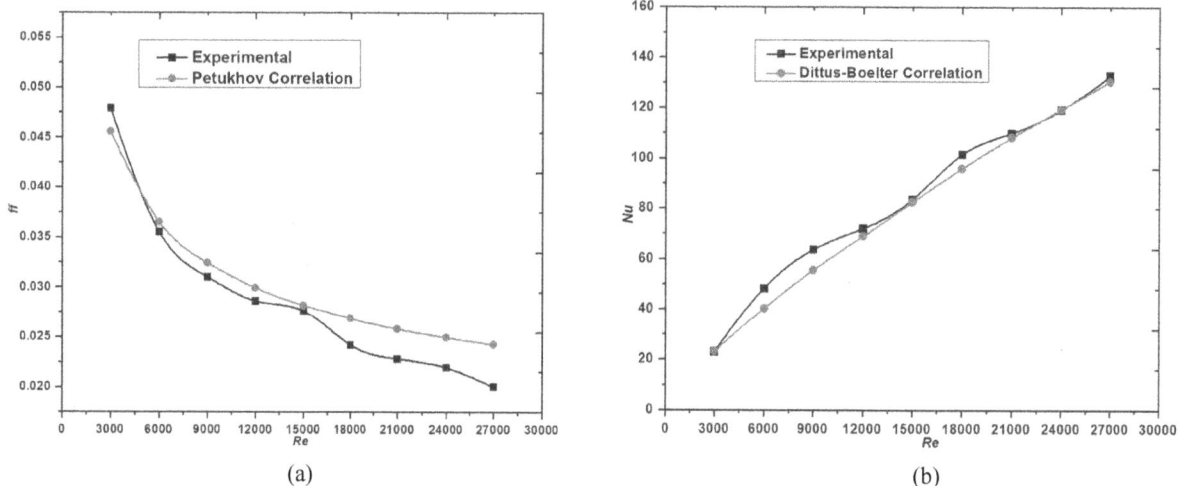

(a)

(b)

Fig. 62.4 Calibration curve in terms of ff v/s Re and Nu v/s Re

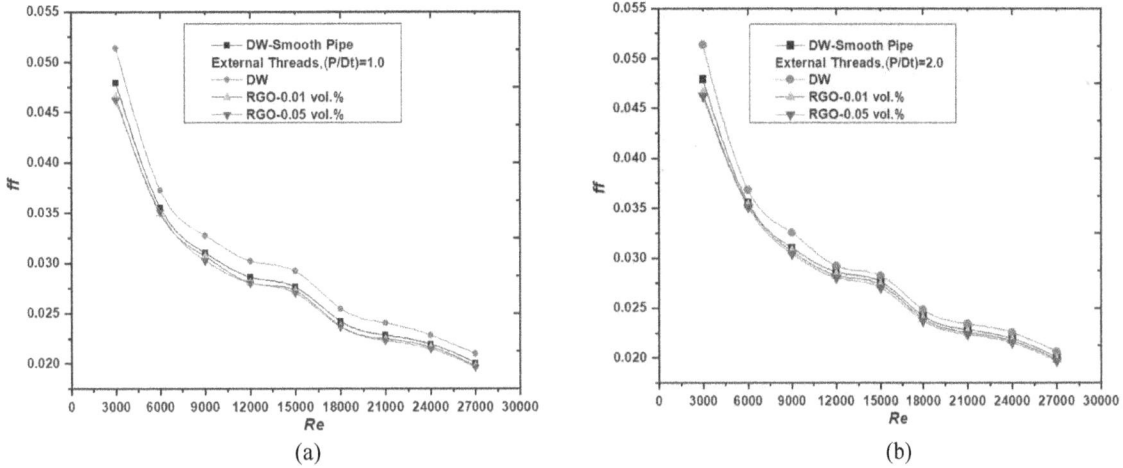

Fig. 62.5 Effect of RGO nanofluid and external threaded pipe on ff

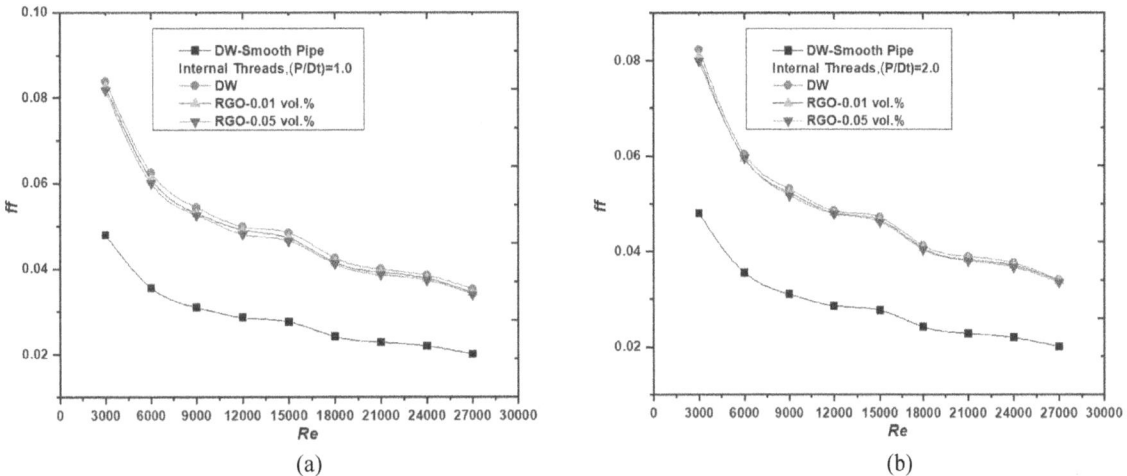

Fig. 62.6 Effect of RGO nanofluid and internal threaded pipe on ff

5.2 Effect of Augmentation on Friction Factor (ff)

Figures 62.5 and 62.6 depict the relationship between the Re and ff, RGO nanofluid concentration, and pitch-to-depth (P/Dt) ratio of threads. Notably, the experimental results reveal that the external V-threaded pipe with a P/Dt ratio of 1.0 exhibits an average 5.5% increase in ff value compared to the smooth pipe, when using distilled water. However, the addition of RGO nanofluids with concentrations of 0.01 and 0.05 volume.% results in an average decrease in ff values of 1.8% and 2.3%, respectively. This trend suggests that higher RGO nanoparticle concentrations lead to lower friction factors. Similarly, for a P/Dt ratio of 2.0, the threaded pipe with distilled water shows an average 3.8% increase in ff, while the nanofluids exhibit average reductions of 1.4% and 2.2% for 0.010 and 0.050 vol.%, respectively. The lubricating properties of RGO nanoparticles are responsible for the decreased friction factor [21].

Figures 62.6 (a) and (b) illustrate the effects of RGO nanofluids at concentrations of 0.010 and 0.050 vol.% in internal V-threaded pipes with pitch-to-depth (P/Dt) ratios of 1.0 and 2.0. The internal threads exhibit higher friction factors compared to external threads, due to increased resistance. Notably, the internal threads with P/Dt ratios of 1.0 and 2.0, using distilled water, show average friction factor (ff) increases of 75.2% and 70.1%, respectively, compared to a plain pipe. The average ff augmentation is 71.7% and 69.4% for 0.010 and 0.050 vol.% nanofluid concentrations, respectively, with a P/Dt ratio of 1.0. Similarly, ff increases of 68.2% and 67% are observed for 0.010 and 0.050 vol.% concentrations, respectively, with a P/Dt ratio of 2.0. According to these results, internal threaded pipes with a higher P/Dt ratio have a lower friction factor, mostly as a result of less surface roughness, mixing, and turbulence. On the other hand, the friction factor decreases when the concentration of RGO nanofluid rises (Sundar et al. 2017).

5.3 Impact of Augmentation on Nusselt Number (Nu)

The heat exchanger's increased Nusselt number (Nu) is seen in Figures 62.7 and 62.8. These graphs unequivocally

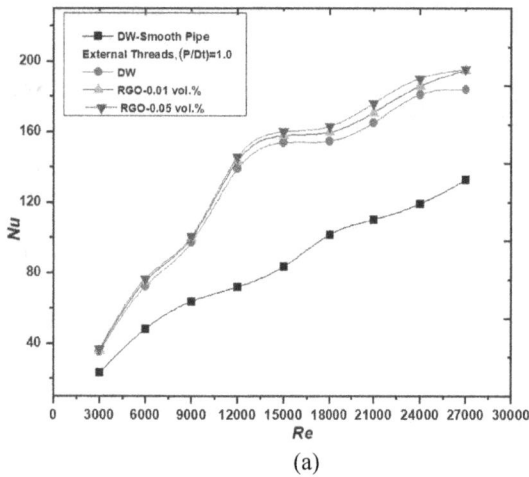

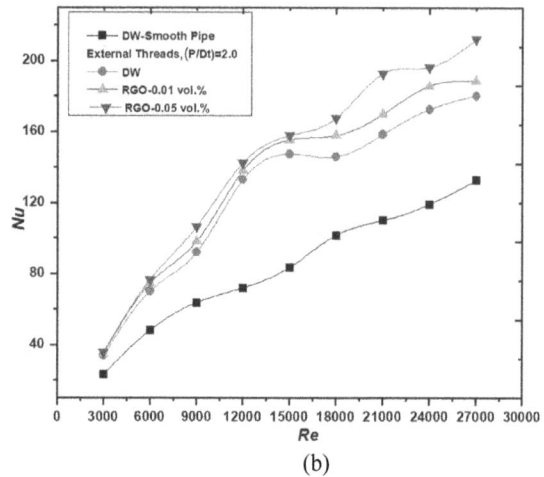

Fig. 62.7 Effect of RGO nanofluid and external threaded pipe on Nu

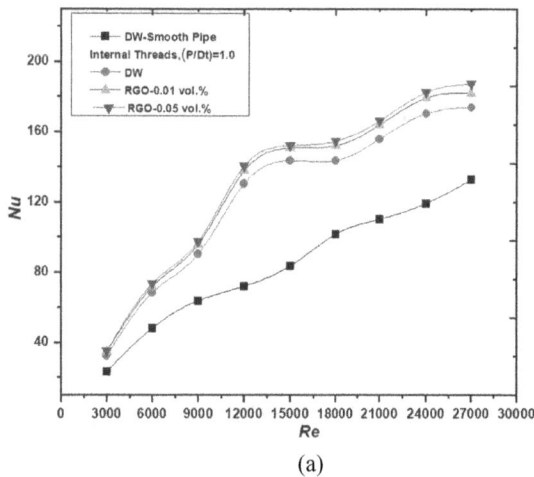

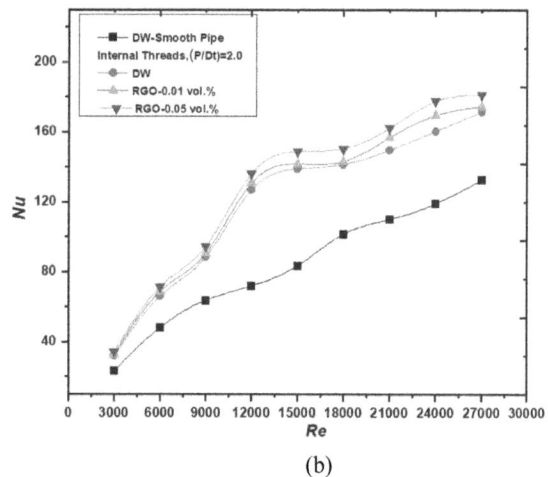

Fig. 62.8 Effect of RGO nanofluid and internal threaded pipe on Nu

show how threaded profiles and nanofluids significantly increase the rate of heat transfer. Figures 62.9(a) and (b) specifically show how Nu for externally threaded pipes grows with Reynolds number (Re), thread pitch-to-depth (P/Dt) ratio, and concentrations of nanofluid (NF). Interestingly, when compared to smooth pipes, the external threads with (P/Dt) ratios of 1.0 and 2.0 show Nu improvements of 56.8% and 50.4%, respectively. Additionally, for (P/Dt) ratios of 1.0 and 2.0, the combination of external threading with RGO nanofluids at concentrations of 0.01 volume percent and 0.05 volume percent results in average Nu improvements of 62.1%, 66.6%, and 59.6%, 64.3%, respectively (Chandra Sekhara Reddy and Vasudeva Rao 2014).

The influence of internal threads and nanofluids on a heat exchanger's Nu was illustrated in Fig. 62.8 (a) and (b). According to the results, as compared to smooth pipe, internal threads with a (P/Dt) ratio of 1.0 and 2.0 show 47.0% and 42.6%, respectively. Nu is improved on average by 54.9% and 57.6% with an internal threaded pipe (P/Dt) ratio of 1.0 and RGO nanofluid of 0.01 and

0.05 vol.%, respectively. Similarly, as compared to a smooth pipe with distilled water, a (P/Dt) ratio of 2.0 and nanofluid concentrations of 0.01 and 0.05 volume. % demonstrate an average improvement of Nu by 47.1% and 53.5%, correspondingly. Based on these experimental findings, it has been determined that, in this specific heat exchanger configuration, externally threaded pipes outperform internally threaded pipes in terms of heat transfer. This is because of a number of factors, including a higher convective heat transfer coefficient, improved annulus turbulence, increased surface area, and decreased internal hydraulic resistance. The rate of heat transfer also increases as the concentration of the RGO nanofluid grows because of the high thermal conductivity of the RGO nanoparticles.

6. CONCLUSION

- External threaded pipes conduct heat more effectively than inside threaded pipes. The v-thread's pitch-to-depth ratio falls, the friction factor rises, and Nu rises.

- Because RGO nanoparticles are lubricating, adding them to distilled water base fluid lowers the friction factor (ff) and raises the Nu.
- The optimal configuration for Nu enhancement (91.5%) is an external threaded pipe with a P/Dt ratio of 1.0 and 0.050 volume.% RGO-DW nanofluid concentration. In comparison, the lowest Nu enhancement (31.5%) occurs with an internal threaded pipe with a (P/Dt) ratio of 2.0 and 0.010 volume% RGO-DW nanofluid concentration.
- Internal threaded pipes with a (P/Dt) ratio of 1.0 and 0.010 vol.% nanofluid concentration result in the highest ff increase (72.6%). External threaded pipes with a (P/Dt) ratio of 2.0 and 0.050 vol.% RGO-DW nanofluid concentration have the greatest ff drop (3.5%).

REFERENCES

1. Ahirwar, B.K., and A. Kumar. 2024. Enhancing Thermal Performance: A Sophisticated Analysis of CuO-Water Nanofluids and Twisted Tape Inserts in Tubular Heat Exchangers-a Numerical Study: Enhancing Thermal Performance: A Sophisticated Analysis of CuO: B. K. Ahirwar, A. Kumar. *Journal of Thermal Analysis and Calorimetry* (December 1).

2. Ahmad, F., S. Mahmud, M.M. Ehsan, and M. Salehin. 2023. Thermo-Hydrodynamic Performance Evaluation of Double-Dimpled Corrugated Tube Using Single and Hybrid Nanofluids. *International Journal of Thermofluids* 17 (February 1).

3. Akhavan-Behabadi, M.A., R. Kumar, M.R. Salimpour, and R. Azimi. 2010. Pressure Drop and Heat Transfer Augmentation Due to Coiled Wire Inserts during Laminar Flow of Oil inside a Horizontal Tube. *International Journal of Thermal Sciences* 49, no. 2: 373–379.

4. Akyürek, E.F., K. Geliş, B. Şahin, and E. Manay. 2018. Experimental Analysis for Heat Transfer of Nanofluid with Wire Coil Turbulators in a Concentric Tube Heat Exchanger. *Results in Physics* 9: 376–389.

5. Chandra Sekhara Reddy, M., and V. Vasudeva Rao. 2014. Experimental Investigation of Heat Transfer Coefficient and Friction Factor of Ethylene Glycol Water Based TiO2 Nanofluid in Double Pipe Heat Exchanger with and without Helical Coil Inserts. *International Communications in Heat and Mass Transfer* 50: 68–76. http://dx.doi.org/10.1016/j.icheatmasstransfer.2013.11.002.

6. Ebrahimi-Moghadam, A., S. Kowsari, F. Farhadi, and M. Deymi-Dashtebayaz. 2020. Thermohydraulic Sensitivity Analysis and Multi-Objective Optimization of Fe3O4/H2O Nanofluid Flow inside U-Bend Heat Exchangers with Longitudinal Strip Inserts. *Applied Thermal Engineering* 164, no. June 2019: 114518. https://doi.org/10.1016/j.applthermaleng.2019.114518.

7. Heyhat, M.M., S. Kimiagar, N. Ghanbaryan Sani Gasem Abad, and E. Feyzi. 2016. Thermal Conductivity of Reduced Graphene Oxide by Pulse Laser in Ethylene Glycol. *Physical Chemistry Research* 4, no. 3: 407–415.

8. Khashaei, A., M. Ameri, and S. Azizifar. 2024. Heat Transfer Enhancement and Pressure Drop Performance of Al2O3 Nanofluid in a Laminar Flow Tube with Deep Dimples under Constant Heat Flux: An Experimental Approach. *International Journal of Thermofluids* 24 (November 1).

9. Khlewee, A.S., Y. Alaiwi, T.A. Jasim, M.A.T. Mahdi, A.J. Hussain, and Z. Al-Khafaji. 2024. Numerical Investigation of Nanofluid-Based Flow Behavior and Convective Heat Transfer Using Helical Screw. *Journal of Advanced Research in Numerical Heat Transfer* 27, no. 1 (November 30): 85–106. https://semarakilmu.com.my/journals/index.php/arnht/article/view/10220.

10. Panja, S.K., B. Das, and V. Mahesh. 2025. Performance Evaluation of a Novel Parabolic Trough Solar Collector with Nanofluids and Porous Inserts. *Applied Thermal Engineering* 258 (January 1).

11. Praveenkumara, B.M., B.S. Gowda, G.L. DushyanthKumar, and M.J.B. Prakash. 2023. A Short Review on Thermal Properties of Nanofluids in Heat Transfer Applications. *Lecture Notes in Mechanical Engineering* 1: 195–200.

12. Praveenkumara, B.M., and B. Sadashive Gowda. 2023. Experimental Study of the Impact of Low Concentration MgO-Distilled Water Nanofluid on Thermal Performance of Concentric Tube Counter Flow Heat Exchanger. *Sadhana - Academy Proceedings in Engineering Sciences* 48, no. 4. https://doi.org/10.1007/s12046-023-02294-x.

13. Praveenkumara, B.M., B. Sadashive Gowda, and K.N. Abhijeet Bharatwaj. 2022. An Experimental Investigation of Study the Combined Effect of Threaded Pipe and Twisted Tap Inserts on Heat Transfer Rate of Double Pipe Heat Exchangers. *Materials Today: Proceedings* 82: 108–117. https://doi.org/10.1016/j.matpr.2022.12.107.

14. Salem, M.R., M.M. Ellaban, R.K. Ali, and A.E. Elmohlawy. 2024. Experimental Investigation of the Performance Attributes of a Double Pipe Heat Exchanger Equipped with Baffles of Conventional or Flower Layouts. *Applied Thermal Engineering* 253 (September 15).

15. Samruaisin, P., V. Chuwattanakul, P. Thapmanee, M. Kumar, P. Naphon, N. Maruyama, M. Hirota, and S. Eiamsa-ard. 2025. Thermohydraulic Performance Evaluation of a Heat Exchanger Mounted with Oval Inclined Twisted Rings. *Engineering Science and Technology, an International Journal* 63 (March): 101981. https://linkinghub.elsevier.com/retrieve/pii/S2215098625000369.

16. Singh, K., D.P. Barai, S.S. Chawhan, B.A. Bhanvase, and V.K. Saharan. 2021. Synthesis, Characterization and Heat Transfer Study of Reduced Graphene Oxide-Al2O3 Nanocomposite Based Nanofluids: Investigation on Thermal Conductivity and Rheology. *Materials Today Communications* 26, no. December 2020: 101986. https://doi.org/10.1016/j.mtcomm.2020.101986.

17. Sundar, L.S., K. V. Sharma, M.K. Singh, and A.C.M. Sousa. 2017. Hybrid Nanofluids Preparation, Thermal Properties, Heat Transfer and Friction Factor – A Review. *Renewable and Sustainable Energy Reviews*. Elsevier Ltd.

18. Thianpong, C., K. Wongcharee, K. Kunnarak, S. Chokphoemphun, S. Chamoli, and S. Eiamsa-ard. 2024. Parametric Study on Thermal Performance Augmentation of TiO2/Water Nanofluids Flowing a Tube Contained with Dual Counter Twisted-Tapes. *Case Studies in Thermal Engineering* 59 (July 1).

Note: All the figures in this chapter were made by the authors.

For Product Safety Concerns and Information please contact our EU
representative GPSR@taylorandfrancis.com
Taylor & Francis Verlag GmbH, Kaufingerstraße 24, 80331 München, Germany

www.ingramcontent.com/pod-product-compliance
Lightning Source LLC
Chambersburg PA
CBHW081043220326
41598CB00038B/6964

* 9 7 8 1 0 4 1 2 0 9 7 0 6 *